Neurobiology

NEUROBIOLOGY

THIRD EDITION

GORDON M. SHEPHERD, M.D., D.PHIL.
Professor of Neuroscience
Yale University

New York   Oxford
OXFORD UNIVERSITY PRESS
1994

Oxford University Press

Oxford New York Toronto
Delhi Bombay Calcutta Madras Karachi
Kuala Lumpur Singapore Hong Kong Tokyo
Nairobi Dar es Salaam Cape Town
Melbourne Auckland Madrid

and associated companies in
Berlin Ibadan

Published by Oxford University Press, Inc.,
198 Madison Avenue, New York, New York 10016-4314

Oxford is a registered trademark of Oxford University Press

Library of Congress Cataloging-in-Publication Data
Shepherd, Gordon M., 1933–
Neurobiology / Gordon M. Shepherd.—3rd ed.
p. cm. Includes bibliographical references and index.
ISBN 0-19-508842-5 (cloth).—ISBN 0-19-508843-3 (paper)
1. Neurobiology. 2. Molecular neurobiology. I. Title.
[DNLM: 1. Nervous System—physiology. 2. Neurons. 3. Membrane
Potentials. 4. Neuronal Plasticity. 5. Behavior—physiology.
WL 102 S549n 1994] QP355.2.S52 1994 612.8—dc20
DNLM/DLC for Library of Congress 94-1679

4 6 8 9 7 5
Printed in the United States or America
on acid-free paper

To my mother
Eleanor Murray Shepherd
for her wisdom, humor, and courage

Preface

The field of neurobiology continues its extraordinary growth, borne on waves of new technology; stimulated by new insights into the brain, mind, and behavior; and motivated by the increasing relevance of basic research to relief of human suffering from nervous and mental diseases. As we make our way through the "Decade of the Brain" toward the Third Millenium, this task grows ever more urgent. The primary aim of this book continues to be to make this vast and expanding world of knowledge intelligible to students by presenting a unified view of the principles of nervous organization.

This edition enters the era of digital electronic publishing by the simultaneous preparation of a tutorial and accompanying disk entitled "Electrophysiology of the Neuron," by John Huguenard and David McCormick (Oxford University Press, 1994). The student can study the different functional properties of neurons both from this textbook and from interactive computer models carefully integrated with the text. Drs. Huguenard and McCormick are well known for their research on these methods. They have produced models which should have broad appeal to introductory students and their teachers as well

as experienced research workers. It has been a pleasure to work with them and support them in this endeavor. Their tutorial realizes a goal I have had, since working with Wilfrid Rall on the first computer models of neurons some 30 years ago, of incorporating realistic models into the way students think about nervous system functions.

This edition emphasizes two of the major themes in current neurobiology, the roles of *synapses* and of *active membrane properties* in shaping the input-output functions of neurons and neural networks. With the combined text and computer models, the student and teacher can together reproduce the classical experiments demonstrating these properties that underlie the information-processing capabilities of different types of neurons. These accounts have been expanded in Section I of the book on Molecular and Cellular Mechanisms to provide a comprehensive introduction to these topics.

A second new area of emphasis in this edition is *development*. Increasingly the analysis of nervous organization is focussing on how it arises as development proceeds. The mechanisms of gene regulation are critical for understanding the basis of development, and these are introduced at an early stage in Chapter 2 on Molecular Biology; I am greatly indebted to my colleague Susan Hockfield for providing most of this new account. Chapter 9 on Developmental Neurobiology has also been extensively revised and expanded to provide an overview of the main steps in the development of the human nervous system, as well as an introduction to the host of factors which control the differentiation, migration, and maturation of the cells of the nervous system. My colleague Pasko Rakic has provided valuable comments and suggestions in writing this account. These principles are demonstrated in different systems throughout the book.

A third area of emphasis is *plasticity*. Just as organization arises out of development, so does it exhibit remarkable abilities to be molded by activity and to recover from injury. We see this in the responses of individual neurons as well as in the organization of systems underlying behavior. These properties are highlighted in a number of the different systems discussed in this book.

In addition to these new areas of emphasis, the main conceptual framework of the book has been strengthened. The first edition emphasized the comparative approach to understanding the principles of nervous organization. The second edition added an emphasis on levels of organization as crucial for understanding these principles. The past few years have seen an increased focus on these ideas. As a Co-Director of the James S. McDonnell Summer Institute in Cognitive Neuroscience (1988–1992), I had the opportunity to develop these ideas as a basis for cognitive functions. I thank John Bruer for this opportunity, and Michael Gazzaniga, Michael Posner, and Steve Kosslyn for stimulating discussions. Thus, in this edition, these twin principles of *levels of organization* and *comparative systems* are made even more explicit than before. Together they constitute a contribution to the theoretical foundations of neuroscience. These principles are introduced

in Chap. 1 and applied throughout the book, culminating in their application to understanding the cerebral cortex and human behavior in Chap. 30. I have benefitted greatly from the advice of Patricia Goldman-Rakic in revising this chapter.

In order to make room for the new material, I have trimmed some of the sections in the previous edition that experience had indicated were of less priority in using the book for teaching. Most of this involved examples taken from invertebrate cells and systems, no longer needed to make general points about the universality of the principles involved. As a result the book is now focused more on the primate and human, which have been of primary interest to most of the students using the book, without sacrificing the importance of the principles illustrated by the comparative approach.

In this revision I have been even more dependent on advice and assistance from my colleagues than previously, due to the increasing complexity of the field and the press of work in the laboratory and various duties. It is a pleasure to express my gratitude to the following people for writing parts of chapters: Susan Hockfield (Chap. 2); Stephen Smith (Chaps. 5, 8); Pietro De Camilli (Chap. 6); Gordon M. G. Shepherd (Chaps. 14, 15); David Corey (Chaps. 14, 15); and Bartley Hoebel (Chap. 26). I am grateful to the following people for suggestions and criticisms regarding specific chapters; Roger Thomas (Chap. 4); Tomas Hökfelt (Chap. 8); Pasko Rakic (Chaps. 9, 30); Karl Pfenninger (Chap. 9); Bernice Grafstein (Chap. 9); David Smith (Chap. 11); John Kaas (Chap. 12); Alan Light (Chap. 12); Peter Matthews (Chaps. 13, 20); Nigel Daw (Chap. 16); Sten Grillner (Chap. 20); Sarah Leibowitz (Chap. 26); Solomon Erulkar (Chap. 27); and Patricia Goldman-Rakic (Chaps. 29, 30). The following neuroscientists have kindly provided specific illustrations: Stephen Hersh, Alan Peters, David McCormick, Paul Trombley, Stuart Firestein, Frank Zufall, Robert LaMotte, Ron Dubner, David van

Essen, Robert Burke, and John Connor.

It is a pleasure to acknowledge my gratitude to these colleagues, and to offer my apologies for any inadequacies of the final text in representing their contributions. I also apologize to colleagues who may feel their contributions or fields are not adequately or accurately represented. This introductory text differs from many others in trying to provide a sense of the work on the frontiers of the field. Some of the examples therefore reflect arbitrary choices of material to serve this purpose.

I should like to emphasize that, as in past editions, this edition attempts to "see the brain steadily and see it whole," with a comprehensive coverage of brain functions that includes material both for the introductory and the more advanced student. Although designed for a one-semester course, it thus contains material that goes beyond that scope. The purpose in doing this is to let teachers and students understand the full range of functions of the nervous system and the richness of methods being used to study them, so that they have a framework within which to select the topics they feel are of primary relevance for their course or their interest. This should continue to make the book adaptable for use in a variety of ways, for neurobiology and psychobiology courses for undergraduates, neurobiology courses for medical school and graduate students, and for people coming from other fields such as physics and computer science.

In a work of this extent it is a great advantage to have readers who have gone through the book carefully for accuracy both in content and typography. I am indebted in this respect to the translators of the first edition into Spanish and Russian, and of the second edition into Japanese, Chinese, and German. Vera Boeckh and Daisuke Yamamoto have been particularly thorough, and I thank them deeply. Miriam LeGare has again supplied me with numerous suggestions gained from experience in teaching with the book.

Charles Greer, Tom Getchell, Doron Lancet, John Kauer, and John Hildebrand have continued to provide wise counsel and advice. I gained much from a brief sabbatical with Hersch Gerschenfeld in the laboratory of Philippe Ascher. I owe a special debt to the colleagues in my laboratory, including Stuart Firestein, Frank Zufall, Anne Williamson, Paul Trombley, David Berkowicz, Rick Jensen, Haiqing Zhao, Michael Singer, Paul Kingston, and Albert Telfeian. By the vigorous pursuit of their work they have contributed in many ways to the knowledge distilled in these pages. I am also deeply grateful to the National Institutes of Health and our funding agencies, including the National Institute for Deafness and Other Communicative Disorders, which supports our work on the olfactory system; the National Institute of Neurological Disorders and Stroke, which supports our work on human cortex and epilepsy; and the National Institute of Mental Health and the National Aeronautics and Space Agency, which jointly support our work on integration of sensory data under the Human Brain Project.

Like its predecessors, this edition would not have seen the light of day without the constant support of Jeffrey House; it is a pleasure to mark the 20th anniversary of our ventures together. Denise Palluoto has provided expert support in the office. Somehow out of all this hectic activity a book emerges, a process my wife Grethe understands much better than I do; I suspect she had it all planned out.

Hamden, Connecticut   G.M.S.
January 1994

Contents

Neurobiology

1

Introduction

In all of modern science, no field is developing more rapidly, attracting more interest, and promising more insight into ourselves as human beings than the study of the nervous system. Whether you are a beginning student who is intrigued by what you read about the brain in newspapers and magazines and see on television programs, or a science student with experience in conducting experiments and reading scientific articles, this book invites you to join in learning about the most interesting and important organ in the world.

Why Study Neurobiology?

There are many reasons for studying neurobiology. First, we all learn from an early age that our behavior, and the behavior of all animal life, depends on the nervous system. As we grow older, we experience the full richness of human behavior—the ability to think and feel, to remember and create—and we wonder, if we have any wonder at all, how the brain makes this possible. How this comes about is the subject matter of neurobiology.

A second reason is simply that the brain is the most complex of all biological organs. Some therefore believe that understanding how it works offers the greatest challenge in all of biological science.

A third reason is that the brain is the organ through which we think. Some therefore believe that this offers the supreme philosophical challenge: understanding how the brain can enable us to understand how the brain enables us to understand. (If you already know the answer to this enigma, proceed directly to page 684!)

A fourth reason is that the brain is the organ that makes us human. Everything we learn about the brain gives us potential insight into the nature of being human. This makes neurobiology more than just a philosophical exercise or intellectual game; it could help to ensure that we human beings have a future. What is the neurobiological basis of racism—the fear and hatred of people who are different? Do terrorism and crime get built into our brain circuits? Why do humans seem bent on self-destruction through environmental pollution and the development of weapons of annihilation? Why do we have this in our brains, and how can we control it? In all of science and medicine, neurobiology is the only field that can ultimately address these critical issues. These could be the most urgent reasons for studying the brain.

In addition to these special reasons, neurobiology is important as a field in biomedicine. Everything we learn about the nervous system, from the most primitive worms and slugs to the human, brings us closer to being able to prevent or relieve the suffering of nervous diseases such as epilepsy, Parkinson's disease, or senile dementia, and mental disorders such as drug addiction, depression, or schizophrenia.

Finally, as in every field of science, neurobiology can be pursued for the fun of using new technologies, the challenge of discovery, and the beauty of new insights into nature that scientific endeavors offer as their own reward.

What Is Neurobiology?

The things we experience about human behavior in our own lifetimes have been pondered ever since the philosophers of ancient Greece first conceived the notion that human beings have a mind and a soul, and began to probe their nature. However, despite the strivings of humankind's profoundest thinkers in the more than 2500 years since then, it is only very recently that we have begun to acquire any realistic idea at all about the true nature of the brain.

The relevant history may be briefly told. Scarcely 100 years ago was it learned that the brain is composed of cells, and that the cells are connected together to form circuits. Scarcely 40 years ago were the main kinds of nerve cell activity directly recorded, and the junctions, called synapses, identified that allow nerve cells to communicate with each other. Only 15 years ago were we provided the means to analyze directly the molecular mechanisms that control the differentiation of nerve cells and the expression of their functional properties. These and other milestones along the way are summarized in Table 1.1.

It should be obvious therefore that neurobiology is a very young science, and is acquiring new knowledge at a very rapid rate. What you will learn about the nervous system in the course of studying this book will far exceed even the wildest dreams that René Descartes might have had as he struggled to formulate the first concepts of the brain as a machine, in the early part of the seventeenth century. But will this vast amount of new and detailed knowledge you learn lead to understanding?—to a coherent view of how the nervous system mediates behavior?

In order for this to occur, it is not enough to learn the facts; the facts must be related to each other in ways that make them meaningful. Scientists recognize that for a fact to have meaning, it must be seen within the context of a general principle. The main aim of neurobiology, therefore, and the main aim of this book, is *to identify the principles underlying the mechanisms through which the nervous system mediates behavior.*

One of the ways in which principles are useful is that they force us to define the elementary units, the basic building blocks, of a field of knowledge. Thus, in physics, we understand matter in terms of elementary atomic particles and forces. Similarly, we can gain a good understanding of the heart in terms of the coordinated actions of cardiac cells. In neurobiology, the problem is much more complicated. We know that nerve cells process information, but we do not understand the different forms that this information takes within the nervous system. We also know that nervous function depends on the coordinated actions of nerve cells, but we do not understand the enormous variety of functions that the nervous system must have in order to mediate the different types of behavior. However, as our brief review of history revealed, some of the structures, substances, and properties have been identified that must contribute to the elementary units that underlie those functions. These elementary units reflect different levels of organization, that range from the single molecule or ion to the complicated systems that underlie the expression of a behavior such as feeding or thinking.

From these considerations we can deduce a basic premise, that an understanding of

Table 1.1 Some steps in acquiring knowledge about the basic mechanisms of the brain

600–400 B.C.	Greek philosophers describe the mind and soul; thinking depends on the brain (or heart?).
1543	Vesalius accurately describes gross anatomy of the human nervous system.
1637	Descartes characterizes the brain as a machinelike mechanism, independent of, but related to, the soul.
1798	Galvani discovers the electrical nature of nervous activity.
1891	Cajal and others determine that the nervous system is composed of independent nerve cells (neurons) connected together to form pathways.
1897	Sherrington proposes that nerve cells form pathways by communicating with each other through junctions called synapses.
1920s	Langley, Loewi, Dale, and others, identify chemical substances (neurotransmitters) that function as messengers which act on receptors at the synapses.
1940s	Shannon, Weaver, and Weiner introduce concepts of information processing and control systems (cybernetics).
1950s	Hodgkin, Huxley, Katz, and Eccles make precise recordings of electrical signals with microelectrodes. Electron microscopy reveals synapses and neuronal fine structure.
1950s	Single cell analyses by Mountcastle, Lettvin, Hubel and Wiesel reveal brain circuits for feature abstraction.
1960s	Integrative functions of dendrites are recognized: synaptic circuits and synaptic interactions without impulses are identified.
1970s	Neuromodulator substances and second messengers are found that greatly extend the duration and complexity of neuronal interactions.
1970s	Computerized imaging techniques permit visualization of brain activity patterns in relation to sensation and cognition.
1970s	Molecular methods are introduced for analyzing genetic mechanisms (recombinant DNA technologies) and single membrane proteins (patch clamping).
1980s	Advances in computers and neural networks provide new models of nervous system functions (vision, language, memory, logic).
1990s	"The Decade of the Brain": new emphasis on combining information from different levels of analysis into integrated models of brain function and nervous disease.

nervous function requires identifying the elementary units at different levels of organization and understanding the relations between the different levels. We can summarize this view with a more precise definition of the subject matter of modern neurobiology, and of this book: *Neurobiology is the study of nerve cells and associated cells and the ways that they are organized into functional circuits that process information and mediate behavior.*

The Levels of Neural Organization

The importance of the different levels of organization for understanding the princi-

ples of neurobiology can be illustrated by an example. Let us consider a simple *behavior,* such as reading this page (see A in Fig. 1.1). We know, by observation, that this act has several component parts. First, there is our sensory *perception* of the page and the symbols printed upon it. Second, there is a process of *comprehension,* which gives those symbols meaning. Third, there is *movement* of our eyes so that we can scan the page.

These parts *describe* the behavior of reading the page. In order to *understand* how this occurs, we need to look inside the brain. Tracing fiber tracts through the brain has been one of the traditional concerns of neurobiologists. From these studies

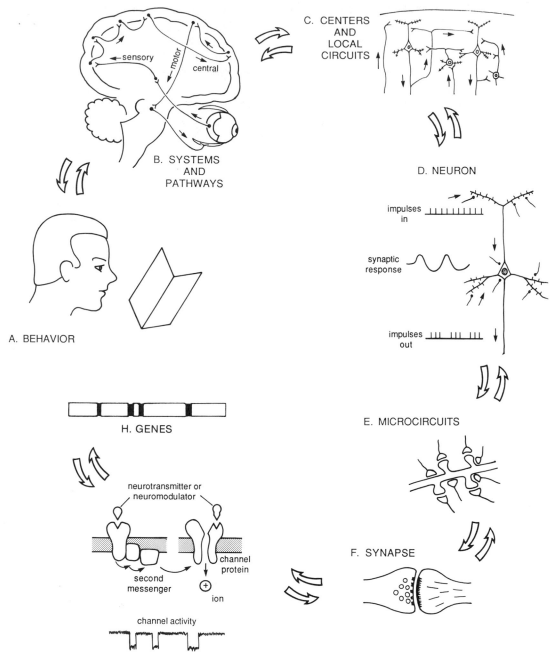

Fig. 1.1 Diagrams illustrating the different levels of organization of the nervous system. **A.** At the highest level is a *specific behavior,* such as reading a page of this book. **B.** *Distributed systems* within the brain mediate a specific behavior. The diagram is of a lateral view inside the brain, showing in schematic fashion the *sensory* (visual) pathway from the eye, the *central* pathway for comprehension, and the *motor* pathway for controlling eye movements. **C.** *Local circuits* provide for the organization of each area within a distributed system. The diagram shows the local circuits in an area of cortex. **D.** The *nerve cell (neuron)* is the basis of local circuit organization. Impulses

6

we can lay out the *pathways* within the brain that are involved in reading this page (B in Fig. 1.1). First, there is the sensory pathway, which starts within the eye and ends within the visual cortex at the back of the brain. Second, there are the central connections reaching forward to the frontal lobes of the brain, that underlie comprehension. Third, there are the descending motor pathways that control the movement of the eyes. In this way we identify the major *sensory, central,* and *motor systems* that are involved in this particular behavior. These systems are at the highest level of brain organization.

To understand how a system works, we need to analyze the organization of the centers. Within each center are populations of nerve cells (also called *neurons*) together with satellite cells called *neuroglia*. There are typically several different types of nerve cell. Neurons characteristically have short branches, called *dendrites;* many neurons, especially those connecting to other centers, also have a long branch called an *axon*. Each cell connects with other cells within the center in specific ways through its axon and/or dendrites. These connections form what are called *local circuits* (C in Figure 1.1). Through these, the center receives information coming from other centers, performs some specific processing operations on it, and sends the output to other centers.

Local circuits are constructed of *neurons,* which represent an important level of organization (D in Fig. 1.1). The main function of the neuron is to generate a characteristic type of activity and integrate it with the activity it receives from other neurons through long-distance and local circuit connections. One common pattern of organization of a neuron is for the dendrites to receive inputs and the axon to send outputs. The output is characteristically carried in a discharge of brief electrical changes, termed *impulses,* which can travel long distances along the outer membrane of the axon without diminishing in amplitude.

Within local circuits are neurons and patterns of connections that are stereotyped and distinctive. These are referred to as *microcircuits* (E in Fig. 1.1), which mediate the specific types of operations that are carried out in that center.

In order for neurons to communicate with each other and form circuits, they need to have points of contact with each other. These are called synapses (F in Figure 1.1). A single neuron may send its output through hundreds of synapses on other neurons; conversely, it may receive inputs through hundreds or thousands of synapses made upon it by other neurons. Each synapse may be considered as a semi-independent input–output unit. Synapses are therefore critical building blocks of microcircuits and local circuits in the same way that transistors are the building blocks of computers.

At the *molecular* level, the synapse is a complicated structural and functional unit. The chemical type of synapse involves release of a neurotransmitter substance from a presynaptic process, which causes an increase or decrease in the flow of ions

---

arriving in output fibers from other areas set up excitatory or inhibitory synaptic potentials, which are integrated within the neuron to control its impulse output to other neurons. E. *Microcircuits* are formed by patterns of synaptic connections onto neurons together with local sites of active properties. They provide the basis for the integrative actions of a neuron. F. The *synapse* is the basic input–output unit for transmission of information between neurons. The organization of synapses into microcircuits provides the basis for complex information processing between neurons. G. *Macromolecules* are organized to form the synapse; different types of macromolecules underlie the generation of impulses and other active properties involved in information processing. H. *Genes* carry the information for synthesis of proteins during development and in response to intercellular signals and intracellular changes in activity.

through a channel protein in the membrane of a postsynaptic process. The substance may act directly on the channel protein, or indirectly, through second messengers, as indicated in G of Fig. 1.1. The molecular structure of the different channel proteins and the ways they are expressed by the genome are elucidated using biochemical methods and recombinant DNA technologies.

Also indicated in G is the fact that the action of a channel protein in allowing ions to flow across the membrane may be recorded by special electrophysiological methods, called patch clamping. Analysis by these methods enables one to show how single-channel events are triggered by the neurotransmitter at a synapse to generate the synaptic potentials which control the firing by the neuron. Similar methods permit analysis of the single-channel events that underlie the nerve impulse.

Molecules are encoded by genes, and the *gene* is therefore the most basic level in the hierarchy of organization (level H in Fig. 1.1). The mechanisms controlling the expression of genes are of critical importance for understanding the development of the nervous system. In addition, it is becoming clear that gene expression is altered in the adult by different states of activity. Modern molecular biology provides the tools for analyzing these mechanisms directly.

Figure 1.1 summarizes most of the levels of organization that are present in the nervous system. If you turn back to the definition of neurobiology, you will see that it embraces most of these levels.

Figure 1.1 helps to show how one starts with a given behavior and works downward through successive levels of organization to identify the units of function underlying that behavior. If we do that for the act of reading this page, we can begin to understand the mechanisms that enable us to do this act. However, the job is not complete until we compare this analysis with the results from other systems, and are able to generalize the levels of organization and their functional units across all

systems. Only then may we say that we can begin to understand the general principles that apply across all systems.

This book is founded on the premise that these general principles are present and are being revealed by modern neurobiological investigations; our task is to identify them and weld them into a coherent framework.

The Concept of Functional Units

A useful framework for this purpose is to combine the ideas of levels of organization and functional units, as expressed in Fig. 1.1, in the following concepts (modified from Shepherd, 1972):

1. The nervous system is organized in terms of functional units. A functional unit is defined as a structural entity with a specific function.
2. Neurons (and functionally related cells such as muscle, glands, and neuroglia) provide the structural basis for functional organization. Neurons have different combinations of structural processes (cell bodies, axons, dendrites, synapses) and physiological properties (synaptic interactions, passive and active properties). For a given site on a given neuron, the synaptic relations and physiological properties depend on the integrative context within the functional units of which it is a part.
3. Functional units are formed at different levels of organization, from genes and gene products through synapses, microcircuits, dendrites, neurons, and local circuits, to pathways and distributed systems. The nervous system is built of overlapping assemblies and hierarchies of such units of increasing extent and complexity.

Several scientists have had similar ideas in thinking about biological organization in general and nervous organization in particular. For example, François Jacob (1974) observed that the complexity of biological organisms arises out of a hierarchy in which, "at each level, units of a relatively well defined size and almost identical structure associate to form a unit of the level above. Each of these units formed by the integration of sub-units may be given the general name 'integron.' " The integron is very similar to the idea of a functional unit. A further attraction of this view is its premise that these integrated units of complex function are the basis of evolution (Jacob, 1974; see also Phillips, 1977).

The question of the fundamental units for nervous function was also discussed by John von Neumann, who played a central role in the development of the digital computer. In his stimulating book *The Brain as a Computer* (von Neumann, 1957), he observed that the neuron is commonly regarded as the basic computational element of the brain. However, he concluded that the neuron is too complicated; the fundamental unit is more likely to be the synapse, since each neuron contains many synapses, and each synapse seems more like the kind of single logic gate equivalent to a transistor. Thus, the neuron and the synapse represent two levels of functional units in the hierarchy of brain organization, a view very much in accord with the ideas expressed above and summarized in Fig. 1.1.

Understanding Brain Function

We thus see that behavior emerges out of the coordinated activity of a hierarchy of functional units (see also Churchland and Sejnowski, 1992). There is a further implication that a functional unit can be understood only within the context of the behavior that it mediates. Thus, no matter how complex a neural circuit may appear, we can be confident that it is designed to mediate specific naturally occurring behaviors. This same principle must apply, no matter how deeply one delves into the hierarchy of organizational levels, even down to the individual molecule gating a flow of ions. We may summarize this principle with the proposition that "Nothing in neurobiology makes sense except in the light of behavior."

From this discussion, you can begin to appreciate that behavior is a product of nervous organization and is also an agent in molding that organization.

We have emphasized the presence of many levels of functional organization between genes and behavior, because it is an important fact to keep in mind when you learn about the results of any particular method in studying the brain and judge how adequate that method is for explaining the neural basis of brain function. As an example, among the most interesting studies of brain functions are those using brain scans, which we will discuss in later chapters. As shown in Fig. 1.2A, these studies are at the highest levels of organization; the scans provide overall maps of the locations of brain activity in relation to behavior, but they give little insight into the neural mechanisms at lower levels that are responsible for generating the maps. This contrasts with the approach that may be summarized as "molecular reductionism" (Fig. 1.2B), driven by the belief that the cloning of a gene or the identification of a neurotransmitter or neurohormone, by itself, can provide the entire explanation for a given behavior. Our ultimate goal, instead, is to carry out studies at all levels of organization underlying different behaviors, so that a balanced view integrating all levels is achieved (Fig. 1.2C). Partial success at filling in the levels is already possible in certain types of diseases, such as Parkinsonism, in which the specific type of behavioral disorder can be correlated with loss of a specific neurotransmitter of a specific cell type and the circuits in which it is involved (see Chap. 21). Figures 1.1 and 1.2 thus provide general schemes for testing the adequacy of our current state of knowledge of neural mechanisms in accounting for different behaviors and cognitive functions.

Although behavior is thus central to an understanding of brain organization, this does not mean that a study of brain organization leads easily to an understanding of the neural basis of behavior. This in fact is

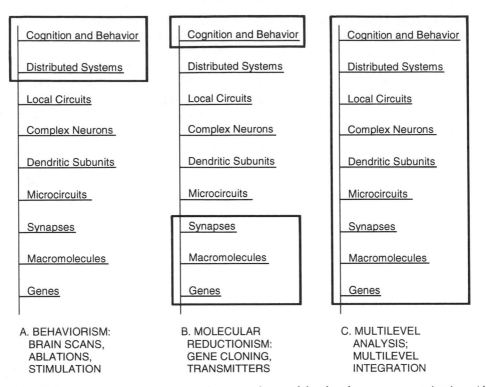

Fig. 1.2   Relations between experimental approaches and levels of nervous organization. (After Shepherd, 1994)

the greatest challenge and most difficult task in the study of neurobiology. We must always keep in mind that, for any complex organism, a given behavior may be due to an infinity of mechanisms; this is as true of the brain as of any other system. As a consequence, it is not enough to show that a given behavior *could* be mediated by a proposed model; the task is to prove that this is in fact *how the brain does it.*

This principle is therefore important in assessing any claim for a particular mechanism underlying a brain function. It is especially important in assessing models of neural mechanisms based on computer simulations. Many recent models, based on principles of parallel distributed processing, are providing a wealth of insights into how complex systems *may* process information (see Rumelhart and McClelland, 1986), but the extent to which these models give insight into how the brain does it is a matter that requires careful assessment. One of the main messages of this

book is that facts about brain function have to be obtained first from actual experiments on animals and humans, and these facts then need to be used to develop more realistic models for simulating brain functions (see Shepherd, 1990a; Churchland and Sejnowski, 1992). We will illustrate this important principle by comparing experimental results and theoretical models at many places in this book. The accompanying tutorial workbook and computer program on computer models of electrical signaling mechanisms (Huguenard and McCormick, 1994) shows the close coordination that is now possible between experiment and theory, not only in carrying out research on the brain, but even at the earliest stages of learning about the brain.

The Plan of This Book

The principles summarized in Figs. 1.1 and 1.2 indicate the plan of this book. Instead of working our way down, we will start at the bottom, at the gene and molecule level, and work our way upward toward behavior.

In the remaining chapters of this section, we will discuss the molecular and cellular organization of the neuron, and how they provide for the basic functions that apply generally throughout the entire nervous system. We will then consider the organization of different systems. We will first study the sensory systems that are responsible for our sensations and perceptions of the world around and within us. Next we will consider the motor systems that enable us to act on our environment. Finally, we learn about the central systems that not only mediate sensorimotor coordination, but also provide the complex circuits that underlie the highest nervous functions such as memory, emotions, and cognition. In fact, we will end by considering experimental evidence for the systems underlying the kind of cognitive functions illustrated in Fig. 1.1.

In this book, special emphasis will be placed on the ways in which nerve cells are connected through their synapses to form functional circuits. The synaptic circuit will be shown to be a key concept in enabling us to understand how nerve cells are organized to process information and mediate behavior. Another key concept is the distribution of excitable properties at many sites throughout the neuron.

We will see that these new findings require certain modifications in the classical concept of the neuron as the main type of functional unit in the brain. In place of the old idea that each neuron is a simple functional unit, receiving synaptic inputs in its dendrites and emitting signals through its axon, is an enlarged view in which the neuron can provide multiple sites for input—output units in both its dendrites and axon. Similarly, the neuron also provides multiple excitable sites for local integration. In this view, synaptic units, organized into multineuronal circuits and assemblies, provide much of the the basis for nervous

Fig. 1.3 An affectionate view of Stephen Kuffler from the early years of modern neurobiology. Armed with his trusty microelectrode, Don Quixote seeks the secrets of the neuron. (From Kuffler, 1958)

organization; see discussion above. One of the main purposes of this book is to identify these circuits that underlie different behaviors.

Emphasis is given in this book on comparing circuits in invertebrates and vertebrates in order to identify basic principles of organization. For those pursuing biology in its broadest aspects, the invertebrates represent solutions to adaptation that are of interest in their own right, quite apart from their relevance to vertebrates. Those who are primarily interested in the vertebrates should heed the warning of E. J. W. Barrington (1979):

Vertebrate studies by themselves . . . tell us little, if anything of the origin of vertebrates, or of the origin of the principles of biological organization that have determined the course of their adaptive evolution. Indeed, the appeal that the vertebrates make to our anthropocentric tendencies can be dangerously deceptive. It can easily lead to over-optimistic generalization from limited data, obtained from some laboratory mammal that has nothing to recommend it for the purposes other than its convenience and its compliant behaviour. If, therefore, we are to evaluate and exploit the dramatic advances of contemporary biology . . . we need as one essential condition the widest possible extension of our understanding of the principles of animal organization.

Those whose interests are confined mainly to humans will see, when we come to consider higher mental functions in the final chapters, that our understanding is drawn from a wide perspective on the principles of organization of neuronal circuits that has been built up in the preceding chapters.

I
MOLECULAR AND CELLULAR MECHANISMS

2

Molecular Neurobiology

Until recently, our knowledge of nerve cells at the molecular level was limited by the difficulty of applying traditional biochemical methods to nervous tissue. The major problem was the incredible diversity of cell types within the nervous system. Our understanding of molecular mechanisms therefore had to be inferred mostly from studies of other cell types, such as blood cells or gland cells, which are more accessible and can be obtained as large, homogeneous populations. Modern methods of molecular biology have changed all this. These methods permit the analysis of any given cell type even though it may constitute only a very small fraction of an entire tissue. Thus, we an now study molecular mechanisms directly in the brain. Because of these advances, we can increasingly build the study of neurobiology on a foundation of the molecular biology of the nerve cell.

This new field of investigation has the name of molecular neurobiology. It is concerned with three fundamental questions. First, *what is the molecular composition of nerve cells?* In order to answer this question, we will review briefly the main types of biological molecules and their functions within cells. We will then consider methods for analyzing how the chemical composition of nerve cells is determined by the flow of information from the genes, taking as examples the development of the hybridoma and recombinant DNA technologies.

Second, *how does the molecular composition of nerve cells provide the basis for their special functional properties?* The most characteristic function of nerve cells is their ability to transmit signals at specialized junctions called synapses. This involves the secretion of signal molecules, called neurotransmitters, which bind to receptor molecules in the membrane of the receiving cell. We will see how application of a variety of modern methods is permitting a unified picture to be built up at the molecular level of how nerve signals are transmitted at different synapses.

The third fundamental question is: *how are nerve cells assembled into functional circuits during development?* For this we need to understand the mechanisms of gene expression, as they relate to the unfolding structure and function of a given neuron, and the way this is orchestrated with similar or complementary events in neighboring neurons with which it interacts. In addition, we must consider how the environment can influence the development of an immature nerve cell. Environmental signals include the contacts between two cells (cell–cell interactions) and the release of neurotrophic factors that influence cell growth. The mechanisms by which signals at the surface of the nerve cell are transformed (transduced) into cellular responses have recently been explored at the molecular level. Insight into signal transduction has helped in understanding how the particular environment and experience of the organism at each stage of development interact with the products of gene expression to determine the neural basis of behavior. We will consider this question later in Chap. 9, and in many other places in this book.

Molecular Biology of the Neuron

The Biomolecular Quartet

One of the beautiful simplifying principles in biology is that all living organisms have been constructed of just four main types of molecules. Each type is composed of carbon atoms, linked together by covalent bonds (see Fig. 2.1). Simplest among these are the *fatty acids;* these consist of straight chains of carbon atoms, ending in carboxyl (—COOH) groups. A second type is the *sugars* (also called carbohydrates); these contain hydroxyl groups (—C—OH) and aldehyde (—CHO) or ketone (—CO—) groups. In larger sugars, such as glucose, a hydroxyl group near one end combines with the aldehyde or ketone group at the other end to form a ring structure. A third type is the *amino acids.* These consist of a carbon atom which is bound to four different moieties: a hydrogen atom (—H), an amino group (—$^+NH_3$), a carboxyl group (—COO$^-$), and a variable side chain (—R). Finally, a fourth main type is the *nucleotides.* These consist of a pentose (5-carbon) sugar linked to a nitrogen-containing ring structure called a base.

Simple versions of these four basic types of molecule were formed at an early stage

---

Fig. 2.1 Four basic types of biological molecules, and the ways they are built into polymers by different types of chemical bonds.

A. In *fatty acids,* the hydrophilic carboxyl group at one end is very reactive. These groups readily interact with hydroxyl groups of other molecules to form esters, which in turn can become covalently linked to phosphate groups and other molecules to build up complicated molecules. As an example in the nervous system, fatty acids are important constituents of myelin, a special membrane that enwraps some nerve axons.

B. *Sugar* molecules in ring conformation, such as glucose, are joined by two hydroxyl (OH) groups to form a glycosidic (—C—O—C—) bond. In this manner disaccharides (two sugars), oligosaccharides (several sugars), and polysaccharides (many sugars) can be built up. Polysaccharides are important constituents of recognition molecules at the surfaces of nerve cell membranes.

C. *Proteins* are formed from amino acids by joining the amino (—$^+NH_3$) group of one amino acid to the carboxyl (—COO$^-$) group of the other to form an amide (—CONH—) bond, also called a *peptide bond.* Two amino acids thus joined are called a *peptide;* longer chains of amino acids are called oligopeptides and polypeptides. Long peptide polymers form proteins; they become folded in complex ways, as a result of weak interactions (such as hydrogen bonds) between different parts of the chains. Small amino acids and peptides can function as neurotransmitters and neuromodulators; larger polypeptides and proteins function as their receptors.

D. *Nucleic acids* are polymers built by phosphodiester bonds between nucleotides. This may be summarized by a block diagram, in which nucleotides are represented by single letters (e.g., U for uracil and C for cytosine). In a chain of nucleotides, the 5′ carbon end is shown on the left, and the 3′ carbon end is on the right. Single-ring bases are called pyrimidines, and double-ring bases are called purines. A sugar plus its base is called a nucleoside; a nucleoside with a phosphate (PO$_4$) group attached to the sugar is called a nucleotide.

A Fatty Acids

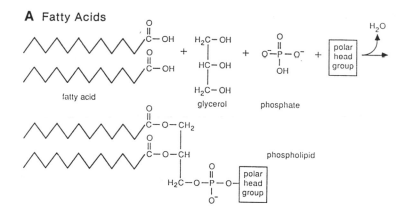

fatty acid

glycerol

phosphate

polar head group

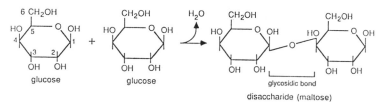

phospholipid

polar head group

B Sugars

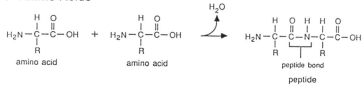

glucose

glucose

glycosidic bond

disaccharide (maltose)

C Amino Acids

amino acid

amino acid

peptide bond

peptide

D Nucleotides

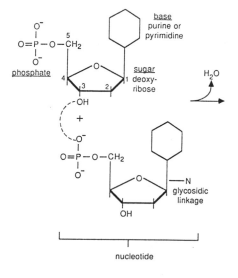

base
purine or
pyrimidine

phosphate

sugar
deoxy-
ribose

glycosidic
linkage

nucleotide

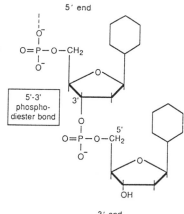

5′ end

5′-3′
phospho-
diester bond

3′ end

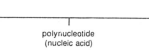

polynucleotide
(nucleic acid)

17

in earth's evolution. The next steps occurred when single units of a given type of molecule combined with each other to form larger units, called polymers. Thus, *lipid* polymers were built of repeated fatty acid molecules. *Carbohydrate* polymers, built of sugar molecules, are called saccharides; glycogen, for example, is a polysaccharide composed of repeated glucose molecules. Amino acids were linked by peptide bonds to form *peptides;* long peptides formed *proteins*. These basic steps in building polymers are illustrated in Fig. 2.1 (we will consider nucleotide polymers later).

Each of these main categories of molecule is suited for a particular range of functions. Within each category, different sizes and compositions of the molecules bestow specific properties. These different properties were the crucial basis for the evolution first of simple cells, and then of multicellular organisms. The different types of molecules in each of the categories are summarized in Table 2.1. With regard to the nervous system, this table conveys two important points. One is that, with the exception of fat storage and the immune response, cells of the nervous system (neurons and their satellite cells, glia) express most of the main types of molecules and functions found in other cells. This is particularly true of such basic cell processes as metabolism, second messenger systems,

and secretion. In addition, they express certain molecules and functions to a greater degree, or uniquely. These lie particularly in the categories of membranes, signaling molecules (neurotransmitters and hormones and their receptors), and recognition molecules that are crucial to the assembly of synaptic circuits during development. The function of a liver cell compared to a lymphocyte, for example, can be understood by the differences in the molecules they make. So, too, the function of a nerve cell can best be understood through its molecular composition. In recent years proteins have received most attention in this regard, for two reasons. First, we understand a considerable amount about how protein structure is encoded by nucleic acids (as discussed below). Second, the techniques of molecular biology provide powerful means to identify and study proteins and their synthesis. The general and special molecular properties of nerve cells are the subject of Chapter 3.

Nucleic Acids Encode Molecular Information

The evolution of lipid, sugar, and protein molecules provided for most of the functions of cells, including nerve cells, but a crucial function is still to be accounted for: how did the different molecules get assembled in a reliable and efficient man-

Table 2.1 Types of biomolecules and their functions[a]

	Lipids	Sugars	Amino acids	Nucleotides
Small molecules	fatty acids energy transfer (acetyl-CoA) hormones	energy (glucose)	metabolism neurotransmitters hormones	energy transfer (ATP) second messengers (cAMP)
Large molecules	membranes fat storage (triacylglycerols)	energy storage (glycogen) molecular recognition (glycolipids, glycoproteins)	cell structure enzymes receptors antibodies	coenzymes DNA and RNA

[a]Note the difference between the functions of small and large molecules (formed by polymerization of small molecules). All of these functions are found in nerve cells and/or their satellite cells (neuroglia), with the exception of fat storage.

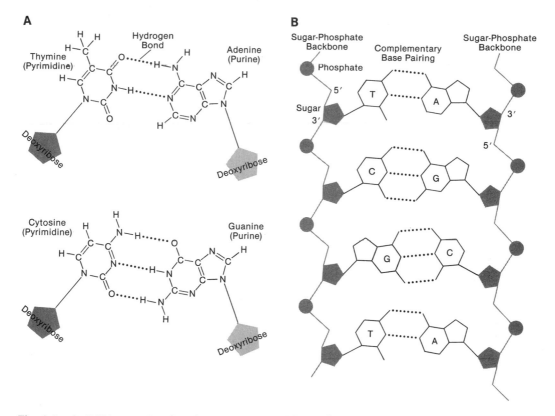

Fig. 2.2 **A.** DNA contains four bases, two pyrimidines (thymine and cytosine), and two purines (adenine and guanine). Through hydrogen bonds, adenine pairs with thymine and cytosine with guanine. **B.** Over the length of DNA, the four bases exhibit complementary pairing. The guanine–cytosine pairing has three hydrogen bonds and the adenine–thymine pairing has two hydrogen bonds. One consequence of this is that stretches of DNA that are rich in guanine and cytosine are more tightly annealed than stretches that are rich in thymine and adenine. (Modified from Watson et al., 1983)

ner? This was especially important for proteins, with their long chains of amino acids arranged in highly specific sequences.

This capacity for assembling polymers was in fact probably one of the first properties of molecules to arise. It was a property well suited to the fourth major category of molecule, the *nucleic acids*. As shown in Fig. 2.1D, nucleic acids are built up of units called nucleotides, consisting of a sugar, a nitrogenous base, and a phosphate group. There are four main nucleotides (see Fig. 2.2A): thymine, adenine, cytosine, and guanine. Two are purines (adenine and guanine) and two are pyrimidines (thymine and cytosine). The phosphate group is quite reactive, and two nucleotides can be readily

joined together by the formation of a bond between the phosphate group of one nucleotide and the sugar of the other. Thus, for the case of deoxyribonucleotide, an oxygen of the phosphate group attached to the number 5 carbon of one deoxyribose molecule is covalently bonded to the hydroxyl group of the number 3 carbon of the second deoxyribose molecule. As shown in Fig. 2.1D, this results in the formation of a phosphodiester bond linking the two. Polynucleotides, consisting of long chains of nucleotides, can thus be easily built up. A polynucleotide containing deoxyribose as the sugar is called *deoxyribonucleic acid* (DNA); one containing ribose as the sugar is called *ribonucleic acid* (RNA). Another difference between DNA and RNA is that in RNA the pyrimidine thymine is replaced by uracil.

In addition to the functions we have already noted, such as serving structural uses and providing reactive sites, nucleic acid polymers also have another inherent property, that of being able to serve as templates for the assembly of additional polymers (see Fig. 2.3). This is due to the ability of each nucleotide unit to form a weak bond with an identical or complementary nucleotide unit that is brought into register with it (thymine pairs with adenine, and cytosine with guanine; see Fig. 2.2). This unit can then form phosphodiester bonds with neighboring units also brought into register, so that a new chain is formed. The key is to have the weak bonds strong enough to bring the new units into register, but weak enough so that the new chain can be easily separated from the first as soon as it is assembled. In early evolution this process of assembly was probably quite slow, but eventually it came to be catalyzed by specific enzymes and could proceed quite rapidly.

The ability of nucleic acids to serve as templates in this fashion conferred upon them the ability to perform three functions which were crucial for the evolution of living organisms (see Fig. 2.4). The first was *replication*, the ability of the nucleic acid polymer to reproduce itself. A single

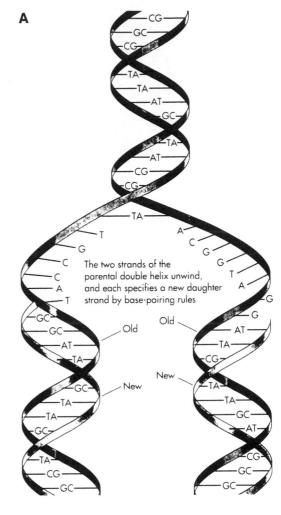

A

The two strands of the parental double helix unwind, and each specifies a new daughter strand by base-pairing rules

Old   Old

New   New

Fig. 2.3 1Complementary base pairing has two important consequences. **A.** DNA can replicate itself with high fidelity. When DNA strands separate (denature), two daughter strands are synthesized from each of the parent strands. Each daughter strand is an exact complement of the parent strand from which it is synthesized. This is called the semiconservative replication of DNA. **B.** The second consequence is that information can be transferred to RNA (translation) and subsequently into protein (transcription). RNA is synthesized from denatured regions of DNA. (Modified from Watson et al., 1983)

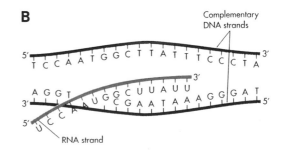

B

Complementary DNA strands

RNA strand

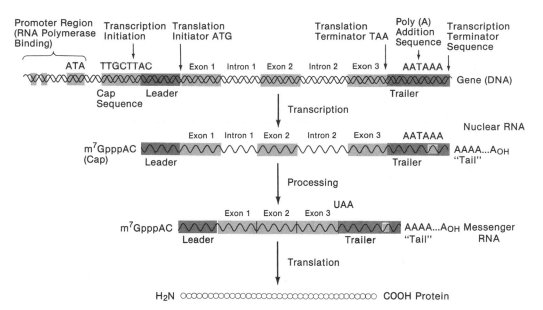

Fig. 2.4 Once synthesized from a DNA strand, RNA undergoes a series of steps in its maturation to become a competent messenger RNA (mRNA) molecule. RNA is first synthesized from DNA and contains sequences that do not encode protein. These are the upstream and downstream untranslated regions, and noncoding regions within the gene itself, called introns. Sequences of the gene that do ultimately encode protein are called exons. RNA processing splices out the introns from the mRNA. (Modified from Alberts et al., 1989)

strand of RNA or DNA could produce a double strand, and a double strand could reproduce itself as another double strand. This made it possible for a cell to divide into two daughter cells, each receiving exact copies of the original RNA or DNA. By this means, cells could increase their populations, and a species could perpetuate itself through successive generations. The second function was *transcription*. A single strand of DNA could serve as a template for assembling RNA, and this RNA could then serve as a template elsewhere in the cell. Thus there could be communication or transfer of the information throughout a cell. The third crucial function that could be served by this template was in the construction, not simply of more nucleic acids, but of proteins. This process, of converting information from nucleic acids to proteins, is termed *translation* (see Fig. 2.4).

DNA and RNA have particular properties which resulted in their being selected during evolution to perform these func-

tions. These properties were first recognized in the model of James Watson and Francis Crick in England in 1953. They showed that the DNA polymer takes the form of a double helix, in which the pentose sugars and phosphates constitute two intertwining backbones and the bases project toward the interior, where the bases of one strand are constrained by hydrogen bonds to be in perfect register with the bases of the other. Because of this perfect register, DNA can act as a reliable repository of the precise information needed to encode for protein synthesis. RNA, on the other hand, has become specialized for receiving the transcribed code and translating it into the proper sequences of amino acids to form proteins; as we shall see, several types of RNA cooperate in this task.

The unit of information for protein synthesis is a triplet of successive nucleotides in the RNA; each successive triplet constitutes a code "name," a codon which specifies a particular amino acid (see Fig. 2.4).

The RNA nucleotide triplet is encoded and stored in the DNA by the complementary nucleotide triplet. The chain of DNA nucleotides that specifies one polypeptide molecule, such as a structural protein or an enzyme, is called a gene; the total amount of DNA constitutes the genome of the organism.

The basic scheme, for replication of DNA, and for transfer of the genetic information from DNA to RNA for the manufacture of proteins, as outlined above, was worked out in only about a dozen years. By 1965, the genetic code had been broken, and the codons for all essential amino acids had been recognized. Synthesis of the amino acids into proteins was shown to require three types of RNA, each specialized for a different task: *messenger RNA* (mRNA), to read specific lengths of the DNA base code and bring it to the cytoplasm; *transfer RNA* (tRNA), to bring the individual amino acids to be connected; and *ribosomal RNA* (rRNA), to serve as the site of synthesis. We will have more to say about these steps in neurons in later chapters (Chaps. 3 and 8).

There is thus a flow of genetic information, from DNA to DNA for replication, and from DNA to RNA to proteins for protein synthesis. These simple paths for flow of genetic information within the cell became established as a "Central Dogma" of molecular biology. This is represented in Fig. 2.4.

The genome of the mammal is believed to consist of approximately 100,000 genes. The complexity of the brain is reflected in the fact that more of the genes are devoted to encoding proteins in the nervous system than in any other organ. The brain may express over 50,000 different mRNA species, whereas the kidney probably expresses only 10,000 or so. As many as 30,000 of the total number of genes may be specific for the brain. An important principle is that many of the proteins crucial for neuronal function are produced by multigene families. These proteins are related through common amino acid sequences, indicating their common evolutionary origins. They can be adapted to different functions through diversity of amino acid sequences generated at different levels of transcription and translation.

There has been debate about the number of gene products that are required for the assembly and function of the nervous system. Some investigators have suggested that, through combinatorial schemes, only a small number of gene products, perhaps fewer than 100, is required. Other investigators have suggested that the number is likely to be many tens of thousands. As in the immune system, the number of different gene products may be nearly as great as or greater than the coding capacity of the entire genome. In the nervous system several different mechanisms are utilized to expand the number of gene products generated from a single sequence of DNA. Among these are alternative splicing of precursor RNA (see Fig. 2.5), alternative polyadenylation sites, and differential glycosylation of protein products. In the immune system additional protein complexity is achieved through recombination of the DNA in precursor cells. Whether DNA recombination is also used in the nervous system has not yet been resolved.

Transcription Factors

Now that we understand how genes code for proteins and have had a preliminary glimpse of how differences in protein expression determine cell and tissue type and function, we are in a position to discuss one of the most important issues in biology: how gene expression is regulated. The principles underlying the development of the nervous system are dealt with in Chap. 9 and in many places throughout the text. Here we wish to orient ourselves to the basic mechanisms acting at the level of the gene that will apply generally to gene expression.

Research in this area has focused on regions of DNA outside the regions that code for proteins; in fact, this accounts

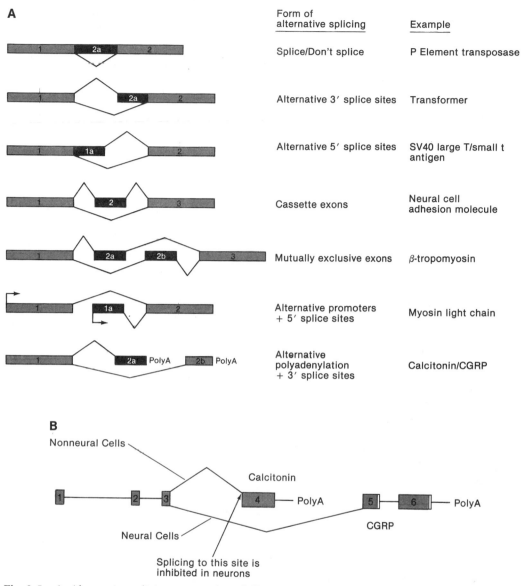

Fig. 2.5 A. Alternative splicing can produce different mRNAs from a single gene. Splicing produces mRNAs that contain (or lack) specific exons in different tissues. This diagram demonstrates different forms of alternative pre-mRNA splicing. **B.** In the nervous system, the gene that produces the mRNA for calcitonin undergoes alternative splicing to produce the mRNA for calcitonin gene-related peptide (CGRP), a neuromodulator. (Redrawn from Alberts et al., 1989)

for most of the DNA in the mammalian genome. Some of the DNA located near coding regions contains sequences that are important for the regulation of cell-specific DNA transcription. DNA sequences that do not themselves code for protein, but that regulate the transcription of mRNA from neighboring stretches of DNA, are called *regulatory elements*. Regulatory elements are often located just upstream (in the 5' direction) from a coding region. (The direction of transcription and translation is from the 5' carbon atom toward the 3' carbon atom of the sugar molecule contained within the nucleotide; see Fig. 2.1D and legend.) However, as will be described further, one class of regulatory elements, called enhancers, can be located in other noncoding regions.

Two different classes of regulatory elements exist: those that are required for the initiation of RNA transcription for virtually all genes, and those that regulate transcription in a cell- or tissue-specific manner. A sequence called the *TATA box* is found at a location about 25 base pairs upstream of the transcription start site of most eukaryotic genes (Fig. 2.6). The name TATA box is derived from the findings that the sequence itself is almost invariant for all genes: GNGTATA (A/T)A(A/T). When the same sequence is found in many different DNA segments and is believed to serve the same function wherever it appears, it is called a *consensus sequence*.

The TATA box is the site at which the protein complex that synthesizes mRNA binds to the DNA (see Fig. 2.6). Before this enzyme (RNA polymerase II) can bind to DNA and initiate synthesis of RNA, several other proteins (called TFIID, TFIIA, TFIIB, and TFIIE) associate with the TATA box. As illustrated in Fig. 2.6, these proteins bind sequentially to form the RNA-synthesizing apparatus that includes RNA polymerase II.

The second class of regulatory sequences is called *enhancer* or *promoter elements*. Unlike the TATA box, these sequences are found associated with only some genes and are responsible for tissue- or cell-specific gene regulation. Promoter elements are usually located within the first several hundred bases upstream of a gene, while the location of enhancer sequences is highly variable. Enhancer regions may be located as far as 10 kilobases(kb) upstream from the transcription start site. Enhancers have also been mapped to introns and to the region downstream of the coding region of a gene.

Like the TATA box, enhancer and promoter elements function in the regulation of DNA transcription by serving as binding

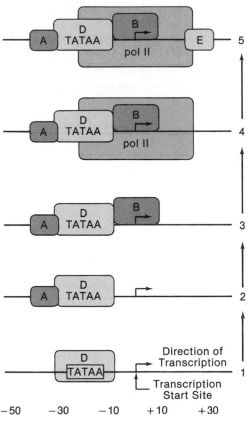

Fig. 2.6 RNA transcription from DNA requires the assembly of a protein complex on DNA. The complex is assembled in an orderly way, where a protein called TFIID binds first to the region of the TATA box, and then TFIIA, TFIIB, RNA polymerase, and TFIIE bind in turn. (Modified from Watson et al., 1983)

sites for specific proteins. The proteins that bind these regulatory sequences modulate the rate of transcription. These proteins are called *transcription factors,* and they can either increase or decrease the rate of transcription. Most transcription factors have two functional domains: one domain is responsible for binding to specific DNA sequences; the other domain is responsible for binding to other proteins (either other transcription factors or elements of the RNA polymerase II complex), and thereby regulating the transcriptional machinery.

Based on DNA sequence and protein structure, four families of transcription factors have been described (see Fig. 2.7). The *helix-turn-helix* (Fig. 2.7A) family includes the homeodomain proteins, which play a role in establishing the identity of neurons in the mammalian brainstem and spinal cord (as discussed later). As shown in Fig. 2.7, helix 3 binds DNA, whereas helices 1 and 2 bind other proteins.

The *zinc finger* family (Fig. 2.7B) of transcription factors is composed of proteins with tertiary fingerlike structures that re-

quire the association of Zn^{2+}. Krox-20, a transcription factor that is expressed in a defined region of the fetal mouse brainstem, is a zinc finger protein. Zinc finger transcription factors contain an alpha helix that binds DNA (in some cases DNA binding is dependent on an association with a Zn^{2+} ion) and domains that interact with other proteins.

The *leucine zipper* family (Fig. 2.7C) includes the transcription factors fos and jun,

Fig. 2.7 Tissue- or cell-type–specific transcription factors fall into five major families. These transcription factors bind to DNA sequences that are associated with specific genes and regulate tissue-specific expression. (A) The helix-turn-helix family includes the homeodomain proteins of *Drosophila* and mammals. Helix 3 binds to DNA whereas helices 1 and 2 contact other proteins. (B) The zinc finger family includes steroid receptors and Krox-20. The helix on the right of the right model is thought to bind DNA. (C) The immediate early genes, fos and jun, encode leucine zipper proteins. The leucine zipper region mediates dimerization between leucine zipper proteins. The positively charged, amino-terminal ends of these proteins are thought to be responsible for DNA recognition. (D) The helix-loop-helix proteins also have regions for dimer formation and other, positively charged, stretches for DNA binding. (Redrawn from Alberts et al., 1989)

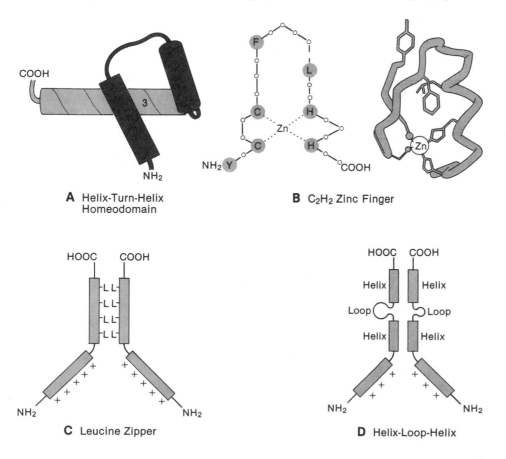

A Helix-Turn-Helix Homeodomain

B C_2H_2 Zinc Finger

C Leucine Zipper

D Helix-Loop-Helix

which have been shown to be upregulated rapidly when neurons are activated. The region of the leucine zipper is alpha helical, with the residue leucine appearing at regular intervals along one side of the helix. This array of leucines is then available to interact in a zipperlike fashion with a similar array of leucines present on a second transcription protein of the leucine zipper class. DNA binding occurs outside the region of the leucine zipper.

The *helix-loop-helix* family (Fig. 2.7D) includes cell fate–determining transcription factors like myoD. MyoD is normally expressed in muscle cells, but when it is introduced into a fibroblast it causes the initiation of muscle cell–like differentiation, including the expression of muscle-specific proteins such as myosin, and the fusion of the myoblast-like cells into functional myotubes. The DNA and protein binding domains of helix-loop-helix transcription factors are similarly organized to those of the leucine zipper family.

The *homeodomain*-containing transcription factors (see Fig. 2.8) are of particular interest because of their pattern of expression and apparent role in neural development. These proteins, which were first identified in *Drosophila,* where mutations cause the transposition of body parts, are believed to encode positional information. The expression of the homeodomain proteins in the rodent nervous system is restricted to well-defined regions. For example, the hindbrain of a fetal mouse is composed of a series of segments, called rhombomeres, each giving rise to a different part of the adult brainstem. Several members of the Hox-B (Hox-2) group of homeodomain proteins have very sharp anterior borders of expression at the boundaries of different rhombomeres (see Fig. 2.8); the expression of Hox-B1 (Hox-2.9) is restricted to rhombomere r4. The expression pattern of Hox-B genes and the identity of neurons in the rhombomeres can be altered experimentally, for example, by exposing a mouse fetus to abnormally high levels of retinoic acid. We will discuss homeodomain gene expression further in Chap. 9.

Molecular Technology for Studying Nerve Cells

In the early years of molecular biology, the main focus was on gaining an understanding of normal genetic mechanisms involved in the processes of replication and protein synthesis in viruses, bacteria, and animal cells. Very soon, however, the results began to reveal specific enzymes which not only provided explanations for these processes, but also put into scientists' hands the means to manipulate the genetic mechanisms themselves.

These methods have been crucial for the development of molecular neurobiology. Their use is now widespread in the analysis of neural systems, and we will study their applications throughout this book. It will therefore be important to review briefly the principles involved. The methods fall into two main groups: recombinant DNA technology and hybridoma technology.

Monoclonal Antibodies

Antibodies are proteins secreted by B lymphocytes in response to infection or invasion of the body by a foreign substance, called an *antigen*. Each antibody molecule recognizes only one specific site *(epitope)* on an antigen. Each B lymphocyte and its progeny make only one kind of antibody. As many as 10^7 or 10^8 different antibodies can be made by the different populations of B lymphocytes in a human, mouse, or rabbit. For example, when a rabbit is immunized with an antigen, the serum of that animal will contain a mixture of all the different B cell antibody responses, each antibody recognizing a single epitope on the antigen. Such a serum can be used to detect the antigen in a number of different ways. It can be used to recognize the antigen from a tissue homogenate run on a polyacrylamide gel and transferred to a nitrocellulose paper sheet (this is called *western blotting*). It can also be used to

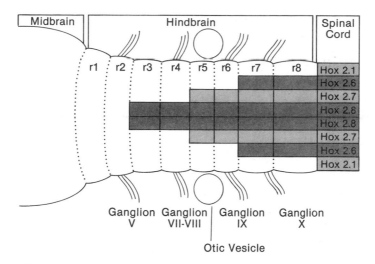

Fig. 2.8 Homeodomain proteins are expressed in segmentally restricted patterns in the embryonic mouse brain. Hox-2 genes are expressed in only some of the rhombomeres of the 9.5-day hindbrain. (From Gilbert, 1991)

detect the antigen in a section of a tissue (this is called *immunocytochemistry*). An antibody attached to a solid support (a column matrix, for example) can be used to purify its antigen (this is called *affinity chromatography*).

Serum antibodies have some practical limitations, due to the heterogeneity in their composition. For example, to obtain a serum antibody with high specificity for a particular antigen, the immunizing antigen should be highly purified. If an antigen is contaminated with other proteins, antibodies will be generated not only to the antigen desired but also to the other proteins present in the immunizing preparation. Furthermore, one cannot easily use a serum antibody to investigate the composition of a mixture of antigens—for example, to look for proteins expressed by only some neurons in one area of the brain.

In theory, if an individual B lymphocyte could be grown in culture, it would be a source of homogeneous antibody molecules, all with the same structure and specificity. However, B lymphocytes do not survive in culture. Methods to propagate antibody-secreting cells in culture had their origins around 1960 in the finding that

certain viruses have the ability to induce cells in culture to fuse together. This produces hybrid cells which contain chromosomes from both parents and can undergo normal mitosis. It was then found that if one of the parents is from a myeloma (a type of lymphocyte tumor), the hybrid cells will retain the ability of the myeloma parent to continue multiplying indefinitely. Hybrid cells with this capacity are termed *hybridomas*. After fusion, the preparation can be exposed to selective media in which the parent cells die. The surviving cells consist of clones of the original fused cells.

By the late 1960s, it was shown that parent cells from different species (for example, human cells and mouse myeloma cells) could be fused. In addition, methods were developed for identification of individual chromosomes. Thus it became possible to construct gene maps, in which specific genes introduced from a parent line could be identified. Of great interest was the possibility that immunoglobulin-secreting hybridomas could be made, and by the early 1970s this had been achieved.

Hybridomas prepared in this way were still heterogeneous in their production of antibodies. The last step was therefore to obtain a homogeneous antibody population. In 1975, George Kohler and Cesar Milstein in Cambridge showed that antibodies of a particular specificity could be obtained by screening the supernatants of individual hybridoma cells with the antigen in question, and isolating and propagating the positive cells. Those cells constituted single clonal lines secreting one type of antibody, which therefore were called *monoclonal antibodies* (see Fig. 2.9A).

These methods together constitute *hybridoma technology*. Like recombinant DNA technology, its power resides in its

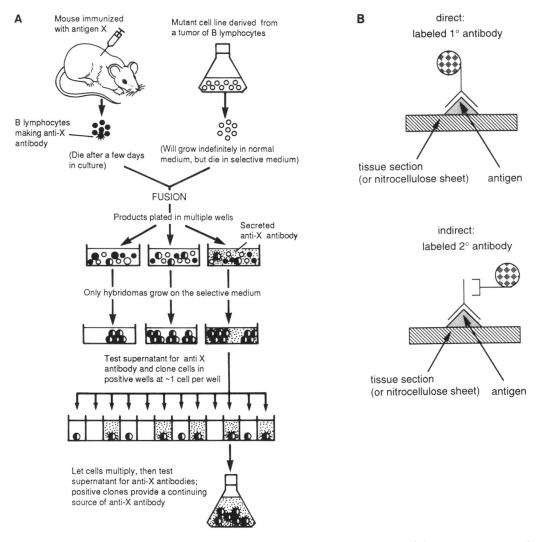

Fig. 2.9 **A.** Preparation of monoclonal antibodies. This shows a summary of the main steps: initial production of polyclonal antibodies; fusion with tumor lymphocytes to produce hybridomas; screening and isolation of specific hybridoma cells; cloning of specific cells to obtain monoclonal antibodies. This method is widely applied in studies of specific molecules in the nervous system. (From Watson et al., 1983) **B.** Once an antibody is available, it can be used to visualize the location of the antigen in a tissue section or on a western blot of an SDS-PAGE gel. If the antibody is labeled (for example, with horseradish peroxidase: HRP), it can be detected directly. Alternatively, an indirect detection method can be employed where an unlabeled antibody to the antigen can be detected with a labeled antibody that is directed to the constant region of the antibody molecule.

ability to make precise identifications of individual genes and gene products. It was quickly realized that monoclonal antibodies (mAbs) are a particularly powerful tool in the investigation of the nervous system. In some of the earliest work, Susan Hockfield and her colleagues (1983) at Cold Spring Harbor showed that mAbs not only can be used to identify specific neuronal types in the brain, but also can provide the means for analyzing the differential expression of specific gene products during development and in relation to different activity states.

Recombinant DNA Technology

The ability to isolate and manipulate individual genes, or parts of a gene, through the techniques of molecular biology has revolutionized the study of biology in general and neurobiology in particular. Central to the analysis of complex tissues like the nervous system are techniques that permit the isolation of one gene (that encodes one protein) from a mixture of other genes.

Cloning the gene for a protein can be accomplished using a variety of different strategies. All require two basic components: (1) information about the structure (amino acid sequence) of a protein and (2) a library of complementary DNAs (cDNAs) that are copies of mRNAs from a tissue that synthesizes the protein.

Information about the structure of a protein can be obtained in several ways. The most direct way is to obtain amino acid sequence data from a purified preparation of the protein. Microsequencing techniques now permit amino acid analysis to be performed on as little as 50 picomoles of protein. An alternative approach to information about a protein's structure is to identify another protein whose function, and presumably structure, is closely related to the protein of interest. This approach has been used for proteins for which purification of even 50 picomoles is too difficult to be practical, a situation encountered for many neural proteins. This approach is also used to identify the genes encoding proteins that are related to a previously identified protein. For example, once the gene for one type of glutamate receptor was cloned, its sequence was used to search for other types (see Chaps. 7, 24).

Amino acid sequence information can be used to infer nucleic acid sequence information because, as mentioned earlier, each amino acid is encoded by a particular nucleotide *codon* (three nucleotides per codon). Oligonucleotide sequence information can then be used to identify the gene encoding the protein. This is accomplished by using the oligonucleotide sequence to search for complementary stretches of DNA in a mixture of DNAs. This is called "screening a library."

DNA libraries are constructed from mRNA obtained from a tissue that is known to produce the protein of interest (see Fig. 2.10). For example, the genes for the nicotinic acetylcholine receptor were first isolated from a library constructed from *Torpedo* electric organ, a tissue in which the receptor protein is especially abundant (see Chap. 7). Messenger RNA is collected and DNA copies (cDNA) of the mRNA are synthesized. These cDNAs are then incorporated into plasmid or bacteriophage vectors. The collection of all the vectors obtained from a single source of mRNA is called a *library*. Vectors can be amplified to many thousands of copies each by growing them in bacteria, their host organisms. Each vector will contain only a single cDNA and the transformation of bacteria is carried out so that each bacterium receives only a single vector. Thus, if each vector-carrying bacterium is grown individually, it provides a virtually unlimited source of an individual cDNA derived from an individual mRNA molecule.

A library can be screened for a specific oligonucleotide sequence encoding a particular protein in a number of different ways. The general strategy is to use oligonucleotide sequences to find complementary cDNAs. The oligonucleotide sequence of interest can be synthesized and then radioactively labeled (so that its presence or

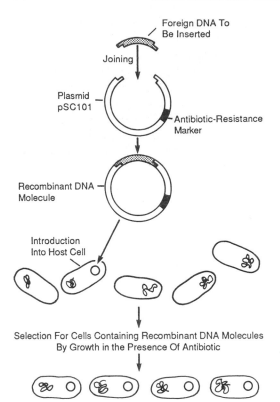

Fig. 2.10 Recombinant DNA procedures. DNA can be cloned, using a plasmid as a vector. (From Watson et al., 1983)

absence can be monitored). Bacteria carrying the library are plated onto dishes and then immobilized on specially prepared membranes. The membranes are then exposed to the labeled oligonucleotide. The oligonucleotide will bind, through complementary base pairing, only to those bacteria that contain complementary cDNA sequences. Those bacteria are then isolated and expanded, and the cDNA is isolated.

An exceptionally powerful strategy for screening libraries (or any source of nucleic acids) for particular sequences uses the *polymerase chain reaction* (PCR) (Fig. 2.11). PCR is a very sensitive technique. As can be appreciated from the preceding discussion, a critical step in isolating a DNA sequence is creating many thousands of copies of the sequence. In PCR, a sequence of DNA is subjected to multiple

rounds of DNA synthesis, denaturation (separation of the two DNA strands from one another), and synthesis. The specific sequence that is amplified by PCR is determined by oligonucleotide sequences used to prime the reaction, called *primers*. Primers are constructed to match, as nearly as possible, sequences that will be contained within the gene of interest. Their sequence can be deduced from available amino acid sequence information of the protein of interest, or from the nucleic acid sequence from a closely related gene. Once the DNA segment of interest has been amplified by PCR, it can be incorporated into a vector and propagated in bacteria.

Another screening strategy uses *expression cloning,* in which the cDNA is engineered to express the protein it encodes. Several vectors are available that contain regions that can direct the synthesis of mRNA from the inserted fragment of DNA. The mRNA is then translated into protein by the bacterial host. The bacteria are then subsequently screened for the presence of a protein of interest. The most commonly used method for identifying an expressed protein is with an antibody. In a manner similar to that described above for screening a library with oligonucleotide probes, bacteria expressing the protein are transferred onto a membrane and the membrane is probed with an antibody. The techniques of western blotting (described earlier for antibodies) are used to visualize bacteria that are synthesizing the protein of interest. This approach can be extended to probing a library with a ligand (rather than an antibody) for the protein of interest. So, for example, a cDNA for an opiate receptor has been identified using a specific ligand for the receptor (see Chap. 8). Many clever variations on the expression screening theme have been used to identify cDNAs encoding neural proteins.

How can one know that a cDNA obtained by screening a library or by PCR represents the gene of interest? The first step in determining whether a cDNA is "correct" is usually to obtain the complete sequence of the cDNA. The sequence of the cDNA should encode an amino acid sequence that corresponds to that of the protein. A second step is to determine whether the temporal or spatial patterns of expression of mRNA recognized by the cDNA match those of the protein. This can be determined by a *northern* blot analysis, that is, using the cDNA to identify RNA prepared from different tissues or different stages in development. It can also be determined using in situ hybridization, where the cDNA is used to visualize the location of mRNA in tissue sections. A third, and perhaps the most definitive, step is to express the cDNA as a protein in vitro and then to generate an antibody to this in vitro protein product. If the cDNA encodes the protein of interest, antibodies to the protein expressed in vitro should recognize the protein of interest.

Often a cDNA does not contain the entire coding sequence of the gene of interest. Once the veracity of a cDNA has been confirmed, that cDNA can then be used to rescreen a library to identify additional clones that may contain the full sequence or that may contain additional segments of the coding region of the gene.

Functional Analysis of Expressed Proteins

How do we assess the function of a protein? There are many ways to do this indirectly; for example, by testing biochemically for the product of an enzymatic reaction or pharmacologically for the action of an antagonist on the binding of a ligand to a receptor. However, the most direct information is gained by recording directly from the protein if it is an ionic

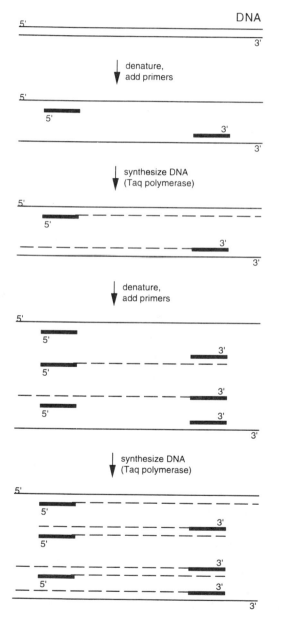

Fig. 2.11 The polymerase chain reaction (PCR) permits the detection of specific sequences of DNA by selective amplification. Primers corresponding to 5′ and 3′ regions of the gene of interest are synthesized. Target DNA (containing the gene of interest) is denatured (by bringing it to high temperature) and then exposed to the primers. When the temperature is lowered, the primers bind to specific sites on the DNA. The addition of DNA polymerase permits the synthesis of new DNA strands from the parent strand. Further cycles of denaturation, primer binding, and DNA synthesis result in the amplification of the sequence of interest. The breakthrough that made PCR possible was the discovery of a DNA polymerase (Taq polymerase) that remains active when heated over 90°C.

channel. The feat of recording from a single molecule seemed beyond the scope of classical experimental methods, but it has in fact become a routine electrophysiological procedure with the advent of the patch-recording technique. Since this method is a basic tool in the analysis of the molecular basis of neuronal function, it will be described briefly here.

Beginning in the late 1960s, scientists found that they could measure the electrical current through single channel-forming proteins that had been inserted into artificial membrane bilayers. This approach was extended to living cells by the method called patch clamping. This method was developed in Germany in the late 1970s by Erwin Neher and Bert Sakmann, who received the Nobel Prize for it in 1991 (see Sakmann and Neher, 1983). In this method, a fine glass tube is pulled out to form a micropipette with a tip diameter of 2–3 μm, and filled with an electrically conducting salt solution. If care is taken to keep the tip clean, and the surface of a cell, such as a muscle cell, is also clean, the tip can be gently placed on the cell surface and it will seal very tightly to the membrane. When this happens, the resistance from the inside of the pipette to ground increases dramatically, from the range of thousands of ohms (kilohms) to billions of ohms (gigohms = 10^9 ohms). The membrane encircled by the tip is called a "patch;" analysis can be carried out with the electrical potential across the membrane clamped at different values in order to obtain the currents that flow through the patch, hence the term *patch-clamp* recording.

Patch recordings can be carried out with different relations between the tip and the cell membrane. Several of the basic configurations are indicated in the diagram of Fig. 2.12. The *whole-cell recording* (C) is similar to the traditional mode of intracellular recording. The *on-cell* (cell-attached) *patch* (A) is the simplest and most direct mode of single-channel analysis. This patch can be removed directly to give an *inside-out patch* (B). Alternatively, the electrode can be gently withdrawn from a whole-cell

configuration, pulling the membrane into an *outside-out patch* (D). Each of these configurations has advantages and disadvantages for analyzing particular properties of channel proteins, as summarized in Table 2.2. The student may refer to these configurations when studying results discussed in later chapters.

New configurations have been developed to supplement the classical ones. An example is the *perforated patch* (see C', Table 2.2). This is made by including in the micropipette solution an agent (such as the antibiotic nystatin) that partially dissolves the membrane. This converts an on-cell patch to a whole-cell patch, with the advantage that the remaining membrane is able to prevent the micropipette solution from entering the cell while still allowing electrical continuity equivalent to a whole-cell recording (Marty and Horn, 1985).

The information that one can obtain about the functional properties of a channel protein is in the form of electrical recordings. Let us take as an example the protein which serves as a combined receptor and ion channel for the neurotransmitter acetylcholine (ACh), which we will discuss in detail when we consider the neuromuscular junction as a model synapse in Chap. 7. As shown in Fig. 2.13, the subunits of this protein are arranged in a circle around a central channel. Here we wish to emphasize those aspects of the recordings that give insight into some of the basic properties of this macromolecule in allowing charged atoms (ions) to pass through the channel. As illustrated in Fig. 2.14, the properties may be summarized as follows. The channel opens abruptly (activates) ①, has a constant open conductance ②, and closes abruptly (inactivates) ③. These are general properties of all channel proteins thus far studied, and suggest that the protein jumps from one conformation to another. There are different open times ④; the shorter open times may be due to binding of a single transmitter molecule, whereas the longer open times may be due to binding of two or more molecules. The open states are often interrupted by brief

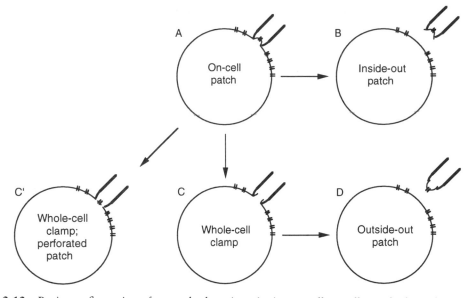

Fig. 2.12 Basic configurations for patch clamping. **A.** An on-cell or cell-attached patch is studied with the cell intact. The short double lines indicate single ionic channels in the membrane. **B.** By withdrawing the patch pipette, one can isolate the patch of membrane under the pipette; this is termed an "inside-out patch" because the inside of the membrane faces the bath. **C.** If additional suction is applied to the cell-attached patch to rupture the membrane under the pipette, the whole-cell membrane potential may be controlled through the pipette, enabling study of the cell with a whole-cell clamp. **C'.** Inclusion of a membrane-permeabilizing agent in the pipette establishes electrical continuity with the cell interior while preventing interchange of fluids between pipette and cell. **D.** From the whole-cell patch, the pipette can be slowly withdrawn to form a tube of membrane, which pinches off to create an outside-out patch. (Modified from Corey, 1983)

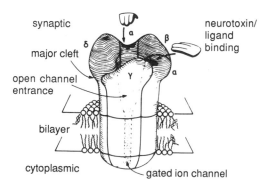

Fig. 2.13 Three-dimensional structure of the protein that serves as the combined acetylcholine receptor–ionic channel. Note that the protein is arranged around a central channel, through which ions flow when the protein is activated by acetylcholine. See text and Chap. 7. (Based on Kistler et al., 1982)

closings ⑤; the molecule thus appears to "flicker" between the open and closed states. This may occur several times while the transmitter molecule unbinds slowly from its binding site. The channel can flicker between the open state and other partially closed states ⑥, indicating that there may be substates between the two. Finally, the noise level during the open state is often higher than at rest ⑦. This channel noise has implied that the open channel "quivers" as a result of the motion of water or neighboring lipid molecules in the membrane.

These remarkable recordings thus give us an immediate sense of looking directly at the action of a single molecule; in the words of the poet Wordsworth, it is as if "we see into the life of things." The protein undergoes abrupt changes in its molecular conformation, which cause abrupt changes

Table 2.2 Advantages and limitations of four configurations for patch clamping

Method	Advantages	Disadvantages
A. On-cell patch (single channels)	no perfusion needed does not disturb cytoplasm modulatory systems intact	cannot change intracellular medium must measure membrane potential with another electrode
B. Excised inside-out patch (single channels)	access to both sides of membrane can change concentration of intracellular ions or regulatory substances, or apply enzymes to inner surface of membrane	cannot change outside medium during experiment must have low-Ca^{2+}-concentration bath perfusion to prevent vesiculation
C. Whole-cell clamp (average current through many channels)	can change internal medium to isolate currents Low access resistance allows single-pipette voltage clamp with good speed (<0.5 msec)	leads to exchange of internal medium hard to clamp cells much bigger than 30–40 μm in diameter
C'. Whole-cell clamp ("perforated patch")	Whole-cell recording without exchanging internal medium	technically more difficult
D. Excised outside-out patch (single channels)	access to both sides of membrane no bath perfusion needed can change concentration of extracellular substances such as neurotransmitters	cannot change inside medium during experiment much have low Ca^{2+} in pipette to prevent vesiculation

Modified from Corey (1983)

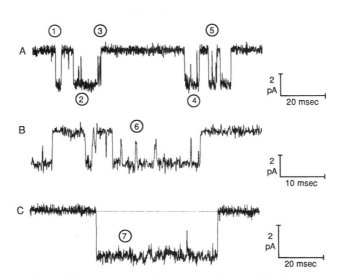

Fig. 2.14 Patch recordings of ionic current through single acetylcholine receptor channels. The main characteristics of the channel activity are labeled ① through ⑦, and are described in the text. **A** and **B** are from a cell-attached patch on a myoball preparation from rat myotubes (embryonic muscle cells); **C** is from an outside-out patch. (From F. J. Sigworth, in Corey, 1983)

in the channel conductance states. This seems to be a fundamental property of channel proteins, whether they are neurotransmitter receptors activated by different ligands or excitable channels activated by changes in membrane voltage, and regardless of what kinds of ions are controlled. Further studies may be pursued, using methods such as site-directed mutagenesis, to alter different parts of the polypeptide chains, to understand the contribution of each part to binding selectivity, ionic selectivity, and channel kinetics. We will discuss examples of these types of analysis in Chap. 7.

3

Neurons and Glia

**The Cellular Basis of Neurobiology:
A Brief History**

Foundations of the Neuron Doctrine

Until about 100 years ago, it was not known whether the cell theory applied to the nervous system. As early as 1836, Jan Purkinje, the great Czech physiologist, had published observations of cells (which later would bear his name) in the cerebellum, but, as can be seen in Fig. 3.1A, these showed little more than the nucleus and surrounding cytoplasm. An important advance was made in 1865, when the observations of Otto Deiters of Bonn were published. In his diagram of a large motor neuron of the spinal cord (Fig. 3.1B), he distinguished between two kinds of fiber arising from the cell body. One kind consisted of a number of branches which appeared to be extensions of the cell body, and which he termed "protoplasmic prolongations," *protoplasm* being the traditional term for the living substance of the cell. The other kind consisted of a single, unbranched, tubular process, or "axis cylinder," which arose from a small, conical mound on the cell body, and in turn became the fiber which left the spinal cord and entered the peripheral nerve that supplied the muscles. The protoplasmic pro-

longations came eventually to be called "dendrites," a term borrowed from botany, meaning, simply, branches. The axis cylinder came to be called "axon." You should be able to distinguish axon from dendrites in Fig. 3.1B.

Despite these advances, a single nerve cell had not yet been seen in its entirety. One could therefore only speculate as to how nerve cells were organized. Many believed that when an axon split up into fine branches within the brain, those branches became continuous with the finest branches of dendrites of other cells, in much the same way that the smallest arteries and veins in the body communicate through capillaries. This became known as the "reticular theory" of nervous organization, in opposition to the cell theory, in which each nerve cell was conceived of as a separate entity whose branches terminate in "free nerve endings."

It seemed almost impossible to resolve this issue, because even if a method could be found that made it possible to stain the finest branches, they would be obscured by the thousands of other branches around them. What was needed was a method that would stain only a small percentage of the cells, but stain them in their entirety. And that is exactly what happened! In 1873,

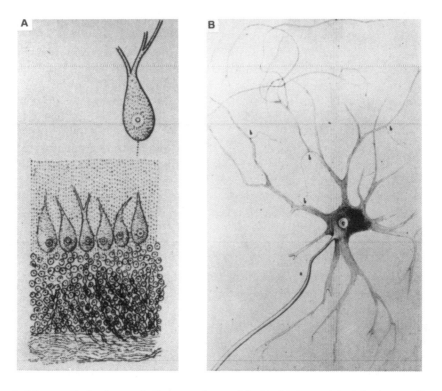

Fig. 3.1 A. Nerve cells in the cerebellum, as observed by Purkinje in 1837. The large cells are now known as Purkinje cells; the small cells packed tightly below are granule cells. The cerebellum is discussed in Chap. 21. **B.** A large motoneuron in the spinal cord, as observed by Deiters in 1865. Note the single long, smooth axon, which is distinctly different in appearance from the branching dendrites. The spinal cord is discussed in Chap. 20. (From Liddell, 1960)

an impoverished doctor, Camillo Golgi of Pavia, was carrying out experiments by candlelight in his kitchen, trying to find a better way to visualize nerve cells. Among the many different methods he tried was a combination of potassium dichromate fixation and silver impregnation. To his astonishment, in nervous tissue this method revealed, here and there, a few cells with their cell bodies and dendrites stained completely black, out to the finest terminal branches. Golgi applied his method to a number of different parts of the nervous system, and published his collected results in 1885, in a comprehensive work in Italian. At first it aroused little interest among anatomists, and the full implications of the results were not realized until a Spanish histologist, Santiago Ramón y Cajal, working in a small laboratory in Barcelona,

stumbled on the method in 1888. The effect of this new vision of the nervous system is best described in his own words (as translated by Sherrington, 1935):

> Against a clear background stood black threadlets, some slender and smooth, some thick and thorny, in a pattern punctuated by small dense spots, stellate or fusiform. All was sharp as a sketch with Chinese ink on transparent Japan-paper. And to think that that was the same tissue which when stained with carmine or logwood left the eye in a tangled thicket where sight may stare and grope for ever fruitlessly, baffled in its effort to unravel confusion and lost for ever in a twilit doubt. Here, on the contrary, all was clear and plain as a diagram. A look was enough. Dumbfounded, I could not take my eye from the microscope.

Cajal worked feverishly, developing the Golgi method and applying it to many parts of the nervous system in many animal species. Figure 3.2 shows some cells in the cerebral cortex. Cajal had the genius to realize that the entity stained by the method was, in fact, the entire nerve cell, and that this procedure provided the long-sought proof that each nerve cell is an entity, separate from the others. He also deduced the basic principles that nervous signals pass through the dendrites as well as the axon of a cell, and that transmission between cells takes place where their axons and dendrites contact each other.

Cajal's outpouring of publications between 1888 and 1891 attracted a number of other anatomists, and most of them agreed with his interpretations. These ideas also fit with conclusions that had been reached from studies of the embryological development of nerve cells by Wilhelm His, of Leipzig, in 1886, and of the way that nerve cells respond individually to injury by August Forel, of Zurich, in 1887. It remained only for someone to assemble all the evidence in a convincing way, and this was done by Wilhelm Waldeyer, a distinguished professor of anatomy and pathology in Berlin, in 1891. Waldeyer's extensive review in a German medical journal finally made a convincing case that, after a 50-year delay, the cell theory applied to the nervous system, too. Waldeyer suggested the term "neuron" for the nerve cell, and the cell theory as applied to the nervous system became known as the "neuron doctrine."

The original papers of these pioneers in neurobiology have been translated, and they are fascinating reading for anyone interested in how we arrived at our modern concept of the neuron as a basic building block of the nervous system (Cajal, 1990; DeFelipe and Jones, 1988; Shepherd, 1991a). Cajal, for his part, never quite forgave Waldeyer for being credited with the doctrine he considered his own. Ironically, Golgi himself never accepted the individuality of the nerve cell, clinging bitterly to the reticular theory, even on the occasion

of his Nobel lecture when he and Cajal shared the award in 1906.

Although the neuron doctrine became widely accepted, final proof required a method that could demonstrate that nerve cell membranes remain everywhere distinct from each other. This is beyond the power of resolution of the light microscope, and the question awaited the advent of the electron microscope. This instrument was first applied to physical materials in the 1940s, and to biological tissues around 1950. Its application to the nervous system was delayed by the same problems of fixing and staining the tissue that we have already noted. However, by the mid-1950s, the investigations of David Robertson in London, Eduardo de Robertis in Buenos Aires, and Sanford Palay and George Palade in New York showed that the nerve cell membrane resembles the basic "unit membrane" of other cells, and that it appears to be continuous around each nerve cell (see Fig. 3.3). This supported the neuron theory in its proposition that each nerve cell is a genetic and anatomical unit like other cells of the body, and its corollary that nervous tissue consists of populations of these units organized into functional systems (cf. Peters et al., 1990; Shepherd, 1991a).

Neuroglia: Essential Components of the Nervous System

Soon after the beginnings of microscopical investigations of cells in the middle of the nineteenth century it was recognized that the nervous system also contained elements which differed from those of nerve cells. This appeared as a kind of amorphous ground substance between nerve cell bodies. In 1860, Rudolph Virchow, a famous German microscopist, suggested the nature of this substance as follows: "These interstitial substances form in the spinal cord and the higher sensory nerves a kind of glue (neuroglia), in which the nervous elements are embedded."

This might imply that neuroglia were not

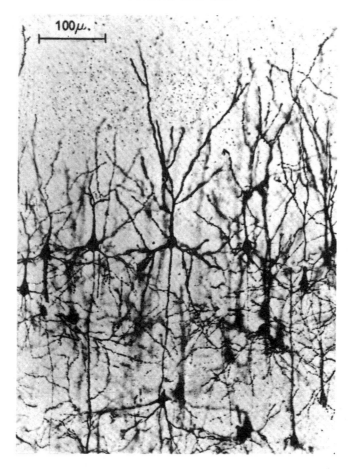

Fig. 3.2 Neurons in the visual cortex of the cat impregnated by the Golgi method. Can you identify the cell bodies, apical dendrites, basal dendrites, and axons of these neurons? (From Sholl, 1956)

cells but only gluelike secreted substances, but Virchow, who is generally regarded as the founder of cellular pathology, further suggested that the neuroglia are cells, some of which have phagocytic actions. From that time on, studies of neuroglial cells, or simply *glia,* developed in parallel with studies of nerve cells. Cajal identified several distinct glial cell types, which were fully characterized by his students in the course of the 1920s. With the advent of the electron microscope, the different types have been further differentiated by their fine structural properties. Modern studies have shown that glial cells have a wide range of functions mediated through inter-

actions among themselves and with nerve cells, which we will discuss later in this chapter and elsewhere in this book.

Cell Biology of the Neuron

Today, the modern study of the cellular structure of neurons and glia is called *cell biology,* which in turn is based on fundamental studies of the molecular biology of the cell (see Alberts et al., 1989). Essential to these studies is the electron microscope together with the biochemical, molecular biological, and electrophysiological techniques described in the preceding chapter. As mentioned previously, the ultimate aim is to understand the cellular functions of

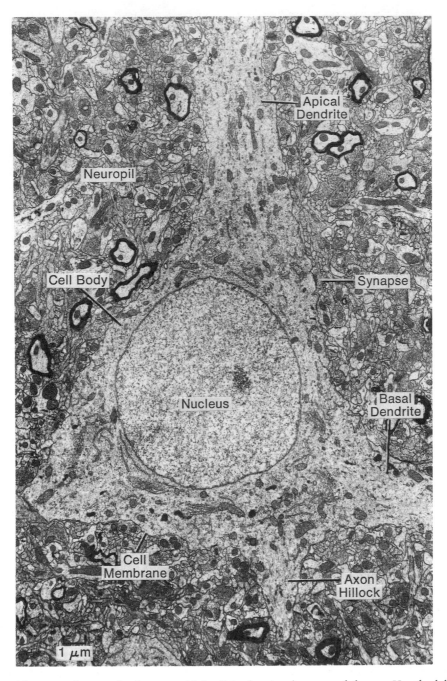

Fig. 3.3 Electron micrograph of a pyramidal cell in the visual cortex of the rat. Use the labels on the micrograph to check your identification of the parts of the neuron as seen with the Golgi stain in Fig. 3.2. Magnification, approximately ×5000. (Courtesy of Steven Hersch and Alan Peters)

gene products. The general theme that has emerged is that the fine structural elements, or organelles, reflect a beautifully orchestrated division of labor within the cell, enabling it to carry out the different functions necessary to sustain its life by means of specific interactions with neighboring cells.

Studies of cellular organelles have traditionally been done in organs other than the brain, such as the liver or pancreas, and one has had to infer their applicability to nerve cells. With the advent of molecular neurobiology, this situation has begun to change, and, as with the study of genes (Chap. 2), it is increasingly possible to obtain information directly from nerve cells themselves.

The neuron is the most fundamental anatomical building block of the nervous system, and the way its functions arise out of the properties of its organelles is essential for grasping the principles of neurobiology. Figure 3.4 is a summary view of the main organelles and their distribution within the different parts of the neuron. We will briefly describe each organelle. It will be seen that the divisions of labor resemble those in other cells of the body, but there are specializations that are adapted for the unique functions of nerve cells. The specializations relating to the synapse will be considered in Chap. 6.

The Plasma Membrane

The nerve cell, like all other cells of the body, is bounded by a plasma membrane. In cross sections of electron micrographs at low magnification (for example, as in Fig. 3.3), the membrane appears as a single dark line. At higher magnification it can be seen that there are, in fact, two dark lines with a light space between them. This gives the membrane a two-layered structure, with inner and outer leaflets.

In the early 1970s, it was proposed that the two layers are two sheets of oppositely oriented phospholipid molecules; the hydrophilic phosphorylated heads lie at the exterior surface of the membrane, and the hydrophobic lipid tails lie at the water-free interior of the membrane. It is an inherent property of phospholipid molecules to form such a membrane (see Chap. 2). The hydrocarbon tails within the membrane do not allow ions and charged molecules to enter the membrane. This prevents the passage of dissociated salts, water molecules, and water-soluble molecules contained within the cell cytoplasm. The lipids of the membrane thus form a natural barrier to retain the cytoplasm. However, the cell has to pay for this by needing special mechanisms for transporting across the membrane the ions and water-soluble molecules essential for metabolism. This is provided by proteins embedded in the membrane.

The lipids form a fluid matrix within which the protein molecules are embedded. This is known as the *fluid mosaic membrane model*. The membrane has been shown to be quite fluid; in some cells the lipids can move laterally at rates of $2\mu m/$sec, whereas protein molecules move about 40 times more slowly (50 nm/sec, or 3 $\mu m/$min). This fluid property of the membrane is of practical use because a nerve cell can be injected with a lipid-soluble dye, such as DiI, which enters the lipid bilayer and can spread throughout the extent of the cell, so that the cell is stained in its entirety much like a Golgi-stained neuron. Moreover, this technique will work even in fixed cells.

Membrane Lipids

The *lipid* moiety imparts an electrical capacitance to the membrane, which stores the charge underlying the membrane potential (Chap. 4). In addition, it has other properties, as indicated in the diagram of Fig. 3.5. The phospholipid bilayer is not, in fact, symmetrical; in the outer layer the head groups contain mostly choline, whereas in the inner layer they contain mostly amino acids. These head groups may interact with the membrane proteins in ways that are essential for the proper functioning of the proteins. Among the

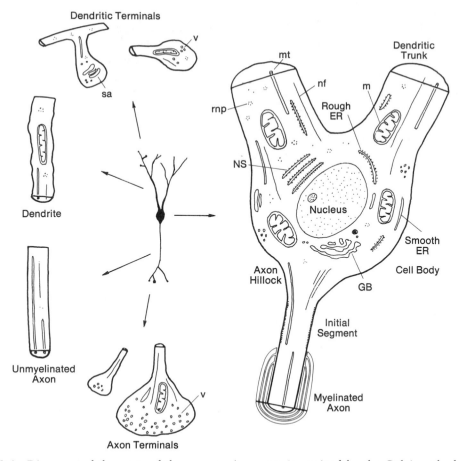

Fig. 3.4 Diagrams of the parts of the neuron. A neuron (as stained by the Golgi method or by intracellular injection of dyes), shown at center, is surrounded by schematic drawings of fine structure (as viewed in the electron microscope) of the different parts. ER, endoplasmic reticulum; GB, Golgi body; m, mitochondria; mt, microtubule; nf, neurofilament; NS, Nissl substance; rnp, ribonuclear protein particles; sa, spine apparatus; v, vesicles. Compare the parts of the neuron with the stained cells in Fig. 3.2, and the fine structure with Fig. 3.3. (From Shepherd, 1979)

phospholipids of the inner layer is phosphatidylinositol, composed of two fatty acid chains linked through glycerol to a phosphorylated six-carbon ring molecule called inositol. Phosphatidylinositol is the source of an important intracellular second-messenger system, as we shall see in Chap. 8.

The outer membrane layer contains fatty acids giving rise to head groups of short chains of sugar molecules. These are called *glycolipids*. The most complex glycolipids are gangliosides, whose head groups contain a complex sugar called sialic acid. They are present in high amounts in neu-rons. These short-chain sugars (called oligosaccharides) protrude into the extracellular space and are believed to function in cell–cell recognition.

Membrane Proteins

The *protein* moiety is crucial to many functions of the cell. These functions depend on the parts of the proteins that lie outside, across, and inside the membrane. Let us consider each briefly.

Surface Proteins. As we saw in Fig. 3.5, the outside surface of the membrane fairly bristles with long molecules arising from,

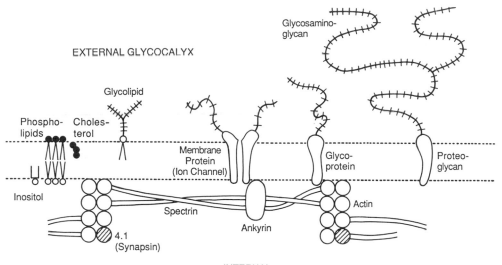

EXTERNAL GLYCOCALYX

Glycosamino-
glycan

Glycolipid

Phospho-
lipids

Choles-
terol

Membrane
Protein
(Ion Channel)

Glyco-
protein

Proteo-
glycan

Inositol

Spectrin

Actin

Ankyrin

4.1
(Synapsin)

INTERNAL
MEMBRANE CYTOSKELETON

Fig. 3.5 Diagram of the structure of the plasma membrane, showing the types of molecules that are characteristic of the membrane and its associated intracellular and extracellular domains. (Adapted from Alberts et al., 1989)

or attached to, membrane proteins. One important group is the *glycoproteins,* which, as its name indicates, has both protein and carbohydrate components (the basic molecular construction of these components should be reviewed in Chap. 2, Fig. 2.1).

This group is divided into at least three main subgroups. The largest to date is the group that belongs to the *immunoglobulin* superfamily. As shown in Fig. 3.6, these contain repeating domains of Ig folds, formed by disulfide ("s") links; most family members also contain fibronectin repeats and have numerous glycosylation sites, which give rise to carbohydrate branches consisting of highly charged sialic acid. In this group is the *neural cell adhesion molecule (N-CAM),* which mediates *homophilic* interactions between cells, that is, interactions between the same molecules on different cells (see Table 3.1). By contrast, other members of the family, such as *TAG-1,* mediate *heterophilic* interactions, that is, interactions between dissimilar types of molecules between cells or between cells and a substrate. Both promote the outgrowth of nervous processes during devel-

opment; application to cultured cells of antibodies to these molecules causes defasciculation of fiber bundles. There has been much speculation on how specific these interactions are in mediating cell–cell recognition. The consensus is that they are important in promoting neurite growth and in enabling fibers to bundle together and grow toward their targets, but specific interactions with target cells appear to be mediated by other types of molecules. This is consistent with the fact that, whereas in antibody molecules the variations in amino acid sequences within the Ig folds are the basis for their specificities for different antigens, such variations are lacking in the neural family members.

A second type of membrane glycoprotein is the *cadherin* family, which is also important in neurite outgrowth and fasciculation. Members of this family consist primarily of fibronectin repeats (see Fig. 3.6). They engage in homophilic interactions (see Table 3.1) that are Ca^{2+} dependent, in contrast with the immunoglobulin-type interactions, which are Ca^{2+} independent. This provides for alternative means of modulating these interactions in relation to

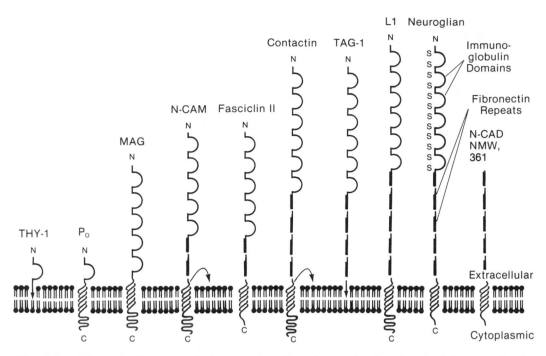

Fig. 3.6 Schematic diagrams of glycoprotein cell surface molecules. On the left are molecules belonging to the immunoglobulin superfamily; on the right is a member of the cadherin family. For abbreviations, see text. (Adapted from Grumet, 1991; Hall, 1992; Grenningloh and Goodman, 1992)

Table 3.1 Cell surface molecules

Types of molecules	Functional characteristics
Membrane glycoproteins	
Immunoglobulins	
NCAM	homophilic interactions; non-Ca^{2+}-dependent; promotes axon growth?
TAG-1	heterophilic interactions; non-Ca^{2+}-dependent; fasciculation; transient expression
Cadherin	homophilic interactions; Ca^{2+}-dependent; fasciculation
N-Cadherin	
Integrin	heterophilic interactions with e.c. matrix; Ca^{2+}-dependent (?); antibodies block neural crest
Extracellular matrix glycoproteins	
Laminin	substrate adhesion; basal lamina in PNS
Fibronectin	substrate adhesion; neural crest migration; axonal regeneration
Tenascin (J1, cytotactin)	
Thrombospondin	

the availability of free Ca^{2+} and calcium-binding proteins.

A third group is the *integrins,* which mediate interactions of cells with the extracellular matrix. These interactions take place through yet another group, the *extracellular matrix glycoproteins,* which are present in the intercellular spaces and lack direct insertion in cell membranes (see Table 3.1). An important member for the nervous system is *laminin,* a trimer containing epidermal growth factor (EGF)-like repeats. Laminin is widely expressed early in development and is believed to promote neurite outgrowth and facilitate extension of axons to their target regions. These properties make it an important ingredient of cell cultures.

These families are growing in members and functions; Table 3.1 should be consulted for additional details. We will discuss some of these in later sections and chapters (for example, P_0 in myelinated axons, below).

Proteoglycans. Some membrane proteins give rise to extremely long sugar chains. These are *proteoglycans;* the carbohydrate moieties are called *glycosaminoglycans.* The carbohydrate may constitute 95% of these molecules. This long branching structure is like a long test-tube brush, or a centipede wriggling its way through the extracellular space. It attracts water, imparting a spongy turgor to the extracellular space. The branching structure contains complex sequences of reactive saccharides, which are believed to play an important role in cell–cell recognition that must take place during development.

Together, the extracellular carbohydrates form a layer around the cell that is called the *glycocalyx;* in addition to being important for structural support, cell adhesion, and cell recognition, it appears also to regulate the diffusion of molecules through the extracellular space.

Intrinsic Proteins. The parts of the proteins that lie within or across the membrane are called *intrinsic membrane proteins;* they have several distinct functions. Some provide sites for reception of neurotransmitters and other neuroactive molecules. Some provide channels for ion movements across the membrane. Some provide carriers for movement of molecules involved in the metabolism of the cell, such as glucose and amino acids, whereas others pump ions inward or outward, or both. Finally, proteins that are attached to molecules in both the extracellular and intracellular spaces serve to anchor the cell. These different types of intrinsic membrane proteins are represented in Fig. 3.5.

Cytoplasmic Proteins. On the *cytoplasmic* side, certain of the membrane proteins attach to other proteins that form a complicated latticework near the inner surface of the membrane. These are called *membrane skeletal proteins.* Most of our information has come from studies of the red blood cell. David Anderson's (1984) comment—

Although the biconcave erythrocyte seems an unlikely model for the neuron with its complex architecture, it has proved to be surprisingly relevant.

—nicely catches the spirit of modern times, when our understanding of molecular biology is drawn from a variety of cells.

The current view of the relations between membrane proteins and membrane skeletal proteins is summarized in the diagram of Fig. 3.5. Several types of proteins, including a short-chain *actin,* form nodal points under the membrane; in addition, a protein called *ankyrin* is bound to an intrinsic membrane protein (in erythrocytes, this is band 3, an anion channel). Forming the links between these nodes is a type of filamentous protein called *fodrin.* There are several types of fodrin. The first to be characterized was found in red cells; called *spectrin,* it is a heterodimer (that is, its peptide subunits are different) consisting of α and β subunits of molecular weights (MW) 240,000 and 222,000 respectively.

The cross-links between spectrin and actin are responsible for giving the red cell its characteristic biconcave shape.

In the brain, spectrin is found only in cell bodies, and only in fully differentiated neurons. There is, in addition, a brain-specific fodrin, a heterodimer of an α and a γ (MW = 235,000) subunit; this fodrin is present in both developing and mature neurons, and is transported into the axon and dendrites. These differing distributions have suggested that the fodrins may be part of the mechanism that generates the different parts of the neuron, as well as microdomains within those parts. The evidence in the red cell implies that they may be important in determining the shape of axonal and dendritic branches. The linkage of spectrin to the red cell anion channel limits lateral diffusion of the channel in the membrane. In neurons, similar linkages to channel proteins appear to play a role in organizing the subsynaptic membrane (Chap. 6), or to be subject to activity-dependent changes underlying memory (see Chap. 29). A role in vesicle exocytosis has also been suggested.

The Nucleus

Each nerve cell contains a nucleus. As in other eukaryotic cells, the nucleus contains the genes, consisting of DNA and its associated proteins. Through its DNA, the nucleus plays two essential roles. One is to provide for DNA replication during mitosis. Mitotic activity is essential to the functioning of many body cells (for example, two million red blood cells are produced every second!). This function is present in the nervous system during development but is lost in most mature neurons (exceptions are the olfactory receptor neuron in vertebrates and certain cells in the avian brain). The other role is to transcribe DNA into RNA in order to synthesize proteins, especially the enzymes, which then synthesize the macromolecules of the cell. By this means the nucleus generates the instructions for controlling the differentiation and growth of the neuron during development,

and for maintaining its integrity during maturity.

The mechanisms of DNA replication and RNA transcription have been described in Chap. 2 and are covered in detail in many textbooks of biochemistry and molecular biology (see Alberts et al., 1989). Our interest here will be in how the gene products expressed in the nucleus are targeted within the cell.

Traffic between the nucleus and the rest of the cell must pass through the nuclear pore. Passing outward from the nucleus are tRNA, mRNA, and various shuttle proteins; passing inward from the cytoplasm are ribosomal proteins and shuttle proteins. There is evidence that these molecules move along filaments, possibly actin within the nucleus connecting the nucleolus to the pore, and microtubules within the cytoplasm (Dingwall, 1992; Meier and Blobel, 1992). It has been speculated that there may be myosin motors for driving transport along the actin filaments and kinesin or dynein for transport along the microtubules (see below). How all these substances move in both directions through the pore without creating a traffic jam is a question under study.

Ribosomal RNA (rRNA) is transcribed from DNA in a special part of the nucleus called the nucleolus (see Chap. 2). A large and a small subunit are immediately joined to special proteins to create a ribonuclear protein particle, called a *ribosome*. Processing of the particle is not completed until the mature ribosome is leaving the nucleus, so that it does not interact prematurely with mRNA.

The molecular machinery contained in the nucleus can be affected in several ways. For example, the nucleus is the prime target of one of the major classes of hormones, the steroid hormones. (See Chaps. 8, 24, and 27). Certain chromosomal aberrations lead to deficits or malfunctioning in nerve cells, as in Down's syndrome (mongolism). Finally, the nucleus may be the site of action of environmental toxins; for example, certain mushroom toxins inhibit the nuclear polymerases that produce mRNA.

These actions on nuclear mechanisms offer the opportunity to neurobiologists for experimental manipulations that can provide insight into the relations between gene expression, neuronal properties, and behavior.

As in other cells, the position of the nucleus defines the location of the *cell body* (also called soma); the cell body, in other words, is the part of the neuron that contains the nucleus. The size of the nucleus varies with the size of the cell body. In the largest vertebrate neurons the cell body may reach 100 μm or more in diameter, and the nucleus may be as large as 20 μm. The smallest nerve cell bodies are around 5μm in diameter. In these cells the nucleus may almost fill the cell body, leaving only a very thin rim of cytoplasm. The fact that processes may extend for long distances poses a special problem for the nucleus in controlling protein synthesis in all parts of the nerve cell (see below).

Ribosomes and the Rough Endoplasmic Reticulum

When ribosomes reach the cytoplasm they are sorted into two populations. Some remain "free" within the cytoplasm, either singly or in clusters ("polyribosomes"). In most cells of the body these ribosomes manufacture proteins that remain within the cell. The other population of ribosomes is attached to the *endoplasmic reticulum* (ER). This extends throughout the cytoplasm of the cell body as a system of sheets, channels, and membrane-enclosed spaces (large, flat spaces called cysternae, small spaces called vesicles). The part of this system that has attached ribosomes is called the *rough ER*. The membrane-bound ribosomes of the rough ER synthesize proteins that in most cells are destined for insertion into the plasmalemma or secretion from the cell.

As described in Chap. 2, synthesis of proteins depends on the third type of RNA, called *transfer RNA (tRNA)*. This combines with a given amino acid to "activate" it, so that it is ready to be assembled into

place. The assembly is directed by mRNA. The mRNA attaches to the ribosomes and moves across them, bringing its template into position to select *(translate)* the right sequence of activated amino acids to form a given polypeptide. By suitable foldings of the chain, the polypeptide becomes a protein.

In nerve cells there is a characteristic accumulation of rough ER near the nucleus that is termed the *Nissl substance* (see Figs. 3.4, 3.7). This is obviously a site of intense protein-synthesizing activity. In large neurons the Nissl substance is large and dense; in small neurons it may consist only of scattered particles. Close examination reveals that many of the ribosomes within the Nissl substance are not attached to the ER membrane, but lie between the membranes as polyribosomes. It has been speculated that these clusters may be involved in synthesizing complex proteins specific for nerve cells, perhaps specific for different types of nerve cells. Some of the protein synthesized in the Nissl substance may be destined for secretion in the form of transmitter substances. However, most proteins are probably used for maintenance of the large expanses of branching processes of the neuron.

The ribosomes manufacture the molecular machinery for most of the characteristic functions of cells: enzymes, regulators, carriers, receptors, transducers, contractile and structural elements, and membranes. As in the case of the nucleus, this machinery is susceptible to modification or disruption by various influences. For example, antibiotics exert their effects by interfering at various specific steps in transcription or translation. Diphtheria toxin inactivates one of the elongation factors responsible for assembly of polypeptides in nerve cells.

Secretion and the Golgi Complex

We have described the rough ER; what about the other part of the endoplasmic reticulum, that lacks attached ribosomes? This is called the *smooth ER*. It is an extremely versatile structure, which is put in

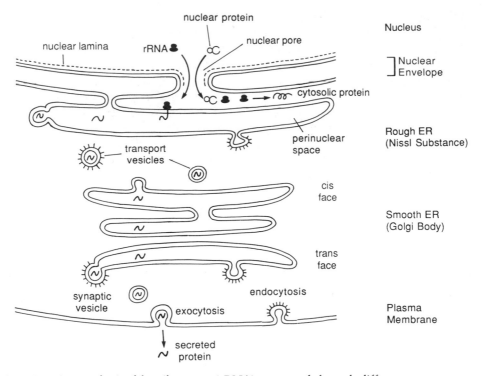

Fig. 3.7 Proteins synthesized by ribosomes (rRNA) are sorted through different compartments of the neuron. *Nuclear proteins* return through the nuclear pore. *Cytosolic proteins* are synthesized within the cytosol and remain there. *Secretory proteins* are synthesized within the endoplasmic lumen, which is continuous with the space between the inner nuclear membrane and outer nuclear membrane of the nuclear envelope (see diagram). Buds from the rough endoplasmic reticulum (ER) form transport (shuttle) vesicles which fuse with the cis face of the Golgi smooth ER. The contents of the smooth ER are transferred by continuity or vesicles to the trans face of the Golgi body, where synaptic vesicles are formed. Secretion occurs by exocytosis. (Based on Palade and Farquhar, 1981, and others)

the service of cells in a variety of ways. With regard to other cells of the body we will note only two examples. At one extreme, the smooth ER may form a rigidly defined inner system of sheets and channels, as in the skeletal muscle fiber. In muscle, it is called the sarcoplasmic reticulum; this forms the T-system, which has as its primary functions the conduction of the impulse into the interior of fiber and the regulation of the calcium level there in relation to muscle contraction. This will be discussed further in Chap. 17.

A second function of the smooth ER is to prepare proteins for insertion into the plasmalemma. This mechanism has been worked out for the muscle acetylcholine receptor; it involves the successive synthesis and maturation of different receptor subunits, their transport in the membranes of transport vesicles to the cell surface, and their insertion into the plasma membrane by vesicle fusion (see Chap. 6).

A secretory cell such as that of the pancreas employs its smooth ER in a different manner. As indicated in Fig. 3.7, the proteins manufactured by the rough ER are transferred (via shuttle vesicles) to the system of flattened pancakelike sacks, or cysternae, called the *Golgi complex*. The membranes here are called *transitional ER*. The Golgi complex has an orientation, there being an internal, *forming* face (cis) and an external, *releasing* face (trans).

From the latter, vesicles bud off to form secretory granules. The secretory granules remain within the cytoplasm of the cell until the cell is stimulated with the appropriate factors. The granules then move to the apical surface where their membranes fuse with the plasma membrane and discharge their contents. This process, called *exocytosis*, requires energy and the presence of free calcium ions.

In nerve cells, the Golgi complex is represented by a smaller accumulations of cysternae that are not clearly polarized and that tend to be dispersed in the cytoplasm and extend out even into the dendrites (but not the axon). In the neighborhood of the Golgi complex are clouds of small vesicles which may well be analogous to the shuttle vesicles in a secretory cell (see Fig. 3.7). However, most neurons do not secrete large granules like those of secretory cells. The relation of the smooth ER and the Golgi complex to the secretion of transmitter and neuromodulatory substances is still being worked out.

Lysosomes

In addition to systems for manufacture and transport of substances, cells also have an internal digestive system comprised of *lysosomes*. They vary in size from small vesicles to large, rounded sacks; and they contain a variety of hydrolytic enzymes that degrade and digest a wide range of substances that originate both inside and outside the cell. *Primary lysosomes* are formed from the Golgi complex by the budding off of vesicles which then acquire a dense matrix about them; these are termed *alveolate vesicles*. They contain acid hydrolases specific for different digestive tasks. When these processes begin, the vesicle becomes a *secondary lysosome*, and may assume a variety of forms as a digestive vacuole. The contents of the vacuoles are disposed of by diffusion of the breakdown products into the cytoplasm, or by exocytosis.

This general scheme appears to apply to nerve cells, but with some interesting differences. Because of the long distances between many nerve terminals and their cell body, the terminals must carry out some functions in semiautonomy. Nerve stimulation causes alveolate vesicles to appear in the terminal, where there is a cycle of pinocytosis and exocytosis that shares some common features with the sequence of steps that occurs through the lysosomal system in the cell body. We will discuss this cycle further in relation to synaptic mechanisms (see below and Chap. 4).

Mitochondria

Most of the cellular functions we have described require energy, and most of this energy comes from *mitochondria*. Apart from the nucleus, this is the most complex organelle within the cell. As indicated in Fig. 3.4, it is cigar-shaped, with a smooth outer membrane and an inner membrane that is thrown into internal folds called cristae. In general, the higher the energy demands in a particular cell, the more tightly packed the cristae. The inner membrane is relatively impermeable, so movement of substances across this membrane requires special transport mechanisms. The outer membrane is freely permeable to ions and water. Electron microscopic observations indicate that the outer membrane is continuous with the smooth endoplasmic reticulum.

The mechanisms for production of energy are summarized in Fig. 3.8. One source of energy is *glycolysis,* the chain of reactions that breaks down glucose by *anaerobic metabolism* to yield, for each molecule of glucose, a net of two molecules of *adenosine triphosphate* (ATP) with its high-energy phosphate bonds. All the glycolytic enzymes in the chain are soluble proteins which exist free within the cytoplasm. In contrast, *aerobic metabolism*, involving the acetylation of pyruvic acid and the reactions of the *citric acid cycle*, yields a net from each glucose molecule of 36 molecules of ATP, obviously a much more efficient source. All of the citric acid cycle

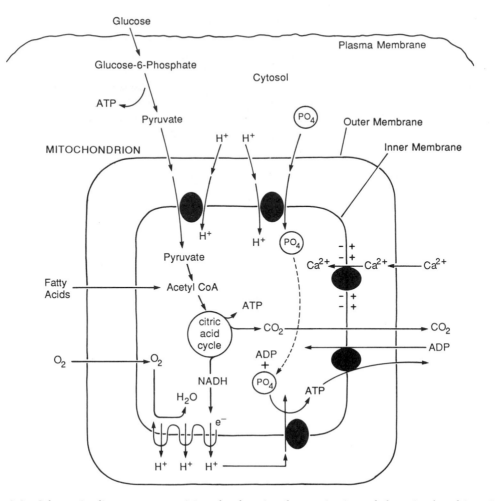

Fig. 3.8 Schematic diagram summarizing the functional organization of the mitochondrion. In normal neurons of the adult vertebrate nervous system, the main and probably exclusive source of energy is glucose. The metabolic pathway for energy metabolism begins with breakdown of glucose in the glycolytic chain to pyruvate, which takes place in the cytosol. This is followed by aerobic metabolism, by the citric acid cycle inside the mitochondrion, yielding 36 ATP molecules for each molecule of glucose. The energy is generated by the proton pump in the inner membrane, as shown in the diagram. Also shown in the diagram are sequestration of Ca^{2+}, utilization of O_2, and production of CO_2.

enzymes are contained within the inner matrix of the mitochondria. The enzymes of the associated electron transport chain are components of the inner mitochondrial membrane. It is believed that electron transfer sets up a gradient of H^+ across the inner membrane, and the potential energy in this gradient is used to form ATP from adenosine diphosphate (ADP) (see Fig. 3.8). The ATP is then available for the various energy-requiring processes of the cell. A primary energy demand in neurons is for running the Na/K ATPase membrane pumps that restore ionic gradients after synaptic transmission.

In most cells of the body a variety of sugars can be taken up by the cell and metabolized to yield energy, or they can be stored within the cell as glycogen. However, nerve cells in the vertebrate brain are special, in that they are almost exclusively dependent on glucose. (In invertebrates the corresponding substance for supplying energy is trehalose.) Other substances are excluded by the blood–brain barrier, formed by tight junctions between the endothelial cells of capillaries in most parts of the brain. Most nerve cells also lack the ability to store glycogen, further increasing their dependence for energy on circulating glucose as well as the oxygen needed for its aerobic metabolism. This is why we lose consciousness if the blood supply to our brains is interrupted for only a few seconds. Among the other functions of mitochondria is the ability to store calcium, which, as we shall see, can be a factor in regulating calcium in nerve terminals. Defects of mitochondrial metabolism may play a role in neurodegenerative diseases. Reduced levels of ATP in the cell would initiate a chain of events, including reduced ionic pump activity; partially depolarized membrane potentials; activation of channels allowing excessive Ca^{2+} entry, with resultant excitotoxic effects on the cell (Beal et al., 1993). We will discuss the role of NMDA receptor channels in these effects later (Chaps. 8, 29).

Mitochondria also contain small ribosomes and even a few strands of DNA. This indicates that the mitochondrion itself has most of the features of a cell, and, indeed, mitochondria probably originated from a prokaryotic microbelike ancestor. It is hypothesized that this aerobic microorganism developed a symbiotic relation with the anaerobic eukaryotic cell and became incorporated into it, to the mutual benefit of both.

The Cytoskeleton

If the neuron contained only the organelles discussed so far, it would be a stationary simple sack of genetic and metabolic machinery. However, the neuron grows in size and moves during development, sends out branches, transports substances and organelles within those branches, and makes synapses with other cells. There are thus demanding requirements for maintaining a complicated structure that yet provides for flexibility and movement.

These requirements are met by three types of filamentous proteins that form an internal network called the *cytoskeleton*. This can be visualized in neural tissue specially prepared for electron microscopy by the quick-freeze, deep-etch procedure. Let us discuss each of the filament types.

Microtubules and Neurofilaments
Microtubules. These are long, unbranched tubes of approximately 20 nm in diameter (see Fig. 3.9). The walls are composed of subunits of tubulin. The globular polypeptide monomer has a molecular weight of approximately 50,000. Assembly into microtubules begins with the formation of a heterodimer of α- and β-tubulin subunits containing similar amino acid sequences. These then polymerize into thin protofilaments of alternating subunits. The microtubule consists of 13 protofilaments arranged around the central core. Microtubules are polarized with plus and minus ends, new guanosine triphosphate (GTP)-tubulin monomers being added at the plus ends. Tubulin monomers are GTPases; when polymerization occurs, GTP is hydrolyzed to guanosine diphosphate (GDP). Thus, microtubules are in a dynamic balance between polymerization and depolymerization (see also Chap. 9).

The assembly and stability of microtubules is regulated by a group of proteins termed *microtubule-associated proteins* (MAPs). In electron micrographs they can be seen as lateral projections protruding from the microtubules and forming crossbridges between adjacent microtubules (see Fig. 3.9). There are two main groups, both of which have been cloned and sequenced. One group, called *MAP2 proteins,* consists of three isoforms produced from a single gene by alternative mRNA splicing. MAP2a and MAP2b are large proteins (288 and 280 kDa respectively) expressed mainly in nerve cell bodies and dendrites.

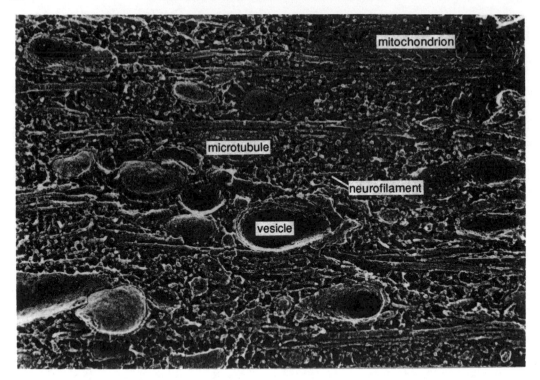

Fig. 3.9   Elements of the cytoskeleton, as seen in a preparation of the frog sciatic axon that has been quick-frozen and deep-etched. (Modified from Hirokawa et al., 1982, in Baitinger et al., 1983)

MAP2c is a smaller protein (70 kDa) expressed transiently in developing axons. The other main group, called *tau proteins,* has at least six isoforms produced by alternative mRNA splicing. They are also developmentally regulated. They are expressed in cell bodies and, in contrast to MAP2 proteins, in axons rather than dendrites. Tau proteins are of additional interest with regard to Alzheimer's disease, which is characterized by a deterioration of mental function that can strike older people. The pathological findings show accumulations of the protein amyloid and neurofibrillary tangles containing tau proteins. It is possible that abnormal phosphorylation of tau proteins contributes to the development of this disease (see Goedert et al., 1991).

From these considerations it is apparent that MAPs have multiple functions, and these extend even further than those already mentioned. In all cells undergoing mitotic division, the mitotic spindle consists of microtubules. In motile cilia, and in the tails of spermatozoa, they are arranged in a rigid pattern, consisting of a ring of nine pairs (doublets) around a central pair. This appears to provide an internal skeleton for the cilium, as well as mediating the forces that move it. In contrast, in many cells the microtubules are structural elements that lack any obvious orientation. Certain plant alkaloids, such as *colchicine,* bind with microtubules to depolymerize them, stopping mitosis in metaphase and also inhibiting the transport of substances in the axon (see below).

Neurofilaments. These are long solid filaments approximately 10 nm in diameter. They belong to the class of structure which in other cells is called *intermediate filament* (because it is intermediate in diameter between microtubules and microfilaments).

In contrast to microtubules and microfilaments, which have been conserved across phyla, intermediate filaments are composed of proteins showing considerable polymorphism, varying with different types of cells, and also within cell types. Thus, the proteins in neurofilaments differ from muscle or epithelial cell intermediate filaments. In neurofilaments three main polypeptides have been identified, with molecular mass of about 70 kDa (NF-L), 160 kDa (NF-M), and 200 kDa (NF-H). each filament is composed of chains of units consisting of an alpha-helical central rod, an amino-terminal head, and a carboxy-terminal tail. The long lengths and special structures of the tails differentiate neurofilaments from intermediate filaments in other types of cells. Glial cells are also rich in these filaments, where they are composed of a polypeptide called *vimentin* and, in some glial cells (see below), a glial fibrillary acidic protein (GFAP) with a molecular weight of 50,000.

Although neurofilaments are very stable in their basic structure, they are nonetheless involved in the continual remodeling of nerve processes that takes place during development and activity. Their role in these processes appears to be mediated to a great extent by phosphorylation (Nixon and Sihag, 1991). Neurofilaments are among the most phosphorylated proteins in the cytoskeleton. Phosphorylation of the carboxy-terminal sidearms on the subunits extends them outward from the subunit rod, enabling the sidearms to link to surrounding cytoskeletal structures (see Fig. 3.9). This is believed to brake the slow transport of the neurofilaments through the cytoskeleton, contributing to the stabilizing influence of neurofilaments on axon length. It also spaces the neurofilaments within the axon, thereby influencing, and perhaps making it a primary determinant of, the diameter of the axon.

Neurofilaments are especially prominent in large axons, where they outnumber the microtubules; on the other hand, in small axons and dendrites the proportions are reversed. Neurofilaments, and their relation to microtubules, change during aging, and in association with Alzheimer's disease, as discussed above.

Intracellular Transport. The presence of microtubules and neurofilaments in axons and dendrites has naturally suggested that they might be involved in the transport of substances, and a variety of biochemical studies, including those with colchicine (see above), have supported this notion. Intracellular transport between cell body and outlying processes is vital to the economy of the nerve cell, and we have already noted one example of this in the transport of substances (including HRP) from nerve terminals to the cell body. This direction is referred to as *retrograde*. Transport from cell body to terminals, on the other hand, is in the *orthograde* direction.

One of the most vivid demonstrations of these movements has been by the use of video-enhanced microscopy. In this method, axoplasm of a large axon, such as the squid giant axon, is spread out in a dish. The light microscopic image is recorded by a video camera under high magnification. The ability to set the level of light and contrast in the image (the same way one can on a television screen) enables one to visualize objects that would otherwise be below the resolving power of the light microscope. In addition, the image can be digitized and fed to a "frame-grabbing" computer for further enhancement. The result is an image in which particles (vesicles) can be seen moving along tubular structures (microtubules). Astonishingly, particles move in either the orthograde or retrograde direction along the same tubule, or change from one tubule to another.

Experiments showed that the motive force for the vesicle movement did not come from the microtubules themselves or from MAPs. A soluble protein was extracted from the axoplasm which, when coated onto a glass slide, could cause movement of overlying microtubules; furthermore, it could be coated onto microbeads, enabling them to be transported along the

microtubules. These investigations culminated in the identification of two main
motor proteins. *Kinesin* is a 350-kDa protein which moves along a microtubule
toward its plus end; *dynein* is a larger,
1200-kDa protein which moves toward the
minus end. Dynein in fact is a member of
a family of motor molecules that also is
involved in movement of the microtubules
within cilia and flagella. The opposite orientation of the motors explains how vesicles or beads can move in opposite directions along the same microtubule (see Fig.
3.10). By virtue of these properties, in the
normal intact axon, kinesin binds vesicles
in the cell body and moves them along
microtubules in the anterograde direction
toward the axon terminals, whereas dynein
moves them in the retrograde direction
from the terminals back to the cell body
(see below).

When radioactively labeled amino acids
are injected in the vicinity of cell bodies,
the amino acids are taken up by the cell
bodies and incorporated into protein,
which is then transported down the axon
to the terminals. These experiments have
identified two general types of *axonal
transport: slow transport* at a rate of about
one millimeter per day, and *fast* (rapid)
transport at rates of several hundred millimeters per day. Many of the substances
transported are intimately related to functions involved in synaptic transmission.
The transport of peptides is illustrated in
Fig. 3.11. The relation of transport to synaptic functions will be discussed in Chap. 8.

Intracellular transport has provided the
basis for a valuable method for demonstrating nerve connections in the central nervous system. The location of labeled substances after an injection can be ascertained
by making sections of the tissue and
exposing them to photographic film, a
method known as *autoradiography* (the tissue takes its own photograph, so to speak).
An example of this method is shown in
Fig. 3.12A.

Equally important is the transport of
substances in dendrites. In 1968, Tony

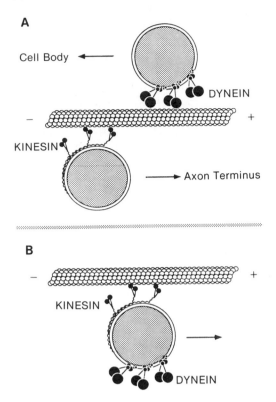

Fig. 3.10 Possible models for the mechanisms
of fast axonal transport. A. Dynein and kinesin
bind preferentially to a vesicle; kinesin moves
toward the plus end of a microtubule, away
from the cell body, whereas dynein moves toward the minus end. B. Alternatively, both dynein and kinesin bind to the same vesicle, but
only one is active. (From Valee and Bloom,
1991)

Stretton and Ed Kravitz at Harvard showed
that a dye (Procion Yellow) injected from
a micropipette into the cell body of a neuron is transported into the dendrites and
can be visualized by fluorescence microscopy. This was a key discovery for cellular
neurobiology, for it enabled an investigator
to see the whole dendritic tree of the cell
that was being recorded. It was like having
a Golgi stain of the very neuron one wished
to study. Since then, a host of substances
have been used for this purpose, including
HRP and dyes such as Lucifer Yellow; an
early study of radioactively labeled amino
acids in motoneuronal dendrites by autora-

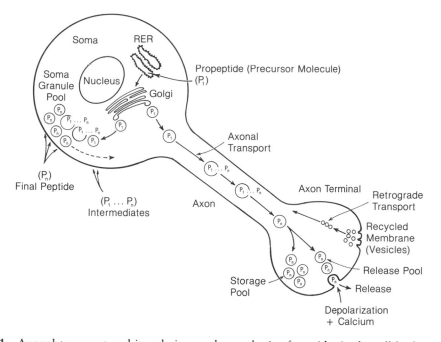

Fig. 3.11 Axonal transport and its relation to the synthesis of peptides in the cell body and their release from terminals. Propeptides (P₁) are synthesized in the rough endoplasmic reticulum and transported through the Golgi body. Transfer of intermediates or final peptide (Pₙ) then takes place to storage (soma granule pool), or by axonal transport to their terminals for release. Not shown is dendritic transport for release from dendritic synapses. (From Mains et al., in Gainer and Brownstein, 1981)

Fig. 3.12 Intracellular transport of radiolabeled compounds, as demonstrated by autoradiography. **A.** Transport of [³H]proline into visual cortex of monkey. Injection was made into one eye; [³H]proline was incorporated into protein in retinal ganglion cells, transported to their terminals in the lateral geniculate nucleus, and transferred to cells therein which project to the cortex. The alternating bands demonstrate ocular dominance columns (see Chap. 16). (From Wiesel et al., 1974) **B.** Transport of [³H]fucose into dendrites of a motoneuron after intracellular injection into the cell body, where it is incorporated into glycoprotein. (From Kreutzberg et al., 1975)

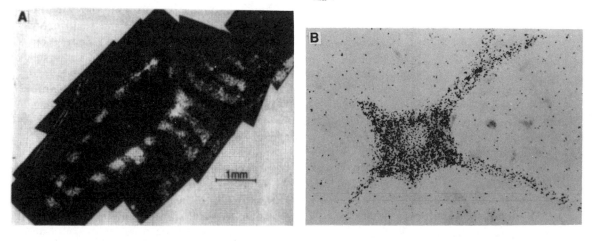

diography is illustrated in Fig. 3.12B. The ability to correlate physiological properties with the morphology of the cell revealed in this way is the basis of much of our understanding of the integrative functions of neurons.

Microfilaments

The third type of fibrillar structure in the cell is the *microfilament*. The best understood of these are in skeletal muscle, where thick filaments (12-nm diameter, composed of myosin) and thin filaments (5-nm diameter, composed of actin) are arranged in a highly geometrical array to provide the mechanism for muscle contraction (Chap. 17). Even in many nonmuscular cells, actin surprisingly accounts for up to 10% of the total cell protein, and most or all of this may be in the form of microfilaments. Microfilaments are abundant in growing nerve processes (Chap 9). They are also abundant in neuroglia, as discussed in the next section, and they are involved in certain kinds of neuronal junctions (Chap. 6). In many freely moving cells they are present just beneath the plasma membrane, where it is believed they control movement of the membrane and fluidity of the underlying cytoplasm. Evidence for this has been obtained by observing the inhibition of movements of macrophages after treatment with cytochalasin B, a substance obtained from fungi that disrupts microfilaments. In the neuron, interactions between fodrin and actin and other elements provide the means for controlling the form and movement of the plasma membrane, the membrane skeletal proteins, and the cytoskeleton.

Neuroglia and Glial Sheaths

Neuroglia are the other main type of cell in the nervous system. They play critical roles in controlling the environment of the nerve cells as well as being intimately involved in many of their functions.

Types of Neuroglia

Neuroglial cells are very numerous; in some parts of the nervous system they out-number the nerve cells by 10 to 1. They have been most closely studied and classified in the vertebrate nervous system. One of the main types is the *astrocyte* (Fig. 3.13A). These have many processes which radiate out in all directions from the cell body, giving the cell a star-shaped appearance. Within the central nervous system some of the processes terminate as end-feet on the surfaces of blood vessels. Astrocytes located in the white matter of the brain are called *fibrous* astrocytes, by virtue of the large numbers of fibrils present in the cytoplasm of their cell bodies and branches. Those located in the grey matter have fewer fibrils, and are called *protoplasmic* astrocytes. In the electron microscope, astrocytes are seen to have a somewhat dark cytoplasmic matrix and large numbers of neurofilaments (these are the fibrils seen in the light microscope) as well as glycogen granules in the cytoplasm, all characteristics that are different from those of nerve cells. Astrocytes also are interconnected by gap junctions (Chap. 6).

The functions of astrocytes are believed to include: (1) providing structural support for nerve cells; (2) proliferation and repair following injury to nerves; (3) isolation and grouping of nerve fibers and terminals; (4) participating in metabolic pathways which modulate the ions, transmitters, and metabolites involved in functions of nerve cells and their synapses, and (5) generating slow waves of K^+ and Ca^{2+} transients. Earlier speculations, that they form part of the blood–brain barrier, or that they are involved in the transport of nutrients from the blood vessels to the nerve cells, seem now to be largely discounted. In the vertebrate central nervous system, a special kind of cell, called *radial glia,* appears only during embryonic life, and provides guidelines for migrating neurons to follow (Chaps. 9 and 30).

Some glial cells have distinctly fewer and thinner branches than astrocytes, and these are termed *oligodendrocytes* (oligo = few, dendro = branches). In the electron microscope, they are seen to contain few neurofilaments and glycogen granules, but nu-

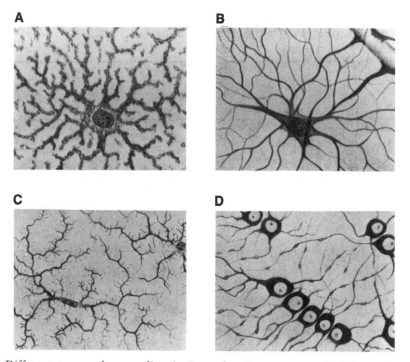

Fig. 3.13 Different types of neuroglia. **A.** Protoplasmic astrocytes. **B.** Fibrous astrocytes. **C.** Microglia. **D.** Oligodendrocytes. (After del Rio-Hortega, in Bloom and Fawcett, 1975)

merous microtubules. The branches are often hard to distinguish from those of nerve cells, but they can be differentiated because they never take part in synaptic connections. The functions of oligodendrocytes include the formation of the myelin around axons in the central nervous system (see below), and it has also been proposed that they have a symbiotic relation with certain nerve cells, involving complex metabolic exchanges.

A third main type of glial cell is termed *microglia*. These are small cells scattered throughout the nervous system. Wherever there is injury or degeneration, these cells proliferate, move to the site, and transform into large macrophages that remove and phagocytize the debris.

Molecular biological studies have added considerably to our knowledge of these main glial types. In cultures from the early developing optic nerve, three types of glia can be identified: oligodendrocytes and two types of astrocytes (types I and II). Martin Raff and his colleagues in London (Miller et al., 1989) have shown that differentiation of these cell types involves a complex sequence of interactions mediated by growth factors. In brief, oligodendrocytes and type II astrocytes originate from a common precursor called *O-2A*. During early development, type I astrocytes secrete a protein identical to platelet-derived growth factor *(PDGF)* that stimulates proliferation of O-2A cells. Later, they secrete a different protein identical to ciliary neurotrophic factor *(CNTF)*; this, together with a factor secreted by mesenchymal cells into the extracellular matrix, induces differentiation of O-2A cells into type II astrocytes; without it, the O-2A cells differentiate into oligodendrocytes. These interactions can be followed in tissue culture with the help of staining with antibodies to the different cell types.

Nerve Sheaths

A very important function of neuroglia is to provide special sheaths around some axons. These sheaths not only protect the axons, but are also intimately involved in molecular and structural modifications of the axons needed to conduct impulse signals rapidly over long distances.

The simplest arrangement is one in which a single axon, or group of axons, is embedded in a glial cell, as shown in Fig. 3.14A. This is the common situation for very small-diameter fibers, in both invertebrates and vertebrates. The cells that provide these sheaths in vertebrate peripheral nerves are modified glial cells, called *Schwann cells.* The point at which the

Schwann cell membranes come together to enclose the axon or axons is called the *mesaxon* (it is analogous to the mesentery that encloses the intestines in the abdomen). The axons ensheathed in this manner are termed *unmyelinated,* for reasons that will become apparent below.

A somewhat more complicated arrangement is found where there are several loose folds of Schwann cell membrane around a given axon. This is characteristic of many larger invertebrate axons (see Fig. 3.14B).

The most complicated arrangement is

Fig. 3.14 Nerve sheaths. **A.** Single glial cell surrounding thin nerve fibers. **B.** Loose glial wrappings. **C.** Tight glial wrapping, forming myelin. **D.** Relation of oligodendroglial (G) cell to myelin folds around nerve fiber and node (N) of Ranvier. An oligodendroglial cell supports 30–40 myelinated segments of different axons in the central nervous system, as indicated in the diagram, whereas a Schwann cell supplies a single myelinated segment in the periphery. (From Bunge, 1968)

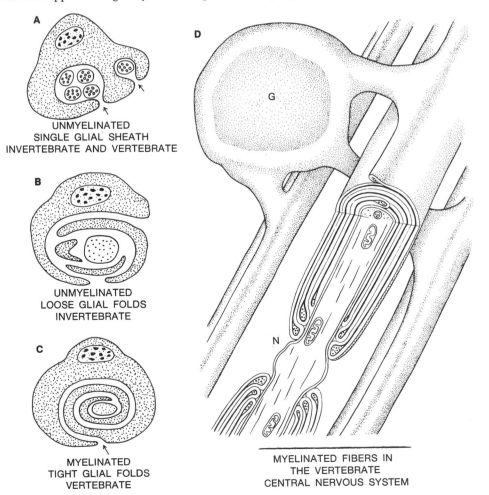

A
UNMYELINATED
SINGLE GLIAL SHEATH
INVERTEBRATE AND VERTEBRATE

B
UNMYELINATED
LOOSE GLIAL FOLDS
INVERTEBRATE

C
MYELINATED
TIGHT GLIAL FOLDS
VERTEBRATE

D

MYELINATED FIBERS IN
THE VERTEBRATE
CENTRAL NERVOUS SYSTEM

one in which there are a number of layers of Schwann cell membranes tightly packed around a single axon. The layers are formed by the wrapping around of the Schwann cell membrane in a spiral manner during development, as indicated in Fig. 3.14C. By virtue of their tight packing and modified composition, these layers form a special tissue called *myelin*. This is such an important structure that all nerve fibers can be generally classified as either *unmyelinated* (see above) or *myelinated*. In the central nervous system, myelin is formed by oligodendrocytes (see Fig. 3.14D).

Myelinated tissue has a fatty consistency and, to the naked eye, a white appearance (as in the white matter of the brain). The fibers are revealed in the light microscope as *black* structures when treated with the common lipid stains. This is why, in many brain atlases, the "white" matter appears black and the "grey" matter appears relatively white.

A single Schwann cell in peripheral nerve supplies myelin for a length of about 1 mm of axon. At the borders of this region, the myelin layers overlap progressively in the manner shown in Fig. 3.14D. There is a gap of 1–2 μm between neighboring myelinated regions that is known as the *node of Ranvier*. Here the plasma membrane of the axon is unsheathed; in its place are hairlike microvilli arising from satellite cells. A myelinated fiber thus consists of *nodes* of naked (unsheathed) nerve membrane alternating with *internodes* of myelinated membrane. The manner in which this imparts special properties for efficient conduction of nerve impulses is explained in Chap. 5.

Our current understanding of the structure of myelin comes from a convergence of many methods. In the electron microscope, myelin is clearly recognized as repeating dark and light layers with a period of about 18 nm, corresponding to two thicknesses of compressed plasma membrane (see Fig. 3.15). Biochemical studies using cell fraction procedures show that myelin is about 80% lipid and 20% protein. Cholesterol is one of the major lipids. Cerebrosides and phospholipids are present to different degrees in different tissues and species. The main protein constituents have been identified, cloned, and sequenced (see Fig. 3.15). They include myelin-associated glycoprotein (MAG) and protein zero (P_0), both members of the immunoglobulin family of cell surface proteins (see above); and peripheral myelin protein-22 (PMP-22). Another important group of proteins is called myelin basic protein (MBP). It may be recalled that the cell surface glycoprotein family plays important roles in cell adhesion (see above); presumably this role is adapted in myelin to mediating adhesion between successive layers of membrane of the same Schwann cell.

Defects in these protein constituents of myelin are strongly implicated in specific nervous diseases (for review, see Waxman, 1992). For example, injection of MBP into rats produces an autoimmune reaction characterized by inflammation followed by demyelination of central nerve fibers. Called experimental allergic encephalomyelitis, it is being explored as a model for the demyelination that occurs in multiple sclerosis in humans. Mutant strains of mice are providing other examples of myelin disease. The shiverer mutant is due to loss of expression of MBP. In the trembler (Tr) mutant, the gene expressing PMP is affected. Studies of trembler indicate that the defect is not only in laying down myelin, but includes effects on axon diameter, slow axonal transport, and neurofilament density and state of phosphorylation (Suter et al., 1993).

The results of these studies are interesting for two main reasons. First, they provide important models for understanding the pathogenesis of hereditary diseases of myelin in humans. Second, they show that Schwann cell functions are not limited to laying down myelin, but include trophic interactions that are critical in determining many structural and functional properties of the axons they envelop. Thus, they provide excellent examples of the ways that

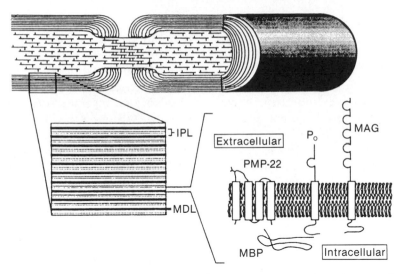

Fig. 3.15   Some of the ultrastructural and molecular properties of myelin. The main diagram shows a cutaway view of the region of a node of Ranvier. Neurofilaments are present in the axoplasm, at a higher density at the node. The myelin consists of alternating layers of major dense line (MDL) and intraperiod line (IPL), which are continuations of the cytoplasm and surface membrane, respectively, of the Schwann cell. The membrane of the MDL in turn contains specific proteins, as indicated. MAG, myelin-associated glycoprotein; MBP, myelin basic protein group; P_o, protein zero; PMP-22, peripheral myelin protein 22. (Modified from Suter et al., 1993)

glial cells play integral roles in the nervous functions of the nervous system.

Myelin is found almost exclusively in vertebrates. It would be possible to think of myelin, therefore, as an essential element in the higher nervous functions of which vertebrates are capable. The main contribution of myelin is probably that it permits rapid signal conduction over long distances. Even over long distances, many axons are thin and unmyelinated, and provide for slower impulse transmission (see Table 12.1). However, rapid and precise integration of information from widely separated regions may be presumed to be necessary for the evolution of higher nervous functions. Loss of this precision has a disastrous effect on brain function, as when myelin undergoes degeneration in multiple sclerosis.

Differentiation of Axon and Dendrite

A major theme in this review of the molecular and cellular biology of the neuron has been the distinction between axon and den-

drite. It will be useful to gather together this information for a coherent synthesis of the structural basis of this important distinction.

To begin with, the cell body is that region of the cell around the nucleus. In nerve cells, as in other cells, the main organelles of the cytoplasm are gathered here in order to interact with the nucleus and with each other. These include the Golgi body and (in nerve cells) the Nissl substance, in addition to large numbers of mitochondria, rough and smooth ER, polysomes, and fibrillar structures.

Next are the branches. One of the cardinal features of neurons is that their branches have widely different patterns. Like trees, neurons acquire different names on the basis of different branching patterns. However, the branching patterns are so diverse that it is sometimes difficult to make the distinction between what is an axon and what is a dendrite. Let us see if we can take some logical steps toward making these distinctions.

Current studies confirm the classical distinctions by showing that axons and dendrites differ in the constituents in all three of the major cell compartments: intracellular, membrane, and extracellular. A summary of the main determinants is provided in Fig. 3.16 and Table 3.1. Gary Banker and his colleagues at Albany and the University of Virginia carried out detailed studies of the mechanisms underlying the differentiation of processes by observing nerve cells in tissue culture. As shown in Fig. 3.16A, the first processes are short and undifferentiated; they contain microtubules with plus ends oriented outward and a protein called *growth-associated protein (GAP-43)*. The critical step in differentiation is the onset of growth of one of these processes at a more rapid rate than the others; this is correlated with restriction of GAP-43 to this process, together with the appearance of the tau protein (tau). This process becomes the axon; all the rest become dendrites. There thus appears to be a molecular switch that commits one, and only one, of the processes to becoming an axon with molecular properties different from the rest. Since all processes at first contain outward-oriented microtubules and GAP-43, one could regard all processes as initially potentially axons, so that the switch may both designate one process to become an axon and turn off the rest from that path. Banker's studies have shown that there is a critical period during which a newly formed axon can be amputated and the cell can still give rise to a new axon from another process. After differentiation of the axon, the other processes begin to extend and take on the branching appearance of dendrites; as they do so, they begin to contain microtubules oriented in both directions (see Fig. 3.16B).

In the maturing and adult neuron, the dendrites and axons continue to take on differentiated properties. Some of these are summarized in Fig. 3.16C. New proteins appear differentially, such as phosphokinase A and calcium/calmodulin kinase in the dendrites. Organelles also appear differentially. Neurofilaments are present in both dendrites and axons, but they are

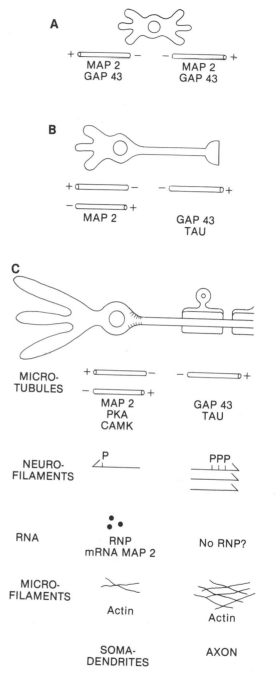

Fig. 3.16 Differences in molecular constituents and cell organelles between axons and dendrites. A. Before differentiation. B. Early differentiation of an axon. C. Maturing and adult neuron, with dendrites and an axon. CAMK, calcium calmodulin kinase; GAP 43, growth-associated protein 43; P, phosphate group; PKA, protein kinase A; RNP, ribonucleoprotein; other abbreviations as defined previously. (Based on Goslin and Banker, 1989, and others)

much more numerous and more heavily phosphorylated in axons. Actin filaments are also more numerous in axons. Ribonucleoprotein particles are found in dendrites; some of these represent mRNA for local synthesis of MAP2. Classically it is believed that protein synthesis is lacking in axons. However, polyribosomes extend at least into the initial segments of some axons (Steward and Ribak, 1986).

In the extracellular compartment are constituents specific for dendrites or axons. Most prominent is the myelin laid down around axons by Schwann cells in the peripheral nervous system and oligodendrocytes in the central nervous system. Such ensheathments are characteristic for large-diameter axons; small-diameter axons, however, may be either myelinated or unmyelinated, and the smallest diameter axons are unmyelinated.

We consider finally the surface membrane. Differences in the molecular composition of the cell membrane are the basis of *cell surface domains,* which are closely correlated with the intracellular and extracellular compartments. In the case of myelinated axons, for example, we will see that the unmyelinated nodes contain a high density of voltage-gated Na^+ channels for impulse generation (Chap. 6). The axon hillock and initial segment also have a high density of Na^+ channels; these regions are characterized in addition by the presence of a dense undercoating seen in electron micrographs (see Fig. 3.4).

Recent work provides strong evidence of the creation of these membrane domains in relation to the differentiation of axons and dendrites. The domains can arise in two main ways. One way is by the random insertion of membrane proteins, which diffuse within the lipid moiety and are trapped at specific sites by elements of the cytoplasmic skeleton. The other way is by intracellular transport and insertion at specific points (Elson, 1993). Studies of cultured axons have shown that membrane channels diffuse more readily in the dendrites than the axons; moreover, there is a barrier to diffusion between the cell body and the axon hillock. The axon hillock appears to be an *anchored domain* (in which membrane channels are inserted by targeting and attached to the cytoskeleton), as does the myelinated axon, whereas the soma-dendrites appear to be *barrier-limited domains* (where membrane proteins are able to diffuse up to limiting barriers). These properties suggest that dendrites provide a relatively flexible substrate for positioning of synapses and for activity-dependent remodeling. We shall have evidence of these properties in many parts of this book.

Neuron Terminology

From these considerations we may feel some confidence that the identification of axons and dendrites can be securely made on a molecular basis. However, in practice one is more often faced with the problem of identifying axons and dendrites in stains of whole cells, such as cells stained by the Golgi method. Neurons vary so widely in form that it has been difficult to be sure that one is using the terms axon and dendrite in a consistent manner for all types of cells. Here we will discuss some simple rules for accomplishing this.

To begin with the simplest case, some nerve cells have long fibers that connect to other regions of the nervous system. These are called *projection neurons, principal neurons,* or *relay cells.* They characteristically have a single long axon that makes the distant connections. This process can be recognized in Golgi-stained material because the fiber arises from a cone-shaped part (axon hillock) of the cell body or a dendritic trunk. It usually (but not always) maintains the same diameter throughout its length, despite giving off branches. The branches usually arise at right angles. Large axons and their branches may have a myelin covering.

Thus, in projection neurons, an axon can nearly always be identified by one or another (often all) of the above criteria. Then, all the other processes of the cell

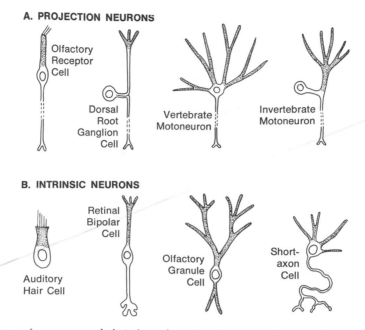

Fig. 3.17 Types of neurons and their branches. Dendrites are shown with stippling, axons with clear profiles.

are dendrites. Thus, *dendrites are all those branches of a nerve cell that do not fulfill the criteria for being an axon.* Figure 3.17A shows several examples of projection neurons in which these definitions have been applied. Note that despite the specializations of the dorsal root ganglion cell and the invertebrate neuron, the definition can be applied with ease. The term "neurite" is sometimes used to refer to both the axonal and dendritic branches of invertebrate neurons.

The other main type of nerve cell is contained wholly within one region of the nervous system. These are called *intrinsic neurons,* or *interneurons.* Examples of these are shown in Fig. 3.17B. The problem that arises here is that many intrinsic neurons do not require an axon for their functions. Thus, some have almost no processes whatever (the hair cell); others have only short processes (the bipolar cell); some have no process qualifying as an axon (the granule cell). The latter are usually called *anaxonal,* or *amacrine* (a = no, macrine = long pro-

cess) cells; their processes may thus all be called dendrites. Only in the case of a cell with a short axon (short-axon cell) does one have an interneuron in which one can make the usual distinctions that apply to the projection neuron.

As noted earlier, dendrites were originally termed "protoplasmic prolongations," and the modern studies with the electron microscope have fully confirmed the correctness of that idea. As indicated in Fig. 3.4, the main organelles of the cell body extend without any sharp boundaries into the trunks of the dendrites. Thus, large dendrites can be clearly distinguished from large axons. However, small axons and small dendrites are not so dissimilar in their fine structure. The terminals of axons are often characterized by a large number of synaptic vesicles (see Chap. 6), but small axon terminals differ little from small dendritic branches and terminals in their fine structure. The sites of synapses on axons and dendrites will be considered in Chapter 6.

4

The Membrane Potential

We have seen that a nerve cell, like other cells of the body, contains a variety of organelles for carrying out different functions. Another property it shares with all cells is an electrical potential difference that exists across the cell membrane. This is termed the *membrane potential*. Most people do not realize that this is so fundamental a property that one can equate it with life itself: to be alive, cells must have membrane potentials; cells that have lost their ability to maintain an electrical potential across their membranes are dead cells.

The membrane potential is thus a necessary condition for the cellular functions carried out by the different organelles. In addition, in nerve cells, and in related muscle and gland cells, the membrane potential is the source of special functions: brief changes in the membrane potential are the basis for electrical signaling. It is therefore obvious that we must begin our study of neural signaling with a clear understanding of the basis of the membrane potential. In this chapter we will consider the mechanisms by which the membrane potential established, the metabolic machinery by it is maintained, and the electrical that allow membrane potential spread passively through the

cell. This will provide us with the foundation for understanding the nature of electrical signals—action potentials and synaptic potentials—in subsequent chapters.

Nerve Cells and Their Ions

The organelles of a neuron are embedded in a cytoplasm that is made up mostly of *water, protein,* and *inorganic salts,* as shown schematically in Fig. 4.1. The proteins range from structural macromolecules and enzymes of high molecular weights, down through smaller subunits like polypeptides and peptides, all the way to the various amino acids. Many of these molecules have terminal groups that are dissociated in the aqueous medium of the cytoplasm, and their net electrical charge makes them *ions*. In the squid giant axon these *organic ions* can be determined simply by squeezing out the axoplasm and assaying it. The main organic ion is isethionate; it has a net negative charge, and is therefore an organic *anion*, represented by A^- in Fig. 4.1. Glutamate, aspartate, and organic phosphates have been suggested to be present in other nerves. Whatever their identity, the net charge on these molecules makes them anions.

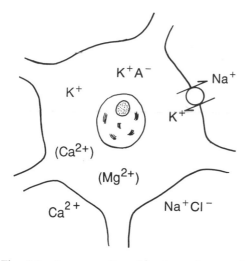

Fig. 4.1 A nerve cell and its ions. A, organic anions; Cl, chloride; K, potassium; Na, sodium; Ca, calcium; Mg, magnesium.

In addition to organic ions there are *inorganic ions*. These are needed for the operation of many enzymes; for maintaining electrical, chemical, and osmotic equilibrium within the cell and between it and the outside; and for other functions. The main intracellular *cation* is potassium, as shown by K^+ in Fig. 4.1. Also present in smaller but significant amounts are calcium and magnesium. The inorganic *anions* include chloride, phosphate, and sulfate. Outside the cell, the main *extracellular cation* is sodium (Na^+) and the main *anion* is chloride (Cl^-).

It can be seen at a glance in Figure 4.1 that the electrolyte composition inside and outside the cell is quite different. This is summarized more quantitatively in Fig. 4.2 and Table 4.1, for the case of an invertebrate cell (the squid giant axon) and a vertebrate cell. The latter is exemplified by frog muscle fibers, which have been well studied, and are believed to be similar to vertebrate neurons in these respects.

The electrical activity of nerve cells is derived from two factors: the unequal distribution of ions across the cell membrane, as shown in Figs. 4.1 and 4.2, and the membrane's different and constantly changing permeabilities to these ions. Let us consider first how the unequal distribution gives rise to an electrical potential.

The Nernst Potential

Ions diffuse through the membrane to achieve a balance of chemical forces, and now we can discuss how this gives rise to an electrical difference. In order to do this, we will delve a bit deeper into the nature of diffusion processes; the diagram of Fig. 4.3 serves as an intuitive guide to the following discussion.

When a substance diffuses in a solution, the force that moves the molecules from a region of high concentration to one of low concentration is a *chemical force*. We say that this force moves the molecules down their *concentration gradient,* like gravity makes marbles roll down an inclined plane. This can be described mathematically in several ways, for example, in terms of how fast the substance moves (called the *flux*), or how much *work* would be necessary to oppose the movement of the substance. If we consider our system of two regions separated by a membrane permeable to K^+, then the unequal concentrations of K^+ on either side mean that there is a force on K^+ that moves it *outward*, down its concentration gradient. In terms of the work (W_c) to oppose this chemical force, we have

$$W_C = 2.3 \, RT \log \frac{[K^+]_{OUT}}{[K^+]_{IN}} \quad (4.1)$$

where R = gas constant (a measure of the energy of the substance), T = absolute temperature (a substance is more active as the temperature is raised), and the concentrations of K^+ inside and outside are in moles. For present purposes we will not worry about the units of these parameters, but only concern ourselves with gaining an intuitive grasp of the relations.

Now, as K^+ diffuses outward through the membrane, Cl^- is diffusing *inward*,

Table 4.1 Ionic concentrations for squid axon and mammalian muscle fiber

Ions	Invertebrate squid axon (seawater≈blood)		Vertebrate muscle (neurons) (interstitial fluid)	
	Internal	External	Internal	External
Cations				
K^+	400	(10)	124	2
Na^+	50	460	10	(125)
Ca^{2+}	(.4)	10	5	2
Mg^{2+}	10	54	14	1
other	—	—	—	—
Total	460	534	153	130
Anions				
Cl^-	40–150	560	2	77
HCO_3^-	—	—	12	27
$(A)^-$	345	—	74	13
other	—	—	(65)	(13)
Total	460	560	153	130

Concentrations in mM. The values for the mammalian muscle fiber are believed to be representative of neurons. ( ) indicates estimates, to give electroneutrality between cations and anions. Note lack of osmotic equilibrium across the membrane (between internal and external medium).

After Aidley (1989)

down *its* concentration gradient (see again Fig. 4.3). This means there is a tendency for K^+ to become separated from its accompanying negative ion. However, since opposite charges attract, there is an *electrical force* tending to pull the K^+ back inside toward the inwardly diffusing Cl^- ions. The work (W_E) required to oppose this electrical force is given simply by

$$W_E = FE \qquad (4.2)$$

where $F =$ Faraday's constant (a measure of electrical charge per mole of substance) and $E =$ electrical potential difference, due to the charge separation across the membrane, measured in volts.

When the system is at equilibrium, there will be no net movement of K^+ or any

Fig. 4.2 Ionic concentrations for an invertebrate neuron (squid axon) and a mammalian muscle fiber. (Based on Aidley, 1989)

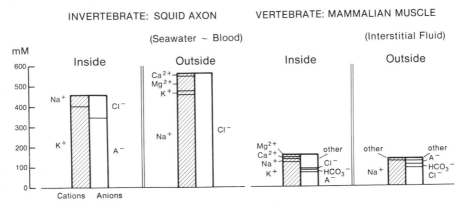

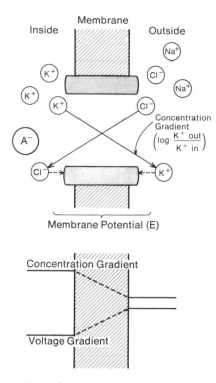

Fig. 4.3 Relations between electrical and chemical gradients across the membrane. (Modified from Woodbury, 1965)

$$E_K = 58 \log \frac{[K^+]_{OUT}}{[K^+]_{IN}} \text{ mV} \qquad (4.4)$$

$$= 58 \log \frac{20}{400} \text{ mV}$$

$$= -75 \text{ mV}$$

If you learn only one equation in your study of neurobiology, the Nernst equation is the one to learn, because it is fundamental to the nature of electrical potentials in all cells, as well as the electrical activity in neurons. For any given ion species, E is the potential at which there is no net flux of ions across the membrane; it is therefore referred to as the *equilibrium potential* for that ion. In other words, it is the potential that the membrane tends toward when the membrane is permeable to that particular ion. Since the potential E exists across the cell membrane, it is called the *membrane potential*. When we take up the specific kinds of electrical activity that are associated with snyapses and with impulses, we will see that they all take the form of changes in the membrane potential.

other substance, and the chemical force tending to move K^+ out will be just balanced by the electrical force tending to move it in. The two forces are therefore equal, and that means we can set Eq. (4.1) equal to Eq. (4.2), thus

$$W_E = W_C$$

$$FE = RT \log \frac{[K^+]_{OUT}}{[K^+]_{IN}}$$

$$E = 2.3 \frac{RT}{F} \log \frac{[K^+]_{OUT}}{[K^+]_{IN}} \qquad (4.3)$$

This is the *Nernst equation*, named for Walther Nernst, a German chemist, who derived it in 1888. We refer to *(E)* as the *Nernst potential*, or the *diffusion potential*. For the case of the squid axon at room temperature (18°C) the constant 2.3 *RT/F* = 58 mV, and, for the concentration of K outside and inside, we have

The Membrane Potential

Sir Arthur Eddington, the British astronomer, once remarked that "You cannot believe in astronomical observations before they are confirmed by theory." Much the same applies to the experiments we do in biology: we can begin to believe in results only if we have an adequate grasp of the theories that seek to explain the nature of the systems we study. That is why some of the theoretical basis of the membrane potential has been presented. With an understanding of how a membrane potential *might* arise, we are ready to set up an experiment that will tell us the actual value of the membrane potential, and how it fits with the theory. You can reproduce this classical experiment yourself using the computer disk in the companion tutorial of Huguenard and McCormick (1994).

The aim of this experiment will be to measure the electrical potential across a nerve membrane, using the giant axon of the squid as our test subject. All electrical measurements involve recording the difference between some quantities of electricity at two electrodes. In this case, one electrode is outside the axon, and the other is inside. The squid axon is so large (up to 1 mm in diameter) that the internal electrode can be a fine wire inserted longitudinally through a cut end. Generally, however, a nerve cell is must smaller, and in order to insert the electrode through the membrane, one uses a microelectrode, fabricated especially for this purpose. A similar approach is used for making electrodes for patch recordings.

Methods for Recording Membrane Potentials

Intracellular Recordings. The method used for making a microelectrode for *intracellular* recording is to take a length of glass capillary or pipette tubing, heat it in the middle, and quickly pull it apart so that one obtains tips which are very fine but still open; they are called *micropipettes.* The first ones, made around 1950, were pulled by hand over a small Bunsen burner flame (it took a steady hand and eye to do it!), but machines (microelectrode pullers) were soon devised to make them automatically. A salt solution is placed in the tubing at the large end; if the tubing contains a fine glass thread, the solution will fill the pipette to the tip by capillary action (Fig. 4.4). The micropipette is now a microelectrode. When the pipette tip is inserted into the axon, the salt solution serves as an electrical conductor between the axoplasm at the tip and a wire in the large end, connected to a suitable electronic amplifier and recorder. The most convenient recorder is a *cathode ray oscilloscope.* The cathode ray tube is constructed on the same principle as a television tube, except that it has only a single beam that travels across the screen, to register electrical changes at different speeds and at different amplifications (see Fig. 4.4).

When both electrodes are outside the nerve, there is naturally no signal registered on the oscilloscope (the beam just keeps moving across and being reset at the same level). However, when the pipette tip is delicately pushed through the axonal membrane (sometimes a gentle tap on the table will help), the oscilloscope beam is abruptly deflected in a direction that indicates that the pipette tip has become electrically *negative* relative to the outside electrode (this occurs at the arrow shown in Fig. 4.5A). If it is a "good" penetration, the membrane seals around the pipette, and, if nobody trips over a cord in the dark or slams a door (!), the recording may be stable for many minutes or even hours. The pipette tip must be small (less than 1 μm diameter) in order to minimize damage to the cell membrane. Even so, the small opening in the membrane around the micropipette creates a small electrical shunt across the membrane.

Under these conditions, the stable negative potential that is recorded is the *resting membrane potential;* we assume that all the potential difference between the two electrodes is due to the potential difference across the membrane. The value of the resting membrane potential in a typical experiment is in the range of -60 to -70 mV. Because of this internal negativity, we say that the membrane has a negative polarity. If the membrane potential moves toward zero, we say that it becomes *depolarized;* if it increases above the resting value, we say it becomes *hyperpolarized.* The significance of such changes in polarization will soon become apparent.

Patch Recordings. Micropipettes for patch recordings are prepared by the same methods, except that the tip is cleaned so that it can adhere to the cell membrane instead of being inserted through it. Recordings of the membrane potential are made in the whole-cell configuration or the variation involving a perforated patch, as we discussed in Chap. 2 (see Fig. 2.13). Similar micropipettes are used for obtaining single-channel recordings in the other configurations illustrated in Fig. 2.13. For patch

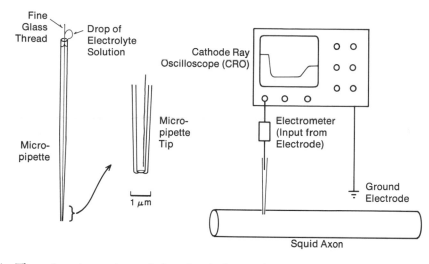

Fig. 4.4 The micropipette is used for electrical recording (extracellular, intracellular, patch), electrical stimulation (current or voltage clamp), or delivery of substances (microionophoresis or pressure ejection). Preparation of an intracellular recording micropipette is shown on the left. The diagram on the right shows the arrangement for recording from a squid axon and observing potentials on a cathode ray oscilloscope (CRO).

recording from cell bodies the tip can be somewhat larger (2–3 μm diameter) compared with the tip of an intracellular micropipette (typically less than 1 μm). Thus, patch electrodes are actually easier to make and use than the traditional intracellular type. Since they also appear to cause less damage to the cell membrane, they are

increasingly replacing intracellular techniques in many recording studies.

Identifying Ionic Conductances

In our example of the squid axon, the recorded membrane potential is close to the value predicted by the Nernst equation for potassium, but it is not exactly the

Fig. 4.5 **A.** The resting membrane potential and its dependence on K^+ outside the membrane. **B.** Graph of experimental results (open circles), and comparison with theoretical curves assuming dependence solely on K^+ (straight line) and including Na^+ (curved line). (From Hodgkin and Horowicz, 1959; see Huguenard and McCormick, 1994)

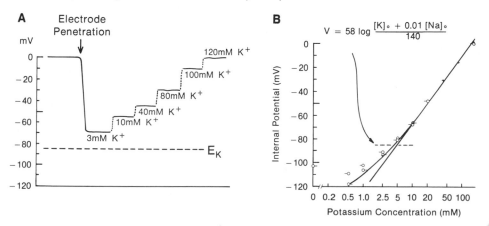

same. To test further the correspondence between theory and experiment, the concentration of $[K^+]_{OUT}$ in the bathing medium can be varied and the effects on the membrane potential observed. This is illustrated in Fig. 4.5A. As $[K^+]_{OUT}$ increases, the concentration gradient across the membrane decreases, as does the membrane potential; the membrane thus becomes progressively depolarized. The results are plotted in B and compared with those expected from the Nernst equation. It can be seen that there is a good fit for higher potassium concentrations, but the experimental curve deviates from the theoretical one at low concentrations, the prevailing situation under natural conditions.

These experiments suggest that normally the membrane potential reflects the presence of other ions besides K^+, such as, Cl^- and Na^+. We can calculate individual Nernst potentials for these ions as well; for Na, we have

$$E_{Na} = 58 \log \frac{[Na^+]_{OUT}}{[Na^+]_{IN}} \qquad (4.5)$$

$$= 58 \log \frac{460}{50}$$

$$= +55 \text{ mV}$$

Thus, the equilibrium potential for Na^+ has a polarity opposite that for K^+, in accord with their opposing concentration gradients (see Fig. 4.2). If the permeability of the membrane for Na^+ were equal to that for K^+, then the two equilibrium potentials would tend to cancel each other out, and the membrane potential would be near zero. However, the permeability to Na^+ has been shown to be only about 1/25 of that of K^+, so that the effect of Na^+ is to decrease (depolarize) only slightly the membrane potential from the K^+ equilibrium potential.

The combined effects on the membrane potential of more than one ionic species can be expressed in a single large equation as follows:

$$V_m = 58 \log \qquad (4.6)$$
$$\frac{P_K[K^+]_{OUT} + P_{Na}[Na^+]_{OUT} + P_{Cl}[Cl^-]_{IN}}{P_K[K^+]_{IN} + P_{Na}[Na^+]_{IN} + P_{Cl}[Cl^-]_{OUT}}$$

in which V_m = membrane potential and P = relative membrane permeability. This equation was derived by David Goldman of Bethesda in 1943. One of the premises on which it is based is that the potential gradient, or electrical field, within the membrane is constant; hence it is referred to as the *constant-field equation*. It lets us take account of the contribution of any ionic gradient to the membrane potential by simply weighting its effect in accord with the permeability.

For present purposes we neglect the contribution of Cl^- (it is near equilibrium across the membrane) and focus on the effect of Na^+. Using the weighting factor of $1/25 = 0.04$ for P_{Na} (and neglecting Cl^-), Eq. (4.5) yields a value of -60 mV for V_m, and the curve for different values of $([K^+]_{OUT})$ now falls very closely on the curve obtained from the experimental data (Fig. 4.5).

The Equivalent Circuit

We thus have a reasonable explanation for the membrane potential, and we would like to have a convenient way to represent it. Representing the various factors with diagrams is cumbersome, and carrying around equations in one's head is too abstract. The most convenient form is an electrical analogue, or model, as shown in Fig. 4.6. This little electrical circuit represents the membrane or, more correctly, a membrane site. Each equilibrium potential is represented by a *battery* across the membrane, which has the appropriate polarity and voltage *(E)* for that ion. In series with the battery is a *resistance (R)*, which is related to the membrane *permeability* for that ion. In order to understand this relation, we begin with Ohm's Law, which states that $E = IR$, where E is the electrical potential in volts (V), I is the current in amperes (A), and R is the resistance in ohms (Ω). The

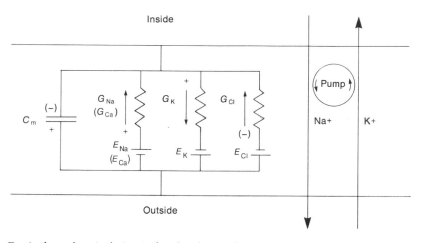

Inside

Outside

Fig. 4.6 Equivalent electrical circuit for the electrical properties of the nerve membrane.

reciprocal of the resistance is *conductance* (G), measured in siemens (S); that is, $R = 1/G$. Conductance is related to membrane *permeability (P)* as follows (using K^+ as the ion in question):

$$G_K \propto P_K \frac{[K^+]_{OUT}}{[K^+]_{IN}} \qquad (4.7)$$

Thus, theoretically the membrane could be quite permeable to K^+, but if there are no K^+ ions outside, the conductance would be zero (there would be no ions to carry the current through the permeability channels). Under physiological conditions, however, Eq. (4.7) is a reasonable approximation.

The "channels" for each ionic species are separate and independent, as shown in Fig. 4.6. In addition, the lipids of the membrane impart an electrical *capacitance* (C) to it; that is, they are poor conductors, and are able to store electrical charges on either side of the membrane. In fact, the membrane potential we record, representing the algebraic sum of the ionic batteries, is the potential due to the separation and storage of charge across the membrane capacitance (C), as illustrated in the circuit of Fig. 4.6. When changes in the membrane potential occur slowly, we can neglect the effects of the storage of charge across the capaci-

tance, but when changes are rapid, the time course over which the capacitance charges or discharges is a very important factor in shaping the signals. We will see later how this comes about during synaptic and impulse activity.

The Membrane Potential and Metabolism

The fact that we can account for our experimental recording of the membrane potential as a combination of diffusion potentials does not mean the end of our interest in how this potential arises. In many ways it is only the beginning. In our theoretical discussion we simply assumed a high concentration of K^+ inside the cell to begin with. But during life, the cell is supplied with nutrients and substances through the blood and interstitial fluid, where the K^+ concentration is, as we have seen, very low. How does the cell actually obtain K^+ and replenish the small amounts that are continually lost by leakage due to outward diffusion through the permeability channels? It cannot be by passive diffusion, because the concentration gradient is only in the direction of *outward* diffusion. A similar problem concerns Na^+; how does the cell prevent a buildup of Na^+ at rest, and particularly as a consequence of impulse activity (see next chapter)? And, in the face of these ion movements, how does the cell maintain osmotic equilibrium, so that it does not swell up and burst?

From these considerations it is clear that the cell is more than a mere bag of salt solution. The cell must contain metabolic mechanisms that can maintain and adjust ion concentrations under resting as well as changing conditions, and against ion concentration gradients. More specifically, there must be mechanisms in or at the membrane to pump ions across the membrane against their concentration gradients, and a source of energy in the cell to keep the pumps going. This process of moving substances in this way is called *active transport,* and the mechanism for doing it is called a *metabolic pump.* There is an obvious analogy with the raising of well-water by a good old-fashioned barnyard pump, worked by an energetic farm kid.

The first step in identifying this mechanism was to show that ion transport across the membrane requires energy. This was revealed in experiments on the squid giant axon, in which the rate of efflux of radioactive Na^+ was measured. Under resting conditions the rate is relatively low, but it can be raised by stimulating the axon repetitively, which loads the axon because of Na^+ influx during the impulse (see next chapter). Following such stimulation, the rate of Na^+ efflux is relatively high (see Fig. 4.7). The efflux is blocked when the axon is poisoned with cyanide, a well-known metabolic inhibitor. Injection of ATP partially restores the efflux (see Fig. 4.7), suggesting that the energy supply comes through ATP.

Direct evidence has subsequently been obtained that the membrane pump produces an actual current flow across the membrane. Such currents are called *membrane currents.* In order to demonstrate them, electrophysiologists use a method called a *voltage clamp,* in which the intracellular electrode is used not to record voltages but to record the currents that give rise to the voltages. The voltage clamp is applied to the membrane so that the membrane potential can be set at a particular level and the underlying membrane currents can be observed in isolation. This can be done with the appropriate electrical instrumentation. The basic features of the voltage clamp are illustrated in Fig. 4.8 and explained in the legend. This method was first applied to the squid giant axon, where it yielded the experimental results used for developing the model of the nerve impulse. It has subsequently become the workhorse method for analysis of currents underlying all the main types of membrane mechanisms in neurons and other cells of the body, as we shall see in subsequent chapters.

An elegant experiment demonstrating pump currents with the voltage clamp was carried out by Roger Thomas in London in 1972. He used a giant cell of the snail *Helix,* in which four electrodes (one a double-barreled electrode) could be inserted. The arrangement is shown in Fig. 4.9. Electrode ① was used to measure the membrane potential. Electrode ② was used to "clamp" the membrane potential at a preset value and record the current necessary to hold it at that value. Electrode ③ was used to inject Na^+ into the cell, and electrode ④ was used to complete the circuit, by injecting K^+, so that this current would not cross the membrane. Electrode ⑤ was an ion-sensitive electrode that monitored the Na^+ concentration in the cell. The results in Fig. 4.9 showed that when Na^+ was injected, it was necessary to pass an inward current to hold the membrane potential "clamped." This implied that there was an equal and opposite outward current associated with the raised Na^+ concentration, due to the action of the pump. The net current represented movement of only about one-third of the injected Na ions, implying that the pump ejects three Na ions for every two K ions that enter the cell. This ratio means that continued action of the pump will restore the ratios of Na^+ and K^+ to their resting levels, thus restoring the membrane potential to its resting

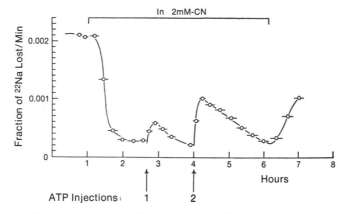

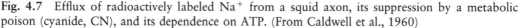

Fig. 4.7 Efflux of radioactively labeled Na$^+$ from a squid axon, its suppression by a metabolic poison (cyanide, CN), and its dependence on ATP. (From Caldwell et al., 1960)

Figure 4.8 **A.** Experimental setup for space clamp and voltage clamp of squid giant axon. A voltage source ① sets the membrane at a given level, which is recorded by an amplifier ②. This amplifier is connected to a feedback amplifier ③, which feeds back current across the membrane that just balances off the ionic current induced by the imposed voltage. The current is measured across a resistance ④. (After Kandel, 1976) **B–D.** Simplified equivalent circuits for flow of ionic currents under conditions of space and voltage clamp. **B.** In the clamped membrane, capacitative currents can be ignored, and the equivalent circuit reduces to conductance pathways for ionic currents driven by their electrochemical voltage sources. **C.** When K conductance is blocked, the membrane current (I_m) is due only to Na ionic flow (i_{Na}). **D.** When Na conductance is blocked, I_m is due only to i_K. (Modified from Hubbard et al., 1969)

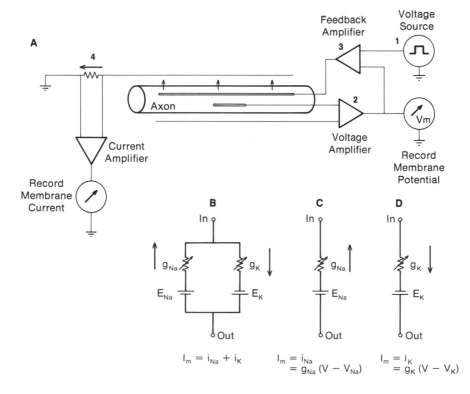

73

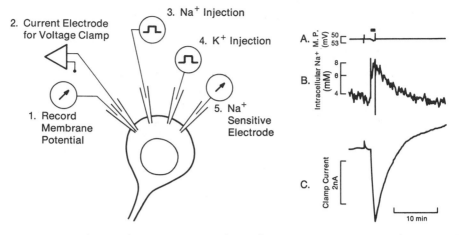

Fig. 4.9 Experimental setup for carrying out voltage-clamp experiments in a snail neuron. To the right are shown (**A**) the steady membrane potential recorded by electrode ① while the membrane is voltage-clamped by electrode ②; (**B**) the injection of Na$^+$ by microionophoresis through electrode ③, measured in milliamperes (mA) of the injection current; (**C**) the current due to the action of the Na$^+$ pump, recorded by electrode ⑤ under the voltage-clamp conditions maintained by electrode ②. Injection of K$^+$ could be done through electrode ④. See Chap. 5 for further details of the voltage-clamp method. (Modified from Thomas, 1972)

level. For this reason the pump is referred to as *electrogenic;* that is, its action contributes to the difference in ion concentrations on either side of the membrane which is the basis of the membrane potential.

The Membrane Pump

Because of the ions it transports, the pump is known as the Na$^+$, K$^+$ *pump;* because it requires high energy from the hydrolysis of ATP to ADP, it is also a *(Na$^+$, K$^+$)-ATPase.* Studies in red cell ghosts, in which the intracellular and extracellular conditions could be precisely controlled, showed that the pump works only when Na$^+$ and ATP are present on the inside and K$^+$ is present on the outside. The glycoside ouabain inhibits the pump by competing for the K$^+$ binding site on the outside.

Biochemical analysis has shown that the purified enzyme consists of a catalytic subunit (α) of approximately 100,000 daltons, coupled to a glycoprotein subunit (β) of approximately 45,000 daltons. The binding sites are on the catalytic subunit; the

function of the glycoprotein is as yet unknown.

The complete amino acid sequence of the α unit has been obtained by gene cloning. Analysis of hydrophobic regions in the sequence has suggested that there are eight transmembrane regions, giving the tertiary structure shown in Fig. 4.10. Between transmembrane regions 4 and 5 there is a large cytosolic domain, which contains the ATPase hydrolysis site. Binding sites for Na$^+$ and K$^+$ have been identified on cytosolic domains (see Fig. 4.10), but not yet on the extracellular domains. As indicated in the diagram, the α subunit shows considerable sequence homologies with Ca^{2+}-ATPase, which pumps Ca^{2+} out of the sarcoplasmic reticulum (see Chap. 17) as well as a bacterial K$^+$-ATPase, suggesting a common gene origin for these membrane proteins with similar enzymatic functions.

The pump functions as an antiport, because it transports two different ions in opposite directions across the membrane. The mechanism as presently understood is represented in Fig. 4.11. In A, the catalytic

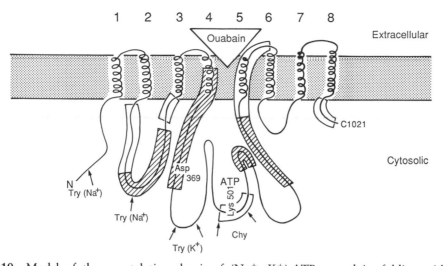

Fig. 4.10 Model of the α catalytic subunit of (Na^+, K^+)-ATPase and its folding within the plasma membrane, based on amino acid sequencing from cDNA clones. Numbers 1–8 indicate the hydrophobic regions that cross the interior of the membrane. Most of the molecule forms loops within the cytosol. Hydrolysis of ATP occurs at the lysine (Lys) residue site 501, coupled with an intermediate step of phosphorylation of aspartate (Asp) residue site 369. Na^+ and K^+ binding sites, determined by trypsin (Try) cleavage, are indicated in the diagram (Chy = chymotrypsin cleavage site). The chemical energy released by ATP hydrolysis is coupled to transport of Na^+ and K^+ across the membrane. The drugs ouabain and digitalis inhibit ion transport, leading to decrease of the membrane potential, by binding to a site(s) on the external surface of the α (or β) subunit. Sequence homologies with the Ca^{2+}-ATPase of the sarcoplasmic reticulum in muscle are indicated by open (clear) boxes, and further homologies with bacterial K^+-ATPase are shown by shaded boxes. (Based on Cantley, 1986)

subunit binds K^+ on the outside; this is followed by dephosphorylation, which causes a conformational change in the protein, resulting in transfer of K^+ into the cytoplasm. This leads to B, where Na^+ binding to its internal site is followed by phosphorylation, which induces another conformational change that transfers Na^+ to the outside. The protein is then ready to bind K^+ on the outside again, and the cycle repeats itself.

The Na^+, K^+ pump is found in all cells. It normally accounts for as much as one-third of the total energy consumed by the cell. This can rise to two-thirds in nerve cells after periods of electrical activity, when ion gradients are being restored. Both nerve impulses and synaptic potentials create these demands. Nerve impulses do so especially at nodes of Ranvier and in small nerve fibers, where there is a large surface-to-volume ratio, hence large movements of ions across the membrane into and out of a relatively small intracellular volume. Synaptic terminals create these demands because ion pumps and vesicle recycling are highly energy dependent.

Membrane Transport Mechanisms

In this discussion we have focused on one possible mechanism for carrier-mediated transport of ions. This is only one of the many ways that substances can move across a membrane. Figure 4.12 summarizes some of the main categories of transport mechanisms that have been identified in biological membranes. Beginning at the left are the simplest cases, of ① passive diffusion of ions and ② simple bulk flow.

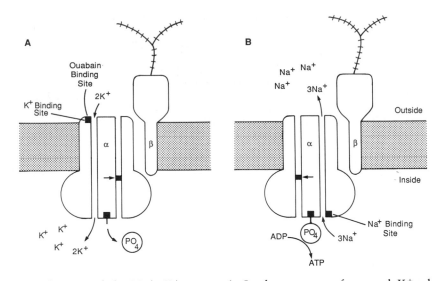

Fig. 4.11 Mechanism of the Na$^+$–K$^+$ pump. **A.** In the presence of external K$^+$, there is an allosteric conformational change which causes dephosphorylation at the ATP binding site. In this conformation, there is transport of K$^+$ to the inside, against the K$^+$ gradient across the membrane. **B.** When Na$^+$ binds to its specific site(s) on the inside, there is a conformational change which favors phosphorylation at the ATP binding site. This favors transport of Na$^+$ to the outside, against the Na$^+$ gradient. The transport occurs in the stoichiometric ratio of 3 Na$^+$ for 2 K$^+$. The molecule alternates between these two phosphorylation states and associated conformations to carry out its pumping function. (Based on Alberts et al., 1989; Cantley, 1986; Shull et al., 1986)

Fig. 4.12 Main categories of transport mechanisms in biological membranes. (Diagram adapted from C. L. Slayman)

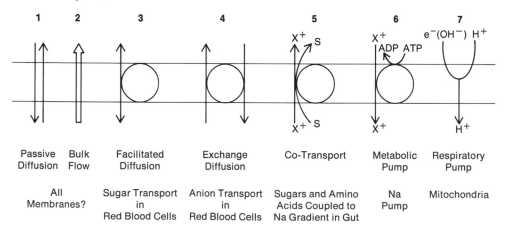

Next is carrier-mediated passive diffusion, in one direction ③ or in both directions ④. A very common passive mechanism is the linking of one substance to another; thus ⑤ sugars and amino acids are cotransported along with Na^+ down its concentration gradient in many cells. Finally, we have the systems requiring energy. These include the active pump just discussed ⑥, driven by high-energy phosphate, and the proton pump ⑦, driven by respiratory enzymes, that is present in the inner membrane of the mitochondrion, as discussed in Chap. 3. While these mechanisms are all present in biological membranes, most of them have also been demonstrated in artificial membranes composed of various organic substances. This has provided a powerful tool for experimental analysis, and has suggested that transport phenomena are to a large extent inherent in the properties of organic molecules and macromolecular complexes, when arranged in monolayers or very thin membranes.

In conclusion, both passive and active properties contribute to the membrane potential. These properties are present in different combinations in different cells. As a reflection of this, it is being recognized that the membrane potential does not have the same value in all neurons, nor do neurons respond in the same ways to change brought about by activity. In some cells or fibers the membrane potential may have a relatively high value, around -80 mV, inside negative. In other cells, the value may be lower, as low as -40mV. A low value is found, for example, in vertebrate retinal receptor cells, where it is associated with a large resting leak of Na ions (see Chap. 16). The metabolic mechanisms for active transport are *temperature dependent*, which means that their contribution to the membrane potential in poikilothermic animals will vary according to temperature changes during the day, or in different seasons. The amount of pumping will also depend on the *size* of a nerve process: the smaller the process, the higher the surface-to-volume ratio and the higher the rate of pumping activity required to maintain internal concentrations. If a neuron has a high level of spontaneous impulse firing, there may not be a "resting" membrane potential between the successive action potentials. Thus, the membrane potential, even at rest, is a very important variable for the different functions of nerve cells.

5

The Action Potential

The Nerve Impulse: A Brief History

The earliest ideas about the nature of the signals in the nervous system, going back to the Greeks, involved notions that the brain secretes fluids or "spirits" that flow through the nerves to the muscles. However, a new era opened in 1791 when Luigi Galvani of Bologna showed that frog muscles can be stimulated by electricity. His postulate of the existence of "animal electricity" in nerves and muscles soon led to a focus of attention almost exclusively on electrical mechanisms in nerve signaling.

In the 1840s, Galvani's countryman, Carlo Matteuci, who also had a distinguished career in Italian government, obtained the first evidence for the electrical nature of the nerve impulse. This was soon followed up and put on a sound and systematic basis by the extensive studies of Emil du Bois Reymond of Berlin. In 1850, Reymond's colleague, Herman von Helmholtz, later to be the famous physicist, was able to measure the speed of conduction of the nerve impulse, and showed for the first time that, though fast, it is not all *that* fast. In the large nerves of the frog, it is about 40 meters per second, which is about 140 kilometers per hour. This was another

landmark finding, for it showed that the mechanism of the nerve impulse has to involve something more than merely the physical passage of electricity as through a wire; it has to involve an *active biological process*. The impulse therefore came to be called the *action potential*.

The ability of a nerve to respond to an electrical shock with an impulse is a property referred to an *excitation*, and we say that the nerve is *excitable*. In the earliest experiments there were no instruments that could record the impulse directly; it could be detected only by the fact that, if a nerve was connected to its muscle, the shock was followed (after a brief period for conduction in the nerve) by a twitch of the muscle. The brief nature of the twitch indicated that an impulse must also occur in the muscle, so that the muscle was also recognized as having the property of excitability.

The electrical nature of the nerve impulse and its finite speed of conduction were important discoveries for physiology in general, and indeed for all science, because they constituted the first direct evidence for the kind of activity present in the nervous system. It appeared that, just as the heart pumps blood and the kidney makes urine, so now one could say that the nervous system produces impulses. In addition, the fact that the impulse moves at only a moderate speed had tremendous implications for psychology, for it seemed to separate the mind from the actions that the mind wills—in effect, it provided empirical evidence to support the idea of dualism—that the mind is separate from the body. It was one of the stepping stones toward the development of modern psychology and the study of behavior, as well as contributing to the debate on the nature of the mind and the body.

The Sodium Impulse

There are several essential facts about the impulse in nerve (or in any other cell, for that matter) that we start with in analyzing the underlying mechanism. First, the action potential is a *membrane event;* it consists of a transient change in the *membrane potential.* This was already suspected in the nineteenth century, and it has been elegantly demonstrated in squid axons, where impulses continue to be conducted even though all the axoplasm has been squeezed out.

The second important fact about the action potential is that it consists of a transient *depolarization* of the membrane potential. This had also been suggested by the experiments of the nineteenth century, but direct demonstration was first obtained with intracellular recordings from the squid axon. The basic experimental setup for recording the membrane potential in the squid axon has been described in Chap. 4 (see Fig. 4.4), and the similar arrangement for recording the impulse is shown in Fig. 5.1. The earliest results, by K. C. Cole and H. J. Curtis at Woods Hole in 1939, showed that not only does the membrane depolarize (in other words, become less negative inside), but it passes zero and actually becomes almost 50 mV positive inside at the peak of the action potential.

What can account for this finding? It cannot be simply a transient breakdown in permeability to allow all ions to move across the membrane, because that would only depolarize the membrane to zero, not beyond. The clue is provided by a third key fact, that in squid axons the action potential depends on the presence of *sodium ions* in the external medium. If they are removed, the action potential is reduced in amplitude, as illustrated by the recordings and the graph in Fig. 5.2. (This experiment may be reproduced by the student in Huguenard and McCormick, 1994.) It may be remembered that changing sodium has very little effect on the resting membrane potential, in accord with its low permeability relative to potassium in the resting membrane (see Fig. 4.5). The results of Fig. 5.2 indicate that, by contrast, the sodium permeability, and hence the sodium conductance, is very high at the peak of the action potential. That the sodium conductance can account for the peak is confirmed by the fourth fact, that the peak is near the Na equilibrium potential of approximately +55 mV, positive inside.

These facts thus suggest that the action potential in the squid axon involves a transient increase in the conductance of the membrane for sodium, providing for an inward rush of positively charged Na ions down their concentration gradient. This reduces the charge difference across the membrane and actually reverses it, as the membrane potential tends toward the sodium equilibrium potential of about +55 mV, positive inside.

What is the mechanism that produces this inward rush of sodium ions in response to a small depolarization of the membrane? The *energy* is provided by the electrochemical gradient of Na^+ across the membrane, according to the principles outlined in the preceding chapter. The path followed by the Na ions is provided by the *transmembrane channel* of a protein that is Na^+ *selective.* The explosive nature of the flow of Na^+ ions, triggered by an initial small depolarization of the membrane, is due to *voltage-sensitive properties* of the Na^+

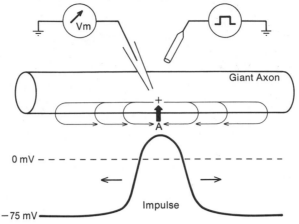

Fig. 5.1 The impulse in the squid axon. The impulse has been triggered by a brief depolarization at A. Note that the impulse has the ability to spread in both directions when elicited experimentally in the middle of a nerve.

channel protein. All of these properties can now be understood at the molecular level.

We will begin with the molecular structure of the sodium channel. Then we will discuss the properties of single-channel events and the synchronous actions of many channels that give rise to the sodium impulse. We will then consider the other two main types of voltage-gated channels that are involved in generating impulses, those that are selective for Ca^{2+} and K^+.

We will see that all three types of channel belong to the same family of membrane proteins, so that the principles derived from one can be applied to all. The distinct properties of each member derive from differences in amino acid residues at specific key sites in the proteins. We will see how these different channels mediate varied functions in neurons. Finally, we will take a perspective on the general significance of excitability and impulse generation in many different types of cells.

Fig. 5.2 Dependence of the action potential on Na ions. **A.** Impulse in normal (100%) seawater, and reduction in amplitude of impulse when external Na^+ is reduced to one-third normal concentration. **B.** Graph of effect of different external $[Na^+]$ on action potential amplitude. (From Hodgkin and Katz, 1949; see Huguenard and McCormick, 1994)

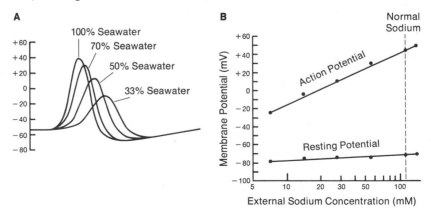

Molecular Structure of the Sodium Channel

The first information about the composition of the sodium channel came from biochemical studies. Certain toxins, such as tetrodotoxin (TTX), act by specifically binding to and blocking the sodium channel and the impulse. Biochemists took advantage of this property of TTX to do binding assays and obtain the channel protein in purified form. This work was first done in the electric eel, *Electrophorus electricus,* the same animal in which the initial work on the acetylcholine (ACh) receptor was done. As we note in Chap. 6, the high density of neuromuscular junctions in the electric organ (electroplaque) provided a rich source of ACh receptors; by the same token, the high density of modified muscle cells, whose summed action potentials generate the electric current which stuns the prey, provides a rich source of sodium channels.

It was found that the *Electrophorus* sodium channel consists of a single continuous polypeptide with a molecular weight of 260,000–300,000. The amino acid sequence of the polypeptide was determined by recombinant DNA techniques resembling those used to sequence the ACh receptor (see Chap. 7); in fact, the first work was carried out in the same laboratory of Shosaku Numa in Japan, by a team of 18 collaborators! The methods included microsequence analysis of purified polypeptide segments of the complete protein; preparation of a cDNA library from electroplaque mRNA; obtaining a cDNA complementary to one of these segments; use of this clone to screen a cDNA library to obtain seven longer overlapping clones; sequence analysis of these oligonucleotides, leading to the complete sequence of the entire cDNA coding for the sodium channel protein (Noda et al., 1986).

The entire cDNA consists of a chain of 7230 nucleotides, containing the triplet codes for a polypeptide chain of 1820 amino acids. Interestingly, the chain lacks the signal peptide at the amino terminal that is characteristic of a transmembrane protein translated by the ribosome on the endoplasmic reticulum (recall Chap. 3). It resembles other exceptions, such as the erythrocyte anion (chloride) channel (Chap. 3), certain ion pumps (Chap. 3), and rhodopsin (Chap. 16), in this regard.

In order to determine how this polypeptide chain forms a secondary structure that is inserted into the plasma membrane, the pattern of amino acid sequences was analyzed by computer. This showed that there is a pattern of about 300 amino acids forming a unit which is repeated four times. This repetition is called *internal sequence homology.* Second, the entire chain was analyzed for the hydrophobic or hydrophilic properties of the amino acids. This showed a pattern of six segments with each homology unit (Fig. 5.3A). Finally, the tertiary structure of the polypeptide was deduced from this analysis. The structure is shown diagrammatically in Fig. 5.3B.

For a channel protein to be the basis for impulse generation it must have three specific properties: (1) the ability to sense changes in membrane potential, in order to initiate voltage-sensitive ion flows; (2) a pore, with selectivity for a given ion species; and (3) a means for inactivation, so that it can be activated again. Much research has been directed toward identifying the parts of the Na^+ channel responsible for these properties. Our present understanding may be summarized as follows.

The Voltage-Sensitive Region. Movement of ions through the channel, which is blocked at resting potential, occurs when the membrane is depolarized to a sufficient amount (called the *threshold potential*). This implies that some part of the molecule must have the ability to sense a change in the transmembrane voltage. Clay Armstrong of the National Institutes of Health suggested a model in which a transmembrane segment with an excess of positive charges could act as a voltage sensor; he speculated that increasing membrane depolarization, placing positive charge on the inside of the channel segment, would exert outward force on the segment, causing a

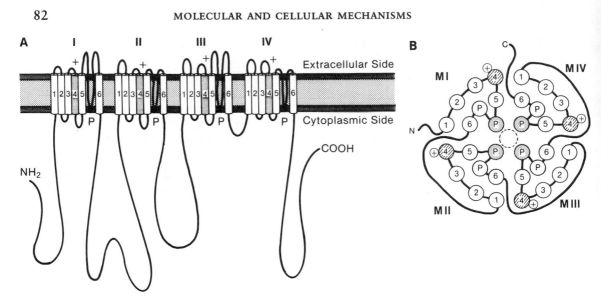

Fig. 5.3 Presumed tertiary structure of the Na$^+$ channel protein based on hydropathicity plots of the primary amino acid sequence. **A.** The channel protein consists of four repeating subunits, each containing six presumed transmembrane segments. Segment 4 contains an excess of positively charged residues and is assumed to be the voltage sensor. A long loop between segments 5 and 6 is believed to dip into the membrane and form the face of the pore. A cytoplasmic loop contains the inactivation gate. **B.** View looking down on the membrane to see the arrangement of the four subunits around the central pore. The Ca^{2+} channel protein is similar in its construction. (Modified from Catterall, 1988, and Stevens, 1991, in Kandel et al., 1991)

screwlike movement of the alpha helix that would give it a voltage-sensing property (Armstrong, 1981). This movement would also transfer positive charge toward the outside of the membrane, creating a virtual tiny current that in fact has been recorded with very sensitive electrical measurements as a *gating current.*

Current evidence points to the S4 segment as the likely site for this property (Catterall, 1988). The sequence analysis shows that this segment is unusual in that arginine and lysine, the most positively charged of amino acids, are arranged in every third position among otherwise neutral residues (see Figure 5.3). The movement of the voltage-sensor segment presumably causes an allosteric change in the rest of the molecule, which opens the channel slightly and allows Na ions to flow through it toward the inside of the membrane, down their concentration gradient. The interaction between the sensor and

the pore thus contains the heart of the impulse mechanism.

The Pore-Forming Region. The structure of the functioning protein is formed by arranging the four repeating subunits around a central pore, or channel. The channel must have a cross section no larger than 3×5 angstroms (= 0.3–0.5 nm) if it is to be selective in letting only Na$^+$ pass through. There may be room for only one segment from each subunit to form the face of the pore.

The identity of the pore-forming segment has been of considerable interest. This has been studied by making successive point mutations of different amino acid residues in the different segments. The results have pointed to a long stretch of residues between the highly hydrophobic transmembrane segments 5 and 6. This stretch is termed the P (for "pore") segment. As indicated in Fig. 5.3A, this is believed to dip into the membrane from the extracellular side. Current research is aimed at relating particular residues in this segment to the key properties of selectivity and permeability for Na$^+$ (see Fig. 5.4); these properties are not as well understood as for Ca^{2+} and K$^+$ channels (see below).

The Inactivation Region. As soon as the channel has been opened to allow a surge of Na ions to flow through it, there must be a mechanism to close it so that it can be activated again. Early studies by Armstrong and Francisco Bezanilla at the University of Pennsylvania showed that inactivation does not occur (the Na current stays on indefinitely) if protease (an enzyme that digests proteins) is injected into the cell. This pointed to a region of the membrane protein exposed to the inside of the cell as the site for the inactivation mechanism. It was speculated that the mechanism could work like a ball and chain, with the ball being a globular part of the protein that could plug the inside mouth of the pore, and the chain being a flexible part that could allow the ball to flop in and out of the mouth, depending on the voltage across the membrane. This mechanism is illustrated in Fig. 5.4. It has been tested most thoroughly at the molecular level in K$^+$ channels, as we will discuss shortly.

Single Channel Function

The physiological action of a single sodium channel polypeptide can be recorded by the patch-clamp technique (cf. Chap. 2). This was first accomplished in immature muscle cells (myotubes) grown in tissue culture. Typical recordings are shown in Fig. 5.5A. In this experiment, the potential across the membrane patch was controlled by a command voltage step from a hyperpolarized level to a depolarized level. The recording shows that the depolarization induced an immediate response, consisting of a square pulse of current. The response to a depolarization shows that these are voltage-gated events, in contrast to a synaptic receptor, which responds only to the presence of its

ligand (see Chap. 6). The responses are immediate, as expected if the voltage sensor induces an immediate conformational change in the channel wall to allow Na ions to begin to flow through the channel. The inward direction of the current across the membrane is in accord with the flow of Na ions down their concentration gradient.

The responses have the general characteristics of single channel events that we noted previously (Chap. 2). These include rapid opening to a constant plateau, rapid closing, and flickering of the open channel. When many of these single channel events are summed and averaged, one obtains, as in Fig. 5.5B, a smooth time course of the Na current. In these summed records, one refers to the onset as *activation* of the sodium channels, and the decay from the peak as *inactivation*. It can be seen that the Na channels undergo rapid activation, immediately followed by inactivation, in response to a depolarization of the membrane. This is exactly what is required for generating an impulse.

The Sodium Channel: Hodgkin-Huxley Model

The information about single channels in a patch of membrane must next be interpreted in relation to the actual situation, in which many sodium channels are distributed in the membrane of an axon. In this analysis, we want to know the time course of the ionic conductance change and current flow across the membrane, and we want to know also if any other ions might be involved. To do this we must set the membrane potential at different levels and determine how the ionic currents vary. This requires a technique for achieving a *space clamp* of the membrane, so that the impulse is held stationary over a length of the axon, together with a *voltage clamp* of that membrane, which enables the membrane potential to be set and held at a particular value and the response of the membrane measured in terms of the transmembrane current. The basic setup was illustrated in Fig. 4.8A.

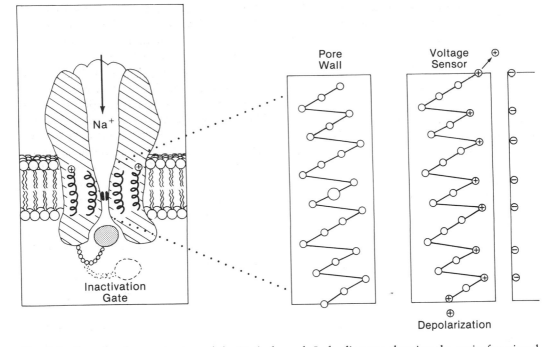

Fig. 5.4 Functional organization of the Na$^+$ channel. Left, diagram showing the main functional domains: pore, voltage sensor, and inactivation gate. The arrangements of the residues within the segments lining the pore and forming the voltage sensor are shown schematically on the right. (Based on Hille, 1992, and Unwin, 1993)

This method was applied by Alan Hodgkin and Andrew Huxley of Cambridge University in a famous series of experiments that was published in 1952. (You can reproduce these experiments with the computer models in Huguenard and McCormick, 1994.) The way they went about the analysis is indicated in Fig. 5.6A. They set the membrane potential at different levels and recorded the membrane currents during voltage clamping; as shown in Fig. 5.6A,a, the currents consisted of an early inward phase and a later outward phase. They then repeated the experiments while replacing the Na in the external medium. This produced recordings, as in Fig. 5.6A,b, consisting of only a later outward phase, which increased in amplitude as the membrane was set at more depolarized levels. They postulated and confirmed that this component of the current is carried by K$^+$; the current is stronger the further the membrane is depolarized away from the

K$^+$ equilibrium potential. When this component was subtracted from the control recording, they obtained an early component, as in Fig. 5.6A,c; this began to appear with small depolarizations, was largest at around zero potential, and reversed to an opposite polarity around +55 mV (Fig. 5.6B,c). This is what would be expected if this component was carried by Na$^+$, driven by the Na$^+$ equilibrium potential of around +55 mV.

Subsequent experiments using poisons that selectively block Na$^+$ and K$^+$ conductances have confirmed these basic results. As shown in Fig. 5.6B,c, when K$^+$ conductance is blocked by adding tetraethylammonium (TEA) to the external medium, only the early, Na$^+$ component remains. When, on the other hand, the Na$^+$ conductance is blocked by adding tetrodotoxin (TTX), a poison found in the ovary of the puffer fish, only the late, K$^+$ component is present.

A. SINGLE CHANNEL CURRENTS

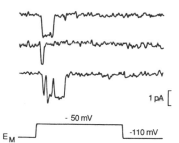

B. SUMMED CURRENTS

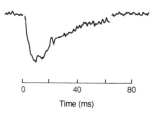

Fig. 5.5 Recordings of single voltage-gated Na channels. **A.** Three traces showing single channel activity in response to a depolarizing step from -110 mV to -50 mV (monitor of clamped membrane potential (E_m) is shown below). Note the abrupt channel openings and closings (cf. Chap. 2). **B.** Smooth record obtained by summing or averaging the responses in 144 traces. These are referred to as macroscopic, as contrasted with microscopic (single channel), currents. They resemble the results seen in intracellular and whole-cell patch recordings. (From Patlak and Horn, 1982)

From measurements of the ionic currents, Hodgkin and Huxley were able to obtain the Na^+ and K^+ conductances by the application of Ohm's law to the reduced circuits for the membrane, as illustrated previously in Fig. 4.8B–D. They then derived equations that describe the turning on of these conductances. The only remaining piece of the puzzle was to account for the fact that, after rising rapidly, the Na^+ conductance quickly falls. This was described by the process of Na inactivation; as we have noted, it is possible to block this selectively by application of the enzyme pronase within the axon. The interrelations between these three factors—*Na$^+$ conductance, K$^+$ conduc-*

tance, and *Na$^+$ inactivation*—are shown in Fig. 5.7. This figure also brings out a crucial property of the Na conductance—that it is involved in a positive feedback relation with the membrane depolarization. When the membrane begins to be depolarized, it causes the Na^+ conductance to begin to increase, which depolarizes the membrane further, which increases Na^+ conductance, and so on. This is the kind of self-reinforcing *regenerative* relation that characterizes various kinds of devices; for example, a similar relation between heat and chemical reaction underlies the explosion of gunpowder. One can say that it is the property that puts the "action" in the action potential! It gives the impulse a *threshold,* below which it fails to fire, above which it is fully successful; one says that it is "all or nothing." This property is due to the interaction between Na and K conductances. Below threshold, the constant outward flow of K across the resting membrane prevents depolarization by small increases in Na conductance; "threshold" is the amount of depolarization at which inward Na flow exceeds the passive outward K flow, so that the regenerative mechanism takes over. The nerve is briefly *refractory* to restimulation, because of the Na inactivation during the peak and the high K conductatnce that keeps the membrane hyperpolarized during the afterpotential. The regenerative property also allows the impulse to propagate at a *constant amplitude* for long distances. This nonlinear regenerative behavior, in which the Na^+ and K^+ conductances occur in *sequence* rather than simultaneously, is the key way in which the mechanism of the action potential differs from the mechanism of the synaptic potential.

Based on their analysis of the experimental results, Hodgkin and Huxley were able to finish their study by deriving equations for the variables controlling the conductances, and, using these equations, construct a model that reproduced the action potential. As implied in Fig. 5.7, the fit of experiment and theory is very close. Fur-

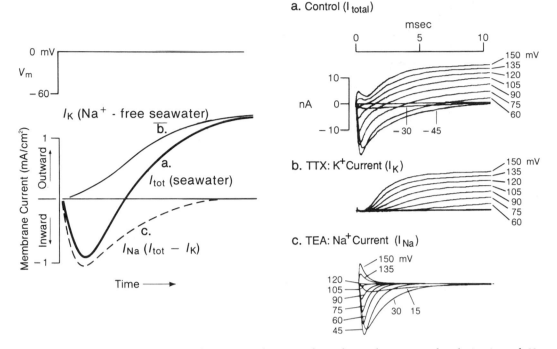

Fig. 5.6 **A.** Response of the squid axon membrane under voltage clamp, to a depolarization of 60 mV. a. Response in normal seawater. b. Response due to potassium current (I_K) when extracellular Na^+ is replaced by choline (a nonpolar molecule that maintains the osmotic pressure of the solution). c. Calculated response due to Na^+ current ($I_{Na} = I_{total} - I_K$). (From the original study of Hodgkin and Huxley, 1952) **B.** Separation of ionic currents by use of nerve poisons. a. Response in normal seawater. Different amplitudes of voltage steps are indicated on the right (in mV). b. Response due to I_K when I_{Na} is blocked by tetrodotoxin (TTX). c. Response due to I_{Na} when I_K is blocked by tetraethylammonium (TEA). (From Hille, 1977; see also Huguenard and McCormick, 1994)

ther details of the model are given in the legend, and in the references.

The Hodgkin-Huxley model of the nerve impulse is a consummate matching of experiment and theory; it is one of the great achievements of modern neurobiology and, indeed, of modern science. Its importance for neurobiology in general, and the study of membrane mechanisms in nerve cells in particular, cannot be overemphasized; it has been the solid touchstone for all subsequent research.

Calcium Channels

In the Hodgkin-Huxley model for the impulse it is Na ions, powered by the Nernst potential, that provide for the inrush of positive charge that depolarizes the membrane. Is there any other cation with a similar concentration gradient that could do the same? Inspection of Table 4.1 and Fig. 4.2 shows that free calcium ions have a concentration gradient in the same direction, with a Nernst potential that is even more positive than for Na.

Evidence for a calcium-based impulse was first revealed by Katz and his colleague Paul Fatt at London, in muscle fibers of the crab (cf. Fatt and Ginsborg, 1958). Since then, a great many studies have shown that virtually every excitable cell—nerve, muscle, and gland—has voltage-gated Ca channels in its membrane (reviewed by Hagiwara, 1983). The reason for this appears to be that Ca ions play critical roles

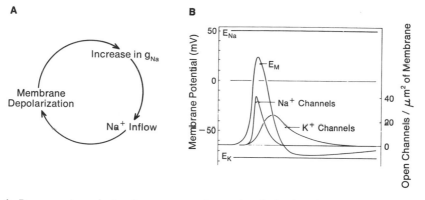

Fig. 5.7 **A.** Regenerative relation between membrane depolarization, increase in Na$^+$ permeability and conductance, and Na$^+$ current that underlies the action potential. **B.** Reconstruction of changes in ionic conductance underlying the action potential according to the Hodgkin-Huxley model; scale for the membrane potential (E_m) is shown on the left. The equilibrium potentials for Na and K are also indicated on the left $(E_{Na} = +50$ mV, $E_K = -77$ mV). Changes in Na and K ionic conductances are scaled on the right in terms of calculated open channels per square micrometer of membrane. The time courses of these changes are controlled by a set of equations that constitutes the Hodgkin-Huxley model. (For further details, see Hodgkin and Huxley, 1952; Hille, 1992; Huguenard and McCormick (1994))

in many fundamental cell processes. We have already had evidence of this in our discussion of the cell biology of the neuron (Chap. 3). Calcium channels are the main basis for impulse generation in the muscles of many invertebrate species, and in smooth muscles and many gland cells of vertebrates. They carry some of the inward current in vertebrate cardiac muscle and in many types of nerve cells across all phyla. However, their importance is not primarily in relation to long-distance impulse propagation, as in the case of Na channels. Instead, they have significance in relation to the functions controlled by Ca ions within the neuron. The most fundamental of these is the control of secretion of neurotransmitter (see next chapter). In addition, Ca ions play a variety of critical roles in shaping the patterns of impulse generation in neuronal cell bodies and dendrites, as we shall discuss in this chapter.

The molecular constitution of Ca channels has been revealed by studies of a type of channel that is found in the T-tubule system of skeletal muscle. As we shall see in Chap. 17, this channel binds dihydropyridines, a property which was used as a means for tagging the molecule and car-

rying out cloning and sequencing. The results have shown that the channel protein has a high sequence homology with the Na channel and a tertiary structure that is very similar, consisting of six transmembrane segments, an S4 voltage-sensor segment, and a P segment presumed to line the pore. Thus, the diagrams of Figs. 5.3 and 5.4 apply to the Ca channel as well.

A characteristic of Ca channels is that they are present at relatively low density in the membrane, so analysis of their properties depends heavily on the patch-recording technique in which individual channels can be isolated and analyzed. Figure 5.8 shows the analysis of two distinctive types of Ca channel. In one type, a relatively modest depolarization results in transient activation of the channel; the summed action of many channels thus consists of a transient peak with a rapid decay. This type is called a *T (for transient) channel,* or an *LVA (low-voltage activation) channel.* The other type requires a relatively strong depolarization but continues to undergo activation and inactivation during the sustained depolarization. This is termed an *L (for long-lasting) channel,* or an *HVA (high-voltage activation) channel.* As you might expect,

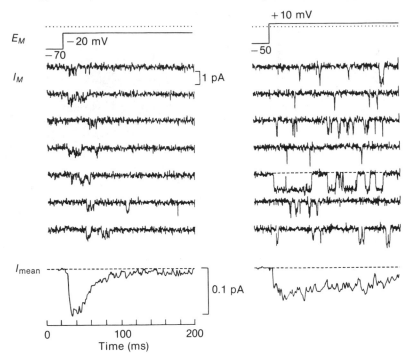

Fig. 5.8 Two types of calcium channels, as recorded from cardiac ventricle cells of the guinea pig by an on-cell patch clamp. **A.** T-type channel. Above, voltage command to the membrane potential (E_M); middle, successive recordings of single channel currents (I_M); below, record of summed traces. Note transient nature of channel openings and summed currents. **B.** L-type channel. Note the larger command depolarization, the persistent channel activity, and the sustained summed currents. (From Nilius et al., 1985)

the properties differentiating these two types are used for distinctive functions within the neuron. The T channel is used to shape patterns of impulse discharge into bursts in many central neurons (see below). The L channel is deployed at synaptic terminals to control secretion of neurotransmitter (see next chapter). Between these two types are many gradations in terms of threshold, activation and inactivation kinetics, amount of unitary conductance, and so on. We will discuss these gradations and the interactions of Ca and K channels in relation to impulse patterns later in this chapter.

Potassium Channels

The K^+ gradient across the membrane is in the opposite direction from that for Na^+ and Ca^{2+}, so that whereas Na and Ca conductance increases are in the depolarizing direction, K conductance increases are in the polarizing direction. We have already seen how the neuron makes use of this property with two different types of membrane conductance. One is maintenance of the resting membrane potential; these channels are activated by very small amounts of depolarization, are essentially noninactivating, and hold the membrane potential at a relatively polarized level (-70 to -80 mV) near the K equilibrium potential of around 95 mV (see Chap. 4). The other is the delayed rectifier K channel (I_K). This is an integral part of the Hodgkin-Huxley model for the nerve impulse. As we have seen (Fig. 5.7), this is activated by relatively strong depolarization, such as occurs during the upswing of the action potential, and acts to restore quickly the membrane potential to its resting level.

It turns out that there are many different types of K channels. Although it was at first bewildering, there seems to be logic behind the diversity. The nerve cell is ceaselessly active in a variety of ways, and the K channels constitute one of the main mechanisms for maintaining the equilibrium of the cell in the face of these demands. Since the potassium equilibrium potential is usually near the resting membrane potential, activation of K channels tends to return the membrane potential to this level during the following periods of Na or Ca impulse activity. Without this mechanism, the cell would tend to go into excessive electrical activity and the organism would have persistent seizures. Buildups of Ca are especially to be avoided because of the critical role in Ca in many intracellular processes; therefore, it would be sensible for the cell to have a mechanism for turning on a K conductance after a Ca conductance. Also, some impulses are quite prolonged, in which case it would be an advantage to turn off the K conductance during the impulse so that it would not oppose the prolonged depolarization.

Molecular Basis of the A Channel

Currently, the best understood K channel is the so-called A channel. This was identified in neurons of the sea slug *Anisodoris* by John Connor and Charles Stevens at the University of Washington as a K conductance that is activated by small depolarizations, but only transiently (Connor and Stevens, 1971). Its role during the action potential is relatively small; its main role is in neurons whose membranes have a resting depolarization that tends to make them fire trains of impulses. In these cells, between impulses, the slow depolarization of the membrane is opposed by the tendency of the A conductance toward repolarization. The role of the A conductance is therefore to control relatively long intervals between impulses in the range of 20–200

msec, for example, at low rates of firing (5–50 impulses per second). As a cell fires faster, its frequency will be increasingly limited by the absolute refractory period set by the time to repolarize after the impulse, which usually takes 1–2 msec (see Fig. 5.7); that is, the limiting impulse frequency for most cells is 500–1000 impulses per second.

The molecular basis of the function of the A channel has been analyzed in the fruit fly *Drosophila*. Larry Salkoff and Robert Wyman at Yale found that A currents in muscle cells were affected in a mutant strain called *Shaker* (because their legs shake under ether anesthesia) (Salkoff and Wyman, 1981). Molecular cloning of the Shaker gene, as well as genes for other voltage-gated K channels, has been carried out. The Shaker channel protein turns out to be equivalent in size to one of the four subunits of the Na and Ca channel proteins, with a high sequence homology to those subunits. Thus, it contains six transmembrane segments, a highly charged S4 voltage-sensor segment, and a long P segment which may form the lining of the pore (see Fig. 5.9). It is presumed that four of these units combine to form the functional A channel. At least four variations on this basic structure are produced from the same Shaker gene locus by alternative RNS splicing. Homomeric A channels (of identical subunits) and heteromeric A channels (of different subunits) are all functional, but with different activation and inactivation properties. It is tempting to speculate that by "mixing and matching" subunits, the cell achieves a fine tuning of channel properties for specific loci within the cell.

The A channel has been an attractive subject for analysis of the molecular basis of channel function. Two functions have been of particular interest so far: inactivation and pore structure. Point mutations of different residues allowed Richard Aldrich and his colleagues at Stanford to carry out an elegant series of experiments in which they could show that the inactivation mechanism takes the form of the ball-and-

K⁺ channel

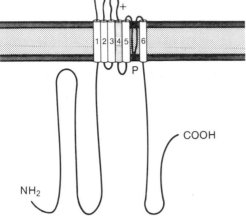

Fig. 5.9 Presumed tertiary structure of the K⁺ channel polypeptide based on hydropathicity plots of the primary amino acid sequence. The polypeptide consists of one unit containing six presumed transmembrane segments. Segment 4 contains an excess of positively charged residues and is assumed to be the voltage sensor. A long loop between segments 5 and 6 is believed to dip into the membrane and form the face of the pore. A cytoplasmic loop contains the inactivation gate. Four units together are believed to form the complete K⁺ channel protein. Note the similarity to the Na⁺ (and Ca²⁺) channel protein (Fig. 5.3A). (Modified from Catterall, 1988, and Stevens, 1991, in Kandel et al., 1991)

chain model, as we discussed in relation to Fig. 5.4 (Hoshi et al., 1990). Pore structure has similarly been investigated by making point mutations and observing the effects on membrane currents. Studies by Rod MacKinnon and his colleagues at Harvard (Heginbotham et al., 1992) have yielded the fascinating result (see Fig. 5.10) that mutations in only one or two critical residues within the P segment have profound effects on channel selectivity, changing from K⁺ to Ca²⁺ to nonspecific cation (K⁺, Ca²⁺, and Na⁺). These results not only support the idea that all three of the voltage-gated cationic channels we have discussed here belong in the same family, but they also extend the family to include

relatives in the family of cyclic nucleotide-gated nonselective cationic (CNGC) channels. Some of these relations are indicated in Fig. 5.10. We will study the CNGC channels in relation to nitric oxide (Chap. 8) and sensory transduction in the olfactory (Chap. 11) and visual (Chap. 16) systems.

Models of Ionic Conductances

The foregoing account should indicate that progress in understanding voltage-gated channels is providing a much clearer view of the principles underlying neuronal functions than we dreamed of even a decade ago. With only the Na mechanism for long-distance impulse communication, the active properties of the rest of the neuron were a mystery. With the addition of the varieties of Ca and K channels, the full range of the cellular functions of the neuron has come into the picture, and the full range of impulse patterns that neurons generate can be appreciated (see Llinás, 1988). It is as if we are listening to a symphony by a full orchestra, with the different combinations of instruments exploring many worlds of sound and tempo.

To understand how the different ionic conductances work together to produce different patterns of impulse firing, we will summarize the functional properties of each of the main channel types (see also the summary in Table 5.1). I am indebted to Stephen J. Smith for providing most of this account. We will show how the individual channel types contribute to specific patterns in the firing of a neuron. To simplify this task, Fig. 5.11 has at its center the simple pattern of impulse discharge based on the Na and K conductances of the Hodgkin-Huxley model; radiating from it are examples of modulations of this firing pattern brought about by adding some of the other types of membrane conductances that are discussed in the text.

A large step forward in analyzing and understanding these properties has been the ability to make specific computer models

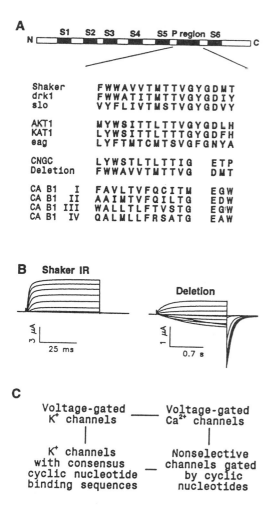

of them. This approach began in the 1960s with the work of Wilfrid Rall and his colleagues (see Rall, 1967; Rall and Shepherd, 1968), and in recent years has blossomed with the development of modeling software and the arrival of powerful workstations. As a result, each membrane conductance can be modeled so that its properties can be explored and its contribution to the integrative actions of individual neurons can be assessed. A complete tutorial on these models has been developed by John Huguenard and David McCormick, entitled *Electrophysiology of the Neuron: An Interactive Tutorial* (1994), as a companion to the present text. Its examples are carefully integrated with examples in this text, so that the reader can simulate each of the main types of ionic conductance and analyze its properties. By correlating the readings here with the tutorial, the student can begin to capture the excitement of doing experiments on real membrane signals.

Sodium (Na) Channels

Activation Characteristics. As discussed above, the classical $I_{Na(t)}$ channels underlying the axonal impulse are closed at the resting potential but open within 1 msec or less upon membrane depolarization. Na channels remain open for a maximum of a few milliseconds, even if depolarization is maintained, because of the characteristic inactivation process inherent in the channels' voltage-gating mechanism. When the

Fig. 5.10 Mutation analysis of the pore region of the Shaker K⁺ channel. **A.** The Shaker channel polypeptide is shown, with the presumed six transmembrane segments and the pore (P) segment. Below is shown the deduced amino acid sequences of the pore segment for Shaker, for a Shaker mutant (Deletion), and for other members of the S4-containing ion channel superfamily. (Drk1 is a mammalian voltage-activated K channel; slo is a Ca²⁺-activated K channel; AKT1 and KAT1 are K⁺ channels from the plant *Arabidopsis;* eag is another K⁺ channel from *Drosophila;* CNGC is the cyclic nucleotide-gated channel from bovine retina; and CA B1 I–IV are homologous segments of a brain Ca²⁺ channel. The group consisting of AKT1, KAT1, and eag are K⁺ selective and have pore sequences more like Shaker, but they resemble CNGC channels in their overall sequence homology and their possession of a cy-

clic nucleotide binding sequence.) **B.** Voltage clamp records of whole-cell currents from *Xenopus* oocytes injected with mRNA coding for the Shaker and mutant (Deletion) K⁺ channel. Note that the mutant shows voltage gating, but the outward (upward) current is slower, and there is loss of selectivity for K⁺, as shown by inward (downward) tail current after the pulse. **C.** Interpretation of the evolutionary relations between different types of channels in the S4 ion channel superfamily. (Modified from Heginbotham et al., 1992)

Table 5.1   Neuronal ionic currents

Current	Description	Function
Na$^+$		
$I_{Na,t}$	transient; rapidly inactivating	action potentials
$I_{Na,p}$	persistent; noninactivating	enhances depolarizations; contributes to steady-state firing
Ca^{2+}		
I_T, low threshold	"transient"; rapidly inactivating; threshold negative to -65 mV	underlies rhythmic burst firing
I_L, high threshold	"long-lasting"; slowly inactivating; threshold around -20 mV	underlies Ca^{2+} spikes that are prominent in dendrites; involved in synaptic transmission
I_N	"neither"; rapidly inactivating; threshold around -20 mV	underlies Ca^{2+} spikes that are prominent in dendrites; involved in synaptic transmission
K$^+$		
I_K	activated by strong depolarization	repolarization of action potential
I_C	activated by increases in [Ca^{2+}]$_i$	action potential repolarization and interspike interval
I_{AHP}	slow afterhyperpolarization; sensitive to increases in [Ca^{2+}]$_i$	slow adaptation of action potential discharge
I_A	transient, inactivating	delayed onset of firing; interspike interval; action potential repolarization
I_M	"muscarine" sensitive; activated by depolarization	contributes to spike frequency adaptation
I_Q, I_h	"queer"; activated by hyperpolarization/ mixed cation current	prevents strong hyperpolarization
$I_{K,leak}$	contributes to neuronal resting "leak" conductance	helps determine resting membrane potential

Adapted from McCormick (1990)

membrane is returned to the resting potential range, Na channels are reprimed for subsequent activation cycles (see Fig. 5.11). The unitary conductance of the channel is small, approximately 10 picosiemens (pS).

Pharmacology. A broad range of naturally occurring toxins block or otherwise modify Na channel function. The guanidinium toxins, represented by tetrodotoxin (TTX) and saxitoxin (STX), are strong blocking agents which have been widely used to block Na currents in physiological studies and to label channel proteins for molecular analysis or determination of density (see Table 5.2).

Diversity. There are subtypes (for example, $I_{Na,p}$) that activate and inactivate more slowly, or show little inactivation; they are found particularly in cell bodies and dendrites in CNS neurons (see Table 5.1).

Calcium (Ca) Channels

Activation Characteristics. Ca channels are normally closed at the resting membrane potential and opened reversibly by depolarization to more positive potentials. The opening and closing occurs within a few milliseconds in most instances. Ca channels are also subject to inactivation during prolonged depolarizations. Some instances of Ca channel inactivation are secondary to intracellular Ca ion accumulation; others are a direct effect of membrane depolarization.

Pharmacology. Agents used to block Ca channel conduction include a number of inorganic cations: Cd^{2+}, Co^{2+}, Mn^{2+}, Ni^{2+}, and La^{2+}. Organic Ca channel blocking agents include verapamil, diltiazem, and the dihydropyridines (nifedipine, nitrendipine, nisoldipine, etc.). A number

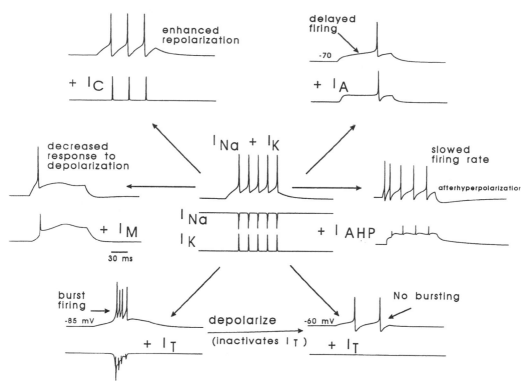

Fig. 5.11 A summary of different types of voltage-gated currents and the impulse firing patterns they produce in a neuron in response to a steady injection of depolarizing current. At the center is shown the repetitive impulse response of the classical Hodgkin-Huxley model (voltage recordings above, current recordings below). Radiating out from it are the changes in this pattern associated with the addition of different types of ionic channels. See text and Table 5.1. Kindly provided by David McCormick. These different patterns are reproduced by the models in the tutorial workbook and computer program of Huguenard and McCormick (1994): see their Fig. 11 (Hodgkin-Huxley); Fig. 12 (I_A); Fig. 14 (I_C, I_L, I_K); Fig. 15 (I_{AHP}); Fig. 16 (I_T); Figs. 17, 18 (I_M).

Table 5.2 Densities per μm^2 of channels, pumps, and receptor proteins in membranes of neurons and other cells

Sodium channels

Garfish olfactory nerves	1–35
Squid giant axon	100–600
Rabbit vagus nerve	100
Rabbit myelinated nodes of Ranvier	12,000

Sodium pumping sites

Garfish olfactory nerve	300
Rabbit vagus nerve	750

Receptor molecules

Muscle endplate: acetylcholine receptors	10,000
Rhodopsin in photoreceptors	50,000
(fat cells; insulin receptors)	1

For references, see Hille (1992)

of protein toxins from snails of the genus *Conus* and some spiders also have been found to block various Ca channels. Certain dihydropyridines, such as Bay-K 8644, have been found actually to promote the opening of some Ca channels. Ba^{2+} and Sr^{2+} ions can permeate through Ca channels and are sometimes substituted for Ca^{2+} ions in studies where investigators wish to eliminate other specific actions of the Ca^{2+} ion, such as intracellular activation of second messengers or vesicle release.

Diversity. Multiple Ca channel subtypes have been distinguished on the basis of activation voltage threshold, inactivation characteristics, single-channel conductance, and pharmacological sensitivities. In many instances, multiple subtypes are present on individual cells.

L type. This is one of several types of Ca^{2+} channel identified in neurons by Richard Tsien and his colleagues (Nowycky et al., 1985). The *L* (long-lasting) channel has a high threshold; it is opened by strong depolarization and is noninactivating (see Fig. 5.8B). Its unitary conductance is 25 pS. It is blocked by dihydropyridines and by omega toxin from *Conus.*

T type. This type, by contrast, has a very small unitary conductance (9 pS). It has a low threshold, around a membrane potential of-70 mV, and is rapidly inactivating (hence *T* for transient) (see Figs. 5.8A and 5.11). Low-threshold Ca spikes of this type may be important in boosting weak signals in dendrites, contributing significantly thereby to subthreshold synaptic integration. T channels play an important role in voltage-dependent burst discharges (see Fig. 5.11 and Chap. 25)

N type. N stands for "neither" (*T* nor *L*). This channel has a high threshold, requiring strong depolarizations (around -30 mV) to activate it. In response to continued depolarization, it slowly inactivates. It has a unitary conductance of 13 pS, and is blocked by omega toxin. There is evidence that it may regulate neurotransmitter release from presynaptic terminals.

I_B *type.* In addition to Ca^{2+} current during the impulse, a separate Ca^{2+} current has been identified that is very slow. This current provides for a slow depolarization that underlies the generation of bursts of impulses in certain kinds of pacemaker neurons and in neuroendocrine cells when they secrete their neurohormones.

Potassium (K) Conductances

The action of all K channels is to stabilize the membrane potential at a relatively polarized level (near E_K), and thus to oppose depolarization, which would lead to increased impulse firing or neurotransmitter output. K channels are thus diversified into many types to control the variety of excitatory influences on the cell.

Leak ($I_{K,leak}$; I_L). This is perhaps the most ubiquitous of the K channels. It activates extremely slowly with small depolarizations. It is thus well suited for maintaining the resting membrane potential, and also contributes to the repolarizing phase of the action potential, impulse frequency coding, and subthreshold synaptic integration. It also serves to link intracellular metabolism to membrane excitability via protein phosphorylation and Ca^{2+}. The unitary conductance is approximately 20 pS. It is blocked by TEA, 4-aminopyridine (4-AP), Ba, and Co.

Delayed Rectifier (I_K; I_{DR}). This is the classical conductance of the squid axon (see Fig. 5.11). It is activated after a brief delay by strong depolarization and inactivates very slowly. The unitary conductance is approximately 10 pS in squid axon, and the blockers are similar to those for I_L. At some sites this late K conductance may be absent (this appears to be the case at the node of Ranvier). At these sites, repolarization of the membrane occurs as a consequence of Na^+ inactivation.

Calcium-Dependent K Channel (I_C). The voltage dependence of this current is mostly secondary to activation of voltage-dependent Ca channels. The inflow and cytoplasmic accumulation of Ca leads to opening of this type of K channel. The resulting slow hyperpolarization is important in shaping action potentials in neuronal cell bodies and controlling slow rates of impulse firing by opposing increases in excitability associated with increase in Ca influences (see Fig. 5.11). The unitary conductance in snail neurons is 10–20 pS. It is blocked by apamin.

Afterhyperpolarization (I_{AHP}). This current is slowly activated by depolarization, and continues to build up during a repetitive spike discharge (see Fig. 5.11). This buildup reflects the sensitivity of this current to the buildup of Ca^{2+}. The effect of I_{AHP} is thus to cause the gradual slowing of a prolonged spike discharge; this slowing is called *spike frequency adaptation*. This property enables a neuron to cease firing to a maintained excitatory input. Central neurons vary in the degree to which they express this important property, and the degree to which it can be modulated.

Early K (I_A). This type is also called fast transient K. It is rapidly activated and inactivated by small depolarizations and is turned on particularly after being deinactivated by hyperpolarizations such as follow an impulse. It therefore is well suited for opposing interimpulse depolarizations, thus contributing to control of slow impulse firing rates (see Fig. 5.11). The unitary conductance is approximately 20 pS, and it is blocked by 4-AP.

Anomalous Rectifier (I_{AR}; I_Q;I_h). All of the K conductances considered thus far turn on with depolarization, to help return the membrane to its resting potential level. There are, however, K conductances that turn *off* with depolarization and turn on only when the membrane is hyperpolarized above its resting level. In this range, the K current is inward (to drive the membrane toward E_K). This type of channel was therefore called the inward rectifier, or anomalous rectifier. Since this conductance is turned off during depolarization, it helps to maintain prolonged depolarization of the membrane, as during the depolarized plateaus of long-lasting impulses. This is important in the cardiac action potential (see Chap. 17), in the prolonged action potential generated by many egg cells in association with fertilization, and during the Na action potentials generated in the electroplaque (in this case their function is presumably to enhance the depolarization by the Na action potential, in order to generate the greatest possible amount of electric current by the electroplaque organ).

M Current (I_M). A final interesting type of current to be noted is I_M. This is activated by depolarization, but, unlike I_K, it does not inactivate. This constant current can be affected by neurotransmitters and appears to provide a mechanism for adjusting the sensitivity of a neuron to synaptic inputs. (See Fig. 5.11).

In the electrical analogue of the membrane, these ionic conductances are present as elements in parallel with the others. Figure 5.12 shows how they can be incorporated into the electrical circuit analogue of the membrane previously shown in Fig. 4.6.

Channel Densities and Local Excitability

Using the fact that the nerve poison tetrodotoxin (TTX) selectively blocks Na^+ channels, studies have been carried out to determine the amount of binding of radioactively labeled TTX (and related poisons) to nerve fibers, in order to obtain an estimate of the density of Na^+ channels per area of membrane. Studies of this type have revealed that the density varies considerably in different neurons and in different parts of the neuron. Table 5.2 summarizes some results, together with estimates for related properties such as pumping sites and receptors. The extremely high density of Na^+ channels in the node of Ranvier (see Chap. 3) presumably reflects the specialization of this small patch of membrane for generation of the action potential during saltatory conduction (see below). The differences among the nerves shown may reflect in part their differing sizes; for example, olfactory nerves are among the thinnest fibers in the nervous system (diameter $= 0.2\mu m$), with correspondingly very high surface-to-volume ratio.

These studies make it clear that the density of voltage-dependent ion channels varies widely, as an expression of the functional specialization of the nerve

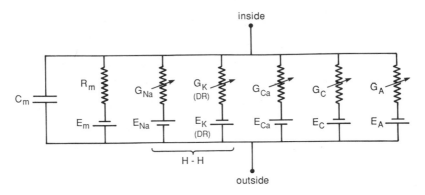

Fig. 5.12 Equivalent circuit of a patch of neuronal membrane which includes different types of voltage-gated ionic channels discussed in the text, and illustrated in Fig. 5.11. This is an expanded version of the single equivalent circuit in Fig. 4.6. Leak resistance of the membrane is R_m. H–H refers to the Na and K conductances of the Hodgkin-Huxley model. Other conductances shown are calcium (G_{Ca}), calcium-activated K ($G_{K(Ca)}$), and transient K ($G_{K(A)}$). For other abbreviations, see Fig. 4.6.

membrane. Local regions of a neuron may thus have different excitable properties; this is an extension of our discussion of membrane microdomains in Chap. 3, and it is an important principle underlying the complexity of neuronal organization.

Voltage Dependence and Neurotransmitter Sensitivity

Thus far we have considered the ion channel to be formed by a macromolecular complex which permits ions to pass through according to the potential difference across the membrane. This is a purely "voltage-dependent" channel. Traditionally it was believed that this property made impulse channels distinct from synaptic channels, which were presumed to be sensitive only to the action of a chemical transmitter substance liberated from another neuron. However, it has been found that in many cases the voltage-dependent channels are also sensitive to neurotransmitters.

A nice demonstration of this was provided by the experiments of Kathleen Dunlap and Gerald Fischbach, at Harvard. They made intracellular recordings from disseminated cell cultures of the chick dorsal root ganglion (Fig. 5.13). These cells have a prolonged action potential, with an early peak that is due to Na^+ current and a later component that is carried by Ca^{2+}. Neurotransmitter substances were ionophoresed onto the cells from extracellular micropipettes. As shown in Fig. 5.13B, several of these substances (serotonin, γ-aminobutyric acid, and norepinephrine) reduced the slow Ca^{2+} component of the impulse. Other substances, such as dopamine or various peptides, had no effect.

These cells lack dendrites and receive no synapses onto their cell bodies. However, their axon terminals in the spinal cord receive synapses from several types of fibers (these belong to the category of axoaxonic synapses). It has been suggested that these fibers may release one or another of the above transmitters to depress the Ca^{2+} component of the impulse and thereby reduce the amount of transmitter released by the dorsal root fibers onto motoneurons and other spinal cord cells. This mechanism of reducing the input to motoneurons is called *presynaptic inhibition.* A similar mechanism has been invoked to explain plastic changes of synapses underlying learning in *Aplysia,* as we will discuss in Chap. 29.

These studies have shown that not only may there be excitable channels at specific

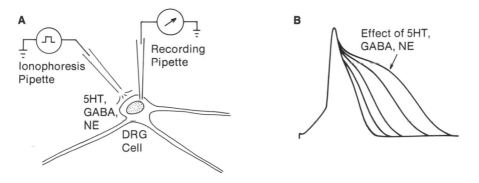

Figs. 5.13 A. Experimental setup for recording intracellularly from a single chick dorsal root ganglion cell in culture while ionophoresing neurotransmitter substances. **B.** Effects of different substances on impulses. Long-duration spikes before, short-duration spikes after, ionophoresis of substances. (Diagram B redrawn from Dunlap and Fischbach, 1978)

sites on a neuron, but these channels may also be sensitive to the actions of neurotransmitters and/or internal second messengers. On the other hand, work on synapses has shown that the response to a neurotransmitter may depend on the level of the membrane potential. Thus, at one extreme, is the traditional voltage-gated channel, such as the Na channel underlying the axonal impulse, and at the other is the traditional rapid synaptic channel, which can be activated either directly by a neurotransmitter or indirectly by a second messenger. Between are channels, such as I_M, that contain mixtures of both properties. These channels provide the nervous system with greater flexibility in modifying trains of impulses or integration of synaptic inputs, depending on the ongoing state of the organism and level of activity in its neural circuits. This should be kept in mind when we discuss the properties of synapses in Chaps. 7 and 8.

Conduction of the Action Potential

We have seen that, during the action potential, positive current carried by Na^+ or Ca^{2+} flows into the cell. Where does it go from there? As shown in Fig. 5.14, the current will have to flow back out through the membrane to complete the circuit. It cannot all return at the point of entry, and

it therefore spreads out along the fiber, seeking pathways of least resistance to get outside. How far it extends down the fiber depends on the ratio of the resistance of the cytoplasm to the resistance of the membrane. The higher the membrane resistance (or the lower the cytoplasm resistance), the further the current will tend to flow along the fiber.

Local Currents

This spread of electric current through the constant resistance and capacitance properties of the nerve is called *electrotonus*. It was first studied in the late nineteenth century. At that time, electric cables for long-distance telephone communication were being laid down, and it was recognized that the equations used to describe the spread of electricity in the cables were similar to those that applied to nerve. From that time, the electrotonic properties of nerve cells have been referred to as *cable properties*.

We will consider the spread of electrotonic potentials, and their crucial role in the integrative functions of dendrites, in Chap. 7. Here we simply point out that whenever an impulse is set up, electrotonic currents flow through the neighboring membrane. These currents, also called *local currents*, spread the depolarization of an impulse to neighboring membrane sites where, if threshold is reached, the impulse

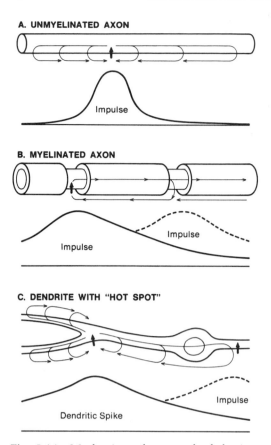

A. UNMYELINATED AXON

Impulse

B. MYELINATED AXON

Impulse

Impulse

C. DENDRITE WITH "HOT SPOT"

Dendritic Spike

Impulse

Fig. 5.14 Mechanisms for spread of the impulse. **A.** Continuous conduction in an unmyelinated axon. Amplitude scale is in millivolts. **B.** Discontinuous (saltatory) conduction from node to node in a myelinated axon. **C.** Discontinuous spread from "hot spot" to "hot spot" in a dendrite. In all diagrams, the impulses are shown in their spatial extent along the fiber at an instant of time. The extent of current spread is governed by the cable properties of the fiber. For an orientation to these properties, see Jack et al. (1975) and Shepherd (1990).

is also generated. The process continues along the fiber, at a rate determined by the size of the fiber: the larger the fiber, the faster the rate, other things being equal. This is a function of the cable properties; the larger the axon, the smaller the internal resistance and the farther the spread of the local currents (see Fig. 5.14). We say that the local currents spread *passively*, whereas the impulse is *conducted*, or *propagates*.

Continuous propagation of the impulse, as just described, occurs in unmyelinated axons. These may be very large, in certain invertebrate axons like the squid giant axon. In vertebrates, they are the smallest fibers, from several micrometers in diameter down to around 0.2 μm.

Saltatory Conduction

Myelinated fibers, as we have seen, have a coating of many membranes that is interrupted at intervals by nodes of Ranvier. The nodes are the sites of impulse generation; here, the Na$^+$ channels are packed together at a density of some 12,000/μm^2 (see Table 5.2), the highest density yet known in the nervous system. By contrast, the intervening membrane under the myelin has few voltage-dependent conductance channels, and spread through the internodal segments is therefore purely electrotonic (see Fig. 5.14B). It is very effective: the high resistance of the nerve membrane coupled with the low capacitance of the myelin wrapping makes the current tend to flow down the fiber to the next node rather than leak back across the membrane. The impulse in effect jumps from node to node, and this form of propagation is therefore called *saltatory conduction*. It is an efficient mechanism that achieves maximum conduction speed with a minimum of active membrane, metabolic machinery, and fiber size. It is especially prevalent in vertebrates, where it has been an important factor in making possible high-speed conduction over many fibers between many nerve centers.

A variation on this type of conduction is present in the dendrites of some neurons. As shown in Fig. 5.14C, there are patches of active membrane in the dendritic tree. These patches are separated from the site of impulse initiation in the cell body and axon by passive dendritic membrane. The active patches are called "hot spots." They are believed to boost the response to synaptic inputs in the dendrites, thereby enhancing the effect of distant dendritic inputs in affecting impulse generation at or near the cell body. As already noted, hot spots may generate impulses by Na$^+$ conductance mechanisms, whereas other parts of the dendritic tree may have slower Ca^{2+} conductances. In addition to their role in conductance, dendritic spikes may have significance for other functions of dendrites; these include synaptic integration and control of output from presynaptic dendrites, as discussed in Chap. 7, and specific types of information processing in dendrites, as discussed in Chaps. 29 and 30.

The Many Functions of the Impulse

Traditionally, the impulse has been regarded as the characteristic functional property of the nerve cell. It has been assumed that this property of peripheral nerves can be generalized to all nerve cells within the brain. This has led to two common beliefs: first, that the nerve cell can be defined as a cell that generates impulses, and second, that impulses are the only means by which nerve cells communicate with each other. These beliefs are widespread today, but neither is correct.

The idea that nerve cells communicate only through impulses was perhaps natural—after all, it is the only means by which signals can travel rapidly over a long stretch of peripheral nerve. It therefore seems reasonable to suppose that rapid signaling over long axons between regions within the brain similarly depends on impulses. However, if we look within brain regions, much of the communication takes place over short distances, and in these cases the electrotonic spread of synaptic potentials within nerve cells can be sufficient, without need of action potentials. We will describe this type of signaling in Chap. 7, where we will see that many nerve cells are able to carry out nervous functions without generating action potentials.

A second reason for doubting that the impulse is the one defining property of the neuron is that the neuron is not the only kind of cell that can generate impulses. As mentioned above, skeletal muscle was recognized to have this property, and so was cardiac muscle. As so often happens in science, these were easily regarded as exceptions to the rule. However, it was soon found that they were not the only exceptions; in the 1870s John Burdon-Sanderson, one of the leading British physiologists, showed that, when a Venus flytrap is touched by an insect (or by an experimenter), impulses can be recorded in association with the rapid closing of its leaves around the prey. This experiment showed that the impulse is not exclusive to nerve cells, or even to the Animal Kingdom.

These results have been confirmed and amplified in recent years, particularly since the advent of the microelectrode and the development of techniques to record intracellularly from cells in a variety of organisms, and, most recently, to make patch recordings of single channel events. The surprising finding has been that many very different kinds of cells show the ability to generate impulses. Some examples are illustrated in Fig. 5.15. In the giant cells of the fungus *Neurospora* (A), impulses are very slow, lasting over one minute. By comparison, a number of higher plants have cells that generate impulses in bursts (B); these cells are found especially in long-stemmed plants like peas and pumpkins, where they appear to be involved in pumping sap through the vasculature. In single-cell organisms like the *Paramecium* (C), impulses are involved in sensory responses as well as in the control of ciliary movement.

Within the Animal Kingdom, a number of studies have shown that oocytes of many species respond to an electric shock with an impulse; the example in D is from a primitive chordate *(Tunicata)*, and similar findings have been reported in invertebrates (for example, the oocyte of annelid worms) and in higher vertebrates (for example, the rat). Impulses have also been recorded in the skin of tadpoles (E). In mammals, a number of types of gland cells have been found to generate impulses; two examples are the cells of the anterior pituitary that

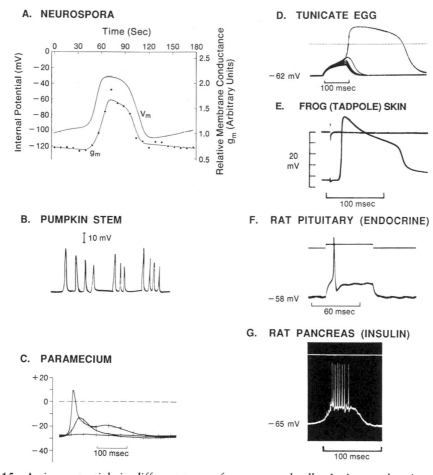

Fig. 5.15 Action potentials in different types of nonneuronal cells. **A.** A very slow impulse in the elongated cells of the giant alga. **B.** Impulse bursts in vascular cells of the pumpkin stem. **C.** Ca^{2+} action potential in a protozoan. **D.** Example of slow action potential characteristic of many types of ova. **E.** Impulse typical of several types of epithelial cell. **F.** Action potential of a pituitary cell that secretes melanocyte-stimulating hormone (MSH). **G.** Characteristic impulse burst associated with secretion of insulin from β cells of the pancreas. (A from Slayman et al., B from Sinyuhkin and Gorchakov, C from Eckert et al., D from Hagiwara and Miyazaki, E from Roberts and Stirling, F from Douglas and Taraskevich, G from Dean et al., all from Shepherd, 1981, which includes full references; see also review in Hille, 1992)

secrete pituitary hormones (F) and the islet cells of the pancreas that secrete insulin (G).

These examples by no means exhaust the list; one might mention, for instance, the finding of voltage-gated channels in cells of the immune system, and impulses in certain kinds of cancer cells. However, they are sufficient to indicate that voltage-gated channels are present in a wide variety of cells throughout the plant and animal kingdoms. This is a good illustration of why it is important to take a broad molecular and cellular approach to the functions of nerve cells. In line with that approach, we would say that *excitability*—the ability to generate impulses by means of voltage-gated channels—*is a property expressed to a greater or lesser extent in many different cells, depending on their particular functions.*

Table 5.3 Functions of excitability in cells

Development	Motility
in fertilization	ciliary
in cell division	vascular
in morphogenesis	muscle
Ion transfers across membranes	Nerve signaling
Bioluminescence	propagation
Secretion	Signal boosting
hormonal	Local information
glandular	processing
synaptic	Plasticity

Modified from Shepherd (1981), Llinás (1988) Hille (1992)

Some of the functions of impulses that have been identified in plant and animal cells are summarized in Table 5.3. Some of these, such as bioluminescence or fertilization, are obviously special for certain types of nonnervous cells. However, some of them have implications for nerve cells as well. These include the possible role of excitability in cell division and morphogenesis; in transfer of ions across membranes; in control of transmembrane movements of various substances; or in motility of nerve processes underlying growth and plasticity.

In addition to these general properties, we have noted certain functions of excitability already identified in neurons, such as prolonged effects on impulse firing, complex integration in dendrites, and control of output from presynaptic dendrites (reviewed in Llinás, 1988, and Shepherd, 1990). Many of these functions are likely to be of increased importance in smaller processes such as axon terminals and dendritic spines, where even small conductance changes may have very large effects on membrane potential, internal ion concentrations, and associated metabolic machinery.

Excitability can thus be seen to play many possible roles in nerve cells. Some of these involve general properties common to many other cells. From this perspective, the impulse can be regarded as an expression of the cell biology of the neuron.

6

The Synapse

The Synapse: A Brief History

In Chap. 1, we saw that the main business of neurobiology is to understand how individual neuronal elements are organized into functional systems. The principal means of organization is through connections between the neurons, called *synapses*. As Sanford Palay, a modern scholar of the nervous system, has expressed it: "The concept of the synapse lies at the heart of the neuron doctrine." Let us take a brief excursion into history to see how our knowledge of synapses has come about.

The origin of the idea of the synapse is particularly associated with one man, Charles Sherrington, an English physiologist. Around 1890, when Cajal and his contemporaries were establishing the anatomical evidence for the neuron as a cell, Sherrington was just beginning his study of the reflex functions of the spinal cord. His work involved a painstaking analysis of the anatomy and physiology of the spinal nerves and spinal cord, and the results provided the foundation for all subsequent concepts of the reflex as a basic unit of function in the spinal cord as well as in other parts of the nervous system, as will be discussed in Chap. 19.

Sherrington's results also set him to thinking about how activity conducted in the sensory fibers to the spinal cord is transferred to the motor cells that innervate the muscles. His studies had convinced him that the transfer involved properties different from those involved in the conducting of signals in the fibers themselves. If Cajal and his colleagues were right, and the sensory nerves arborize and terminate in free endings, then these different properties must be associated with some kind of special contact between those endings and the motor cells. And so it was, when Michael Foster came in 1897 to revise his standard physiology textbook of the day, and asked Sherrington to contribute the chapters on the spinal cord, that Sherrington advanced the following simple proposal:

So far as our present knowledge goes, we are led to think that the tip of a twig of the arborescence is not continuous with but merely in contact with the substance of the dendrite or cell body on which it impinges. Such a special connection of one nerve cell with another might be called a *synapse*.

The term *synapse* is derived from the Greek, meaning to clasp, connect, or join. Sherrington thought of it, anatomically, as a site of "surfaces of separation," but he always emphasized that it was first and foremost a *functional* connection. He conceived of the possible functions very broadly, as shown in a passage from his famous book, *The Integrative Action of the Nervous System,* published in 1906:

Such a surface might restrain diffusion, bank up osmotic pressure, restrict the movement of ions, accumulate electric charges, support a double electric layer, alter in shape and surface-tension with changes in difference of potential . . . or intervene as a membrane between dilute solutions of electrolytes of different concentration or colloidal suspensions with different sign of charge.

We will see that a broad framework of this kind is very much needed to embrace the varieties of interactions between nerve cells shown by modern studies.

One of the important properties of spinal reflexes is that they always proceed from sensory to motor, never in the reverse direction. Sherrington suggested that this is imparted by a one-way valvelike property of the synapses. This seemed to fit with an-

other idea, that the dendrites and cell body of a neuron are its *receptor* parts, where signals are received, and the axon and its terminals are the *effector* parts, where signals are emitted. This had been deduced around 1890 by Cajal and a Belgian anatomist, Arthur van Gehuchten, who had taken up the Golgi method very soon after Cajal, and it was christened the *Law of Dynamic Polarization.* Figure 6.1 illustrates how this concept applied to the flow of activity in neurons in invertebrates (A) and vertebrates (B).

This "law," soon accepted as a corollary to the neuron doctrine, provided an attractive and logical framework for understanding how individual nerve cells could be connected into groups and chains for transmitting nerve signals. However, in the 1960s, methods for analyzing single nerve cells showed that in many parts of the nervous system, axons interact with other axons, and dendrites interact with other dendrites or with axons. Thus, the Law of Dynamic Polarization describes the overall flow of information through a nervous center, but at any site on any neuron within

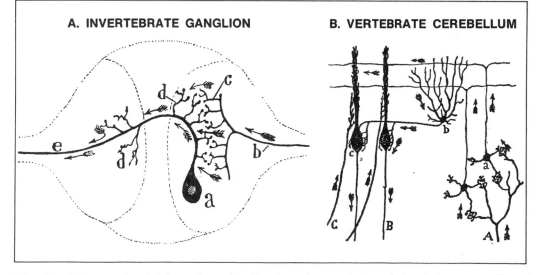

Fig. 6.1 Diagrams by Cajal, to show the direction of transmission of signals in nerve cells and nerve circuits according to his law of dynamic polarization. **A.** Invertebrate ganglion. a, cell body of ganglion cell; b, incoming axon; c, axon terminal; d, neurite (dendritic) terminal; e, projection axon of ganglion cell. **B.** Vertebrate cerebellum. A, incoming axon (mossy fiber); B, output axon of Purkinje cell; C, second type of incoming axon (climbing fiber); a, granule cell; b, basket cell; c, Purkinje cell. (From Cajal, 1911)

that center there are opportunities for complex interactions (with neighboring axons, cell bodies, or dendrites) that enhance the ability to process information (Bodian, 1972; Shepherd, 1972).

Modern research in neurobiology has focused much attention on the synapse and its role in the functioning of the nervous system. This is not by accident, for it is possible to regard the synapse as the essential and defining property of the neuron. Thus, one can suggest that a *neuron is a cell connected to other cells by synapses that mediate specific signals involved in behavior.* Note how this parallels rather closely our definition of the subject matter of neurobiology in Chap. 1.

Having stated the importance of the synapse, we also want to recognize that it is not the only means by which nerve cells can interact with each other. Interactions, in fact, take many forms, and the organization of neurons is built on these as well. Our study therefore requires a broad view of the nature of interneuronal relations.

Volume Transmission

In some situations, a terminal or other part of a neuron may release substances that diffuse through the intercellular clefts and affect neurons not actually in contact with it. Cells in the midbrain of vertebrates, for example, project their axons diffusely to many parts of the nervous system, where they ramify and terminate, in many cases, without making definite contacts on target dendrites. The branches and terminals nonetheless contain vesicles and neurotransmitter substances, and it is therefore believed that such axons release these substances to act on the target dendrites. The range of action of such substances is limited by diffusion, as well as other factors such as inactivating enzymes. These substances may act as neuromodulators as well as specific neurotransmitters, as we will discuss in Chap. 8.

This type of transmission is sometimes referred to as *volume transmission*. It has been of particular interest in regions in which the site of release of a neurotransmitter substance is in a different layer from the site of highest density of receptors for those transmitters, as shown by receptor binding studies (see Fuxe and Agnati, 1991). In such cases it has been concluded that only volume conduction can enable the transmitter to act on the receptors. This may well be true in many cases. This interpretation assumes that the density of a receptor in a histological section is a direct measure of strength of effect of the neurotransmitter that acts on that receptor. It has been pointed out (Dryer, 1985) that the size of a process bearing a receptor is also an important factor; the smaller the process, the greater may be the electrical response generated by the receptor in that process. Thus, a few receptors on a small process may give a bigger electrical response than many receptors on a large process. Similar considerations apply to biochemical responses within small compartments such as dendritic spines. Thus, it is necessary to understand the electrical and biochemical properties of nerve cells in order to interpret neurochemical findings regarding the densities and locations of neurotransmitter release sites and neurotransmitter receptors (see Chaps. 7 and 8).

Membrane Juxtapositions

Let us next consider the situation in which the membranes of two neurons are situated close together, separated only by the ubiquitous extracellular space, or cleft, of about 20 nm. We term this a *juxtaposition* (juxta = next to) of the two membranes. This relationship is a very common one in the nervous system; in fact, one can say as a general rule that neurons usually are crowded against each other, with their membranes in juxtaposition to each other, except where glial membranes are intruded in order to keep particular processes apart. As indicated in Fig. 6.2, certain types of

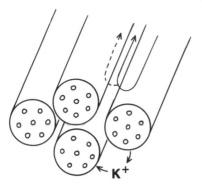

Fig. 6.2 Membrane juxtapositions, as exemplified by a bundle of unmyelinated axons, which provide for interactions through ions (K$^+$) or electric current (--). The current flows indicated in the diagram occur during an impulse, increasing the excitability of neighboring axons; accumulation of extracellular K$^+$ following an impulse peak depolarizes neighboring axons, also increasing their excitability.

fine, unmyelinated axons (for example, the axons of the olfactory nerve, and the parallel fibers of the cerebellum) have this membrane-to-membrane relationship with each other. It also occurs throughout the neuropil of the local regions of the nervous system, between the terminals of axons and dendrites. The membranes of neurons and glia are juxtaposed nearly everywhere they occur together.

This juxtaposition of membranes provides for several possible functions. Any movement of substances, such as ions or metabolites, out of one cell into the intervening cleft may have effects on that same cell as well as on all the juxtaposed processes of other cells. Uptake of substances may occur by this route, as for example the uptake of K$^+$ (see Fig. 6.2) or the uptake of the neurotransmitter γ-aminobutyric acid (GABA) by glial cells. Juxtaposed membranes also provide for electrical interactions between neighboring processes under some conditions (see dashed line in Fig. 6.2); the sites at which such interactions occur are called *ephapses* (see Chap. 7).

Membrane Junctions

The closest stage of relatedness between neurons is a specific contact of their membranes. This occurs at sites where (1) the two membranes come close together or are fused, and/or (2) the membranes appear more dense. Such sites are found between cells throughout the body. Depending on details of structure, they are called occluding junctions, desmosomes, tight junctions, gap junctions, septate junctions, or zonulae adherens. They vary widely in size and form, ranging from small spots to long strips or patches. Such junctions provide for several possible functions: simple adhesion; transfer of substances during metabolism or embryological development; restriction of movement of substances in the extracellular compartment.

Tight Junctions

An instance of the latter function is provided by the *tight junctions* between the cells that line the blood vessels and ventricles of the brain. The two outer leaflets of the unit membrane of these junctions, as illustrated in Fig. 6.3, are completely fused, to form a five-layered complex. These tight junctions restrict the movement of substances in the extracellular space and are responsible for the so-called *blood–brain barrier*, which limits transport from brain capillaries and thereby protects the brain from circulating substances.

Gap Junctions

An important type of membrane junction is the *gap junction*. Here, the outer leaflets are separated by a gap of 2–4 nm, to form a seven-layered complex (Figs. 6.3 and 6.4). In many cases, the presence of these junctions has been correlated with the physiological finding of a low-resistance electrical pathway between two neurons. On this basis they have been categorized as *electrical synapses* (we will study these electrical interactions in Chap. 7). The junction varies in diameter from 0.1 to 10

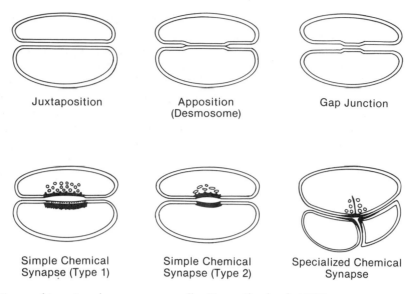

Juxtaposition

Apposition
(Desmosome)

Gap Junction

Simple Chemical
Synapse (Type 1)

Simple Chemical
Synapse (Type 2)

Specialized Chemical
Synapse

Fig. 6.3 Types of junctions between nerve cells. (From Shepherd, 1979)

Fig. 6.4 Diagram of a gap junction (electrical synapse). Channels provide for intercellular exchange of low-molecular-weight substances and electric current. Some gap junctions pass current in only one direction (rectifying junctions); this is regulated by several factors as indicated in the diagram. Channel walls are composed of six protein subunits (referred to collectively as a connexon) which span the lipid bilayer of each plasma membrane. Because of the gap between the membranes, extracellular substances can percolate between the channels. (From Makowski et al., 1977)

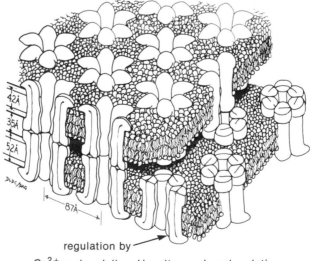

regulation by
Ca^{2+}, calmodulin, pH, voltage, phosphorylation

μm. At high resolution, dense material is seen beneath each apposed membrane, and it can be shown that the membranes are part of two systems of channels, the one continuous with the extracellular space, the other connecting the two cells. This arrangement is shown in Fig. 6.4A. Recently it has been possible to make patch recordings from two adjacent cells (B), and observe the single channel currents (D). The single channel conductance is relatively large (120 pS) compared with voltage-gated (Chap. 5) and ligand-gated channels (see below). As indicated in the diagram, the gap junction channels may be gated (activated or modulated) by several factors: the concentration of intracellular free Ca^{2+}; intracellular pH; membrane voltage; calmodulin; or phosphorylation.

Gap junctions are widely distributed throughout the body, and are the most common type of junction between cells in most body organs. Through these junctions multicellular functional units are formed for interchanges of small molecules that are important during embryonic development and into adult life, as was mentioned in Chap. 1. Some of the many functions proposed for gap junctions and related cell-to-cell channels are listed in Table 6.1. In the nervous system, gap junctions are especially common between glial cells. They are common between neurons in invertebrates and lower vertebrates, and have been found between certain types of neurons in the mammalian brain (see Chap. 7). The relatively low incidence of gap junctions in the mature mammalian brain is noteworthy. This reduction could reflect a mechanism to increase the metabolic and functional independence of individual neurons, in order to permit more complex information processing.

Chemical Synapses

The most complicated type of junction in the nervous system, and the type considered to be the most characteristic, is the chemical synapse (Fig. 6.3). It differs morphologically from other types of membrane appositions in being oriented, or polarized, from one neuron to the other. This polarization is determined mainly by two features: (1) there is usually a group of small vesicles near the site of contact, and (2) there is an increased density associated with the apposed membranes, which in some synapses is particularly marked opposite the vesicle-containing process. As was

Table 6.1 Physiological roles of cell-to-cell channels

A. Tissue homeostasis
  1. Equilibration of individual cell differences in electrical potential
  2. Equilibration and buffering of individual cell differences in small molecules
  3. Transport of nutrient substances from cell to cell
B. Regulating signal transmission
  1. Signals affecting cytoplasmic processes
    a. Chemical signals: cyclic nucleotides, various metabolites
    b. Electrical signals: propagation of electrical activity in heart, smooth muscle, electrical synapses
  2. Signals affecting genetic processes
    a. Cellular differentiation: ubiquity of cell-to-cell channels in embryonic tissue
    b. Cellular growth: channels are necessary for control of growth and prevention of unregulated (cancerous) growth
C. Cellular organization
  1. Amplification of cell responses through diffusable messenger modules (such as cyclic nucleotides) or voltage-sensitive electrical responses
  2. Hierarchical interactions: driving of secondary cells by a primary pacemaker, as in the heart and some neural circuits

Adapted from Loewenstein (1981)

first shown at the neuromuscular junction, transmission is from the vesicle-containing process to the other process, so that one can identify the *presynaptic* and *postsynaptic* process with confidence.

Neuromuscular Junction

The best understood synapse in the nervous system is the *neuromuscular junction* between motoneuron axon terminals and skeletal muscle fibers. Situated as it is in the muscles of the body, it is readily accessible to experimental study. Since the 1940s, it has been studied as a model synapse, and in recent years it has been a primary focus for the application of the new tools of molecular neurobiology. We will therefore summarize briefly the ways that these methods have been applied, and the model they have produced for the way that a chemical signal brings about a response at the molecular level of a synapse.

Overview of the Neuromuscular Junction

When we move our muscles to perform a motor act, the signal to do so has to pass from our nervous system to the muscle fibers to be contracted. The signals are conveyed by nerve cells called motoneurons, whose cell bodies are in the spinal cord and brainstem. The signals are in the form of nerve impulses carried in the axon of the motoneuron (see Fig. 6.5A). Each muscle in our body is composed of many individual muscle fibers. At the muscle, the motor axon divides a number of times, and each terminal branch finds its way to an individual muscle fiber. There the terminal runs along a length of the muscle fiber, forming the specialized region of contact with the underlying muscle membrane called the neuromuscular junction, or *muscle endplate* (see Fig. 6.5B).

Like every other synapse, the neuromuscular junction is made up of a presynaptic process (the axon terminal, in this case) and a postsynaptic process (the muscle). The nerve impulse traveling down the motoneuron axon spreads into the terminals,

as shown in B, where it triggers the release of a chemical substance called a neurotransmitter, in this case, a small molecule named acetylcholine (ACh). Most of the release is associated with tiny spheres, called *vesicles,* that fuse with the terminal membrane facing the narrow space, or cleft, between the two processes (Fig. 6.5C). The ACh diffuses the short distance (approximately 80 nm) across the cleft to reach receptor molecules in the specialized endplate region of the muscle membrane. Activation of these receptors causes ions to flow through them across the membrane, thereby initiating the electrical response of the muscle, called the endplate potential (EPP). This in turn spreads from the endplate region to the surrounding muscle membrane, initiating the impulse response of the muscle that finally leads to muscle contraction.

The release and action of the neurotransmitter substance on the molecular receptor is the crucial link that provides for synaptic communication of information between cells. Because of its importance, and because it is accessible to the full range of molecular biological and physiological methods, the ACh receptor at the neuromuscular junction has become the most thoroughly studied and best understood of all receptor proteins. This is called the nicotinic ACh receptor (nAChR), because the action of acetylcholine on this receptor is mimicked by the action of nicotine. We will see later that ACh can act on another type of receptor, the muscarinic AChR (mAChR), with a different effect that is mimicked by another drug, muscarine (Chap. 8). Let us therefore focus on the nicotinic AChR and ask: what is its molecular composition, how is it encoded in the genome, and how does it function at the molecular level?

Amino Acid Sequence of the Acetycholine Receptor Molecule

The first question to ask about the ACh receptor is, what kind of macromolecule is it? Over the years, there were speculations

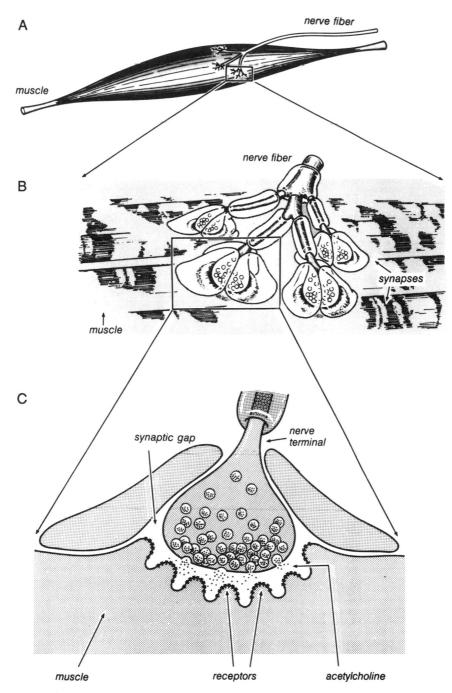

Fig. 6.5 The vertebrate neuromuscular junction. **A.** A muscle is innervated by a motor nerve fiber. **B.** The nerve fiber branches to form synaptic junctions with individual muscle fibers. **C.** Each neuromuscular junction consists of a presynaptic nerve terminal from which neurotransmitter (acetylcholine) is released; a synaptic cleft; a postsynaptic area on the muscle containing the acetylcholine receptors; and a surrounding envelope of glia. (From Miller, 1983)

that it was a lipid or an enzyme. However, by the 1960s, it was generally agreed that it was a receptor protein. By around 1980, it was established that it is a pentameric protein, that is, made up of five separate polypeptide chains, or subunits. One chain, labeled α (alpha), is double, and there are single copies of β (beta), γ (gamma), and δ (delta). The total molecular weight, as determined biochemically by polyacrylamide gel electrophoresis (PAGE), was approximately 250,000; the current estimate (see below) is approximately 270,000, with molecular weights for the four subunits of approximately 40,000, 48,000, 58,000, and 64,000 respectively. It has been known for some time that the α subunits are the sites of binding of the ACh.

In order to characterize this protein biochemically, it was an advantage to find a tissue where it was present in high concentration; this is a common strategy for any biochemical analysis. As noted previously, such a tissue is present in the electric ray, *Torpedo*. This species of cartilaginous fish has an electric organ, composed of flattened modified muscle cells arranged in parallel columns called electroplaques. Each cell has a neuromuscular junction that covers much of one face of the cell. Transmission takes place at the junction in the usual manner described above. However, because of the flattened geometry of the cells and their innervation on only one side, there is an electrical asymmetry of the cell so that there is a net electrical potential between the two faces. The stacked arrangement of the cells further means that the voltages of the cells add together like batteries connected in series. Thus, an electroplaque containing a column of 500 cells can generate a potential of 500 times 150 mV = 75 volts. The fish uses this to stun its prey. This is a good example of how cells and cell mechanisms are adapted for specialized functions; the neuromuscular junction in this tissue functions in the normal manner, but it activates muscle-like cells that have lost their contractile apparatus and have become specialized for voltage

generation. Since as much as half of the innervated membrane of the cell may be occupied by ACh receptors, the electroplaque has been an excellent source of this protein for many of the biochemical studies that have been carried out to date. It has also provided a rich source of voltage-gated Na channels (see Chap. 5).

Using this preparation, recombinant DNA technology was used to clone and sequence the DNA that codes for each of the polypeptide subunits of the ACh receptor. This was first accomplished by Toni Claudio and her colleagues (1983) at the Salk Institute in California for the γ subunit. At about the same time, Shosaku Numa's laboratory at Kyoto University worked out the sequences for the α subunit, and shortly thereafter for the other subunits. The main principles involved in this technology have been summarized in Chap. 2.

From this analysis it was concluded that the α-subunit precursor is composed of 461 amino acids, with the sequence indicated in Fig. 6.6. This subunit is referred to as "precursor" because it represents the molecule before it is processed and joined to the other subunits and inserted into the postsynaptic membrane. After an initial 5′ amino-terminal untranslated sequence of 160 nucleotides, there is a 24-residue sequence containing mostly hydrophobic amino acids, thus resembling the signal peptide of secretory proteins (see Chap. 8). This hydrophobic sequence is cleaved and may be involved in translocation of the subunit into the synaptic site at the plasma membrane.

In further work, the primary sequence of the *mammalian* ACh receptor α subunit was obtained, using a cDNA fragment for the *Torpedo* α subunit to probe a library of cloned cDNA from newborn calf muscle. This showed exactly the same number of amino acids, and over 80% sequence homology. In addition, a *human* genomic DNA library was screened by hybridization with a calf α-subunit cDNA probe. By nucleotide analysis of the cloned human

```
α  MILCSYWHVGLVLLLFSCCGLVLGSEHETRLVANLL-EN-YN-KVIRPVEHHTHFVDITVGLQLIQLISVDE-VNQIVETNVRLRQQWIDVRLRWNPADY
β  MENVRRHALGLVVMMALALSGVGASVMEDTLLSVLF-ET-YNPKV-RPAQTVGDKVTVRVGLTTLTNLLILNEKIEE-MTTNVFLNLAWTDYRLQWDPAAY
γ  MVL---TLLLIICL-ALEVR---SENEEGRLIEKLL-GD-YDKRII-PAKTLDHIIDVTLKLTLTNLISLNEK-EEALTTNVWIEIQWNDYRLSWNTSEY
δ  MGNIHFVYLLISCLYYSGCS---GVNEEERLINDLLIVNKYNKHV-RPVKHNNEVVNIALSLTLSNLISLKE-TDETLTSNVWMDHAWYDHRLTWNASEY

α  GGIKKIRLPSDDVWLPDLVLYNNADGDFAIVHMTKLLLDYTGKIMWTPPAIFKSYCEIIVTHFPFDQQNCTMKLGIWTYDGTKVSIS-------PESDRP
β  EGIKDLRIPSDVWQPDIVLMNNNDGSFEITLHVNVLVQHTGAVSWQPSAIYRSSCTIKVMYFPFDWQNCTMVFKSYTYDTSEVTLQ---HALDAKGERE
γ  EGIDLVRIPSELLMLPDVVLENNVDGQFEVAYYANVLVYNDGSMYWLPPAIYRSTCPIAMVTYFPFDWQNCSLVFRSQTYNAHEVNLQL--SAEEGE---A
δ  SDISILRLPPELVWIPDIVLQNNNDGQYHVAYFCNVLVRPNGYVTWLPPAIFRSSCPINVLYFPFDWQNCSLKFTALNYDANEIITMDLMTDTIDGK-DYP

α  ------DLSTFMESGEWVMKDYRGWKHWVYYTCCPDTIPYLDITYHFIMQRIPLYFVVNVIIPCLLFSFLTGLVFYLPTDSG-EKMTLSISVLLSLTVFL
β  VKEIVINKDAFTENGQWSIEHKPSRKNW-----RSDDPSYEDVTFYLIIQRKPLFYIIVYTIIPCILISILAILVFYLPPDAG-EKMSLSISALLAVTVFL
γ  VEWIHIDPEDFTENGEWTIRHRPAKKNYNWQLTK-DDTDFQEIIFFLIIQRKPLFYIININIIAPCVLISSLVVLVYFLPAQAGGQKCTLSISVLLAQTIFL
δ  IEWIIIDPEAFTENGEWEIIHKPAKKNI-YPDKFPNGTINYQDVTFYLIIRRKPLFYVINFITPCVLISFLASLAFYLPAESG-EKMSTAISVLLAQAVFL
                                          ├───────── M1 ─────────┤           ├──── M2 ────┤

α  LVIVE-LIPSTSSAVPLIGKYMLFTMIFVISSIII--TVVVLNTHHRSPSTHTMPQWVRKIFIDTIPNVEFFS---------------------TMK
β  LLLADKV-PETSLSVPIIIRYLMFIMILVAFSVIL--SVVVLNLHHRSPNTHTMPNWIRQIFIETLPPFLWIQRPVTT-PSPD----------SKPTIIS
γ  FLIAQKV-PETSLNVPLIGKYLIFVM-FV-SMLIVMNCVIVLNVSLRTPNTHSLSEKIKHLFLGFLPKYLGMQ-LEPSEETPE---KPQPRRRSSFGIMI
δ  LLTSQRL-PETALAVPLIGKYLMFIMSLV-TGVIV-NCGIVLNFHFRTPSTHVLSTRVKQIFLEKLPRILHMSRADESEQ-PDWQNDLKLRRSSSVGYIS
                          ├───── M3 ─────┤

α  RASKEKQENKIFADD-IDISD-----ISGKQV---TGEVIFQ------TPLIKNPD--------VKSAIEGVKYIAEHMKSDEESSNAA-EE---WKYVAM
β  RANDEYFIRK-PAGDFVCPVDNARVAVQOPERL---FSEMKWHLNG--LTIQPVTLPQ------DLKEAVEAIKYIAEQLESASEFDDLK-KD---WQYVAM
γ  KAE-EYILKK--PRSELMFEEQKDRHGL--KRVNKMTSDIDIG-----TTVDLYKDLANFAPEIKSCVEACNFIAKSTK--EQ--NDSGSENENWVLIGK
δ  KAQ-EYFNIK--SRSELMFEKQSERHGLVP-RV---TPRIGFGNNNENIAASDQLHD------EIKSGIDSTNYIVKQIK--EK--NAYDEEVGNWNLVGQ
                                                               ├───────────── A ──────────────┤

α  VIDHILLCVFMLIC-IIGTVSVF-AGRLIELSQ---EG-------------
β  VADRLFLYVFFVIC-SIGTFSIFLDASHNVPPDNPF-A-------------
γ  VIDKTACFWIALLLFSIGTLAIFLTGHFNQVPEFPFPGDPRKYVP------
δ  TIDRLSMFIIITPVM-VLGTIFIFVMGNFNHPPAKPFEGDPFDYSSDHPRCA
   ├─── M4 ───┤
```

Fig. 6.6 Amino acid sequences of the α, β, γ, and δ subunits of the acetylcholine receptor (AChR) or *Torpedo californica*, is deduced from cDNA clones. Amino acids that are identical in at least three of the subunits at a given position are enclosed in solid lines. Amino acids at a given position which have opposite hydrophilic or hydrophobic properties are enclosed in dotted lines. The four transmembrane α helices originally proposed by Claudio et al. (1983) and Numa et al. (1983) are indicated by heavily lined boxes. The amphipathic α helix (see text) is indicated by a dashed line box. The cleavage site of the signal protein is indicated by S. Recent improved crystallographic data at 9 = Å resolution suggest that the M2 segment is an α helix lining the channel, and the other hydrophobic segments are β barrels (see Fig. 6.9). (From Numa et al., 1983, in Changeux et al., 1984)

DNA, it was possible to show the structure of the human gene for the ACh receptor α-subunit. As shown in Fig. 6.7 this has the common form of a split gene, with eight introns separating nine exons. The exons appear to correspond to some degree with different functional domains of the subunit (see below). The complete sequence of the human α subunit consists of 457 amino acids, showing 97% homology with the α subunit of the calf. This presumably reflects the closeness of the cow and human on the evolutionary scale. The other subunits (β, γ, δ) have also been sequenced, and show considerable homologies with the α subunits (see Fig. 6.7).

Structure of the Receptor

A variety of methods have been used to obtain evidence regarding the tertiary structure of the ACh receptor. Four hydrophobic intramembranous sequences were originally identified, labeled M1–M4 (see Figs. 6.6 and 6.8). Each sequence was 19–27 amino acids in length. Assuming that they are arranged in an α-helix, this is sufficient to span the membrane (approximately 4 nm thick). Subsequently, evidence

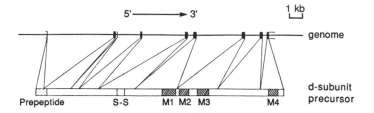

Fig. 6.7 Arrangement of the gene that codes for the ACh receptor α-subunit peptide in the human. The 5' to 3' orientation of the chromosomal DNA is indicated above; the scale shows 1 kilobase (kb). M1–4 represent the membrane-spanning segments of the α-subunit peptide as originally proposed; S–S, disulfide bonds near the ACh binding site. Note that nine exons code for the α-subunit precursor peptide; several exons appear to code for different structural and functional domains. (From Numa et al., 1983)

Fig. 6.8 The nicotinic ACh receptor (a) and the GABA$_A$ receptor (b) belong to the same family. I–IV, the four membrane-spanning segments; C, cysteine residues; MA, an amphipathic segment; filled triangles, sites of N-glycosylation; P, site of phosphorylation. Note the relative excess of negatively charged residues near the ends of the membrane domains of the nAChR and of positively charged residues for the GABA$_A$ receptor, which contribute to the ion selectivity of the channels (see the channel model in Fig. 6.9). (Modified from Barnard, 1992)

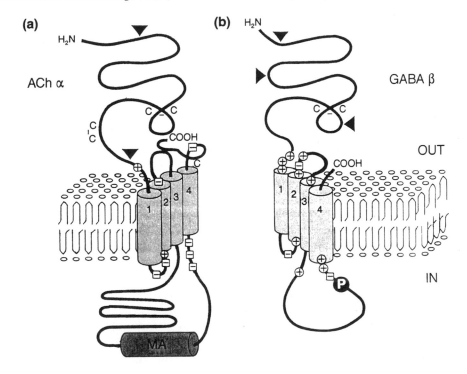

was obtained in the α subunit for an additional sequence, M5, between M3 and M4, which may be in the cytoplasm or in the membrane (see Figs. 6.6 and 6.8).

Recent studies have shown that the nicotinic AChR is a member of a superfamily that includes the receptors for amino acid transmitters, which mediate most of the fast synaptic actions in the nervous system. Within this superfamily, the nAChR is paradoxically most closely related to specific types of receptors for GABA and glycine, the main transmitters for inhibitory synaptic actions. More distantly related are the receptors for glutamate, the main transmitter for excitatory synaptic actions in the nervous system. We will have more to say about these amino acid transmitters and their receptors below and in Chap. 7. The similarities between the nAChR subunit and the GABA$_A$ receptor subunit are shown Fig. 6.8 and discussed in the legend.

With regard to the nicotinic AChR, each subunit is encoded by a separate gene. Gene expression therefore must be coordinated within this gene family in order for the entire ACh receptor to be assembled. It is presently envisioned that each subunit is translated from its mRNA by rRNA in the endoplasmic reticulum (see Chap. 3), where the M1–M3 segments anchor the newly formed molecule within the endoplasmic membrane. Posttranslational processing includes the infolding of segments M4 and M5 within the membrane. The five subunits then come together and are inserted into the postsynaptic plasma membrane in a circular arrangement, in the sequence α-β-α-γ-δ or α-γ-α-β-δ in clockwise rotation as one looks down on it from outside. The resulting three-dimensional structure was depicted in Fig. 2.13. Mature nicotinic AChRs in the central nervous system have a different subunit composition, consisting of α-β-α-β-β.

One reason why the nAChR is such an important model is that it is the only receptor for which there is X-ray crystallographic data on its three-dimensional struc-

ture (see Unwin, 1993; Sansom, 1993). This has been possible because of the electric organ, where the receptors are so tightly packed that they can be obtained in an almost completely crystalline form.

A schematic representation of the nAChR is shown in Fig. 6.9. By combining the crystallographic data of the outward form with the sequence data on the different segments it has been possible to guide experimental mutations of individual residues and observe their effects on different properties of the channel. These studies are yielding a picture of the functional organization of the channel which is equivalent in its level of detail to what we have seen for voltage-gated channels. Let us summarize this picture briefly.

Functional Organization of the Acetylcholine Receptor Molecule

The sites of binding of ACh are in the α subunit. The main binding site is believed to be in a pocket in the external part of the peptide chain near the disulfide bridge (residues 192–193; see Figs. 6.6 and 6.9). The ACh interacts with residues within the pocket by means of weak bonds. This presumably brings about allosteric changes in the configuration of the receptor molecule, which are communicated extensively throughout the molecule in order to open gates in the central pore to allow ions to flow through in order to generate the electrical response. Allosteric changes in molecular structure are the mechanism by which many proteins carry out their functions.

Mutation experiments have provided fascinating clues to the functional organization of the pore. As we shall discuss in the next chapter, the channel is nonselective for cations (Na^+, Ca^{2+}, K^+), producing a net inward flow of positive charge when activated by ACh binding. External ions are collected in the vestibule of the receptor and are attracted to charged residues in the wall of the vestibule and more toward the narrower opening of the pore within the plane of the membrane. The pore is be-

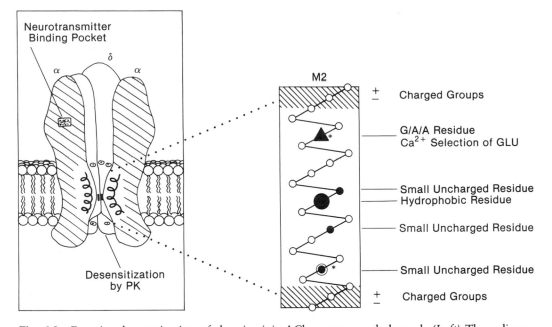

Fig. 6.9 Functional organization of the nicotinic ACh receptor and channel. *(Left)* Three-dimensional structure based on crystallographic analysis and sequence data. The coiled structures indicate α-helix M2 segments lining the channel. Behind them are other segments presumed to be β barrels. *(Right)* Detail of M2 residues lining the channel surface. Large filled circle, leucine residue that is highly conserved in AChR family; small filled circles, small uncharged residues that are conserved in AChR family; hatching, charged groups at either end of pore; asterisks, mutated residues in the study of Galzi et al. (1992), which changed the selectivity of the channel from cations to anions. (Adapted from Unwin, 1993)

lieved to be lined by the M2 segments of each of the five subunits, whereas the β-barrels provide a supportive framework. As shown in the expanded view of Fig. 6.9, at each end of the pore are charged residues that help to determine the ionic selectivity of the channel: negative residues attract positive ions (cations) and repel negative ions (anions) in the case of the AChR, whereas positive residues attract anions and repel cations in the case of the other members of this channel family that bind GABA and glycine (see below). The most selective region is within the pore, which an ion can penetrate only if its energy of attraction to the residues lining the pore is greater than its energy of hydration (its attraction to water molecules).

Site-directed mutagenesis experiments suggest that several specific residues within the pore are important in determining ion *selectivity* (see Fig. 6.9). Other residues are implicated in determining the *permeability* of the ions, that is, the ease with which the selected ions can pass through, which is reflected in the *conductance* of the channel for that ion. A third property is involved in *shutting* the channel; specific residues within the channel are also implicated in this. Each of these properties is exquisitely sensitive to the presence or absence of specific residues; for example, the mutation of a particular hydrophobic residue can change the channel from closed to open (Reval et al., 1991), and the mutation of a different residue can change the channel from conducting cations (as in the case of the nAChR) to conducting anions (as in the case of the GABA$_A$ and glycine receptors) (Galzi et al., 1992).

Our present understanding of how the pore is organized to provide for each of these functions is shown in Fig. 6.9 and discussed further in thc legend. As Nigel Unwin of London has emphasized, the strategy of channel organization appears to be that the pore lining surfaces

provide opportunities for numerous weak interactions, at various levels, with the diffusing ions. The interactions presumably work in a coordinated way, adding up over the length of the structure. Such multiple influences, rather than just one or two at particular levels, probably underlie the high ion selectivity combined with high transport efficiency that neurotransmittergated channels are able to achieve. (Unwin, 1993)

The adaptations of this basic plan for the receptors for GABA, glycine, and glutamate will be discussed below and in the next chapter.

Synapses in the Central Nervous System

The concept of the synapse as a site of small *synaptic vesicles* (about 40 nm in diameter) and *membrane-associated densities* originated in the first electron microscopical studies of the neuromuscular junction and was quickly extended to the central nervous system by Palay, Palade, De Robertis, Bennett, and others, in the 1950s (see above). Since that time, this concept has been universally accepted as the morphological definition of the synapse when viewed at relatively low magnification. The accumulation of synaptic vesicles is mediated by a cytoskeletal matrix to which the vesicles are anchored. Additionally, the vesicles closer to the surface appear to be directly anchored to the plasma membrane. When special stains are used, dense projections radiating into the cytoplasm from the presynaptic membrane are visible. These form a gridwork in the presynaptic terminal and synaptic vesicles appear to move between these projections to fuse with the membrane during neurotransmitter release (Fig. 6.10). The points at

which the vesicles actually contact the plasma membrane to undergo exocytosis are called vesicle attachment sites, or *active zones*.

Besides synaptic vesicles, nerve terminals may contain another population of secretory organelles, the so-called *large dense-core vesicles*: these have a larger diameter than synaptic vesicles, have an electron-dense core, are present in much lower number, are not accumulated under the presynaptic plasmalemma, and are often excluded from synaptic vesicle clusters. In most nerve terminals, the large dense-core vesicles are so infrequent that they can be seen only after serial sectioning. Their name distinguishes them from a type of small vesicle with dense cores, which are synaptic vesicles containing catecholamines. Following certain fixation conditions, the catecholamine content of such vesicles forms an electron-dense precipitate, making the core appear dense. Other membranous organelles present in nerve terminals are mitochondria, which provide the energy required for neurotransmitter release, and endosomes and coated vesicles, which are involved in membrane recycling (see Fig. 4.7). Elements of the smooth endoplasmic reticulum act as intracellular Ca^{2+} stores; gold labeling studies have shown that these stores are present in both presynaptic (Fig. 6.11A) and postsynaptic (Fig. 6.11B) terminals.

The major feature of the postsynaptic site is a membrane thickening and the presence at many synapses of electron-dense material (*postsynaptic density*, or *PSD*) lining the cytoplasmic face of the membrane. Elements of the endoplasmic reticulum are often located in close proximity to the postsynaptic membrane (the so-called *spine apparatus* in the case of dendritic spines). The pre- and postsynaptic sides are connected within the *synaptic cleft* by a dense matrix. At the neuromuscular junction this matrix is relatively thick and continuous with the basal lamina surrounding the muscle fiber.

Further details of synaptic structure have been revealed by a variety of methods. One

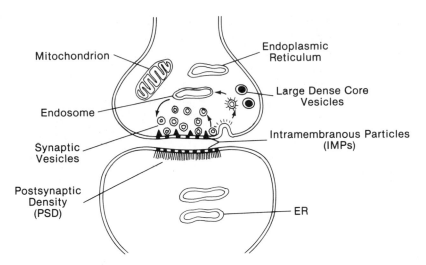

Fig. 6.10 Some of the morphological and biochemical properties of a chemical synapse. See text.

Fig. 6.11 Immunogold labeling for a calcium channel. The black particles are gold precipitates which mark the presence, on membranes of the endoplasmic reticulum, of the ryanodine receptor, a Ca channel first identified in the sarcoplasmic reticulum of skeletal muscle (see Chap. 17 for further description). This receptor is a target for the second messenger IP_3 (see Chap. 8). *(Left)* Presynaptic terminal containing labeled membranes. *(Right)* Purkinje cell spine containing a labeled spine apparatus, postsynaptic to a parallel fiber containing synaptic vesicles. (From Takei et al., 1992)

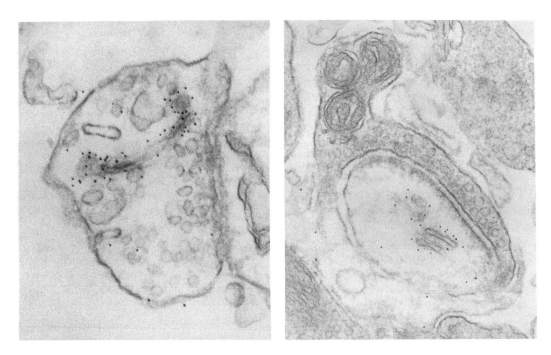

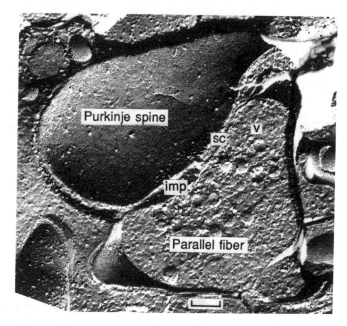

Fig. 6.12 A freeze-fractured specimen of synaptic terminals in the cerebellum. The line of cleavage is such that one sees the inner surface of an outer leaflet of a Purkinje cell dendritic spine; it also cuts across a synaptic terminal of a parallel fiber. Note the vesicles (v) in the presynaptic terminal, the widened synaptic cleft (sc), and the accumulation of small particles (imp) in the postsynaptic terminal membrane. Bracket, 0.1 μm. (From Landis and Reese, 1974)

of the most useful has been the freeze-fracturing technique, in which a tiny block of tissue is frozen and then fractured by a swift blow with a sharp blade. Subsequently, a thin metal layer is deposited on the fractured surface to produce a metal cast, which can be examined in the electron microscope. Figure 6.12 is a micrograph of a specimen prepared in this way. The lines of cleavage are not between the membranes of the two neighboring neurons, but between the inner and outer leaflets of the same membrane. As can be seen, the fracture line, in its course through the tissue, jumps from one membrane to the next or cuts entirely through a process. In this view, the fracture line has cut through a synapse between a pre- and a postsynaptic process. There is a collection of *intramembranous particles (IMPs)* on the inner surface of the outer leaflet of the postsynaptic membrane. IMPs are found widely in the membranes of cells, where they appear to

be the electron microscopic images of membrane proteins. When IMPs are present in high density, they serve as an additional morphological component for defining a synapse. Clusters of homogeneously sized IMPs in the postsynaptic membrane are thought to represent ligand-gated channels. However, they are present only in certain types of synapses, mainly those believed to have an excitatory action (see below) Special arrays of IMPs in the presynaptic membrane are thought to be channels for Ca^{2+}, which plays a critical role in synaptic transmission and cell metabolism (Fig. 6.10; Chaps. 7 and 8). In a modification of the freeze-fracture technique, tissues can be quick-frozen, fractured, and deep-etched before metal shadowing. The etching step has permitted investigators to visualize not only intracellular organelles (see Chap. 3) but also the cytoskeletal matrix which connects synaptic vesicles with each other.

Our knowledge of the synapse has been

greatly expanded by molecular, biochemical, and immunocytochemical studies. Several receptors present in the postsynaptic plasma membrane (plasmalemma) have been cloned, as we have already noted. Progress has been made in elucidating the molecular composition of synaptic vesicles, and several proteins present in the presynaptic plasmalemma have been identified and characterized. We will summarize these studies; Pietro De Camilli has kindly provided the following account (see De Camilli, and Jahn, 1990; Jessell and Kandel, 1993).

Molecular Components of Synaptic Vesicles

As mentioned, nerve terminals generally contain two classes of secretory vesicles, *synaptic vesicles* and *large dense-core vesicles*. These two classes differ in a variety of properties. Synaptic vesicles are highly specialized secretory vesicles which store and secrete nonpeptide neurotransmitters, such as acetylcholine, GABA, glycine, and glutamate. Their exocytosis takes place selectively at active zones and follows nerve terminal depolarization with extremely short latency. After exocytosis, synaptic vesicle membranes are reinternalized by endocytosis and reused for the generation of synaptic vesicles. At each cycle of exo–endocytosis they are reloaded with neurotransmitter by a neurotransmitter transporter which is present in their membrane and which derives its energy from the proton electrochemical gradient generated by an ATPase of the vacuolar type.

Large dense-core vesicles contain peptide neurotransmitter and may also contain amines. They are very similar in properties to secretory granules of endocrine cells, which store peptide hormones and amines. Their exocytosis (see Fig. 6.10) is not restricted to active zones and is preferentially triggered by trains of action potentials.

Synaptic vesicles and large dense-core vesicles have different roles in neuronal signaling. Secretion via synaptic vesicles is responsible for fast point-to-point signaling, whereas secretion via large dense-core vesicles is involved primarily in modulatory signaling and in distant signaling between neurons and between neurons and other cells.

Little information is available on the protein composition of large dense-core vesicle membranes. By contrast, significant progress has been made in elucidating the protein composition of synaptic vesicle membranes. This progress was greatly aided by the enormous abundance of synaptic vesicles in brain tissue, an abundance which has made possible the concentration of synaptic vesicles in high yield and purification by subcellular fractionation. The brain is homogenized to obtain suspensions of cell fragments and cell organelles, which are then separated by means of centrifugation, chromatography, or immunoisolation. Information on the protein composition of synaptic vesicles and of the nerve terminal plasmalemma is summarized in Fig. 6.13. All proteins depicted in the figure are represented by protein families. The ratios of the different isoforms of a given protein vary greatly from one neuron to another, but at least one member of the protein family is thought to be expressed by virtually all neurons. We will briefly summarize these families.

Synapsin, which was discovered by Paul Greengard and his collaborators at Yale and Rockefeller Universities, was the first synaptic vesicle protein to be identified and thoroughly characterized (see Fig. 6.13). It is a peripheral membrane protein which binds both synaptic vesicles and actin. It may act as a cross-link between vesicles and the actin-based cytomatrix of nerve terminals. It is a major endogenous substrate for cAMP and Ca^{2+}/calmodulin-regulated protein kinases, and it is thought to play a crucial regulatory role in neurotransmitter release by controlling the availability of synaptic vesicles for exocytosis.

Synaptobrevin is a small protein that is anchored to the vesicle membrane by its hydrophobic C-terminal region. A key role

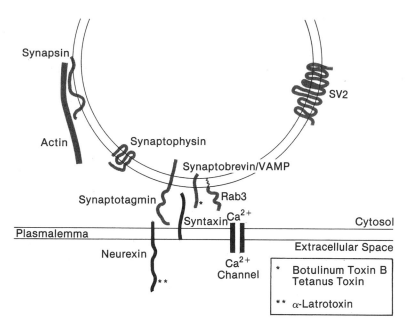

Fig. 6.13 Schematic drawing summarizing some of the types of proteins associated with synaptic vesicles and the presynaptic membrane to which they fuse to release their contents of synaptic transmitter. Sites of action of various toxins are indicated by asterisks. See text for full description. (Drawing by P. De Camilli)

of this protein in the exocytotic process was suggested by the finding that it is the target for some of the clostridial neurotoxins which act as potent inhibitors of synaptic vesicle exocytosis. Clostridial neurotoxins are Zn^{2+} endoproteases: tetanus toxin and botulinum toxin type B appear to produce their effects by selectively cleaving synaptobrevin.

Rab3 is a member of a large superfamily of *rab* GTPases which share a structural homology to the RAS oncogene. Members of the rab family are thought to be involved at all stations of the secretory and endocytic pathway where membrane fusion reactions take place.

Synaptophysin is a membrane glycoprotein with multiple transmembrane spans. It has a structural, but not primary sequence, homology to the connexons, the unit elements of gap junction. It has been reported to form ion channels in lipid bilayers, but its function remains elusive.

Synaptotagmin, an integral membrane protein with a single transmembrane span, has been shown to interact with Ca^{2+} and phospholipids via its cytoplasmic domain. For this reason it was proposed to play a role in Ca^{2+}-dependent exocytosis. However, recent studies have suggested that synaptotagmin may not be essential for synaptic vesicle exocytosis. It may participate in the regulation of synaptic vesicle fusion.

SV2 has the domain structure of a membrane transporter, but its precise function is unknown.

In addition to the proteins depicted in Fig. 6.13, synaptic vesicle membranes contain a proton pump of the vacuolar type which pumps protons into the vesicle lumen.

Molecular Components of the Presynaptic Membrane

Some important components of the presynaptic plasmalemma include the following proteins.

Transporters are involved in the reuptake of neurotransmitters from the synaptic cleft into the cytosol. These differ from their counterparts in the synaptic vesicles, which transport neurotransmitter from the cytosol into the vesicle.

Syntaxin is a protein that may play a role in the docking or fusion of synaptic vesicles.

Ca^{2+} channels are involved in the depolarization–secretion coupling.

Neurexins are a family of cell surface receptors which may participate in cell-to-cell recognition or in the interaction of the axonal surface with the extracellular matrix. This family of proteins is thought to include the receptor for α-latrotoxin, the main component of black widow spider. This toxin acts as a potent stimulus to synaptic vesicle exocytosis, but its mode of action remains unclear.

With this information at hand, the elucidation of the molecular machinery for synaptic vesicle exocytosis is proceeding rapidly and may be solved in the near future. An interesting convergence is occurring between studies on synaptic vesicle fusion and studies on membrane fusion in other systems. Genetic studies carried out in yeast have led to the identification of a number of proteins which appear to participate in membrane fusion at various stations of the secretory pathway and during exocytosis in particular. Mutations affecting yeast homologues of syntaxin, synaptobrevin, and rab3 have been found to impair exocytosis, suggesting a key role of these molecules in the docking/fusion process.

The emerging picture indicates that a common basic machinery is used in all types of membrane fusion along the secretory and endocytic pathway. Several proteins in synaptic vesicles and the presynaptic plasma membrane represent nerve cell–specific variants of proteins which have general roles in cellular functions. The Ca^{2+} regulation of synaptic vesicle exocytosis may be conferred by some specific variants of these proteins which regulate this basic machinery.

Molecular Components of the Postsynaptic Membrane

On the *postsynaptic side,* the main elements identified thus far are the membrane receptors that bind the neurotransmitter molecules and allow ion movements through channels, the postsynaptic density (PSD; Fig. 6.10), and the enzymes that phosphorylate and dephosphorylate the PSD proteins. Let us consider each of these briefly.

Membrane receptors are of two types. One type consists of channel proteins that themselves bind the neurotransmitter molecule, inducing an allosteric change in the channel to allow ions to flow; the receptors for ACh, glutamate, GABA, and glycine belong to this class (see above and Chaps. 2 and 7). Alternatively, the neurotransmitter may bind to a nonchannel protein that is coupled to enzymatic cascades which usually involve generation of intracellular second messengers (see Chap. 8). The first type of receptor transduces the chemical signal into an electrical signal over a brief time scale and is responsible for classical synaptic transmission. The second type of receptor plays primarily a modulatory role in synaptic transmission and is often involved in distant types of communication between neurons. The same nonpeptide neurotransmitter, such as glutamate, may act both on channel receptors (also called ionotropic receptors) to produce a fast electrical response and on second messenger–linked receptors (such as glutamatergic metabotropic receptors) to produce modulatory changes. Peptide neurotransmitters are thought to act only on the second type of receptor. Activation of both types of receptor may lead to further reactions, such as phosphorylation of nearby enzymes and proteins, which ultimately affect the membrane properties, cytoskeleton, or metabolism of the terminal or the rest of the neuron. These receptor mechanisms thus allow for a wide range of brief or long-lasting effects on the postsynaptic terminal or neuron, as we shall see in Chap. 8.

The postsynaptic density has been the

subject of numerous investigations, using subcellular fractionation in combination with other biochemical methods. Much of the density is due to the membrane cytoskeleton, whose basic organization we have already discussed (Chap. 3). In addition, the PSD is enriched in certain components. Microtubules and neurofilaments and their associated proteins are anchored in the PSD. Proteolytic enzymes are present; they act to degrade components of the membrane cytoskeleton, particularly fodrin and microtubule-associated proteins. The possible role of one such enzyme, calpain, in a mechanism for memory is discussed in Chap. 29. Recent work has shown that the major PSD protein, accounting for up to 50% of the total PSD protein, is virtually identical to the 50-kD subunit of Ca/calmodulin-dependent protein kinase II (Camkinase II). These elements are summarized in Fig. 6.10.

Protein phosphorylation, as already mentioned, is important in presynaptic release mechanisms, and it is also important in postsynaptic responses. PSDs are enriched in several types of phosphoproteins which are substrates for the Camkinase II, and the number is growing. In addition, PSDs are enriched in protein phosphatases, as they must be if the proteins are to be able to undergo dephosphorylation. In Chap. 8, we will learn how these phosphorylation and dephosphorylation reactions control the physiological responses of neurons to their synaptic inputs.

Additional components of the postsynaptic membrane include receptors for components of the specialized cytomatrix in the synaptic cleft which keep pre- and postsynaptic membranes in register. Two such components at the neuromuscular junction are S-laminin and agrin, which we will discuss in Chap. 9. Both proteins represent highly specialized forms of extracellular matrix proteins.

Molecular Mechanism of the Synapse

How do these molecular components function together as a coordinated mechanism to provide for synaptic transmission? Figure 6.14 provides a summary of the basic steps involved. Although this scheme appears daunting at first, it can easily be broken down into the several main types of mechanisms discussed previously. For the case of rapid synaptic transmission, these are as follows:

1. *Depolarization.* The presynaptic membrane is depolarized ①.
2. *Activation of voltage-sensitive Ca^{2+} channels.* Depolarization produces an influx of Ca^{2+} into the presynaptic terminal at the site of active zones ②. This results within hundreds of milliseconds in a very transient elevation of cytosolic Ca^{2+} in close proximity to these sites. Given the clustering of Ca^{2+} channels at active zones, such an elevation may be of the order of 100 μM Ca^{2+}.
3. *Vesicle exocytosis* (③–⑥). The high Ca^{2+} induces fusion of synaptic vesicles already docked at the plasmalemma, possibly by removing an inhibition of the fusion process already engaged. Ca^{2+} elevations are rapidly dissipated by dilution, Ca^{2+} buffering proteins, and Ca^{2+} extrusion mechanisms. Depolarization-induced Ca^{2+} entry may also trigger exocytosis of large dense-core vesicles with much slower activation kinetics. The best stimulus for large dense-core vesicles is a train of action potentials (see Chap. 8).
4. *Presynaptic restitution.* Compensatory mechanisms are activated to restore basal conditions in the presynaptic terminal. These include transmitter reuptake ⑧ₐ and synaptic vesicle membrane recycling ⑧ᵦ.
5. *Cleft mechanisms.* The transmitter diffuses across the cleft to act on its postsynaptic receptors. Diffusion ⑦ is controlled by the cleft extracellular matrix. The concentration of the transmitter is brought back to basal levels by diffusion, and in some cases by hydrolysis ⑧ or by reuptake ⑧ₐ. Neuropeptides may diffuse over long distances.
6. *Postsynaptic receptors.* Nonpeptide

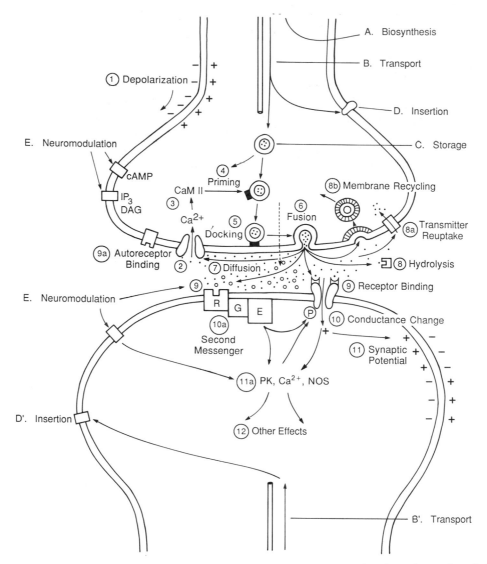

Fig. 6.14 A summary of some of the main biochemical mechanisms that have been identified at chemical synapses. **A–E.** Long-term steps in synthesis, transport, and storage of neurotransmitters and neuromodulators; insertion of membrane channel proteins and receptors; and neuromodulatory effects. ①–⑫. These summarize the more rapid steps involved in immediate signaling at the synapse. These steps are described in the text, and are further discussed for different types of synapses in Chapter 8. Abbreviations: IP_3, inositol triphosphate; CaM II, Ca/calmodulin-dependent protein kinase II; DAG, diacylglycerol; PK, protein kinase; R, receptor; G, G protein; E, effector.

transmitters act directly on channel proteins in the postsynaptic membrane ⑨, causing a conductance change ⑩ and a consequent postsynaptic potential response ⑪ which spreads along the membrane in decrementing fashion (see next chapter). In general these receptors have relatively low affinity for their transmitter ligands. Other types of transmitters (peptide or nonpeptide) may bind to receptors ⑨ linked to second messenger systems ⑩ₐ. These typically have higher affinities for their ligands and may be activated even at some distance from the release site. These receptors may in turn act directly through G proteins and second messengers on membrane channels, or indirectly through kinases ⑪ₐ or other enzymes. These enzymes may also have many other effects on the molecular machinery of the neuron ⑫ related to motility, growth, and metabolism.

7. *Gaseous neurotransmission.* Calcium influx or mobilization in the presynaptic (step 2) or postsynaptic (step 5) terminal may also activate cytosolic enzymes, with diverse effects. These include activation of enzymes that produce gaseous signal molecules, such as NO and CO, which diffuse into neighboring terminals to activate target enzymes and membrane channels. These mechanisms may thus be characterized as nonvesicular, and cytosol-to-cytosol.

In addition to these specific steps in rapid neurotransmission, many related mechanisms act over longer periods of time. As indicated in Fig. 6.14, these include the following:

A–D. *Metabolic machinery.* These include biosynthesis, transport, and storage of transmitters and vesicles; insertion of membrane proteins; and assembly of elements of the terminal. Corresponding events take place in the postsynaptic terminal.

E. *Presynaptic modulation.* The release process is subject to modulation by neuromodulators, acting either directly on membrane channels or indirectly through internal second messengers. The release process can also be modulated by feedback action of the released transmitter or autoreceptors. In addition to its rapid effects on the fusion process, Ca^{2+} may also produce longer term modulatory changes.

E'. *Postsynaptic modulation.* The postsynaptic site may be modulated in several ways. First, peptides released from the presynaptic terminal (see small open circles in the diagram)

may bind to postsynaptic receptors, affecting binding of their neurotransmitter ligand ⑨. Second, neuroactive substances from other sources (neighboring terminals, blood-borne hormones, and so on) may bind to membrane channels ⑨ or second messenger receptors ⑩ₐ. Third, peptides may act internally on protein kinases ⑪ₐ and other enzymes or components of the postsynaptic process.

This provides only a brief summary of the main steps involved in synaptic transmission. The student should work carefully through this diagram to understand each step as a basic type of molecular mechanism as well as a part of the whole synaptic system. In Chap. 7 we will discuss the physiological actions of different transmitters involved in rapid synaptic actions, and in Chap. 8 we will discuss the properties of synapses involved in slower actions mediated by transmitters and neuropeptides as well as the properties of diffuse interactions mediated by gaseous signal molecules.

Enlarged Concepts of the Synapse

The foregoing account has considered the synapse as a mechanism for unidirectional transmission between cells, as envisaged in the classical studies. The complexity revealed by modern research has widened this view to include several new concepts.

First, as we have already seen, the presynaptic terminal may have receptors for its own released products; thus, the presynaptic terminal may be postsynaptic as well to its own transmitter. This can be seen as an expression of a general principle of regulation of enzymatic activity by feedback of the output.

Second, the postsynaptic terminal may send retrograde signals to the presynaptic terminal. These may be rapid signals, as in the case of gaseous molecules such as NO, which mediate activity-dependent effects (Chap. 8), or they may be slowly acting growth factors, which play roles in determining the types of neurotransmitters expressed (Chap. 9). By these actions, the synapse can be viewed as having a bidirectional nature (Jessel and Kandel, 1993).

Third, it is obvious that the synapse is not a simple link between two neurons, but rather is a complex organelle in its own right. Furthermore, it is unique among cell organelles in being constructed by more than one cell; it is in fact a *multicellular organelle*. Like other organelles, such as the mitochondrion or the nucleus (see Chap. 3), it satisfies certain criteria: it is an anatomical entity, it has a distinct biochemical composition, and it has specific functions. As Fig. 6.14 demonstrates clearly, this organelle has multiple points of control, a property that is essential for coordinated function. This view helps us to appreciate the synapse as a building block of nervous circuits, as we discussed in Chap. 1. It will also help us to understand how synaptogenesis is a main concern of neuronal development (Chap. 9), and how the synapse is adapted for different information-processing functions in different nervous pathways.

Two Types of Synapses

In 1959, E. G. Gray of London, working on the cerebral cortex, obtained evidence for two morphological types of synapses. There is a consensus that, despite many local variations and gradations between the two, this division has some validity. The two types are illustrated in Fig. 6.3. The distinguishing features may be summarized as follows:

Type 1: synaptic cleft approximately 30 nm; junctional area relatively large (up to 1–2 μm in diameter); prominent accumulation of dense material next to the postsynaptic membrane (that is, an asymmetric densification related to the two apposed membranes).

Type II: synaptic cleft approximately 20 nm; junctional area relatively small (less than 1 μm in diameter); membrane-associated densifications modest and symmetrical.

Following the recognition of these types, evidence was obtained in 1965 by Uchizono of Japan that, in many parts of the brain, type I synapses are associated with spherical vesicles (diameter approximately 30–60 nm) which are usually present in considerable numbers. Type II synapses, on the other hand, are associated with smaller (diameter 10–30 nm) vesicles, which are less numerous and which, significantly, take on various ellipsoidal and flattened shapes. The distinction between round and flat types of vesicles is by no means a sharp one; in many synapses a vesicle simply tends to the one shape or the other.

In addition to the membrane-related densities and the vesicle shapes, a third distinction is the distribution of intramembranous particles. Type I synapses tend to have a dense collection of IMPs in the postsynaptic membrane, as already shown in Fig. 6.12; by contrast, this is lacking at many type II synapses, at least in the vertebrate nervous system.

The recognition of the two types of synapse has provided anatomists with a useful tool to unravel the synaptic organization of local brain regions. Much of this usefulness has been based on the premise that all the synapses made by a given neuron onto other neurons are either of one type or the other. This is commonly called the *morphological corollary of Dale's Law*, Dale's Law being usually understood as stating that a given neuron has the same physiological action at all its synapses. As we will see in Chap. 8, this is neither what Dale, in fact, put forward nor what electrophysiology reveals. Nor has it been proven that the morphological corollary has universal validity.

Many neuroanatomists have been skeptical of the validity of the two types of synapse on the basis of the fact that the flattening of vesicles has been shown to depend on the osmolarity of solutions used in preparing the tissue for electron microscopy.

But, in a sense, everything the electron microscopist sees is a distortion of the true dynamic living state. The interpretation of electron micrographs, and of any preparations of anatomical specimens, must be made with this constantly in mind. Despite these reservations, the recognition of the two types of synapse has been the basis for remarkable progress in the understanding of the synaptic organization of the brain.

Synapses and Terminals

Types I and II synapses provide relatively small areas of contact between neurons. They may be characterized as *simple* synapses. They are typical of the contacts made by small terminals, both axonal and dendritic, and they are also the type of contact made by most cell bodies and dendrites when those structures are presynaptic. It is probably fair to say that they make up the majority of synapses in the brain. This, in itself, bespeaks an important principle of brain organization, that the output of a neuron is fractionated, as it were, through many synapses onto many other neurons, and, conversely, that synapses from many sources connect to a given neuron. This is referred to as *divergence* and *convergence*, respectively. It is an essential aspect of the complexity of information processing in the brain.

In addition, there are, in many regions, much more extensive contacts with more elaborate structure that may be characterized as *specialized synapses*. The neuromuscular junction is an example in the peripheral nervous system. In the vertebrate central nervous system, we find an example in the retina, where the large terminal of a receptor cell makes contact with several postsynaptic neurons; within the terminal, the synaptic vesicles are grouped around a special small dense bar. This arrangement is shown very schematically in Fig. 6.3 and is described further below and in Chap. 16.

One may also categorize the terminals that bear the synapses. A terminal may be small and have a single synapse onto a single postsynaptic structure, as shown in most of the diagrams of Fig. 6.3. These may be characterized as *simple terminals*. On the other hand, a large terminal, with complicated geometry, may be characterized as a *specialized terminal:* examples are the neuromuscular junction and the basket cell endings around the Purkinje cell (Chap. 21). In many regions of the brain, large terminals have synapses onto more than one postsynaptic structure. The receptor terminal in the retina mentioned above is an example; another example is the large terminal rosette of the mossy fiber in the cerebellum, which has as many as 300 synaptic contacts onto postsynaptic structures (see below).

Within the brain are all possible combinations of synapses and terminals. Simple synapses may be established by any of the parts of the neuron: terminals, trunks, or the cell body. Simple synapses may also be made by specialized terminals, as in the case of the mossy fiber of the cerebellum. On the other hand, specialized synapses may be made by small terminals, as in the spinule synapses of the hippocampus, and, finally, specialized synapses may arise from specialized terminals, as in the case of the retinal receptor.

Patterns of Synaptic Connections

Synapses are also categorized by the kinds of processes that take part in the synapse. Thus, for example, a contact from an axon onto a cell body is termed an *axosomatic* synapse, whereas that onto a dendrite is termed an *axodendritic* synapse (Fig. 6.15A). Similarly, a contact between two axons is termed an *axoaxonic* synapse (C, E), and a contact between two dendrites is termed a *dendrodendritic* synapse (B).

A single synapse seldom occurs in isolation in the brain; it is usually one of a number of synapses that together make up a larger pattern of interconnecting synapses. In Chap. 1, these were referred to as *microcircuits*. The simplest of these pat-

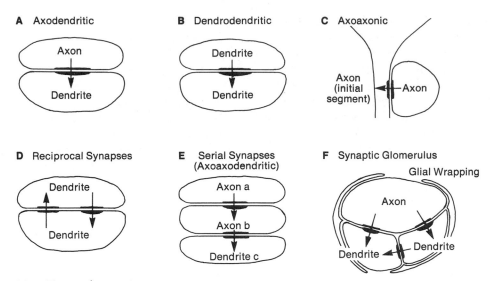

Fig. 6.15   Types of synaptic arrangements.

terns is that formed by two or more synapses situated near each other and oriented in the same direction; they may for example be all axodendritic. A more complicated pattern is one in which there is a synapse from process (a) to process (b), and another from (b) to (c). Such a situation is diagramed in Fig. 6.15E. These are referred to as *serial* synapses; examples are axoaxodendritic sequences and axodendrodendritic sequences.

Another pattern has a synapse from process (a) to process (b), and a return synapse from (b) to (a). This is diagramed in Fig. 6.15D. These are referred to as *reciprocal* synapses. If the two synapses are side by side, they are called a *reciprocal pair*. If the two synapses are far apart, a *reciprocal arrangement* results. Finally, there are patterns of synaptic connections between tightly grouped clusters of terminals, called *synaptic glomeruli* (Fig. 6.15F).

The first synapses identified by electron microscopists were simple contacts made by simple terminals, of the axosomatic and axodendritic type. Since these simple arrangements were in accord with the idea of a "polarized" neuron, they came to be regarded as "classical" synapses. The axoaxonic and dendrodendritic types were

identified later, as were the serial and reciprocal arrangements and the various types of specialized synaptic contacts and terminals. Since these synapses, terminals, and patterns did not fit classical concepts, the practice grew of referring to the simple synapses as "conventional" and to all the other synapses as "unconventional" or even "nonusual."

In recent years, so many examples of complex synaptic arrangements have come to light that we are in danger of believing that most parts of the nervous system, in both invertebrates and vertebrates, are organized in unconventional ways! This of course is absurd. It simply shows that nature does not always work according to our simple preconceptions; certainly the nervous system does not put these labels on its synapses. We may conceive that, in any given region, it is faced with specific tasks of information processing, and it assembles the necessary circuits from the available neuronal components. Thus, far from being unconventional, the complex synaptic arrangements, in fact, give expression to the extraordinary flexibility of the nerve cell as a fundamental anatomical unit of the nervous system. This has undoubtedly been crucial to the evolution of synaptic circuits that meet the needs placed on them by adaptive pressures for processing information, sometimes with competing priorities, in the most efficient ways possible. After considering the multitude of synaptic patterns that have emerged, David Bodian (1972) summarized the situation in the following eloquent manner:

In synaptic systems . . . we see not a stereotyped mechanism for the transfer of information from cell to cell, but another display of the fact that every conceivable capability of living organisms to solve adaptive problems is likely to be put to the test in the evolution of life.

Identification of Synaptic Connections

If the synaptic connections between neurons are so complex, how do we identify which neuronal processes contribute to a given synapse or synaptic cluster? Neuroanatomists have developed many methods for doing this. Sometimes the fine structure differs sufficiently to identify the processes in single sections, as in the case of the retina shown in Fig. 6.16A. In some cases, three-dimensional reconstructions from serial sections are necessary, as illustrated in the olfactory bulb in Fig. 6.16B. Another method is to treat the tissue with antibodies to an enzyme that is involved in the synthesis of a neurotransmitter. Figure 6.16C shows that, in olfactory bulb tissue treated with antibodies to glutamic acid decarboxylase (the enzyme that synthesizes GABA), the granule cell spines are positive, whereas the mitral cell dendrites are not. This is consistent with other evidence that the granule cell dendrites inhibit the mitral cell dendrites by the action of GABA at the dendrodendritic synapses. A fourth method is to inject a cell with a dye or with HRP which under the electron microscope (EM) is electron-dense. A fifth method is to transect a bundle of input fibers and identify the synapses made by degenerating terminals. A final method is to impregnate tissue by the Golgi method and then partially deimpregnate and examine the synapses made by or onto partially deimpregnated cells and terminals.

The examples above are all taken from vertebrates. An example from invertebrates is shown in Fig. 6.17. Here (A) the cell bodies of two neurons in the stomatogastric ganglion of the lobster were injected with the dye Procion Yellow. The neurons were visualized under light microscopy in whole mounts of the ganglion and precisely reconstructed from serial sections made through the cells. In the EM, serial reconstructions of ultrathin sections show that the dendritic trees that arise from the axons of these cells have numerous expansions (varicosities) where synapses are present. A typical reconstruction is shown in Fig. 6.17B. The results have been summarized by Allen Selverston, Don Russell, John Miller, and David King of San Diego as follows (Selverston et al., 1976):

Each synaptic varicosity is functionally bipolar . . . ; it both projects synapses onto and receives synapses from many other processes. These bifunctional varicosities are found on all of the dendrites of the neuron. Hence input and output are each distributed over the entire dendritic arborization. Although neuronal input and output are traditionally thought of as segregated onto polarized regions of the neuron (onto "dendrite" and "axons"), this does not seem to be the case in stomatogastric ganglion neurons. In this, the stomatogastric ganglion may show functional similarity to those regions of vertebrate central nervous system (such as olfactory bulb and retina) where dendrodendritic interactions are important.

From Synapses to Circuits

From these considerations it is obvious that the flow of information through a neuron and between neurons is much more complicated than depicted in the diagrams of Cajal in Fig. 6.1. Although this seems perplexing at first, some relatively simple general principles about how synapses are organized into circuits have emerged.

A useful generalization to begin with is that synapses are usually made by one of

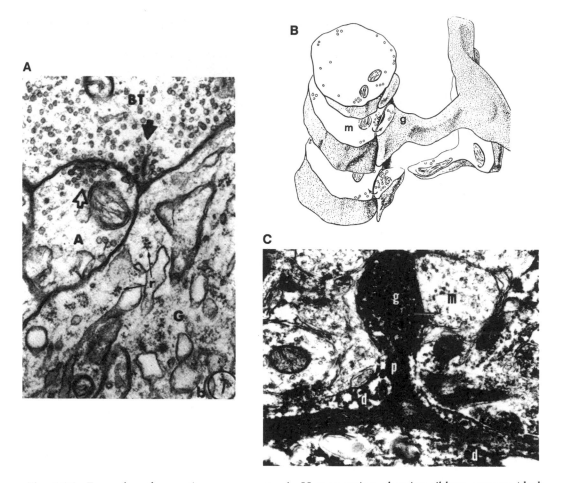

Fig. 6.16 Examples of synaptic arrangements. **A.** Human retina, showing ribbon synapse (dark arrow) between bipolar terminal (BT) and amacrine (A) and ganglion (G) cells. Also shown is a reciprocal synapse from amacrine to bipolar cell (open arrow). r, ribosomes. (From Dowling and Boycott, 1966) **B.** Olfactory bulb, reconstruction in three dimensions from serial EM sections, showing reciprocal synapses between mitral (m) and granule (g) cell dendrites. (From Rall et al., 1966) **C.** Rat olfactory bulb, treated with antibodies for glutamic acid decarboxylase (GAD), the enzyme that synthesizes GABA. Reaction product is found in granule cell spine (g), pedicle (p), and dendrite (d), but not in mitral cell dendrite (m). Arrow indicates site of granule-to-mitral dendrodendritic synapse. (From Ribak et al., 1977)

three types of neuronal element. First are the *input* fibers coming from other regions, usually ending as axon terminals. Second are the *relay neurons* of the given region. Third are the *intrinsic neurons* within that region. Thus, as shown in Fig. 6.18, a given synapse can be made by any one of these elements, and a complex synaptic arrangement generally involves some specific way in which the three elements are interrelated.

We call these three elements the *synaptic triad*. Sometimes the elements are very tightly organized, as in serial or reciprocal synapses. Sometimes they are loosely organized, as in spread-out regions like the ventral horn of the spinal cord or the cerebral cortex. Sometimes the interneuronal elements may be lacking, as in some simple relay structures. Or perhaps the output neuron is lacking, if the output is humoral,

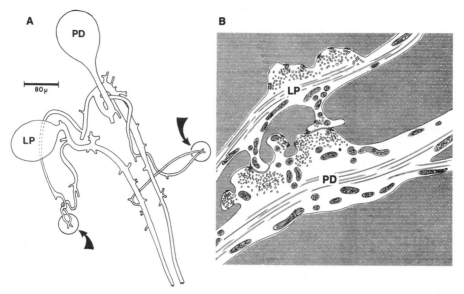

Fig. 6.17 Lobster stomatogastric ganglion. **A.** Two cells [pyloric dilator (PD) and lateral pyloric (LP)] have been reconstructed from serial sections. Circles show sites of synaptic connection between the cells. **B.** Synaptic connections, as reconstructed from serial EM sections, at right-hand site in A. (From Selverston et al., 1976)

or the cell body is situated elsewhere. In the face of all this diversity, it is useful to use the synaptic triad as a general framework for identifying the main kinds of connections present in a region, and comparing them with those of other regions.

With the identification of patterns of synaptic connections, we are in a position to

begin to identify circuits at different levels of organization (see Fig. 6.19). We will take as an example the pathway for the sense of smell, and see if we can identify its levels of organization in vertebrates and invertebrates. We will apply the same principles introduced in Chap. 1.

At the first level of synaptic organization

Fig. 6.18 Diagram showing how the triad of synaptic elements—input terminals, projection neuron, and intrinsic neuron—provides the basis for synaptic organization of local circuits in vertebrates and invertebrates.

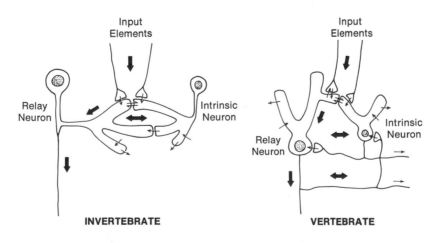

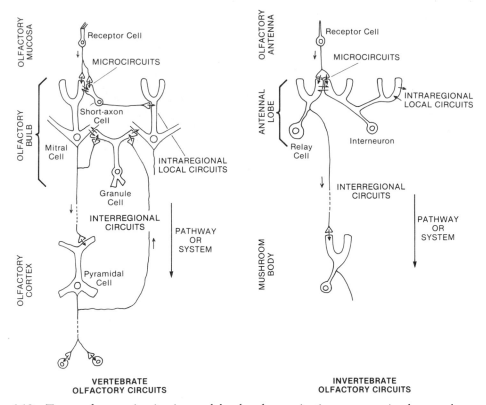

Fig. 6.19 Types of synaptic circuits, and levels of organization, as seen in the vertebrate and invertebrate olfactory pathways.

is the arrangement of synapses at a local site on a cell body, dendrite, or axon terminal. This may involve simple *convergence* of several inputs onto that site, or simple *divergence* to several output sites. In addition, it may involve *serial* relays of information, or *reciprocal* interactions. In all these cases, there is a set of synaptic connections that acts as an *integrative unit*. These local patterns of connections have been termed microcircuits (Shepherd, 1978). It is very common for a particular type of microcircuit to be repeated throughout a given layer or on a given cell type, thus acting as a *module* for a specific kind of information processing or memory storage. The microcircuits on a computer chip represent the same operating principle.

At a higher level of organization is the circuit that connects different neurons over longer distances within a given region. This

transmission may take place through a dendritic branch or dendritic trunk, or it may take place through the axon of an interneuron, or the axon collateral of an output neuron. The key point about all of these pathways is that they remain within a given region (see Fig. 6.19). The term introduced for all these types is *local circuit* (Rakic, 1976). The most restricted local circuits include microcircuits; the most extensive involve interlaminar and intraregional axonal connections. The functions of these different types vary widely. Some provide for reexcitatory spread of activity from neuron to neuron; others provide for antagonistic interactions between neighboring integrative units within a region.

The next highest level of organization involves connections of one region with another; we may refer to this as a *projection circuit*. Usually a region receives input from more than one other region, and usually a region projects output to more than one other region. Thus, the same principles of convergence, divergence, and integration of different kinds of information operate at this level, too. It is also common for there to be feedback from one region to another. Note that feedback loops are present at all levels. The more local feedback loops can be regarded as *nested* within the more extensive projection feedback loops (see Fig. 6.19).

At a still higher level are sequences of connections through several regions. These are said to constitute a *pathway* or *system*. The function is usually to transmit information from the periphery into the central nervous system (as in a sensory system), or from central to periphery (as in a motor system). However, in any pathway there are often connections running in the opposite direction to provide for *descending*, *ascending*, or *centrifugal* control.

Finally, at the highest level (at least the highest thus far identified) are sets of connections between a number of regions, which together mediate a behavior that involves to some extent the whole organism. These are called *distributed systems*. They are characteristic of higher functions of motor and sensory systems, and of many central systems.

7

Synaptic Potentials and Synaptic Integration

Synaptic Potentials: A Brief History

By the early years of this century, the idea of the synapse was firmly established. Cajal had demonstrated that nerve cells are individual entities, requiring that transmission between them takes place, as he phrased it, "by contiguity, not continuity." Sherrington had made contiguity explicit in his concept of the synapse. Sherrington also provided evidence for some of the physiological properties of the synapses in his studies of transmission through reflex arcs in the spinal cord. He showed that reflex discharges were graded in strength, showed summation without refractoriness, displayed inhibition as well as excitation, and often long outlasted the stimulus, all properties that are clearly differentiated from those of impulses in the nerves.

We have seen that during this time, biochemists like Dale and Loewi were laying the foundations for the view that synaptic transmission takes place through chemical messengers. However, the opposite view, that synaptic transmission occurs by means of electrical current passing from one neuron to the next, was held by many neurophysiologists. They objected that many of the biochemical experiments involved collecting substances in perfusates of isolated organs that were stimulated at high rates, so that the results admitted of more than one interpretation. It was also difficult to generalize from these peripheral organs to synapses in the central nervous system where experimental methods, both physiological and biochemical, were at that time almost completely lacking. It was a situation that gave rise to much heated debate, from the 1920s to the early 1950s; some sense of it can be gained from the remarks of Alexander Forbes (one of the few who was able to maintain his good humor) of Harvard, summing up a symposium on the synapse that was held in 1939:

So goes the controversy. Dale in discussing it remarked that it was unreasonable to suppose that nature would provide for the liberation in the ganglion of acetylcholine, the most powerful known stimulant of ganglion cells, for the sole purpose of fooling physiologists. To this Monnier replied that it was likewise unreasonable to suppose action potentials would be delivered at the synapses with voltages apparently adequate for exciting the ganglion cells merely to fool physiologists.

All this confusion was swept away by the advent of the microelectrode and the electron microscope in the 1950s. The old controversy was supplanted by the clear evidence that some synapses are chemical and some electrical (and some are mixed).

The first synapses studied with intracellular methods were the neuromuscular junction, by Bernard Katz and his colleagues in London, and the motoneuron, by John Eccles and his colleagues in New Zealand and Australia. Since that time, neurophysiologists have characterized the action at a synapse in terms of the electrical responses which they could record when a presynaptic process acts through the synapse on a postsynaptic process. The postsynaptic electrical response usually takes the form of a transient change in membrane potential, and this change is referred to as a *postsynaptic potential,* or simply *synaptic potential.*

Electrical Fields

The simplest means for effecting a change in the membrane potential of a postsynaptic cell is through the flow of current from a neighboring cell. We have already considered this situation in discussing an *ephapse* in the preceding chapter (Fig. 6.3). Electrical interactions of this type are limited in effect because the high resistance of the nerve membrane limits the amount of current that can cross the membrane. They are also diffuse in spatial extent, unless there are special glial wrappings to limit the extracellular spread of current. Finally, the postsynaptic activity is rigidly locked to the presynaptic activity. Interactions through electrical fields may be important in some situations in synchronizing activity of cell populations, but they provide limited means for specific types of neuronal interactions.

Electrical Synapses

Effective electrical coupling between cells is achieved through gap junctions. As described in detail in the preceding chapter (see Fig. 6.4), the intercellular channels at these junctions have a very low resistance to current passing between the two neurons, and at the same time they prevent loss by leakage to the extracellular space. Thus, a potential change in a presynaptic terminal may be transmitted to a postsynaptic terminal with little attenuation. Experimentally this is the direct test for the presence of an electrical synapse, and it was first reported by Ed Furshpan and David Potter in 1959 at a synapse between two nerve fibers in the crayfish. The synapse they studied is made by the lateral giant fiber onto the giant motor fiber in the abdominal ganglion; it mediates rapid flip movements of the tail that are used for defensive reflexes (see Chap. 19). As shown in Fig. 7.1, an action potential in the presynaptic fiber spreads with little attenuation into the postsynaptic fiber (A), whereas spread in the reverse direction (B) is very limited. This directionality of current flow is called *rectification.*

Electrical synapses (and their morphological substrates, gap junctions) have been found between neurons at many sites in the nervous systems of invertebrates and lower vertebrates.

Several sites have been found in the mammalian brainstem. In the mesencephalic nucleus of the fifth cranial nerve, there are electrical synapses between cell bodies and between cell bodies and initial axonal segments. In the vestibular (Deiters') nucleus the synapses occur between cell bodies and axon terminals. A spike initiated in one cell is transmitted to a neighboring cell as a short-latency depolarization by current flow through the axon terminals and branches. In the inferior olive, dendritic spines are interconnected by electrical synapses. The spines also receive chemical synapses, and it has been suggested that when they are active they shunt current away from the electrical synapses, thereby uncoupling the cells.

The salient features of electrical synapses derive from the nature of the direct connections. They operate quickly, with little or no delay. They can provide for current flow

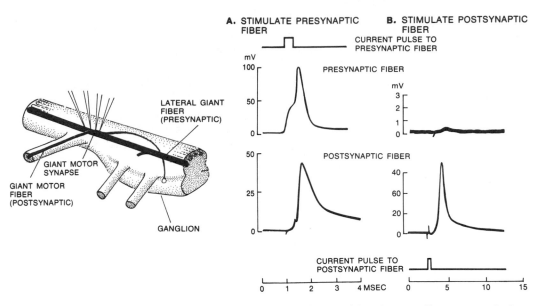

Fig. 7.1 Transmission through an electrical synapse in the crayfish. There is effective spread of an impulse from the lateral giant fiber into the giant motor fiber (**A**). The lack of spread in the opposite direction is called *rectification* (**B**). (After Furshpan and Potter, 1959, in Nicholls et al., 1992)

in both directions, or alternatively they can offer more resistance in one direction than the other (rectification). They provide a means of synchronization of populations of neurons. Their actions can be fixed and stereotyped in the face of repeated use, and are less susceptible to metabolic and other effects than chemical synapses. However, this does not mean that electrical synapses cannot be influenced by events within the cytoplasm. Experiments in salivary gland cells have shown that increases in internal Ca^{2+} block the ability of a small molecule (fluorescein) to pass through gap junctions. This uncoupling effect may depend on a concomitant fall in intracellular pH. There is evidence from freeze-fracture studies that the uncoupling involves changes in the geometrical arrays of intramembranous particles at the gap junction. Whether this mechanism is operative in the modulation of electrical coupling between nerve cells has not been determined. In addition to mediating electrical transmission, gap junctions are also important in other ways for intercellular communication and the organiza-

tion of cells into multicellular ensembles as discussed in Chap. 6.

Chemical Synapses

Chemical synapses, operating through the release of a neurotransmitter, are the predominant type throughout the nervous system. Let us begin our discussion of the physiological mechanisms of chemical synapses with the model of the neuromuscular junction, and then consider the two main types of central synaptic actions: excitation and inhibition. As a preparation, the student should review the biochemical steps involved in synaptic actions that are summarized in the preceding chapter.

Neuromuscular Junction

Our understanding of the process for release of ACh from the nerve terminal and the subsequent events that give rise to a response in the muscle membrane is based on the brilliant investigations of Bernard Katz and his co-workers, beginning in the early 1950s. We can incorporate these classical findings with the newer results by starting with the molecular events and building up to the muscle response in the following way. (This will illustrate the principle of levels of organization within the synapse.)

At the molecular level, the effect of ACh is to open conductance channels in the endplate membrane, as shown in Fig. 7.2D. When channels open, they change the electrical potential across the endplate membrane. The smallest change is an increase in "noise," due to slow "leaking" of ACh from the nerve terminal. With a patch electrode (Chap. 2) this can be recorded as the opening and closing of individual AChR channels. As discussed in Chap. 6, the channel allows cations to flow according to their Nernst potentials: most of the current is carried by Na^+ flowing in and K^+ flowing out, with a net inward current that depolarizes the membrane. The student at this point should review the summary of

the patch-recording technique in Chap. 2 and the relations between neurotransmitter binding and channel opening and closing.

The next larger change is caused by a number of channels opening in synchrony. This is called a *miniature endplate potential* (MEPP in Fig. 7.2C). It was discovered by del Castillo and Katz in 1954. The unitary nature of the MEPP indicates that it is due to a packet, or quantum, of ACh. Approximately 10,000 ACh molecules are contained in each quantum. The quantal nature of transmitter release and action has been fundamental to our understanding of synaptic transmission. It has been tempting to equate one quantum with one vesicle.

A single MEPP is due to a quantum released at a single active zone. When several quanta occur at or about the same time, their effects summate to produce a larger potential, as shown in Fig. 7.2B. When the motoneuron wants to signal the muscle to contract, the impulse it sends

Fig. 7.2 Synaptic properties exemplified in different types of recordings from the neuromuscular junction. **A.** Intracellular recording of endplate potential (EPP) giving rise to an action potential (AP) in the muscle cell: experimental setup shown at left. **B.** High-gain recording, showing summation of miniature endplate potentials (MEPPs). **C.** Very high gain recording, showing noise induced by ionophoresis of ACh (compare with control trace below). **D.** Extracellular patch clamp recording from junctional site, showing currents passing through single AChR channels. This was one of the first reports of single-channel activity, using a subgigohm ($<10^9$ ohms) recording seal. (A–C from Katz, Miledi, and colleagues (see text); D from Aidley, 1989)

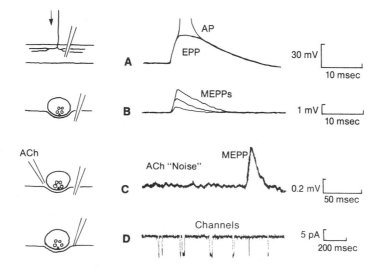

through the axon invades the terminal and depolarizes the membrane of the nerve terminal. There are approximately 1000 active zones in a typical ending. They do not all automatically release their quanta when depolarized by the impulse. Each one has a certain probability of release, so that the total number of quanta released, m, is equal to the number, n, of possible quanta (i.e., roughly the number of active zones) times the average probability of release, p, so that one has the following relationship:

$$m = np$$

This simple equation summarizes the *quantal hypothesis,* and was used by del Castillo, Katz, Martin, and subsequent workers to develop the idea that quantal release is a statistical, probabilistic process, not a deterministic process. An impulse releases 100–200 quanta, implying a p of 0.1–0.2. The quanta are released in synchrony by the impulse, and their effects summate to give a large potential, called the endplate potential (EPP), which can be recorded by an electrode with its tip inserted inside the muscle fiber near the endplate. As shown in Fig. 7.2A, the EPP spreads from the endplate to initiate the muscle impulse, which leads to muscle contraction.

The release of transmitter molecules in quanta has been found at a number of synapses, and may apply to most chemical synapses. The release process is controlled by the amount of depolarization of the nerve terminal membrane; as shown in Fig. 7.3A, the more the applied depolarization in the terminal, the more the depolarizing response in the endplate membrane. This is an expression of a generalization linking depolarization with Ca^{2+} entry and transmitter release (Fig. 7.3B). Special note should be made of the fact that the nerve terminal depolarization increases the frequency (probability of occurrence) of the MEPPs; the individual amplitudes remain the same. The experiment in Fig. 7.3A shows clearly how the depolarizing response is due to the summation of MEPPs overlapping in time.

Grading of postsynaptic responses with the amount of presynaptic depolarization is a crucial property for synapses in many central regions. A synapse from an incoming axon terminal is normally activated by an impulse invading the terminal, but a synapse from a dendrite may be activated by the graded depolarizations of synaptic potentials within that dendrite. The grading of depolarizations and the differing probabilities of release are properties that add to the complexity of operation of synaptic circuits in a number of regions of the nervous system, and need to be kept in mind as we apply the neuromuscular junction model to other synapses.

Isolated Preparations for Studying Central Synapses

In order to analyze the mechanisms of central synapses, one needs suitable preparation. In the case of invertebrate ganglia, one can excise the intact ganglion, maintain it in a recording chamber, stimulate selected input and output pathways, and record the synaptic responses using intracellular electrodes. This ease of preparation, together with the large sizes of many of the cells, is why invertebrate ganglia are so attractive to electrophysiologists.

Studies in the mammalian brain were traditionally carried out in the intact anesthetized animal, with all its attendant problems of maintaining the proper levels of anesthesia, allowing for the depressant effects of the anesthetic agents, and controlling the respiratory and vascular pulsations that dislodge the electrode from its intracellular position. A big step forward was the discovery by Chosaburo Yamamoto and Henry McIlwain in London in 1966 that slices of the mammalian brain can be prepared and maintained in a recording chamber. Since that time, many parts of the brain have been studied in vitro, including even slices of human cortex obtained in neurosurgical operations. Other prepara-

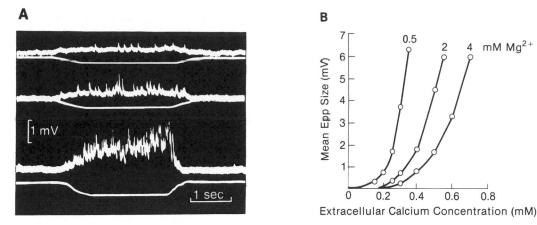

Fig. 7.3 Presynaptic control of the frequency of miniature endplate potentials (MEPPs). **A.** Recordings at three levels of depolarization of terminals. Recordings (upper traces) were made as in Fig. 7.2B, while current (bottom traces) was applied to the presynaptic nerves through an extracellular electrode. (From del Castillo and Katz, in Katz, 1962). **B.** Graph of the steep dependency of the amplitude of the endplate potential on the concentration of Ca^{2+}. The experiment also shows that Mg^{2+} depresses the effect of Ca^{2+}. (From Dodge and Rahamimoff, 1967)

tions include the mammalian brainstem perfused through its blood vessels, and the entire brain of the turtle, from medulla to olfactory nerves, simply bathed in Ringer's solution.

These in vitro preparations allow investigation of individual synapses and synaptic circuits under conditions in which the region is completely stable, and pharmacological manipulations can be carried out with different drugs in the bathing medium. One of the first and best known preparations is the hippocampal slice, which we will describe briefly.

A summary diagram of the synaptic organization of the hippocampus is shown in Fig. 7.4. As can be seen, the cell layers form two C-shaped sheets, facing and overlapping each other. One of these sheets is the hippocampus proper, containing large pyramidal neurons, which are the main output cells. The other sheet is the dentate fascia, whose output neurons are called granule cells. By virtue of the main fiber pathways, there is a natural sequence of activity in these regions. The sequence actually begins in the entorhinal cortex, a region which receives and integrates multi-

sensory inputs from the touch, auditory, olfactory, and visual pathways, as well as inputs from the cingulate gyrus as a part of limbic systems (see Chap. 28). The entorhinal cortical output fibers ① traverse ("perforate") the surrounding cortex and terminate mainly in the dentate fascia. The dentate granule cells have relatively short axons ("mossy fibers") which connect to the nearest part of the hippocampal pyramidal cell population ②. The axons of these pyramidal cells ③ project to the septum through the fornix; in addition, they send a collateral branch ④ to connect to the long apical dendrites of pyramidal cells in another part of the hippocampus. These cells project to a nearby cortical area, the subiculum ⑤, which in turn projects to the septum through the fornix ⑥.

The hippocampus is well suited for preparing in a slice, because it is constructed as a series of lamellae, each of which contains the circuit indicated in Fig. 7.4. In the chamber under the microscope, stimulating electrodes can be placed on each of the layers or pathways and recordings made from each of the types of cell with microelectrodes. These studies have provided di-

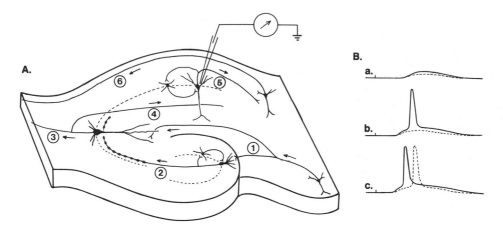

Fig. 7.4 A. Synaptic organization of hippocampus. The circuits are shown as they would be present in a slice of the hippocampus prepared for recording in an isolation chamber. Different parts of the circuits ①–⑥ are described in the text. **B.** Intracellular recordings from pyramidal cell showing plasticity of synaptic actions. a. Response to weak volleys in radiation fibers ④ before (dotted trace) and after (solid trace) tetanization of radiation fibers to produce long-term potentiation (LTP). b and c. Same, stronger volleys. (After Andersen et al., 1977)

rect confirmation of previous results in the intact animal, and have enabled more precise analyses to be carried out on neuronal properties and synaptic actions. Both intracellular and patch electrode recordings can be obtained, and pharmacological manipulations can be carried out. The slice preparation approaches the neuromuscular junction and other peripheral preparations in its accessibility for modern experimental approaches to membrane mechanisms.

All of the input connections of the hippocampus are believed to operate with the transmitter glutamate, whereas the interneuronal synapses operate mostly via GABA. The hippocampus offers the opportunity to carry out detailed analyses of these two primary transmitter systems for excitatory and inhibitory synaptic actions, respectively. Moreover, these synaptic systems show considerable activity-dependent plasticity (see Fig. 7.4, right), which is of great interest as a model for memory mechanisms in the mammalian brain. In this chapter we confine our attention to basic synaptic mechanisms; we take up their possible roles in memory and learning in Chap. 29.

In addition to slice preparations, *isolated cell cultures* are providing another very important means for analyzing synaptic properties. The disadvantage of primary cell cultures is that the cells initially regress to earlier developmental states before growing out their processes in the culture dish, and this takes place in an artificial environment lacking the neighboring structures that constrain normal growth and provide for normal synaptic connections. However, many of the properties of normal function are retained, as we saw in discussing the outgrowth of processes in Chap. 3. Cell cultures have provided important information on properties of synaptic release and reception, as we shall see.

Central Excitatory Synapses

The primary type of excitatory synapse in the central nervous system operates by the release of glutamate to act on glutamate receptors. Drawing on our discussion in the previous chapter, we may summarize the main steps at a glutamate synapse with the diagram of Fig. 7.5. Glutamate is synthesized from glutamine ①; it is stored

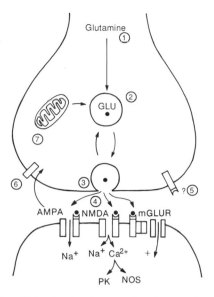

GLUTAMATE

Fig. 7.5 Molecular mechanisms of gluta-matergic synapses. ① synthesis of glutamate (GLU) from glutamine; ② transport and storage; ③ release of GLU by exocytosis; ④ binding of GLU to receptors identified by specific agonists [α-amino-3-hydroxy-5-methyl-4isoxazole (AMPA); N-methyl-D-aspartate (NMDA)]. The AMPA receptor gates Na^+ and K^+ flux; the NMDA receptor regulates a Ca^{2+}-permeable conductance state which is normally blocked by Mg^{2+} and high resting membrane potential; when this block is relieved by membrane depolarization Ca^{2+} flows in to depolarize the membrane further and activate other second messenger systems. The receptor mechanisms illustrated in the figure may represent different conductance states of the same receptor-channel complex. ?⑤ binding to presynaptic receptors; ⑥ reuptake. A similar sequence is involved at synapses utilizing aspartate. mGLUR indicates metabotropic glutamate receptor. (Based on Cooper et al., 1993; Jahr and Stevens, 1987; Cull-Candy and Usowicz, 1987)

② and released ③ in a Ca^{2+}-dependent manner (similar to ACh at the neuromuscular junction). On the postsynaptic membrane, glutamate acts on two main types of receptors, which have been distinguished on pharmacological grounds. One type binds quisqualate and kainate; this type

also binds a compound called AMPA (α-amino- 3- hydroxy- 5- methyl- 4- isoxazole propionic acid), and is therefore called the *AMPA receptor*. The other type binds a compound called N-methyl-D-aspartate (NMDA) and is therefore referred to as the *NMDA receptor*. Both receptors are parts of a protein molecule that forms the ionic channel that is gated to produce the post-synaptic effect. They are similar to the nAChR receptor in this important respect; in fact, as we have seen, they belong to the same superfamily of receptor-channel proteins (Chap. 6).

The physiological actions of the two types of glutamate receptors can be analyzed in slice preparations or in cell cultures with the application of substances that selectively block each component. This analysis is summarized schematically in Fig. 7.6 for an experiment in which recordings are made from a pyramidal neuron in a hippocampal slice (position ⑤ in Fig. 7.5). Let us start our experiment with the cell at a normal resting potential of −80 mV. At this potential, activation of excitatory synaptic input, by delivering an electrical shock to the Schaffer collaterals ④, sets up a depolarizing excitatory postsynaptic potential (EPSP) in our neuron. We then test to see if the EPSP contains subcomponents by introducing into the bath APV (2-amino-5-phosphonovalerate), which is a selective blocker of the NMDA receptor. This has a small effect of reducing the decaying phase of the EPSP. Subtracting the two traces shows that most of the EPSP is due to the AMPA receptor, with only a very small late contribution by the NMDA receptor.

The next step in the analysis of an EPSP is to determine the equilibrium potential of the response; according to our discussion of the Nernst potential (Chap. 4), this will tell us which ions are likely to be carrying the current. In our experiment, we find that the AMPA response vanishes around 0 holding potential; like the endplate potential at the neuromuscular junction, this equilibrium potential (reversal potential) is

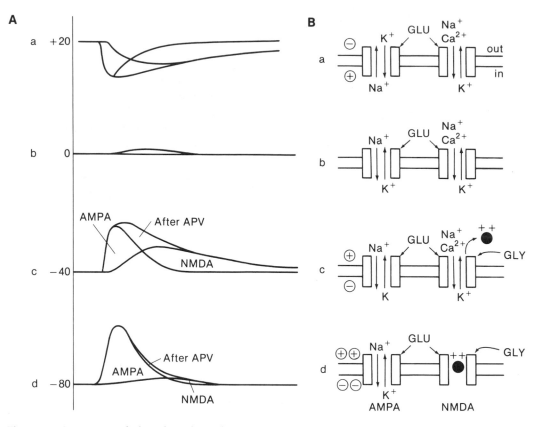

Fig. 7.6 Summary of the physiological properties of glutamatergic synapses. **A.** Diagrams of intracellular recordings from a neuron responding to excitatory synaptic input at different holding potentials (a–d); the responses are shown before and after exposure to antagonists of the AMPA and NMDA receptors. **B.** Diagrams showing the AMPA and NMDA channels and the current flows through them at the different holding potentials. See text.

intermediate between the equilibrium potentials of Na^+ and K^+, and the membrane conductance is likely to be a nonspecific cation conductance. However, the NMDA component reverses at slightly more positive potentials, suggesting that there is a larger permeability to Na^+ and/or Ca^{2+}.

When we examine the response at intermediate holding potentials, as at -40 mV in Fig. 7.6, we find that the AMPA response is smaller than at -80 mV, as we would expect, since the holding potential is closer to the reversal potential at 0 mV. However, unlike the endplate potential, the NMDA component is actually larger in this intermediate range. How can this be? The answer came from testing for the effects of

other ions in the bathing medium. In patch clamp recordings of single channels from cultured brain cells of the mouse, Philippe Ascher and his colleagues in Paris found that if they carried out their experiments in the absence of external Mg^{2+}, glutamate activated the channels similarly at both negative and positive holding potentials (see Fig. 7.7A), in accord with the behavior of the AMPA potentials in Fig. 7.6. However, if Mg^{2+} was added to the bath in its normal extracellular concentration, the single-channel events were much reduced at negative holding potentials. From this kind of data one can construct a current–voltage (I–V) plot, which shows (Fig. 7.7B) that in Mg^{2+}-free solution the cur-

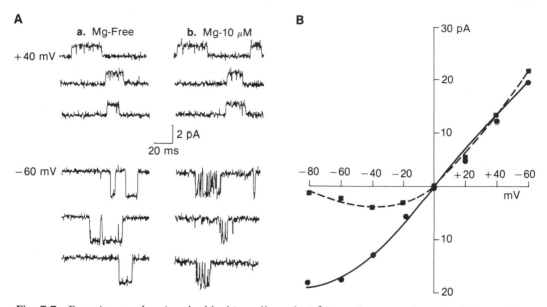

Fig. 7.7 Experiments showing the blocking effect of Mg^{2+} on glutamate channels. **A.** Recordings of single glutamate channels in cultured brain cells of embryonic mouse exposed to glutamate. a. The channel openings in Mg-free Ringer are relatively symmetrical at positive ($+40$ mV) and negative (-60 mV) holding potentials (the directions of current flow may be seen in the diagrams of Fig. 7.6). b. The channel openings with Mg^{2+} in the bath are similar at positive ($+40$ mV) holding potentials, but show frequent interruptions, due to blocking of the channel, at negative (-60 mV) holding potentials. **B.** Graph of the relation between glutamate-induced membrane currents and different holding potentials in the absence (continuous line) and presence (dashed line) of Mg^{2+} (these current responses underlie the NMDA component of the voltage responses at different holding potentials shown in Fig. 7.6). (From Nowak et al., 1984)

rent–voltage relation is nearly linear over most of the physiological range, whereas in Mg^{2+}-containing solution, the relation is nonlinear at negative potentials. In fact, this curve recalls the nonlinear property of the Na^+ conductance underlying the action potential (see Chap. 5).

Based on these and many other experiments, the proposed mechanism of the glutamate-sensitive receptor-channel protein is summarized in Fig. 7.6B. Near the normal resting potential of -80 mV, activation by glutamate induces nonspecific cationic current through the AMPA channels but not through the NMDA channels, which are blocked by Mg^{2+} due to attraction of Mg ions to the negatively charged inner end of the channel. At depolarized membrane potentials, the increased internal positivity, combined with in-

creased external negativity, removes the Mg^{2+} from the blocking position, allowing current to flow. The NMDA channel is much more permeable than the AMPA channel to Ca^{2+}, which enters the postsynaptic region of the cell. The slower time course of the NMDA response is due to the fact that the receptor is only slowly desensitizing, and the channels are activated in long-lasting bursts.

In summary, the NMDA channel has several critical properties. There is a voltage dependence, due to the voltage sensitivity of the Mg^{2+} block; note that this is different from the voltage sensitivity of a transmembrane segment in the case of the Na^+ channel underlying the impulse. There is an associated threshold depolarization for activation, which a highly nonlinear property. There is prolonged, nondesensitizing activation of the channels. And there is a high permeability for Ca^{2+}; in fact, most of the current in the NMDA channel is carried by Ca^{2+}. This is highly significant, because Ca^{2+} serves many functions in the nervous system. This makes the NMDA channel a prime candidate for Ca^{2+}-mediated second messenger functions (see Chap. 8) and functions related to plasticity and memory mechanisms (see Chap. 29). It also makes it a prime suspect for causing cytotoxicity and neuronal death if prolonged membrane depolarization, such as may occur in traumatic injury to nerve cells or in stroke, should lead to prolonged NMDA channel activation.

Molecular Biology of the Glutamate Channel. A subunit of the non-NMDA receptor was first cloned by Steven Heineman and his co-workers in 1989. The first NMDA receptor subunit was cloned by Nakanishi and co-workers in 1991. As noted in the previous chapter, these glutamate receptors belong to the same superfamily as the ACh receptor, with four presumed transmembrane segments but with low sequence homology to the AChR. The single channel conductances of the neurotransmitters-activated channels are summarized in Table 7.1.

Studies in expression systems have confirmed most of the properties of the native channels. In addition to the properties noted above, there appear to be an unusually large number of regulatory sites on the NMDA receptor which can produce allosteric changes that affect channel function. The glutamate receptors as a class also show an unusually high degree of sequence diversity. There are already more than a half-dozen types of glutamate receptors, and the number is growing (summarized in Barnard, 1992). It is speculated that this large number of types and subtypes reflects the widespread distribution of glutamate synapses in the nervous system and the way that they can be fine-tuned in different regions for specific functional properties.

Central Inhibitory Synapses

Rapid inhibition in the central nervous system is mediated by two main types of synapses with directly gated receptors, operating through the amino acid transmitters α-aminobutyric acid (GABA) and glycine. We will focus on GABA, with a briefer consideration of glycine.

GABAergic Synapses

The main steps at a GABAergic synapse are summarized in Fig. 7.8. Like glutamate, GABA arises from common pathways of intermediary metabolism; in fact, GABA is synthesized from glutamate ①, by the enzyme glutamic acid decarboxylase (GAD), which can be used to localize GABAergic synapses by immunocytochemical methods (cf. Fig. 6.17). After storage ② and release ③, GABA acts on postsynaptic receptors ④. The receptors are of two types. One is a directly gated channel that is selective for Cl^-; this is called the $GABA_A$ *receptor.* The other is a receptor that acts on membrane channels indirectly through a G protein; this is called the $GABA_B$ *receptor.* A common action is to open channels selective for K^+. $GABA_B$ receptors are also found on presynaptic terminals ⑤, where they may modulate transmission. There are also high-affinity uptake systems in the presynaptic terminals ⑥ and glia ⑥. Degradation of GABA is by transamination in mitochondria ⑦. Note the active participation of glia in the uptake and resynthesis of GABA.

The physiological properties of these two types of GABA receptors are illustrated in Fig. 7.9. Let us imagine we are recording again from a pyramidal neuron at ⑤ in Fig. 7.4 while stimulating the input pathway at ④. Let us further assume that, in addition to setting up EPSPs in our cell, these fibers also activate interneurons which then make inhibitory synapses onto our cell (location 5A). If our cell is relatively active and somewhat depolarized, for example, to a resting potential of around -50 mV, the response

Table 7.1 Single-channel conductances of neurotransmitter-activated channels

Preparation	Agonist	Method	γ (pS)	T (°C)
Cation-permeable excitatory channels				
Amphibian, reptile, bird, and mammalian endplate	ACh	SF	20–40	8–27
Bovine chromaffin cells	ACh	UC	44	21
Aplysia ganglion	ACh	SF	8	27
Locust muscle	Glutamate	UC	130	21
Mammalian neurons	Glutamate (AMPA)	UC	15	21
Mammalian neurons	Glutamate (NMDA)	UC	50	21
Chloride-permeable inhibitory channels				
Lamprey brainstem neurons	Glycine	SF	73	4
Cultured mouse spinal neurons	Glycine	SF	30	26
Cultured mouse spinal neurons	GABA	SF	18	26
Crayfish muscle	GABA	SF	9	23

Abbreviations: NMDA, *N*-methyl-D-aspartate; SF, stationary fluctuations; UC, unitary currents.

Adapted from Hille (1992) and C. F. Stevens (personal communication)

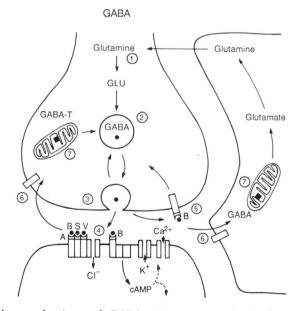

Fig. 7.8 Molecular mechanisms of GABAergic synapses. ① Synthesis of γ-aminobutyric acid (GABA) from glutamine (catalyzed by glutamic acid decarboxylase); ② transport and storage of GABA; ③ release of GABA by exocytosis [corelease with a neuropeptide such as enkephalin (Enk) or somatostatin (SOM)]; ④ binding to a $GABA_A$ receptor blocked by bicuculline (B), picrotoxin, or strychnine (S), which are coupled to a chloride channel; the GABA receptor also has a site for binding of benzodiazepines, such as Valium (V); $GABA_B$ receptors, by contrast, are linked via a G protein and/or cAMP to K^+ and Ca^{2+} channels: these are blocked by baclofen; ⑤ binding to presynaptic receptors; ⑥ reuptake in presynaptic terminal, and uptake by glia; ⑦ transamination of GABA to α-ketoglutarate (catalyzed by GABA transaminase, $GABA_T$, regenerating glutamate and glutamine; glial glutamine then reenters the neuron. (Modified from Cooper et al., 1993; Aghajanian and Rasmussen, 1988; Nicoll, 1982)

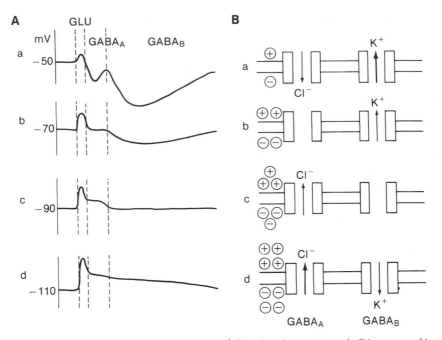

Fig. 7.9 Summary of the physiological properties of GABAergic synapses. **A.** Diagrams of intracellular recordings from a neuron responding to a sequence of excitatory and inhibitory synaptic inputs at different holding potentials (a–d). Note the different phases of the response, and the different reversal potentials for the GABA$_A$ and GABA$_B$ responses. **B.** Diagrams showing the GABA$_A$ and GABA$_B$ channels and the current flows through them at the different holding potentials. See text.

might look like trace A in Fig. 7.9. The sequence consists of an initial depolarizing EPSP, due to the glutamatergic synapses, followed by two phases of hyperpolarization. These are termed inhibitory postsynaptic potentials (IPSPs).

How do we reveal the mechanisms of these potentials? To answer this question, we follow our usual procedure of finding the equilibrium potentials for the conductances underlying the two phases, by observing the response at different holding potentials. Setting the holding potential successively to more polarized (negative) levels reveals the reversal potential first for the early phase (b, at −70 mV) and then for the later phase (c, at −90 mV). This is consistent with equilibrium potentials for Cl$^-$ and K$^+$, respectively. Thus, the early phase is due to activation of GABA$_A$ receptors whose channels conduct Cl$^-$ ions, and the later slower phase is due to activation

of GABA$_B$ receptors, which leads to activation of channels that conduct K$^+$ ions. These mechanisms and the directions of the ion flows are indicated in the diagrams to the right in Fig. 7.9. It may be noted that patch recordings of the activity of single GABAergic channels give results at different holding potentials that are in accord with the directions of current flows shown in the diagrams.

Molecular Biology of the GABA$_A$ Channel. The GABA$_A$ channel was first cloned and sequenced in 1987 (Schofield et al., 1987). It became evident that it belongs to the same family as the nACh receptor. At this point the reader should turn to Fig. 6.8 to review the topology of the GABA$_A$ receptor in comparison with the nACh receptor. Within the framework of similarities in transmembrane segments and extracellular domains, there are specific differences related to the different functional properties. In the GABA receptor there is an excess of positively charged residues near the extracellular ends of the segments, which is presumed to contribute to the selectivity of the channel for the negatively charged Cl ions that flow through the channel when it is open (we saw, in contrast, that the nicotonic ACh receptor has a preponderance of negatively charged residues at this site, which contributes to its selectivity for positively charged cations). In addition, there are distinct differences between the cationic and anionic channels in the residues in membrane segment M2, which is consistent with various evidence we previously discussed that this segment lines the channel surface.

By molecular cloning, five different types of subunits have been recognized, so it is believed that the channel is formed, like the AChR, by an arrangement of five subunits around a central pore, similar to the arrangement of the ACh receptor. Like its cousin, the glutamate receptor, multiple isoforms have been found for most of the subunits. These multiple isoforms could produce hundreds of different combinations, suggesting that as in the case of glutamate channels, natural GABA channels are probably composed of different subunits with different specific types of properties in different local regions of the brain.

In addition to this diversity of subunits, the GABA receptor, like the glutamate receptor, is remarkable for the multiple ways in which different ligands can bind to it to bring about allosteric changes in its function. This is indicated schematically in Fig. 7.10, which summarizes the results of many types of experiments, including analysis of receptor binding by GABA agonists and antagonists, electrophysiological studies of the actions of these agents, and behavioral studies of the effects of these agents in animals and humans. The effects of these agents may be summarized as follows (based on Sieghart, 1992).

GABA is the natural ligand at the synapse, opening the channel to the flow of

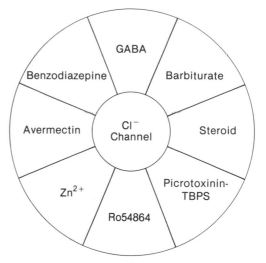

Fig. 7.10 Schematic representation of the multiple allosteric binding sites on the GABA$_A$ receptor for different ligands. The effects of GABA, benzodiazepines, barbiturates, picrotoxin, and steroids are discussed in the text. Avermectin, a drug with potent insecticidal and anthelmintic actions, increases Cl$^-$ conductance. Ro54864 is a convulsant that inhibits Cl$^-$ conductance. Zn^{2+} is present at high concentrations in synaptic terminals; it is released during synaptic activity and inhibits GABA receptors at some synapses. (From Sieghart, 1992)

Cl$^-$ ions which polarize the membrane and reduce neuronal excitability. Paradoxically, GABA acts at the low-affinity binding sites. Bicuculline is a competitive inhibitor at this site.

Benzodiazepines are used clinically for a variety of purposes: to reduce anxiety, oppose convulsions, relax muscles, and act as sedatives and hypnotics. These agents bind to high-affinity sites to enhance the actions of GABA. Single-channel recordings show that this is by increasing the frequency of channel openings during GABA binding.

Barbiturates are important clinically as sedatives and hypnotics. They achieve their effects by enhancing the actions of GABA and, at high concentrations, they enhance Cl$^-$ conductance themselves. Patch recordings show that their action is to pro-

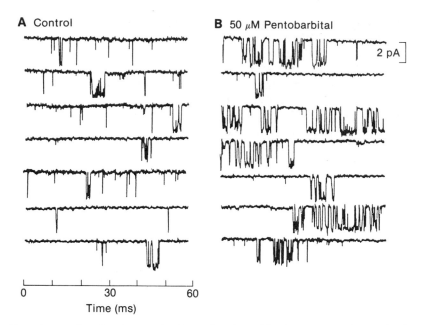

A Control   **B** 50 μM Pentobarbital

2 pA

0   30   60
Time (ms)

Fig. 7.11 Experiment showing the mechanism of action of barbiturates at the single channel level. Recordings are from GABA$_A$ receptors expressed in an embryonic kidney cell line, isolated in outside-out patches, and exposed to GABA. **A.** Control recordings in normal Ringer, showing brief channel openings. **B.** Recordings with pentobarbital in the bath, showing prolonged bursts of repeated openings, thereby increasing the total open time of the channel. The direction of the (positive) current is inward at this holding potential of −70 mV because the Cl⁻ concentrations on either side of the membrane were equal (normally, with the Cl concentration gradient inward, the positive current would have been outward, that is, negative Cl ion flow inward, at this potential). (From Puia et al., 1990)

long the bursting of the channels (see Fig. 7.11).

Steroids of certain types have anesthetic, sedative, or anxiolytic effects that are mediated by enhancing the action of GABA. There is evidence that this is due to prolonging the open time of the Cl⁻ channel.

By contrast with these enhancers of GABA action, which reduce activity in the nervous system, *picrotoxin* is well-known for its ability to induce convulsions; in accord with this, its action is to antagonize the actions of GABA. It appears to do this not by inhibiting GABA or benzodiazapine binding, but by binding to the receptor to bring about a steric change that blocks Cl⁻ entry into the channel.

These are the main types of ligands of clinical interest that have allosteric binding sites and documented actions on the GABA receptor. Other agents and their effects are noted in the legend to Fig. 7.10.

Glycinergic Synapses

Glycine is the other major inhibitory neurotransmitter in the nervous system. Glycine synapses traditionally have been regarded as less prevalent than GABA synapses. The traditional view was that GABA synapses are found mainly in the forebrain, whereas glycine synapses are confined to the spinal cord and brainstem. However, this separation has broken down (see Fig. 7.12A and B), and we will discuss the widespread distribution of glycine synapses in Chap. 25.

Glycine synapses operate by a mechanism similar to that of GABA synapses, acting directly to increase the conductance of a postsynaptic receptor to Cl⁻. Traditionally the two types could be distin-

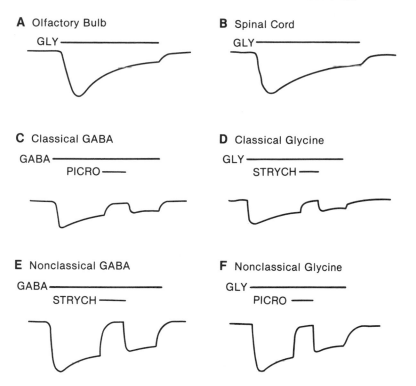

Fig. 7.12 Experimental analysis of GABA and glycine receptors. Drawings of typical patch recordings from cultured rat neurons stimulated by glycine, GABA, and antagonists. Stimulation was by miniature flow pipes, which allowed brief controlled applications of the compounds. **A, B.** Both forebrain (olfactory bulb) and spinal cord neurons respond with inhibitory currents to glycine. **C, D.** In olfactory bulb neurons, responses to GABA are blocked selectively by low concentrations of picrotoxin; responses to glycine are blocked by strychnine. **E, F.** At high concentrations of picrotoxin and strychnine, there is cross-reactivity between GABA and glycine responses. (After Trombley and Shepherd, 1994)

guished on pharmacological grounds. GABA synapses were selectively blocked by bicuculline and by picrotoxin (see above), whereas glycine synapses were blocked by strychnine (see Fig. 7.12C, D); in both cases, blockade led to increased neuronal excitability and convulsions. However, this pharmacological distinction is also breaking down; as shown in Fig. 7.12E and F, there can be cross-reactivity with higher doses, and this is seen particularly at earlier ages. These findings suggest that glycine receptors are more complex than originally thought.

The glycine receptor was first cloned and sequenced by Schofield et al. and by Glenningloh et al. in 1987. The receptor shows high sequence homology with the GABA receptor and thus belongs with it in the nACh receptor family. Like the other members of the family, it is composed of five subunits, each composed of a protein with four transmembrane segments, a long extracellular domain where ligand binding takes place, a long intracellular loop where phosphorylation can occur, and an M2 segment believed to form the surface of the pore. The receptor pentamere has a stoichiometry of two α and three β subunits. Molecular cloning is revealing an increasing variety of subunits. Receptors show different subunit construction in relation to development and to different regions of the brain, as we shall discuss in Chap. 25.

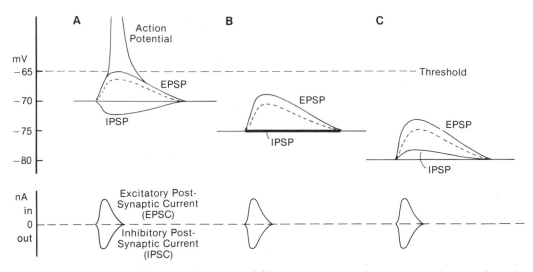

Fig. 7.13 Integration of EPSP and IPSP at different resting membrane potentials. **A.** When the resting MP is below (less negative than) the inhibitory equilibrium potential *(E_i)*, the conductance increase during the IPSP causes a hyperpolarization of the membrane. **B.** When the resting MP is equal to *E_i*, the inhibitory conductance increase causes no change in potential. **C.** When the resting MP is above (more negative than) *E_i*, the IPSP causes a depolarization of the membrane. In all cases, integrative summation of the EPSP and IPSP causes a decrease in amplitude of the EPSP (dashed line), making the neuron less excitable.

Synaptic Integration

It is largely through the interaction between excitatory and inhibitory synapses that the competition for control of the membrane potential in different parts of the neuron is carried out. This competition lies at the heart of the study of the dynamics of synaptic organization. The principle goes back to Sherrington; following him, the process by which different synaptic inputs are combined within the neuron is termed *synaptic integration.*

The interaction of a single EPSP and IPSP serves as a paradigm for synaptic integration in neurons, and it will be useful to grasp certain essentials. Let us assume an excitatory synapse and a nearby inhibitory synapse, the activation of which individually produce an EPSP and IPSP, respectively, as shown in Fig. 7.13A. Assume now that the two are activated simultaneously. The effect of the IPSP is to reduce the amplitude of the EPSP, away from the threshold for impulse initiation, as is

shown in Fig. 7.13A. The dotted line traces the resulting transient; it represents the "integrated" result of the two synaptic potentials.

Now it is commonly thought that this process of integration is a matter of simple algebraic addition of the two opposed synaptic potentials; to wit, "depolarization plus hyperpolarization equals membrane potential." However, this simple formula does not have general validity. As shown in Fig. 7.13B, when the resting potential is at the inhibitory equilibrium (reversal) potential, no IPSP is recorded, but there is still a reduction of a simultaneous EPSP, due to the shunting effect of the increased inhibitory conductance. And when the resting membrane is more polarized (Fig. 7.13C), the IPSP is in fact depolarizing (toward the inhibitory equilibrium potential), yet its effect is still to reduce the EPSP by virtue of the increased conductance. The essential inhibitory action is therefore not a hyperpolarization of the membrane, but rather an increase in ionic conductance, which drives the membrane potential toward the equilibrium potential for those ions.

It is thus the opposition of synaptically activated conductances and ionic currents that controls the relative amounts of depolarization and hyperpolarization of the membrane potential. In addition, one must consider the geometrical relations between excitatory and inhibitory synaptic sites in a dendritic tree, and the electrotonic flow of current through the dendrites. Synaptic integration thus involves a complex interplay between ionic conductances and neuronal geometry (see below).

Ionic Currents

It will be useful at this point to consider more closely the relation between an ionic conductance change and the resultant change in membrane potential. For an example we take the case of a brief increase in conductance to Na ions. As shown in Fig. 7.14, Na^+ moves inward through its conductance channel at the active site (A). In the electrical circuit, the current reaches a point on the inside where it can travel in two directions. Some current passes onto the inner surface of the membrane capacitance, where it deposits positive charge that depolarizes the membrane. Some passes along the inside of the nerve cell to the next patch of membrane (site B), where it can follow three paths: onto the membrane capacitance, through the membrane resistance, or further along the fiber. Ultimately all the current must pass out across the membrane and pass back along the outside of the cell to the negative pole of the Na battery.

Careful study of the diagram and current flows will help answer two questions that are often puzzling to the student. The first is, how can inward and outward current both depolarize the membrane? As can be seen, the reason is that inward current at an active site and outward current at a neighboring site both have the same effect, of putting positive charge on the inside of the membrane capacitance. The same reasoning applies to the relation between oppositely directed current flows and hyperpolarization.

The second questions is, what are the time relations between the current flows and the potential changes? When the flow is rapid the potential response is slower, because charge is transiently stored on the membrane capacitance. The amount of slowing depends on the time constant (τ_m) of the membrane, given by the product of the membrane capacitance (C_m) and membrane resistance (R_m): $\tau = RC$. The relation between the rapid synaptic current flows and the slower synaptic potentials is shown in Fig. 13. For very slow or constant changes in conductance, a steady state exists in which the capacitance becomes an open circuit and can be ignored, and the spread of current to the neighboring site is determined solely by the resistance along the paths.

These relations between current and potential underlie most of the electrophysiological properties of nerve cells. For instance, the diagram of Fig. 7.14 applies equally to the case of activation of an action potential and its propagation by local currents, as we discussed in Chap. 5. The slowing of potential responses and the decay of potential spread are both governed by electrotonic properties. The diagram also emphasizes that at any site on a nerve cell there can be two pathways for interactions: internally with other parts of the same cell, by means of electrotonic spread or impulse generation, or externally through synapses onto neighboring cells.

Conductance-Decrease Synapses

Thus far we have considered synaptic responses as being due to the opening of conductance channels. By contrast, there are responses that are produced by the closing of conductance channels. These are called conductance-decrease synapses. They were first described by Forrest Weight and his colleagues in sympathetic ganglion

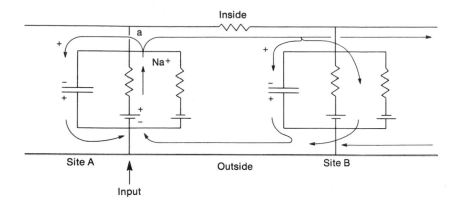

Fig. 7.14 Current flows underlying the depolarization of membrane patches. Initial input (as from an EPSP, applied current, or local potential) at site A causes inward current flow of positively charged Na ions. At (a), current can flow in two directions: outward, to depolarize membrane capacitance; or longitudinally and then outward, to depolarize capacitance of neighboring membrane patch (site B). Thus membrane depolarization is brought about by both inward ionic current and outward capacitative current. Note that external current flow completes the circuit.

cells, and have been analyzed extensively in recent years by Paul Adams and his colleagues.

An excitatory response can be generated at a conductance-decrease synapse in the following manner. At rest, the membrane is somewhat depolarized (say, at -60 mV), representing a balance between a resting K conductance (g_K) and a significant amount of Na conductance (g_{Na}). When the synapse is activated (stimulated), the effect of the released transmitter is to turn off the resting g_K. The membrane moves toward the equilibrium potential of the remaining ion (Na^+) to which it is permeable, thus producing a depolarizing response.

This type of synaptic response has been demonstrated in sympathetic ganglion cells, in the slow EPSP and the late, slow EPSP (we will discuss these in Chap. 18). The K current which is turned off is in fact the M current (I_m), previously discussed in Chap. 5.

You probably realize by now that the opposite effect could be produced by turning off a Na current, and you are right; Nature has not missed this opportunity. If both Na and K currents are present at rest, the membrane will be partially depolarized.

The effect of turning off I_{Na} is thus to hyperpolarize the membrane toward E_K. This type of mechanism has been seen most clearly in the response of vertebrate photoreceptors to light; we will study this further below and in Chaps. 10 and 16.

Conductance-decrease synapses have several properties that are of interest. The decreased conductance raises the resistance of the membrane, thereby increasing the amplitude of any other synaptic responses nearby. However, this also increases the time constant of the membrane, and therefore slows any nearby responses. Conductance-decrease synapses thus help to increase the strength of neighboring responses, and, by slowing them, increase their chances of summation. They thus can contribute significantly to enhancing the integrative properties of neurons.

Integrative Organization of the Nerve Cell

We are now in a position to ask: how do synaptic potentials carry information through the cell and provide for transfer of information to other cells? There are some underlying principles, but with many adaptations in different nerve cells. An introduction to the synaptic organization of neurons and circuits in different regions may be found in Shepherd (1990b). Let us consider three examples that cover some of the basic properties.

Stretch Receptor Cell

It will be instructive to begin with one of the first model neurons, the stretch receptor of the crayfish. This preparation was introduced to intracellular electrophysiology by Stephen Kuffler and Carlos Eyzaguirre in a classic series of experiments in the 1950s.

A schematic diagram of the stretch receptor cell is shown in Fig. 7.15A. The cell has several large dendritic trunks, which enter the muscle and terminate in fine branches. When stretch is applied to the muscle (Fig. 7.15B), a depolarization is set up in the dendritic branches, graded with the amount of stretch. This depolarization is due to a nonspecific increase in permeability to cations, which moves the membrane potential toward an equilibrium potential around zero. This *receptor potential* is thus similar to an EPSP. We will discuss mechanisms of transduction of the stimulus in this cell in later chapters (Chaps. 10 and 13).

An intracellular electrode inserted into the cell body records the receptor potential together with the discharge of impulses that arise from it (Fig. 7.15B). In this situation, as in all recording experiments, the electrode is in one particular spot in the nerve cell, and we must deduce the sites where the different types of activity are actually initiated. With regard to the receptor potential, the site of generation is clearly known; it is in the dendritic terminals, which are 100–300 μm away from the cell body. We therefore know that our recording must represent an attenuated version of the true receptor potential, because of the leakage of current across the membrane of the dendritic trunk as the current flows toward the cell body and out into the axon, according to the cable properties of the dendrites. But what about the impulses? Where do they arise—in the dendrites, in the cell body, or in the axon? An elegant experiment by Charles Edwards and David Ottoson, working in Kuffler's laboratory in 1957, showed that, contrary to expectation, the site of impulse generation is in the axon, at a considerable distance from the cell body. To complete the picture, there are inhibitory nerves which make synapses on the dendrites of these cells. Stimulation of these fibers produces IPSPs, as indicated in Fig. 7.15C. How do these responses interact with the other activity in the cell?

In order to appreciate these relationships we need to think about the nerve cell in its spatial dimension. We need to take a point in time during an impulse response (see dashed line in Fig. 7.15B), and ask: what is the distribution of potentials throughout the cell at this point in time? We can answer this question in an intuitive way with the help of the diagram in Fig. 7.16. We will follow the methods of Wilfrid Rall, of the National Institutes of Health, who founded the study of the flow of electrical activity in dendritic systems (see Rall, 1977; also Jack et al., 1975).

Beginning with the dimensions of the cell in Fig. 7.16A, we construct a model of the cell in which its electrical cable properties are incorporated in a simple equivalent cylinder (B). In C, we plot the distribution of the different types of potentials in this equivalent cylinder. To begin with, the *receptor potential,* as we have already mentioned, is generated in the fine terminals and spreads electrotonically through the cell, as indicated in the figure. The action potential, by contrast, is generated in the axon; it propagates in the axon toward the central nervous system and also spreads electrotonically back into the cell body and dendrites. The IPSP is generated in the distal dendrites, and spreads, also by electrotonic means, as indicated in the diagram.

The normal sequence of events is thus: stimulus→setting up of receptor potential in terminals→electrotonic spread through dendrites and cell body to axon→generation of action potential in axon when depolarization reaches threshold→propagation

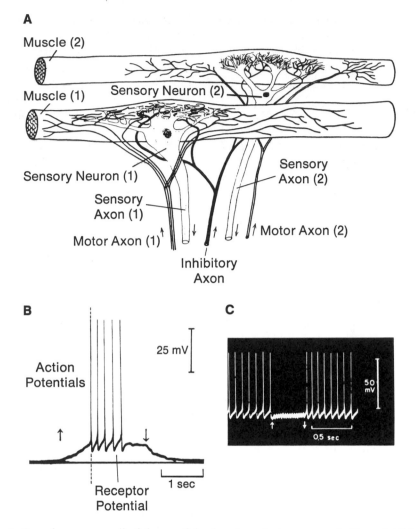

Fig. 7.15 **A.** Stretch receptor cell of the crayfish, showing relation to muscle fibers, inhibitory axon, and motor axons to muscles. **B.** Excitatory response of receptor cell to stretch of the muscle (arrows), as recorded by intracellular electrode inserted into cell body of sensory neuron (2). Dotted line indicates point in time for display of potentials in Fig. 7.16. **C.** Inhibition of the excitatory response by stimulation of the inhibitory axon (between arrows). (A from Burkhardt, B from Eyzaguirre and Kuffler, in Aidley, 1978; C from Kuffler and Eyzaguirre, 1955)

of action potential forward (orthodromically), and electrotonic spread backward (antidromically) into cell body and dendrites. The IPSP, acting during a response to stretch, opposes the response by making the membrane tend toward the relatively hyperpolarized inhibitory equilibrium potential.

Several questions often arise at this point in students' minds; let's see if we can anticipate and answer them.

Question 1: "It seems to me a funny way to organize a cell; how can the potentials spread so far?" *Answer:* As we noted in Chap. 4 and in connection with Fig. 7.14 above, electrotonic spread depends on just three factors: the resistance of the cytoplasm, the resistance of the cell membrane, and the diameter of the dendrite or axon. The two main ways that neurons control the effectiveness of spread is by means of the membrane resistance and the diame-

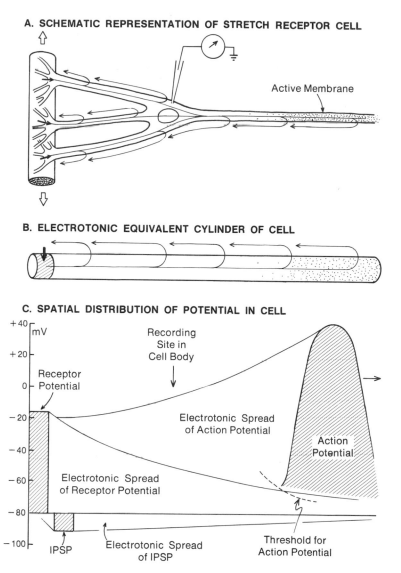

Fig. 7.16 Diagrams illustrating the analysis of spread of electrical activity in the stretch receptor cell. See text.

ter. *The greater the membrane resistance,* the less current leaks out across the membrane, and the more effective is the spread of a receptor potential and synaptic potential (and also, as we saw earlier, the local currents spreading in front of an action potential). *The greater the diameter,* the easier it is for the current to spread through the interior (cytoplasm) of the cell. Thus, stretch receptor neurons have a relatively *high* membrane resistance, and sufficiently large diameter dendrites and axon, to pro-

vide for effective electrotonic spread of potentials within them.

Different neurons vary greatly in this respect. In many cases, there is effective spread throughout much of a dendritic tree, and this provides the basis for the integration of many synaptic inputs. In other cases, there is limitation of spread; one sees this, for example, in the thin necks of some spines that arise from dendrites, and in very thin processes that connect parts of cells. In these cases, it appears that there is an isolation of the activity at a given site from the ongoing activity in the rest of the cell, so that different parts of the cell can operate as independent or semi-independent functional units.

Question 2: "Is the electrotonic potential active or passive?" *Answer:* Recall that the electrotonic potential is a *passive* potential. It is the spread that occurs along a process when all of the electrical properties of that process remain constant, at their resting values. Strictly speaking, "active" is used only in reference to the regenerative property of the action potential. Any time there is a change in the membrane potential at some site (due to a receptor potential, synaptic potential, or action potential), current always flows electrotonically through neighboring regions to equalize the distribution of charge. Part of the strategy of the functional organization of a neuron is to distribute synaptic sites and action potential sites in strategic parts of the neuron, linking them through passive electrotonic spread.

Question 3: "I don't understand; is the electrode at the cell body recording directly the receptor potential and the action potential, or isn't it?" *Answer:* Strictly speaking, the electrode is recording the *electrotonically spreading* receptor potential and action potential. If we wanted to record the receptor potential directly, we would have to put the electrode tip into the terminals. We would then record a very big depolarization, as indicated in the diagram. Similarly, the action potentials recorded in the cell body are attenuated versions of the action potentials in the axon, as shown by the lower amplitudes in the recordings of Fig. 7.15. This serves as a reminder of the fact that *every recording gives a selective view of what is going on in a neuron,* being weighted for events happening near at hand.

Question 4: "I thought there was a general rule in the nervous system that inhibition occurs at the cell body. Why does it occur in the dendrites in this cell?" *Answer:* The rule has many exceptions. It is true that many neurons receive inhibitory inputs to their cell bodies, and even initial segments. It is presumed that this provides for very effective control of impulse initiation at these sites. However, the stretch receptor cell shows that another effective placement is near the site of excitatory input. At this site, there is maximum opportunity for shunting of excitatory currents by the increased conductances of the inhibitory channels, thus depressing the receptor potential. Thus, the IPSP at this site tends to gate the receptor potential near its site of origin, rather than gating the impulse near *its* site of origin. The nervous system provides variations on these two themes in different regions, presumably reflecting needs for the interaction of excitation and inhibition in the processing of different kinds of information.

Mitral and Granule Cells

For our next examples we turn to the olfactory bulb and the two main types of cells within it. One is the principal neuron, the mitral cell, and the other is the granule cell, an inhibitory GABAergic interneuron.

As we discussed in Chap. 6, the mitral and granule cells interact by means of synapses between their dendrites. The nature of these interactions has been revealed by experimental analysis combined with computer simulations of the mitral and granule cells. The sequence of activity in the microcircuits formed by these synapses is illustrated in Fig. 7.17. The sequence begins with the olfactory nerve input to a mitral cell, which sets up an EPSP (see open arrow in A) that spreads through the cell to activate the action potential (AP) in the initial segment. This much is similar to the stretch receptor. As in that cell, there is also antidromic electrotonic spread of the action potential into the dendrites, as indicated by the arrows in A. The novel element in the mitral cell is that this depolarization activates numerous dendrodendritic synapses (A), each of which sets up an EPSP in a spine of a granule cell dendrite (B). It is as if the mitral cell dendritic tree func-

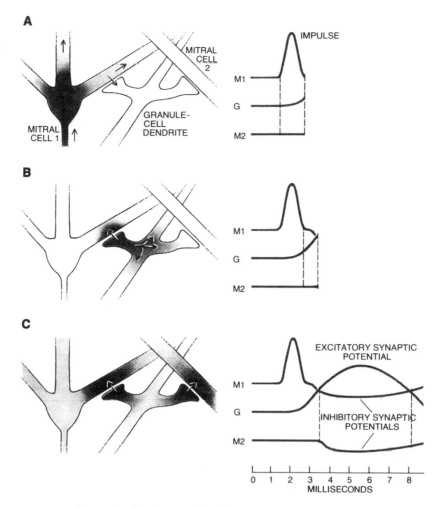

Fig. 7.17 Diagram illustrating local interactions between mitral and granule cells in the vertebrate olfactory bulb, by means of synapses between their dendrites. Sequence begins (A) with impulse in mitral cell (1); in (B) an EPSP has been initiated in the granule cell spine; in (C) the spine feeds back an IPSP onto mitral cell (1), and sends inhibition laterally onto mitral cell (2). All interactions are by dendrodendritic synapses. (From Shepherd, 1978, based on Rall and Shepherd, 1968, and Shepherd and Brayton, 1979)

tions as a large presynaptic terminal, similar to the multiple active zones of a nerve terminal at a neuromuscular junction. The depolarization of the spine in turn activates the reciprocal synapse of the spine onto the mitral cell dendrite (B); this synapse is inhibitory, which brings about feedback inhibition, which hyperpolarizes the mitral cell dendrite (C; M1). In addition, the EPSP spreads within the granule cell dendritic tree to depolarize neighboring spines, which brings about lateral (surround) inhibition of the dendrites of neighboring mitral cells (C;M2). These interactions were first postulated from computer models based on close matching of data from electron microscopical (Rall et al., 1966) and electrophysiological data (Rall and Shepherd, 1968); they have been tested and confirmed since then by numerous anatomical, electrophysiological, and neuropharmacological studies.

Several new principles of synaptic organization are exemplified by these interactions.

First, input and output through the granule cells can be mediated solely by synaptic potentials in the granule cell spines and electrotonic spread between them, without need of transmission by an impulse. There are by now many examples of neurons that perform complex input–output operations without generating propagating impulses—that is, *neurons without impulses* (see Roberts and Bush, 1981). This does not rule out the contribution of voltage-gated channels to nonlinear functions in these neurons; it means only that a propagating impulse is not needed for communication between nearby parts of the neuron.

Second, granule cells lack axons; thus, neuronal output can take place without need of either impulses or axons. As noted in Chap. 6, there are many examples in the nervous system of *neurons without axons* (anaxonal neurons), such as the amacrine cells of the retina.

Third, dendrites can be sites of synaptic output as well as synaptic inputs: that is, *dendrites can be presynaptic as well as postsynaptic*. As we have already noted, this evidence, together with the evidence for axoaxonic synapses, requires a revision of Cajal's doctrine of the dynamic polarization of the neuron. These do not appear to be synapses with unique properties radically different from the synapses made by axon terminals; rather, they appear to be synapses of the same basic types, excitatory or inhibitory, but placed in positions on the neuronal somadendritic surface where they can carry out specific input–output operations.

Fourth, models of the granule cell spines provided evidence that *the dendritic spine is an anatomical, metabolic, and functional entity*. A primary function is to *generate large synaptic depolarizations* that mediate feedback and lateral inhibition of the mitral cells. By virtue of small size and thin neck of the spine, the amplitude of the EPSP is much larger than it would be if the same

conductance change took place on the dendritic branch or soma (Shepherd and Brayton, 1979). Generation of large synaptic potentials is a near-universal function of dendritic spines, whether they are both pre- and postsynaptic, as in the granule cell, or only postsynaptic, as in cerebellar Purkinje cells and cortical pyramidal neurons (see below). The thin neck also contributes to limiting the metabolic communication of the spine head with the rest of the neuron (Shepherd, 1974); thus, the large synaptic potentials may be associated with *high local concentrations of ions, second messengers, and target proteins*. The neck also limits effects on the head from the rest of the neuron (see Chap. 29).

Fifth, the models of granule cell spines further indicate that spines can act as *semi-independent input–output units*, by virtue of the long, thin neck that connects the spine head to the parent dendritic branch. This has drawn attention to the possibilities for semi-independent subunits at many levels of dendritic organization (see also Ralston, 1971). Thus, the neuron is not one homogeneous integrative unit; it is potentially *many integrative subunits*, each with the possibility of mediating a local synaptic output to another cell or a local electrotonic output to another part of the same cell. In this local activity, backward (antidromic) spread of activity is as functionally important as forward (orthodromic) activity.

Finally, the fact that mitral cells take part in these interactions shows that relay neurons that generate impulses and have long axons can have presynaptic dendrites and take part in local dendrodendritic microcircuits as well as axonless interneurons such as granule cells.

Cortical Pyramidal Neurons

As our last example we consider the pyramidal neuron, the principal (output) neuron of the cerebral cortex. We have already seen these cells in Golgi stains, with their long apical dendrites and short basal dendrites covered with numerous dendritic spines (see Fig. 3.2). We will have further occasion to study their properties with regard to the highest levels of processing in sensory and motor systems in Sections II and III, in memory mechanisms in Chap. 29, and in higher cognitive functions in Chap. 30. It is manifest that an understanding of the functional organization of this type of neuron must be critical for an understanding of human brain function.

Although cortical pyramidal neurons come in various shapes and sizes, we can identify a "canonical" pyramidal neuron in the same way we identify a gene family by certain common features. Figure 7.18A schematically represents the minimum architecture needed to capture the integrative structure of a cortical pyramidal neuron. This includes an apical dendrite with distal branches, several branching basal dendrites, and dendritic spines at different levels of the apical and basal dendritic tree. Only four spines are shown, in order to signify the minimum number needed to generate some essential types of processing that may occur through spine interactions. Not shown are additional features such as the long axon that gives off recurrent collaterals.

To make a computational model of this cell, we represent each main anatomical subunit by a compartment, as in Fig. 7.18B. Thus, there are compartments for the spines (A_1, A_2; B_1, B_2), distal dendritic branches (A, B), distal dendritic branch points (C), apical dendritic trunk (D), basal dendrites (E), and soma (F). Each of these subunits plays a critical role in information processing within the neuron through its particular combination of synaptic inputs, active properties, and electrotonic linkage to neighboring structures. Thus, the spines are the principal sites of excitatory synaptic inputs; located nearby are less frequent but strategically placed inhibitory synapses. There is evidence for voltage-gated properties of distal dendrites and spines, which boost the amplitude of local EPSPs. Computer simulations have shown that, by the interactions of these combined local synap-

tic and action potentials, spines on distal dendrites can general specific types of information processing, such as all the basic types of logic gates that underlie the operation of a digital computer (discussed futher in Chap. 30). Cortical dendrites thus provide a rich substrate for local information processing.

A larger type of subunit is formed by clusters of spines on distal dendritic branches, whose summed activity is necessary to spread effectively to more proximal branches. Thus, branches A and B would feed into the main dendritic stem, D. This means that the branch point C would be critical as a gate in determining the effectiveness of this spread. Spread to more proximal dendritic sites is promoted by synchrony of distal activity and boosting by local voltage-gated membrane in dendrites and spines. This can bring about a kind of saltatory conduction, as in the case of axons, but with intervening passive membrane without myelin.

These several levels of dendritic processing thus put a new perspective on the soma. In contrast to the common view that inputs to the soma should carry specific information and distal inputs carry slower, more modulatory types of information, we can see that the organization of the pyramidal neuron favors a great deal of computational complexity in the distal dendrites. The soma (F) may be seen as the final summing point for this cascade of local operations in both apical (A–D) and basal (E) dendrites. As far as is known, there are no local synaptic outputs through presynaptic dendrites, as in the case of the mitral cell. The local operations in pyramidal neurons thus affect the final output of the cell only when they reach threshold for impulse generation at the initial segment-axon hillock.

It remains to note the significance of inhibition in this cell. We have already commented on the fact that local inhibition in distal dendrites (A, B) can contribute to specific information processing. Another key location for inhibitory synapses is at

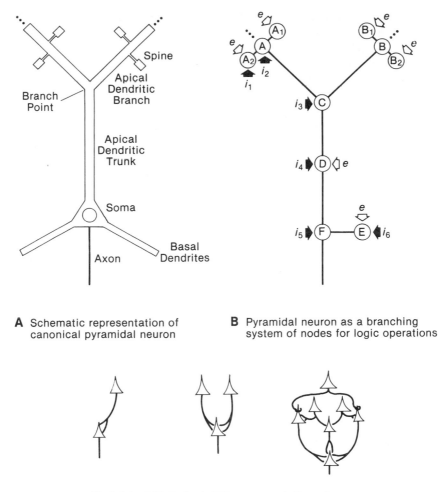

A Schematic representation of
canonical pyramidal neuron

B Pyramidal neuron as a branching
system of nodes for logic operations

C Original McCulloch-Pitts representation of neurons
as oversimplified nodes for logic operations

Fig. 7.18 Integrative organization of the cortical pyramidal neuron. **A.** Representation of the basic architecture of a "canonical" pyramidal neuron. Dotted lines indicate extensions of the dendritic branches and their spines to varying lengths in different neurons. Branches and spines on basal dendrites are not shown, nor is the axon with its recurrent collaterals. **B.** Compartmental representation of the pyramidal neuron. Each compartment contains an equivalent cylinder representing passive, synaptic, and active properties, as in Fig. 4.6. See text for full explanation of symbols and significance. (Modified from Shepherd et al., 1989) **C.** Example from the representations of neurons as simple nodes and of their interconnections for generating logic operations. (Modified from McCulloch and Pitts, 1943)

branch points (C), where they can gate summation of input from two branches. A third key location is on the apical dendritic trunk (D), where inhibitory inputs can isolate the processing in distal branches and gate its effects in determining impulse output. Finally, inhibition at the soma (F) determines the effectiveness of summation of apical and basal dendritic inputs to the soma and can directly counteract the depolarizing pressure underlying impulse output. Thus, the neuron provides for many gradations of inhibitory actions, from the most local to the most global.

put. Thus, the neuron provides for many gradations of inhibitory actions, from the most local to the most global.

In assessing these actions it is important to take into account not only the density of synapses but also the size of the dendrite on which they are located. A small spine or dendrite has a high input resistance, which gives a large-amplitude synaptic potential for a given amount of synaptic current, whereas a large dendrite or cell body has a low input resistance, giving a small synaptic potential for a given synapse, another demonstration of the principle that "small is large, and large is small" (recall the importance of this concept in our discussion of volume transmission in Chap. 6).

In conclusion, the original concept of the functional organization of the neuron at the dawn of the computer age was as a simple summing node, in which two functions were represented: the strength of synaptic inputs and the binary generation of an impulse. This is represented in the simple diagrams of the neuron and its interactions with its neighbors in Fig. 7.18C, from the paper of Warren McCulloch and Walter Pitts of MIT in 1943. By contrast, we have seen that most neurons have an extensive somadendritic surface and a complex branching geometry, which provide multiple sites for local processing of synaptic and active properties. In this view, the neuron is not one node but many; it is itself an extensive computational system, equivalent in computer terms to an integrated, multifunction chip. Incorporating this rich complexity into network simulations of neural circuits is one of the greatest challenges facing theoretical neuroscience today.

8

Second Messengers
and Neuromodulators

Biochemistry of the Synapse:
A Brief History

In contrast to studies of electrical activity in the nervous system, studies of biochemistry had a slower beginning. The laboratory apparatus was primitive, and, throughout most of the nineteenth century, studies were limited to characterizing the presence of fats, proteins, and carbohydrates in ground-up samples of the brain.

A decisive step forward was taken around 1900 by a school of English physiologists under John Langley studying the autonomic nerves to the internal organs of the body. They found that electrical stimulation of these nerves produced characteristic bodily changes (increase in heart rate, increased blood pressure), and that these changes were mimicked by the injection of extracts of the adrenal gland. According to Masanori Otsuka and Zach Hall (1979):

The idea that chemicals might mediate transmission between excitable cells is thought to have originated in 1903 during a conversation between two young scientists: Otto Loewi, a 29-year-old Austrian pharmacologist, and Thomas R. Elliot, a 25-year-old British medical student and physiologist.

In a preliminary communication to the Physiological Society in 1904, Elliott, a student under Langley, postulated that impulses in the autonomic nerves cause the release of an epinephrine-like substance from the nerve terminals onto the effector cells in the adrenal gland. In the same year, Langley further postulated that the cells in the gland have excitatory and inhibitory "receptive substances" which determine what the response and action will be. These were indeed far-reaching suggestions, for in their postulates of a *chemical link* between cells, its dependence on the amount of *electrical impulse activity,* and the presence of specific *molecular receptors,* they presaged many of the essential properties of synaptic transmission.

The culmination of this line of work was the demonstration in 1921 by Loewi that the vagus nerve inhibits the heart by liberating the substance acetylcholine. The work of Henry Dale and his collaborators in England in the 1930s provided evidence for acetylcholine as the transmitter substance in autonomic ganglia as well as at the junctions of nerves onto skeletal muscles.

As biochemists tackled the molecular mechanisms of the nervous system in the

1950s, they faced two main problems. One was that methods were lacking for getting at the molecular structure of accessible synapses such as the neuromuscular junction; this awaited the arrival of recombinant DNA technology, as we saw in Chaps. 2 and 7. The other problem was how to study inaccessible synapses tucked away within the central nervous system. The main traditional method was to homogenize the brain and analyze the biochemical constituents of different fractions. Most of these whole-brain studies made little connection with the cellular anatomy and physiology of the time. It was not until the development of a method for visualization of biogenic amines by their fluorescence around 1960 that one had a tool for identifying neurotransmitter substances in individual neurons and neuron terminals. On the receptor side, it was realized by the 1970s that the response to a transmitter is not necessarily limited to an immediate change in membrane conductance, but can involve activation of second messenger systems with many effects within the neuron. A third main development, beginning in the 1970s, has been the identification of a wide variety of small peptides, which can have trophic actions on neuronal growth, modulate synaptic transmission or second messenger systems, or have other effects within the neuron.

We are now in a fourth stage, in which the function of the synapse is being explored on a molecular basis.

As a result of this work, we are coming to regard neuromuscular transmission and the rapid actions of excitatory and inhibitory central synapses as *simple synaptic actions*. In contrast are a variety of *complex synaptic actions* that include one or more of the following features: different first messengers (neurotransmitters or neuropeptides), acting at different receptors, activating different second messengers, which act on different target proteins and channels. A given synapse may mediate both simple and complex actions; indeed, a given transmitter may do both, as we

Table 8.1 Second messenger systems in neurons

Free calcium ions (Ca^{2+})
G-binding proteins
Phosphoinositide hydrolysis (IP_3, DAG)
Cyclic nucleotides
Channel gating
Protein phosphorylation
Protein carboxymethylation
Phospholipid methylation
Arachidonic acid metabolites (prostaglandins, leukotrienes, thromboxanes)
Gases (nitric oxide, carbon monoxide)

Abbreviations: DAG, diacylglycerol; IP_3, inositol triphosphate.

Adapted from Cooper et al. (1993)

have seen in the case of the NMDA glutamate receptor. Therefore, we distinguish between simple and complex actions, which in turn may define a given synapse as simple, complex, or mixed in its effects. The more complex and longer lasting mechanisms are grouped under the term *neuromodulation*. We will first consider the main types of second messenger systems, and then summarize the main types of synapses.

Second Messenger Systems

Second messengers are defined as molecules that serve as functional links between receptors of external (first) messengers, and effector mechanisms in the receptive cell (metabolic processes, genes, ion channels). The number of these links is growing, and it is becoming obvious that there are third, fourth, and higher order messengers in many reaction chains. In addition, the same molecule can be active at different steps in different chains. Thus, Ca^{2+} can be the trigger for a reaction chain, or the final objective that is modulated; as another example, G proteins can be steps in modulation of cAMP levels, or they themselves can act directly on ion channels.

A list of second messenger systems in neurons is shown in Table 8.1. We will discuss several of the best understood of these.

Calcium as a Second Messenger

The most ubiquitous messenger in cells in general and neurons in particular is intracellular free ionized Ca. Several methods have been developed for determining calcium levels. One measures the luminescence of aequorin, a protein isolated from jellyfish, which reacts with Ca^{2+} to emit light. Another method uses metallochromic dyes, such as arsenazo III; when this complexes with Ca^{2+} it changes its absorption spectrum, which can be measured by differential spectroscopy. Ion-exchange microelectrodes that are selective for Ca^{2+} have been developed, which give a direct measure of intracellular free Ca^{2+}. Electron microscopy combined with X-ray or protein microprobe spectroscopy has permitted localization of several ion species, including Ca^{2+}, in relation to cell organelles.

All of these methods, together with traditional biochemical methods of measuring radioactive Ca^{2+} fluxes, are in agreement in showing that the levels of free ionized Ca in nerve cells are extremely low. The levels are in the range of 10^{-6} to 10^{-8} M. This may be compared with estimates of about 10^{-4} M of total Ca^{2+}/kg of axoplasm for the case of the squid giant axon (and about 10^{-2} M in seawater; see Table 4.1). Thus, most of the Ca in a neuron (or any other body cell for that matter) is in the bound form, and only a very small proportion is free and ionized within the cytoplasm. This is essential for the second messenger functions of Ca, for it means that the cell can use small changes in local Ca^{2+} levels to promote large or significant effects. This is the basis for the critical role that Ca^{2+} plays in such diverse functions as secretion, axoplasmic flow, motility, contraction, enzymatic reactions, and membrane permeability.

With regard to neurotransmitter release, the key event is the influx of Ca^{2+} (see Chaps. 6, 7). As shown in Fig. 6.8, binding of Ca^{2+} to Ca^{2+}/calmodulin-dependent protein kinase II is considered to be an essential step in priming of synaptic vesicles, and Ca^{2+} may also play a more direct role in bringing about vesicle fusion. The Ca ions enter the presynaptic terminal through voltage-gated channels when the terminal is depolarized by invading nerve impulses. The resultant transient elevation of cytosolic Ca concentration has been visualized at the giant synapse of the squid through combined use of indicator dye and video-enhanced microscopy. An experiment using the dye fura-2 is shown in Fig. 8.1. These results indicate that synaptic Ca^{2+} channels are highly localized to the region of synaptic contact, but that opening of the channels leads to transient increase of Ca^{2+} concentration throughout much of the presynaptic terminal. After Ca^{2+} dependent release of neurotransmitter has occurred (see Chap. 7), the Ca^{2+} must be cleared from the cytoplasm. The most important mechanisms for doing this are binding to calmodulin, binding to endoplasmic reticulum and other organelles, sequestration in cisternae, uptake by mitochondria, and efflux by pumping. These mechanisms controlling the availability of free Ca are in turn linked in many ways to second messenger systems, protein phosphorylation steps, pathways for transmitter synthesis, and the general metabolism of the cell. Through these means, Ca^{2+} plays important roles in longer term processes underlying synaptic plasticity, development, and memory and learning; these will be discussed in later chapters. Monitoring of Ca^{2+} in dendritic spines is discussed in Chap. 29.

Cyclic Nucleotide Systems

Cyclic AMP. The best understood of the second messenger systems is the type that uses a cyclic nucleotide as the intracellular signal molecule. Most prevalent is the adenosine $3',5'$-cyclic monophosphate (cAMP) system. As shown in Fig. 8.2 (top), cAMP is synthesized by the enzyme adenylate cyclase, which removes the two outer-

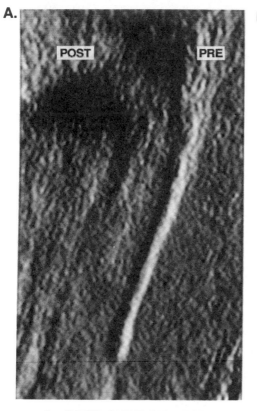

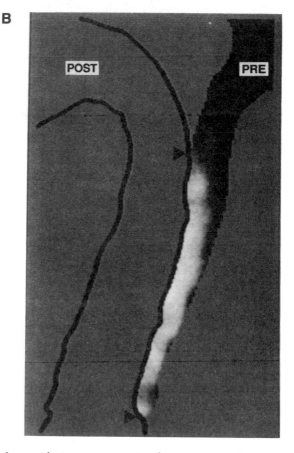

A. SQUID GIANT SYNAPSE
B. PRESYNAPTIC Ca IMAGE

Fig. 8.1 Changes in intracellular Ca^{2+} at the squid giant synapse. **A.** The synaptic region was visualized by video-enhanced microscopy, using a combination of brightfield and fluorescence illumination. The tapered presynaptic (PRE) ending of the secondary giant fiber makes synaptic contact where it envelops a segment of the tertiary giant fiber (POST), which forms an inverted U in this view. The presynaptic fiber was injected with a fluorescent Ca^{2+} indicator dye (fura-2), and appears lighter. **B.** The spatial distribution of elevated Ca^{2+} during a train of 50 impulses stimulated at 100 Hz in the presynaptic fiber. The brighter areas of the presynaptic process indicate raised Ca^{2+} concentration. The gray-scale values reflect changes in fluorescence of fura-2 during stimulation divided by resting levels of fura-2 fluorescence. The physiology of the squid giant synapse is further considered in Fig. 8.6 and Figs. 19.3–19.5. (From Smith et al., 1987; illustration and description kindly provided by Dr. S. J. Smith)

most phosophate groups and creates an additional ester bond between the remaining phosphate group and the No. 3 carbon of the ribose molecule. The cyclic AMP is in turn hydrolyzed by phosphodiesterase (PDE) to the inactive AMP.

In order to function as a second messenger, the basal concentration of cAMP in the cytoplasm must be held at a low level. Normally this level is less than 10^{-6} M. When adenylate cyclase is stimulated, the

synthesized cAMP has only a brief period during which it can act before it is degraded by PDE.

How is the synthesis of cAMP stimulated or inhibited? This takes place through a three-component system of membrane-associated proteins that consists of a receptor, a linking protein, and the adenylate cyclase. This triumvirate operates in the following manner (see Fig. 8.2, bottom). In step 1, the transmitter, hormone, or other

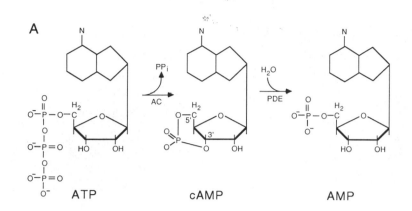

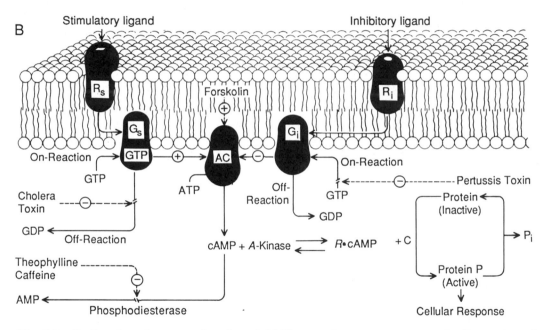

Fig. 8.2 Cyclic adenosine monophosphate (cAMP) second messenger system. **A.** Conversion of adenosine triphosphate (ATP) into cAMP, catalyzed by adenylate cyclase, and hydrolysis by phosphodiesterase (PDE) to adenosine monophosphate (AMP). **B.** Signal pathways related to cAMP. Receptor proteins (R) act through GTP-binding proteins (G) to activate or inhibit adenylate cyclase (AC). The cAMP produced binds to the regulatory subunit of protein kinase A, releasing its catalytic unit to phosphorylate target proteins that produce cellular responses. The actions of drugs at various sites in the pathways are indicated.

External signal molecules that act at excitatory receptors (R_s) include epinephrine (at β-adrenoceptors), thyroid-stimulating hormone (TSH), vasopressin, glucagon, serotonin, and dopamine. Molecules that act at inhibitory receptors (R_i) include epinephrine (at α_2-adrenoceptors), acetylcholine (M_1 receptors), opioids, angiotensin II, and dopamine (D_2 receptors). (Modified from Berridge, 1985)

first messenger binds to a receptor protein (R). This causes an allosteric change which activates the linking protein (called a *GTP-binding protein,* or *G protein*) to bind preferentially GTP instead of GDP. When this occurs, the G protein moves laterally in the membrane to associate with the adenylate cyclase molecule. This has two effects: it first activates adenylate cyclase to synthesize cAMP (step 2) and it also activates the GTPase activity of the G protein to hydrolyze GTP to GDP (step 3), returning the G protein and the membrane receptor to their base configurations (step 4), ready for another activation cycle.

All of the components of this second messenger system have been cloned and sequenced. As in the case of the other systems we have studied, of voltage-gated and ligand-gated channels, this has shown that each component represents a molecular family based on a canonical residue sequence, with different isoforms related to particular types of synapses and neurotransmitter ligands. The *receptor* is exemplified by the β-adrenergic receptor shown in Fig. 8.3. According to the hydrophobicity plots, it consists of seven putative transmembrane domains, and therefore it is referred to as a 7TD receptor (cf. Fig. 2.8). The seven transmembrane domains form a pocket, within which the neurotransmitter or neuropeptide acts. Differences in residues lining the pocket are believed to be the basis for the different affinities of receptors for their specific first messengers at different types of synapses. On the external side, the long sequence of residues at the *N*-terminal end provides additional possibilities for pocket formation and ligand-binding and modulation sites. On the cytoplasmic side, there are sites for binding to the G protein and also sites at which phosphorylation can take place to regulate the sensitivity of the receptor. This basic pattern is adapted for each of the neurotransmitter-mediated 7TD receptor synapses we shall discuss below. It is also adapted for transduction of sensory stimuli, as we shall see in discussing the

senses of smell and taste (Chap. 11) and vision (Chap. 16).

The *GTP-binding proteins* are similarly a marvellously adaptable class of signal molecule. The membrane-associated type participates in membrane signaling mechanisms as indicated in Fig. 8.2. The traditional idea of a simple linear sequence, from a receptor through a G protein to a target enzyme (that is, adenylate cyclase), has been replaced by the growing appreciation of multiple G-protein pathways. G proteins may be either stimulatory or (G_s) or inhibitory (G_i). Many of them have been cloned and sequenced. Within a single cell there may be several isoforms, each activated by more than one receptor and acting on more than one target. Overall, it has been estimated by Birnbaumer (1990) that there are over 40 different first messengers and 90 different 7TD receptors acting through 20 or more different types of G proteins to regulate most of the main types of second messenger systems summarized in Table 8.1.

A similar variety characterizes the *G-protein targets*. The traditional target, as shown in Fig. 8.2, is adenylate cyclase, which catalyzes the synthesis of the second messenger cyclic AMP (cAMP). In most cells of the body, the typical action of the cAMP is to activate a protein kinase, an enzyme that phosphorylates a target protein. In rapid membrane signaling systems, such as synapses, the sites of phosphorylation are usually serine or threonine residues (see below). The addition of a phosphate group activates the protein so that it can carry out a specific function. The number and variety of phosphoproteins are enormous, involving nearly all of the cellular machinery we have discussed thus far. Some representative phosphorylation effects are summarized in Table 8.2.

Why are several steps interposed in the linkage between the first messenger and the final effect of the second messenger? One reason is the multiple control points that this provides. As Elliot Ross (1992) has pointed out, this provides the opportunity

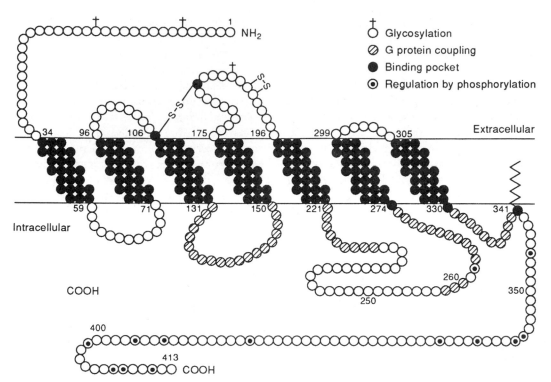

Fig. 8.3 The β-adrenergic receptor as an example of the seven-transmembrane domain (7TD) receptor superfamily. Probable secondary structure as indicated by hydrophobicity plots. Possible sites for *N*-glycosylation, G-protein binding, and phosphorylation are indicated. In the tertiary structure, there is a pocket formed by the transmembrane domains, within which the neurotransmitter ligand is presumed to act; a model is shown in Fig. 11.9. (Modified from Watson and Girdlestone, 1993)

Table 8.2 Some types of proteins that are targets for phosphorylation

1. Regulatory proteins (G proteins)
2. Cytoskeletal proteins (microtubule-associated proteins or neurofilaments)
3. Synaptic vesicle proteins (synapsin I)
4. Neurotransmitter-synthesizing enzymes (tyrosine hydroxylase)
5. Neurotransmitter receptors (ACh receptor or β-adrenergic receptor)
6. Ion channel proteins (Na$^+$ or Ca^{2+} channels)

Adapted from Nestler and Greengard (1984)

for setting up metabolic logic gates in these signaling pathways, such as AND gates through simultaneous convergence of actions, as well as AND-NOT gates when activation and inhibition converge. Another

reason is the amplification that is achieved. A single receptor protein activates not just one but many G proteins; each adenylate cyclase molecule synthesizes many cAMP molecules. The enzyme cascade thus achieves effects that are specific, powerful, and under exquisite control.

Cyclic GMP. Another cyclic nucleotide that acts as a second messenger is guanosine 3′,5′-cyclic monophosphate (cGMP). In general, it is present in much lower concentrations than cAMP, and traditionally it has been thought to be much less important. Indeed, it has been called the "wayward child" of the cyclic nucleotide family (Guy, 1991). However, in the 1970s it became recognized as the second messenger in transduction of light into electrical signals in photoreceptors (see Chap. 16). In recent years there has been increasing evidence for its role as a second messenger in a number of other cell types, where it may mediate a variety of actions, some of which are summarized in Fig. 8.4. One of the main actions is on a membrane channel to modulate its conductance, as in photoreceptors and odoreceptors. There is now evidence for this type of action in other types of cells, too, for example, rod bipolar cells, retinal ganglion cells, and cells of the central nervous system (see Ahmad et al., 1994). This action is direct, without the intervention of an intervening phosphorylation step. As indicated in Fig. 8.4, the guanylate cyclase may be membrane associated, where it mediates transduction of light or of peptide first messengers, or it may be cytoplasmic, where it mediates the action of gaseous signals (discussed further below).

G Proteins as Second Messengers

In addition to acting on second messenger–generating enzymes such as adenylate cyclase, G proteins can also act on other targets, including direct actions on ion channels themselves. This first came to light in experiments on single cultured heart cells. In whole-cell patch recordings, the muscarinic ACh effect of activating a K^+ conductance was found to depend on G proteins and GTP analogues. ACh stimulation of a whole cell did not activate channels contained within an on-cell patch, suggesting that the activation did not depend on a diffusible substance (such as cAMP), but rather that the G protein was acting directly on the K^+ channels. We will discuss these mechanisms and their significance in regulating the heartbeat in Chap. 18.

These mechanisms in heart cells have turned out, as has happened so often in the past, to have their counterparts in nerve cells. G-protein–stimulated opening of K^+ channels, coupled in some cases to inhibition of adenylate cyclase, has been reported at several types of synapses in the brain; direct actions on Ca^{2+} channels have also been reported. Through these actions, to-

Fig. 8.4 Summary of some of the pathways involved in controlling GMP as a second messenger. See text. (From Goy, 1991)

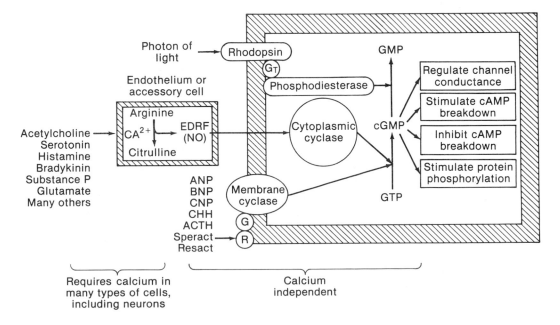

gether with their role in controlling cAMP and cGMP, G proteins are crucial constituents of most synapses that have neuromodulatory properties. It is therefore not surprising that G-protein dysfunction is implicated in several types of mental disorders and is a prime target for development of psychotropic therapeutic drugs; progress in this area is reviewed by Aghajanian and Rasmussen (1988).

Cyclic Nucleotide–Gated Channels

Just as G proteins may act directly on membrane channels, so may the second messengers, cAMP and cGMP. As noted above, this class of cyclic nucleotide–gated (CNG) channels was first recognized in photoreceptors, where cGMP regulates the dark current (Chap. 16). It was subsequently identified in olfactory receptor cells (Chap. 11) and recently has been found in retinal neurons and brain cells. The cAMP appears to be generated by a G-protein–linked 7TD receptor, whereas the cGMP may be controlled by a receptor-mediated activation of a phosphodiesterase (photoreceptor) or by a gaseous interneuronal messenger such as nitrous oxide (see below).

In all cases thus far the native channel conducts a nonspecific depolarizing cationic current. In the heart, activation of cardiac pacemaker I_f channels by direct action of cAMP has been demonstrated (see Chap. 18). It will be interesting to see whether CNG channels are present at synapses in the central nervous system. The direct action of the cyclic nucleotides provides for rapid onset of control of channel gating. Patch clamp analysis has shown that the single channels show prolonged bursting activity in the presence of agonist, a behavior similar to that of NMDA channels (reviewed in Zufall et al., 1994). The properties of CNG channels will be discussed in detail in Chaps. 11 and 16.

Gaseous Messengers

Nitric Oxide. Ever since the discovery of the anesthetic effects of ether, it has been known that gases can have effects on brain cells. However, the possibility that such gases might be produced within the brain and have physiological effects on neuronal signaling has been recognized only very recently. The story began with the analysis of the mechanisms of vasodilation of blood vessels. Around 1980 it was found that substances such as ACh, which cause vasodilation, do so by acting on endothelial cells lining the blood vessels; these cells release a substance that was initially named endothelial-derived relaxing factor (EDRF), which relaxes the adjacent smooth muscle cells of the vessel wall by stimulating production of cGMP within them. It soon emerged that EDRF is in fact the gas nitric oxide (NO).

The first evidence for this system in the brain was brought forward by John Garthwaite and his colleagues in Liverpool in 1988 (reviewed in Garthwaite, 1991). Synthesis of NO takes place when an increase in Ca^{2+} stimulates *NO synthase* to convert arginine to citrulline, with production of NO (see Fig. 8.4). The target for NO is cytoplasmic guanylate cyclase. However, increased Ca^{2+} inhibits guanylate cyclase, which means that in order to have an effect, NO must act on other cells. This it does by readily diffusing through the lipid plasma membrane (as do other gases such as anesthetics) to enter other cells, where it activates guanylate cyclase, producing cGMP, which can then act directly on CNG channels or other targets, as summarized in Fig. 8.4. Although it might seem from these properties that NO would act too diffusely to be useful as a second messenger, it has an extremely short half-life of only a few seconds, so that its effects are limited to other cell processes in the immediate vicinity within a few micrometers of the site of origin.

The distribution of NO synthase in different regions and pathways within the nervous system is described in Chap. 25. The possible role of NO in mechanisms of synaptic plasticity at glutamatergic synapses is discussed in Chap. 29.

Carbon Monoxide. Although carbon monoxide might seem to be an unlikely candidate to be a physiologically useful signaling molecule, in fact there is increasing evidence that it represents a second category of gaseous intercellular messenger substances. It shares many of the properties of NO, in being freely diffusible between cells and having an extremely brief half-life.

Membrane Lipid System

In Chap. 3, we pointed out that membrane lipids play more roles than simply providing a fluid medium for proteins to float in. One of these additional roles is an important second messenger system. This begins with phosphatidylinositol, one of the phospholipids present in the inner membrane (cf. Chap. 3). Phosphatidylinositol (PI) is phosphorylated in two steps to phosphatidylinositol-4,5-bisphosphate (PIP_2), which is a substrate for a phosphodiesterase called phospholipase C (PC). This enzyme can be activated by a membrane receptor protein when it binds to its appropriate lipid; recent evidence suggests that this activation is also mediated by a G protein (see Fig. 8.5).

The key effect of PC is to cleave the lipid chains, producing inositol triphosphate (IP_3) and diacylglycerol (DAG). Both of these are second messengers.

IP_3 is water soluble and can diffuse into the cytoplasm. There it acts on receptors in the endoplasmic reticulum to increase efflux of Ca^{2+} from inside the ER to the cytoplasm (Fig. 8.5). The Ca^{2+} (which when mobilized in this way is really functioning as a third messenger) then exerts its various effects as indicated in Fig. 8.5.

The second compound produced by phospholipase C, diacylglycerol (DAG), consists of glycerol joined to two fatty acids (stearic and arachidonic acid). DAG appears only briefly in membranes, being incorporated again into phosphatidylinositol, or being enzymatically degraded (in which case the arachidonic acid can contribute to synthesis of prostaglandins). Associated with the appearance of DAG is activation of protein kinase C (PKC). This phosphorylating enzyme is found widely in cells and

is present with the highest specific activity in brain. In most cells it is present in the cytoplasm, whereas in the brain it appears to be preferentially associated with synaptic membranes, as previously noted in Chap. 6.

Protein kinase C has been shown to phosphorylate several proteins in the brain, including B50 (F1) protein associated with presynaptic membranes; synapsin 1 associated with synaptic vesicles; microtubule-associated proteins; and other cytoplasmic and membrane-associated proteins whose functions are still unknown. In many of its actions, PKC acts synergistically with the Ca^{2+} mobilized by IP_3; it may also act synergistically in other ways, as with cAMP and Ca/calmodulin-dependent protein kinase II on synaptic vesicles (see Chap. 6).

The mechanism of phosphorylation is considered in more detail in Fig. 8.6. A protein kinase catalyzes the transfer of the terminal phosphate group of ATP to a hydroxyl group on a substrate protein. As shown in Fig. 8.6A, the usual site of transfer is the hydroxyl group of a serine or threonine residue. Phosphorylation at these sites induces conformational changes which bring about short-term physiological changes. In addition to these sites, phosphorylation can also occur at tyrosine residues, through a tyrosine kinase (Fig. 8.6B). This type of phosphorylation is involved in control of normal growth, as well as the abnormal growth responses to oncogenes (see Chap. 9). Some examples of common phosphoproteins are listed in Table 8.2.

What kind of experiments can we do to study the effects of protein phosphorylation? A direct approach was taken by Rudolfo Llinás and Paul Greengard and their colleagues, who wished to test for the effects of protein phosphorylation on synaptic transmission. They used the giant synapse of the squid, in which the pre- and postsynaptic processes are large enough that one can insert microelectrode tips into them. (This is the same synapse used for the Ca^{2+} measurements shown in Fig. 8.1.) By injecting Ca/calmodulin-dependent protein kinase (Ca/CaM PK) from one elec-

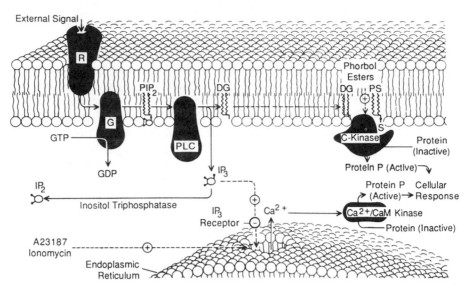

Fig. 8.5 Inositol-lipid second messenger system. Receptor proteins (R) activate a GTP-binding protein (G) which stimulates phospholipase C (PLC) which hydrolyzes phosphatylinositol-4,5-bis-phosphate (PIP_2) into two signal molecules: inositol triphosphate (IP_3) and diacylglycerol (DAG). IP_3 acts on the endoplasmic reticulum to release Ca^{2+}, which functions as a third messenger to activate Ca^{2+}/calmodulin-dependent protein kinase (Ca^{2+}/CaM kinase); this kinase, in turn, phosphorylates target proteins to produce cellular responses. DAG, on the other hand, activates a protein kinase C, which phosphorylates target proteins to produce cellular responses. Drugs can act at the sites indicated in the pathways. External signal molecules that activate the membrane receptors include acetylcholine, vasopressin, thyrotropin-releasing hormone (TRH), serotonin, thrombin, and antigens. (Modified from Berridge, 1985)

trode and recording the response from another electrode, they were able to test for the effect of this enzyme and its site of action. As shown in Fig 8.6C, injection of Ca/CaM PK into the postsynaptic terminal produced no physiological response, whereas injection into the presynaptic terminal induced a large increase in the postsynaptic response. The interpretation of these results was that the injection had enhanced the normal action of Ca/CaM PK in phosphorylating synapsin I, detaching it from the synaptic vesicles and allowing the vesicles to undergo exocytosis and release their transmitter. This has been part of the evidence supporting the sequence of events at the synapses we outlined in Chap. 6 (Fig. 6.8).

This is only one example of the kinds of proteins that are phosphorylated as a result of second messenger actions (see Table 8.2).

Main Types of Modulatory Synapses

We are now in a position to consider the functional organization of different types of modulatory synapses. These are commonly grouped into types according to the transmitter released by the presynaptic process. They are further divided into subtype on the basis of binding by different pharmacological agonists or antagonists, and the different types of postsynaptic responses associated with each. These criteria have given rise to several traditional categories of synapses, and we will consider each of these in turn. Building on the common framework for all synapses introduced in Chap. 6, we will show how it is adapted

A. Phosphorylation of threonine (or serine) residue

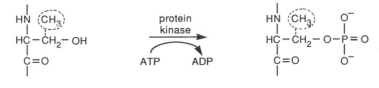

B. Phosphorylation of tyrosine residue

C. Test for protein phosphorylation at squid giant synapse

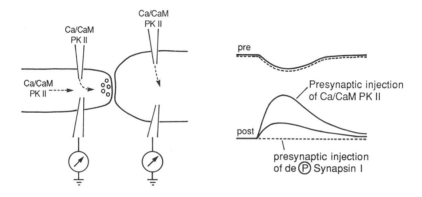

Fig. 8.6 Molecular mechanisms of protein phosphorylation. **A.** Some protein kinases catalyze the transfer of a phosphate group to a threonine or serine (additional carbon in threonine indicated by dashed lines) residue of a substrate protein. This induces a conformational change which produces a rapid cellular response. **B.** Some protein kinases catalyze transfer of a phosphate group to a tyrosine residue of a substrate protein; this leads to a slower cellular response. Examples of such kinases are those involved in normal and abnormal (tumor) growth (see Chap. 9, Fig. 9.9). (After Nestler and Greengard, 1984) **C.** Experimental test for the functional role of protein phosphorylation in synaptic transmission. On the left, the experimental setup involved injecting Ca/calmodulin-dependent protein kinase (Ca/CaM PK) into either the pre- or postsynaptic terminal of the giant synapse of the squid, while recording intracellularly from those sites. The recordings, on the right, showed that neither presynaptic (dashed line) nor postsynaptic (solid line) injection of Ca/CaM PK had an effect on the presynaptic potential; however, pre- (but not post-) synaptic injection greatly enhanced the postsynaptic potential. Presynaptic injection of dephosphorylated synapsin I (deP) had no effect on the postsynaptic potential. (Adapted from Llinás et al., 1985)

to achieve a wide variety of modulatory effects between neurons.

Acetylcholine

Acetylcholine (ACh) is unique as a small transmitter molecule in that it does not appear to belong to a larger clan of related compounds.

Ever since the work of Langley, Dale, and Loewi early in this century, it has been known that synapses that use ACh as a transmitter can be divided into two main classes, depending on whether they are activated by the application of the substance nicotine or muscarine. The two classes are therefore called nicotinic and muscarinic cholinergic synapses. The neuromuscular junction, as we have seen, is a nicotinic synapse.

By now we have become well acquainted with the function of ACh as a transmitter of motoneurons at the neuromuscular junction (Chap. 7). The basic mechanism at a central nicotinic synapse is summarized in Fig. 8.7A. ACh is synthesized in the terminal from choline and acetyl-CoA by choline acetylase ①; this is followed by packaging and mobilization ②, release ③, and action on postsynaptic ④ and presynaptic receptors ⑤. Brief action of the ACh is ensured by the presence in the cleft of acetylcholinesterase, an enzyme that rapidly hydrolyzes ACh into acetate and choline ⑥; the choline is taken up again by the presynaptic terminal ⑦ for resynthesis into ACh. It can be seen that direct binding of ACh to the channel protein and the rapid hydrolysis of ACh in the cleft are adaptations for quick action, as is needed for control of skeletal muscles.

In contrast to nicotinic synapses, muscarinic synapses are characteristically found where the synaptic actions are slower and more subject to metabolic and other factors; examples are the synapses of motor nerves onto autonomic ganglia, glands, cardiac and smooth muscle, and certain sites in the central nervous system. At these synapses, the presynaptic mechanisms involve the same basic steps (see Fig. 8.7B, ① to ③), but the postsynaptic mechanisms are different. ACh binds to a muscarinic receptor (mAChR) molecule that is distinct from the channel protein. The muscarinic receptor is not soluble in detergents, like most other membrane-associated proteins, so progress in understanding its molecular structure by traditional biochemical methods was slow. However, in 1986 a cDNA for a muscarinic cholinergic receptor in the pig brain was cloned and expressed by Shosaku Numa's laboratory (Kubo et al, 1986). The polypeptide consists of 460 amino acid residues, with a molecular weight of 51,000. Analysis of hydrophobicity suggests the presence of seven membrane-spanning segments. A pleasant surprise was that this pattern is similar to that of the visual protein rhodopsin (see Chap. 16) and the β-adrenergic receptor (see above), a result supported by a high degree of amino acid sequence homologies. A common ancestor for the genes encoding these proteins of diverse function is thus implied. It is now known that the muscarinic receptor belongs to the 7TD G protein–linked receptor superfamily (see Table 8.3).

The binding sites for ACh are not yet identified, but there are several types of receptors (M_1, M_2) based on affinities for different ligands and the postsynaptic effects they produce. Muscarinic responses are slow because the receptor molecule is coupled to ion channels by second messenger systems. In different types of nerve, muscle, and gland cells, responses to receptor stimulation have been found to be associated with activation of G proteins (brain), inhibition of cAMP formation (heart), stimulation of cGMP formation, and increased turnover of inositol phosphates (glands). The second messenger generated by these responses may lead to one or several of the actions discussed previously, including phosphorylation of a membrane channel protein to produce an electrical response and phosphorylation of a membrane or cytosolic protein to produce a longer term metabolic response. As shown

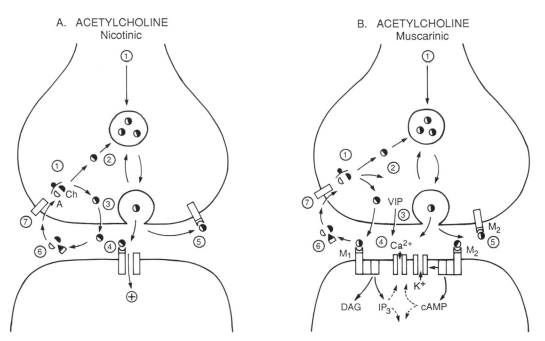

A. ACETYLCHOLINE
Nicotinic

B. ACETYLCHOLINE
Muscarinic

Fig. 8.7 Molecular mechanisms of cholinergic synapses. **A.** Nicotinic cholinergic synapse. The main steps are indicated by numbers ① to ⑦ in this and the following diagrams: ① synthesis of acetylcholine (ACh) from acetate and choline (catalyzed by choline acetylase) in the cell body or the synaptic terminal; ② transport and storage in vesicles; ③ release by exocytosis, or from the cytosol (recycling of vesicle membrane indicated by dashed arrow); diffusion in the synaptic cleft (release is blocked by botulinum toxin or high Mg^{2+}); ④ binding of ACh to the nicotinic cholinergic receptor molecule, which opens the channel for net inward cationic flow, depolarizing the membrane (nicotine is an agonist at this receptor); ⑤ binding of ACh to the presynaptic receptor; ⑥ hydrolysis of ACh into acetate and choline by acetylcholinesterase; ⑦ reuptake of choline into presynaptic terminal. (Modified from Dunant and Israel, 1985) **B.** Muscarinic cholinergic synapse: ① synthesis; ② transport and storage; ③ release (corelease with a peptide, such as vasoactive intestinal peptide, VIP); ④ binding to M_2 postsynaptic receptor activates a G protein which directly modulates K^+ channel. Additional G protein modulation of cAMP second messenger system leads to complementary effects on Ca^{2+} channels. Other types of mAChRs are coupled to IP_3 second messenger system. ⑤ binding to presynaptic M_2 receptor; ⑥ hydrolysis by acetylcholinesterase; ⑦ reuptake of choline.

in Fig. 8.7B ④, a common mechanism at many muscarinic cholinergic synapses is for the mAChR to activate a G protein which modulates a K^+ channel; opening the channel polarizes the membrane and inhibits the cell, whereas closing the channel makes the cell more excitable (see M current in sympathetic ganglion cells: Chaps. 5 and 18). The G protein may also modulate adenylate cyclase production of cAMP, which in turn modulates Ca^{2+} channels, usually in a way that complements the effects on K^+ channels (i.e., closing Ca^{2+} channels

would reinforce the inhibitory effect of opening K^+ channels).

We have considered nicotinic and muscarinic synapses as separate entities for convenience of explanation, but the two types of receptors may be intermingled so that the neuron can give a more complex response. Additional degrees of complexity are provided by the co-release at some muscarinic synapses of a peptide, such as vasoactive intestinal peptide (VIP; see below), which can modulate the ACh response, and by the presence of autoreceptors on the

Table 8.3 Neurotransmitter and neuromodulator receptors and subtypes

Neurotransmitter	Receptor	Type	Second messenger	Action
glutamate	GLU_{AMPA}	4TD	—	E (fast)
	GLU_{NMDA}	4TD	—	E (fast)
	mGluR	[7TD]	PL	E (slow)
GABA	$GABA_A$	4TD	—	I (fast)
	$GABA_B$			I (slow)
ACh	nAChR	4TD	—	E (fast)
ACh	mAChR	7TD	cAMP/PL	E, I (slow)
NE	α_1	7TD	PL	postsynaptic
	α_2	7TD	PL	presynaptic
	β	7TD	cAMP	postsynaptic
DA	D_1	7TD	cAMP	increase
	D_2	7TD	cAMP	decrease
5HT	$5HT_1$	7TD	cAMP	regulated
	$5HT_2$	7TD	cAMP	regulated
Histamine	H	7TD	cAMP	regulated

presynaptic terminal, which can modulate the presynaptic release processes (see Fig. 8.7B).

Biogenic Amines

Several types of small molecules have an amine group and can affect a variety of cells, from platelets to neurons; they are classified as monoamines, or biogenic amines. They form several subgroups: (1) the catecholamines (dopamine, norepinephrine, and epinephrine), all of which contain a catechol nucleus and share a common synthetic pathway; (2) serotonin, or 5-hydroxytryptamine (5HT); and (3) histamine. Let us consider the most important of them as far as the nervous system is concerned.

Adrenergic Synapses. This term can apply to synapses that use either norepinephrine (NE) or epinephrine (E); of these, NE is the most prevalent in the nervous system, being the main neurotransmitter of ganglion cells in the sympathetic nervous system (Chap. 18) and of locus ceruleus cells, which project widely throughout the vertebrate brain (Chap. 24).

As indicated in Fig. 8.8A, synthesis of catecholamines ① begins with the amino acid tyrosine (Tyr), which is taken up by nerve cells from the blood. NE synthesis takes place in the cell body or in the synaptic terminals themselves; it is packaged into vesicles that enter a storage pool as shown in ②. With depolarization and Ca^{2+} influx, the NE is released by exocytosis ③, and acts on postsynaptic receptors ④, termed α_1-adrenergic receptors. The primary effect, mediated through second messengers, is on Ca^{2+} channels. In addition, the released NE acts on α_2 receptors, located primarily on the presynaptic terminal ⑤. Through second messengers and protein phosphorylation, α_2 receptors exert diverse controls on the state of the synapse, including changes in the gating of K^+ channels. This may take place by the direct action of a G protein. Termination of NE action occurs by reuptake ⑥ into the presynaptic terminal, where the level of NE is controlled through enzymatic degradation or inactivation ⑦.

A second type of adrenergic receptor is the β receptor. As shown in Fig. 8.8B, this differs from the α receptor mainly in being primarily postsynaptic and being linked to

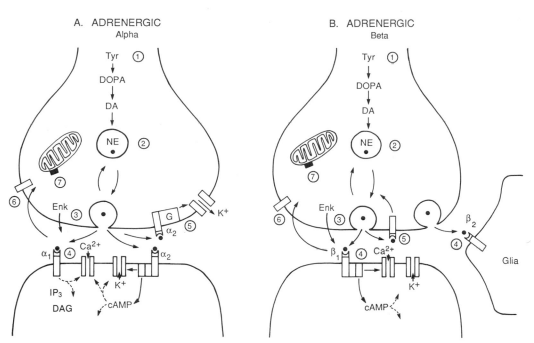

Fig. 8.8 Molecular mechanism of adrenergic synapses. **A.** α-Adrenergic synapse: ① synthetic pathway is through tyrosine (Tyr) to 3,4-dihydroxyphenylalanine (DOPA; catalyzed by tyrosine hydroxylase) to dopamine (DA; catalyzed by DOPA decarboxylase) to norepinephrine (NE; catalyzed by dopamine-β-hydroxylase); ② transport and storage (storage is blocked by reserpine); ③ release by exocytosis (increased by amphetamine); co-release with neuropeptides such as enkephalin, Enk; ④ binding of NE to postsynaptic receptors. Examples are shown of binding to α_1 receptor, which leads to molulation of Ca^{2+} channels, and binding to α_2 receptors, which are linked to adenylate cyclase and modulate K^+ channels; there may also be direct actions of G proteins on K^+ channels; ⑤ binding of NE to presynaptic α_2 receptors; ⑥ reuptake, which terminates NE action (blocked by tricyclic antidepressant drugs); ⑦ degradation by monoamine oxidase (MAO) [there may also be inactivation by catechol-O-methyltransferase (COMT)]. **B.** β-Adrenergic synapse: ①–③ synthesis, transport, storage, and release as in A above; ④ binding of NE to β_1 receptor, which leads to phosphorylation of ionic channels through cAMP; β_2 receptors are also found on glia; ⑤–⑦ presynaptic receptors, reuptake, degradation, and inactivation, as in A above. (Based on Cooper et al., 1993; Aghajanian and Rasmussen, 1988; Tsien, 1987; and others)

the cAMP second messenger system. One effect of this system is to decrease the membrane conductance of the postsynaptic cell and inhibit its firing. This was first shown in 1971 by George Siggins, Barry Hoffer, and Floyd Bloom, then in Washington, D.C. They used a multibarrel pipette which allowed them to record the impulse activity of a Purkinje cell in the cerebellum (Chap. 22) from one barrel while ejecting different drugs from the other barrels (a technique called microionophoresis). They showed that ionophoresis of NE or cAMP produced a suppression of impulse firing simi-

lar to that produced by stimulation of the noradrenergic fibers of the locus ceruleus that innervate the cerebellum. These actions were blocked by ionophoresis of β-receptor blockers such as antipsychotic drugs (chlorpromazine) and prostaglandins, or by drugs that block activation of adenylate cyclase.

Subsequent studies have shown that β receptors act through the cAMP second messenger system in many cells in the nervous system. The actions are different in different cells; for example, in heart cells NE and E act mainly through β receptors and cAMP, whereas in dorsal root ganglion neurons they act through α receptors and DAG (Tsien, 1987). A β_2 receptor has been characterized which, in the brain, appears to be more localized in glial cells. Also indicated in Fig. 8.8A, B is the fact that adrenergic endings may co-release neuroactive peptides such as enkephalin (see below). As already discussed (see Fig. 8.3), the β-adrenergic receptor has been cloned and sequenced and is a member of the 7TD receptor superfamily.

Dopamine Synapses. In neurons that utilize dopamine (DA) as their transmitter, the biogenic amine synthetic pathway stops with DOPA decarboxylase, the enzyme that synthesizes dopamine (DA) (see ① in Fig. 8.9A). Storage ② and release ③ mechanisms appear to be similar to those for NE discussed above. DA acts on two types of receptors ④, both of which are linked to adenylate cyclase–cAMP second messenger systems. Stimulation of D_1-receptors causes an increase in cAMP levels, whereas stimulation of D_2 receptors causes a decrease in cAMP, in the postsynaptic neuron. The DA receptor belongs to the 7TD receptor superfamily. There are also autoreceptors ⑤ on the presynaptic terminal, which are thought to exert an inhibitory feedback effect on the DA neuron. The action of DA is terminated largely by reuptake ⑥. Mechanisms for degradation and inactivation ⑦ are similar to those for NE.

Dopaminergic synapses have been of great interest because of the evidence, beginning in the 1960s, that antipsychotic drugs that are effective in alleviating the symptoms of schizophrenia have the common effect of interfering with transmission at dopaminergic synapses in the brain. Recent studies indicate that the primary mode of action of antipsychotic drugs is to bind to, and block, the D_2 receptors. In contrast to these blocking actions of antipsychotic drugs, it is known that drugs that are DA agonists such as amphetamine, which enhance or mimic the actions of DA, can induce schizophrenia-like behavior. It has

been tempting therefore to believe that schizophrenia is due to overactivity of DA neurons in the brain. Although malfunction of DA synapses is likely to be important in schizophrenia, there is general agreement that other mechanisms and other transmitter systems must be involved, too. We will return to this question in Chap. 24 when we consider the location of transmitter pathways in the brain.

Serotonin Synapses. Serotonin (5-hydroxytryptamine: 5HT) differs from the catecholamines in having, in addition to the catechole ring, an indole ring, from which the alkyl group with its terminal amine group arises. The precursor for 5HT is tryptophan, which is taken up from the bloodstream, as indicated in ① of Fig. 8.9B. The 5HT synthesized is stored in vesicles ② and released ③ by mechanisms similar to those for other biogenic amines. The serotonin receptor has been cloned and sequenced, and it too belongs to the 7TD receptor superfamily. Serotonin receptors are divided into two types, depending on whether they preferentially bind 5HT (5HT-1) or a central stimulatory drug (neuroleptic) called spiperine (5HT-2). The 5HT-1 receptors are linked to G-protein cAMP second messenger systems which regulate K^+ and Ca^{2+} channels. We will study an example of this in a molluscan neuron (Chap. 29), where a serotonin receptor can mediate sensitization of a synapse. There is little evidence for 5HT autoreceptors ⑤. Reuptake ⑥ is believed to be the main mechanism for terminating 5HT action, and degradation occurs in the presynaptic terminal ⑦.

Interest in the behavioral actions of 5HT began around 1950 with the realization that its molecular structure resembled that of lysergic acid diethylamide (LSD), a hallucinogenic drug, and the finding that the stimulatory action of 5HT on the smooth muscles of the gut was antagonized by LSD. From this observation arose the theory that the hallucinations associated with LSD are due to blockade of 5HT receptors

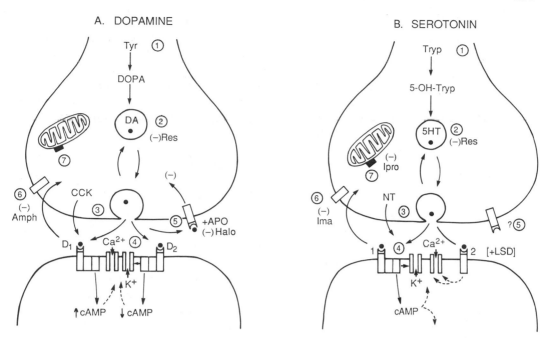

Fig. 8.9 **A.** Molecular mechanism of dopaminergic synapses: ① synthesis by enzymatic pathway from Tyr to DOPA to DA (see legend, Fig. 8.8); ② transport and storage (storage inhibited by reserpine, Res); ③ release of DA by exocytosis; co-release of a neuropeptide such as cholecystokinin (CCK); ④ binding to D_1 receptor, acting through stimulatory G protein (G_s) to increase levels of cAMP, or to D_2 receptor, acting through inhibitory G protein to lower levels of cAMP (antipsychotic drugs such as butyrophenones block D_2 receptors); G protein can also have a direct action on K^+ channels at some synapses; ⑤ binding of DA to presynaptic receptors [typically D_2; DA receptors are stimulated by psychoactive drugs such as apomorphine (APO), blocked by haloperidol (Halo)]; ⑥ reuptake terminates DA action; ⑦ degradation by MAO and inactivation by COMT. **B.** Molecular mechanism of serotonergic synapses: ① synthetic pathway is from tryptophan to 5-hydroxytryptophan (catalyzed by tryptophan hydroxylase) to 5-hydroxytryptamine (5HT, or serotonin) (catalyzed by 5-hydroxytryptophan decarboxylase); ② transport and storage (blocked by reserpine, Res); ③ release of 5HT by exocytosis; co-release with a neuropeptide, e.g., neurotensin, NT; ④ binding to a $5HT_1$ postsynaptic receptor (coupled to G protein and cAMP), or to a $5HT_2$ receptor (LSD is an agonist/antagonist); ?⑤ possible binding to presynaptic receptors; ⑥ reuptake terminates 5HT action (blocked by tricylic antidepressant drugs such as imipramine, Ima); ⑦ degraded by MAO. (Based on Cooper et al., 1993; Aghajanian and Rasmussen, 1988; and others)

in the central nervous system (see step ④ in Fig. 8.9B), and that this receptor blockade might be a mechanism underlying some psychoses. As in the case of the DA theory of schizophrenia, this theory has been considerably modified. We will discuss these matters further in Chap. 24.

Other Biogenic Amines. Closely related to norepinephrine is epinephrine (E), which is synthesized by carboxylation of the terminal nitrogen group of norepinephrine. Epi-

nephrine is produced mainly in the adrenal medulla, and is the main circulating hormone released during "fight or flight" stress reactions. It has a limited distribution in the brain (Chap. 24).

Histamine is another monoamine, synthesized from histidine which has been taken up from the blood. Histamine is found primarily in mast cells, which release it as part of their reactions to allergens. If you suffer from hay fever, you know that an antihistamine tablet not only relieves your stuffy nose, but also produces drowsiness; this may be due to an action on histamine receptors in the brain, which have been found mostly in the hypothalamus. Different types of histamine receptors have been identified; one type is found in the neocortex and hippocampus, where stimulation leads to changes in neuronal metabolism believed to be mediated by cAMP. The histamine receptor also belongs to the 7TD receptor superfamily.

The biogenic amines discussed above are found in invertebrates as well as vertebrates. In addition, a common transmitter in some invertebrates is octopamine, a biogenic amine formed by an alternative synthetic pathway from tyrosine. In the lobster, octopamine is believed to act through cAMP and protein phosphorylation to prime extensor muscles to respond more vigorously when stimulated. This underlies the extensor posture assumed by the lobster in defensive behavior, in contrast to the flexion posture of aggressive behavior that is mediate by 5HT (as we will learn in Chap. 19). Both of these priming actions are characteristic of neuromodulatory transmitter actions (see below).

Amino Acids

We have already discussed the amino acid transmitters in the previous chapter, where we learned that glutamate is the main transmitter for rapid excitatory synaptic actions and γ-aminobutyric acid (GABA) is the main transmitter for rapid inhibitory actions. The fastest actions of glutamate are mediated by activation of the AMPA receptor; slower synaptic actions take place through activation of the NMDA receptor.

In addition to these glutamate receptors, which are integral parts of the regulated ionic channel, there is a third type called the *metabotropic glutamate receptor* (mGluR). This is coupled to G proteins and has its action through the inositol trisphosphate (IP_3) second messenger pathway. This receptor has recently been cloned and sequenced. The hydrophobicity plots suggest the presence of seven transmembrane segments, but otherwise there is no sequence similarity to the members of the 7TD superfamily of G-protein–coupled receptors that we have discussed so far. The mGluR has an extensive hydrophilic extracellular amino terminal domain which shares some sequence homology with AMPA receptors and may be the site of glutamate binding. There is also an extensive hydrophilic intracellular carboxy terminal domain. RNA blots show that mGluR is most concentrated in the cerebellum and olfactory bulb. In situ hybridization shows prominent mGluR expression in cerebellar Purkinje cells, hippocampal pyramidal cells, dentate granule cells, olfactory mitral/tufted cells, and thalamic cells (Masu et al., 1991). Since hippocampal pyramidal cells and cerebellar Purkinje cells are known to be rich in IP_3 receptors, it is speculated that mGluR may be involved in mediating long-term potentiation (LTP) in the hippocampus and long-term depression (LTD) in the cerebellum. We will discuss these mechanisms in Chaps. 29 and 22, respectively.

Purines

In recent years, evidence has grown that nucleosides may function at some synapses as transmitters. Most attention has been focused on adenosine, which is composed of a nitrogenous base (in this case, the purine adenine) and a ribose sugar. A adenosine is already familiar to us; it is one of the building blocks of the genetic code, has a role through ATP in energy metabolism, and has a role through cAMP as a second messenger.

Evidence that adenosine or ATP may function as a first messenger comes from several sources. Various studies have indicated that adenosine or ATP may exert excitatory or inhibitory influences on neurons. Binding studies of adenosine analogues have permitted identification of adenosine receptors in the brain. Two types of receptor, one that inhibits adenylate cyclase (A_1) and one that stimulates adenylate cyclase (A_2), have been identified. We will discuss the distribution of purinergic systems further in Chaps. 18 and 24.

Dale's Principle

Although the nervous system as a whole can use different substances at different synapses, this is not necessarily true for an individual neuron. The metabolic unity of the neuron would seem to require that it release the same transmitter substance at all its synapses. This is *Dale's Principle* and, since it can be easily misunderstood, it is well to quote the original formulation. In a review of synaptic transmitter in the autonomic nervous system many years ago, Dale (1935) wrote,

. . . the phenomena of regeneration appear to indicate that the nature of the chemical function, whether cholinergic or adrenergic, is characteristic for each particular neurone, and unchangeable. When we are dealing with two different endings of the same sensory neurone, the one peripheral and concerned with vasodilatation and the other at a central synapse, can we suppose that the discovery and identification of a chemical transmitter of axon-reflex dilation would furnish a hint as to the nature of the transmission process at a central synapse? The possibility has at least some value as a stimulus to further experiment.

The principle implies that during development some process of differentiation determines the particular secretory product a given neuron will manufacture, store, and release (see Chap. 9). The usefulness of the principle in the analysis of synaptic circuits is explicit in Dale's statement, for, if a substance can be established as the transmitter at one synapse, it can be inferred to be the transmitter at all other synapses made by that neuron.

Dale's law applies only to the presynaptic unity of the neuron; it does not apply to the postsynaptic actions the transmitter will have at the synapses made by the neuron onto different target neurons. These actions may be similar, or they may be different. Several possibilities for diversity of action exist for the transmitter released from a single neuron. Such neurons have been termed *multiaction cells,* and have been particularly well studied in invertebrates. Eric Kandel (1976) summarized the conclusions from this work as follows:

1. The sign of the synaptic action is not determined by the transmitter but by the properties of the receptors on the postsynaptic cell.

2. The receptors in the follower [postsynaptic] cells of a single presynaptic neuron can be pharmacologically distinct and can control different ionic channels.

3. A single follower cell may have more than one kind of receptor for a given transmitter, with each receptor controlling a different ionic conductance mechanism.

As a result of these three features, cells can mediate opposite synaptic actions to different follower cells or to a single follower cell.

These may be regarded as corollaries to Dale's Principle.

What of the possibility of cells with multiple transmitters? Among invertebrates, four putative transmitters have been reported in single neurons of *Aplysia*. In the vertebrate, there is increasing evidence, especially from the extensive studies of Tomas Hökfelt and his co-workers in Stockholm, for the presence of more than one transmitter substance in single nerve terminals in many parts of the nervous system. As we have already discussed above, a common pattern appears to be the presence of one or more peptides within a monoamine-containing terminal (an example of the way this has been demonstrated is shown in Fig. 8.10). The fact that some synaptic terminals contain more than one type of synaptic vesicle is also suggestive in this regard.

These findings do not negate the idea expressed by Dale that the neuron has a metabolic unity; the fact that the unity includes the ability of a neuron to manufacture and release more than one kind of neuroactive substance may therefore be recognized as Dale's Modified Principle. However, if different substances could be released at different sites (for example, one substance from dendrites and another from

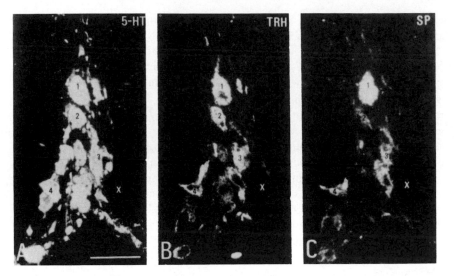

Fig. 8.10 Immunocytochemical demonstration of neuroactive peptides. Parts A–C are consecutive thin sections of the raphe nucleus in the medulla oblongata, treated with antisera to serotonin (**A**), thyroid hormone-releasing hormone (**B**), and substance P (**C**). Several cells (1–4), identifiable in all three sections, can be seen to be immunoreactive to all three antisera. X marks a blood vessel, used to orient the slides. Arrows point to a process emanating from cell 1. (From Johansson et al., 1981)

axons), this would violate the *functional* unity of the neuron implied by Dale's Principle. Since the functional unity of the neuron has already been disproved by the evidence for semi-independent synaptic input–output units within the dendritic trees of many neurons (see Chap. 6), it would not be unreasonable to expect that some of these units might release different neuroactive substances, or combinations of substances. The possibility could at least serve as a stimulus to further experiment.

Neuropeptides

One of the most important developments of recent years in neurobiology has been recognition of the widespread distribution of *neuroactive peptides* in the nervous system. Many of these compounds were identified in other organs by biochemists working over the past 50 years or so, and were only recently identified in the nervous system. The names of these compounds, such as vasopressin or prostaglandin, reflect the organ in which they were first found or the

physiological action that was first studied. Similarly, some have names, like luteinizing hormone releasing hormone (LHRH), reflecting one specific function of the nervous system. A number of peptides previously identified in the gut have turned up in the central nervous system. Thus, the nomenclature for these substances can be somewhat confusing.

Like neurotransmitters, peptides can be identified as neuroactive substances only by meeting certain criteria. This begins with a bioassay, which involves procedures for extraction and testing which are actually quite stringent. Separation and purification of the peptide have traditionally depended on a number of chromatographic techniques and other standard and advanced biochemical methodologies. However, these have been largely replaced by recombinant DNA technology, because it is now much faster to sequence the cDNA and synthesize the protein from it than to sequence the protein itself. The localization of peptides in neural tissue relies on immunological methods; the high sensitivity and high specificity of monoclonal antibodies have been of great advantage in this work. The ability of a substance to evoke a response when experimentally applied to a neuron is a final, and important, criterion in the identification of a peptide as being involved in the normal function of a given type of synapse or neuron.

Some of the best known neuroactive peptides are summarized in Fig. 8.11. This represents only a modest proportion of the neuroactive peptides in the brain, which probably total over 50, and may reach 100 or more. There is, in fact, no definitive list; a review of peptides (Cooper et al., 1987) prompted the comment: "We write with pencil in one hand and eraser in the other!" It can be seen that many of the peptides contain 2–10 amino acids, and thus overlap in size with small amino acid transmitters, on the one hand, and hormones, on the other.

The large numbers of peptides, and their idiosyncratic names, give the impression that this is a heterogeneous collection of substances. However, the more that is learned about them, the more it is possible to discern some powerful unifying principles underlying their functions. First, neuroendocrine cells, secreting peptides, were among the first kinds of neurons to appear in the evolution of primitive nervous systems. Second, neuropeptides are strongly conserved in phylogeny, so that similar substances or similar amino acid sequences appear in different species, both invertebrate and vertebrate. Third, many peptides of the nervous system are also found in other tissues of the body, such as the gut, as already mentioned. This has suggested that many peptidergic neurons may derive embryologically from a common type of neuroectodermal precursor cell, a cell characterized biochemically by "amine precursor uptake and decarboxylation: APUD." Finally, neuroactive peptides as a class share certain biochemical and physiological properties. Table 8.4 summarizes some of these, and contrasts them with properties of neurotransmitters.

With this as a general orientation to neuroactive peptides, let us consider in more detail the main aspects of their synthesis and mechanism of action.

Synthesis

In comparison with the transmitter substances discussed above, synthesis of neuroactive peptides is more complicated. Let us take the opioid peptides as an example. Synthesis of these molecules (see Fig. 8.12) follows the plan for peptide hormones in which amino acids are first assembled by the ribosomes into a large polypeptide *prehormone*. This is then reduced in the Golgi body to a somewhat smaller *prohormone*, which in turn is cleaved into fragments that are the active peptides, and secreted in vesicles. This general sequence has been shown to apply to the manufacture of most neuroactive peptides. Note that, as a general rule, synthesis of neuropeptides requires gene activation, DNA transcription, and RNA translation, with the final peptide being transported from soma to release sites, as previously reviewed in Chap. 3. This is an important distinction, as compared with most transmitter substances, which can be synthesized immediately by cytosolic enzymes, often at synaptic sites, more under the influence of local activity states than global genetic control.

Receptors

We have seen that most neurotransmitters can bind to and activate more than one kind of receptor; the same applies to neuropeptides. The opioid receptors illustrate very nicely this principle of multiple receptor types.

In 1975, several laboratories reported biochemical studies showing that there are receptors in the brain that specifically bind morphine, the substance known since time immemorial as both pain-killing and addicting. Did the brain evolve receptors just to bind this substance, or are there neurons that secrete morphinelike substances (called "endorphins")? This was soon answered, still in 1975, by the finding in brain

CARNOSINE

(Ala)(His)

THYROTROPIN RELEASING HORMONE (TRH)

p(Glu)(His)(Pro) NH₂

Met-ENKEPHALIN

(Tyr)(Gly)(Gly)(Phe)(Met)

Leu-ENKEPHALIN

(Tyr)(Gly)(Gly)(Phe)(Leu)

ANGIOTENSIN II

(Asp)(Arg)(Val)(Tyr)(Ile)(His)(Pro)(Phe) NH₂

CHOLECYSTOKININ-LIKE PEPTIDE

(Asp)(Tyr)(Met)(Gly)(Trp)(Met)(Asp)(Phe) NH₂
              |
            SO₃H

Ala	ALANINE
Arg	ARGININE
Asn	ASPARAGINE
Asp	ASPARTIC ACID
Cys	CYSTEINE
Gln	GLUTAMINE
Glu	GLUTAMIC ACID
Gly	GLYCINE
His	HISTIDINE
Ile	ISOLEUCINE
Leu	LEUCINE
Lys	LYSINE
Met	METHIONINE
Phe	PHENYLALANINE
Pro	PROLINE
Ser	SERINE
Thr	THREONINE
Trp	TRYPTOPHAN
Tyr	TYROSINE
Val	VALINE

OXYTOCIN

(Tyr)(Cys)
(Ile)
(Gln)(Asn)(Cys)(Pro)(Leu)(Gly) NH₂

VASOPRESSIN

(Tyr)(Cys)
(Phe)
(Gln)(Asn)(Cys)(Pro)(Arg)(Gly) NH₂

LUTEINIZING-HORMONE RELEASING HORMONE (LHRH)

p(Glu)(His)(Trp)(Ser)(Tyr)(Gly)(Leu)(Arg)(Pro)(Gly) NH₂

SUBSTANCE P

(Arg)(Pro)(Lys)(Pro)(Gln)(Gln)(Phe)(Phe)(Gly)(Leu)(Met) NH₂

NEUROTENSIN

p(Glu)(Leu)(Tyr)(Glu)(Asn)(Lys)(Pro)(Arg)(Arg)(Pro)(Tyr)(Ile)(Leu)

BOMBESIN

p(Glu)(Gln)(Arg)(Leu)(Gly)(Asn)(Gln)(Trp)(Ala)(Val)(Gly)(His)(Leu)(Met) NH₂

SOMATOSTATIN

(Ala)(Gly)(Cys)(Lys)(Asn)(Phe)(Phe)(Trp)
(Cys)(Ser)(Thr)(Phe)(Thr)(Lys)

VASOACTIVE INTESTINAL POLYPEPTIDE (VIP)

(His)(Ser)(Asp)(Ala)(Val)(Phe)(Thr)(Asp)(Asn)(Tyr)(Thr)(Arg)(Leu)(Arg)(Lys)(Gln)(Met)(Ala)(Val)(Lys)(Lys)(Tyr)(Leu)(Asn)(Ser)(Ile)(Leu)(Asn) NH₂

β-ENDORPHIN

(Tyr)(Gly)(Gly)(Phe)(Met)(Thr)(Ser)(Glu)(Lys)(Ser)(Gln)(Thr)(Pro)(Leu)(Val)(Thr)(Leu)(Phe)(Lys)(Asn)(Ala)(Ile)(Val)(Lys)(Asn)(Ala)(His)(Lys)(Lys)(Gly)(Gln)

ACTH (CORTICOTROPIN)

(Ser)(Tyr)(Ser)(Met)(Glu)(His)(Phe)(Arg)(Tyr)(Gly)(Lys)(Pro)(Val)(Gly)(Lys)(Lys)(Arg)(Arg)(Pro)(Val)(Lys)(Val)(Tyr)(Pro)(Asp)(Gly)(Ala)(Glu)(Asp)(Glu)(Leu)(Ala)(Glu)(Ala)(Phe)(Pro)(Leu)(Glu)(Phe) NH₂

Fig. 8.11 Neuroactive peptides, arranged in order of increasing number of carbon atoms. (Modified from Iversen, 1979)

Table 8.4 Properties of neurotransmitters and neuropeptides

Properties of neurotransmitters	Properties of neuropeptides
medium to high concentration	extremely low concentration
high-affinity binding to receptors	low-affinity binding to receptors
low potency	extremely high potency
high specificity	high specificity
moderate rate of synthesis	low rate of synthesis (in vitro)
small molecules (2–10 carbons)	small to medium-size molecules (4–100 carbons)

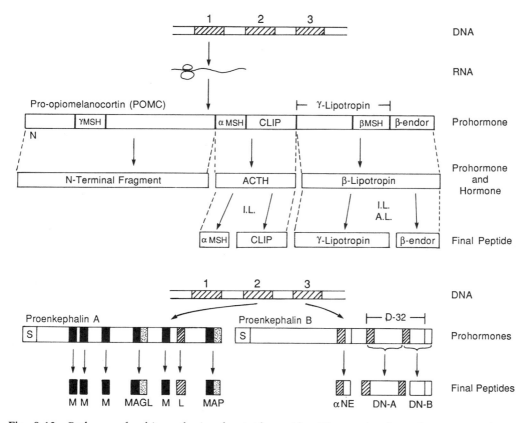

Fig. 8.12 Pathways for biosynthesis of opioid peptides. These arise from three genes. Gene 1 *(above)* gives rise to a pecursor molecule termed pro-opiomelanocortin (POMC). Posttranslational processing produces the peptides shown: α-, β-, and γ-melanophore stimulating hormone (MSH); β-endorphin; ACTH; and corticotropin-like intermediate lobe peptide (CLIP). The enkephalins *(below)* arise from two other genes that code for proenkephalins A and B: M, met-enkephalin (Tyr-Gly-Gly-Phe-Met); L, leu-enkephalin (Tyr-Gly-Gly-Phe-Leu); MAGL (M-Arg-Gly-Leu; MAP (M-Arg-Phe); α-neoendorphin (OLNE); dynorphin A (DN-A); dynorphin B (DN-B); dynorphin 32 (D-32). I.L., intermediate lobe of pituitary; A.L., anterior lobe of pituitary. (Based on Marx, 1983)

extracts of two small pentapeptides, called enkephalins, that had appropriate biochemical specificity in binding assays and analgesic effects when injected into animals. Both enkephalin and endorphin amino acid sequences are contained within the precursor molecule β-lipotropin. It is only recently with the application of recombinant DNA methods that we know that the enkephalins and endorphins and related peptides arise not from one but from three separate genes, in the manner already indicated in Fig. 8.12.

It did not take long after the identification of different opioid peptides for different types of receptors to be postulated and analyzed. Receptors were characterized by the physiological responses of different peripheral tissues, such as the vas deferens, to morphine agonists and antagonists. These tissues then served as bioassay systems for testing for opioid peptides in extracts of different brain regions. From such studies it became apparent that there are several types of receptors. These types have been further analyzed in brain slice preparations, in which single-neuron activity can be recorded during application of different pharmacological agents. Opioid receptors have been cloned and sequenced, and found to be members of the 7TD family.

A summary of three main types of opioid receptors and their mechanisms of action is provided in Fig. 8.13. The mechanism at this peptidergic synapse begins in the presynaptic neuron with the synthesis ①, storage ②, and release ③ of the opioid peptide. The different types of receptors ④ are shown as if they are all on one postsynaptic neuron, though of course in actuality each is specific for different types of postsynaptic neurons.

The mu (μ) receptor is defined by the fact that it preferentially binds morphine and its antagonist, naloxone, with a dissociation, or equilibrium, constant (K_d) of around 1 nM. Activation of this receptor leads, through a cAMP second messenger system, to opening of K^+ channels, which reduces the excitability and the impulse

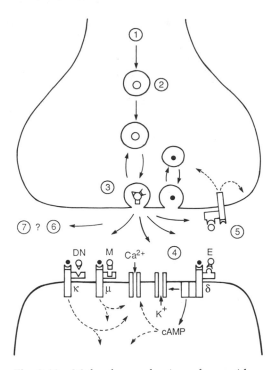

Fig. 8.13 Molecular mechanism of a peptidergic synapse, as exemplified by opioid peptides: ①, synthesis in cell body; ②, transport and storage in vesicles (see Fig. 3.11); ③, release by exocytosis; co-release with neurotransmitters at many synapses; ④, binding to one of several types of receptor. M (μ) receptor preferentially binds morphine; it activates cAMP to open K^+ channels and close Ca^{2+} channels. Delta (δ) receptor is similar but binds enkephalin preferentially. K (κ) receptor preferentially binds dynorphin (DN), and acts through second messenger to close Ca^{2+} channels. Not shown are two other types, γ and ε. (Based on Snyder, 1984; Cooper et al., 1993; Aghajanian and Rasmussen, 1988; and other sources)

firing rate of the postsynaptic cell or neuron. The sigma (σ) receptor similarly opens K^+ channels; it differs from the μ receptor in having different binding characteristics for agonists, particularly enkephalins (see Fig. 8.13 and legend). The kappa (κ) receptor preferentially binds the endorphin substance dynorphin (DN). Activation of this receptor leads to a closing of Ca^{2+} channels and a reduction in Ca^{2+} current. This has the effect of shortening Ca^{2+} impulses and reduces the amount of transmitter released from the postsynaptic cell. Thus, different receptor types can achieve similar effects on cell excitability and transmitter release by controlling different ionic channels. These actions are not unique to opioid peptides; they are also produced by acetylcholine and monoamines in different cells. The opioid peptides may also act on presynaptic receptors at some synapses, to bring about negative feedback through modulation of presynaptic Ca^{2+} channels, as indicated in Fig. 8.13.

The distribution of the different types of opioid receptors in the brain, and their behavioral effects, will be discussed in Chap. 24.

The mechanisms illustrated in Fig. 8.13 are examples of the variety of mechanisms in which neuropeptides are involved. In these examples, the effect on the receptors is to inhibit the postsynaptic cell; by contrast, the actions of other peptides bring about a decrease in the K^+ currents, which renders the postsynaptic cell more excitable (see Chaps. 17 and 18).

Multiple Messenger Mechanisms

From our discussion of Dale's Modified Principle, we have learned that a single neuron may secrete both a neurotransmitter and one or more neuroactive peptides. In recent years, it has been realized that co-release combined with multiple receptors provides the means for greatly amplifying the complexity of actions and the levels of control at a single synapse.

Some of these possibilities are summarized in the diagrams of Fig. 8.14. In A is shown the classical view: this synapse functions through a single transmitter that acts at a postsynaptic receptor and an autoreceptor, when driven by depolarizing input (arrow). Now take the case of a terminal in which there is coexistence of a peptide. Assume, as in B, that at a given level of input (say, a slightly higher frequency of impulses traveling down the axon and invading the terminal) there is co-release of the peptide. There are several

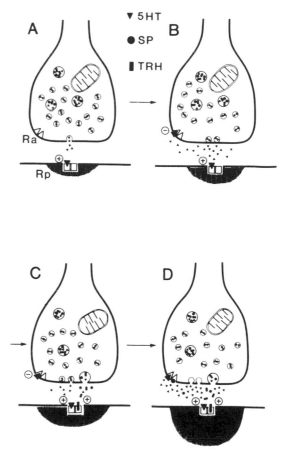

Fig. 8.14 Schematic representation of a nerve ending in the ventral horn containing 5HT, TRH, and substance P (SP), whereby the small vesicles (diameter about 500 Å) contain only 5HT and the larger vesicles (diameter about 1000 Å) contain, in addition, substance P and TRH. An attempt is made to illustrate a sequence of events occurring hypothetically when presynaptic nervous impulse activity is increased. **A.** At low frequency, small numbers of 5HT molecules, released from small synaptic vesicles, cause a small activation ⊕ of the postsynaptic receptor (R_p). **B.** Increased release of 5HT results in activation of the presynaptic autoreceptor (R_a) ⊖, which inhibits further 5HT release and prevents a larger postsynaptic response. Further increase in impulse traffic may activate large vesicles and release TRH and substance P, whereby (**C**) TRH acts together with 5HT on postsynaptic receptors, and (**D**) substance P blocks the inhibitory 5HT autoreceptor, resulting in a profound postsynaptic activation. (Figure and legend from Hökfelt et al., 1984)

sites at which this peptide may act. As shown in B, it may act on the presynaptic side, to block the autoreceptors for the transmitter, thus freeing the terminal from feedback inhibition, and leading to increased transmitter release and increased postsynaptic response. Alternatively, it may act on the postsynaptic terminal, as shown in C. Here it could change the affinity of the transmitter receptor, or it could act on a separate receptor that is linked to a common second messenger system. Finally, as shown in D, the peptide may have independent actions on membrane receptors, or on cytosolic receptors which in turn act on the nucleus (this applies especially to steroid hormones), and there may also be direct actions in the nucleus.

These cover only a few of the possible actions of peptides and their relations with transmitters. An important point is that the action of a peptide may not be manifest by itself, but rather in its interaction with the neurotransmitter. The student should be sure to see this point in working through the examples of Fig. 8.14B and C. This means that experimentally it tends to be difficult to demonstrate the physiological function of a peptide unless the experimental conditions are precisely right. Also, the action of a peptide may involve effects on metabolism or growth, as implied in D, and therefore be on a time scale different from that being examined in an acute experiment in which one is looking at a neuronal response to rapid application of a substance.

Time Courses of Action

As is obvious from our discussion above, the time courses of action of neuroactive substances vary over a wide range. This is summarized schematically in Fig. 8.15A. The time courses of action of the classical neurotransmitters, such as acetylcholine, represent the briefest types. The slower actions of neurotransmitters overlap with the periods of facilitation and depression that occur in the aftermath of activity, and with

the effects mediated by peptides and hormones. These in turn overlap with trophic effects, and with the neuronal interactions that underlie such processes as development and plasticity.

Figure 8.15A illustrates further the problem that can arise in defining a neurotransmitter. Thus, to the traditional criteria mentioned previously, one must now add another dimension: the time course of action. Brief actions are more characteristic of neurotransmitters, whereas long-term effects are more representative of neuromodulators. For substances such as the monoamines, whose actions generally fall between these two extremes, the term neurotransmodulator has even been suggested!

Although this profusion of time courses of action complicates the terminology, it is nonetheless telling us something very important about how the nervous system works. Animal behavior involves a wide range of time courses of action. Sensory systems are organized to receive and process information extremely quickly, often within milliseconds. Motor systems similarly provide for very rapid movements, also in the millisecond range, such as those made in typewriting or playing a piano. On the other hand, some movements are very slow; examples are the maintenance of a standing posture, or the prolonged contraction of muscles of a clam that close the shell. The same applies to central systems, from the quick initiation of a voluntary movement, to the slow modulation of states of sleep or arousal during the day and night, to the gradual changes that occur during development, maturation, and aging. One of the most profound questions about the nervous system is how it coordinates all these activities taking place simultaneously within the same neurons and neural circuits. At least part of the answer appears to be that it utilizes different substances with actions on different receptors with different time courses, as we have discussed above.

An example of different time scales of

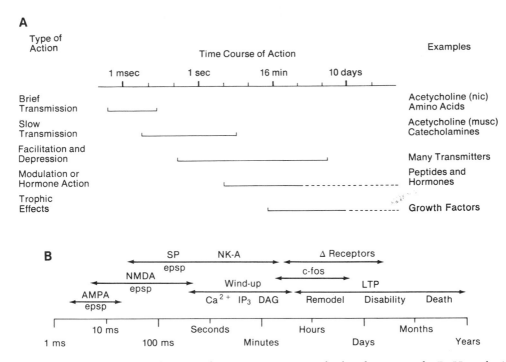

Fig. 8.15 **A.** Time courses of action of neurotransmitters and related compounds. **B.** Hypothetical temporal arrangement of various neural events that might contribute to the synaptic transmission and long-term modulation of pain. AMPA, NMDA, and SP mediate progressively longer excitatory postsynaptic depolarizations. Various second messenger systems may contribute to temporal amplification events such as windup. Early immediate gene (c-*fos*) induction may contribute to longer term alterations in cellular constitution, expressed cell surface receptors, and survivability. See Chap. 12. (Modified from Wilcox, 1991)

action in a specific system is illustrated for the pain pathway in the spinal cord in Fig. 8.15B. We will discuss the neural basis for these different actions in Chap. 12.

Transport of Substances

Closely related to synaptic transmission and its associated metabolic processes is the transport of substances within the nerve cell. Far from being the static structure visualized in microscopic sections, the neuron at the molecular level is in constant motion. As noted in the discussion of cell organelles in Chap. 3, there is ongoing synthesis of transmitter molecules, macromolecules, and vesicle membranes in the cell body, and movement out into the axon and dendrites (see Fig. 8.16). Some of these

substances pass out of axon terminals and are taken up by postsynaptic cells, as shown by transneuronal transport of labeled amino acids incorporated into protein. Proteins and small enzymes also are taken up by axon terminals and move in the axon toward the cell body; this is the basis of the mapping of axonal projections by the horseradish peroxidase (HRP) technique. A similar movement of substances takes place in dendrites, involving transmitters, enzymes, and even such large molecules as nucleoside derivatives. Some of these substances are those taken up from neighboring terminals by transneuronal transport. Ions and small molecules move directly between cells through the channels of gap junctions. Thus, there is constant biochemical transport and communication

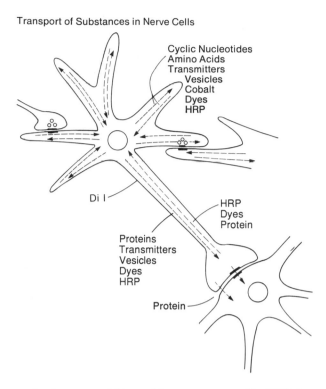

Transport of Substances in Nerve Cells

Cyclic Nucleotides
Amino Acids
Transmitters
Vesicles
Cobalt
Dyes
HRP

Di I

HRP
Dyes
Protein

Proteins
Transmitters
Vesicles
Dyes
HRP

Protein

Fig. 8.16 Transport of substances within and between nerve cells. HRP, horseradish peroxidase.

between all parts of the neuron and between neighboring neurons.

These movements take place at different rates. Axoplasmic transport, in mammals, as noted in Chap. 3, ranges from slow (about 1 mm/day) to fast (100–400 mm/day). What do these rates tell us about how quickly events are taking place at the molecular level? This can be answered if we plot distance against time on logarithmic scales, as in Fig. 8.17. The fastest transport in axons, of 400 mm/day, thus translates to a rate of about 5 μm/sec. Surprisingly, this is similar to the rate that has been estimated for diffusion of phospholipids in the cell membrane, as shown in the graph. Slow transport is at a rate of 0.01 μm (10 nm)/sec; this is even slower than the estimated rate of diffusion of proteins in the cell membrane. These values serve to emphasize that the membrane, as well as the internal cell substance, is in dynamic flux.

With regard to transport of synaptic transmitters across the vesicular membrane, this would take many minutes even by fast transport in the shortest axons ($\leqslant$1 mm in length), and would be a matter of hours in the longest axons. This helps to explain why axonal terminals contain some of their own metabolic machinery for transmitter synthesis and reuptake. In the case of output synapses from dendrites, however, the distances from the cell body are characteristically less than 1 mm, and the synapses could seem to be able to draw more directly on the metabolic resources of the cell body for sustaining their activity. For output synapses from the cell body itself, this of course becomes obvious.

Other rates are also shown in Fig. 8.17, for comparison. Note that the slowest rate of nerve conduction is more than five orders of magnitude faster than the fastest axonal transport; even the slowest impulses travel a micrometer in less than a microsec-

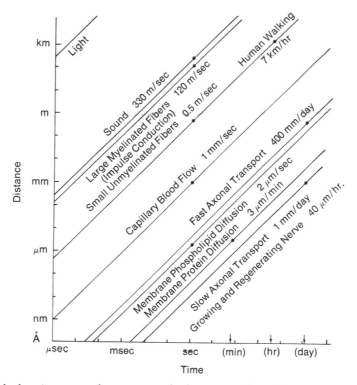

Fig. 8.17 Graph showing rates of movement of substances and conduction of activity for different dimensions of time and space. Small dots indicate values used for common expressions for rates.

ond. Synaptic transmitter diffusion through the cleft (see earlier in this chapter) works out to a rate about 1 μm/ms, similar to the rate of capillary blood flow. The reader can insert other rates and relations as well. The general principle, that physiological processes at molecular levels take place in incredibly short times, is readily apparent. It is also the reason why the time domain for microelectrode and single-channel analysis of synaptic functions commonly falls into the range of milliseconds.

Energy Metabolism and 2-Deoxyglucose Mapping

The vertebrate brain is virtually completely dependent on glucose for energy metabolism. Glucose is taken up by neurons and phosphorylated by hexokinase to glucose-

6-phosphate. As was discussed in Chap. 3, it is then metabolized in the cytosol through the glycolytic chain to pyruvate, which enters the mitochondria and undergoes oxidative metabolism by the Krebs cycle to yield high-energy phosphates. The initial steps in this sequence are indicated in Fig. 8.18. The high-energy phosphate is incorporated into adenosine triphosphate (ATP) and made available for the ongoing metabolism of the neuron and for the immediate demands related to nervous activity.

What types of activity require energy? We have seen that ions move passively through their conductance channels in the membrane; however, the concentration gradients are maintained by the metabolic pump, which requires energy. The squid giant axon can continue to generate action potentials for hours after metabolic poison-

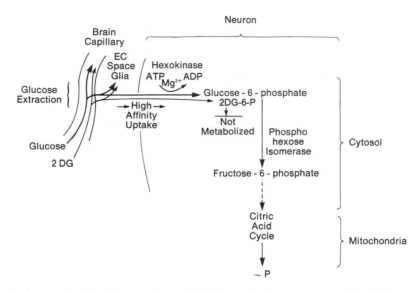

Fig. 8.18 Pathways involved in uptake and initial metabolism in nerve cells of glucose and the analogue, 2-deoxyglucose.

ing, because the passive ion flows are small compared with the large amounts of available ions. The ion flows are proportionately larger in smaller fibers, with their larger surface-to-volume ratios; in the finest unmyelinated fibers, impulse activity therefore places immediate demands on the metabolic pump. Similar factors are involved at synapses; the ion flows themselves may be passive, but the restoration and maintenance of ion concentrations require energy, as does the synthesis of transmitters and the recycling of membrane. Also, higher rates of ion pumping may be expected at sites where the resting membrane potential is relatively low because of an increased permeability to Na (as at nodes of Ranvier, and in retinal receptors). The high density of mitochondria in the small processes of axons and dendrites reflects to a large extent the energy demands of synapses at those sites.

These types of activity thus have immediate energy demands, and there is a method that uses this property to map the distribution of activity in the brain during different functional states. The method was introduced by Louis Sokoloff and his col-

leagues at the National Institutes of Health, and makes use of an analogue of glucose, 2-deoxyglucose (2DG), which simply lacks an oxygen on the second carbon atom. As shown in Fig. 8.18, 2DG is taken up like glucose and is phosphorylated by hexokinase. However, the resulting 2DG-6-P is not a substrate for phosphoglucose isomerase and cannot be metabolized further; it is trapped in the tissue. Sokoloff and his colleagues reasoned that if 2DG was labeled with radioactive carbon (^{14}C) and injected in tracer amounts into an animal, the sites of increased [^{14}C]2DG-6-P could be marked by exposing sections of the brain tissue to X-ray film. As many sections can be made as desired, so that the activity pattern associated with a particular functional state can be mapped throughout the entire brain.

The method has been applied to many systems. Among the most dramatic results are those that have been obtained in the visual system. A monkey is injected a day after one eye has been removed. The autoradiograms of the visual cortex show alternating dark and light stripes, which represent the ocular dominance columns that

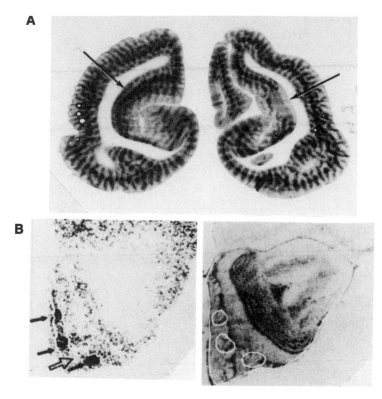

Fig. 8.19 Autoradiograms obtained using the $[^{14}C]$2-deoxyglucose method. **A.** Monkey visual cortex after removal of one eye. Arrows indicate blind spots. (From Sokoloff, 1977) **B.** Rat olfactory bulb after exposure to odor of amyl acetate. *(Left)* Autoradiogram showing three small dense foci (arrows) and intervening light region (open arrow). *(Right)* Outlines of dense foci fall precisely on small groups of glomeruli when superimposed on Nissl-stained section of bulb. (From Stewart et al., 1979)

had previously been demonstrated by electrophysiology and anatomical methods (see Chap. 16). These results are shown in Fig. 8.19A. For comparison, the method has also been successfully applied to the olfactory system, where relatively little was known about spatial activity patterns. Surprisingly, after an injection into a rat breathing an odor, small intense foci of activity are found in the olfactory bulb over the glomerular layer, where the axons from the olfactory receptor cells terminate. These results are shown in Fig. 8.19B (see Chap. 11 for further discussion of olfactory bulb function).

These results indicate the power of this method for confirming and extending our previous knowledge about particular systems, and for providing new insights into systems in which information about spatial activity patterns has not been obtainable by other methods. The applications of this and related methods range from the analysis of local circuit organization to the identification of activity patterns underlying cognition, as we shall see in Chap. 30.

9

Developmental
Neurobiology

How do the structures and functions of the nervous system develop? This is a fascinating subject that addresses the age-old question of how the organism arises out of the interplay between nature and nurture, between genes and environment. In addition, it is of practical importance, because knowledge of how the nervous system is assembled often gives insight into the functional organization of the mature system, and it also tells us what goes wrong in nervous and mental diseases that have a genetic basis. The tools of molecular biology make developmental mechanisms accessible as never before, and provide thereby the hope not only of understanding normal development, but also of intervening to correct or prevent genetic disorders.

Neuroscientists and developmental biologists have been alive to these opportunities; the pace of research has been increasing rapidly, and developmental neurobiology is already a vast field. Since developmental mechanisms are so important, they are integrated into most of the chapters of this book; for example, the student has already been introduced to gene regulation underlying development (Chap. 2) and to cell surface molecules involved in neuronal recognition (Chap. 3). Here we present a brief summary of the main principles.

Because of the special interest of the human for most introductory students, we will focus on the development of the vertebrate nervous system in general, and the human brain in particular, to illustrate these themes. We will also emphasize those aspects that relate specifically to the establishment of synaptic connections and organization of neuronal circuits. More extensive treatments of development may be found in several excellent texts: see Gilbert (1991) and Purves and Lichtman (1985).

Brain Development: An Overview

The nervous systems of all animals develop through a sequence of stages. It is a reflection of the importance of basic principles that we can use almost any vertebrate species to illustrate the main stages of development of the vertebrate brain, and many of the principles of invertebrates as well. Let us begin with an overview of the main stages, using as our model the human brain. We will first orient ourselves to the main anatomical regions before studying the molecular and cellular mechanisms.

Regional Differentiation of the Brain
Our knowledge of the development of the human brain was laid down by anatomists

in the latter part of the nineteenth century and put on a systematic basis early in the twentieth century. As we will study later, the brain starts as a simple tube in the dorsal midline of the head region. The earliest stage at which this tube begins to display different regions is around 4 weeks, when one can discern a most rostral (anterior) part, called the prosencephalon (pro = front, encephlon = head), a middle part called the mesencephalon (mes = middle), and a caudal (posterior) part called the rhombencephalon (rhomb = lozenge-shaped). At about this time a bulge called the optic vesicle appears; this will become the neural part of the retina. These details are shown in diagram A of Fig. 9.1A, and the developmental sequence is summarized in the flow chart of Fig. 9.1B.

By a week later (B in Fig. 9.1A) several new regions have differentiated: the telencephalon has given rise to the cerebral vesicle, which will become the cerebral cortex and basal ganglia, and to the diencephalon, while the rhombencephalon has given rise to two parts, the metencephalon (met = between) and the myelencephalon (myel = fibers). The hypothalamus has also appeared, with its connection to the rudimentary pituitary gland. These early structures are starting to outgrow their place in line, bulging out to produce the cervical flexure at the isthmus (I). These changes continue through the sixth week (C in Fig. 9.1A) until the seventh week. By this time the thalamus, the main structure connecting the telencephalon to the rest of the central nervous system, has appeared, as well as the medulla, pons, and cerebellum (see Fig. 9.1B), and all the regions have become much more clearly demarcated.

From this time forward the human brain begins to diverge from the relatively balanced pattern of regional growth in lower vertebrates and follow the mammalian pattern with an increasing overgrowth of the telencephalon. This can already be appreciated by 2 months (see E in Fig. 9.1A). By 3 months the neocortex of the telencephalon is beginning to cover the more caudal brainstem (F and G in Fig. 9.1A) and is beginning to show evidence of a division into different lobes. Later stages are shown in diagrams H–J at the bottom of Fig. 9.1A. The continued overgrowth of the neocortex is associated with infoldings of the cortical surface, beginning around 6 months, and this process becomes quite extensive by the time of birth. By then, all one can see from the side is this mantle of neocortex, except for a glimpse of the cerebellum (whose growth is also associated with considerable infolding) and the caudal brainstem (see J in Fig. 9.1A).

This gives only a brief overview of the main aspects of the development of the brain. There are many associated changes as well, which include the elaboration of ventricles inside the telencephalon, mesencephalon, and rostral metencephalon, as well as the development of the cerebral vasculature, spinal cord, ascending and descending tracts connecting the brain to the spinal cord, and the cranial nerves, which provide the sensory and motor pathways to and from the brain.

Genetic Basis of Brain Segmentation

From this overview of brain development it is obvious that specialization of different regions is a hallmark of brain organization. Ever since the classical studies it has been known that these regions reflect the basic pattern of segmentation of the body. Recently, with the discovery of mechanisms of gene regulation in complex organisms, we are beginning to understand the genetic basis of the segmentation.

The basic body plan emerges during development by a cascade of sequential gene regulation. This was first worked out in *Drosophila* and appears to apply widely to most segmented organisms (see Gilbert, 1991, for review). The cascade begins with *maternal mRNAs* within the blastoderm; these code for proteins which interact with zygotic genes, called *gap genes*, to lay down broad domains along the anterior–posterior axis within the embryo. Gap gene pro-

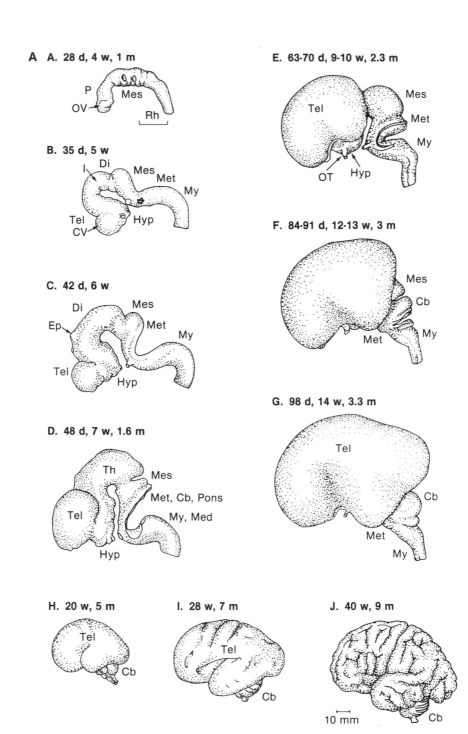

A. 28 d, 4 w, 1 m

P
OV
Mes
Rh

B. 35 d, 5 w

I
Di
Mes
Met
My
Tel
CV
Hyp

C. 42 d, 6 w

Di
Ep
Mes
Met
My
Tel
Hyp

D. 48 d, 7 w, 1.6 m

Th
Mes
Met, Cb, Pons
My, Med
Tel
Hyp

E. 63-70 d, 9-10 w, 2.3 m

Tel
Mes
Met
My
OT
Hyp

F. 84-91 d, 12-13 w, 3 m

Mes
Cb
Met
My

G. 98 d, 14 w, 3.3 m

Tel
Cb
Met
My

H. 20 w, 5 m

Tel
Cb

I. 28 w, 7 m

Tel
Cb

J. 40 w, 9 m

10 mm
Cb

194

B

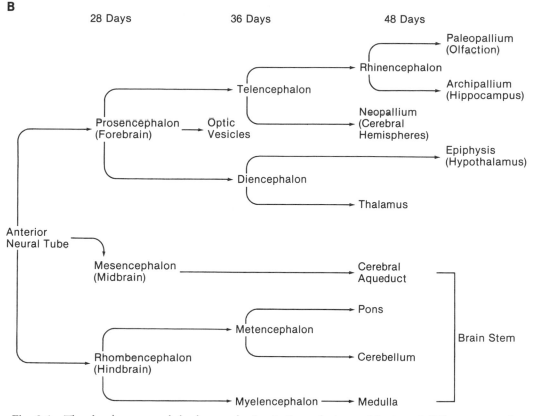

Fig. 9.1 The development of the human brain. **A.** Lateral views of brains of different ages after conception. For convenience, the times are indicated in days (d), weeks (w), and months (m). Scale bar is A: 0.5 mm in A, increasing to 6 mm in G; for H–J, scale bar in J is 10 mm. Cb, cerebellum; Di, diencephalon; Ep, epiphysis; Hyp, hypothalamus; I, isthmus; Med, medulla oblongata; Mes, mesencephalon; Met, metencephalon; My, myelencephalon; OT, optic tract; ov, optic vesicle; P, prosencephalon; Rh, rhombencephalon; Tel, telencephalon; Th, thalamus. **B.** Summary of the timing of regional specialization of the human brain during the first 7 weeks after conception. (A adapted from Hochstetter, in Sidman and Rakic, 1982; B modified from Gilbert, 1991)

teins cause transcripton of *pair-rule genes,* and their products cause transcription of *segment polarity genes,* which increasingly define transverse segments along the anterior–posterior axis. All of these gene products act in a coordinated fashion to control *homeotic genes,* which specify the types of tissues and organs within each segment. Mutations in gap genes cause gaps (hence the name) in the segmentation pattern, whereas mutations in homeotic genes cause bizarre transpositions in which a body part such as a leg can grow out of an inappropriate segment, such as one related to the head. These have long been recognized in *Drosophila,* where they were first named

"homeotic mutants" by William Bateson in 1894.

As we learned in Chap. 2, homeotic genes control a complex regulatory system. Each gene contains a segment called a *homeobox,* a 183 base pair sequence that encodes an amino acid sequence which functions as the DNA-binding region, called a *homeodomain,* in a DNA-binding protein. The homeodomain has the helix-turn-helix motif (see Chap. 2). Homeodomains can bind to regulatory domains on their own proteins as well as to other homeobox-containing genes, providing for autoregulation and combinatorial interactions. In addition, homeobox-containing *(hox)* genes can also be activated by growth factors, as we will discuss later in relation to neurulation.

The basic work in identifying hox genes has been carried out in *Drosophila.* These genes have then been used as probes to identify homologous genes in mammals such as the mouse. As summarized in Fig. 9.2 (see also Fig. 2.8), the hox genes in the mouse fall into families that have their counterparts in *Drosophila,* and the order of the genes within the chromosomes is similar, as is their pattern of expression. This basic pattern in turn applies to the human as well.

This remarkable similarity across species indicates that segmentation of the body plan is fundamental to complex organisms and, in particular, that different brain regions reflect adaptations of this basic pattern. It is a beautiful demonstration of how investigations of many types of animals, which may not seem related at first, are absolutely necessary in order to understand the principles of biological organization. It also demonstrates how studies of organisms distantly related to humans can nonetheless, by their attractiveness as experimental models, give insight into human function and disease.

Cellular Mechanisms

Origins of the Nervous System

The study of the cellular basis of brain development began with microscopic investigations during the nineteenth century. One of the greatest of the early pioneers was Wilhelm His, a Swiss who worked in Leipzig. Much of the work, with many young colleagues, was carried out in his home. Since the methods of preserving (fixing) brain tissue were still primitive, His's microscopic material was of inferior quality, but his ideas were clear and profound. In the 1880s his description of the axon as an outgrowth from the immature nerve cell was an important step toward the concept of the neuron as a cell and the formulation of the neuron doctrine (see Chap. 3). To him also we owe the terms *dendrite* (see Chap. 3), for the branches from the cell body, and *neuropil,* the cell-free regions containing connections between axons and dendrites.

By His's time some general ideas about development had begun to emerge. This includes the *germ layer theory,* which recognized that the early embryo first differentiates into three distinct layers of cells: an outer layer called *ectoderm,* an inner layer called *endoderm,* and a layer between called *mesoderm.* As covered in introductory biology texts, all organs of the body are derived from one or more of these layers, the nervous system being derived mainly from the ectoderm. Since the work of His and his contemporaries, there has been agreement at the descriptive level on the cellular basis of how this comes about.

The main early steps in forming the nervous system are summarized in Fig. 9.2. The process begins with the *neural plate* (see A), a midline strip of ectoderm on the surface of the embryo just after gastrulation has occurred. As the strip thickens, the edges roll up (B); the edges then meet, forming a tube which gradually sinks into the underlying mesoderm (C). This is the *neural tube,* which forms the spinal cord and brain. Continued growth in thickness takes place (D), and in the rostral end the tube begins to differentiate into the regions we observed in Fig. 9.1A. This whole process is called *neurulation.* Associated with neurulation are changes in surrounding tissues; of these, the most relevant to note are the appearance of *somites,* which will become the segments of the body trunk and limbs, and the *neural crest,* which gives rise to structures of the peripheral nervous system (see below).

With the application of methods of biochemistry and molecular biology, modern work has begun to reveal at a molecular level the mechanisms underlying these remarkable changes. One would expect that these kinds of cellular rearrangements would depend heavily on cell surface recognition molecules to mediate interactions be-

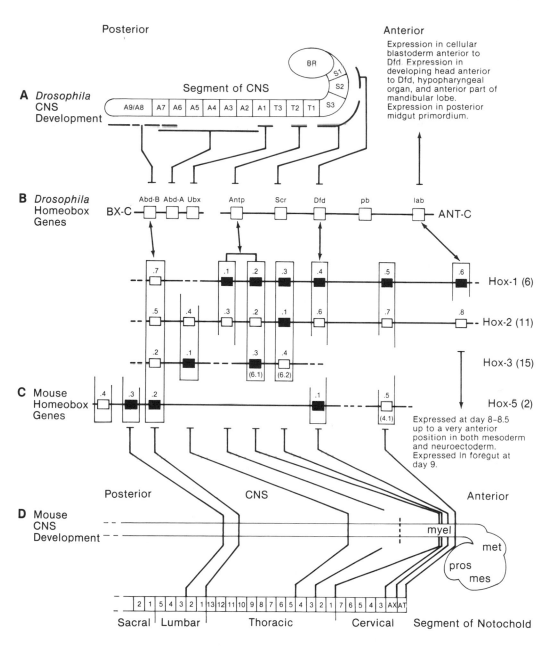

Figures in parentheses refer to chromosome number. Abd-B, abdominal B; Abd-A, abdominal A; Ubx, ultrabithorax; Antp, Antennapedia; Scr, sex combs reduced; Dfd, deformed; zen, zerknült; pb, proboscipedia; lab, labiopedia.

Fig. 9.2 Summary of the organization and expression of homeobox-containing genes in *Drosophila* and mouse. **A.** The segmental organization of the central nervous system of *Drosophila*. **B.** Bithorax (BX-C) and Antennapedia (ANT-C) homeobox genes, with their expression domains indicated in A. **C.** Mouse genes belonging to the Hox-1, 2, 3, and 5 families; those whose expression domains have been identified by in situ hybridization are indicated by filled boxes. **D.** Expression domains of the mouse hox genes in the mouse spinal cord and brainstem. (Adapted from Duboule and Dolle, in Schindler, 1990)

197

tween a cell and its environment and with other cells. In Chap. 3 we learned that there is a family of such molecules of the cadherin type. Once such a protein has been identified and its function postulated, a standard procedure is to prepare antibodies to the protein in order first to relate it to different types of cells, and then to see if it can be blocked and its function deduced. The first step is illustrated in Fig. 9.3, where the localization of three different types of cadherins (E for epithelial, N for neural, and P for placental) has been studied As we learned in Chap. 3, cadherins are a family of integral membrane glycoproteins which are involved in Ca^{2+}-dependent cell–cell adhesion interactions. Each of the three types is mainly expressed in one type of tissue but transiently expressed in others during development. During formation of the neural tube, the ectoderm of the neural plate expresses E-cadherin, which mediates adhesion between epithelial cells at special junctions called *zonulae adherens*. During neurulation, this changes to N-cadherin (see Fig. 9.2B–D), suggesting that N-cadherin may play a critical role in the morphogenetic events underlying the origin of neural structures from epithelial precursors. At this time, cells of the neural tube also begin to express N-CAM, another member of the family (see Chap. 3). Note also the appearance of combinations of N- with E- and P-cadherins in the neighboring structures.

Stages of Neuronal Development

In his comprehensive investigations of the nervous system, Cajal was fascinated by the development of nerve cells and their connections. He combined his studies of Golgi silver-stained neurons in the adult with studies of their forms and movements in embryonic tissue, and thus helped to lay the basis for the modern approach to the subject at the cellular level. Figure 9.4A is from his study in the 1890s of the development of granule cells in the vertebrate cerebellum.

From this early date it has been recog-

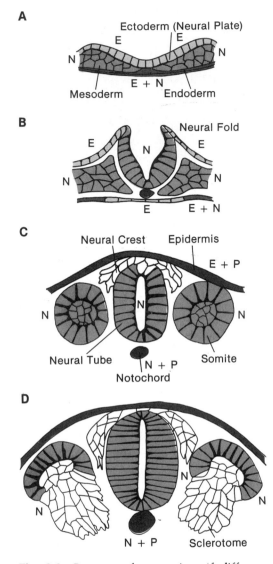

Fig. 9.3 Patterns of expression of different cadherin molecules in different cellular structures at different stages of neurulation in a chick or mouse embryo. The different cadherins are E (ectodermal), N (neural), and P (placental), named for their primary site of expression. (Adapted from Takeichi, modified from Alberts et al., 1989)

nized that, just as in the emergence of the gross anatomical structure of the nervous system, at the cellular level, neuronal development *(neurogenesis)* involves a series of steps. Let us use Cajal's study of the cere-

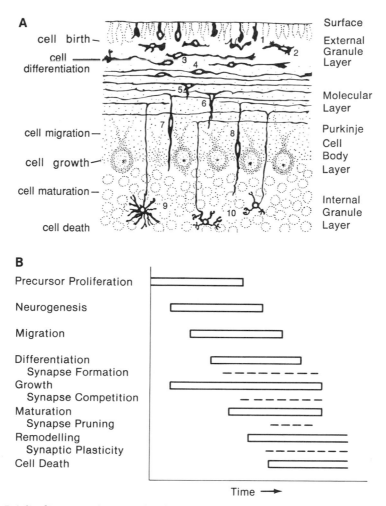

Fig. 9.4 **A.** Cajal's diagrams showing the development of granule cells in the mammalian cerebellum. Onset of developmental steps: cell birth and lineage, ①–②; cell differentiation and synaptogenesis, ③–④; cell growth, ⑤–⑥; cell migration, ⑦–⑧; cell maturation, ⑨–⑩; cell death (not shown). After cell birth, the other steps overlap considerably in duration (see text). **B.** Summary of the main steps in neuronal and synaptic development, to emphasize the overlapping time periods for the different mechanisms. (Modified from Gilbert, 1991)

bellum as an example (Fig. 9.4A), while placing it within the somewhat expanded framework provided by modern research (Fig. 9.4B). The first step is the *proliferation* of precursor cells through a sequence of cell divisions up to the birth of the neuron. After this final division the cell is destined to become a neuron of a given type or limited range of types. Soon after, the cell begins to *migrate* from its site of origin to its final location (5–9 in A), during which it undergoes *differentiation*

(3–10) toward its final form, with extension of axon and dendrites and with *growth* in size (9–10). Modern research also demonstrates the expression of functional membrane properties and the formation of synapses during this time (see Fig. 9.2B). All of these mechanisms characteristically have well-defined time periods that overlap considerably. *Maturation* of the cell into its final form and function (9–10 in A) is a process that in many animals, especially humans, may last over considerable periods of time (see B). In addition to production of cells and cell processes, *pruning* of synaptic connections and *cell death* are important factors in early maturation as well as in old age (see B).

Early studies were carried out within the field of histology (the light-microscopic study of tissues), and the events in the formation of different tissues were called *histogenesis*. Today, our focus is on the molecular and cellular mechanisms that are involved, and the study of neurogenesis is therefore a branch of molecular and cell biology. Neurogenesis was originally thought to occur primarily during embryonic life, and was therefore regarded as a part of *embryology*. We now realize that events related to neurogenesis and the laying down of neural circuits continue into early life; moreover, as noted above, many of the properties underlying the development of neurons and their functional capacities continue to be expressed throughout adolescence and even after, and can be changed under normal conditions or in response to injury or aging. We therefore study all these aspects within the broad field of *developmental neurobiology,* and regard brain development as a process that extends throughout most of the life of the animal.

In this chapter, we will briefly consider the mechanisms underlying each of the steps indicated in Fig. 9.4A and B.

Neuronal Birth

The first step in building a nervous system is to generate nerve cells. An important rule that applies, with few exceptions, is that *neurons once generated cannot be replaced;* in this regard, the nervous system differs from the liver, skin, or immune system, and resembles other more highly differentiated organs like the heart and lungs. Why should this be? One possibility is that complex organisms, particularly mammals and primates, need stable populations of neurons to preserve learned behaviors and

memories in their functional networks over decades of life (see Rakic, 1985). This rule means that the determination of whether a cell is to become a neuron, and what type of neuron in which part of the nervous system, must be completed, once and for all, relatively early in development. This determination depends on the genes, by the selective process called *gene expression.*

Since most cells in the body carry in their DNA the same complete instructions for the entire body, how is gene expression in neuron precursors selectively controlled to yield cells different from those destined to build other organs? The *regulation of gene expression* is thus central to understanding neurogenesis.

Neurogenesis of Central Neurons

Our knowledge of events leading up to neurogenesis of central neurons begins with the studies of F. C. Sauer in 1935. He observed the cells in the wall of the neural tube of the chick embryo and noted changes in them that were related to mitotic activity. As illustrated in Fig. 9.5A, these are bipolar cells with two processes which reach to the inner ventricular surface and outer marginal surface of the neuroepithelium. During the interphase period of DNA synthesis (S) the nucleus of the cell moves outward toward the marginal surface and then inward toward the ventricular surface. Mitosis occurs with the nucleus near the ventricular surface, with loss of the superficial process. After mitosis, the daughter cells send a process to the marginal surface and the cycle repeats itself.

Very early in development the mitotic activity produces new germinative cells, and the cells are therefore referred to as *neuroepithelial germinal cells.* The rate of mitotic activity is controlled by a number of hormones and growth factors, such as *growth hormone* and somatomedins, which include *insulinlike growth factors* (IGFs); *platelet-derived growth factor* (PDGF), and *epidermal growth factor* (EGF). These act in different combinations

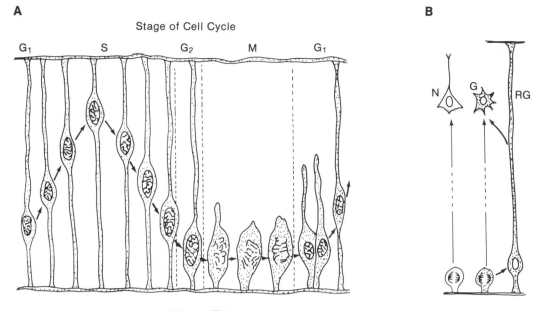

Fig. 9.5 A. Diagram of the neuroepithelium of the neural tube of the chick embryo summarizing the migration of the nucleus of a germinal cell in relation to the cell cycle. Each cell represents an interval of approximately 0.5 hr. Phases of the cell cycle: G_1, prereplication gap; S, DNA synthesis; G_2, premitotic gap; M, mitosis. **B.** Diagram illustrating the origins of neurons (N) and glia (G), including the dual-cell-line theory of glial origin through radial glia (RG). Based on studies in the primate cortex of staining for glial fibrillary acidic protein (GFAP); shading indicates GFAP positive. See text. (A from Sauer, in Jacobson, 1989; B from Cameron and Rakic, 1991)

at different stages of the cell cycle. They also act at later stages of neuronal differentiation, as we shall see later.

What determines the final mitosis, at which time the germinal cell becomes a primitive neuron? And when is the final neuronal identity determined, before or after the final mitosis? At present there are no answers to these questions that can characterize neurogenesis in general in all species. However, some insight into the second question has been obtained in studying the origins of neurons and glia in the cerebral cortex (see Chap. 3). By combined methods, including monoclonal antibodies to antigens specific for neurons or glia, it has been possible to analyze whether neurons and glia begin to differentiate before or after the final mitosis. As shown in Fig. 9.5B, there is evidence that glial fibrillary acidic protein (GFAP) is pres-

ent in a subpopulation of germinal cells. Pasko Rakic at Yale postulated that the GFAP negative germinal cells give rise to neurons, whereas the GFAP positive germinal cells can give rise either directly to various types of specialized mature glia, or to a transient type known as radial glia, which in turn can give rise to some of the other types. Radial glia provide a scaffolding for outward migration of newly formed neurons to their final places in the neural wall. In Chap. 30 we will discuss further this role, as well as the determination of different neuronal phenotypes in the human cerebral cortex.

Model Systems for Studying Neurogenesis and Development

The problems of working with complex organisms such as the primate have induced many workers to seek simpler models among invertebrates. We have already mentioned *Drosophila,* which has all the advantages of rapid reproduction and small size. However, analysis of central developmental mechanisms is difficult because of the very large populations of neurons. A tiny roundworm, *Caenorhabditis elegans,* has therefore attracted attention. Because of the small size and small number of neurons (309), it has been possible to make serial electron microscopic sections through the entire nervous system at different embryonic stages and trace the lineage of every neuron back to the zygote. This work has been carried out by Sidney Brenner and a growing cadre of colleagues (see Sulston et al., 1983). A new initiative of cloning the genome of this animal, which has recently been undertaken, will provide a foundation for analyzing the molecular basis of neural development and function.

The results of the studies in invertebrates have been summarized as follows (Purves and Lichtman, 1985):

. . . these studies of lineage underscore the relative stereotyped and programmatic development of invertebrates. Invertebrate cells seem to differentiate nearly immediately—after all, they have only a few divisions in which to create the finished product. . . . Consistent with rapid differentiation is the . . . limited ability of invertebrate embryos to replace missing cell lines following ablation. A variety of experiments . . . on developing vertebrates . . . indicate that cells at early stages have a much broader range of fates. This raises the interesting question of whether vertebrates and invertebrates are fundamentally different in this respect.

Because of these concerns, there is increasing interest in simpler models among vertebrates. An example is the zebrafish, a common pet swimming in many home fish aquaria. Geneticists are attracted by the brief generational time (3 months) and the large number of eggs. Experimental biologists are attracted by the rapidly developing embryos, which in addition are transparent, so that stained cells can be observed in vivo (in the living animal) (see Kimmel

and Varga, 1986). Thus far, cell lineages have been followed and several interesting mutants have been identified. The next steps are to construct genetic linkage maps of the chromosomes and make mutants for each step of neural development, as is being done for *Drosophila* and other species.

Summary. A general conclusion arising from studies of simpler organisms is that larger brains require many successive cell divisions in order to generate the requisite large number of neurons. This protracted series of cell divisions gives an opportunity for many factors to be involved in determining gene expression and ultimate neuronal phenotype. As pointed out by Easter et al. (1985), these factors include competition between neurons, trophic interactions, and modification by functional activity. One may postulate that these dynamic factors are present to different degrees in the development of most nervous systems, invertebrate and vertebrate, and reach their greatest importance in the human.

To construct the large human brain there must be mechanisms to control the numbers of neurons and glial cells generated for each region (see Williams and Herrup, 1988). Within each region, there are multiple divisions which over time gradually restrict the eventual fate a cell can assume. These fates are products of cell interactions with environmental cues that include the various growth factors and DNA binding proteins described above, and which we will discuss further below.

Thus, at every developmental stage there is interaction of the genome with the internal state and environment of the cell, which in turn depend on interactions with neighboring cells. An important general conclusion to be drawn from this phenomenon is that development is not simply a reading out of the genetic code. The code probably contains an economical minimum of instructions; as Gunther Stent (1981) has pointed out, they probably serve mainly to set neurons along general paths, with final forms and functions determined by cellular interactions (See also Edelman, 1988).

Cell Migration

A general rule in the vertebrate nervous system is that neurons do not remain at their site of origin, but rather migrate to their final position. This is a necessary consequence of the fact that the nervous system starts as a thin tube of ectoderm within the embryo (the neural tube), and the final product is a much larger structure (the nervous system). Also, the initial relations between the sites of origin of neurons may be very different from their final relations, as already indicated in the diagrams from Cajal (Fig. 9.4A).

Migration of Central Neurons

Within the neural tube we have seen (Fig. 9.5B) that after the last division of a precursor cell the new neuron migrates toward the surface to its final position. Pasko Rakic, then at Harvard, postulated in 1972 on the basis of electron microscopical studies that central neurons migrate to their target layers along special *radial glia,* which are transiently present during development to play this role. Many experiments have supported this fundamental mechanism. A vivid demonstration has been provided by U. Gasser and Mary Beth Hatten at Rockefeller University. They prepared mosaic cultures containing radial glia from the cerebellum and neurons from the hippocampus and, using videomicroscopy of the combined cultures, showed directly that the neurons "climb on" the glial fibers and migrate along them. They concluded that, since the neurons and glia come from different regions (are "heterotypic"), the radial glia are a *permissive* pathway (they permit the neurons to migrate along them) rather than an *instructive* pathway (they do not instruct the neurons to head for a specific target). Thus, "changing patterns of gene expression in the young neuron . . . are likely to control cell migrations in the developing brain" (Hatten, 1993).

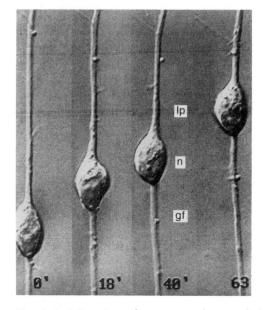

Fig. 9.6 Migration of a neuron along radial glial fiber during early development. Videomicroscopy of a combined culture of radial glia from the cerebellum and neurons from the hippocampus of the rat. The neuron (n) migrates along the glial fiber (gf) by extending a leading process (lp), which actively searches along the fiber with numerous extensions (lammelopodia and filopodia) similar to growth cones. Elapsed time is shown below in minutes. (From Gasser and Hatten, 1990)

Further studies have shown that some migration can take place across the radial glia (O'Rourke et al., 1992). We will return to this question in studying the development of the cerebral cortex (Chap. 30).

Migration of Peripheral Neurons

Neurons of the peripheral nervous system have a different origin. They arise largely from the neural crest, so called because it is a clump of cells which first appears on the dorsal surface of the neural tube (as we saw in Fig. 9.3, these cells are negative for cadherin molecules). The neural crest has attracted attention for several reasons. One is the amazing variety of types of cells derived from the neural crest (see Table 9.1), which invites analysis of differentiation mechanisms. A second is the long mi-

Table 9.1 Neural crest derivatives

Neurons
Spinal ganglia (dorsal root ganglion cells: substance P)
Autonomic ganglia (adrenergic and cholinergic ganglion cells)
Intestinal ganglia (serotonergic and peptidergic cells)
Glia
Schwann cells

Nonneural cells
Melanocytes
Chromaffin cells (adrenal medulla)
Cartilage
Some facial bones
Blood-forming cells
Meninges

gratory pathways, which make experimental manipulations easier. A third is the accessibility of these peripheral structures to experimental analysis. All these things gladden the heart of a developmental biologist, and explain why the neural crest is a favorite model for analysis of cellular mechanisms of development.

How does this structure give rise to different cell types? One possibility is that the neural crest cells are *multipotent,* each capable of giving rise to the full range of different cell types (see Fig. 9.7A). Most studies have been compatible with this postulate. However, it is hard to rule out the possibility that when the neural crest forms it already consists of a mixture of unipotent cells, each committed to a single *phenotype.*

One approach to this problem is to label the cells at different times of development. An ingenious technique for doing this depends on the fact that, in the Japanese quail, chromatin granules are clumped within the nucleus in a characteristic fashion, making it possible to recognize these cells when they are transplanted into a host animal, such as the chicken. This results in a *quail-chick chimera.* Taking advantage of this technique, Nicole Le Douarin and her colleagues in Rouen and Paris have made transplants at different ages and followed the migration of the quail cells in the chick from the neural crest to their final locations (Le Douarin, 1982). This method can also be applied to analysis of development of cells in the neural tube. Even more astonishingly, whole regions of quail brains can be transplanted into chick brains to produce adults which vocalize with quail-like crowing sounds (Le Douarin, 1993; see Chap. 24).

Another approach to analyzing differentiation of neural crest cells is to use in vitro systems. A recent study (Stemple and Anderson, 1992) used binding to a fluorescently labeled antibody directed at the nerve growth factor (NGF) receptor, and a fluorescently activated cell sorter, to isolate neural crest cells. The study showed that when these cells were subjected to different types of culture conditions they could be made to differentiate into neurons (by growing on fibronectin plus polylysine), Schwann cells (fibronectin alone), or other cell types on other substrates (see Fig. 9.7B). Thus, neural crest cells appear to be multipotent. Furthermore, it was possible to maintain this multipotency after serially subcloning the neural crest precursor cells through 6–10 generations, which should permit further analysis of the ability of these cells to continue to divide in vivo as they migrate. How does a precursor cell decide to stop dividing repeatedly into daughter precursor cells and divide for the last time into its differentiated phenotype? Previous work, as well as these experiments, suggest that factors within the mesoderm, through which the cells migrate, play a large role in this determination. This method may hold promise for analyzing these mechanisms (see Bronner-Fraser, 1993).

Growth Cones

During migration neurons begin to elaborate the axons and dendrites that will eventually form neural pathways and networks. Axons in particular have to make long journeys to form their synaptic connections. How does the axon make this jour-

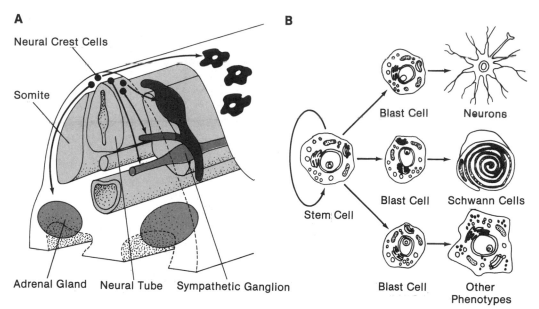

Fig. 9.7 **A.** Schematic diagram of some of the migration pathways and fates of neural crest cells. This diagram assumes that the neural crest cells are multipotent, differentiating into their final phenotypes during migration. **B.** Diagrams summarizing an in vitro system of neural crest stem cells which differentiated into different phenotypes under different culture conditions, and which retain the ability of self-replicate. See text. (Adapted from Bronner-Fraser, 1993; B from Stemple and Anderson, 1992)

ney and find the correct "address" of its target? There was keen interest in this question among early investigators. Cajal, with his genius for preparing the right material and drawing the right conclusions, studied single growing fibers in Golgi-impregnated specimens. He saw enlargements at the ends of the fibers, and called them "growth cones." He imagined that the growth cone is endowed with ameboid movements, which enable it to push aside obstacles in its way until it reaches its destination.

The evidence that this in fact is true soon came from the experiments of Ross Harrison at Yale in 1907. Harrison introduced the technique of tissue culture to biology. In pieces of neural tissue excised from the developing nervous system and maintained in artificial media in a dish, he observed the growth cones and the movements that Cajal had hypothesized. The results, illustrated in Fig. 9.8A, have been repeated and confirmed many times since.

Detailed studies have begun to reveal the special properties of the growth cone. The molecular organization of the growth cone, as revealed by recent findings, is summarized in Fig. 9.8B. Clockwise, starting at left, we see the nerve process (axon or dendrite: ①) that extends from the cell body. Within are microtubules (MT) and microtubule-associated proteins (MAP-2), which provide the "rails" for transporting vesicles to the cone; the MT are assembled from tubulin monomers (T). The process also contains neurofilaments (NF). Associated with the cytoskeleton ② are actin and myosin, which provide contractility for moving the growth cone. Neurotransmitters such as glutamate (GLU) may be released by nearby or target cells to act as morphogens ③; they also can affect growth cone membrane excitability, which is mainly due to voltage-dependent Ca^{2+} channels ④ (see below). Recepters ⑤ for other neuromodulators (NM) are coupled

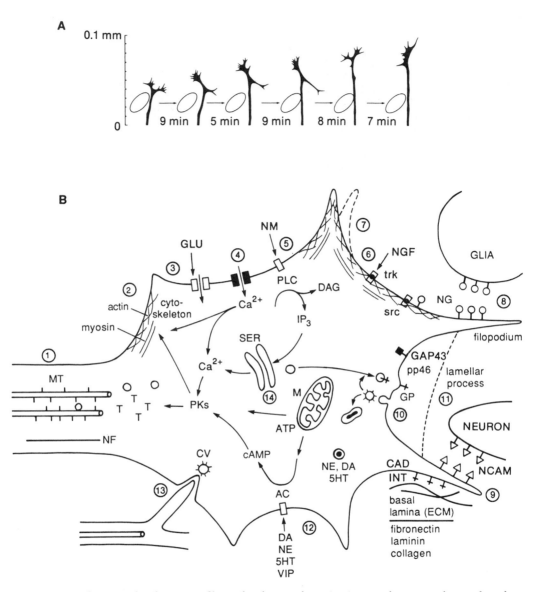

Fig. 9.8 A. The growth of a nerve fiber of a frog embryo in tissue culture, as observed under a microscope at intervals of time indicated. Ellipse indicates red blood cell, which did not move. **B.** Molecular organization of the neuronal growth cone, as summarized from recent research. Numbers ① to ⑭ indicate different functional domains and molecular mechanisms. See text. Clockwise from ①: MT, microtubules; MAP-2, microtubule-associated proteins; NF, neurofilaments; NM, neuromodulators; NGF, nerve growth factor; components of the phospholipid second messenger system (PLC, phospholipase C; IP₃ inositol trisphosphate; DAG, diacylglycerol;)); NG, neuroglian; GAP43, growth associated protein 43; GP, glycoprotein; NCAM, neural cell adhesion molecule; CAD, cadherin; INT, integrin; ECM, extracellular matrix; AC, adenylate cyclase; VIP, vasoactive intestinal peptide; CV, coated vesicle; M, mitochondria; SER, smooth endoplasmic reticulum; PK, protein kinases. Abbreviations for neurotransmitters (GLU, NE, DA, 5HT) as in Chap. 8. (A from Harrison, 1907; B based on Pfenninger, 1986; Lockerbie, 1987; Rakic et al., 1994; K.H. Pfenninger, B. Grafstein personal communication; and many others)

to phospholipase C (PLC). Their activation thus results in changes in concentration of the second messengers inositol trisphosphate (IP_3) and diacylglycerol (DAG). These in turn exert their actions through changes in Ca^{2+} and protein kinase C (PKC), respectively. Tyrosine kinases ⑥ act as receptors, such as *trk* for NGF; others, such as *src*, associate with cell adhesion molecules. Changes in contractility of the cytoskeleton, bringing about movement of the filopodia ⑦, may be produced by these second messengers.

The growth cone membrane interacts with its environment through multiple types of cell-surface molecules. One type is neuroglian (NG) ⑧; another is N-CAM ⑨. Other glycoproteins such as the cadherins (CAD) and the integrins (INT) interact with constituents of the extracellular matrix (ECM), such as laminin, fibronectin, and collagen. These interactions control the movement of the growth cone through its immediate environment, and the recognition of its target. The various types of membrane glycoproteins (GP) are delivered to the plasmalemma by exocytosis, recycled by endocytosis, and degraded in lysosomes.

Forward movement of the growth cone occurs by extension of filopodia on or through the substrate, followed by formation of lamellar processes which fill in the spaces between the filopodia ⑪. Various kinds of neurotransmitters and neuromodulators act on receptors that use cAMP as a second messenger ⑫. Interactions with other processes appear to include invaginations ⑬ and budding off of coated vesicles (CV).

Within the growth cone ⑭, organelles perform the basic cell functions described in Chap. 3: mitochondria supply ATP, and the smooth endoplasmic reticulum (SER) sequesters Ca^{2+} and packages vesicles. Various phosphokinases (PK) activate target substrates. Protein pp46 was isolated by Karl Pfenninger (1986) at Colorado; this is a major growth-associated protein (also known as GAP43) that is phosphorylated by Ca^{2+}/calmodulin-dependent ki-

nase and PKC. It appears to be identical to phosphoprotein F1, which has been implicated in synaptic plasticity and memory mechanisms (see Chap. 29).

Many studies have suggested that the growth cone membrane generates Ca^{2+} action potentials ④. Several demonstrations of this have been provided by treating cells in culture with calcium- or voltage-sensitive dyes. Optical recordings and electrical recordings of presumed Ca^{2+} action potentials are very similar. It is believed that the Ca^{2+} that flows into the cell during the action potential controls the contractility of the actin filaments and the shape and movement of the growth cone and other parts of the neuron.

Cell Growth and Growth Factors

It is clear from the preceding sections that interactions between neurons and their target cells are critical for many aspects of development. We next inquire how these interactions take place.

Nerve Growth Factor

The first evidence for a specific chemical substance that could stimulate or control neuronal development came from experiments conducted around 1950, which showed that implantation of a type of soft-tissue tumor (sarcoma) from a mouse on one side of the vertebral column of a chick embryo caused enlargement of the spinal ganglion on that side. It was soon established, by Rita Levi-Montalcini, working with Viktor Hamburger in St. Louis, that a humoral agent was involved, and that it specifically stimulated the growth of cells derived from the neural crest.

This substance was named nerve growth factor (NGF). NGF is one of a family of neurotrophins, so called because of their trophic influence on neurons. It has been found that NGF's influences come after the mitotic phase of cell proliferation and potentially can be exerted on phases of growth, differentiation, elaboration of processes, maturation, or survival. For example, sympathetic ganglion cells grown in tissue culture show almost no growth in a medium lacking NGF, and ultimately die. If, instead, NGF is added, there is vigorous growth of cell bodies and neurites beginning within a few hours.

These kinds of experiments showed that NGF is produced by target cells, such as gland, muscle, or skin cells. Evidence regarding the specific effect of NGF on a developing neuron can be obtained by analyzing the timing of NGF production in a target population and expression of NGF receptors on ingrowing nerves. Hans Thoenen and his colleagues (Davies et al., 1987) studied this problem, using as a model the developing sensory innervation of the skin surrounding the whiskers in the snout of a fetal rat. First they used in situ hybridization for NGF messenger RNA and found that the onset of transcription in the target skin cells does not precede, but is precisely coincident with, the arrival of the sensory fibers of the trigeminal nerve. Then they assessed the time of expression of NGF receptors by observing binding of radioactively labeled NGF to the sensory neurons; they found that binding does not occur until after the fibers reach the whisker skin. These results thus suggest that other factors must be primarily responsible for controlling outgrowth and guidance of fibers to their targets, and that, in this case, NGF is mainly important in the survival of axons during the period in which they compete for target innervation.

NGF was purified and synthesized in the early 1970s and has subsequently been cloned and sequenced. The native peptide consists of three pairs of subunits, of which the active one is a 118 residue β subunit. In recent years it has become evident that NGF is not an only child but a member of a very large extended family of *cytokine's*. Table 9.2 provides a snapshot of the current known members of the family. NGF's siblings are two other neurotrophins: brain-derived neurotrophic factor (BDNF) and a subfamily of neurotrophins referred to as *NT-3, 4, 5*. Its first cousins

include several neuropoietic factors, such as ciliary neurotrophic factor (CNTF), which is necessary for survival of ciliary ganglion cells. A diverse group of additional cousins includes epidermal growth factor (EGF), one of the best studied of growth factors, and insulinlike growth factor (IGF), which relates NGF to some of the diverse actions of insulin. This latter group is the subject of intense interest by many investigators for insight into mechanisms controlling growth and differentiation of many cell types. Many *neuropeptides* also exert diverse effects in growth. Finally, there is increasing evidence that classical *neurotransmitters* can act as "morphogens", influencing many of the steps involved in neuronal development (see above and Chap. 30).

Growth Factor Receptors and Oncogenes

Our understanding of the receptors for growth factors has its origins in the study of genes that control cell growth, which in turn originated in studies of viruses, particularly retroviruses. These consist of RNA and reverse transcriptase; when the virus infects a cell, the reverse transcriptase makes first a single complementary DNA followed by a second DNA strand to form a double helix, which is then integrated into a host chromosome (recall that reverse transcriptase is an important tool in recombinant DNA techniques: Chap. 2). The virally derived DNA functions as an oncogene (v-onc); its gene products stimulate unrestrained cell divisions, transforming the normal cell into a cancer cell. The action of oncogenes is therefore mainly on *cell proliferation*. The use of molecular probes against v-onc has permitted demonstration of the existence of similar DNA sequences in normal cells. These are therefore called proto-oncogenes (or c-onc for cell oncogene), and have control sequences and DNA associated proteins that provide for the orderly and self-terminating gene expression that characterizes normal cell proliferation and growth. We also know that the activity of some of these proto-

Table 9.2 Different types of growth factors

Classification	Abbreviation	Definition
Cytokines		
Neurotrophins	NGF	Nerve growth factor
	BDNF	Brain-derived neurotrophic factor
	NT-3, 4, 5	Neurotrophins 3, 4, and 5
Neuropoietic factors	CDF/LIF	Cholinergic differentiation factor/leukemia inhibitory factor
	CNTF	Ciliary neurotrophic factor
	ONC	Oncostatin M
	GPA	Growth-promoting activity
	MANS	Membrane-associated neurotransmitter-stimulating factor
	SGF	Sweat gland factor
Hematopoietic factors	GCSF	Granulocyte colony-stimulating factor
	IL-1, 2, 6, 11	Interleukins
Growth factors	EGF	Epidermal growth factor
	FGF	Fibroblast growth factors, acidic (a), basic (b), and others
	TGF	Transforming growth factors, α and β
	TNF	Tumor necrosis factors, α and β
	IGF	Insulinlike growth factor
Neuropeptides	ACTH	Adrenocorticotropic hormone
	CGRP	Calcitonin gene-related peptide
	CAP2	Cardioacceleratory peptide 2
	CCK	Cholecystokinin
	CRF	Corticotropin-releasing factor
	ENK	Enkephalin
	GAL	Galanin
	LHRH	Luteinizing hormone releasing hormone
	NPY	Neuropeptide Y
	NT	Neurotensin
	PHI	Peptide histidine isoleucine
	SOM	Somatostatin
	SP	Substance P
	VP	Vasopressin
	VIP	Vasoactive intestinal polypeptide
Neurotransmitters	ACh	Acetylcholine
	5HT	Serotonin
	CA	Catecholamine

Adapted from Patterson and Nawa (1993)

oncogenes may be related to differentiation and maturation in postmitotic cells.

What is the molecular basis for the mechanism of proto-oncogenes? Of the 100,000 or so functional genes in the human genome, perhaps 100 are involved in the control of growth. This small number suggests that many proto-oncogenes may function like *regulator genes,* exerting their control by throwing developmental switches or activating families of genes that express different gene products which, acting together, are necessary for coordinated growth.

One of the important developments in recent years is the discovery of relationships that exist between the products of *proto-oncogenes, homeotic genes,* and various types of *growth factors* (see Bender, 1985). A key link was forged by experiments to obtain the DNA sequence for a homeotic gene in *Drosophila* (called

Notch) and the nematode *Caenorhabditis* (called lin-12). The surprising result was that the predicted amino acid sequences of the two gene products from these unrelated animals (showing quite different phenotypes for these gene loci) were similar. Even more surprising, each contained a number of repeated sequences that showed homology with epidermal growth factor (EGF).

EGF is well-known as a peptide of 53 amino acids. It is derived, in the manner of most peptides (see Chap. 8), from a longer precursor of some 1200 amino acids, which resembles the precursors for Notch and lin-12 peptides in having multiple EGF-like repeats. EGF binds to an external receptor site on an integral membrane protein that acts as a tyrosine kinase. As we noted in discussing second messengers in Chap. 8, phosphorylation of tyrosine residues appears to be specifically involved in cell growth. Binding of EGF leads to phosphorylation of tyrosine residues in the receptor itself, as well as in other proteins; this acts as a signal to bring about proliferation of several cell types, including fibroblasts and glia. Virally derived oncogenes bring about proliferation of transformed cells by phosphorylating tyrosine residues in these same proteins.

The receptors for many components of receptor-activated systems have been cloned and sequenced. A general scheme for relating these various gene products to the sequence of activation from receptors to DNA synthesis is shown in Fig. 9.9A. One of the best models for a specific system is provided by the recent work of a number of laboratories showing the relation between tyrosine kinase phosphorylation and ras activation. As shown in Fig. 9.9B, receptor activation by a growth factor peptide produces autophosphorylation of the receptor. The activated receptor binds to a special linking protein called Grb2, which in turn binds to a ras activator protein known by various names, such as (in *Drosophila*) Son of sevenless (Sos1). Receptor-activated Sos stimulates GDP–GTP exchange on ras, which triggers phosphorylation of serine-threonine residues (see Chap.

8) in a cascade of kinases that leads to second messenger signals to the nucleus.

Cell Differentiation

The third step in neuronal development is the differentiation of specific structures, properties, and connections. This overlaps greatly with other steps; in some cases differentiation is well underway before migration starts. The term is also used by some workers to refer to the whole process of generation of the sequence of neuroblast precursors as well as the events that bring about the final neuronal form. Let us consider several examples that illustrate principles underlying neuronal differentiation.

Cell Fate and Pioneer Fibers

The central nervous system of the grasshopper has been favorable for studies of neural development. The embryonic nerve cord is relatively thin and translucent, which has allowed individual cells and even their growth cones to be visualized under the microscope in the living animal, and intracellular electrodes have been introduced for recording and dye injection. The entire sequence of steps, from precursor cell to mature form, has been analyzed for several types of neuron. As illustrated in Fig. 9.10A, B, a ganglion, such as the third thoracic ganglion, contains a specific set of precursor cells. There is a lateral group of 30 neuroblasts (NB), and a midline group of medial precursors (MP) plus a single medial neuroblast (MNB). To the right in the figure, the pattern of cell divisions is indicated. Each MP divides only once. Each NB divides several times to give rise to a chain of ganglion mother cells (GMC); each of these divides into a ganglion cell (GC), which then differentiates to reach its final form as a mature neuron (N). Each NB generates from 10 to 100 progeny according to this fixed pattern, after which it degenerates and dies.

By making intracellular injections of fluorescent dyes at different stages, it has been possible to study the development of axons and dendrites of individual identified

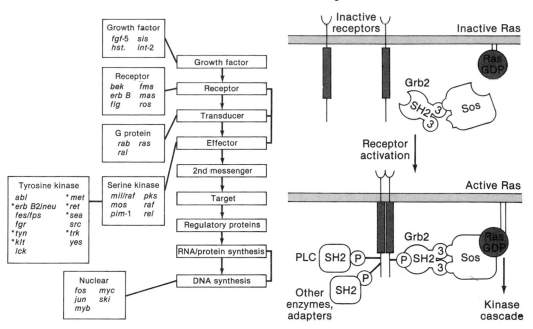

A Oncogenes

B Linking mechanism

Fig. 9.9   A. The canonical sequence from growth factor receptor to DNA synthesis, with different oncogene products and their positions in the sequence. B. Proposed mechanism whereby a linking protein (Grb2) is bound, on the one hand, to a phosphorylated receptor and, on the other, to another linking protein Sos, to cause activation of ras GTP binding protein, which initiates a cascade leading to DNA synthesis. (From McCormick, 1993)

neurons. The earliest axonal pathways, connecting neighboring ganglia, are made by "pioneer" cells. One of these is the MP2 cell, as illustrated in Fig. 9.10C. Similarly, sensory neurons in the periphery (PN1) send pioneer fibers to the ganglion. Within the ganglion, later differentiating cells send their axons along the paths laid down by the pioneer axon. As an example, the MP3 cell, which initially lacks any processes, first differentiates an axon at day 6 (see Fig. 9.10D). From day 6 to day 12, there follows an elaboration of the dendritic branches, to reach the final form characteristic of the mature neuron, called an H neuron.

Excitable Properties

Correlated with the differentiation of neuronal form is the acquisition of specific physiological properties. This has been studied in a variety of systems, including muscle cells, neuroblastoma cells in tissue culture, large identified neurons such as Rohon-Beard cells in the frog, and several types of cell in the grasshopper.

A common finding in many studies has been that cells are excitable at early stages of development, and that the inward (depolarizing) current is carried by Ca^{2+}. This has already been pointed out in Chap. 5. At these early stages, the cells are characteristically coupled by gap junctions. The Ca^{2+} spike may be localized in the cell body, or in the growth cone, as discussed above. In some cells the Ca^{2+} spike persists into maturity (for example, in muscle cells). In many systems there is a change to a spike with both a Na^+ and a Ca^{2+} component, and finally one with only a Na^+ component. This particularly applies to projection neurons with long axons. By contrast, many nonspiking cells have inexcitable membranes at all stages of develop-

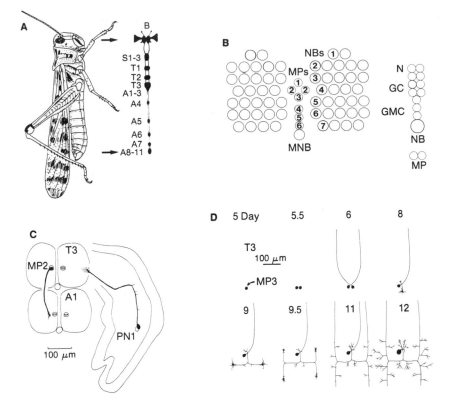

Fig. 9.10 Differentiation of neurons in the grasshopper central nervous system. **A.** Lateral view of grasshopper. Central nervous system, showing brain (B) and associated anterior segmental (S), thoracic (T), and abdominal (A) ganglia. **B.** Identified precursor cells in T3 ganglion, including lateral neuroblasts (NB), medial neuroblast (MNB), and midline precursors (MP). *To the right:* Sequence of cell divisions: MP cells give rise to neurons by a single division; NB cells give rise to sequence of cells, from ganglion mother cells (GMC) to ganglion cells (GC), which differentiate into neurons (N). **C.** "Pioneer" axons laid down by the MP2 cell and a peripheral pioneer neuron (PN1). **D.** Differentiation of MP3 cell into an H neuron from 5 to 12 days of age; drawn from cells injected with Lucifer Yellow. (A–C from Goodman and Bate, 1981; D from Goodman et al., 1981)

ment: alternatively, some pass through earlier stages of excitability.

It should be emphasized that any of these properties may be localized to a particular site in a neuron, a particular stage of development, or in relation to a time of neuron origin. Thus, there is a population of dorsal unpaired median (DUM) neurons in grasshoppers, which contains subpopulations of cells depending on the sequence of their origin, starting with the median neuroblast. The earliest cells to be born have both soma and axon spikes; later come cells with spikes in the dendrites and axon, then cells with only axon spikes, and finally nonspiking local-circuit neuron. This finding is consistent with the pattern, in many vertebrates, that small interneurons differentiate last, and are thus most susceptible to shaping by environmental and other factors (see below).

The presence of Ca^{2+} spikes at early stages of development may be important for the control of the motility of growth cones, as already noted above. In addition, Ca^{2+} must be important in relation to the neurosecretory activity of the developing neuron, and it could also play a role in controlling the insertion of neurotransmitter receptor proteins into the surface membrane. These and other possible functions of Ca^{2+} in the developing neuron may be reviewed in relation to the discussion of the role of Ca^{2+} as a second messenger in Chap. 8. The importance of impulse activity in the establishment of topographic maps will be discussed below.

Transmitter Determination

When and how does a neuron become committed to the neurotransmitter it will use?

Experiments on sympathetic ganglion cells have thrown some light on this question. The background for these studies was the demonstration that when "trunk" regions of the neural crest, which normally give rise to sympathetic ganglion cells, were transplanted to anterior regions that normally give rise to vagal (parasympathetic) cells, the normal adrenergic character of the trunk cells was lost. These divisions of the autonomic system are considered in Chap. 18. When sympathetic ganglion neurons are maintained in culture in the absence of any other cell type, they develop adrenergic properties; that is, they take up, store, synthesize, and release norepinephrine from their terminals. Furthermore, they form morphological synapses that contain small granular vesicles, like normal adult adrenergic synapses.

In contrast, when the neurons are maintained together (co-cultured) with other kinds of cells such as cardiac cells, they develop cholinergic properties. They synthesize up to 1000 times as much acetylcholine as the tiny amounts found in isolated cultured neurons; they make cholinergic synapses onto other neurons or onto muscle cells; and the amount of NE and the number of granular vesicles are greatly reduced. This can occur solely in the presence of medium that contained the cardiac cells or other nonneuronal cells. The medium is thus called *conditioned medium,* and is believed to contain a substance or substances that mediate the effect. By varying the amount of exposure to other cells or conditioned medium, a balance between adrenergic and cholinergic properties in the same cell can be achieved.

These results show that neural crest cells are potentially "dual function" neurons. Very early in development, these cells may actually synthesize and release both types of transmitter simultaneously. They retain this ability to be adrenergic or cholinergic past the last mitosis, and the expression of one trait or the other is determined by chemical factors in the environment of the cell. These experiments therefore provide evidence at the single-cell level for the ways in which the cell genotype interacts with the environment to produce the cell phenotype.

Development of the Synapse

Making the correct synaptic connections requires a precise coordination between the synaptic nerve process and its postsynaptic target. The factors involved in this coordination have been best studied in the neuromuscular junction, and we will therefore consider this first as a model for the individual synapse. We will then take up the mechanisms for formation of synaptic circuits in the central nervous system.

Studies of the development of neuromuscular junction and of its reaction to nerve transection have made it possible to identify some of the molecular mechanisms that guide the nerve to its target and control the differentiation of the junctional complex. There are four main types of interaction that take place between nerve and muscle, and these are summarized in Fig. 9.11.

Presynaptic Factors

The most obvious factor is the presynaptic nerve. As indicated in Fig. 9.11 A and B, this may have several effects on the postsynaptic target. It may contribute to the differentiation of the myotube and development into a mature muscle cell, either by release of transmitter or through electrical junctions; it may also help to shape other membrane properties of the muscle.

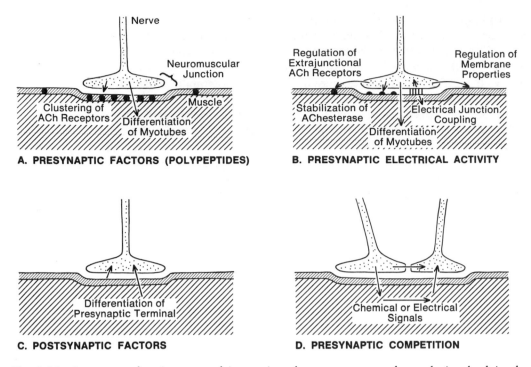

Fig. 9.11 Summary of main types of interactions between nerve and muscle involved in the development of the vertebrate neuromuscular junction. (Based on Lømo and Rosenthal, 1972; Thompson et al., 1979; Frank and Fischbach, 1979; Landmesser, 1980; Hopkins and Brown, 1984; Hall and Sanes, 1993; and many others)

At the junction, the release of transmitter (acetylcholine) by presynaptic activity affects the distribution of extrajunctional ACh receptor molecules and encourages stable deposition of the degradating enzyme acetylcholinesterase.

The dense clustering of ACh receptor molecules that occurs in the postsynaptic membrane depends on the presence of the presynaptic terminal. Although it is attractive to think that the clustering might be induced by presynaptic motoneuron activity, in fact it has been shown to be independent of this. It appears instead that clustering is induced by release of several types of molecular signals (summarized in Hall and Sanes, 1993). One of these is *agrin,* which is synthesized in the motoneuron, transported in the axon, and secreted into the basal lamina. Molecular cloning has shown that it shares homologous regions with EGF and laminin. It is believed to act,

together with FGF, on tyrosine kinase receptors which stimulate tyrosine phosphorylation; this sends second messenger signals to the local nuclei in the myotube to synthesize ACh receptors and insert them into the endplate region. Other factors (see Table 9.2 above) contribute to activation of the genes expressing ACh receptors. Finally, impulse activity leading to synaptic transmission at the endplate (B in Fig. 9.11) contributes to decreasing the density of extrasynaptic ACh receptor, leaving only the clusters at the endplate regions.

Postsynaptic Factors

If the presynaptic axon is crucial for the development of the postsynaptic muscle, the same can be said of the importance of the muscle for the nerve (C in Fig. 9.11). When the motor axon terminal first arrives at the muscle it has a very primitive form, and most of the morphological differentiation of the terminal occurs after a synaptic contact has been established, under the influence of the postsynaptic muscle. This appears to be mediated by several factors, which include specific molecules from three possible areas: molecules (1) contained within the extracellular matrix around the terminal, (2) projecting from postsynaptic membrane, or (3) released from the muscle cytoplasm.

Competition Between Axon Terminals

A fourth factor is that the axons interact among themselves. A given muscle first receives axons from several motoneurons (polyneuronal innervation); in the course of development, all the synapses except those from one motoneuron are eliminated. This appears to reflect competition between the axons. The signals mediating this competition may be either chemical or electrical. They may pass directly between the axons, or indirectly by first affecting the muscle, as shown in Fig. 9.11D.

It thus seems clear that the development of specific synaptic sites, as exemplified by the neuromuscular junction, depends not on one factor, but on multiple factors. The presynaptic process, extracellular matrix, and postsynaptic process each contribute in a carefully programmed sequence of mutual interactions. Both chemical and electrical factors are involved. Axons from several presynaptic cells compete for input to one postsynaptic cell. In these basic mechanisms, the neuromuscular junction exemplifies many of the properties of development of synapses, and many of the plastic properties involved in regeneration and learning (see Chap. 29).

Establishment of Synaptic Connections

The essence of nervous organization is the establishment of synaptic circuits. Normal behavior requires precision in assembling these circuits, and it is not surprising, therefore, that many of the disorders of behavior that occur in humans and other animals are due to abnormalities in the development of the circuit connections (see Chap. 30). In development, as in politics, timing is all. From the experimental evidence it is possible to begin to identify the intricate sequence of steps that is involved in establishing specific connections. Table 9.3 summarizes the main steps. As can be seen, the entire process spans the time from the birth of a neuron, through migration and differentiation, to the final maturation of each part of the cell.

Retinotectal Pathway

The ways in which neuronal populations interact in forming circuits have been studied in many systems, under many experimental conditions. Present thinking has been much influenced by the pioneering experiments of Roger Sperry, of the California Institute of Technology, who received the Nobel Prize in 1981. This work was carried out in amphibians, in the retinotectal pathway, which is composed of

Table 9.3 Steps in establishing specific synaptic connections

1. The neuron must leave the cell cycle at a set time.
2. It must migrate to the appropriate region.
3. It may develop a spatial identity with respect to its neighbors.
4. Dendrites must develop in a characteristic shape and orientation.
5. The axon must leave the cell body and grow in the right direction toward its region of termination.
6. The axon must direct its branches to the appropriate side of the brain.
7. The axon must direct branches to the right region or regions.
8. Within a region, the axon must ramify in the right subdivision or layer.
9. The terminal field of the axon must be ordered in a particular topographic relationship with the cell bodies in the regions of its origin and termination.
10. The axon terminals may end only on certain cell types within the terminal distribution area.
11. The axon terminals may end only on certain parts of these cells (parts of the dendritic surface, for example).

After Lund (1978)

the axons of retinal ganglion cells that project to the optic tectum, the main relay center for visual information in the lower vertebrate brain (see Chap. 16). Sperry cut one optic nerve and rotated the eye through 180°. When regeneration of the optic nerve was complete, he found that the axons grew back to their previous target sites; the map of the retina onto the tectum was preserved, despite the rotation of the eye and the disorganization and regrowth produced by the transection (Fig. 9.12). He therefore postulated that individual axons and their individual target cells have matching biochemical "identification tags," so that they establish synaptic connections by a *chemical affinity* between them. He suggested that this affinity is responsible not only for the reestablishment of connections during regeneration, but also for the establishment of connections during normal development.

Because of the precision with which the retina is mapped onto the tectum, this system has been the object of considerable study. Subsequent work has greatly extended Sperry's original finding and has somewhat qualified the interpretation. The original hypothesis stressed rigid genetic mechanisms in laying down the patterns of synaptic connections in the nervous system. Current research is focused on signal molecules that are expressed in a position-dependent fashion in the retina, where the axons originate, and the tectum, where they terminate. Two candidates are the cell surface proteins called TOP, and a 33-kDa protein. Monoclonal antibody staining of TOP shows that this molecule is expressed in steep and complementary gradients in both retina and tectum (see C and D in Fig. 9.12), suggesting that it could provide position cues for relating the two structures.

The importance of impulse activity in the establishment of synaptic connections has been studied in the visual system by making injections of tetrodotoxin (TTX) into one eye at early stages of development in kittens. The TTX blocks Na^+ impulses in the axons (see Chap. 5), thereby disrupting the normal impulse activity from the retinal ganglion cells to their target cells in the lateral geniculate nucleus and the optic tectum (see Chap. 16). This procedure impairs the normal development of laminae in the lateral geniculae nucleus and distinct ocular dominance columns in the cortex, and retinotopic maps in the optic tectum are degraded. The general conclusion from these experiments is that coarse topographic organization first takes place, but impulse activity is required for fine-tuning of the connections (reviewed in Shatz, 1990).

Further evidence of the importance of synaptic activity during development comes from a preparation in which a third eye is grafted onto a frog embryo; the ganglion cell axons grow into the optic tectum and form alternating stripes of projection between the normal projection from the contralateral eye. Martha Constantine-Paton and colleagues at Yale (see Constantine-Paton et al., 1990) showed that treatment with TTX abolishes these stripes, in accord with the experiments in kittens cited above. They then showed that treatment with APV, which blocks glutamatergic NMDA receptors (see Chap. 7), has a similar effect. This indicates that synaptic activity and postsynaptic responses similar to those involved in long-term potentiation (LTP) appear to be a part of developmental processes as well (see Chaps. 7 and 29).

In conclusion, the retinotectal system continues to be a useful model for exploring the mechanisms underlying the formation of neural maps. The lessons learned thus far have been summarized by Scott Fraser (1992) of Irvine as follows:

> . . . the simple pattern of the retinotectal projection cannot be . . . the product of a single, simple patterning mechanism. Experiments, computer simulations, and common sense all suggest that several processes collaborate. . . . The cooperative action of multiple cues . . . makes the system robust. Unfortunately, it also makes decisive test experiments difficult to perform, as the elimination of an

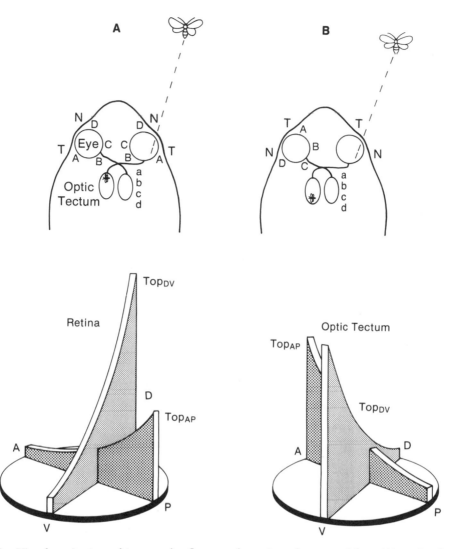

Fig. 9.12 Visual projection of image of a fly onto the retina of a normal frog (**A**) and a frog in which the eyes have been rotated through 180° (**B**). Note retinal points A–D and their projections onto the optic tectum. **C.** A current view of the retinotopic map based on the expression of two cell surface proteins TOP$_{DV}$ and TOP$_{AP}$. The relative intensity of staining of monoclonal antibodies for these two antigens is indicated by the height of the graphs. A, anterior; D, dorsal; P, posterior; V, ventral. (A, B from Lund, 1978; C adapted from Sanes, 1993)

important guidance cue might generate only a subtle defect in the projection pattern. . . . Clearly, the challenge for the future is to develop integrative experimental designs adequate for exploring the cooperative and competitive actions of multiple cues on neuronal patterning.

The formation of visual maps will be further discussed in Chaps. 16 and 30.

Genetic Mutations

The formation of synaptic circuits has also been studied under abnormal conditions induced by genetic mutations. In invertebrates, the most studied species from this point of view are the nematode roundworm *Caenorhabditis elegans* and the fruit fly *Drosophila melanogaster*. The strategy of these experiments is to produce a single mutation by a chemical means or X-irradiation, characterize the behavioral abnormality produced, and then identify the underlying abnormality in synaptic connection.

In vertebrates, mutant species have been particularly useful in studying the ways that cortical structures are formed. The earliest and most complete studies were carried out in mutant strains of mice by Richard Sidman and Pasko Rakic and their colleagues. Most of these strains have been identified by the effects of the mutation on locomotion, and the cellular mechanisms have been analyzed especially in the cerebellum. Depending on the kind of locomotor disorder, the mutant strains have acquired names of reeler, staggerer, weaver, nervous, jumpy, and so on. Each type of disorder is associated with a particular way in which the cerebellar cortex is disorganized.

The results are summarized in Fig. 9.13 for three mutant strains—weaver, reeler, and staggerer—and compared with the normal. In brief, the normal cerebellar cortex (A) consists of an orderly layering of the main neuronal types, and sets of specific connections between them (see Fig. 9.3A above, and more detail in Chap. 21). In the homozygous weaver (B), most of the granule cells degenerate prior to the time of their migration from the surface layer to the deeper position they occupy in the normal adult. Also affected is a type of radial glial cell called a *Bergmann glia,* which normally serves as a guide for the migration of the granule cell bodies during development. A recent study has shown that weaver granule cells transplanted into normal mice migrate normally, suggesting that the defect is in the cellular milieus, perhaps in the glia, as originally proposed.

In the reeler mutant (C), the main abnormality is a lack of granule cells, which remain in a layer at the surface, so that the normal relations between the layers in the adult are reversed. As a consequence, there are few parallel fibers, and the Purkinje cells lack their characteristic branching pattern. The effects on the Purkinje cell are taken to indicate that the maturation of dendritic branches and dendritic spines into their final form is under control of the granule cell axons in their normal arrangement as parallel fibers.

As a final example, in the staggerer mutant (D), the main site of gene action appears to be the Purkinje cells, which fail to develop their mature dendritic tree, and retain several embryonic characteristics. The granule cells migrate, leaving behind their axons as parallel fibers in the normal manner. However, in the absence of their normal Purkinje cell dendritic targets, the granule cells degenerate. This occurs despite the fact that postsynaptic specializations (submembranous densities) are present in the malformed Purkinje cell dendrites.

These studies thus give evidence of the multiple factors involved in the establishment of neuronal circuits. A problem in interpretation is that there is rarely an isolated site of gene action; on the contrary, each mutant exhibits a mixture of effects. The effects are not confined to the cerebellum, but involve a variety of abnormalities throughout the nervous system. Despite these complications, some of the gene actions are surprisingly specific, as already discussed. In addition, electrophysiological recordings from the Purkinje cells in weaver and reeler indicate that these cells, despite their grossly abnormal morphology, have relatively normal excitable membrane properties. The picture one obtains clearly shows the resourcefulness of the developmental process in attempting to assemble its circuits despite the abnormalities which may occur. The functions of the cerebellum will be discussed in Chap. 21. The "knock-out" method for gene deletion will be discussed in Chap. 29.

Maturation

Maturation refers to the process by which neurons and circuits achieve their final form. It is difficult to specify precisely as a distinct process because it is continuous

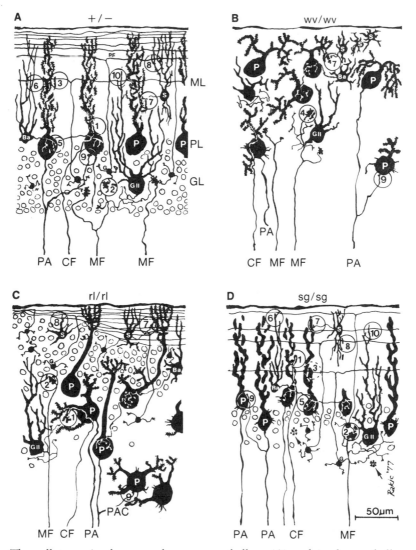

Fig. 9.13 The cell types in the normal mouse cerebellum (**A**) and in the cerebellum of several mutant strains (**B–D**). Ba, basket cell; CF, climbing fiber; G, granule cell; GII, Golgi type II cell; MF, mossy fiber; P, Purkinje cell; PA, Purkinje cell axon; PAC, Purkinje cell axon collateral; PF, parallel fiber; S, stellate cell. Layers in A; ML, molecular layer; PL, Purkinje cell body layer; GL, granule cell layer. Circled numbers indicate corresponding types of synapses: ① climbing fiber synapses on P cell dendrites; ② mossy fiber synapses on G cells; ③ parallel fiber synapses on P cells; ④ mossy fiber synapses on GII cell basal dendrites; ⑤ climbing fiber synapses on P cell bodies; ⑥ parallel fiber synapses on Ba cell dendrites; ⑦ S cell synapses on P cell dendrites; ⑧ parallel fiber synapses on S cell dendrites; ⑨ P cell axon collateral synapses on P cell bodies; ⑩ parallel fiber synapses on GII cell apical dendrites. See text for explanations. (From Caviness and Rakic, 1978)

with the preceding steps of differentiation and involves the finalization of many of the processes listed in Fig. 9.4B. Of these various steps, we will discuss three of special interest.

The first point is that the elements that make up a synapse may be assembled first in immature form before being "fine-tuned" to their mature form by posttranslational mechanisms. As noted above, this appears to be the case for the ACh receptor at the neuromuscular junction. Maturation may also involve a process of "stabilization" (see Changeux et al., 1984). It is possible that many synapses do not reach a final form, but are subject to a continuous process of activity-dependent remodeling; this has been postulated to be the synaptic basis for learning and memory (Chap. 29).

The second point is that the neurons in a given region do not all differentiate, migrate, and achieve maturation at the same time. We have seen in the cerebellar cortex, for example, that the granule cells migrate to their final positions and take on their final form long after the Purkinje cells have been in place. Studies in other regions have indicated that small interneurons similarly reach their final positions and shapes after the projection neurons. It has been postulated that this may be a general rule in the nervous system, making small neurons and local synaptic circuits more open to influence by the experience of the animal.

An important step in maturation is the acquisition of myelin. The ability to stain for myelin led the early histologists to focus their studies of development on the onset of myelination, and use this as a criterion of maturation. However, most axons in the brain are thin and unmyelinated or only thinly myelinated, and it is now recognized that the myelination of larger axons is only one among many steps in maturation. Nonetheless, it is true that different regions vary considerably in their times of myelination. It is one of the later stages of maturation, beginning usually late in embryonic life or early in postnatal life after the projection neurons are well in place, and continuing for considerable periods of time (into childhood, in the case of humans).

Cell Death

We tend to think of death as the end point of old age, but that is not the view from the perspective of development. From this perspective, degeneration and death of spe-

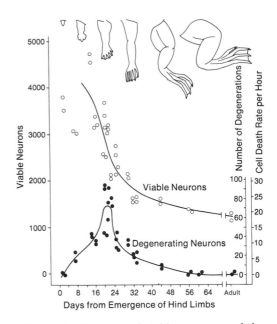

Fig. 9.14 Numbers of viable neurons and degenerating neurons in the ventral horn of the lumbar spinal cord of the frog at different times in relation to development of the hindlimbs. (Modified from Hughes, in Jacobson, 1978)

cific cells, fibers, and synaptic terminals are integral parts of the process of development. This was shown in the vertebrate nervous system in 1949 by Viktor Hamburger and Rita Levi-Montalcini, at Washington University (see Purves and Lichtman, 1985, for a review). They found that in dorsal root ganglia and in the spinal cord motor regions, large numbers of cells degenerate during brief and specific time periods early in embryonic life. It was observed that this occurs approximately at the time when the peripheral fibers establish their connections in the periphery.

Like many important discoveries, this lay dormant for a time, but after a decade or so the question was taken up and pursued in a number of other regions. The results have shown that cell death is a common phenomenon in many regions. The amount of loss is considerable, reaching as high as 75% in some cases (see Fig. 9.14). In many cases there is a coincidence of cell death with the time at which the cells of that region are innervating their targets. From this observation, it has been hypothesized that there is a competition for innervation of neuronal targets, and the cells that die are the ones that lose out in the competition. This, in turn, implies that the cells that survive receive some signal or sustaining trophic factor from the cells they innervate. There is recent evidence that NGF and related growth factors can play this role at sites in the periphery (see above). Thus, the establishment of synaptic connections appears, at least in some cases, to involve competition for targets, validation of successful connections, and elimination of unsuccessful or redundant connections.

These properties relating to cell death are seen at the level of individual terminals and synapses, too. For example, in spinal motoneurons there are synapses on the initial segment of the axon that disappear later in embryonic life. In the cerebellum, there is a time during early development when the Purkinje cell bodies bristle with small spines which receive climbing fiber synapses; later, these spines, and their synapses, completely disappear. Perhaps the early connections help to guide other synapses to their sites, or perhaps they provide for some control of excitability that is necessary at a particular stage of development. Overproduction of synapses in the developing cerebral cortex will be described in Chap. 30.

New Neurons in Old Brains?

Some tissues of the body retain the ability to form new cells from existing precursors during the life of the animal. For example, the cells in the skin are constantly turning over, and the liver can regenerate much of its substance. The nervous system is severely limited in this respect. It appears to be a general rule, in both invertebrates and vertebrates, that once the processes of development are complete there is little or no further generation of new nerve cells. This, of course, is the main reason why

injuries to the nervous system have such devastating effects.

Despite this general rule, there are exceptions. One example is the vertebrate olfactory receptor neuron. As illustrated in Fig. 9.15A, precursor cells ① in the olfactory epithelium within the nasal cavity normally undergo a slow cycle of mitotic activity that generates immature neurons ②, which gradually mature ③ as they migrate toward the surface. Maturation consists of sending out a dendrite with cilia at the tip (the site of reception of odorous molecules: see Chap. 10), and an axon which establishes synaptic contacts in the olfactory bulb (the first relay station of the olfactory pathway; see Chaps. 6 and 11). The mature receptor neuron eventually dies ④, sending an as yet unknown signal to the precursor cells ⑤ to undergo mitosis again. This constant remodeling goes on at a low rate throughout adult life. The new neurons perform the amazing double feat of generating the correct receptors for receiving the odor-carrying molecules in the dendrites, and sending their axons to the right synaptic targets in the olfactory bulb.

This process can be massively activated by cutting the olfactory nerves. As shown in Fig. 9.15B, transection causes immediate (within a few days) degeneration of all neurons (③) whose axons have been cut. This in turn sends a powerful signal to all the precursor cells ⑤ to differentiate new neurons. As a consequence, as shown in Fig. 9.15C, the olfactory bulb is reinnervated by the neurons with uncut axons ② and the axons of the newly differentiated neurons (⑥ and ⑦).

The physiological properties of the neurons change at different stages of development. The ability to generate impulses appears already in young rats before birth; Robert Gesteland (1986) in Cincinnati has obtained evidence that immature neurons are first relatively nonspecific (responding to many different odors), showing more specific responses (to fewer odors) just before term, perhaps when their axons establish synaptic contacts in the olfactory bulb.

A. NORMAL NEUROGENESIS AND MATURATION

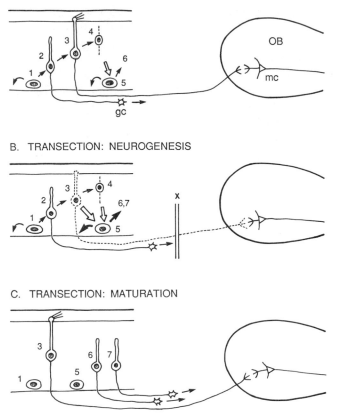

B. TRANSECTION: NEUROGENESIS

C. TRANSECTION: MATURATION

Fig. 9.15 Neurogenesis of the vertebrate olfactory receptor neuron. **A.** Normal sequence of neurogenesis: differentiation from stem cells (1), maturation (2,3), cell death (4), and stimulation of a new cycle (5,6). gc, growth cone of the receptor axon; mc, mitral cell; OB, olfactory bulb. **B.** Transection (×) of the olfactory nerves causes die-back of receptor neurons (3), sending a powerful mitotic signal to stem cells to proliferate new neurons (6,7). **C.** Maturation of unaffected (3) and new (6,7) neurons leads to "regeneration" of the olfactory nerve and reestablishment of synaptic connections to the olfactory bulb. (Based on Graziadei and Monti-Graziadei, 1979; Farbman, 1986; and others)

Immature neurons (such as ② in the diagrams) lack the ability to generate sustained repetitive discharges, as is required to transmit specific information about stimulating odors. Analysis of growth regulation and differentiation of these cells is only just beginning. An understanding of these mechanisms could tell us how these genes might be expressed in other neurons, to permit neuron replacement during aging, degenerative diseases, or after injury.

A second example of the generation of new neurons in the adult vertebrate is in the brains of birds. We will consider this in Chap. 23.

Regeneration and Plasticity

The inability to generate new neurons might imply that the adult nervous system is a static, "hard-wired" machine. This is far from the truth. Although new neurons cannot be generated, each neuron retains the ability to form new processes and new synaptic connections. Thus, although the nerve cell body is a relatively fixed compo-

nent within each center of the adult nervous system, the synaptic circuits it forms
with the processes of other neurons are
subject to ongoing modification. We learn
new skills and new facts, remember them,
and use them in different ways, which in
itself argues strongly that our neural circuits are modifiable. Research at the cellular level has provided considerable evidence
for changes of cellular properties that are
dependent on use. These mechanisms will
be discussed in several chapters, particularly in regard to cellular and molecular
properties underlying memory of learning
in Chap. 29, and functions of the cerebral
cortex in Chap. 30.

What of the ability of nerve cells to respond to injury? Although the nervous system cannot generate new neurons to replace lost ones, each cell can proliferate
new processes to replace those that have
been lost or damaged. The experiments
cited above on regeneration of the optic
nerve demonstrate this capacity, as well as
the ability to reestablish specific synaptic
connections.

Nerve cells are primed to make new connections when other cells are lost. A key
experiment was the study of William
Chambers and John Liu, of the University
of Pennsylvania, in 1958. They transected
the pyramidal tract in the spinal cord, and
several years later examined the fields of
termination of dorsal root fibers within
both sides of the cord. They found that the
field of termination was larger on the side
of the tract lesion, suggesting that the dorsal root fibers give off collateral sprouts
that take over the vacated synaptic sites.

An elegant demonstration of this process
at the synaptic level was provided by Geoffrey Raisman at Oxford in 1969. He studied the septal nucleus, a region in the forebrain that receives two well-defined inputs:
one from the hippocampus, the other from
the median forebrain bundle (MFB). Each
makes characteristic synaptic connections
onto septal neurons, as shown in Fig. 9.16.
When the hippocampal input is cut, there
is an increased incidence of large terminals
making double synaptic contacts, presum-

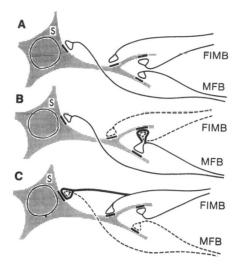

Fig. 9.16 Remodeling of synaptic connections
in the septal nuclei of the rat. **A.** Normal inputs
to septal (S) cell. Fimbria (FIMB) fibers terminate on dendrites, whereas medial forebrain
bundle fibers (MFB) terminate on both dendrites
and soma. **B.** Lesion of FIMB (degenerating
fibers shown by dashed lines); MFB terminals
occupy vacated synaptic sites on dendrites. **C.**
Lesion of MFB; FIMB terminals occupy vacated
sites on cell body. (From Raisman, 1969)

ably from MFB fibers. On the other hand,
when the MFB is cut, the synaptic sites on
the soma are taken over by hippocampal
terminals.

These results reflect the same kind of
competition for synaptic sites that occurs
during normal development. Loss of a particular type of synaptic terminal stimulates
the production of axonal *sprouts*, which
have the *mobility* to move to the vacated
sites, and the necessary *chemoaffinity* to
establish new synaptic connections. In
many respects the process involves reacti-
vation of mechanisms that operated during development.

Similar experiments have now been carried out in a number of regions, demonstrating that when one input is transected, another input expands its terminal field into
the vacated sites on cell bodies or dendrites.
These include such regions as the hippocampus (Chap. 29), red nucleus, olfactory
cortex, and superior colliculus in the verte-

brate. In addition to specific pathways, nonspecific pathways also exhibit these properties. Thus, when the cerebellum is ablated, the amount of histofluorescence for norepinephrine increases in the forebrain. Cells in the locus ceruleus send noradrenergic fibers to both cerebellum and forebrain (see Chap. 24), so it appears that loss of the cerebellar branches induces sprouting of the forebrain branches. This kind of *compensatory sprouting* has been seen in a number of cells, and it suggests that a cell is programmed to make a certain number of synapses and reacts to injury in a way to compensate for the loss and try to restore its appropriate number of connections, even though those connections may be in uncharacteristic or inappropriate places.

Within the cerebral cortex, a vivid demonstration of regeneration has been provided by the experiments of Patricia Goldman-Rakic and her colleagues. They made injections of tritiated amino acids into the frontal association cortex of monkeys on one side, and found transport through fibers of the corpus callosum to the same region of the frontal cortex of the contralateral side (Fig. 9.17A). The experiment was then repeated in animals in which the contralateral site was ablated. The callosal fibers whose terminals were removed by these ablations regenerated; when they reached the contralateral side and found their normal target absent, they swerved and terminated in a neighboring region of cortex (Fig. 9.17B). Although it is not known whether these connections are functional or not, the results attest to the tremendous "force" operating within each neuron to grow out new axons and make new connections even when the normal targets for the axon are not available. A further point of interest in these experiments is the arrangement of the fiber terminations in columns, in both the normal and experimental situations. The column is one of the organizing principles of cortical synaptic circuits, as we shall discuss in later chapters.

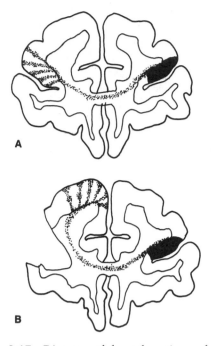

Fig. 9.17 Diagram of frontal sections of the prefrontal lobes of the rhesus monkey. **A.** Injections of tritiated amino acids in the right hemisphere, showing pattern of projection through the corpus callosum to the left hemisphere. **B.** In this animal, the contralateral region of the prefrontal cortex was ablated two months previously. The regenerating fibers revealed by the injected amino acids innervate a new region of cortex. Note columnar patterns of innervation in both cases. (From Goldman-Rakic, 1981)

These experiments thus give evidence of the ongoing competition to make synaptic connections that each neuron engages in during development and throughout adult life. This competition is the basis of much of the plasticity that is inherent in neural circuits.

Brain Transplants

Neurobiologists are only beginning to understand the potentials of this plasticity. For example, recent studies show that a region of the mammalian brain can be excised and grafted onto the brain of another animal. Two kinds of grafts have been

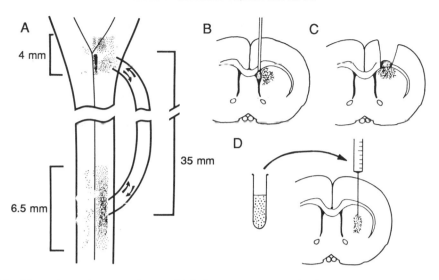

Fig. 9.18   Transplantation procedures which demonstrate growth of grafted nerve cells. **A.** Diagram of the brainstem and spinal cord of a rat, showing a graft of a bridging section of peripheral nerve. The dots indicate cell bodies of neurons labeled retrogradely by HRP applied to the nerve bridge. The labeling shows that regenerating axons can move in both directions through the grafts. **B–D.** Procedures for transplantation of catecholamine-containing neurons into the striatum of the rat brain. **B.** Frontal section of the rat brain, showing placement of pipette tip in the striatum. A piece of substantia nigra, containing dopaminergic neurons which normally innervate the striatum, is deposited in the lateral ventricle and grows as indicated. **C.** A piece of substantia nigra is placed in an artificial cavity created on the surface of the striatum, from which it grows into the striatum. **D.** A suspension of fetal substantia nigra cells, or adrenal medulla containing chromaffin cells, is made (test tube insert) and injected into the striatum. (A from Aguayo, 1985; B from Olson, 1985)

tried. In one, a length of peripheral nerve is placed as a bridge across a site of transection (see Fig. 9.18A). Neurons whose axons have been cut by the transection regenerate new axons, which normally would be unable to penetrate the scar tissue at the site of transection; the graft gives them an alternative pathway through which they may grow. It has yet to be shown that such axons can establish functionally meaningful synaptic contacts on the other side of the transection. These experiments give hope of reestablishing some degree of neural function in patients with severe spinal cord trauma.

A second type of graft consists of introducing new neural tissue to replace neurons and transmitters lost due to disease. One aim of this kind of research is to replace the neurons whose degeneration causes Parkinson's disease. These neurons use do-

pamine as a transmitter at their synapses in a part of the brain called the striatum (see Chap. 21). Dopamine-synthesizing neurons (such as chromaffin cells: see Chap. 18) are introduced into the striatum (see Fig. 9.18B–D) with the hope that they will compensate for the degenerated neurons. Trials using this procedure are presently being pursued in a few selected patients. The interpretation of these experiments is complex; for example, dopamine released by the transplanted neurons may act not only as a transmitter but also to stimulate growth (see Table 9.2 above).

These kinds of experiments thus carry the hope that we can not only reveal more of the plastic capabilities of nerve cells that are normally at work, but also eventually use them to compensate for the effects of injuries and disease in the human brain.

II
SENSORY SYSTEMS

10

Introduction: From Receptors to Perceptions

From early childhood we are aware of the sensations and perceptions that occur as a part of the business of living. Those that are particularly painful or pleasurable become powerful factors in molding the way we develop, the kinds of personalities we acquire, and the goals we work toward. Since sensory experience is so immediate, it is much more readily understandable than many other aspects of nervous function. On the other hand, our senses not only inform us but may often fool us; they constantly test our abilities to make judgments about things.

The fact that sensory perceptions are so accessible to our introspection means that when humans first acquired the ability to think and speculate about their own nature, they were very much aware of the importance of sensory experience. The early Greek philosophers of the sixth century B.C. were able to make the distinction between reason on the one hand and the senses on the other. This is exemplified in the statement of Heraclitus, that "knowledge comes to man through the door of the senses." It was realized that different senses are mediated by different sense organs, but also that the different sensory impressions are united in our minds. Some of these

philosophers even suspected that the site of this integration is in the brain. In these ideas lay the origins of physiology and psychology.

Sensory Modalities

It would be fascinating to trace the development of concepts about the senses since then, but to understand the foundation of modern concepts we need go back only to the nineteenth century. In the 1830s, Johannes Müller of Berlin published a monumental *Handbook of Human Physiology,* which served as the definitive textbook of physiology in Europe and America for many years. In it he summarized the work on sensory physiology and promulgated the *law of specific nerve energies.* This states that we are aware, not of objects themselves, but of signals about them transmitted through our nerves, and that there are different kinds of nerves, each nerve having its own "specific nerve energy." The kinds of nerves considered by Müller corresponded to the five primary senses that Aristotle had recognized: seeing, hearing, touch, smell, and taste. The specific nerve energy represented the *sensory modality* that each type of nerve transmitted. The key point is that the nerve transmits this modality no matter how it is stimulated. Thus, an electric shock or a blow on the head may stimulate the nerves of hearing, eliciting sounds in our ears.

The doctrine was applied in a famous law case in which Müller was called for expert testimony. A man had been assaulted at night, and had accused someone. When asked how he could identify the assailant since it was pitch dark, he replied that he caught a glimpse of him in the light caused by the blow to his head! Müller pointed out that pressure on the eye does indeed cause a light sensation—a phosphene—but this is an expression of the fact that the eye responds to any stimulation with a light sensation, and that this is an entirely internal phenomenon.

In modern terms, we recognize that there are specific *receptor cells* tuned to be sensitive to different forms of *energy* in the environment. The forms of energy serve as *stimuli* for the receptor cells. A summary of the main types of receptor cells in the human body, the organs in which they are located, and the forms of energy to which each is sensitive, is provided in Table 10.1. Note that there are many more modalities than the five senses that we think of in everyday life. Among the conscious senses, we must include pressure, temperature, and pain under "touch," and also joint position and the sense of balance. It is also evident from the list that we have specialized cells sensitive to many stimuli within our bodies that never reach consciousness. Particularly important in this group are stretch receptors in the vasculature, which are involved in the control of blood pressure and heart rate, and in skeletal muscles, which are involved in muscle reflexes. There is also a variety of receptors sensitive to different kinds of chemical factors. The receptors in the viscera and other internal organs are often called *interoceptors* or visceroceptors, to distinguish them from the olfactory, auditory, and visual *exteroceptors* that receive signals from outside the body (also called teleceptors: tele = distant).

Taken as a whole, our array of receptors provides information that ranges from the minutest changes in the internal milieu of our bodies to the faintest signals that waft our way from the furthest reaches of our external world.

Within a modality there may be different submodalities, or *qualities*. Thus, we perceive different tastes and smells; we describe temperature sensations in terms of warmth and cold; we see different wavelengths of light as different colors. In general, just as a modality is determined mainly by a type of receptor, so is a sensory quality based on a differentiation of receptors into subtypes. Each receptor cell subtype is tuned to a more narrow spectrum of the stimulus band. Thus, odor qualities depend on receptor cells in the nose being differentially sensitive to different airborne molecules, and different colors depend on photoreceptors in the eye having different sensitivities to wavelengths of light.

A Comparative Perspective

The sense organs and modalities listed in Table 10.1 are those that are found in the human. If we look back over the course of evolution, we see that in general the main categories of modalities are present in most of the main animal groups. Perhaps the most basic are the chemical and touch modalities, which are necessary for the mere existence of an animal, together with some capacity for light sensitivity, which permits a sense of the cycles of night and day.

At the level of molecules and cell membranes, the basic receptor mechanisms within a given modality share many common features across different phyla and species. However, the receptor cells and sense organs show diversity of forms, just as there is a diversity of body plans and nervous systems throughout the Animal Kingdom. Thus, it is not surprising that the eye of a flatworm or an insect should be different from our own. We could further expect that some species have sense organs that we lack, such as electric fish with their organs for sensing electric currents, or the

Table 10.1 Main types of sensory modalities

Sensory modality	Form of energy	Receptor organ	Receptor cell
Chemical			
common chemical	molecules	various	free nerve endings
arterial oxygen	O_2 tension	carotid body	cells and nerve endings
toxins (vomiting)	molecules	medulla	chemoreceptor cells
osmotic pressure	osmotic pressure	hypothalamus	osmoreceptors
glucose	glucose	hypothalamus	glucoreceptors
pH (cerebrospinal fluid)	ions	medulla	ventricle cells
Taste	ions and molecules	tongue and pharynx	taste bud cells
Smell	molecules	nose	olfactory receptors
Somatosensory			
touch	mechanical	skin	nerve terminals
pressure	mechanical	skin and deep tissue	encapsulated nerve endings
heat and cold	temperature	skin, hypothalamus	nerve terminals and central neurons
pain	various	skin and various organs	nerve terminals
Muscle			
vascular pressure	mechanical	blood vessels	nerve terminals
muscle stretch	mechanical	muscle spindle	nerve terminals
muscle tension	mechanical	tendon organs	nerve terminals
joint position	mechanical	joint capsule and ligaments	nerve terminals
Balance			
linear acceleration (gravity)	mechanical	vestibular organ	hair cells
angular acceleration	mechanical	vestibular organ	hair cells
Hearing	mechanical	inner ear (cochlea)	hair cells
Vision	electromagnetic (photons)	eye (retina)	photoreceptors

Modified from Ganong (1985)

mollusc with its osphradial organ for sensing water entering the mantle cavity. Nor is it surprising that many sense organs may be absent, as with the extreme adaptations of the tapeworm to a parasitic experience in the gut of its host. Finally, despite this diversity, we should not be surprised to learn that some sensory cells and organs are such successful adaptations that, like the vertebrate receptor cells for smell, or the vertebrate eye, they are present relatively unchanged across many different species.

The great diversity of stimuli and cell types (see Table 10.1) might seem to indicate that the principles underlying processing in the different systems would be equally diverse. However, a confluence of work in many fields—molecular biology, membrane physiology, biochemistry, pharmacology, systems neurophysiology, and neural network analysis—has provided undeniable evidence for unifying principles. In fact, this work provides some of the most persuasive arguments supporting the comparative and hierarchical approaches to principles of brain organization, as discussed in Chap. 1. In sensory systems, these principles can be conveniently classed in relation to three main levels of organization: the *sensory receptors,* where the interface between the stimulus energy and the nervous system occurs; the *sensory circuits and pathways,* which transmit the sensory information to the brain; and the higher level operations within the brain that underlie *sensory perception.* A brief overview of these principles of sensory processing will be useful as a preparation for understanding and appreciating the individual sensory systems.

Sensory Receptors

In order to detect and discriminate different stimuli in the environment or within the body, the stimuli have to be converted from their different forms of energy into the common currency of nervous signals. The initial conversion is termed *transduction,* and the cells in which it takes place are, by definition, *sensory receptor cells.*

A Superfamily of Transduction Mechanisms

Figure 10.1 represents schematically some of the main types of sensory receptor cells in the vertebrate. The sites at which transduction occurs are indicated in shading. As a general rule, the stimulus acts on membrane receptors. These may be distributed all over the plasmalemma of an entire cell, as in some chemical receptor cells such as those sensitive to oxygen tension in the blood (to the far left in the figure). In most cases, however, transduction takes place in the membrane of a specialized site in the sensory cell. The specializations may have many different forms. In some cases they may be microvilli (as in taste); in others they may be cilia (smell or vision). In most receptor cells of the skin, viscera, and muscles, the sites of transduction are in the terminals of nerve fibers. The terminals may be free (naked nerve endings), as in the skin, or embedded in special structures, as in corpuscles or muscle spindles. Finally, the sites may be in special intracellular membranes, as in visual receptors.

The diversity of receptor cell types summarized in Fig. 10.1, together with the diversity of types of sensory stimuli, has

seemed to defy the search for basic principles underlying the transduction processes. After all, why should the reception of odor molecules be similar in any way to the reception of photons? However, as scientists have compared their results in recent years, they have been gratified to find a surprising number of similarities between the systems, particularly at the molecular level. Some of these results are summarized in Table 10.2. Here it can be seen that in vision, olfaction, and some modalities of taste, transduction involves variations on basic types of G-protein-linked second messenger systems, such as we have already encountered in reviewing neurotransmitter and hormone actions in Chap. 8. On the other hand, mechanoreception and other modalities of taste involve direct gating of a membrane channel. In general, all the mechanisms can be understood as special adaptations of general membrane signaling mechanisms found widely in the cells of the body.

One suspects that Johannes Müller would be delighted with this new information and would seize the opportunity to recast his doctrine of specific nerve energies into a molecular form. We might assist him by suggesting that *each sensory modality is based on an adaptation of a superfamily of membrane signaling mechanisms which converts the stimulus energy into an allosteric molecular change, leading to the gating of ionic current in a membrane channel* (see Shepherd, 1994).

Common Operations in Sensory Transduction

We next inquire whether there are common principles underlying the steps in information processing carried out by these different transduction mechanisms. Here, too, a common set of operations that applies to all sensory systems is beginning to emerge (Shepherd, 1992; Block, 1992; see also Corey and Roper, 1992). As illustrated in Table 10.3, these start with *detection,* which depends on the sensitivity of the transduction mechanism in individual cells

Oxygen   Taste   Smell   Somatosensory   Muscle   Hearing   Vision

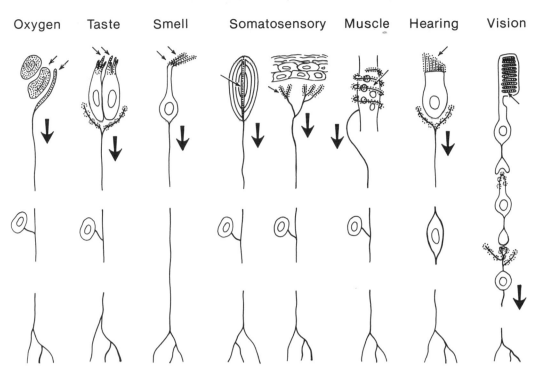

Fig. 10.1 Different types of sensory receptor cells in vertebrates. Small arrows indicate sites where sensory stimuli act. Stippling indicates sites for transduction of the sensory stimuli, and also for synaptic transmission; both of these sites mediate graded signal transmission. Heavy arrows indicate sites of impulse initiation. (Adapted from Bodian, 1967)

and the summation of responses across cell populations. Most systems *amplify* weak signals through positive feedback and active processes which enhance the signal-to-noise ratio. *Discrimination* between signals must then take place; this includes intensity discrimination (as embodied in the Weber-Fechner-Stevens laws; see below) and quality discrimination (which typically involves interactions between populations of receptor cells). In response to prolonged or repeated stimuli most receptor cells *adapt* or desensitize, at rates reflecting their specific modality or submodality. All of these processes converge on the *gating* of the sensory membrane conductance to generate the *receptor potential,* which ultimately leads to encoding of the response in an *impulse discharge,* which is sent to the higher centers in the brain.

In summary, recent work suggests that a

basic set of operations, comprising detection, amplification, discrimination, and adaptation, applies across all sensory systems in transducing sensory stimuli. The precise neural mechanism for a given operation varies in the different receptor cells, but each basic step is essential in the sequence for transducing a local stimulus into a neural response in a single cell and mapping the stimulus field into an ensemble of cells. These operations at the level of stimulus transduction set constraints on the operations at the levels of circuit organization and perception discussed below.

Receptor Potential

It is evident from Tables 10.2 and 10.3 that in all sensory receptors, the common result of transduction is to produce a change in conductance of a membrane channel. The resulting change in membrane

Table 10.2 Steps in sensory transduction

| | | | Taste | | |
Transduction step	Vision	Olfaction	Sweet/bitter amino acids	Salt/sour	Mechanoreception (hair cells)
Energy	Photons	Molecules	Molecules	Na^+, H^+	Displacement
Membrane receptor	7TD family: rhodopsin	7TD family: olfactory	7TD family: gustatory		
G protein	Transducin	G_{olf}	G_{gust}		
G-protein target	Phosphodiesterase	Adenylate cyclase III; phospholipase C	AC; PLC		
Second messenger	cGMP	cAMP; IP_3	cAMP; IP_3		
Protein kinase			Protein kinase Â?		
Membrane channel	Cationic; inward	Cationic; inward Anionic; inward	K^+	$\overline{Na^+}$; $\overline{K^+}$	Cationic; inward
Sensory response	Close channel	Open channel	Close channel	Open; close	Open channel
Adaptation mechanism	Ca^{2+}; phosphorylation?; arrestin	Ca^{2+}; protein kinases ?	?	?	Myosin/actin motor; Ca^{2+}?
Cell body output	Synapses	Impulses	Synapses	Synapses	Synapses

7TD family: 7 transmembrane domain receptor family.

From Shepherd (1991b)

Table 10.3   Common operations in sensory transduction

Transduction operations	Operations in single sensory cells	Operations in cell populations
Detection	Perireceptor mechanisms: filters: carriers: tuning; inactivation Sensitivity Rapidity	Perireceptor mechanisms: filters; carriers; tuning; inactivation Different thresholds
Amplification	Positive feedback Active processes Signal/noise enhancement	Positive feedback Signal/noise enhancement
Encoding/ discrimination	Intensity coding Quality coding Temporal differentiation	Different dynamic ranges Quality independent of intensity Center–surround antagonisms Opponent mechanisms Construction of maps
Adaptation and termination	Desensitization Negative feedback Temporal discrimination Repetitive responses	Temporal discrimination
Sensory channel gating	Open or close conductance gating	
Electrical response	Depolarization or hyperpolarization	
Transmission to brain	Electrotonic spread Active properties Synaptic output or impulse discharges	Spatial patterns: maps and image formation Temporal patterns: directional selectivity, etc.

From Shepherd (1991b)

current produces a change in membrane potential, called the *receptor potential,* also sometimes called the *generator potential.*

The receptor potential shares similarities with a synaptic potential. The reader should review at this point the discussion of the receptor potential in a model receptor neuron, the crustacean stretch receptor cell, in Chap. 7. The main point to recall is that, like a synaptic potential, the receptor potential does not give rise directly to an impulse discharge. The site of receptor potential generation and the site of impulse generation are usually separated. In the stretch receptor cell, and in many other sensory receptors, the separation is some distance through the dendritic branches and axon. In other cases, the sites may be on different cells, requiring one or more synaptic relays in between (see Oxygen, Taste, Hearing, and Vision in Fig. 10.1).

Electrotonic Potential

As we have already discussed in Chap. 7, the spread of a receptor potential, like that of a synaptic potential, is accomplished by means of *electrotonic potentials.* The diagrams in Fig. 7.16 remind us of this fact. In the case of the stretch receptor cell, stretch increases the inward positive current across the nerve terminal membrane, setting up the receptor potential in the terminals. This spreads by electrotonic currents through the cell to the site of im-

pulse initiation in the axon. In the case of
the vertebrate photoreceptor, the action of
the photon on the disc membrane leads
to blockade of the dark current, and the
resulting membrane potential change
spreads electrotonically through the cell to
the site of synaptic output at the receptor
terminal. For the different receptor cells
depicted in Fig. 10.1, the reader may assess
the importance of electrotonic current
spread between the sites of transduction
(stippled) and the sites of impulse generation (thick arrows). For further details on
these mechanisms, the reader may refer to
Chap. 7 and to Chaps. 11–16.

Impulse Generation

The final step at the receptor level is the
encoding of the electrotonically transmitted
receptor response into an *impulse discharge*
in the afferent nerve fiber that carries the
information to the rest of the nervous system. This process has been studied in the
crustacean stretch receptor cell and the vertebrate muscle spindle, and the results are
very similar (Fig. 10.2). The stimulus in
both cases is a stretch applied to the muscle. An important distinction is between the
dynamic (phasic) period of the stimulation,
when the stretch is increasing, and the
static (tonic) period, when it is maintained
constant. If impulses are blocked artificially
(as by tetrodotoxin) in order to observe the
receptor events, it is seen that the receptor
potential rises to a peak at the end of
dynamic stretch, and then falls to a lower,
slowly declining level during static stretch.
When the impulse discharge is recorded,
the impulse frequency also rises sharply
during dynamic stretch, and declines to a
lower level during static stretch. The close
correlation between receptor potential and
impulse frequency can be seen in the graph
at the top of Fig. 10.2.

An important point is that the receptor
potential is graded smoothly and continuously in amplitude in relation to the intensity of the stimulus. Sensory reception thus
involves the transformation, or mapping,

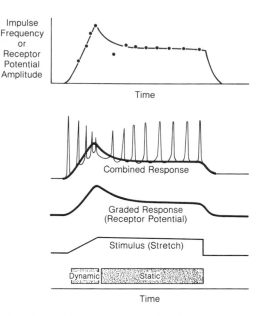

Fig. 10.2 Stimulus encoding in the crustacean
stretch receptor and the frog muscle spindle.
Diagrams show the relations between an applied
stretch (containing dynamic and static phases),
the graded receptor potential, the impulse discharge, and the impulse frequency. (After Ottoson and Shepherd, 1971)

from a continuously varying domain of
sensory stimuli into a neural domain of all-
or-nothing impulses. One can view it as
converting from analog signals to digital
signals. As we noted in Chap. 7, this is also
what takes place during transmission at
many types of synapses. The impulse discharge can function in this way, because
the intervals between impulses (hence the
impulse frequency) vary continuously in
relation to the underlying depolarization
level of the receptor potential and its rate
of change.

From results such as these it is concluded
that the impulse discharge faithfully encodes the parameters of the applied stimulus. However, we can see that it does more
than this; in the case of the stretch receptor
and muscle spindle, it tends to heighten the
response when the stimulus is increasing.
This property is termed *dynamic sensitiv-*

ity. Receptors vary in how rapidly the response declines from the dynamic peak during the ensuing static stimulation. The decline is termed *adaptation.* Receptors that signal slow and prolonged changes are *slowly adapting,* or *tonic,* receptors; those that signal brief changes are termed *rapidly adapting,* or *phasic,* receptors. Our ability to maintain positions of our muscles over long periods of time depends on the slowly adapting properties of our muscle receptors, whereas the rapid fading of our appreciation of a pressure stimulus is due to the rapid adaptation of pressure receptors (Pacinian corpuscles). Adaptation may also be a synaptic property of the transmission of signals through sensory pathways (see below).

In summary, we have seen that there are four main steps (transduction, receptor potential generation, electrotonic spread, impulse generation) in the transfer of information from the domain of the sensory stimulus to the domain of the impulse discharge. We have also seen how the receptor determines the basic properties of the sensory response. Thus, *specificity* resides in the molecular mechanisms of the sensitive membrane. *Intensity* is mapped from graded receptor potentials into an impulse frequency code. *Adaptation* determines the profile of the response in relation to the dimension of time; there is often a tendency to heighten the sensitivity to stimulus change. The distribution of the whole population of receptors determines the *spatial organization* of the incoming information, as we shall soon discuss.

Sensory Circuits

The stimulus has been converted into a frequency code and the impulses are on their way to the central nervous system. Within the central nervous system, sensory information characteristically is relayed through a series of centers. In each center there is opportunity for processing of the signals and integration with other types of information. A *sensory pathway* thus consists of a series of modality-specific neurons connected by synapses. All the circuits within and related to this pathway constitute a *sensory system.*

Basic Principles

The circuits of different sensory systems share some common properties, as illustrated in Fig. 10.3. The axons of primary sensory nerves divide to supply more than one neuron; this is referred to as *divergence.* A neuron in turn is contacted by more than one axon; this is termed *convergence.* These features apply to connections both within and between different centers; thus, incoming axons diverge to more than one center, and different sources converge onto a given center. The chains of connections mean that a pathway consists of connections in *series,* in which there is obviously a temporal *sequence* of events. However, because of the divergence and convergence of connections at successive levels, there are connections in *parallel,* so the different forms of information can be transferred and combined at the same time.

Parallel processing is thus inherent in sensory pathways, and in neural systems in general. This fact has been recognized in recent years by computer scientists engaged in building new generations of more powerful computers. Some of the principles discussed here are finding their way into these computers.

Some central pathways are primarily concerned with transmitting the input from one type of receptor; these are termed *specific sensory pathways.* Other pathways, by divergence of their fibers and convergence with other inputs, become increasingly *multimodal,* or *nonspecific.* Finally, *centrifugal* connections provide for *feedback* of information from one level to the next. In general, specific sensory pathways provide for precise transmission of sensory information, while nonspecific pathways provide for sensory integration and adjustments in behavioral status of the whole

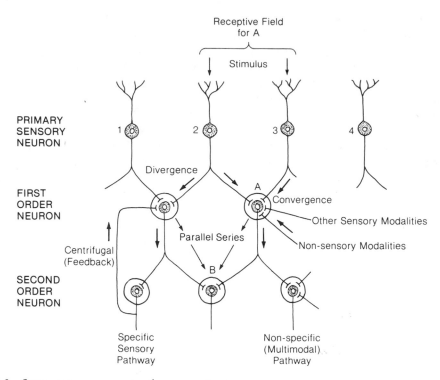

Fig. 10.3   Some common aspects of sensory circuits.

organism. Both can be seen to be necessary for the analytic and synthetic functions of the organism.

Receptive Fields

An important concept in sensory physiology is the *receptive field*. For any neuron in a sensory pathway, the receptive field consists of *all the sensory receptors that can influence its activity*. Thus, cell A in Fig. 10.3 has a receptive field consisting of the two receptors (2 and 3) which connect to it. Cell B, at the second level in this system has a receptive field consisting of receptors 1, 2, and 3. The connections to a cell may be excitatory or inhibitory, and they may be mediated by interneurons at a given level as well as relay neurons connecting levels. As we will see in the following chapters, the properties of receptive fields generally reflect the increasing degree of information processing and feature extraction that occurs in neurons at successively higher levels in sensory pathways

Microcircuit and Local Circuit Organization

Within any center in the nervous system, the types of neurons and synaptic connections have many forms. However, as noted in Chap. 3, the microcircuits of a center are usually built up by patterns of connection between three elements: input fibers, output cells, and intrinsic neurons. These principles of organization are demonstrated very clearly in the centers of many sensory pathways.

Two examples are illustrated in Fig. 10.4. Part A is a schematic diagram of the neurons in the vertebrate retina and their main patterns of synaptic interconnection. There are input elements (the receptors) and output neurons (the ganglion cells). There are interneurons for straight-through transmission (the bipolar cells), and there are interneurons for horizontal interactions. The horizontal connections are organized at two levels, the first (horizontal cells) at the level of receptor input, and

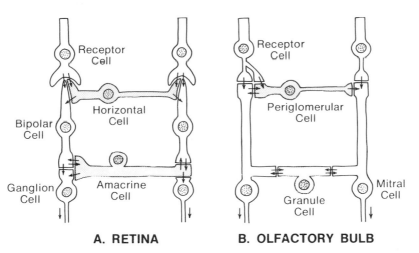

A. RETINA   **B. OLFACTORY BULB**

Fig. 10.4 Comparison between simplified basic circuit diagrams of the vertebrate retina and olfactory bulb. (After Shepherd, 1978)

the second (amacrine cells) at the level of ganglion cell output. The amacrine cells take part in a variety of reciprocal and serial connections that correspond to the complexity of processing which occurs at this level (see Chap. 16).

Figure 10.4B is a similar diagram of the neurons and connections in the vertebrate olfactory bulb. Note that this is an extraction of the design representing intrinsic olfactory bulb circuits in Fig. 6.19. Here, too, are input elements (the olfactory receptors) and output neurons (the mitral cells). In this case the straight-through pathway is provided by the primary dendrite of the mitral cell. Horizontal connections are organized at two levels. The first, through periglomerular short-axon cells, is at the level of receptor input; the second, through granule cells, is at the level of mitral cell output. The granule and mitral cells interact through reciprocal synapses.

From this comparison it can be seen that, although the retina and olfactory bulb process two very different types of sensory information, their microcircuit and local circuit organization have many points in common. In both cases there is provision for straight-through transfer of signals to the output neuron, and local processing of signals through the interneurons. Recall in

addition that olfactory and visual receptors bear certain similarities in their sensory transduction mechanisms (see above). Clearly, there are common principles underlying the functional organization of these two systems.

Lateral Inhibition

We have mentioned that in addition to transmitting faithfully certain aspects of the stimulus, the transduction mechanisms of sensory receptor cells also enhance some aspects. This appears also to be a function of the intrinsic synaptic circuits at successive levels in a sensory pathway. The best known example of this is the lateral inhibition that enhances spatial contrast in the visual system. This was first revealed by Keffer Hartline and his colleagues at Rockefeller University in their studies of the compound eye of the horseshoe crab, *Limulus*. Let us consider this important model system in some detail.

The compound eye of *Limulus* consists of some 800 receptor units, called ommatidia. Each *ommatidium* has at its surface a corneal lens. Behind it are 10–15 receptor *(retinula)* cells arranged in a circle (Fig. 10.5). At the center of the receptor group is the dendrite of a cell whose cell body is deeper and placed to one side, which has

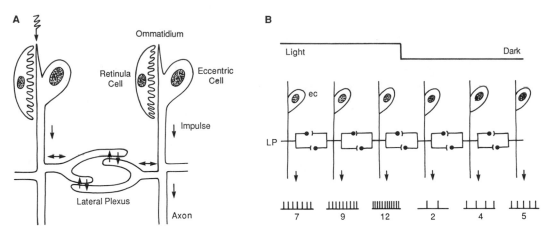

Fig. 10.5 Functional organization of lateral eye of *Limulus*. **A.** Synaptic organization, showing site of light transduction in ommatidium, impulse conduction in eccentric cell axon, and lateral interactions through dendrodendritic synapses in the lateral plexus. **B.** Pattern of activity in population of eccentric cells (ec) in response to light–dark edge stimulus (above). The impulse discharges recorded from the axons at the sites indicated by the arrows are shown below, with the numbers of spikes per unit time (arbitrary) as indicated. The enhancement of response on the light side of the edge and depression on the dark side are due to the inhibitory interactions mediated by the synaptic connections in the lateral plexus (LP). See also Fig. 10.6. (A, B based on Fahrenbach, 1985; Shepherd, 1986).

given it the name *eccentric cell*. This is a modified receptor cell: it has few microvilli, but a long thick axon, and is the principal projection neuron from the ommatidium (see Fig. 10.5).

Intracellular recordings have shown that the photoreceptors are electrically coupled to the eccentric cell dendrite. By this means, the output from the eccentric cell represents the summation of inputs to all the photoreceptors. The electrical synapses are rectifying; they pass current from the receptors to the eccentric cell dendrites, but not in the opposite direction. This prevents the electrical activity of the eccentric cell from interfering with the responses of the photoreceptor.

At rest, the eccentric cell shows background activity consisting of occasional, small depolarizing waves, first called "Yeandle bumps" after their discoverer. These represent photoreceptor responses to single photons, transmitted to the eccentric cell through the electrical synapses. With

light stimulation, these quantal events fuse into a graded receptor potential which depolarizes the membrane. Note that this differs from the response of vertebrate photoreceptors; this difference will be discussed further in Chap. 16.

The receptor potential gives rise to action potentials, which encode the intensity and time course of the stimulation (along the lines indicated in Fig. 10.2), and transmit this information to the central nervous system. The impulses arise in the axon at a distance from the cell body. The mechanism of electrotonic spread of receptor potential into the axon is similar to that discussed for the crayfish stretch receptor in Chap. 7. The axon gives rise to a number of collaterals which form a plexus beneath the ommatidia. Within this plexus, the collaterals are interconnected by reciprocal and serial synapses. These form the circuits which mediate the lateral inhibitory interactions (see Fig. 10.5).

To analyze the way in which the whole

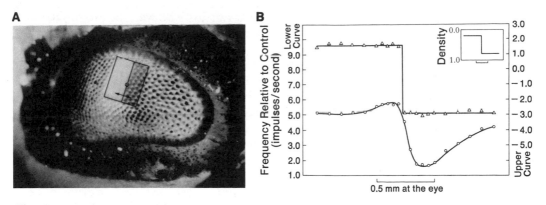

Fig. 10.6 Enhancement of spatial contrast in *Limulus* eye. **A.** Surface of *Limulus* eye, with superimposed rectangular stimulus pattern; pattern is divided into lighter (left) and darker (right) regions. Pattern is centered on test ommatidium (×). Arrows show directions in which the test pattern was displaced, to produce lower curve in graph in B. **B.** Recordings of spike frequency in axon from test ommatidium in A. *Lower curve:* responses to rectangular test pattern in A. *Upper curve:* responses to small spot of light, at high and low intensities corresponding to those of test pattern (see insert). The differences between the two curves illustrate that lateral inhibition enhances the response on the light side of an edge (because there is less inhibition from the more darkly lit neighbors to the right) and depresses the response on the dark side of an edge (because there is more inhibition from the brightly lit neighbors to the left). (From Ratliff, 1965)

ensemble of eccentric cells encodes information about a spatial pattern over the eye, Hartline and his colleagues first used small spots of light that could excite only one or a few eccentric cells. The responses faithfully reflected the relative intensities of the spot, as when changing from a high to a low intensity; this is shown by the square step in the graph of Fig. 10.6B. They then repeated the experiment, using a test pattern containing a region of light and dark (Fig. 10.6A). Under these conditions, the impulse discharges of the eccentric cells near the light–dark boundary were modified so that they did *not* faithfully reflect the real luminosity of the stimulus; the eccentric cells on the light side of the border fired faster, whereas those on the dark side fired slower, than expected. There is thus an enhancement of contrast in the activity of the eccentric cells near the border.

The reason for this enhancement lies in the extensive inhibitory interconnections between the eccentric cells through their collateral branches within plexus lying just below the eye. With relatively even illumination of the eye, the branches mediate relatively similar levels of inhibition onto each other. With a sharp border of contrast, however, the eccentric cell on the light edge receives less inhibition from its neighbor on the dark edge, and it fires faster; conversely, this feeds more inhibition onto the darkly lit neighbor, whose already low level of impulse discharge is suppressed even more. This mechanism is illustrated by the diagram of Fig. 10.5B, and the envelope of firing intensities recorded experimentally is shown in Fig. 10.6B.

The classical studies of *Limulus* emphasized the importance of lateral inhibition for contrast enhancement and feature extraction, but recent work suggests that it also has other functions. In invertebrates, lateral inhibition appears to aid in image reconstruction by contributing to compensation for blurring of the image due to dispersion of light as it passes through the lens. Another function is as a gain control.

Limulus, for example, responds to intensities over 11 log units. In order to cover this enormously wide range, and still have mechanisms that enhance sensitivity at low levels near threshold, there must be a reduction of sensitivity (that is, a gain compression) as the intensity of stimulation increases. One mechanism for achieving this kind of gain compression is through feedback inhibition (see Chap. 16).

Sensory Perception

The study of the quantitative relations between stimulus and perception constitutes the field of *psychophysics.* One of the aims of sensory neurobiology is to understand the neural mechanisms underlying these relations. The ultimate aim is to identify the *building blocks of perception*—the functional mechanisms used to construct our perceptual representation of the world about us.

Let us briefly review the main aspects of a stimulus that contribute to sensory perception as an introduction to the following chapters. These properties of perception are closely related to the transduction processes discussed above.

Detection

The simplest aspect of a perception is the ability to detect whether a stimulus has occurred. The level of intensity is called the *behavior threshold.* We have seen previously that each receptor has its characteristic threshold for responding to some minimum amount of its specific stimulus. As a general rule, several receptor responses must summate in order for transmission to occur in the sensory pathway. This was first shown in a classic study of the visual system by S. Hecht, S. Shlaer, and M. H. Pirenne in 1942. They calculated that a single photon is adequate for stimulating a single photoreceptor in the human retina, but the simultaneous activation of about seven receptors is necessary to perceive that stimulation has occurred. The behavioral threshold is therefore somewhat higher than the individual receptor threshold. In

some systems, however, the thresholds may be very similar (see Chap. 12).

Magnitude Estimation

The next important property of a stimulus is how much of it there is. Primitive visual receptors of invertebrates, and the eyes of primitive vertebrates, are examples of receptors that are concerned mostly with this property. More advanced sensory systems are exquisitely tuned to registering stimulus magnitude over a wide range of intensities, in addition to other sensory qualities.

The study of magnitude estimations involves varying the stimulus in a quantitative manner and determining the physiological, behavioral, or perceptual response along some quantitative scale. This was first attempted and formalized by Ernst Weber (1934) and Gustav Fechner (1860) in Germany, and was the foundation stone of psychophysics as a science.

These studies suggested that the response varies with the stimulus according to an exponential relation (see A and B, Fig. 10.7). This "law" was widely believed until around 1960, when S. S. Stevens at Harvard obtained evidence that in many systems the relation can better be described by a power law (C and D, Fig. 10.7). The beginning student does not have to be concerned about the relative merits of these laws. Over limited magnitude ranges, both are reasonable approximations (Fig. 10.8E). What is important is that in most sensory systems the psychological perception varies in strength with the intensity of stimulation in a quantitative manner.

In an attempt to get physiological evidence for this relation, neurobiologists have made recordings at various levels in several of the sensory pathways. Thus, as shown in Fig. 10.8, the concentration of a taste substance on the tongue was varied while recordings were made from the nerves from the tongue. The results showed that the magnitude of stimulation, of nerve response, and of perceptual estimation were all closely correlated. This experiment could be carried out in humans because of the accessibility of the nerve from the

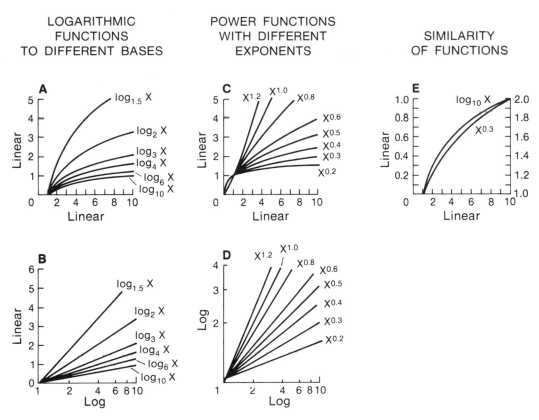

Fig. 10.7 Hypothetical relations between stimulus and response. Stimulus intensity is plotted on the abscissa, response intensity on the ordinate. Logarithmic relation (according to Fechner) is plotted on linear (**A**) and logarithmic (**B**) scales. Exponential relation (according to Stevens) is plotted on linear (**C**) and logarithmic (**D**) scales. The curves may be very similar, as shown in E. (From Somjen, 1972)

tongue for recording as it passes through the chamber of the middle ear. Another approach has been to study the behavioral responses and physiological recordings in primates as a model for humans. In the somatosensory system, behavioral responses are closely correlated with the impulse discharge of neurons at successive levels, as we shall see in Chap. 12.

Spatial Discrimination

In sensory systems, natural stimulation of the receptors occurs in some spatial pattern. The ability to identify the site or pattern of stimulation is termed *spatial discrimination*. This applies to the visual and somatosensory system, and also to the auditory system, in which different sounds affect different parts of the receptor popu-

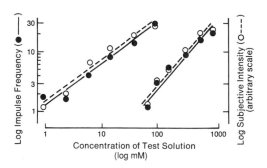

Fig. 10.8 Stimulus–response relations in a sensory system. Graph plots subjective intensity of taste sensation and frequency of impulse discharge in chorda tympani nerve in response to stimulation of the tongue with citric acid and glucose. Recordings were made in humans undergoing middle ear surgery, which exposes the chorda tympani nerve. (Modified from Borg et al., 1967)

lation. A common way to study the somatosensory system for this quality is to test for *two-point discrimination,* to see how close together two points on the skin can be stimulated and still be perceived as two points rather than one. The comparable tests in the visual system are with two points of light (a measure of *visual acuity*), and with two tones in the auditory system.

The general finding is that at low intensities of stimulation, discrimination is poor, and is not much in evidence until some level of intensity above threshold. This is taken to indicate that weak stimulation mainly activates straight-through pathways in sensory systems, to enhance detection, and that only with increased intensities do horizontal interactions, such as those indicated in the diagrams of Figs. 10.3 and 10.4, come into play, to enhance spatial discrimination. This region, between threshold and discrimination, is called the *atonal area* in the auditory system; it is the region between the point at which we say "yes, I hear something" and "yes, I hear a different tone." The student can verify the phenomenon by weak and strong stimulation with two pencil points on his or her skin. That will also demonstrate that two-point discrimination varies widely in different parts of the body surface, as will be described further in Chap. 12.

Feature Abstraction

Natural stimulation does not usually consist of spots of light or points jabbing into the skin. Rather, it involves complex interplays of several stimulus properties. The strategy in studying the components that contribute to perception of a natural stimulus is to begin with spots of light or a simple grating, in the case of the visual system, for example, and then change to increasingly more complex stimuli, such as moving spots and edges with different orientations, as we reach higher levels in the cerebral cortex. The results imply that a unit of perception involves a set of neurons and their connections that is tuned to a coordinated set of several stimulus properties, such as, in this case, light, movement, shape, orientation, and size. The set of properties may be said to constitute a feature, and the mechanisms whereby a neuron or circuit is tuned to this feature in preference to others is called *feature abstraction* or feature extraction. In the somatosensory system, the comparable process relates to the way we feel the texture of a surface by moving our hands over it ("active touch"), or the way we feel the texture of food with our tongues, which makes important contributions to the overall perception of palatability of food.

Both two-point discrimination and feature extraction, as they relate to spatial features, involve mechanisms of lateral inhibition and other types of interactions within sensory pathways. As we have seen, they enhance contrast between stimulated and unstimulated regions, and they enhance changing stimuli over stationary ones. Analysis of these mechanisms has been one of the main achievements of sensory neurobiologists, as we will see in succeeding chapters.

Quality Discrimination

As pointed out previously, a given sensory modality characteristically contains several submodalities or qualities, and the discrimination of these as distinctly different is one of the main attributes of sensory systems. In general, the discrimination of qualities is regarded as being of two types, either analytic or synthetic. Consider a stimulus containing a mixture of two submodalities. In *analytic* discrimination, each submodality retains its individual character. Thus, in the taste modality, there are four basic qualities—sweet, salt, sour, and bitter. When we taste a mixture of, say, sugar and salt, the individual qualities can still be discerned; they do not merge to form a new sensation. We all have this experience when we eat "sweet and sour pork" at a Chinese restaurant. The perception is analytic; it can be analyzed into its components. By contrast, in the perception of color, there are primary colors—red, yellow, green—which when mixed together yield most other colors. The other colors are, in effect, *synthesized* from the primary colors, and have their own qualities, distinct from those of the primaries. We cannot see blue and yellow in a green object.

Pattern Recognition

In studying sensory mechanisms, we do experiments to analyze the system into its components, and then infer the process whereby the system uses those mechanisms to build up its behavioral response or conscious perception. In some cases, sensory systems may actually operate in that way, building up a perception from individual discriminations. However, one of the most vivid of mental experiences is the ability to take in a scene around us and instantly recognize a familiar pattern, or an unfamiliar one, or one that has some special significance. This is a capability that is prevalent throughout the Animal Kingdom. The actions of specific visual objects in releasing innate behavior patterns in many lower animals are striking examples of this capability, as we shall discuss in Section III.

This property of sensory perceptions was first recognized by the Gestalt psychologists in the early part of this century. Gestalt means form, or shape, but as used by these workers it means the patterns we perceive and recognize as unitary wholes. As Edwin Boring (1950) put it:

In perceiving a melody you get the melodic form, not a string of notes, a unitary whole that is something more than the total list of its parts or even the serial pattern of them. This is the way experience comes to man, put up in significant structured forms. . . .

This idea is sometimes thought to be opposed to the concept that perceptions are built up out of neural units, such as we have been discussing, but in fact there is no inconsistency. There are different aspects of the same problem, much as a table is made up of atoms, yet looks like a table. The important point is that many of our sensory experiences consist of complex patterns,

either in space or in time, and we tend to perceive them as wholes more than in terms of their individual parts. This implies that such perceptions involve very large populations of neurons and very extensive sets of circuits. A related point is that although two patterns may be very similar, we can perceive them as distinctly different. The traditional example of this is the ability we all have to recognize our own grandmother in a crowd. Thus, although patterns may be similar, or may grade into each other, our perceptions are separate, distinct, discontinuous entities. This is illustrated in Fig. 10.9, just one of the many examples of this property in the visual system.

These considerations indicate that perceptions involve extensive sets of neurons

Fig. 10.9 In this figure, the black areas appear as faces, the white area as the outline of an urn. Our perception alternates between these two interpretations. This illustrates that we perceive patterns as consistent wholes, each distinct from the other. It also illustrates that perception involves making a "decision" about what is the figure ("signal") and what is the background ("noise"). Finally it shows that perception is not just a passive reception of individual sensory signals, but rather involves an active interpretation by the brain of the meaning of the stimulus patterns it receives. (Figure and legend from Gregory, 1966)

and circuits, and that these sets, though overlapping, nonetheless mediate distinct responses. The analysis of individual neural components is the starting point, but we will obviously need information from many approaches, both experimental and theoretical, before a final understanding is reached.

Toward a Molecular Biology of Perception

How do these principles of perception relate to the membrane mechanisms of signal transduction we discussed at the beginning of this chapter? On the one hand, the billions of ions and molecules that are involved in the response of a photoreceptor cell to light, and the many hundreds of thousands of cells that are activated when we view an object in bright light, would suggest that perceptions are solely the emergent properties of networks. However, we have noted that in the dark a single photon can stimulate a single receptor, and the behavioral threshold for light detection is the simultaneous activation of only seven photoreceptors. In Chap. 16 we will learn how retinal circuits are adapted for passing on the information from these few receptors so that it reaches the level of consciousness. So we must recognize the astonishing likelihood that individual molecules and cells can matter in perception. In addition, we should also recognize that some sensory cells give responses that are determined by specific types of molecules, for example, the cone photoreceptors for specific wavelengths of light, and the olfactory receptors for specific odor molecules, and that these trigger specific behaviors and perceptions.

Thus, in addition to traditional psychophysics, which studies the relation between overall levels of sensory stimulation and perception, there is the possibility of a *molecular biology of perception*, which will study the relation between activation of individual molecules and types of molecules and a consequent behavior or percept. Neural circuits obviously provide the initial link between the two. This promises to give insight, on the one hand, into defects in genetic diseases of human sensory pathways, such as degeneration of specific retinal cells, which causes different types of blindness (see Chap. 16). On the other hand, knowledge of the molecular basis of perception will give deeper insight into the nature of the brain mechanisms underlying higher cognitive functions of humans.

11

Chemical Senses

From an evolutionary point of view it is convenient to start our study of sensory systems with the chemical senses. The first organisms to emerge from the primordial brine defined themselves as organisms by the degree to which they could sustain their own metabolism, and this required the ability to sense the appropriate nutritive constituents in their environment. The chemical senses are thus among our most primitive; on the other hand, they provide us (as well as our animal cousins) with some of our most powerful experiences. If you think about it, the ability of cells to "sense," or respond, to specific chemicals runs as a thread through much of neurobiology: the responses to chemical neurotransmitters and hormones, the development of neural connections, and the sensory responses of chemoreceptors all depend on properties that are similar at the molecular level. In this respect, no other class of sensory receptors better illustrates the study of nerve cells as a part of molecular and cell biology.

The chemical senses may be divided into four categories: common chemical, internal receptors, taste, and smell. The *common chemical sense* includes all those cells that are sensitive to specific molecules or other chemical substances and which respond in ways that are communicated as signals to the nervous system. *Internal receptors* are a subclass of the above-mentioned receptors, which are specialized for monitoring various aspects of the chemical composition of the body that are vital for life. *Taste* and *smell* are familiar to everyone as distinct chemical modalities: taste, for sensing substances within our mouths; smell, for sensing airborne substances originating both in the outside environment and also from ingested foodstuffs.

Internal receptors for a variety of substances have been identified (see Table 10.1, Chap. 10). We will have more to say about the common chemical sense when we discuss pain receptors (Chap. 12), and we will later study glucose receptors and receptors for circulating toxins (Chap. 26). The main outlines of the olfactory system in the insect are covered in Chap. 27. Here we focus on the major systems for taste and smell in the vertebrate.

The Taste System

Cells that are preferentially sensitive to ions or molecules in food materials may be designated as *taste receptor cells,* and the sensory modality is called taste, or *gustation.*

Taste Receptors

In aquatic vertebrates, as in many invertebrates, taste receptors are not always limited to the mouth. For example, in some species of bottom-dwelling fish, fingerlike downward-pointing projections from the anterior (pectoral) fins carry taste receptors at their tips. This arrangement seems well designed to detect foodstuffs in the muddy bottom where these species live.

In most vertebrates, taste receptors are found on the tongue, back of the mouth, pharynx, epiglottis, and upper esophagus. At the cellular level, taste receptor cells are characteristically grouped together in *taste buds.* Taste buds are arranged on *papillae,* which appear as blunt pegs on the surface of the tongue. There are several types of papillae, and these have different distributions on the tongue surface, as shown in Fig. 11.1A, B.

The taste bud contains several cell types. Types 1 and 2 (see Fig. 11.1C) are *support-ing cells;* they have microvilli at their tips, and appear to secrete substances into the lumen of the taste bud. Type 3 is believed to be the *sensory receptor cell;* it has peg-like extensions into the lumen that are probably sites of sensory transduction. The fourth type of cell is the *basal cell.* Radioactive labeling studies indicate that basal cells arise by inward migration of surrounding epithelial cells. The basal cells, in turn, differentiate into new sensory receptor cells. This process continues throughout adult life, because the receptor cells have lifetimes of only about 10 days. Thus, there is continual turnover of sensory cells in the taste buds. We will see that a similar process occurs in vertebrate olfactory receptor cells (next section). It is not known how the sensory specificity of a taste bud is maintained in the face of this continual replacement of the sensory cells.

Taste Transduction: Membrane Mechanisms

When we eat natural foods, our taste sensations are complicated mixtures of qualities.

Fig. 11.1 Basic organization of the human tongue. **A.** Distribution and pattern of taste buds. **B.** Main types of taste papillae, containing taste buds. **C.** Fine structure of a taste bud. See text for explanation. (From Murray, 1973)

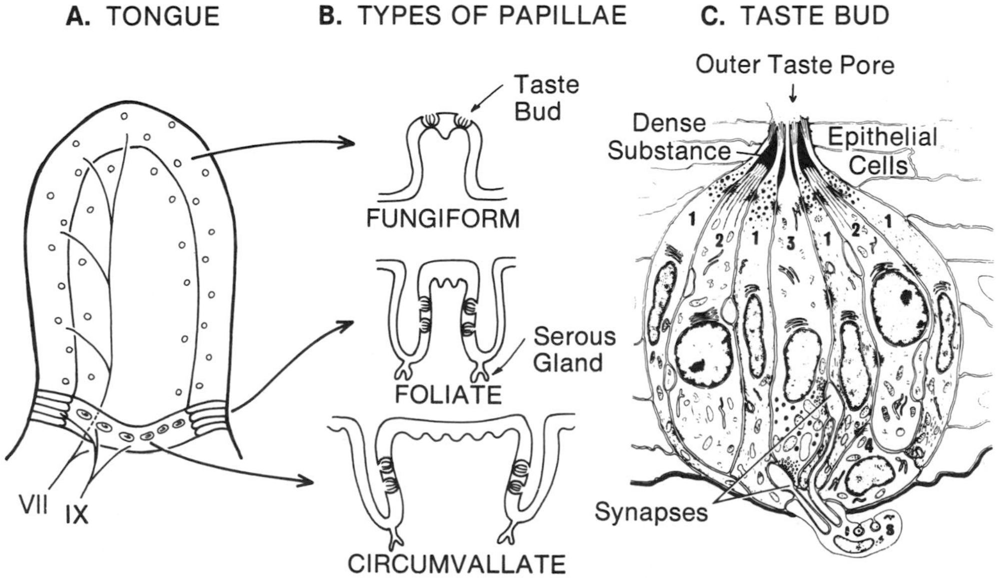

A. TONGUE   **B. TYPES OF PAPILLAE**   **C. TASTE BUD**

However, when humans are tested with pure chemical compounds, the taste sensations can all be grouped into four distinct qualities: *sweet, salt, sour,* and *bitter.* These qualities have seemed to have an obvious logic to them for the behavior of the organism, a logic which may be summarized succinctly as follows: *sweet* is generally associated with nutritious foods, *salt* is essential for fluid balance, *sour* is harmful in excess, and *bitter* is a reliable warning of substances that are injurious or deadly.

What is the molecular basis of these qualities at the receptor level? Since the nineteenth century the possibility has been recognized that each taste quality may have a distinct chemical basis with its own distinct receptor mechanism: hydrogen ions (pH) for sour, NaCl for salt, sugar molecules for sweet, and certain complex molecules for bitter. Until recently it was not possible to isolate and record from individual taste cells due to the tough tissue that surrounds them in the taste bud, so the mechanisms had to be inferred indirectly from recordings from the nerves that supply the taste buds. However, improved cell isolation techniques, and especially the advent of the patch-recording technique, have enabled direct analysis, and in a surprisingly brief time a characterization of each modality at the membrane level has been achieved.

Figure 11.2 summarizes the main types of mechanisms. The mechanisms for sour and salt are the simplest. *Acid* transduction is mediated by a pH-sensitive K^+ channel that is of a type found in many cells in the body. In taste cells these channels are localized in the apical microvilli, so that when they are blocked by protons (H^+), channels (Na^+ and Ca^{2+}) present elsewhere in the cell membrane can generate a depolarizing response of the cell. This is a good example of how a basic channel type is used for a special purpose within a cell by (1) localization to a particular region, through differentiation of the membrane cytostructure, and (2) interaction with other types of channels in the cell.

Salt is mediated by passive movement of Na^+ through Na^+-selective channels in the microvilli down its concentration gradient between outside and inside, thereby depolarizing the cell directly. Like many similar channels in other types of epithelia (for example, kidney tubule cells), these are blocked by the drug amiloride. In some experiments, a pipette with tip diameter of 100 μm has been pressed onto the taste pore, through which recordings of steady-state currents as well as action currents can be made while perfusing the taste pore with different taste solutions. These results suggest that the amiloride-sensitive Na^+ channels are also permeable to H^+ and may therefore contribute to sour taste in some animals.

Completely different mechanisms are involved in sugar and bitter taste. *Sugar* appears to be mediated by a second messenger system in which the sugar molecule first binds to a membrane receptor, which leads to production of cAMP. In experiments utilizing either patch recordings from isolated cells or taste pore recordings, stimulation with an artificial sweetener such as sodium saccharin causes a reduction of voltage-activated K^+ currents; this depolarizes the cell and leads to generation of action potentials. Furthermore, bath application of membrane-permeable cAMP analogues mimicks these effects. It has been inferred that both sugar molecules and taste sweeteners cause an increase in cAMP, which leads to closure of K^+ channels on the basolateral surface of the cell, causing membrane depolarization and consequent generation of action potentials.

Bitter is the most complex and least understood of the taste qualities. Many different types of substances produce a bitter sensation; these include salts, acids, amino acids, peptides, complex sugars, alkaloids, and carbamates. The transduction pathways appear to be correspondingly diverse. One type that has been identified is the phospholipid second messenger pathway. Other mechanisms appear to involve direct actions of these compounds on membrane

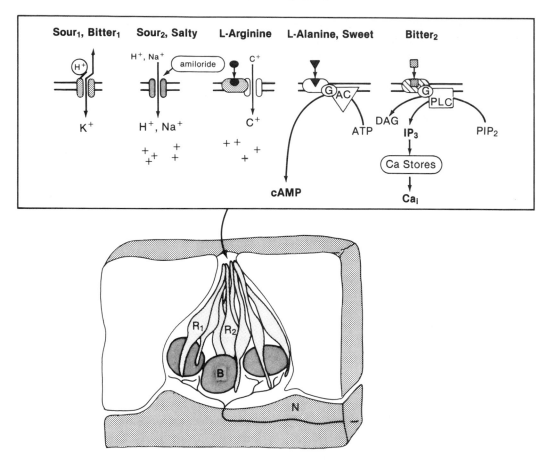

Fig. 11.2 Summary of chemosensory transduction mechanisms occurring in taste buds. Sensory mechanisms illustrated at the *apical membrane* of taste receptor cells include: protons (sour) and bitter compounds blocking K channels; protons and Na⁺ passing through apical, amiloride-sensitive sodium channels; L-arginine binding to a ligand-gated, nonselective cation channel; L-alanine and sweet compounds activating a G protein by binding to appropriate membrane receptors and stimulating adenylate cyclase; bitter compounds activating a G protein that is coupled to phospholipase C and resulting in an increase in inositol triphosphate. Synaptic mechanisms illustrated at the *basolateral membranes* of taste receptor cells include depolarization of the membrane consequent to an influx of cations at the apical membrane; reduction of potassium conductance due to cAMP buildup, which also depolarizes the membrane; generation of an action potential when the depolarization reaches threshold; influx of Ca^{2+} ions; release of neurotransmitter stores consequent to increased $[Ca^{2+}]_i$. AC, adenylate cyclase; B, basal taste cell; C^+, cations; DAG, diacyl glycerol; G, G-protein; IP_3, inositol 1,4,5-trisphosphate; N, nerve bundle; PIP_2, inositol 4,5-bisphosphate; PLC, phospholipase C; R, taste receptor cell. (Figure and legend adapted from Avenet et al., 1993)

and intracellular constituents. A common end effect of all of these mechanisms is a blockade of K⁺ channels and release of intracellular Ca^{2+}.

Since both sweet and bitter seem to involve second messenger pathways, it has been obvious to postulate that they are G-

protein mediated. Starting with a taste–cell enriched cDNA library prepared from mRNA from rat taste papillae and using PCR (Chap. 2), seven distinct G-protein cDNAs have been isolated, one of which appears to be specific to taste cells. This has been termed *G gust*, and it has been postulated to be involved in transduction of sweet or bitter.

Sensory Processing

The taste sensory cells have no axons; information is transmitted from them through synapses onto the terminals of sensory fibers within the taste bud (see Fig. 11.1C). The fibers arise from ganglion cells of cranial nerves VII (facial nerve) and IX (glossopharyngeal nerve) (see Fig. 11.1A). Most of the fibers from the anterior part of the tongue run in the chorda tympani, a branch of VII. In addition, fibers of the superior laryngeal nerve, a branch of the vagus (the tenth cranial nerve), innervate taste buds on the surface of the epiglottis and the upper part of the esophagus. These taste buds, which are distributed widely, total as many as the taste buds on the anterior part of the tongue.

The fact that we distinguish four basic tastes, and that they can be preferentially elicited from different areas of the tongue, suggests that each area might be the basis for a *"labeled line"* carrying information about its specific submodality to the brain. However, the first unit recordings from single fibers in the chorda tympani by Carl Pfaffman, at Brown University in 1941, showed that this could not be the case. A single fiber might respond best to one stimulus, but it also characteristically showed varying degrees of response to other types of stimuli (cf. Fig. 11.3). This implies that a given fiber receives synapses within the taste bud from several receptor cells with differing response specificities. Pfaffman et al. (1976) suggested that "in such a system, sensory quality does not depend simply on . . . activation of some particular fiber group alone, but on the pattern of others active." This became known as the *"across-fiber pattern"* theory of taste quality.

In recent years there has been much debate among workers in the field of taste as to the relative merits of the labeled-line vs. the across-fiber pattern theory. As usual in such cases, there are elements of the truth in both theories. Receptor cells and sensory fibers have their "best" stimulus, but at each level there is graded responsiveness to other stimulus types. Taste information thus appears to be encoded by means of interactions between many elements of different specificities; in other words, elements that have overlapping response spectra. If this is what is meant by "across-fiber patterns," it is the same principle that underlies the encoding of visual information, and possibly of odor information by generalist receptors, as discussed below.

Taste Pathways

The fibers carrying taste information make their synapses centrally in the medulla, in a slender line of cells called the *nucleus of the solitary tract* (see Fig. 11.4). These cells give rise to ascending pathways which differ in different species. The best known examples are the rat and the monkey, which are summarized in the diagrams of Fig. 11.4. Note, in the rat, the relay in the pons, and the bifurcating connections from there to the basal forebrain and the thalamocortical projections. By contrast, in the monkey, the specific taste pathway connects only to the thalamocortical system. This specific parallel pathway through the pontine taste nucleus pathway to somatosensory cortex presumably mediates the conscious perceptions of taste quality, whereas the pathway to hypothalamus, amygdala, and insula carries taste information to the limbic system. This pathway may be important for the affective qualities of taste stimulation, and it might also play a role in affective and memory processes that underlie learned taste aversions in feeding behavior, as we will discuss further in Chap. 29.

We have described these taste pathways with a view to understanding their importance as a basis for our perceptions of taste. However, taste pathways are also important for many other types of behavior. Some of these are summarized in the diagram of Fig. 11.5. On the left are indicated the responses to the main types of

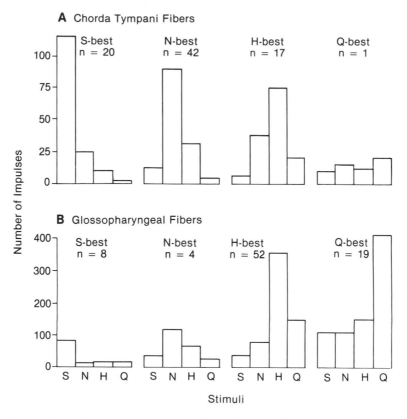

Fig. 11.3 Comparison between response profiles of single fibers in the chorda tympani and glossopharyngeal nerves of the hamster. (From Hanamori et al., 1988)

taste stimuli in the three main taste nerves; note the differing sensitivities of the responses, particularly the preference for acid and bitter in the glossopharyngeal nerve response, as was indicated also in Fig. 11.4. On the right are shown the main types of behavior; in addition to perception, there is a set of reflexes involved in control of motor and secretory actions during food ingestion, and a main class of protective reflexes against ingestion of bitter, toxic, or unpleasant substances. These reflexes are critical for the survival of the organism and, not surprisingly, are built into the brainstem neural circuits early in development. Interestingly, the emotional responses to taste stimuli are also built into the subcortical limbic circuits early in development as well, as we will discuss with regard to feeding mechanisms in Chap. 26.

The Olfactory System

A Comment on Noses

Although we think we smell with our noses, this is a little like saying that we hear with our ear lobes. In fact, the part of the nose we can see from the outside serves only to take in and channel the air containing odorous molecules; the actual sensing is done by receptors lying deep within the nasal cavity. Thus, in analogy with the ear, there is an *outer nose* (what we see from the outside), a *middle nose* (the nasal cavity), and an *inner nose* (the olfactory sensory organ).

In a lower vertebrate like a fish or frog, the nasal cavity may be a relatively simple sac, and the stream of water or air passes directly over the receptor sheet (see Fig. 11.6A). In mammals, the situation is more

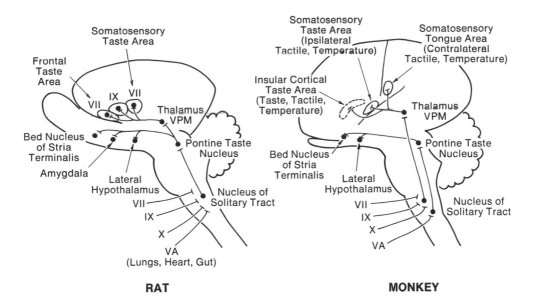

RAT

MONKEY

Fig. 11.4 Taste pathways in the central nervous system of the rat and monkey. VA, visceral afferents; VPM, ventral posterior medial nucleus of the thalamus. (Based on Norgren, 1980, and personal communication)

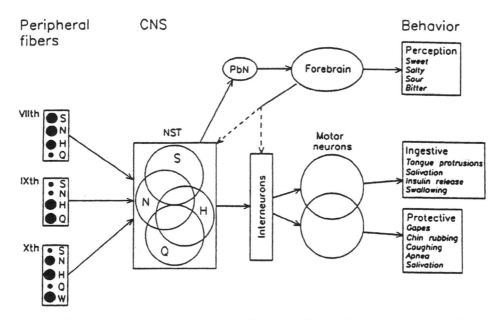

Fig. 11.5 Schematic diagram summarizing *(left)* the preferential taste responses mediated by different sensory nerves; *(middle)* their convergence onto the nucleus of the solitary tract (NST) and various central connections (PbN, parabrachial nucleus); and *(right)* the different types of behavior evoked by taste stimuli. S, sweet; N, salt; H, sour; Q, bitter. See text. (From Smith and Frank, 1993)

253

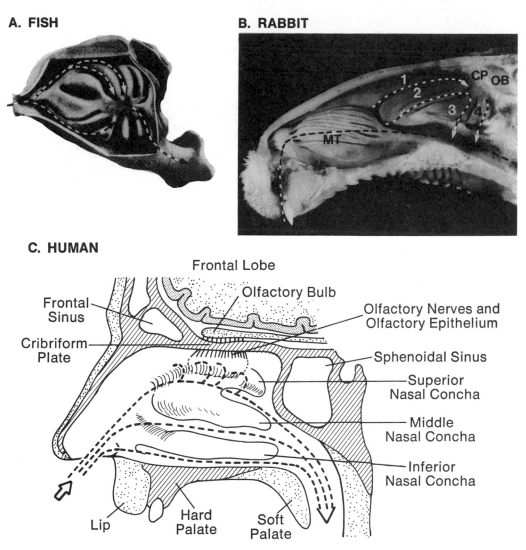

A. FISH

B. RABBIT

C. HUMAN

Fig. 11.6 Comparison of the olfactory organs in several vertebrates. **A.** Fish. Dashed line indicates pathway for flow of water over the folds of the receptor sheet. **B.** Rabbit. Dashed lines indicate air flow. MT, maxillary turbinates, where the incoming air is warmed and humidified; (1–4), nasal turbinates, containing the receptor sheet; CP, cribriform plate, the bone through which the olfactory nerves pass; OB, olfactory bulb, where the olfactory axons terminate. (Photograph courtesy of L. B. Haberly) **C.** Human. Dashed lines indicate air flow over the turbinates and eddy currents over the olfactory epithelium in the dorsal recess of the nasal cavity.

complicated. The keenest sniffers, like opossums, rabbits, or dogs, have wondrously complex nasal cavities (Fig. 11.6B). In these species, the inhaled air first passes through a kind of "air conditioner," which consists of many folds of mucus membrane that warm and humidify it. The air then passes into the pharynx, but it also gives rise to eddy currents that circulate through a complex system of turbinates at the back of the nasal cavity, which are lined with the olfactory receptor neurons. There is as yet no idea of precisely how the air is drawn through these turbinates to reach all the receptors. In the human (Fig. 11.6C), the turbinates are relatively simple, and the olfactory receptor neurons are confined to a patch of membrane in the most dorsal recess of the nasal cavity.

Olfactory Receptor Cells

As noted in Chap. 9 (Fig. 9.15), the olfactory receptor neurons lie in a thin sheet. The mature cell has a long, thin dendrite that terminates in a small knob at the surface (see Fig. 11.7). This knob gives rise to several cilia, which may be up to 200 μm in length but are only 0.1–0.2 μm in diameter. The cilia contain microtubules arranged in the 9 pairs $+2$ pattern typical of cilia elsewhere in the body. When observed under the microscope at high power, the cilia can be seen to be waving, slowly but asynchronously, in contrast to the coordinated, rapid beating of respiratory cilia. The cilia lie in a thin layer of mucus secreted by supporting cells and by Bowman's glands (see diagram, Fig. 11.7). The generation of new olfactory receptor neurons from basal cells has been described in Chap. 9.

Olfactory Transduction: Membrane Mechanisms

Olfactory molecules carried in the air stream within the nasal cavity are absorbed into the mucus and diffuse to the olfactory cilia and terminal knob, where they initiate the transduction process. For many years it was thought that this might involve special kinds of mechanisms peculiar to odors, but it has emerged in recent years (see Chap. 10) that olfactory transduction involves adaptations of general mechanisms of signal transduction. The way this happened makes an interesting story of scientific sleuthing.

The story begins with the development of a cilia-enriched, cell-free preparation of the olfactory epithelium. This enabled Doron Lancet and his co-workers at the Weizmann Institute (Pace et al., 1985) to assay the cilia for components of common second messenger systems that might be involved in odor transduction. They found a highly active adenylate cyclase that was very sensitive to exposure to odorous substances; pharmacological manipulations indicated it was linked to a G protein. This suggested that odor reception involves the binding of odor molecules to a G-protein-linked receptor molecule, in analogy with neurotransmitter and hormone receptors and with rhodopsin in the eye. It further suggested that the cAMP produced by adenylate cyclase might act to phosphorylate sensory membrane conductance channels, in analogy with neurotransmitters and hormones, or it might act directly (Shepherd, 1985), in analogy with the recently discovered direct action of cGMP on the membrane conductance in photoreceptors (see Chaps. 10 and 16).

Physiologists soon tested these ideas, using patch-recording techniques (Fig. 11.7, A, B). Tadashi Nakamura and Geoff Gold at Yale showed that cAMP acts directly on the membrane conductance by exposing inside-out patches from cilia and dendrites to cAMP in the absence of ATP and recording a nonspecific cationic inward current (Nakamura and Gold, 1987). They showed that cyclic nucleotides act directly on membrane conductances in olfactory receptor cells, as they do in photoreceptors. From that time, there has been little doubt that odor transduction involves adaptations of signal transduction mechanisms found in other cells. The next critical step was the demonstration that isolated sensory neurons can respond to odor pulses, and the next, that the cyclic nucleotide–gated conductance and the odor-gated conductance are identical; these steps were accomplished by Stuart Firestein and Frank Zufall at Yale, both for whole-cell currents and at the single-channel level (see Fig. 11.8).

Experiments in several laboratories have shown that odor responses are found only in ciliated neurons; no neurons lacking cilia respond to odors; maximal responses are evoked by pulses applied to the cilia. This has provided direct physiological evidence

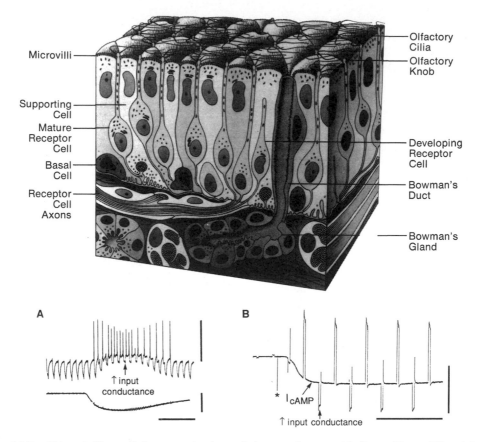

Fig. 11.7 *(Above)* The cellular organization of the vertebrate epithelium. (From Warwick and Williams, 1973) *(Below)* **A.** The response of an olfactory receptor neuron in the salamander to a puff of odor (camphor). The intracellular response consists of a slow depolarizing receptor potential giving rise to a burst of spikes (upper trace). Repeated brief injections of hyperpolarizing current produced smaller downward voltage deflections during the response, showing that the response was associated with an increased input conductance. The summed extracellular voltage response recorded from the epithelial surface is shown in the lower trace. Calibrations: voltage scale, 40 mV for upper trace, 5 mV for lower trace; time scale, 1 sec. **B.** Whole-cell patch-clamp recording from a salamander receptor neuron, showing the action of cAMP. The tip of a recording electrode containing cAMP was sealed onto the membrane of a freshly dissociated neuron. When the membrane under the tip was ruptured by suction (asterisk), cAMP diffused into the cell, eliciting a large inward depolarizing current (slow downward shift of trace), accompanied by an increase in membrane conductance (shown by large responses to brief voltage pulses compared with before rupture; these responses increase, rather than decrease as in A, because the cell is under voltage clamp). Calibrations: current scale, 5 nA; time scale, 5 sec. (A from Trotier and MacLeod, 1986; B from D. Trotier, personal communication)

supporting the biochemical findings that the cilia are the main site of olfactory transduction.

Meanwhile, molecular biologists were identifying the molecular components of the second messenger pathway. Within a

few years, Randy Reed and his collaborators at Johns Hopkins isolated and cloned an olfactory-specific G protein (G_{olf}) and a distinct form of adenylate cyclase (AC III); they and workers at two other laboratories then identified a ciliary membrane channel

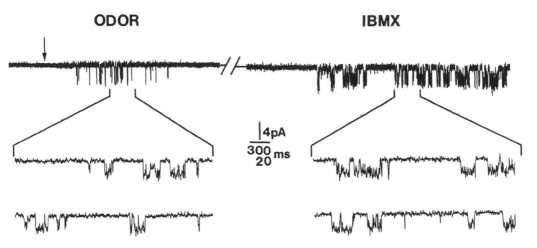

Fig. 11.8 Experiments showing that the odor-sensitive channel is activated by cyclic nucleotides. *(Left)* A cell-attached patch recording from an isolated olfactory cell is made during brief (150 msec) application of a microjet of odor at the arrow. The odor stimulus activates a second messenger system, which induces the single-channel activity shown on a slow sweep above and on two fast sweeps below. *(Right)* the cell was then exposed to IBMX (100 μM), which inhibits phosphodiesterase, causing an increase in cAMP concentration and increased single-channel activity similar to that due to odor stimulation. (From Firestein et al., 1991)

and showed that it shares a high degree of sequence homology with the photoreceptor channel (reviewed by Reed, 1992; see Fig. 16.9). The final challenge was to track down and identify the G-protein-linked receptors. This was first accomplished by Linda Buck and Richard Axel at Columbia, using an ingenious strategy for PCR and molecular cloning, drawing on the prediction (Lancet, 1986) that the receptors would be encoded by a large gene family with sufficient diversity to interact with a broad range of odor molecules. The proteins they isolated are members of the seven transmembrane domain (7TD) superfamily of G-protein-linked receptors, which includes neurotransmitter receptors (serotonergic, β-adrenergic, muscarinic, and histaminergic) as well as rhodopsin (Buck and Axel, 1991). Subsequent work in many laboratories has added to the numbers of members of the olfactory gene family. A hypothetical view of the way an odor ligand may interact within the pocket of the 7TD odor receptor is shown in Fig. 11.9. The reader should compare this receptor with its cousins, the neurotransmitter receptors (Chap. 8) and rhodopsin (Chap. 16), to appreciate the kinship within the superfamily for these different receptor functions.

These different lines of evidence have converged on a consensus model for the second messenger pathways in olfactory transduction (Fig. 11.10). Note the evidence for phospholipid as well as cAMP pathways in some species; the presence of a Ca^{++}-activated Cl^- conductance that contributes to the sensory current; the possible involvement of the nitric oxide-cGMP second messenger pathway; and the multiple mechanisms for desensitization and adaptation. The reader should also review Chap. 8, where there is a more complete discussion of the general properties of these different second messenger systems and a discussion of the special properties of the cyclic nucleotide–gated channel.

Processing of Odor Information

Sensory Neurons. The receptor potential spreads from the cilia through the dendrite

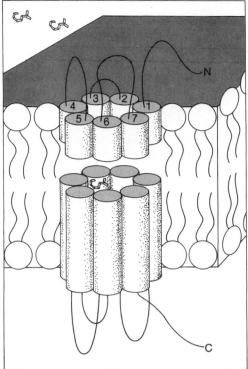

Fig. 11.9 The olfactory receptor molecule is a member of the seven-transmembrane domain (7TD) gene family. The diagram shows a model in which odor molecules (cineole) are proposed to bind within the pocket of the receptor, at a depth similar to that for the binding of ligands to the β-adrenergic receptor (see Chap. 8), and in a position where the benzene head of retinal is bound to rhodopsin (see Chap. 16). (From Shepherd and Firestein, 1991a)

to the cell body by essentially the same process of electrotonic spread that we discussed in the crustacean stretch receptor cell (Chap. 7). In the cell body the depolarization triggers action potentials, which are conducted in the axon to the first relay station, the olfactory bulb.

In response to stimulation with a battery of odors, a given receptor cell characteristically shows a broad spectrum of sensitivity. As shown in Fig. 11.11 (left), it may respond briskly to odor A, weakly to odor B, and not at all to odor C. A cell may be sensitive in this way, in varying degrees, to a number of different odors. These re-

sponses thus fall into the category of "odor generalist," characteristic of all vertebrate olfactory sensory neurons thus far examined, and many insect olfactory receptor neurons. Odor "specialists," such as those in insects that are narrowly responsive only to pheromones, have not been seen in the vertebrate, though the presence of neurons narrowly tuned to mammalian pheromones seems likely.

Several studies have shown that olfactory receptor cells respond to prolonged or repeated stimulation with a slowly adapting, prolonged discharge of impulses (see Fig. 11.11). This is in contrast of course to our behavioral experience that odor sensations fade rapidly: when entering a closed, stifled room with an unpleasant odor, we are initially disgusted, but after a minute or two we may cease to be aware of the odor unless we inhale more deeply. It appears that this adaptation is not due to fading of the receptor response, but rather is a property of inhibitory interactions in central neural circuits in the olfactory pathway.

The Olfactory Bulb. In the olfactory bulb, the olfactory axons terminate in rounded regions of neuropil called glomeruli. These are a constant feature of the bulb throughout the vertebrates, as well as most invertebrates. This suggests that glomeruli play an essential role in processing information carried in odor molecules, a point we will discuss further below. The reader should review the synaptic organization of the olfactory bulb in Chaps. 6, 7, and 10. The olfactory axons make synapses within the glomeruli onto the dendrites of mitral and tufted cells, which are the output neurons of the olfactory bulb.

In single-unit recordings, mitral cells show a range of responsiveness to different odors. There are three basic patterns. As shown in Fig. 11.11 (right) a cell may respond with excitation: a slow, prolonged discharge at threshold, changing to a brief burst followed by suppression at higher concentrations. Or a cell may respond with suppression throughout the stimulation pe-

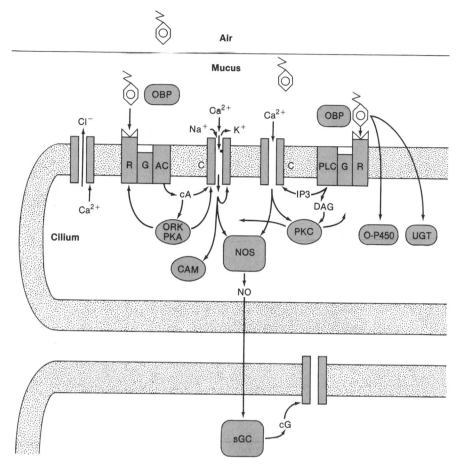

Fig. 11.10 Summary diagram of sensory transduction mechanisms in olfactory cilia. Odor ligands (O) are absorbed from the air into the mucus and act on membrane receptors (R). This activates the G protein-mediated stimulation of adenylate cyclase (AC), producing cyclic AMP (cA), which then activates the cyclic nucleotide–gated channel (C), allowing cations (Na^+, Ca^{2+}, K^+) to diffuse down their electrochemical gradients. The Ca^{2+} acts externally on the channel pore to reduce its conductance, and internally to oppose activation, thereby contributing to adaptation. A Ca^{2+}-activated Cl current also contributes to the sensory response. Other mechanisms contributing to olfactory adaptation incude olfactory receptor kinase (ORK), phosphokinase A (PKA) and calcium/calmodulin (CAM). In some species certain odor ligands also activate a second messenger pathway through phospholipase C (PLC), producing IP_3 and diacylglycerate (DAG), which act on the channel and phosphokinase C (PKC), respectively. Interciliary interactions may occur through activation of nitric oxide synthase (NOS), producing nitric oxide (NO), which diffuses to nearby cilia, activating cytosolic guanylate cyclase (cGC), producing cyclic GMP (cG), which activates the cyclic nucleotide–gated channel. Olfactory binding protein (OBP) in the mucus may be involved in either carrying odor molecules to receptor sites or assisting in removing them. Inactivation of odor molecules may occur by binding to olfactory P450 and uridyl glucuronic transferase (UGT) enzymes.

riod, and at all concentrations. Or a cell may not be affected at all by a given odor. One interpretation has been that the suppression is due to inhibition mediated by the dendrodendritic synapses of granule

cells onto mitral cells (see circuit diagrams in Figs. 6.16 and 7.17). These extensive connections in effect lay down a curtain of inhibition within the bulb, which the excitatory responses punch through, as it

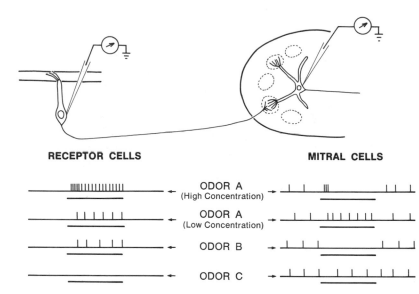

RECEPTOR CELLS   MITRAL CELLS

ODOR A
(High Concentration)

ODOR A
(Low Concentration)

ODOR B

ODOR C

Fig. 11.11 Extracellular single-unit recordings of responses to odors of receptor cells *(left)* and mitral cells *(right)* in the salamander, showing different types of responses and different temporal patterns of activity. (After Kauer, 1974, and Getchell and Shepherd, 1978)

were, carrying specific information about a given odor and its concentration.

Spatial Organization

In order to understand how odor information is processed in the olfactory pathway, we next must inquire how the pathway is organized. In other sensory systems, such as visual, auditory, and somatosensory, one of the important functions is to construct a map of the spatial location of a stimulus. This requires a correspondingly well-organized pathway, with precise topographical projections from points at one level of the pathway to the next. In the sense of smell, however, we do not keep track of the location of an odor source by its location on the sensory sheet, so the space on the sensory sheet, and at higher levels in the olfactory bulb and beyond, is available to process other aspects of the stimulus. Therefore, neural space in this system can be used to process nonspatial information, such as that carried in odor molecules.

The first suggestion that this might be the case was made around 1950 by the physiologist Lord Adrian and the neuroanatomist Wilfrid le Gros Clark, two pioneers in the rise of modern neuroscience. Equally tenable, however, was the opposite view, that the olfactory pathway could function by means of purely diffuse projections from one level to the next. The problem was finally attacked by applying tract-tracing methods, such as axonal transport of radioactively labeled amino acids and retrograde transport of HRP. These results left no doubt that in fact there is a considerable degree of spatial organization. For example, microinjection of HRP into the glomerular layer of the olfactory bulb of the cat results in retrograde labeling of sensory neurons in a limited part of the nasal cavity, and there is a systematic shift of the labeled neurons with injections around the circumference of the bulb (see Fig. 11.12). The projections are not point-to-point, as at the lower levels of the visual or somatosensory pathways, but they show varying degrees of convergence of sensory axons onto single glomeruli or small groups of glomeruli.

The spatial organization of the epithe-

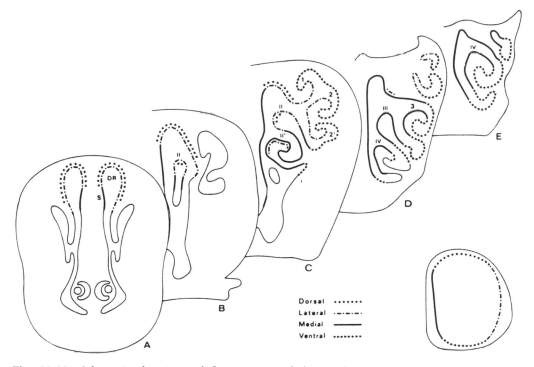

Fig. 11.12 Schematic drawings of five sections of the nasal cavity (A–E, rostral to caudal), illustrating the epithelial projections related to the different surfaces of the bulb, as shown by retrograde transport of HRP from local injections in the olfactory bulbs of rats. In the lower right, a frontal section of the bulb is shown with symbols representing each bulbar surface. DR, dorsal recess; S, septal wall; (1–3), ectoturbinates; (II–IV), endoturbinates. (From Astic and Saucier, 1986)

lium has also been studied by in situ hybridization, using probes for members of the olfactory receptor gene family described above. In the mouse and rat these probes localize to distinct domains within the epithelial sheet. There is thus the possibility that the cells within these domains project to specific glomeruli, as suggested by the HRP results.

Spatial Activity Patterns

How might this spatial organization come into play during processing of odor information? To answer this question, one needs to be able to map odor-induced activity at the different levels in the pathway. The means to do this was first provided by the 2-deoxyglucose (2DG) method of Sokoloff. As already discussed in Chap. 8, odor stimulation elicits different degrees of activity in different glomeruli. This is presumably

due to different degrees of activity in the receptors converging onto the glomeruli. Stimulation with one odor, such as amyl acetate (which has a fruity smell), activates glomeruli in certain regions of the olfactory bulb of a rat. Stimulation with a different odor, such as camphor, activates glomeruli in regions which overlap those of amyl acetate, but nonetheless have a different distribution. Maps for these and several other odors that have been tested are summarized in Fig. 11.13.

These results, together with those from studies using electrophysiological recordings and anatomical tracing, have supported the suggestion that the spatial distribution of activity in the olfactory bulb may carry part of the neural code for different odors. These results are supported by other types of activity markers. For example, expression of mRNA for c-*fos*, an immediate

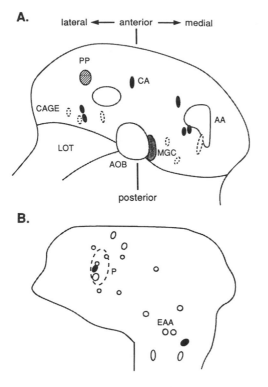

A.

lateral ◄── anterior ──► medial

PP

CA

CAGE

AA

LOT

MGC

AOB

posterior

B.

P

EAA

Fig. 11.13 Summary maps of 2DG activity foci in the glomerular layer of the rat olfactory bulb in response to different odors. AA, amyl acetate; CA, camphor; CAGE, air from the home cage of the rat; EAA, ethyl acetoacetate; PP, peppermint. (From several laboratories, summarized in Shepherd, 1991)

pression of mRNA for c-*fos*, an immediate early gene (see Chaps. 2 and 9), during stimulation with a given odor is localized to cells in specific regions of the olfactory bulb similar to the 2DG domains (Sallaz and Jourdan, 1993). It appears that space can be used for encoding nonspatial aspects of the stimulus, such as the stereochemical configurations of the odor molecules or specific binding properties. These studies thus give insight into how the nervous system maps nonspatial stimulus parameters into neural space.

A Consensus Model for Odor Processing

How do we bring all this information together into a coherent model for the neural basis of olfactory processing? There is by

now an extensive literature on the organization of the projection from the olfactory sensory neurons to the glomeruli of the olfactory bulb, based on anatomical, functional, monoclonal antibody, and molecular biological data. The simplest working hypothesis for the organization of the projection is shown in Fig. 11.14 and may be summarized as follows (Shepherd, 1992).

The key aspects of the model are that sensory neurons with similar ligand specificities project to common glomeruli. The simplest hypothesis is that each neuron expresses a single receptor molecule type; this is undoubtedly true for specialist neurons and one, or only a few, is a reasonable expectation for generalists as well. Narrowly tuned specialist neurons (ON1 in Fig. 11.14) project exclusively to specialized glomeruli (G1 in the diagram); conversely, these glomeruli receive input only from their appropriate specialist neurons. Generalist neurons, on the other hand, show a relaxation of this connectivity rule; there is a preponderance of neurons with similar ligand specificity (for example ON3) projecting to one glomerulus (for example, G3), but there is also a proportion going to other glomeruli as secondary projections. Thus, each glomerulus has a primary input from a set of neurons of similar ligand preference and secondary inputs from sets of neurons with related preferences. This provides a simple algorithm for generating sets of neurons by mitotic division from common stem cells and establishing their connections in the olfactory bulb. Each glomerulus thus has a set of primary and secondary connections which may be characterized as a complex labeled line.

On the right (B) in Fig. 11.14 are shown some of the properties critical for the functioning of the model. At the sensory neuron level (top), response spectra for specialist and generalist types are indicated. The considerable overlap of response spectra of individual neurons is similar to the overlap of response spectra of different cone photoreceptors; in analogy with color vision (see

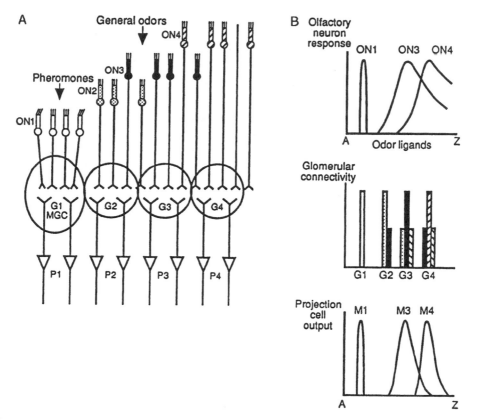

Fig. 11.14 Principles underlying the organization of olfactory glomeruli. A. Sets of olfactory sensory neurons, either specialist (ON1) or generalist (ON2–4), project their axons to the glomeruli. Specialist axons all converge on one target (G1), the macroglomerular complex (MGC) as identified in male insects. By contrast, a given set of generalist sensory neurons sends a primary projection to one or a few primary glomeruli (G2–4), but secondary projections to related glomeruli. Processing within a glomerulus takes place by dendrodendritic microcircuits, whereas interactions between glomeruli are mediated by interglomerular axons or neurites. Outputs from glomeruli are carried by projection (P) neurons (mitral or tufted cells in vertebrates) to central regions (including mushroom bodies in insects, olfactory cortex in vertebrates). B. Functional properties. *(Top)* Specialist sensory neurons (ON1) have membrane receptors with narrow affinities for specific pheromone molecules. Generalist sensory neurons (ON2–4) have receptors with broad overlapping affinities for diverse odor molecules (food, fragrances, repulsive, and so on). By virtue of the convergence patterns on the glomeruli (glomerular connectivity, *middle*), and intra- and interglomerular processing, the projection neurons *(bottom)* extract discriminatory features of the odor molecules and relay this information to central regions. Based on Matsumoto and Hildebrand, 1981; Boeckh et al., 1990; Lancet, 1991; Kauer, 1987; and Shepherd, 1992)

Chap. 16), the overlap is presumably the basis for discriminating odor ligand type independently of odor ligand concentration. At the glomeruli (middle) the preferential connectivity of the primary projection further enhances the response preference of a given glomerulus for a given ligand or related ligands. Finally, at the bulbar output (bottom), narrowed or more specific response spectra of the output neurons (mitral and tufted cells in vertebrates) would reflect lateral contrast enhancement and odor opponency mechanisms within intrabulbar circuits. Analogous mechanisms may operate within the antennal lobe of insects.

This schema may be considered only as a working hypothesis for further testing by experimental analysis and computational modeling. Its main merit is that the properties of sensory transduction in a single neuron can be assessed within a framework of their significance for the properties of the populations of neurons that constitute the olfactory projection. It is only by having a clear concept of that framework of olfactory sensory neuron populations with their differentiated properties that we will be able to understand how the molecular information transduced by the sensory neurons provides the basis for odor perception and discrimination.

Plasticity

The sense of smell is important in controlling early behavior in many species. During this early period, the system displays a remarkable degree of plasticity. Evidence for this plasticity has come to light from studies of rat pups. In one study, Patricia Pedersen and Eliott Blass (1982) at Johns Hopkins wished to know whether an artificial odorous substance could substitute for the normal pheromone that mediates suckling of rat pups. They injected citral (a lemony substance) into the amniotic fluid to expose the fetuses to this substance, delivered the pups by cesarian section, and found indeed that the presence of citral on the mother's nipples was required for suckling to occur. This showed that the olfactory system can adapt to different molecular stimuli for eliciting this crucial behavior.

The neural basis of this type of plasticity has been analyzed by Michael Leon and his colleagues at Irvine. They exposed rat pups to a specific odor (peppermint) for 10 minutes a day during the first 19 days of life, and then examined the 2-deoxyglucose patterns induced in the olfactory bulb by this odor. A distinct focus was seen in the autoradiograms, which was correlated with a recognizable glomerulus not present in control animals (see Fig. 11.13; Leon et al., 1987). It appears that repeated stimulation enhances the response and may induce a morphological rearrangement of the receptor axon terminals in the olfactory bulb. The results are of further interest for supporting the idea that a glomerulus may operate as a functional unit in processing odor information, in analogy with cortical barrels (Chap. 12) or columns (Chap. 16).

Neurotransmitters and Neuromodulators

The olfactory bulb is a rich repository of neuroactive substances. It has among the highest levels of taurine (an amino acid), carnosine (a dipeptide), thyroid hormone releasing hormone (a tripeptide), and opioid receptors in the entire brain. It is also the recipient of many centrifugal fibers from the brain. These arise from the olfactory cortical areas, the basal forebrain (horizontal limb of the diagonal band), and the midbrain (locus ceruleus and raphe).

Through these fibers the olfactory bulb is modulated by central limbic centers, so

---

Fig. 11.15 Schematic diagram of the mammalian olfactory system. This view shows the three levels of the system, proceeding from the olfactory epithelium at the top, through the olfactory bulb, to the olfactory cortex (bottom). There is an attempt to emphasize the different subsystems as they are seen at each level. AMYG, amygdala; AOB, accessory olfactory bulb; AON, anterior olfactory nucleus; epl, external plexiform layer; gc, granule cell; gcl, granule cell layer; glom, glomerular layer; mc, mitral cell; MD thalamus, medial dorsal nucleus of the thalamus; MGC, modified glomerular complex; ml, mitral layer; NHLDB, nucleus of the horizontal limb of the diagonal band; on, olfactory nerve layer; orn, olfactory receptor neuron; OT, olfactory tubercle; PC, piriform cortex; pg, periglomerular cell; SO, septal organ; tc, tufted cell; TEC, transitional entorhinal cortex; VNO, vomeronasal organ. (From Shepherd et al., 1987)

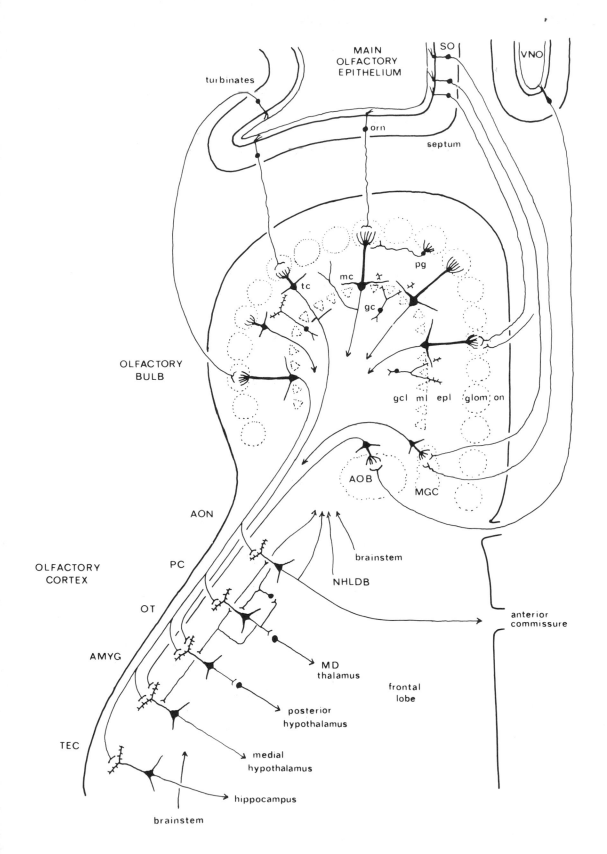

that a given odorous substance has a different meaning, depending on the behavioral state of the animal. Thus, the aromas of food are perceived quite differently, depending on whether we are hungry or sated. Through these centrifugal connections, the bulbar microcircuits are an integral part of central limbic circuits as they mediate nonolfactory functions as well. These multiple controls of olfactory bulb synapses and circuits presumably reflect the sensitiveness of olfactory functions to different developmental and behavioral states, as well as the pervasive influence of olfactory inputs on the life of the animal. This appears to be as true for the lives of most vertebrate species as for invertebrates, and we shall see ample evidence of this when we discuss feeding (Chap. 26) and mating (Chap. 27).

Olfactory Cortex

From the olfactory bulb the mitral and tufted cells send their axons to the olfactory cortex. As shown in Fig. 11.15, there is a parallel pathway, from the vomeronasal organ to the accessory olfactory bulb, and from there to a specific cortical site in the amygdala. The vomeronasal organ is a narrow tube lined with olfactory receptors; it is well developed in certain reptiles and mammals. The receptors appear to be tuned to specific kinds of substances in these species. For example, in snakes, scents of prey are detected by flicking the tongue into the air and drawing it back over the inlet of the organ at the base of the nose. In hamsters, the vomeronasal organ mediates part of the response of a male to the vaginal odor emanating from a receptive female, and is thus an important link in reproductive behavior (see Chap. 27). The vomeronasal organ has been considered vestigial in humans, but that question is beginning to be re-examined.

The olfactory cortex is divided into five main areas. Each has distinct connections and different functions, which we will briefly describe. The *anterior olfactory nucleus* is an integrative center connecting the

two bulbs through the anterior commissure. The *piriform cortex* is the main area involved in olfactory discrimination. The *olfactory tubercle* is also the recipient of ascending dopaminergic fibers from the midbrain (Chap. 26); it has been implicated in various functions of the limbic system (Chap. 28). It has been hypothesized that malfunction of the olfactory tubercle (the anterior perforated substance in the human brain) may contribute to certain kinds of schizophrenia. The *corticomedial* parts of the *amygdala* receive inputs from both the main and accessory olfactory bulbs. Finally, a part of *entorhinal cortex* receives olfactory input, and projects to the hippocampus.

Some of the central projections of the cortical areas are indicated in Fig. 11.15. The piriform cortex projects to the mediodorsal thalamus, which in turn projects to the frontal lobe. This presumably is the circuit for conscious olfactory perception and discrimination. The amygdala, by contrast, projects mainly to the hypothalamus. Through those connections the amygdala is more concerned with the emotional and motivational aspects of odor stimuli (see Chap. 28).

Despite the varying connections and functions of these different cortical regions, they are all constructed on the same general plan. The reader may review in Chap. 6 how synaptic circuits are built of triads of neuronal elements: input, principal, and intrinsic. As shown in Fig. 11.15, in olfactory cortex the afferent input is to the distal dendrites of pyramidal neurons. In addition to their output axon, pyramidal neurons give rise to two types of intrinsic circuit: recurrent excitation through long association fibers, and recurrent inhibition through activation of inhibitory interneurons. This is a basic circuit for initial cortical processing of olfactory information, which is adapted in each of the subregions for differing specific functions. In Chap. 30, we will see how this basic circuit underlies the organization of other parts of the cerebral cortex.

12

The Somatic Senses

Every organism has an external skin or other covering that encloses its body and separates it from the environment. Through this covering, the animal receives information about the presence of objects, other organisms, or physical changes in its environment. The Greek word for body is *soma,* and the sensory modalities that are signaled by receptors at the body surface are referred to collectively as the *somatic senses.*

Before identifying the different components of the somatic senses, it is well to be reminded that the body covering in which receptors are embedded is not a simple structure. Table 12.1 lists some of the functions of the covering (integument) in different animals. It can be seen that the integument has a complex structure which provides for multiple functions. Sensory reception is only one among many special functions. Thus, the sensory receptors in any given animal have structures which reflect the particular functional adaptations of the integument in that species.

Because of the enormous variations in the structure and function of the integument in different species, it is not always easy to assign particular receptor cells to distinct categories. By the same token, it is

Table 12.1   Functions of the integument

Protection against:
  physical harm
  chemical damage
  infection
  water gain or loss
  excessive sunlight
  heat gain or loss
Camouflage (pigmentation)
Structures for:
  locomotion (cilia, feathers)
  aggression and defense (claws, horns, antlers)
  social displays
Exchange of substances:
  respiration of O_2 and CO_2
  excretion of metabolites
Glandular excretions
Special sensory structures

hazardous to assume too readily that a particular sensory modality is exactly equivalent in widely separated species. These considerations make it difficult to apply the terms "modality," "submodality," and "quality" in a consistent fashion. Despite such misgivings, however, it is common to consider that the somatic senses are comprised of four main modalities. These were summarized in Table 10.1. Perhaps the most basic and primitive is the *noxious* sense, the reception of stim-

267

uli that are harmful or signal potential harm to the organism. A second modality is the ability to sense the ambient *temperature.* A third category may be referred to as *crude touch;* this includes *light touch* and *pressure,* which are distinct in some organisms and mixed in others. Finally, there is the fine *tactile* sense, the ability to make precise surface discriminations in space and time.

These sensory modalities can also be grouped according to the nature of the stimulus energy. Thus, stimuli that affect the body may be classed as chemical (some kinds of noxious stimuli), radiant (temperature, sometimes noxious), and mechanical (some noxious stimuli, crude touch, and tactile). Chemical stimuli require chemical transduction by *chemoreceptor* mechanisms in the sensory membrane; temperature is transduced by *temperature receptors;* and the various kinds of mechanical stimuli are transduced by *mechanoreceptors.* The molecular bases of the transduction mechanisms have already been discussed in Chap.10.

Regardless of how one classifies the sensory modalities, the receptors share some basic properties, even across different phyla. Our focus will be on the properties of the receptor cells, and the organization of the sensory pathways, using the mammal, and, where possible, the human, as our model.

Human Skin

Vertebrates differ from invertebrates in that they do not have cuticles and most do not molt, but in other respects their integument is similar in that it can take on a variety of structures and perform many different functions. The structural adaptations include a toughening of the surface to form scales, as in fish, or plates, as in turtles. The integument also may give rise to auxiliary structures such as feathers, as in birds, and hair, as in mammals—structures that are crucial for a species' motor abilities or its ability to withstand extremes of climate and temperature. There are also special adaptations, such as hoofs and claws, for particular kinds of locomotion or manipulation.

There are many fascinating adaptations of the skin senses to these specialized integumentary organs. However, our concern here will be with the reception of stimuli over the whole body surface. By far the greatest amount of information about the general somatosensory system in vertebrates has been obtained in the mammal, much of it in primates and humans. This is partly because the mammalian skin is a relatively general, unspecialized covering. Being soft, it is easy to dissect and experiment upon. An important advantage is that in humans one can stimulate different receptors selectively and test the sensory perception that is aroused. Let us therefore focus our attention on the mammalian skin in general, and the special properties of the human skin where information is available.

The human skin is a complex and fascinating structure. As we all recognize, there are two main types of skin. One type, found on our palms and fingertips, is called *glabrous,* or hairless, skin. The other type, found over most of the rest of our body, is called *hairy* skin. Of course, hair on humans varies widely in amount, from a little to a lot, and in the kind of hair, from peach fuzz to coarse. These variations do not affect the division of skin into these two general classes.

The structures of glabrous and hairy skin are shown in the diagrams of Figs. 12.1 and 12.2. First, we see that both types of skin are divided into two main layers, the epidermis and the dermis. The *epidermis* is the true outer skin, being derived from the ectodermal germ layer of the embryo. It also gives rise to the various specialized structures (hair, feathers, claws, and glands) that become especially prominent in higher vertebrates. The epidermis is composed of a *stratum germinativum,* where cells undergo continual mitosis and migrate toward the surface. As they reach the surface, they undergo degenerative changes,

GLABROUS (HAIRLESS) SKIN

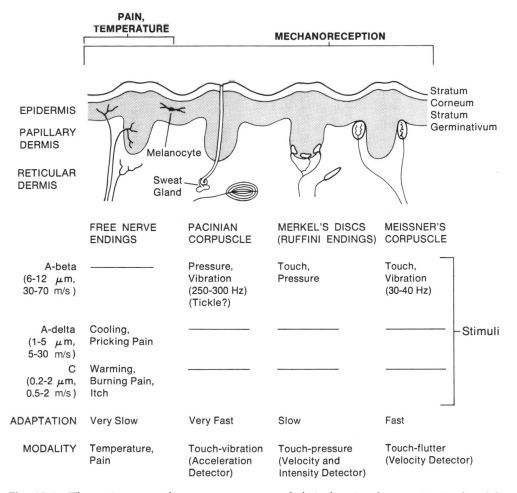

Fig. 12.1 The main types of sensory receptors and their functional properties in the glabrous (hairless) skin of the human. *(Above)* Diagrams showing the structure and location of each type. *(Below)* List of the specific best stimuli, correlated with the type of sensory nerve fiber innervating the receptor (medium-thick myelinated A-beta fibers, thin myelinated A-delta fibers, and thin unmyelinated C fibers; for further information on fiber types, see Table 13.2).

so that at the surface they form a layer of dead, flattened cells, the *stratum corneum*. This is rich in *keratin*, a very stable fibrous protein, resistant to water, most chemicals, and enzymatic digestion. Also present in the epidermis are *melanocytes*, which contain melanin, derived metabolically from tyrosine. Melanin is responsible for the pigmentation of the skin, and it protects the skin and deeper layers from ultraviolet light.

Beneath the epidermis is the *dermis*. This is actually derived from the mesoderm, and it becomes connected to the epidermis during embryonic development. The dermis is the layer that provides for the thick, bony scales that are so characteristic of lower vertebrates like fish. It is only in higher vertebrates that the dermis has evolved a soft, flexible structure, due to a thick, tightly interwoven layer of *connective tissue*. Rich in collagen fibers, this layer also

HAIRY SKIN

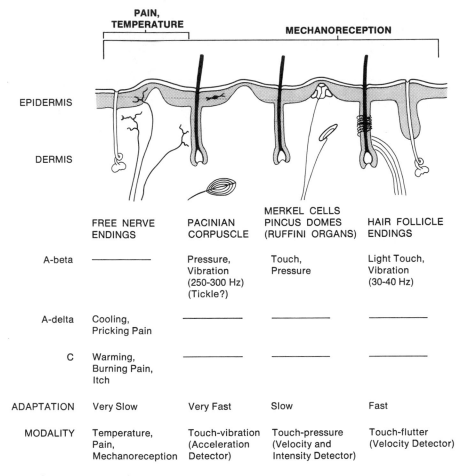

Fig. 12.2 The main types of sensory receptors and their functional properties in the hairy skin of the human. Compare with Fig. 12.1

contains elastic fibers and, in deeper parts, fat cells. These components not only permit the freedom of movement so characteristic of mammals, but also are important for other functions such as protection, insulation, and temperature regulation.

Sensory Receptors

We are now in a position to identify the sensory elements within the skin. By way of brief history, this work began with the earliest microscopic studies of the skin in the middle of the nineteenth century. By

around 1900, a number of specific, small end organs, each attached to a sensory nerve fiber, had been identified. The main interest of the physiologists of the time was in correlating each end organ with a specific sensation. To do this, the skin was stimulated with very small probes: a fine hair or bristle; a pin; a thin wire, heated or cooled. The results showed that sensitivity is not uniform over the skin surface; instead, there appeared to be a mosaic of sensitive "spots." There were "touch spots," "warm spots," "pain spots," and so forth. Furthermore, a given spot seemed to be sensitive to only one modality. These results implied that a sensitive spot is the site of a specific sensory end organ for that modality. A number of correlations between sensory modality and sensory end organ were suggested; thus, free nerve endings for pain, Ruffini organs for warmth, Krause's end bulbs for cold, Pacinian corpuscles for pressure, Meissner's corpuscles and hair follicle endings for touch. These correlations seemed to be a logical realization of the doctrine of specific nerve energies set forth by Müller in the 1830s, as we discussed in Chap. 10.

Our modern correlations differ in certain details, and are not as strict as originally envisioned. Nonetheless, the idea that nerve endings, especially the specialized end organs, have a *preferential* sensitivity for certain types of stimuli is widely accepted, and is a useful framework for our beginning study. Let us therefore consider each of the main types of sensory ending. We will start with free nerve endings, and move toward the more differentiated organs. Although this reverses the sequence in which they are usually described, it is more in line with the evolutionary development from simple to complex, and leads naturally to the description of central pathways later in the chapter.

Free Nerve Endings

The simplest type of sensory receptor in the skin is the free nerve ending. This is just what its name implies: a nerve fiber divides into branches and terminates in naked, unmyelinated endings in the dermis and deeper layers of the epidermis. The modes of termination are similar in glabrous and hairy skin, as indicated in Figs. 12.1 and 12.2.

Free nerve endings respond to mechanical stimuli, heating, cooling, or noxious stimuli. Some endings respond to only one modality; others respond to two or three modalities—these are called *polymodal* receptors. The endings arise from thin fibers, either thinly myelinated axons (*A-delta*

[Aδ] *fibers)* or unmyelinated axons (*C fibers).* The sensations evoked in these two types of fiber are distinctive. When we touch a hot stove, the immediate sharp pain (*pricking pain)* is mediated by the Aδ fibers; the subsequent constant aching (*burning pain)* is mediated by the C fibers.

The mechanisms of activation of these two types of pain fiber are illustrated further in Fig. 12.3. Some nociceptors of C fibers respond to histamine, serotonin, or other chemical mediators that contribute to the inflammatory response to injury. Those that respond to histamine (a potent itch-producing chemical released from mast cells) may contribute to the sensation of itching, whereas others, responding to algesic (pain-producing) chemicals such as bradykinin, produced from a blood-borne precursor, contribute to pain. These responses may be regarded as part of the *common chemical sense.* Among nociceptors responsive to mechanical stimulation, some with thinly myelinated fibers contribute to mechanically evoked pricking pain, whereas some with unmyelinated axons contribute to *aching pain,* produced by heavy, blunt pressure (the student may pinch the web of skin between two fingers to experience this). Endings with combined thermoreceptor and nociceptor sensitivity signal *pricking pain,* a sensation akin to that produced by jabbing a pin into the skin.

Temperature reception is also apportioned to different fibers in primates, *cooling* being sensed mainly by a subpopulation of Aδ fibers and *warming* being sensed by certain C fibers. Cooling and warming receptors are actually defined by the fact that they have peak sensitivities that are cooler and warmer, respectively, than the body temperature. As shown in Fig. 12.4, each may fire faster or slower with increases or decreases in temperature, depending on where in the whole temperature range the change is taking place. Presumably, the overlapping ranges of warming and cooling receptors are part of the mechanism for enhancing the ability to discrimi-

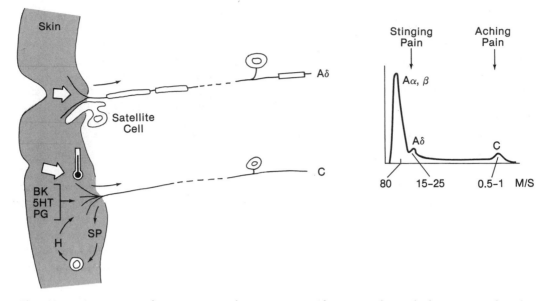

Fig. 12.3 Activation of receptors mediating pain. *(Above)* Mechanical damage to the skin (scratching, sticking, pinching, or pulling) or intense heat activates the endings of moderately fast conducting myelinated fibers (Aδ range), which mediate the sensation of sharp stinging pain. In the human, most Aδ fibers respond to both these modalities. *(Below)* Unmyelinated C fibers have a slow conduction velocity of 0.6–0.8 m/sec. The endings of these fibers respond to a variety of harmful stimuli, including mechanical damage, excessive heat, and noxious chemicals such as bradykinin (BK), serotonin (5HT), and prostaglandins (PG); these fibers are therefore termed "polymodal nociceptors." In the human, nearly all unmyelinated fibers are nociceptors. A common mode of stimulation is by release from the nerve endings of the neuropeptide substance P (SP), which activates mast cells to release histamine (H), which excites the endings. *(Right)* Recording of the compound action potential from a cutaneous peripheral nerve in response to a single electric shock, showing the different components of fast conducting (Aα, β), medium conducting (Aδ), and slow conducting (C) fibers. M/S: meters per second. (Based on Light, 1992)

Fig. 12.4 Graphs of impulse firing frequencies in relation to applied temperature for "cold" and "warm" receptors. (Modified from Kenshalo, 1976, in Schmidt, 1978)

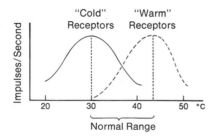

The free nerve endings thus provide for a rich variety of polymodal sensory reception. All of these fibers are slowly adapting; they continue to discharge impulses as long as the stimulus is present. This seems a necessary property for signaling the relatively slow changes that take place in the modalities of pain and temperature, and for signaling the organism to remove the noxious stimulus or adjust itself to the ambient temperature.

Pacinian Corpuscle

nate small changes in temperature near body temperature; in this range, an increased response in one type is accompanied by a decreased response in the other.

The other sensory receptors in the skin are associated with special end organs, and all are endings of medium-sized myelinated fibers, the A-beta (Aβ) type (6–12 μm in diameter). For convenience we start with the Pacinian corpuscle, one of the largest of the end organs, situated deepest in the dermis. The Pacinian corpuscle has a widespread distribution, in the connective tissue of muscles, the periosteum of bones, and the mesentery of the abdomen. It is relatively easy to isolate single corpuscles from the mesentery, and much of our information about Pacinian corpuscles, and indeed about many basic properties of sensory receptors in general, has come from these studies.

The Pacinian corpuscle, like the onion, is composed of concentric layers of cellular membranes alternating with fluid-filled spaces. The picture that has emerged from electron microscopic studies is the naked ending of the nerve fiber surrounded by an inner core of incomplete shells of cell processes and collagen fibers. Making up the bulk of the corpuscle are outer complete lamellae.

The Pacinian corpuscle has been studied physiologically by pressing on it with a carefully controlled probe while recording the response from the nerve where it exits from the corpuscle. If impulse activity is blocked selectively with such agents as local anesthetics or tetrodotoxin, we can record the receptor potential as it spreads into the nerve fiber. As shown in Fig. 12.5A, the receptor response takes the form of a very brief potential wave at the onset of the pressure pulse, and a similar brief wave at the termination. There is no response during the stationary plateau of the applied stimulus. This is therefore an extremely rapidly adapting receptor response. Each wave gives rise to a single impulse in the nerve fiber.

How does this type of response arise? In the isolated Pacinian corpuscle preparation, it was possible for Werner Loewenstein and his colleagues (1971), then at Columbia University, to dissect away very carefully the onion-skin lamellae, so that the stimulus could be applied directly to the naked nerve ending. When this was done, the receptor potential produced by a step pulse was a slowly adapting response (see Fig. 12.5B). This showed that the lamellae act as a filter, to absorb slow changes impressed upon them, while still passing on rapid changes to the nerve endings. Interestingly, the nerve fiber, at the first node of Ranvier, still responds with a single impulse despite the maintained receptor potential, a very nice matching of receptor and nerve properties.

The Pacinian corpuscle is thus constructed to signal rapid changes in *touch-pressure*. This organ is well suited for signaling rapid vibratory stimuli; as indicated in Fig. 12.5C, the maximum sensitivity is in the range of 100–300 Hz. This form of stimulation may be important in our tactile perception of finely textured surfaces or objects. Note the extremely high sensitivity of the corpuscle; less than a 1-μm displacement at its surface is sufficient to give a threshold response. Transduction is believed to be mediated by "stretch-activated" channels in the sensory membrane (cf. Chap. 10).

Tactile End Organs

Most of the other end organs are located superficially in the skin, near the junction between the dermis and epidermis. These are specialized to be sensitive to different types of tactile stimuli.

A systematic study has been carried out in humans to identify different receptor types and relate them to sensory perception. This has been based on a method developed by Åke Vallbo in Sweden for inserting a fine tungsten microelectrode through the skin to record impulse discharges from nerves innervating the galbrous skin of the hand (see Fig. 12.6). Using this method, called microneurography, Vallbo and Roland Johansson have been able to study different types of responses in the awake human and relate them to psychophysical functions.

Two of the basic response properties of any receptor are its size of receptive field and its rate of adaptation. With regard

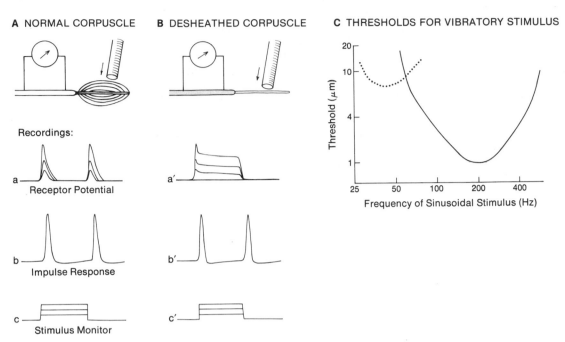

A NORMAL CORPUSCLE   **B** DESHEATHED CORPUSCLE   **C** THRESHOLDS FOR VIBRATORY STIMULUS

Recordings:

a   Receptor Potential

b   Impulse Response

c   Stimulus Monitor

a'

b'

c'

Threshold (μm)

Frequency of Sinusoidal Stimulus (Hz)

Fig. 12.5   Experimental analysis of transduction in the Pacinian corpuscle. **A.** Diagram showing probe for stimulating the intact corpuscle, and recording from the nerve. *(Below)* recordings of the receptor potential and impulse discharge. **B.** Repeat of experiment after removal of lamellae. **C.** Sensitivity of Pacinian corpuscle to vibratory stimulation at different frequencies. Sensitivity of Meissner's corpuscle is shown by dotted line. (A, B based on Loewenstein, 1971; C modified from Schmidt, 1978)

to adaptation, the tactile receptors in the glabrous skin of the hand fall into two groups, fast (FA) or slowly, (SA) adapting. The rapidly adapting type includes the Meissner corpuscle and the Pacinian corpuscle (refer to Figs. 12.4 and 12.6). Within this type, the adaptation shows interesting differences. The Meissner corpuscle (FA I) is not quite as rapidly adapting as the Pacinian corpuscle (FA II), and therefore has a lower range for signaling vibratory frequency. In addition, the threshold for activation of the Meissner corpuscle is much higher than the threshold of the Pacinian corpuscle.

The slowly adapting type of response arises in Merkel's discs (SA I) and Ruffini endings (SA II) (Fig. 12.1). Here also there are differences. Both receptors respond to a steady indentation of the skin with a sustained discharge, but the Merkel's disc shows an overshoot during the initial pha-

sic part of the indentation. It can thus provide information about changes in stimulus intensity as well as steady-state values (Fig. 12.6).

In addition to rates of adaptation, receptors also differ in receptive field sizes (Fig. 12.6). The receptive fields in the hand have been mapped with light tactile stimuli, and the results for the fast-adapting receptors are summarized in Fig. 12.7. The presumed Meissner's corpuscles have very small receptive fields, as shown in the diagram above and the graph below. This small diameter is part of the basis for our ability to make fine spatial discriminations with our fingertips (see below).

In contrast to FA I receptors, FA II receptors have very broad receptive fields. As can be seen in Fig. 12.7B, these fields can cover all of a finger or a large part of the palm. The graph in B shows that the field is almost flat, in comparison with the sharp

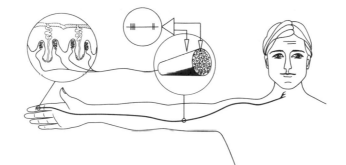

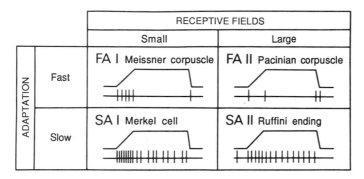

12.6 Different types of mechanoreceptors, as determined by experiments in humans. *(Above)* Experimental setup, in which mechanical stimuli are delivered to the palmar (glabrous) skin of the hand while recordings of impulse responses in single axons are made using thin microelectrodes inserted through the skin into the median nerve bundle. *(Below)* Summary of the four main functional types of response, on the basis of adaptation and receptive field size, and their correlation with receptor type. FA, fast adapting; SA, slow adapting. For each type is shown the stimulus monitor (ramp and hold) and the impulse discharge. Each receptor type is illustrated in Fig. 12.2. (Modified from Vallbo and Johansson, 1984)

Fig. 12.7 Comparison of receptors with narrow (**A**) and broad (**B**) receptive fields. The two broad fields shown in B have their centers of highest sensitivity at the sites marked by the dots. In the graphs, threshold (T) is plotted in multiples of the lowest threshold at the center; the absolute thresholds, in micrometers (μm) of skin indentation, are indicated on ordinates on the right. (From Vallbo and Johansson, 1984)

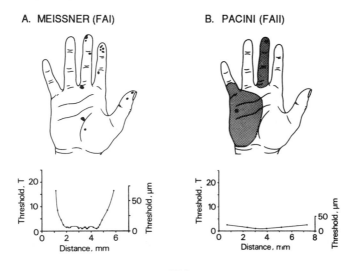

boundaries of the FA I receptors over this same distance.

Pacinian and Meissner corpuscles thus provide an interesting comparison. The Pacinian corpuscle has an extremely low threshold, high temporal resolution, and low spatial resolution; by contrast, the Meissner corpuscle has a higher threshold, lower temporal resolution, but higher spatial resolution. Similar differences have been found between the two types of slowly adapting (SA) receptors: the SA I receptors have small receptive fields, whereas the SA II receptors have very broad receptive fields.

How do these receptor properties relate to the attributes of sensory perception, as tested in these same human subjects? Let us consider three main attributes discussed previously in Chap. 10: threshold, intensity of sensation, and spatial discrimination.

Threshold. The threshold for perceptual detection has been determined and compared with the threshold for eliciting an impulse response. For many years, there had been a debate about this comparison. Psychophysicists believed that the behavioral threshold was higher than the single receptor threshold, indicating that several receptor responses had to summate in central pathways and overcome "noise" in higher centers in order to give a perceptual response. Neurophysiologists, on the other hand, believed that the behavioral detection threshold was set only by the peripheral receptor threshold. Surprisingly, it turned out that both were right. In the regions of highest sensitivity, such as the digits, a single impulse elicited in a sensory nerve can indeed reach consciousness. In regions of lower sensitivity, such as the palm, the behavioral threshold is higher. It appears, therefore, that the most direct pathway to consciousness is reserved for specialized areas of highest sensitivity. (See the discussion of this issue in the visual system in Chap. 10, under *Detection*.)

Intensity of Sensation. In Chap. 10, we saw that there is generally a close correla-

tion between the intensity of sensory stimulation and the magnitude of the perceived sensation, and we learned that this can be described by a simple power law, such that equal ratios between stimulus intensities are correlated with equal ratios between perceived intensities (see Fig. 10.8). The question then arises, similar to the question regarding thresholds, as to whether the intensity of the sensory perception is set by the receptor response.

This was directly tested in the neurographic studies in humans by comparing the stimulus–response functions of the nerve responses with the magnitude estimation functions in the same subjects. Typical results are shown in Fig. 12.8A, for the case of three SA units with small receptive fields at three different sites on the hand. In the graphs, the nerve response functions (N) and the psychophysical response functions (P) are plotted. In all cases, the N response functions rose rapidly and then leveled off; this implied an average exponent in the power law of around 0.7 (see Chap. 10, Fig. 10.8). By contrast, the P response function began slowly and increased in slope, implying an average exponent of around 1.0.

These results suggest that the magnitude of a sensation is not determined solely by the rate of impulse discharge in the afferent fibers. As summarized by Vallbo and Johansson (1984):

. . . it appears that the central nervous system plays a major role in shaping the psychophysical functions. Moreover, different subjects seem to experience relative intensities in a different way probably because the intrinsic properties of their brains differ.

Later in this chapter we will see how the sensory pathways and the microcircuits within them can contribute to shaping the incoming information.

Spatial Resolution. As mentioned above, the size of receptive fields is related to our ability to make spatial discriminations. This ability is tested in humans by two-point discrimination, measured as the dis-

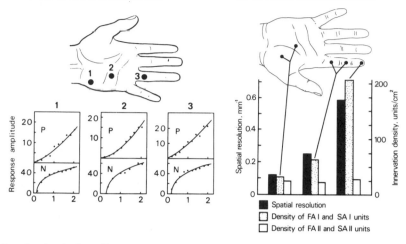

Fig. 12.8 Psychophysical studies comparing neural responses and perception. **A.** Intensity discrimination was measured for three SA receptors whose receptive field centers are at the sites indicated (1–3). For each receptor are plotted the increase in neural firing frequency (N) and perceptual intensity (P) for increasing stimulus intensity. The ordinate scales are in arbitrary units of magnitude estimation (P) and impulse firing frequency (N). Note that the P curve is not a simple function of the N curve. **B.** Spatial discrimination. Two-point discrimination tests were carried out in three skin regions, as shown above. The highest spatial resolution was found in the fingertip, and the lowest in the palm. The graph shows that the spatial resolution was found in the fingertip, and the lowest in the palm. The graph shows that the spatial resolution is correlated with density of receptors with small receptive fields (FA I and SA I receptors). (From Vallbo and Johansson, 1984)

tance between two points that can just be perceived as separate. If you carry out this test yourself with two pencil points, you will find that this ability varies widely, from about 2 mm on the fingertips, to 30 mm on the arm, to 70 mm on the back.

Is the fineness of two-point discrimination set solely by the size of the receptive field? This hypothesis was tested by first determining the behavioral thresholds at different sites on the hand. As expected, this showed the highest resolution near the fingertip, with lower resolution in the rest of the finger, and lowest resolution on the palm (Fig. 12.8B). One might anticipate that this would be correlated with corresponding sizes of receptive fields of the receptors in those areas, but we have already seen in Fig. 12.7 that small receptive fields have similar sizes everywhere on the hand (receptors with large receptive fields do not need to be considered, since they could not mediate fine discrimination in any case).

How then does one achieve increasing spatial resolution without decreasing receptive field size?

The answer lies in the densities of the receptors. As shown in Fig. 12.8B, the density of small field receptors varies widely, and is closely correlated with the two-point discrimination thresholds. (By contrast, the density of large receptive field receptors is constant.) The high density ensures that a restricted stimulus, such as a fine point, will have maximum chance of stimulating one or more receptors.

Active Touch

A vibrating probe is a very effective stimulus for Meissner's corpuscles, at 30–50 Hz (compare in Fig. 12.5C). In life, this kind of receptor is probably most often activated by our fingers moving over rough or irregular objects. If two points on an object are 2 mm apart, and our finger moves past them at a rate of 80 mm/sec, the second point will excite a given site on the finger at a frequency of 40/sec—just right for a Meissner corpuscle. Although as physiologists and neurologists we tend to apply a given stimulus at a single site to study receptor responses or sensations, in natural behavior the fingers and hand take an active part by moving over and exploring surfaces in order to give rise to our sensory perceptions. This is called *active touch*. The tactile sense arising in the distal extremities thus illustrates vividly the close interrelationship that exists between the sensory and motor systems. We will return to the subject of active touch in discussing movement of the hand in Chap. 22.

In real life, stimulation of the skin characteristically activates more than one type of mechanoreceptor in the hand. Later in this chapter we will see how higher centers provide for the synthesis of sensory information that is the basis for our perception of surfaces and objects.

Spinal Cord Circuits

The information transduced by the sensory receptors is transmitted by impulse codes in the sensory nerves to the spinal cord. Within the spinal cord the information has two destinations. First, it is involved in local reflexes at the *spinal cord* level. Especially important in this regard are the circuits for withdrawing a limb from a painful stimulus, the so-called *flexor reflex*. We will study these circuits in Chap. 19. The second destination is the *ascending pathways*, which relay the information to higher brain centers.

Dorsal Horn Microcircuits

Some of the basic aspects of organization of the dorsal horn are summarized in Fig. 12.9. The sensory fibers terminate within the dorsal horn, which is the sensory region of the spinal cord. The dorsal horn is arranged in layers (I–V), as indicated in Fig. 12.9B. The fine, unmyelinated fibers (C fibers) terminate mostly in one of the superficial layers (substantia gelatinosa). The Aδ

fibers terminate in the most superficial layer, called the *marginal zone*. The large myelinated fibers (Aβ) sweep around the dorsal horn, giving off collaterals which ascend in the posterior columns and then terminate on the dendrites of cells in *layers III and IV*.

When we describe the entire somatosensory pathway (see below), the dorsal horn will be represented by a single relay site, but as Fig. 12.9B indicates, this is not a single site but rather a complicated set of microcircuits. The main point to realize is that through these connections, processing of somatosensory information begins at this first relay station.

Some of these microcircuits can be traced in the diagram of Fig. 12.9B and its inset. First, there are connections for *straight-through transmission* of each specific modality (e.g., large myelinated fibers to projection cell [PC] neuron). As shown in the inset, these include not only axodendritic synapses (for forward transmission), but also dendroaxonic synapses (for immediate feedback inhibition) and dendrodendritic synapses (for lateral interactions between responding cells). Second, there are connections that mediate *interactions between modalities* (e.g., INT in Fig. 12.9B, which mediates interactions between responses to the medial division and lateral division inputs). Third, there are connections from descending axons that provide for *modulation* of incoming sensory information by higher brain centers, as will be discussed below.

Transmission and Modulation of Pain

One of the main findings of modern research on pain is that pain is mediated by different fiber systems with different but overlapping properties. A second important finding is that the pain pathway involves integration and modulation by widely diverse neural mechanisms, acting over different overlapping periods of time. We will discuss the pain pathways and several types of modulation that relate to our everyday experience.

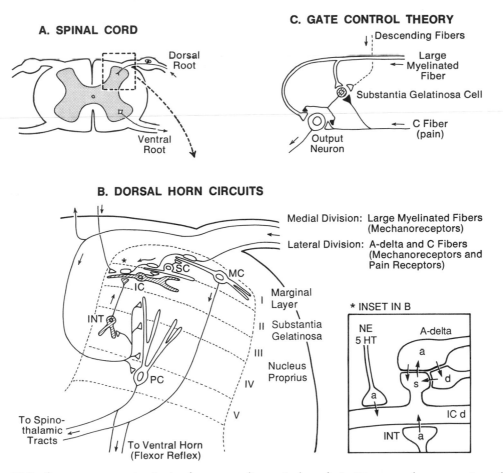

Fig. 12.9 Somatosensory circuits in the mammalian spinal cord. **A.** Diagram of cross section of spinal cord. **B.** Some of the main neuron types and synaptic connections that have been identified in the dorsal horn. I–V, laminae of the dorsal horn; IC, islet cell; INT, interneuron; MC, marginal cell; PC, projection cell; SC, stalked cell. The large myelinated fibers are believed to be glutamatergic; pain fibers contain substance P, somatostatin, vasoactive intestinal peptide, or possibly cholecystokinin; descending fibers contain norepinephrine or serotonin. (Based in part on Carol LaMotte, personal communication; Gobel et al., 1980) *(Inset in B)* Synaptic organization of microcircuits in the marginal layer (site marked by asterisk in B). a, axon; d, dendrite; s, spine; INT, interneuron; IC, islet cell; NE, norepinephrine; 5HT, serotonin. (Based on Gobel et al., 1980, and others) **C.** Simplified diagram illustrating "gate-control theory" of Melzack and Wall (1965). Postulated excitatory terminals shown by open profiles, inhibitory by shaded profiles.

Gate Control Theory of Pain. The idea that there can be summation of inputs on spinal cord cells to modulate pain transmission originated with the observation that pain in the skin frequently can be relieved by gently stimulating skin around the hurt area by light brushing, massaging, or tickling. This suggested that the tactile pathway can have an inhibitory action on the pain pathway.

In a celebrated theory published in 1965, Ronald Melzack and Patrick Wall, then at M.I.T., suggested a circuit within the dorsal horn that could account for this action. At the core of the circuit are the C fibers that transmit pain through excitatory inputs to dorsal horn relay neurons, aided by inhibitory inputs that suppress an inhibitory neuron to the relay neurons, thus helping to maintain their excitability. The essence of the theory is that the inhibitory interneuron is activated by tactile fibers, thereby suppressing pain transmission (Fig. 12.9C). In general, the theory envisaged that transmission through the dorsal horn would be dependent on the activity of the dorsal horn interneurons, which in turn could be set by their own intrinsic resting activity, the tactile and pain inputs, and control by descending fibers from the brain. The interneurons thus act as "switches," or "gates," to enhance or depress pain transmission—hence the name *gate control theory* of pain.

This theory had an invigorating effect on the whole field of pain research. The theory is undoubtedly incorrect in details; for example, the postulate of both excitatory and inhibitory outputs from C fibers has always seemed unlikely. However, the idea of a specific neuronal circuit for pain transmission and modulation was a significant step forward, and has been the focus for much work since then. It also stimulated clinical interest in therapeutic approaches to pain relief, especially the use of electrical stimulation of tactile pathways in peripheral nerves and the spinal cord for reducing pain perception.

Pain Intensity and Hyperalgesia. When the skin is stimulated in a noxious manner, peripheral pain fibers are activated in the way discussed above. How does the frequency of impulse firing relate to the intensity of the perceived pain? This has been investigated by Robert LaMotte and his colleagues at Yale, who compared perceptions of heat stimuli to the forearm of human subjects with the activity evoked by similar stimuli in nerves to that area in monkeys. As shown in Fig. 12.10, bottom panels, the pain ratings in the humans and the nerve activity in the monkeys increased in parallel with hotter stimuli. Then, studying the effect of prior episodes of heat stimuli, they found a large increase in both the perceived intensity and nerve activity. This is called *primary hyperalgesia,* or *sensitization,* and it involves both a lowering

of threshold (weaker stimuli elicit pain responses) and an increase in response intensity. The experiments indicated that activity in polymodal C fibers was primarily responsible for normal pain sensation and for at least a part of the hyperalgesia. However, the subjective sensitization in humans was greater than the neural activity increases, suggesting (1) a species difference; (2) a contribution of mechanoheat-sensitive Aδ fibers; (3) recruitment of neighboring fibers by spread of sensitizing substances from the terminals of activated fibers; or (4) central summation in the dorsal horn of the spinal cord or in higher centers.

Tenderness around a Local Injury: Secondary Hyperalgesia. A related pain sensation is tenderness around the site of a local injury to the skin, called *secondary hyperalgesia.* Experimentally, this can be produced and studied using a small injection of an irritant into the skin. LaMotte and colleagues showed that capsaicin (the active ingredient of hot chili peppers) injected into the skin in the monkey locally activates the endings of low- and high-threshold heat fibers and mechanoheat fibers. In addition, as shown in Fig. 12.11, it activates the endings of neighboring chemonociceptor fibers, which branch widely in the skin. These fibers make connections in the spinal cord onto interneurons in the low- and high-threshold mechanoheat pathway, which in turn connect to the relay neurons conveying pain information from the local inflammation site to the spinothalamic tract. In this way, remote mechanical stimuli enhance activity in the cells transmitting pain information, giving rise to secondary hyperalgesia that can develop in a wide area of skin surrounding a local injury.

Intrinsic Neuromodulation. It should be clear by now that a great deal of processing takes place through the synaptic circuits of the dorsal horn. Investigators are interested not only in the specific circuit connections, but also in the types of actions that differ-

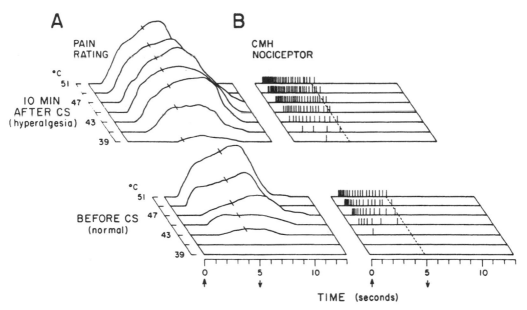

Fig. 12.10 Correlation between the perception of pain in a human subject and impulse firing in a monkey C multimodal heat receptor under normal conditions *(below)* and during hyperalgesia *(above)*. In both cases, heat stimulations of the hair skin were applied (between arrows) in increasing steps before *(below)* and after *(above)* a conditioning mildly injurious stimulation of 50°C and 100 sec duration. (From LaMotte et al., 1991)

ent transmitters and modulators may have in mediating the different types of pain.

The hyperalgesia induced by local chemical injection in the skin in fact is accompanied by profound changes in synaptic activity, synaptic plasticity, and neuromodulatory processes in the dorsal horn. Ronald Dubner and his colleagues at the National Institutes of Health have studied these changes related to hyperexcitability of dorsal horn neurons with a variety of methods, and Fig. 12.12 summarizes their results. Excitatory amino acids (glutamate, aspartate) appear to be the primary neurotransmitters of pain afferents. Activity in the afferent terminals leads to release of Substance P and CGRP, both of which enhance the EAA excitatory action, causing activation of NMDA receptors and resulting persistent depolarization of local circuit neurons. Intense pain inputs activate immediate early genes (c-*fos*) and expression of dynorphin, which can be blocked by NMDA antagonists. It is therefore pro-

posed that dynorphinergic interneurons are activated by these mechanisms and have excitatory inputs to projection neurons in the pain pathway. Inhibitory interneurons may gate transmission through the dorsal horn by several types of feedforward and feedback connections, as shown in the diagram.

It is important to realize that these different mechanisms act over different time scales (see Fig. 8.16). The fastest EPSPs, over time scales of up to 100 msec, are mediated first by glutamate, acting through the AMPA receptor followed by the NMDA receptor, and then by SP. At an intermediate range are second messenger—mediated processes, acting over minutes. Then come the expression of c-*fos* and other immediate early genes, and regulation of various types of neuropeptides, such as dynorphin, and their receptors, acting over many minutes or hours. Next are the slower processes of long-term potentiation and structural plasticity, and, finally,

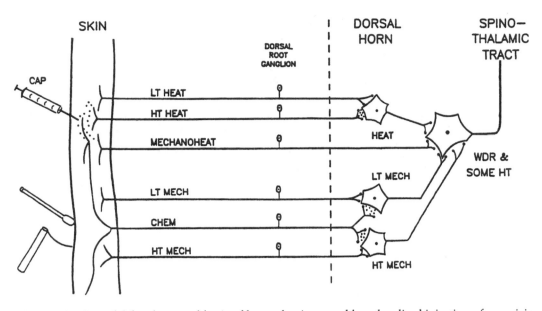

Fig. 12.11 A model for the neural basis of hyperalgesia caused by a localized injection of capsaicin (CAP). The injurious substance activates the endings of locally branching low- and high-threshold (LT, HT) heat fibers as well as mechanoheat fibers. In addition, it activates the endings of widely branching chemonociceptive fibers (CHEM), which release onto dorsal horn interneurons a modulating substance that sensitizes them. This causes increased transmission in distant pain pathways mediating activity in LT and HT mechanonociceptive fibers, enhancing the transmission through wide-dynamic range (WDR) and some HT spinothalamic neurons. This model thus accounts for local hyperalgesia to heat stimuli and remote hyperalgesia to mechanical but not heat stimuli. See text. (From LaMotte et al., 1991)

if activation is so intense as to be excitotoxic, cell death. These processes thus represent specific instances of the general point about overlapping transmitter actions that was presented in Chap. 8.

Descending Modulation. Thus far, we have focused primarily on intrinsic modulation within the dorsal horn. However, there is a vast field of study of descending pathways from many higher centers which exert their own modulatory influences on sensory transmission through the dorsal horn.

The most important regions, shown in Fig. 12.13, may be summarized as follows (see Light, 1992). *Corticospinal tract* fibers have been shown to send collaterals to the periaqueductal grey (PAG) and the raphe nucleus (NRM), as well as to the superficial layers of the dorsal horn, where they pre-

sumably can modulate nociceptive transmission. Connections with the *hypothalamus* enable the effects of painful stimuli to be coordinated with hormonal and autonomic responses (see Chaps. 18 and 27). The *periaqueductal gray* is a complex integrative center for inputs from the autonomic nervous system, the limbic system, and from sensory and motor pathways; it in turn has an inhibitory effect on nociceptive transmission mediated by direct descending connections and by a relay through the NRM. The PAG cells are both glutamatergic and enkephalinergic. The *locus ceruleus* (LC) sends noradrenergic fibers to the spinal cord and elsewhere. There is evidence that both norepinephrine and serotonin mediate the inhibitory effect of morphine on pain transmission through the spinal cord. The *A5 group* of noradrenergic cells is involved in central regulation of

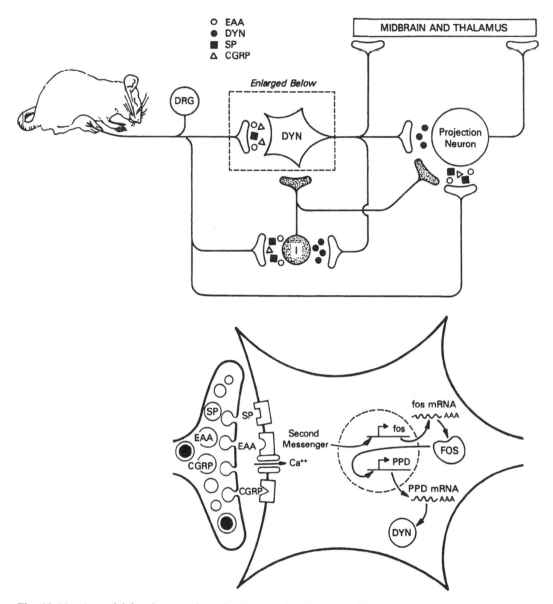

Fig. 12.12 A model for the neural mechanisms underlying central hyperexcitability and excitotoxicity in the dorsal horn. Activity in nociceptive dorsal root ganglion (DRG) cells causes release of excitatory amino acids (EAAs) such as glutamate, facilitated by substance P (SP) and calcitonin gene-related peptide (CGRP). This activates NMDA receptors on local interneurons (I), some of which express and release dynorphin (DYN). Persistent depolarization of DYN interneurons leads to activation of immediate early genes (e.g., c-*fos*), whose protein product (FOS) may regulate preprodynorphin (PPD) expression leading to DYN (see inset). Potentiation of excitatory synaptic actions by DYN, SP, and CGRP may thus be the basis for expansion of pain-receptive fields in hyperalgesia but also hyperexcitability leading to excitotoxicity. See text. (From Dubner and Ruda, 1992)

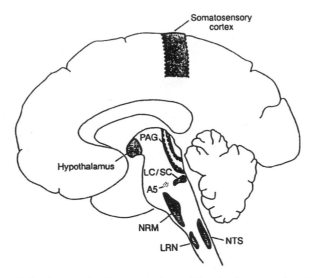

Fig. 12.13 Drawing of the human brain seen in lateral (sagittal) view, showing regions involved in processing and modulation of pain signals. LC/SC, locus ceruleus/subceruleus; LRN, lateral reticular nucleus; N RM, nucleus raphe magnus; NTS, nucleus of the solitary tract; PAG, periaqueductal grey. (From Light, 1992)

blood pressure and heart rate; these cells interact with serotonergic cells of the *raphe nucleus* to inhibit pain transmission through the spinal cord, presumably in coordination with cardiovascular control. Microinjection of morphine into these regions inhibits pain reflexes. The *solitary nucleus* (NTS) is classically involved in taste reception, but it also is an integrative center for inputs from many of the other pain centers and outputs that mediate inhibition of pain reflexes elicited through stimulation of the vagal nerve (see Chap. 18).

The main point to be remembered from these findings is that pain is a heavily modulated sensation, and that the modulation does not occur in isolation but is integrated with many other systems of the brain, particularly those involved in skin reflexes, autonomic regulation, emotion, alertness, and attention. This is fully in accord with our everyday experience that pain hurts; can make us sweat, feel sick to our stomachs, feel anxiety and distress; and can distract our attention away from everything else. In all these ways the pain pathway is fulfilling its role of protecting the organism from harm.

Ascending Pathways

Sensory information reaches the rest of the brain by two main pathways. Since these rise from lower centers to higher centers, they are called ascending pathways. The oldest, phylogenetically, is made up of fibers that arise from dorsal horn cells, cross the midline, and form a tract in the anterolateral part of the white matter of the spinal cord. These fibers ascend all the way through the spinal cord and brainstem, and terminate in the thalamus (see Fig. 12.14). They are thus referred to as the *spinothalamic* tract. These fibers mediate mainly *pain* and *temperature* sensation, but there are also fibers that convey some *tactile* and *joint* information. Along the way in the brainstem they give off numerous collaterals to the reticular formation. The reticular neurons, in turn, form a system of polysynaptic ascending connections, which eventually also feeds into the thalamus. These neurons are part of the *ascending reticular system*, which is involved in *arousal* and *consciousness* (see Chap. 25).

The phylogenetically newer ascending pathway is made up of collaterals of large,

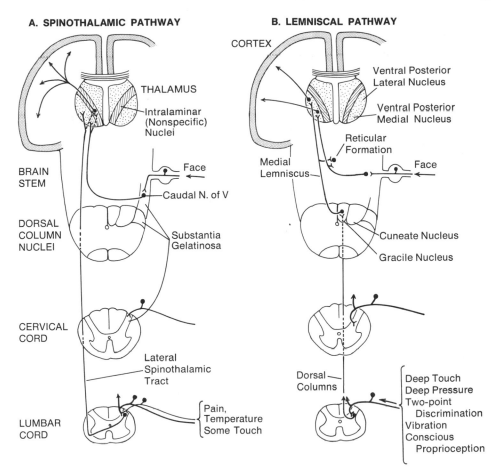

Fig. 12.14 Ascending pathways of the somatosensory system. (Modified from Carpenter, 1976, and Brodal, 1981)

myelinated sensory axons. As shown in Figs. 12.9 and 12.14, these collaterals gather in the posterior, or dorsal, part of the white matter of the cord, and form the *posterior columns* or *dorsal columns*. As part of the somatosensory pathway, this system first can be delineated along the phylogenetic scale in reptiles. These fibers ascend only as far as the lower margin of the brainstem, where they terminate and make synapses in the *dorsal column nuclei*. From there, fibers sweep across the midline to form the *medial lemniscus,* and ascend to terminate in the thalamus. This pathway is often referred to as the *lemniscal system;* it is mainly concerned with conveying the most precise and complex information

about touch and pressure. Lemniscal fibers give off collaterals to the reticular formation, and thus also contribute to arousal mechanisms.

Somatosensory Cortex

The ascending fibers in the somatosensory pathways terminate in the thalamus, where they make synapses with relay cells that project to the cerebral cortex. The thalamus is thus the gateway to the cortex, and it serves this function for all pathways ascending from the spinal cord and brainstem. The somatosensory fibers terminate in the group of cells called the *ventral posterior nucleus* (VP), the lemniscal and spinothalamic fibers in the lateral part (VPL), and the fibers from the trigeminal nucleus, relaying inputs from the face, in the medial part (VPM). In subprimate mammals, like cats, the whole group of cells is referred to as the ventrobasal complex (VBC).

Much of the work on the somatosensory cortex has been concerned with three main questions. First, what is the topographical representation of the body surface? Second, how many cortical areas are there, and how specific are they? Third, what is the intrinsic organization within an area of cortex; what are the basic functional units? These questions are also relevant to the other parts of the cerebral cortex that we will study.

Topographical Representation

Within the ventroposterior nucleus of the thalamus, the fibers terminate in an orderly geometrical arrangement that preserves the relations of the body surface. This arrangement is called *somatotopy,* or *topographical representation;* it is as if the body surface were projected onto the nucleus. This arrangement, in turn, is preserved in the projection of the relay cells onto the cortex. The area within which those fibers terminate defines the somatosensory cortex.

Within this cortex, the relations of the body surface are preserved, but the relative areas are modified. This was established by Wilder Penfield and his colleagues at Montreal in their studies, from the 1930s to the 1950s, of patients undergoing neurosurgical operations. With punctate electrical stimulation, Penfield elicited descriptions from the patients of the tactile sensations (numbness, tingling, pressure) and their apparent sites on the body surface. From this emerged the well-known drawing called a "homunculus" depicted in Fig. 12.15, with its characteristic distortions, particularly the large areas given over to the lips, face, and hands. The area of cortex varies with the acuity of perception, and the large cortical areas reflect the high sensitivity and fine discrimination possible in these parts of the body (a direct measure

of this is provided by the two-point discriminations in Fig. 12.8).

Two particularly striking examples of the precision of cortical representation deserve mention. If you have ever observed a raccoon eating food, you know that it has remarkable forepaws, which, though lacking an opposable thumb, are nonetheless capable of dextrous manipulations resembling those of human hands. In the raccoon, the part of the somatosensory cortex representing the forepaw is greatly developed. It has been possible to identify five individual, small gyri (folds), one for each finger, and another five gyri for each of the volar pads of the paw. A summary of the maps obtained is shown in Fig. 12.16. The elaborate nature of this representation is taken to reflect the acute tactile sensitivity of the raccoon hand. This sensitivity is expressed both in responses to passive stimulation, and by "active touch" during the dextrous movements of the forepaw (see Chap. 22).

A second example concerns the vibrissae (whiskers) of the snout of rodents; these function as sensitive detectors of the environment around the snout. The vibrissae project through VPM in the thalamus. Thomas Woolsey and Henrik van der Loos at Johns Hopkins discovered in 1970 a regular series of five rows of cell groups with hollow interiors, looking much like the cell clusters surrounding olfactory glomeruli in the olfactory bulb. They called each cluster a "barrel," and showed that each barrel represents a vibrissa on the animal's snout (Fig. 12.17).

When a vibrissa is removed early in life, anatomical studies show that the corresponding barrel disappears from the row. When a vibrissa is stimulated continuously, a single, dense focus of 2-deoxyglucose uptake can be demonstrated in the corresponding barrel. The fibers that innervate a vibrissa are tuned to several different submodalities, and these are transmitted to the barrel for that vibrissa. A barrel is thus a *morphological unit,* and it is also a *functional unit* within which *multisensory integration* takes place.

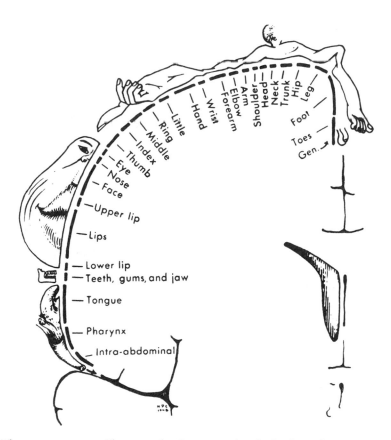

Fig. 12.15 The somatosensory "homunculus," representing the body surface as projected onto the postcentral gyrus of the human cerebral cortex. (From Penfield and Rasmussen, 1952)

Fig. 12.16 Somatosensory cortex of the raccoon. There is a fold (gyrus) for each finger (1–5) and each palm area (volar pad: A–E) of the paw. (From Welker and Seidenstein, 1959)

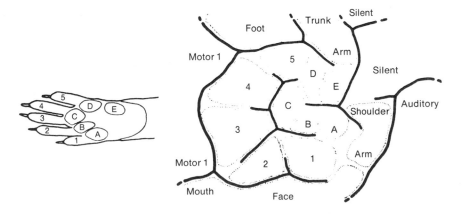

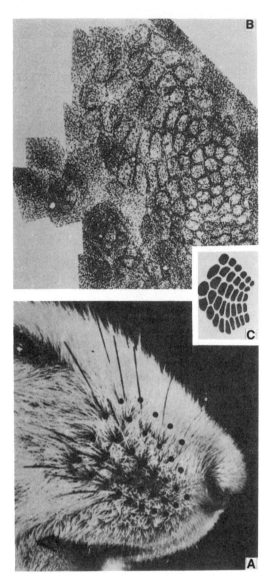

Fig. 12.17 A. Snout of a mouse; the vibrissae (whiskers) are marked by dots. **B.** Section across the somatosensory cortex receiving input from the snout. Note the rings of cells ("barrels," or glomeruli), each corresponding to a vibrissa (the barrel map is summarized in C). (From Woolsey and van der Loos, 1970)

Cortical Areas

The maps for somatosensory cortex were soon shown to apply in their general outline to many species of mammals. In addition, these studies showed that there is a second, smaller area where the body sur-

face is represented. The main area was called S I (primary somatosensory cortex, and the second area S II (secondary somatosensory cortex). With the methods available, it appeared that S I was the only region that receives the thalamic input. The belief therefore arose that *serial processing* takes place at the cortical level, beginning with analytical mechanisms in primary cortex, and proceeding to more integrative mechanisms in secondary cortex. The final step was believed to occur in neighboring regions called *association cortex,* which appeared to lack topographical representations of the body surface, and therefore could be concerned with the synthetic mechanisms that seem to be necessary for perception to occur. This traditional view is depicted in Fig. 12.18A.

This tidy sequence has not been borne out by recent experiments. Instead, it has been found that there are not one but several subdivisions of the thalamus that relay somatosensory information, and these each have their specific inputs to one or more of several somatosensory areas. Furthermore, with more refined microelectrode recording techniques, it has been found that each cortical area is selective in the submodalities it processes. This new evidence is summarized in Fig. 12.18B.

These results have emphasized the importance of *parallel processing* of different aspects of somatosensory stimuli at the cortical level. They have indicated that there is much more detailed mapping present in the cortex, particularly in parts believed to be "associational," than previously suspected. This has required rethinking the question of where and how our unified perceptions are formed. We will examine similar evidence as it relates to other sensory systems in the following chapters, and return to this question in our final discussion of the cortex in Chap. 30.

Cortical Maps Are Both Detailed and Plastic

The cortical maps of the body surface have become increasingly detailed, progressing from the homunculus (Fig. 12.15), through

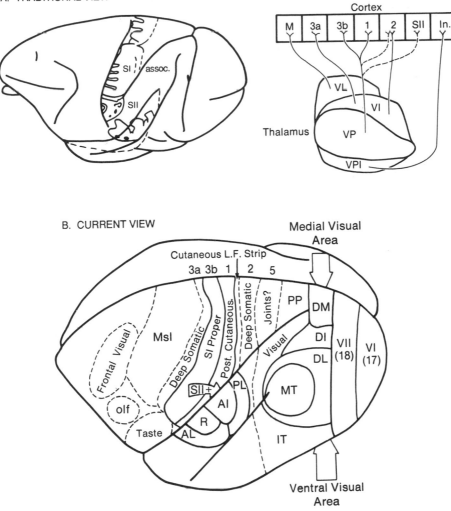

Fig. 12.18 **A.** Traditional view of the monkey somatosensory cortex. **B.** Current view of subdivisions of monkey somatosensory cortex, together with cortical areas for other sensory modalities. **C.** The main thalamocortical connections of the somatosensory system. AI, primary auditory; AL, anterior lateral auditory field; DI, dorsointermediate visual area; DL, dorsolateral visual area; DM, dorsomedial visual area; IT, inferior temporal; MsI, motor-sensory area I; MT, middle temporal visual area; PL, posterior lateral auditory field; PP, posterior parietal visual area; R, rostral auditory field; SI, primary somatosensory; SII, secondary somatosensory; VI, primary visual; VII, secondary visual; VI, ventralis intermedius nucleus; VL, ventralis lateralis nucleus; VP, ventroposterior nucleus; VPI, ventroposterioinferior nucleus. (From Merzenich and Kaas, 1980)

the digits of the raccoon (Fig. 12.16), and finally to the tips of the digits in maps of the somatosensory cortex of the monkey (see Fig. 12.19 A–C).

The extraordinarily fine resolution of these maps suggested initially that the thalamocortical connections underlying them would have to be hard-wired, but studies under a variety of experimental manipulations have shown that in fact the maps have a great deal of plasticity (summarized in Kaas, 1991). For example, repeated tactile stimulation of digits D2 and D3 over a period of several weeks (see Fig. 12.19D) results in a much larger representation of those digits, not only pushing aside the

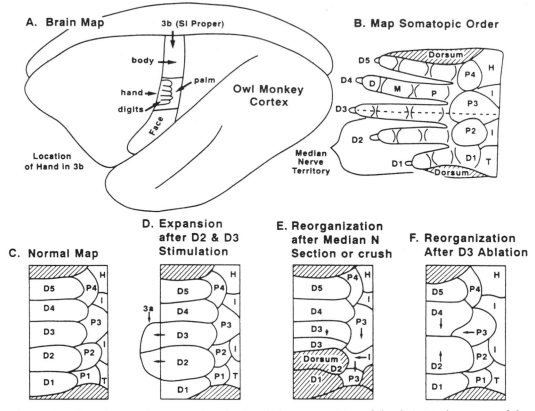

Fig. 12.19 Experiments demonstrating the detailed representation of the digits in the cortex of the owl monkey (A-C), and the plasticity of this representation after different procedures (D-F). See text. (Adapted from Kaas, 1991)

representations of neighboring digits but also extending into the neighboring cortex normally occupied by area 3a. By contrast, when the median nerve is cut and prevented from reconnecting (see Fig. 12.19E), thereby depriving the lateral part of the hand of its normal innervation (see median nerve territory in B), the normal representation of that part is replaced by expansion of the representation of the dorsal skin of those digits and by encroachment of the neighboring digit (3). If the nerve is allowed to regenerate, however, the normal representation is completely reestablished. Finally, if a digit is removed (see Fig. 12.19F), the glabrous skin of the neighboring digits and palm region expand into that territory. Similar changes in cortical representations have been seen for the paws of cats and the digits of raccoons.

These rearrangements have several interesting and important features. First, they show that thalamic inputs to the cortex are both extremely precise and also significantly plastic. Second, these changes take place over varying time scales; in some cases the shifts in representations are slow, developing over weeks, but in other cases they may be surprisingly rapid, beginning within a day or so, or even a few hours. Third, these changes are not limited to primary cortex; ablations of hand areas in primary somatosensory cortex produce similar shifts in representation of the hand in secondary somatosensory cortex.

What is the mechanism of this plasticity? One possibility is that there is sprouting of new collateral fibers in response to persistent stimulation or to denervation of a neighboring region. The other possibility is that there is a shift in the effectiveness of already existing overlapping projections to neighboring areas. The rapid changes seem most likely to be mediated by the latter mechanism. This could occur by decreasing or removing lateral inhibitory connections; in fact, there is evidence that GABA expression in the cortex is reduced following nerve section. It could also occur by changes in the effectiveness of excitatory synapses; the role of NMDA glutamatergic synapses is of special interest in this respect. However, the lateral spread of these changes must be limited by the extent of terminal arborizations; these arborizations cannot be too extensive, or they could not provide for the highly detailed maps. For more extensive and slower shifts, some degree of sprouting may therefore be involved.

These results leave no doubt that somatosensory maps show both high resolution and considerable potential for plasticity. Similar results have been found in other sensory cortical areas, particularly auditory and visual cortex. These results give hope that ways can be found to enhance these mechanisms in order to minimize the functional losses due to brain injury.

Cortical Columns

In the early years of studying the cortex with the new microelectrode recording methods that had become available during the 1950s, Vernon Mountcastle at Johns Hopkins University carried out an analysis of cell responses in the somatosensory cortex of the cat to various types of stimuli. He found that when his electrode penetrated the cortex perpendicular to the surface, all the units he encountered tended to respond to the same sensory submodality (for example, light touch, or joint movement), but when he made an oblique electrode track, he encountered a series of units with different submodalities. From this he deduced that the cortex contains columns of cells with similar functional properties. Figure 12.20 illustrates some typical results from one of the early experiments on the monkey, performed in collaboration with

T. P. S. Powell of Oxford, in which functionally characterized units were localized anatomically.

The concept of the column as a basic functional unit has turned out to be widely applicable in the cortex. The characterization of columnar organization has been one of the dominating forces in studies of sensory areas, motor areas, and even association areas, as we shall see in subsequent chapters. From this and related work, it is beginning to appear that the cortical column expresses a fundamental tendency of nerve cells and circuits to be organized in more or less discontinuous groups, or modules. We have already seen a clear example of this in the olfactory glomeruli, as well as in the somatosensory barrels. A point of some interest is that olfactory glomeruli and somatosensory barrels have dimensions of 100–300 μm, which is similar to the widths of several hundred micrometers for columns in the rest of the somatosensory cortex. Thus the smallest unit of anatomical representation in the cortex yet identified—the barrel—appears to be equivalent in size to the basic functional unit, the column. Other work has similarly revealed fine-grained representations that approach the dimensions of single columns.

Is the cortical column a building block of perception? This was an obvious possibility from the moment of its discovery. Mountcastle early stressed that events within the column probably are involved in the initial steps in cortical processing of sensory information. An expanded view is that columns within both primary and related association areas of cortex provide the neural basis for perception through multiple and reciprocal connections, forming a distributed system as described in Chap. 1. This has been worked out most fully for the visual system, as we shall see in Chap. 16.

Perception of Surface Form and Texture

Earlier in this chapter we discussed the analysis of individual types of skin recep-

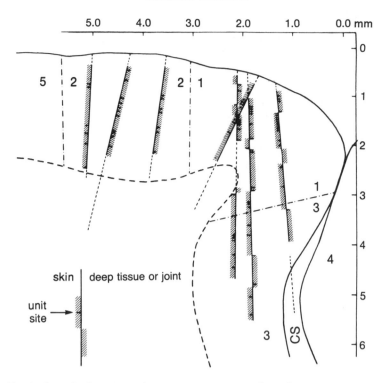

Fig. 12.20 Classical study showing columnar organization of monkey somatosensory cortex. Cross section containing electrode tracks, showing sites of unit recordings (short horizontal bars) and modality of units (shading on left for units activated by stimulation of the skin, on right for units activated by stimulation of joint, periosteum, or deep tissue; see inset). Numbers refer to cortical areas. CS, cortical sulcus. (From Powell and Mountcastle, 1959)

tors and the contribution they may make to the perception of touch. We are now in a position to relate that information to the organization of the higher pathways in order to have an understanding of the neural basis of touch perception.

The pioneering experiments toward this end were undertaken by Mountcastle and his colleagues in the 1960s and 1970s. They carried out psychophysical studies of specific touch perceptions in humans, and then obtained electrophysiological recordings at different levels in the somatosensory pathway of experimental animals using the same stimulus procedures. The approach is similar to that used in the human studies already described, except that in animals one can obtain recordings not only of the peripheral nerve fibers but also, crucially, at every stage up to and

within the cerebral cortex. In this work the activity of a single neuron is taken to be representative of the population response at that particular level. Although this assumption has come under critique, as experimenters have sought to develop multiple electrode methods to represent simultaneously the activity of multiple cells within the population, the results obtained with the single electrode have remained unmatched for their precise data, often from identifiable members of the active populations.

With either approach, it is essential to deliver precise stimuli, and experimenters have developed ingenious devices capable of selectively activating the different types of receptors; we have already discussed examples earlier in the chapter. These devices include rotating drums with embossed characters, which simulate the movement of a digit across a surface, as would be involved in active touch. These stimuli, combined with single-unit recordings from cells in the somatosensory cortical areas in the monkey, have enabled the relative contributions of different receptors to different perceptions to be assessed.

Out of this work has come a consensus that the SAI, RA (FA I), and Pacinian corpuscle (PC) (FA II) afferent systems play complementary roles in the perception of different aspects of surface form and texture. As summarized by Johnson and Hsiao (1992), the SAI system is the primary system involved in the perception of form and roughness when we explore a surface with our fingertips. The PC system comes into play in signaling high-frequency vibrations that occur when we move our fingertips over structures with very fine texture. The RA system is tuned to signal localized movement between the skin and the underlying surface, and also is more sensitive than the SAI system to fine form and texture. Thus, as we use our fingertips to explore and manipulate the objects around us, there is a constantly shifting mixture of signals that is integrated in the cerebral cortex to give our unified perception of the physical properties of those objects. This integration involves the organization principles of cortical columns, detailed somatotopic maps, and multiple representations of the body surface. If we take this as a model, we would expect that other sensory systems would provide for similar types of integration, and we will see that expectation fulfilled in other chapters, particularly in our discussion of visual perception in Chap. 16.

13

Muscle Sense and Kinesthesia

The tactile systems studied in the previous chapter tell us about the world around us. What about the world within? How do we know *when* we move our muscles; how do we know *how much* to move our muscles, limbs, and joints? The precise control of movements has a high priority in the behavior of most organisms, and increasingly so in higher animals. Two examples requiring this precision are shown in Fig. 13.1. Different kinds of motor behavior will be discussed in later chapters; the examples in the figure illustrate that motor control is the outcome of a complex interplay between genetics, motivation, training (especially in humans), and sensory factors. Sensory information is needed to help control posture and sequences of movements and to make adjustments to changes in the environment.

Information about movement is signaled by several different types of receptors. The receptors are located at almost every possible site in the musculoskeletal system at which movement can take place. As indicated in Fig. 13.2, the main sites are in the muscles, tendons, and joints. In addition, tactile receptors are also involved when movements are associated with movement of the skin.

It should be obvious that the central nervous system wants as much information as possible about ongoing movements. An important principle is that it does not entrust that task to just one information channel, but spreads the task among as many complementary kinds of channels as it can. As usual, this makes the subject more interesting, but it makes definitions of terms more difficult. Table 13.1 lists some of the common terms and classifications, which deserve a brief discussion.

Muscle sense usually means the sensory information arising in the muscles and tendons. *Proprioception* is a term introduced by Sherrington to refer to all sensory inputs from the musculoskeletal system, which therefore includes inputs from joint receptors. *Kinesthesia* is the sense of the position and movement of the limbs. Contributing to this are skin receptors as well as proprioceptors. This sense includes conscious sensations—what Charles Bell, in the early nineteenth century, referred to as our "sixth sense"—and therefore applies primarily to higher vertebrates, especially humans. Kinesthesia also includes our sensations of effort, of force, and of weight; contributing to these sensations are signals in descending pathways within the central nervous system (see Chap. 21).

In this chapter we will focus on the pro-

Fig. 13.1 Examples of motor behavior requiring precise sensorimotor coordination. *(Left)* A male three-spined stickleback fish assumes a characteristic threat posture toward its own reflection in a mirror. According to Tinbergen, "this activity is innate, dependent on internal (motivational) and external (sensory) factors. It has an intimidating effect on other males. . . . Historically, it is displacement sanddigging, changed by ritualization." (From Tinbergen, 1951) *(Right)* A ballerina, Jane Brayton, executing an arabesque. Ballet is a typical higher human activity, in which sensory factors, motivation, and innate capacities are molded by instruction, practice, and individual expression. (Photograph by Curt Meinel, courtesy of Ruth and Robert Brayton)

Fig. 13.2 Sites of sensory receptors for muscle sense and kinesthesia, as illustrated by the knee joint.

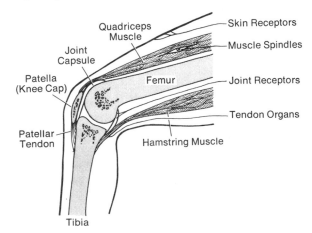

Table 13.1 Nervous substrates for the muscle sense

Classification	Relative importance
Muscle receptors (muscle spindles)	Primary contribution
Tendon receptors	Primary contribution
Joint receptors	possibly in some cases
Skin receptors	perhaps
Central (corollary) discharges	Sense of effort; sense of weight
	Needed for decoding in kinesthesia

Adapted from Matthews (1982)

prioceptors in the musculoskeletal system. We will discuss briefly their central pathways and, in humans, their possible contributions to conscious sensations.

Crayfish Abdominal Stretch Receptor

Proprioceptors are present in many invertebrate species, and are especially important in arthropods. In concert with the development of specialized muscle groups to control the body segments and long, articulated appendages was the development of various kinds of specialized sensory receptors. Among the simplest of these is the stretch receptor cell of the crayfish. We have discussed the spread of electrotonic potentials in this cell in Chap. 7, and the basic properties of the sensory response in Chap. 10. Here this cell will further be used as a model for the control of excitability of muscle receptors.

The long abdomen of the crayfish is composed of a number of segments (see Fig. 13.3). The dorsal cuticular plates of each segment are hinged on each other so that when the segmental muscles contract, the abdomen is flexed. The abdomen can thus be flipped like a tail, which is the method a crayfish uses for a quick retreat from danger (this escape response is described in Chap. 20). The muscles contain two types of fibers, the regular muscle fibers, which move the plates when they contract, and modified muscle fibers, which contain the

terminals of the sensory cells. The modified fibers contribute little to the movement of the plates, but indicate to the sensory cells the state of tension or lengthening of the muscle. The modified muscle fibers occur in two types of bundles, depending on whether they are slow fibers (slow summation of contraction) or rapid fibers (individual contractions). There are correspondingly slowly adapting and rapidly adapting types of receptor cells (see Figs. 7.15 and 13.3).

Each muscle bundle receives a motor innervation in the form of collaterals from the motoneuron axons to the main muscle fibers. Thus, whenever the motoneurons send impulses in their axons to signal the main muscles to contract, they also produce contractions of the sensory muscle fibers (Fig. 13.4A, B). Now, the question arises: what is the reason for having the sensory muscle fibers contract at the same time as the main muscle fiber? The answer is that if the sensory fibers did not contract, they would fall slack as the rest of the muscle contracted and shortened. By contracting at the same time, they, in effect, adjust themselves to the new length of the main muscle, and can therefore signal any departure from that length, such as occurs if an obstacle is encountered. The increase in sensitivity produced by motor stimulation is shown in Fig. 13.4B. Similar operating principles apply to the vertebrate muscle spindle, as we shall see below.

Each sensory cell also receives an inhibitory axon. Stimulation of this axon produces an inhibitory postsynaptic potential (IPSP) in the sensory cell, as we saw in Chap. 7, and the effect of the IPSP in depressing and interrupting an impulse discharge in the receptor cell is shown in Fig. 13.4C, D. The inhibitory axons arise from cells in the segmental ganglion. Because they carry impulses *away from* the central nervous system and its higher centers, they are referred to as *centrifugal fibers*.

It can now be seen that the stretch receptor cell, even though it is located in the periphery among the muscles it innervates,

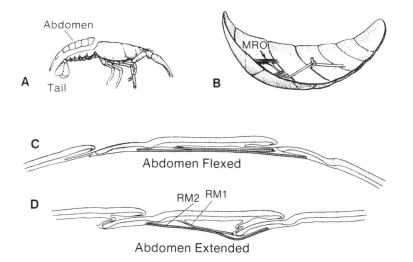

Fig. 13.3 The stretch receptor cell of the lobster, showing relation to muscles and segments of the abdomen. MRO, muscle receptor organ; RM 1, 2, receptor muscles (compare with Fig. 7.15). (A, B from Florey, in Bullock, 1976; C, D from Alexandrowicz, 1951)

Fig. 13.4 Centrifugal control of sensory responses. **A.** Setup for stimulating the excitatory motor axon to the receptor muscle. **B.** During the receptor response to a maintained stretch, stimulation (at arrow) of the motor nerve produces contraction of the receptor muscle, further stretching the receptor cell membrane and increasing the receptor response. **C.** Setup for stimulating the inhibitory motor axon to the receptor cell dendrites. **D.** High-frequency electrical shock (150/sec) to the inhibitory axon (between arrows) inhibits the ongoing receptor response to a maintained stretch. (Adapted from Kuffler and Eyzaguirre, in Kuffler et al., 1984)

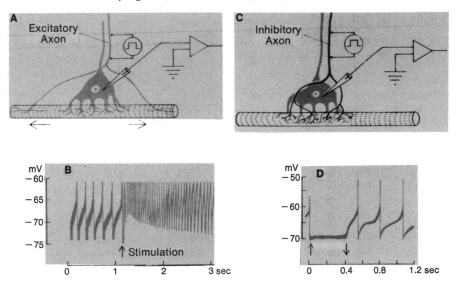

is under exquisite control by the central nervous system. The nervous system can either increase or decrease the sensitivity of this receptor. It is interesting that the output in the motoneurons themselves, in fact, functions as part of the centrifugal control, through the collaterals to the sensory muscle fibers. In the vertebrates, centrifugal control of sensory input occurs mainly in the spinal cord. This is taken to reflect the *encephalization* of nervous control, and the shifting of nervous integration to higher centers, an important feature in the evolution of higher organisms. A similar shift can be seen in motor systems (Chap. 17).

Evolution of Vertebrate Muscle Receptors

Specialized muscle receptors are rather late to appear in vertebrate evolution. There are apparently no sensory nerve endings within the body musculature of fish. In some speices of fish there are sensory fibers, including free nerve endings and corpuscle-like organs, within the connective tissue surrounding the muscles to the fins (Fig. 13.5A). These fibers signal extension and compression produced in the connective tissue by lengthening and contraction of the muscles that control the bending of the fins. A simple type of muscle spindle, consisting of a single modified muscle fiber innervated by a sensory axon, has been recently reported in a jaw-closing muscle in salmon.

Fins, of course, were the evolutionary forerunner of the limbs of terrestrial animals. The amphibians are the first animals in the phylogenetic scale to possess muscle spindles; it is believed that they evolved here to provide the sensory input required for the limb muscles to oppose the force of gravity and maintain posture. Within the muscles are modified muscle fibers, gathered in small bundles and surrounded by a capsule. Their widened midregions reminded early histologists of the spindle used in spinning wool, and they were therefore named *muscle spindles*. There is a single sensory fiber innervating each muscle

spindle (Fig. 13.5B). Each spindle also receives a motor innervation via collaterals from motor axons to the muscles. In the frog, the limb muscles characteristically are composed of different types of muscle fibers with different speeds of contraction. There are fast-twitch, slow-twitch, and nontwitch fibers, with corresponding motor nerve axons to each type of fiber, and corresponding axon collaterals to the muscle spindles (see Fig. 13.5B). The fibers that make up the main mass of the muscle and do all the work are called *extrafusal fibers*, and the modified muscle fibers within the spindle are called *intrafusal fibers* (these names being derived from the fusiform appearance of the spindle). The basic similarities in the arrangements of motor and sensory innervation of the frog spindle and the crayfish stretch receptor may be seen by comparing Fig. 13.5B with Fig. 13.4

Further steps in evolution are seen in birds and mammals. In these species, several differences are evident, as illustrated in Fig. 13.5C. The first difference is that there are two types of intrafusal fibers. One type, which has a collection of nuclei within a saclike central portion, is called a *nuclear bag fiber*. The other type, called a *nuclear chain fiber*, contains a chain of nuclei. In the central portion, each type receives the spiral ending of a large sensory nerve fiber. The ending is called a *primary ending*. It arises from a group Ia axon, the largest of all the peripheral nerve fibers.

Table 13.2 summarizes the diameters and conduction velocities for the Ia fibers and for other types in peripheral nerves.

Near its central portion, the nuclear chain fiber receives smaller spiral terminals. This is called a *secondary ending*. It arises from a group II axon. There may also be a small twig from a group II axon that supplies secondary endings to the nuclear bag fibers (see below).

A second difference is that, rather than the motor innervation being derived from the motor nerves to the extrafusal muscles, the intrafusal muscles receive their own motor axons. These have small diameters,

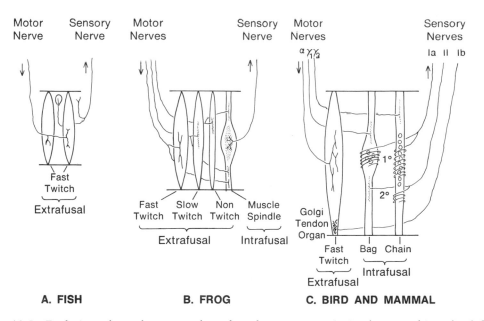

Fig. 13.5 Evolution of vertebrate muscle and tendon receptors. A simple type of intrafusal fiber has been described in certain fish muscles. The innervation pattern is more complicated in mammals, as discussed later. (Based on Barker, 1974)

and are called *gamma (γ) fibers* to distinguish them from the large *alpha (α) fibers* to the extrafusal muscles. Another name for them is *fusimotor fiber*. The γ fibers to bag and chain fibers are distinct from each other.

A third difference is that a new type of sensory organ makes its appearance. This was first described by Golgi in the late nineteenth century, and is therefore named the *Golgi tendon organ*. As indicated in the diagram of Fig. 13.5C, this organ is embedded in the tendon at the end of the muscle. It is innervated by a group Ib axon, only slightly smaller in diameter than the Ia axons.

Nearly all muscles in the mammalian body contain muscle receptors organized along the lines just described. This is rather remarkable, considering how radically such muscles as those in the legs, fingers, tongue, esophagus, and eye differ from each other. According to David Barker of England, who has carried out many detailed studies of muscle receptor anatomy, muscle spindles are present in highest densities in the

hand, foot, and neck, where they are believed to be important in controlling *fine movements*, and also in certain leg muscles (such as the soleus), which are important for *maintaining posture*. By contrast, spindles are fewest in shoulder and thigh muscles, and in muscles (such as the gastrocnemius) involved in initiating gross movements. Spindles are also present in high density in the *extraocular muscles* of the eye in most mammals (such as primates, horses, and pigs). In these muscles, the muscle fibers are divided into subclasses, and the spindles receive some of their motor innervation from motoneuron axon collaterals. The situation thus appears intermediate between the frog and the common mammalian pattern. Muscle spindles have been reported absent from the extraocular muscles in several species (such as rat, cat, and dog). There is as yet no explanation for this puzzling difference.

In summary, muscle receptors are neural sensors that, once appearing in evolution, have been strongly conserved. Like other types of neural modules (e.g., the rhabdom-

Table 13.2 The fiber spectrum of the peripheral nerves in the mammal

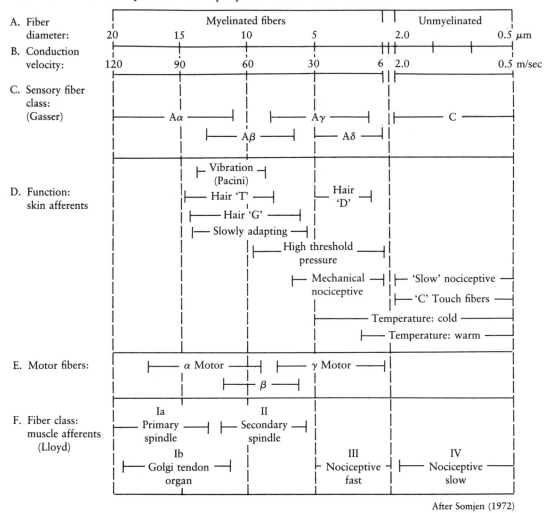

After Somjen (1972)

ere of the insect eye; the cortical column in the mammalian brain: see Chap. 16), they have been adapted to a variety of functions; these include differential sensing of dynamic and static changes in muscles and control of different types of motor reflex responses. Let us next consider the physiological properties of muscle receptors, as exemplified by the frog and the mammal.

Frog Muscle Spindle

The frog muscle spindle, like the Pacinian corpuscle, has played an important part in

the development of our knowledge about sensory receptor mechanisms. It was the first receptor from which single unit recordings, from the sensory axons, were made; this was accomplished by Yngve Zotterman and Edgar Adrian at Cambridge University in 1926. Its response properties were analyzed in a quantitative manner by Brian Matthews in Cambridge in the 1930s. It was the first receptor from which a receptor potential was recorded; this was obtained by Bernard Katz in London in 1950.

A careful analysis of the frog spindle

was carried out by David Ottoson and his colleagues in Stockholm. With fine forceps Ottoson dissected out a single spindle with its single afferent axon and mounted it between two nylon threads in a recording chamber. The setup is shown in Fig. 13.6. Below the setup are schematic diagrams of the ultrastructure of different regions of the spindle: the muscle; a transition zone; and the central, main sensory zone. In the central zone the regular striations of the muscle are partially replaced by connective and reticular tissue, within which are embedded the sensory terminals. The terminals consist of bulbous varicosities several micrometers in diameter, alternating with thin linking processes only a few tenths of a micrometer in diameter. It is believed that sensory transduction occurs in the membrane of the bulbs by mechanically gated channels, as was discussed in Chap. 10.

With the setup shown in Fig. 13.6, stretch is applied very close to the sensory endings, and the receptor properties can be studied with great accuracy. Examples of typical recordings are shown in Fig. 13.7. Note the extremely regular nature of the impulse discharge shown in A, even when the stretch is applied very slowly. The receptor potential shown in B is finely graded in rate of rise and amplitude in correlation with the applied stretch. The greater sensitivity of the spindle during the dynamic phase of the applied stretch is seen in the recordings in A and B and is plotted in the graphs of C and D. These studies leave little doubt of the ability of the spindle to signal the rate of change and the steady amplitude of an applied stretch with exquisite precision.

Mammalian Muscle Receptors

As we have seen, mammalian muscle receptors are characterized by differentiation into spindles and tendon organs, and the spindle fibers are further differentiated into bag and chain fibers, each with primary and secondary sensory endings and with separate γ motor supply.

Motor Supply

The motor supply to the spindle fibers has been a subject of some controversy. Until the 1970s there was a belief that dynamic γ axons supply bag intrafusal fibers and static γ axons supply chain fibers, as shown in Fig. 13.5. This gave way to a more complicated view, depicted in Fig. 13.8, in which there are two types of bag fibers. As summarized by Peter Matthews (1981), in functional terms dynamic action appears the simplest, because it appears to depend on the contraction of one type of intrafusal fiber (bag 1). Static action is more varied; it can be brought about by static γ axons activating bag 2 fibers, chain fibers, or bag 2 and chain fibers together (see Fig. 13.8). It is interesting that chain fibers actually contract more rapidly than bag 2 fibers; this may be related to the fact that bag fibers (both 1 and 2) are activated by electrotonically spreading junction potentials, whereas chain fibers support all-or-nothing impulses. Gradations of local contractions appear to be involved in the responses of bag 1 fibers, as shown by direct microscopical observations.

A further complication to the motor innervation of the mammalian muscle spindle was the finding from the laboratory of Yves Laporte in Toulouse and Paris of motor fibers belonging to the "β" category, between the large α and small γ axons. There are dynamic β axons that supply bag 1 fibers and static β axons that supply chain fibers. These fibers also have branches to the extrafusal muscle fibers, providing for co-activation, as in amphibia (see Fig. 13.5). It has been suggested that dynamic β axons are brought into play during weak muscle contractions, enabling the brain to adjust the sensitivity of the muscle during activity of slow postural muscles (Matthews, 1981), whereas static β axons are active during fast muscle contractions to maintain the sensitivity of the spindle dur-

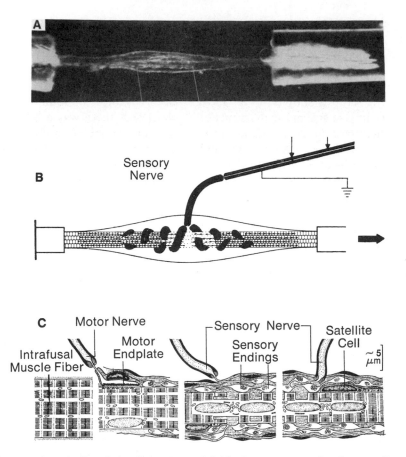

Fig. 13.6 The muscle spindle of the frog. **A.** Darkfield photomicrograph of a spindle mounted between two nylon rods (thickness, 300 μm). The two thin lines indicate the central sensory region. **B.** Diagram of a spindle mounted as in A, showing the recording setup. **C.** Diagram of the fine structure of the spindle, showing the relation of motor and sensory terminals to the intrafusal muscle nerve fiber. (A, B from Ottoson and Shepherd; C from Karlsson et al.; for complete references, see Ottoson and Shepherd, 1971)

ing movement. These actions are supplementary to similar differential actions of the dynamic and static γ motor innervation (see below).

Sensory Responses

These considerations indicate that the sensory and motor innervation patterns make for a rather complicated situation, which permits different combinations of sensory signaling under different states of muscle activity. A number of these combinations have been studied, mostly using the muscles of the hindlimb of the cat (for example, the soleus). Many workers have contributed to this study, and the diagrams in Fig. 13.9 summarize much of this work. Since there are so many permutations of sensory and motor activity, the student should examine this figure slowly—one diagram at a time!

Passive Stretch. The simplest situation is stretch applied to a passive muscle. As shown in Fig. 13.9A, the primary spindle endings give a brisk response, especially to dynamic stretch, while the secondary endings give a slowly adapting response with little dynamic sensitivity. From this it

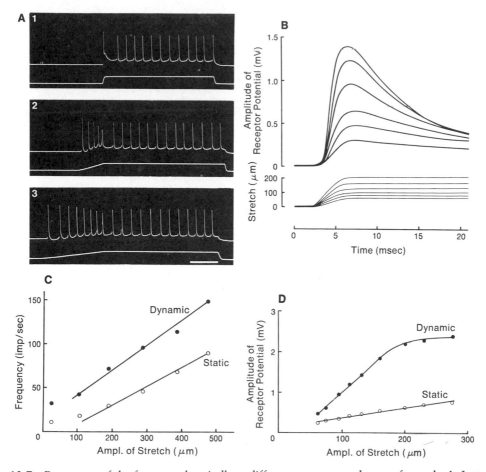

Fig. 13.7 Responses of the frog muscle spindle to different amounts and rates of stretch. **A.** Impulse response to stretches of decreasing rate of extension (1–3), to the same steady extension. Note higher impulse frequencies during dynamic stretching in (2) and (3). (1) 130 mm/sec; (2) 13 mm/sec; (3) 5 mm/sec. Time bar, 50 msec. **B.** Receptor potential recorded from the sensory nerve after blocking impulses with local anesthetic in the bathing medium. Note dynamic peaks followed by decline toward static plateau. **C.** Comparison between impulse firing frequencies at the dynamic peak and the static plateau of the response. Dynamic stretch rate was 2.6 mm/sec; static plateau frequencies were measured 200 msec after the dynamic peak. **D.** Comparison between amplitudes of the dynamic and static receptor potentials for rapidly applied stretches to different static levels. (From Ottoson and Shepherd, 1971)

has been concluded that the primary endings are the main channel through which information about changing stretch of a muscle is communicated, and secondary endings are more specialized for transmitting information about position. Tendon organs show a high threshold and low sensitivity to passive stretch, and thus contribute little information under these conditions.

Passive Stretch with Intrafusal Muscle Contraction. In the normal animal, there is ongoing activity in the γ motor fibers, and the effects of this activity on responses to passive stretch are shown in Fig. 13.9B. In general, there is increased background firing of the sensory fibers, due to the background contractions of the intrafusal muscle fibers, and an increased sensitivity to an applied stretch. There is no effect of the

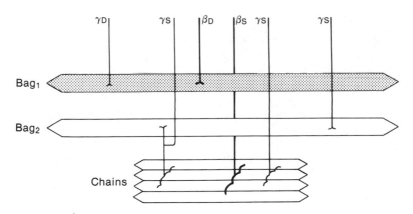

Fig. 13.8 Summary of some of the motor innervation of mammalian muscle spindle fibers by γ axons (D, dynamic; S, static) and by β axons. (Modified from Matthews, 1981)

intrafusal contractions on tendon organs, however.

Extrafusal Muscle Contraction. The activity of the muscle receptors during active contraction of a muscle is shown in Fig. 13.9C, D. During a brief muscle twitch (C), both primary and sensory endings of muscle spindles show a "pause"—an interruption in their ongoing discharge. Tendon organs, in contrast, give a high-frequency burst of impulses during the contraction. In a brief paper in 1928, Pi-Suñer and John Fulton, then at Harvard, speculated that this difference arises because the tendon organ is in series with the contracting muscle, whereas the spindles are in parallel. Because of this arrangement, the tendon organ is subjected to increased tension, while the spindle tends to fall slack. Note the high sensitivity of the tendon organ to active contraction, which stands in contrast to its low sensitivity to passive stretch.

Extrafusal and Intrafusal Muscle Contraction. Finally, we consider the case of active contraction with γ innervation intact, as in the normal cat (Fig. 13.9D). As in part B of the figure, the background discharge is increased, but, in addition, the sensory endings continue to give rise to impulses during the twitch, so that there is no pause. This role of the γ fibers was first indicated by

the work of Lars Leksell in Stockholm in 1945, and was established in a classical series of papers by Carlton Hunt and Stephen Kuffler at Johns Hopkins in the 1950s. By this means, the muscle spindle remains under tension during a muscle contraction, and can thus signal changes in load when they occur. The separate γ innervation of the bag and chain fibers (see D) allows the dynamic and static responsiveness of the sensory endings to be controlled independently, thereby adding to the precision of the sensory signals. The actual situation is more complicated than this, by virtue of the differing innervation patterns of the bag 2 and chain fibers by dynamic and static γ axons and the parallel innervation by dynamic and static β axons, as described earlier.

In summary, different types of motor activity give rise to different patterns of sensory signals from the muscles, as is necessary if the central nervous system is to receive accurate information about the ongoing actions of the muscles. The central nervous system itself contributes to setting the sensitivity of the spindle receptors by ongoing biasing through the γ motor system. The way that these motor and sensory fiber systems are integrated into the neural machinery for muscle control will be explained in the discussion of motor systems in Chap. 19.

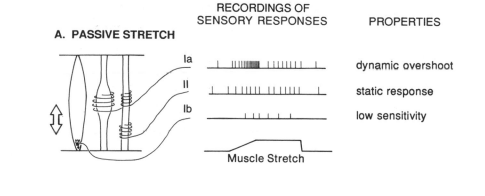

A. PASSIVE STRETCH

Ia — dynamic overshoot

II — static response

Ib — low sensitivity

Muscle Stretch

B. PASSIVE STRETCH WITH INTRAFUSAL MUSCLE CONTRACTION

γ γ

increased background
and response

increased background
and response

low sensitivity

Muscle Stretch

C. EXTRAFUSAL MUSCLE CONTRACTION

α

pause during twitch

pause during twitch

high sensitivity

Muscle Twitch

**D. EXTRAFUSAL MUSCLE CONTRACTION
WITH INTRAFUSAL MUSCLE CONTRACTION**

α γ_D γ_S

discharge during twitch

discharge during twitch

high sensitivity

Muscle Twitch

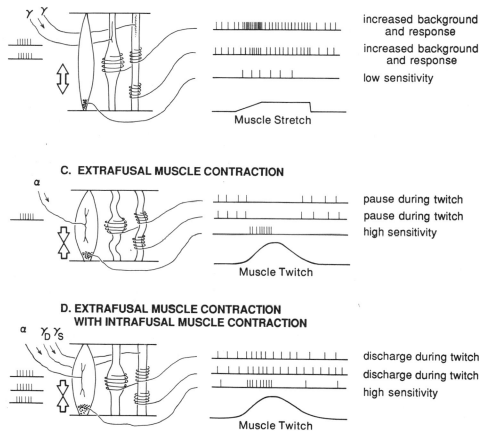

Fig. 13.9 Responses of mammalian muscle spindles and tendon organs under different conditions of muscle stretch, muscle contraction, and centrifugal (gamma) control. **A.** Passive stretch alone causes a dynamic response in Ia spindle endings, a mostly static response in II endings, and little response in Ib (tendon organ) endings. **B.** Stimulation of γ axons *(left)* increases the background activity of both types of spindle endings and their responses to stretch, but has little effect on tendon organ responses. **C.** When impulses in the nerves of a motoneuron *(left)* cause the muscle to contract and shorten (see muscle twitch), the muscle spindles become slack, while the tendon organs are activated. **D.** Stimulation of the dynamic (D) and static (S) γ axons *(left)* causes the corresponding intrafusal fibers to contract, so that they are not slack during a muscle twitch, and can continue to signal the amount of stretch. α, axon of alpha motoneuron; γ, axon of gamma motoneuron [dynamic (D) and static (S)]. (Based on Pi-Suñer and Fulton; Leksell, Hunt, and Kuffler; Matthews, Gordon, and others; references in Kuffler et al., 1984)

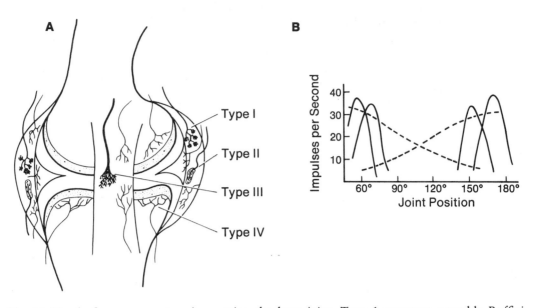

Fig. 13.10 **A.** Sensory receptors innervating the knee joint. Type 1 receptors resemble Ruffini endings in the skin. Type II receptors have the form of flattened Pacinian corpuscles. Type III receptors resemble Golgi tendon organs. Type IV receptors are unmyelinated nerve endings, resembling pain fiber terminals. (From Brodal, 1981) **B.** Tuning curves of joint receptor nerves for different joint positions in the cat. Population of fibers covering only extremes of the range of joint position. Dashed lines show recordings from primary endings of muscle spindles in antagonistic muscles around the ankle joint. Muscle receptors may contribute to sensation of joint position through this activity, which Burgess et al. have termed an opponent frequency code. (Based on Burgess et al., 1982)

Joint Receptors

Joints are typically encased in tough connective tissue capsules and extensions of ligaments of muscles. Embedded in the joint capsules are several kinds of sensory receptor. Each appears to be a modified version of a corresponding receptor in the skin. They have been classified as follows (see Fig. 13.10). The type I receptor consists of small corpuscles around the branches of thin, myelinated fibers. These receptors resemble Ruffini end organs in the skin. They respond to stretch with slowly adapting discharges. The type II receptor is a large corpuscle supplied by a medium-thick myelinated fiber. These receptors resemble Pacinian corpuscles and, like them, are rapidly adapting. The type III receptor consists of a large, dense arborization of a large, myelinated fiber. They

are found in ligaments near the capsule, and they resemble a Golgi tendon organ. They have high thresholds and are slowly adapting. Type IV receptors are free nerve endings of fine, unmyelinated fibers, resembling those in the skin.

This rich innervation suggests that sensory information from the joints might contribute to position sense. The early experiments provided evidence that type I receptors are tuned to respond over narrow angles with the range of joint movement (Fig. 13.10). It was proposed that these slowly adapting discharges signal joint position, whereas the rapidly adapting type II receptors are acceleration detectors. This suggested a beautifully tuned system for signaling joint position, and seemed all the more persuasive in view of other behavioral evidence that muscle receptors did not seem to contribute to position sense.

In recent years, however, the relative significance of the roles played by joint and by muscle afferents has been reversed. Reinvestigation of the slowly adapting fibers indicates that most of these fire impulses only at the extremes of joint movement, but rarely over the middle range, where signals in the normal physiological range of joint position and movement are needed (see Fig. 13.10). On the other hand, there has been increasing evidence for the contribution of muscle receptors to kinesthesia. We will return to these questions when we discuss cortical mechanisms below.

Ascending Pathways

We have seen that information from muscles and joints is carried to the spinal cord in an array of different axons, ranging from the largest myelinated axons (type I and II from muscle receptors) to the several types of fibers from joint receptors. The array of fibers is similar from both the hindlimb and forelimb, but in the spinal cord the connections and ascending pathways are different.

The fibers from the *hindlimb* bifurcate within the cord. One branch terminates within the cord, to take part in segmental reflex circuits (Chap. 19) or to connect to cells of a nucleus called Clarke's column, which projects to the cerebellum via the spinocerebellar pathway (Chap. 21). These fibers also give off collaterals to a nucleus with the somewhat mysterious name of "nucleus Z," which in turn relays through the medial lemniscus to the thalamus. These connections are summarized in Fig. 13.11.

The fibers from the *forelimb* also bifurcate on entering the cord, but their destinations are simpler. One branch takes part in local reflex circuits, whereas the other ascends to the dorsal column nuclei at the anterior end of the cord. This ascending pathway, through the medial lemniscus to the thalamus, is thus similar to that for cutaneous afferents, as discussed in the previous chapter.

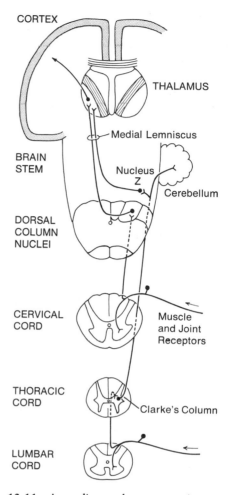

Fig. 13.11 Ascending pathways carrying sensory information from muscles and joints. Compare with somatosensory pathways shown in Fig. 12.14. (Adapted from Carpenter, 1976, and Brodal, 1981)

The different central connections presumably reflect the different functions of hindlimbs and forelimbs. The spinocerebellar pathway, for example, provides for integration of muscle and joint information with cerebellar mechanisms that are essential for sensorimotor coordination and maintenance of muscle tone and posture. This is particularly relevant to the functions of the hindlimbs in standing posture and in locomotion. By contrast, the forelimbs are closely related to the neck and head; their direct connections to lemniscal pathways presumably reflect the more discriminative functions, including manipulation, by the paw or hand, of the forelimb. The separation of these two pathways may therefore reflect functional specialization that was important in the evolution of primates.

The Cortex and Kinesthesia

The ascending pathways carrying muscle and joint information enter the medial lemniscus and terminate in the VP nucleus of the thalamus, as indicated in Fig. 13.11. The fibers terminate topographically, similar to the cutaneous fiber terminations. From here, muscle and joint information is relayed to the cortex. These submodalities have their own specific areas in the cortex, which are closely related to the multiple representations of the somatosensory system. At this point, the reader should refer to Fig. 12.18B. The main cortical representation for muscles is area 3a, which receives input relayed especially from Ia fibers from muscle spindles. There is a representation of some muscle afferents in area 5 of the

parietal lobe, and of various deep tissues in area 2 (see Fig. 12.18B).

These cortical representations are all recent findings, made possible by application of modern techniques such as horseradish peroxidase (HRP) tracing and unit recordings in unrestrained animals. They have brought about a revision in our thinking about perception of movement. As mentioned above, the traditional view had been that muscle afferent pathways do not reach the cortex, and that our perceptions of joint position and the movements of our joints and muscles are mediated only by joint receptors. Proprioceptors were supposed to take part only in subcortical and subconscious muscle reflexes. This seemed to receive support from behavioral studies in humans, in which finger joints were infiltrated with local anesthetic, producing loss of position sense.

This conception began to topple with the finding that relatively few joint receptors are tuned to the physiological range of midjoint ankle position, as mentioned above (Fig. 13.10B). Then, in 1972, G. M. Goodwin, Ian McCloskey, and Peter

Fig. 13.12 A. Site of recording from a single pyramidal neuron in the primary motor area (MI) of an awake, behaving monkey. Also shown are primary (SI) and secondary (SII) somatosensory areas. (From Woolsey, in Henneman, 1980a) Response of this neuron to flexion of the joint at the base of the middle finger (metacarpophalangeal [MC-P] joint) is shown in **B.** The neuron did not respond to other types of movement, as shown in (**C**) flexion of proximal interphalangeal joint (P-IP), (**D**) ulnar (lateral) deviation of MC-P joint, and (**E**) radial (medial) deviation of MC-P joint. (From Lemon and Porter, 1976)

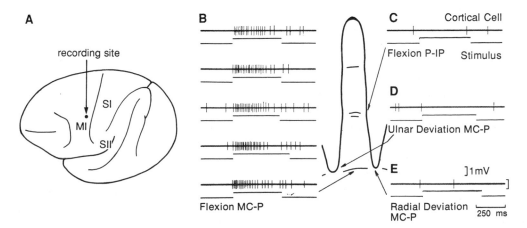

Matthews at Oxford carried out careful behavioral studies which showed that, with local anesthesia of all cutaneous afferents from the arm, position sense of the fingers still persisted. This showed that muscle afferents have access to mechanisms of perception, and it implied that there must be projections of the muscle afferent pathways to the cortex for this to occur. Recent work, as we have seen above, has confirmed the presence of these connections.

Muscle and joint afferents not only have access to somatosensory cortex, they also are relayed to the motor cortex. This is illustrated in Fig. 13.12, which shows results from an experiment in an unanesthetized monkey by Robert Porter and Roger Lemon in Australia. The recording is from an identified projection neuron (a pyramidal cell) in motor cortex (MsI, Fig. 12.18B). The neuron, as can been seen, responds briskly to movement at a single joint of the finger. Results such as these extend the specific parallel and serial pathways underlying perception to include sensory mechanisms within motor areas as well. We shall discuss how this may relate to kinesthesia and "sense of effort" in Chap. 21.

14

The Sense of Balance

All animals exist in a physical environment, and must therefore be able to orient appropriately within it in order to carry out their functions. In a few species, simple contact with the environment is sufficient. This is true, for example, for a sessile animal like the tapeworm, which lives its life attached to the gut wall. However, active organisms are always changing in their relations to the environment, and therefore require constant monitoring of those relations. In general, the more active they are, the more important it is to obtain precise information about different aspects of position and movement. This information is the basis for maintaining the *balance,* or *equilibrium,* of the animal, either in anticipation of motor tasks that involve changes in body position or movement, or through reflexes responding to disturbances from the environment.

The different kinds of sensory information that are used in maintaining balance are indicated in Fig. 14.1. Proprioceptive inputs are a constant source of information about the relative positions and movements of different parts of the body. Cutaneous inputs also contribute. Visual information is important, as anyone can discover by seeing how long it is possible to stand on

one foot with eyes closed! However, these inputs from other sensory systems are usually not enough, and most organisms have therefore evolved sense organs (vestibular apparatus in Fig. 14.1) that are specially adapted for this function.

Sense Organs for Balance

In general, these special organs fall into two categories. One is the *statocyst*. This characteristically takes the form of a fluid-filled pocket that has, in its wall, a patch (called a *macula*) of sensory cells (see Fig. 14.2). The cells have fine hairs which support, at their tips, some dense crystals glued

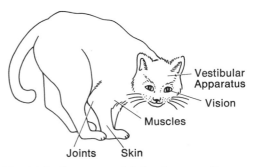

Fig. 14.1 Sensory modalities that contribute to maintaining balance.

310

A. STATOCYST - MACULA

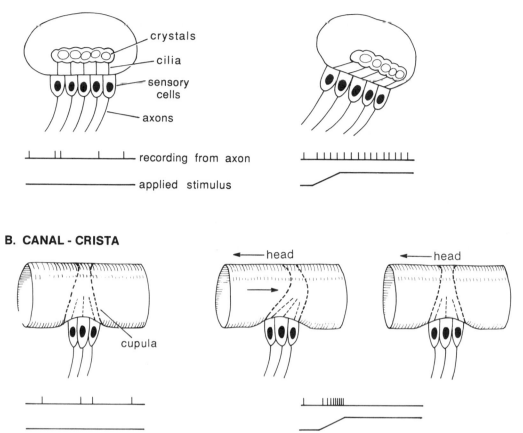

B. CANAL - CRISTA

Fig. 14.2 The main type of organs specialized for sensing balance. **A.** The statocyst, or macula, for sensing the force of gravity and linear acceleration. **B.** The canal, with crista, for sensing angular acceleration of movement.

together with a jellylike material. When the statocyst is tilted, the heavy crystals weigh on the hairs, making them bend (see figure), which leads to an increased discharge of impulses in the sensory fibers. This arrangement is sensitive to velocity. Since the mechanism is sensitive to the force of gravity, the statocyst is a *gravireceptor*. Gravity is a universal force on all organisms, and it is not surprising that nearly all active organisms should have gravireceptors. The statocyst must be an effective organ for this purpose, because, with the conspicuous exception of the insects, most invertebrate as well as vertebrate animals have gravireceptors constructed along the general lines shown in Fig. 14.2.

The other type of organ contributing to the sense of balance is the *canal*. As shown in Fig. 14.2, it is fluid-filled tube with a patch of sensory cells in the wall. These cells also have hairs, which project into the lumen and are embedded in a structure called the *cupula*, a jellylike matrix composed of glycoprotein, which stretches across the lumen of the canal. The patch of cells forms a raised protuberance which is called a *crista*. When the head rotates, the fluid in the affected canal is displaced, which causes a shearing force on the hairs projecting into the cupula, and this is converted into a burst of impulses. As long as the body movement is changing (either accelerating or decelerating), the cupula will be displaced, but when constant velocity is attained, the fluid of the canal moves at the same rate as the body, and the cupula returns to its original position. Thus, the canal type of organ is specially adapted to detect angular *acceleration*. There are a few examples of this type of organ in invertebrates (e.g., lobster, octopus) but they are a constant feature of the vertebrates, where they are called the *semicircular canals*.

Although the principles expressed in Fig. 14.2 are intuitively clear, the details of the functioning of the organs of balance are exceedingly complicated. This is partly because we are not accustomed to thinking in terms of the properties of movement. It is relatively easy to think about *static* responses of statocysts to the force of gravity. The relations of mechanical forces to constant *velocity* are more difficult to grasp. Properties related to *acceleration* are even more difficult, involving as they do concepts of mathematical differentiation, phase lag, and the like. This study is becoming increasingly sophisticated, and interested students will find appropriate references in the legends to the figures.

In this chapter we will be concerned mainly with the peripheral sensory receptor cells and the functional circuits that transmit and process the sensory information within the central nervous system. We will focus especially on the hair cell. As already mentioned in Chap. 10, it is the best understood example of a mechanoreceptor. We will see how its molecular mechanisms serve the sense of balance in the vestibular system and, in the following chapter, how these same mechanisms are adapted for the sense of sound in the auditory system.

The Labyrinth

The fundamental nature of the statocyst as a gravity detector is reflected in the fact that this type of structure, called an *otolith organ,* is present in all vertebrates. Attached to it are one or more canals for detecting rotation, which, because of their form, are called *semicircular canals*. The otolith organ and semicircular canals form the *vestibular organ*. In the course of evolution the otolith organ gives rise to an outpocketing which becomes the organ of hearing, the *cochlea*. Together this makes for a very complicated geometry, which the early morphologists (perhaps in bewilderment) dubbed a labyrinth; since it is all enclosed in membranes, the whole structure is called the *membranous labyrinth*.

The key steps in the evolution of the labyrinth are illustrated in Fig. 14.3. These diagrams also indicate the sites of the sensory receptor cells. The otolith organ is divided into two sacs, utricle and saccule, and the receptor cells are grouped in a macula in each sac. Similarly, each semicircular canal has an enlargement *(ampulla)*, within which the sensory cells are grouped in a crista.

Structure of the Hair Cell

The sensory elements of the vestibular organ are hair cells (Figs. 14.4 and 14.5). In few cells is structure more intimately linked to function; part of the fascination of the modern study of these cells is that each detail of their cellular and molecular structure can be assessed for its role in converting the mechanical stimulus into an electrical signal.

As described above, the hair cells are arranged in sheets within local regions of the labyrinth. Each cell spans the sheet, with a base at one of its ends and an apex at the other. The apex is crowned by an array of slender hairs collectively termed the *hair bundle* (Figs. 14.4 and 14.5). One of these hairs, called the *kinocilium*, is a true cilium, containing nine pairs of microtubules encircling the cilium core. It is placed eccentrically toward the edge of the apex. The rest of the hairs are called *stereocilia;* they actually have a structure resembling microvilli like those lining the wall of the intestine. As shown in Figs. 14.4 and 14.5, they number 40–50 and form an orderly array across the apex of the cell. The array is oriented in several

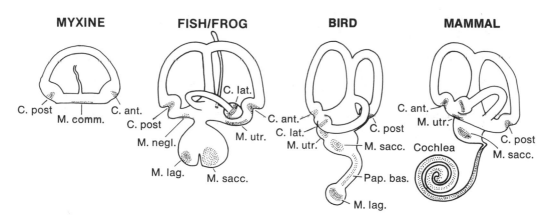

Fig. 14.3 Evolution of the labyrinth in vertebrates. C. ant., anterior crista; C. lat., lateral crista; C. post., posterior crista; M. comm., macula communis; M. lag., macula lagenae; M. negl., macula neglecta; M. sacc., macula sacculi; m. utr., macula utriculi; Pap. bas., papilla basilaris. (From Wersäll and Bagger-Sjöbäck, 1974)

Fig. 14.4 Cellular organization of the sensory cells in the mammalian labyrinth. Note the two types of cell, depending on the size of the afferent terminal. Inset shows relation of single kinocilium to rows of stereocilia. (From Wersäll and Bagger-Sjöbäck, 1974)

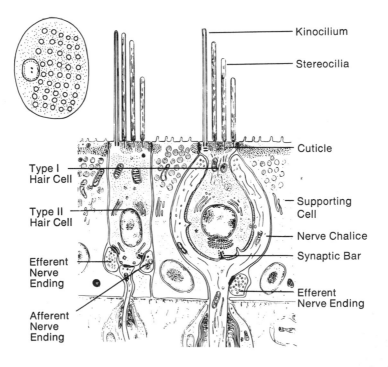

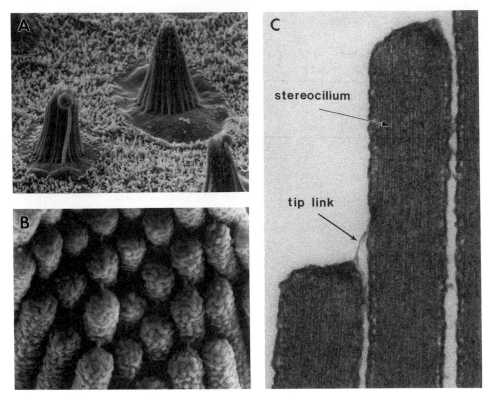

Fig. 14.5 A. View of a typical otolith macula, the bullfrog sacculus. The two hair bundles arise from cells on either side of the striola. The characteristic bulb of the kinocilium demarcates the positive edge of the bundle. **B.** Scanning electron micrograph of the tips of the stereocilia, showing the tip links between them. **C.** Transmission electron micrograph of the tips of two stereocilia. A tip link inserts into the taller stereocilium in an electron-dense plaque. (Micrographs kindly provided by G. M. G. Shepherd and David Corey)

ways: in relation to the kinocilium offset to one side; by the rows orthogonal to the axis set up by the kinocilium; and by the decreasing height of the cilia with increasing distance of the rows from the kinocilium.

The stereocilia are packed with actin filaments, to the exclusion of other internal organelles. The filaments appear early in development and are added as the cilia grow, up to 3000 in the adult. Extending the length of the cilium, they are tightly packed in a paracrystalline array, with the helices of each filament bound to their neighbors by fimbrin. It is assumed that the stiffness of the cilium in resisting imposed movement is due to the fimbrin cross-bridges, which constrain the movement of the actin filaments.

The cilia are constricted at the base, where the outer actin filaments insert into the plasmalemma. The central filaments pass through this constriction to insert into a cuticular plate that spreads across the apex of the cell. The constriction is the hinge point for ciliary movement. It is also the site most sensitive to damage; in auditory hair cells, for example, loud sounds produce deafness by causing local depolymerization of the actin and loss of fimbrin cross-bridges within the basal constriction.

In the extracellular space around the cilia, several types of connecting molecules are attached to the stereocilia plasma membrane, maintaining the structure of the bundle as it moves back and forth, and allowing the individual cilia to slide next to each other as they pivot on their bases.

This generates shear along the tips of the stereocilia, which is essential to the mechanisms of transduction. A critical element in all current theories of hair cell transduction is the so-called *tip link* (Fig. 14.5). This is a very fine connective strand which connects from the side of a cilium near its tip to the side of a neighboring taller cilium. Proposals for its function will be discussed below.

Mechanism of Transduction

The site of transduction and details of its mechanism have been studied in great depth, by A. J. Hudspeth, David Corey, and James Pickles and their co-workers. The simple model emerging from these investigations has recently gained considerable experimental support.

The kinetics of the transduction channels offer many clues. The channels open astonishingly fast, within tens of microseconds. The kinetics are sufficiently fast as to argue strongly against an enzymatic intermediate. Rather, a direct mechanical activation of the channels was posited, involving a *gating spring* which pulls directly on the channels. For increasingly larger positive deflections the channels open faster and faster, but for increasing negative deflections they close at a constant rate. The gating springs were assumed to go slack for large negative displacements, accounting for the asymmetric kinetic behavior.

Next, the site of transduction was localized to the tips of the stereocilia, by measuring where extracellularly recorded receptor potentials were greatest and where focally applied pharmacological blockers of the transduction channels were most effective. Scrutinizing the tip region with the electron microscope, Pickles (1984) discovered fine extracellular filaments which emanate from the tip of each stereocilium and insert into the side of the tallest neighboring stereocilium. These *tip links* occur only along the axis of morphogical and physiological polarization (Fig. 14.5).

This finding immediately inspired a model for transduction (Fig. 14.6) in which tip links correspond morphologically to the gating springs: tip links would be stretched by positive deflections, slackened by negative ones, and unaffected by lateral movements. In this model, tip links are directly coupled to mechanically sensitive channels (as discussed in relation to Table 10.2); increased tension favors channel opening and reduced tension allows channel closure. Evidence for direct mechanical gating of channels came from the observation by Howard and Hudspeth (1988) of minute (4 μm) movements of the bundle associated with channel opening. The most direct support for the involvement of tip links is the finding by Assad et al. (1991) that reducing the Ca^{2+} concentration around the bundle to submicromolar levels abolishes transduction and causes a virtually complete loss of tip links, as seen in electron micrographs. Also consistent with this model are several estimates of the number of transduction channels: there appear to be only one or two per stereocilium. It is still unclear whether the channels are at the upper, lower, or both ends of the tip link; estimates of the numbers of transduction channels agree with any of the possibilities.

It thus appears that deflection of the bundle causes a receptor potential by opening special transduction channels in the tips of stereocilia. Receptor current flows through stereocilia into the cell body, depolarizing it by up to 20 mV (Fig. 14.6D). The transduction channels select poorly among small cations, passing Na^+, K^+, and Ca^{2+} with ease. The high K^+ in endolymph causes this cation to be the dominant charge carrier in vivo. Ca^{2+} is in low concentration in endolymph and thus contributes a small fraction of the current, but it plays an important regulatory role in adaptation.

Adaptation in Hair Cells

Vestibular nerve recordings often show a transient response to maintained stimuli. In the frog sacculus, much of this adaptation mechanism appears to reside in the tips of the stereocilia, and it involves a remarkable mechanical adjustment process. The pro-

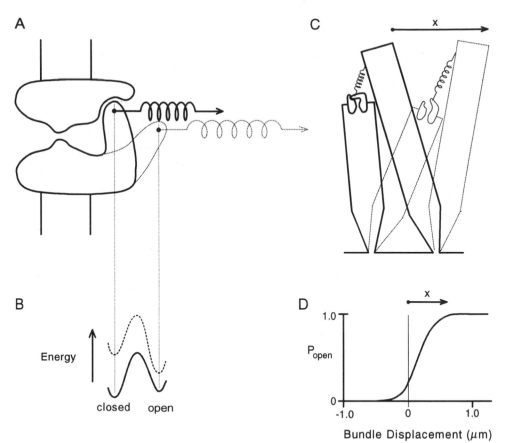

Fig. 14.6 The tip-links model for hair cell mechanotransduction. **A.** A hypothetical ionic channel in the plasma membrane is subjected to distortion by stretch of the tip link during sinusoidal stretch (**B**). **C.** Relative positions of two neighboring stereocilia and the postulated change in length of tip link during applied stretch. **D.** Sensitivity curve of the hair bundle, showing the relationship of channel open probability and hair bundle displacement. See text. (From Pickles and Corey, 1992)

cess is revealed in the hair cell's response to a maintained deflection (Fig. 14.7). First, some fraction of the maximum transduction current, usually around 10%, is on at rest, indicating some amount of resting tension in the gating springs. Second, after adaptation the peak current can still be elicited with a sufficiently large stimulus, so the channels are not being inactivated or desensitized as many voltage- or ligand-gated channels are. Adaptation can be seen as a shift in the sensitivity curve along the deflection axis (Fig. 14.7B), as if the tension on the channels is changing despite the maintained deflection.

A model for adaptation has been pro-posed by Howard and Hudspeth (1988) that is fully compatible with the tip links model for transduction (Fig. 14.7C). It places a force-generating "motor" element at the upper insertion point of each tip link; the motor always climbs upward along the stereocilium, until tension in the tip link causes it to slip.

A positive deflection that stretches the tip links opens channels at first, but as the insertion point slips down, the channels close. Similarly, a negative deflection re-laxes the tip links and closes channels, but they reopen as the motor climbs to restore tension.

Thus, the motor never rests; at steady

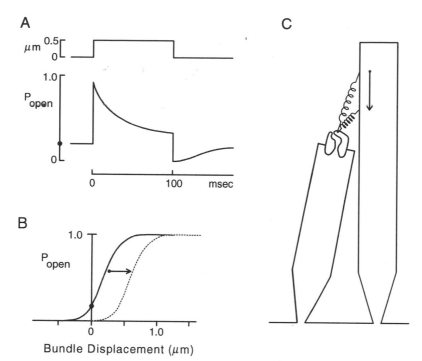

Fig. 14.7 Adaptation in hair cells. **A.** Deflection of the hair bundle toward the tallest stereocilia causes a large increase in the open probability of the transduction channels, followed by a rapid decline toward the resting value. Upon cessation of the stimulus, the channels initially close, then return to the resting value for open probability. **B.** The sensitivity curve shifts during adaptation to accommodate the shift in bundle position. **C.** A model for the mechanism of adaptation, consistent with the tip-links model for sensory transduction. The changes in tension in the tip links are effected by the movement of the upper insertion point of the tip link. (From Pickles and Corey, 1992)

state a resting tension is generated by the balance of climbing and slipping. While the similar labors of Sisyphus went unrewarded, the benefit for the hair cell is that the transduction mechanism is biased to its most sensitive point. This also endows the hair bundle with sensitivity to negative as well as positive deflections.

What is the motor? Recent physiological and biochemical evidence implicates myosin, moving along the actin filaments in the stereocilia. The mechanism of myosin-actin interactions in skeletal muscle contractions is discussed in Chap. 17.

While adaptation of this sort has been found in a variety of hair cell types, it remains to be seen how the extent and rate of adaptation may vary among organs and species.

Ion Channels in the Basolateral Membrane

The only channels observed in the apical surface and bundle are the transduction channels. This arrangement resembles that in other sensory receptors (such as photoreceptors and olfactory receptors), in segregating the functions of transduction and transmission. Along the cell's basolateral walls, however, are voltage- and ion-sensitive channels which contribute to maintaining the resting potential, to shaping and tuning the receptor potential, and to synaptic transmission at the afferent and efferent synapses (see Fig. 14.8).

Inwardly rectifying potassium channels help set the resting potential near the potassium reversal potential. As the receptor current depolarizes the cell, voltage-gated calcium channels open, amplifying the receptor potential. Ca^{2+}-activated K^+ channels then open, limiting further depolarization. In some cells, a delicate interplay of Ca^{2+} channels and Ca^{2+}-activated K^+ channels creates an electrical resonance of the hair cell membrane that strongly amplifies the cell's response to a characteristic frequency. This frequency can vary with channel number and gating kinetics.

The hair cell has no axon and communicates its fluctuations in membrane potential to the central nervous system by means of an afferent synapse. The graded release of an excitatory neurotransmitter, probably a glutamate-like dipeptide, is stimulated by depolarization. A tonic release occurs even without bundle stimulation; it is reduced by hyperpolarization, endowing the synapse with sensitivity to negative bundle deflections. Each of the presynaptic terminals is characterized by a dense body like the presynaptic ribbon of photoreceptors and bipolar cells. The work of Hudspeth and W. M. Roberts has localized both the Ca^{2+} and Ca^{2+}-activated K^+ channels to the presynaptic active site. Their dual participation in electrical resonance and synaptic transmission marks a fascinating economy in design (see Fig. 14.8).

Organization of the Sensory Sheet

The features thus far described are common to all hair cells in the vestibular organs, and in the auditory and lateral line organs as well. The accessory structures that convey and filter mechanical stimuli to the hair bundles are, by contrast, specialized for each type of organ and the sensory stimulus it receives. In the maculae of the sacculus and utricle, the sensory epithelium is overlain with a sheet of dense extracellular matrix, the *otolithic membrane,* directly coupled to the hair bundles via the tips of the kinocilia (Fig. 14.9A). Embedded in this structure are *otoconia,* rocklike crystals of calcium carbonate. The greater density of

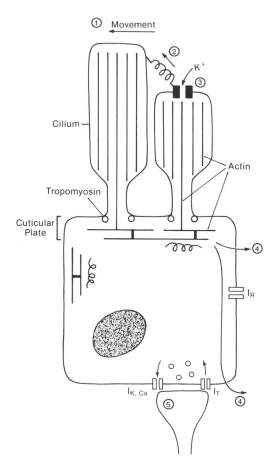

Fig.14.8 Mechanism of transduction in the vestibular hair cell. Movement ① in the direction of increasing cilia height stretches thin interciliary strands ②. This causes an increase in membrane conductance to K^+ ③, which moves into the cilium down its concentration gradient (extracellular K^+ concentration is very high in the endolymph). The resulting depolarization spreads into the cell ④, triggering transmitter release at the hair cell synapse onto vestibular nerve sensory terminals ⑤. See text.

otoconia compared with the surrounding fluid (endolymph) gives them a greater inertia, giving them differential motion when subjected to the linear accelerations of gravity or vibration. In the cristae of the semicircular canals, on the other hand, the hairs project into a dome of jelly, the *cupula,* which spans the lumen of the ampul-

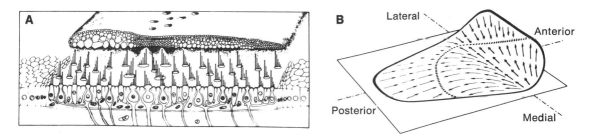

Fig. 14.9 Functional organization of the labyrinth. **A.** Macula, showing covering layer of otolithic membrane. **B.** Different orientations of hairs within the macula. (A from Spoendlin; B from Lindeman, in Wilson and Melvill-Jones, 1979)

lae and detects angular accelerations of the endolymph induced by rotation (cf. Fig. 14.2).

Establishment of the orientation of the ciliary bundle during development represents a remarkable feat of orderly gene expression and cellular targeting of gene products. Experiments have shown that it can take place during differentiation of the otic placode in the absence of the vestibular nerve. Jeff Corwin and his colleagues at Virginia have found that at an early embryonic age the stereocilia emerge in full orderly array as "nubbins" protruding from the apex of the cell, and that differential elongation proceeds beginning in the stereocilia nearest the kinocilium (reviewed in Corwin and Warchol, 1991). Thus, the orientation of the stereocilia is established early. However, within the whole population of hair cells the cilia bundle orientations at first are random, so there is a process of realignment during development to give the orderly patterns of stereocilia orientation seen in the adult.

The vestibular sensory regions display different patterns of hair bundle orientations. As shown in Fig. 14.9B, in the otolith organs the hair bundles are oriented toward (orthogonally to) a curving midline, the *striola*. This arrangement allows detection of all possible stimulus orientations within the plane of the macula. The situation is nicely described in the following quotation from Otto Loewenstein (1974):

In his theoretical exposition of labyrinth function, Mygind . . . compared a macula to the curved palm of a hand holding an irregularly shaped heavy object. . . . We are aware of a changing pattern of contact with the complex contours of the object and localize these accurately as the object rolls in the palm. . . . In this situation the tactile sense, depending chiefly on an apparently random network of . . . nerve endings, performs a feat of pattern recognition which the orderly arrangement of polarized sensors in a macula is bound to surpass. The difference is that we become aware of tactile sensation but remain ignorant of labyrinthine afference.

Similar considerations are involved in the arrangement of the semicircular canals. There, the hair bundles in the cristae are oriented along the axis of each canal. Collectively, the three orthogonally related canals on each side of the head permit detection of all possible angular accelerations of the head (cf. Fig. 14.3).

Sensory Terminals

Although we have considered hair cells as one type, they are in fact divided into two types. As shown in Fig. 14.4 and more clearly in Fig. 14.10, the two types of hair cell reflect different synaptic relations with their afferent nerve terminals. Type II synaptic terminals resemble the small, simple postsynaptic bouton that is common throughout most of the nervous system (cf. Chap. 6). These number approximately 20

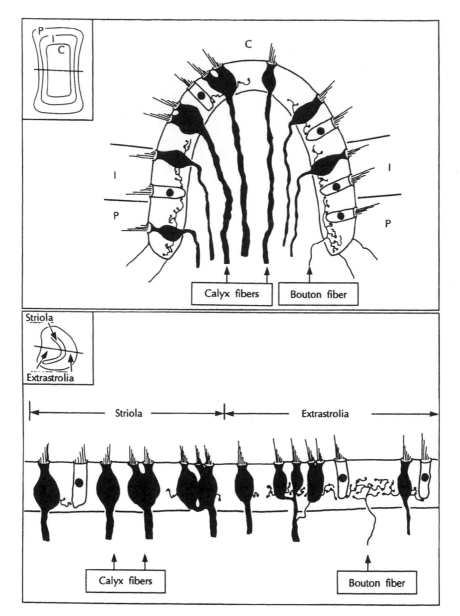

Fig. 14.10 Patterns of sensory innervation in the chinchilla vestibular organs. *(Above)* The sensory epithelium of a crista is divided into central (C), intermediate (I), and peripheral (P) zones. Thick fibers have calyx endings around type I hair cells in zone C. Thin fibers end in branches with boutons on type II hair cells in zone P. A few dimorphic fibers supply both types of cell. *(Below)* The sensory epithelium of the utricular macula contains the striola and extrastriola regions. Calyx fibers supply the striola, bouton fibers supply the extrastriola region, and dimorphic fibers supply both. See Table 14.1 for the physiological properties of these fibers. (From Goldberg, 1991)

per type II hair cell. Type I terminals, by contrast, are very large, enclosing the entire hair cell body in a cup, or calyx. A single such terminal supplies a single type I hair cell.

The distinct morphology of these sensory terminals and their differing synaptic relations to the hair cells are associated with different functional properties of the sensory nerves. As summarized in Table 14.1, there are two types of impulse firing patterns. Calyceal endings are associated with phasic, high-sensitivity responses, which are also relatively transient in response to maintained sensory input. The synapses at these endings thus make a further contribution to sensory adaptation. The small bouton endings are associated with lower sensitivity, more tonic, responses. The two types of endings thus serve to differentiate the responses of the hair cells to different aspects of the sensory inputs to the vestibular organ.

Sensory Signaling

During natural behavior, the transduction mechanisms described above are activated by normal movements of the head. Electrophysiological studies have elucidated the relation between head movements and the neural responses transmitted in the nerves from the semicircular canals. This has required elaborate rotating animal holders, complete with recording instruments. Typical results obtained by Cesar Fernandez and Jay Goldberg (1976) at Chicago are illustrated in Fig. 14.11. Note first the high resting rate of impulse activity in these fibers, in the range of 80 spikes/sec; this high "set point" means that inhibition as well as excitation can be accurately signaled. During slow, constant, angular acceleration (sloped part of monitor trace), the impulse frequency rises to a relatively steady level. This tonic response, presumably reflecting the properties of type II (bouton) endings (see above), is due to the press of the displaced endolymph against the cupula of the crista. The amplitude of the displacement (which is transduced into the receptor potential of the hair cell) is determined by the balance between the inertial force of the fluid and the elastic restoring force of the cupula. When constant velocity is reached, the discharge returns (with an undershoot) to the resting level. Mathematically, one says that this response represents an integration of the angular acceleration to give a signal proportional

Table 14.1 Comparison of the characteristics of mammalian vestibular-nerve afferents

Irregularly discharging	Regularly discharging
Thick and medium-sized axons	Medium-sized and thin axons
Calyx and dimorphic units in the central (striola) zone	Dimorphic and bouton units in the peripheral (extrastriola) zone
Phasic-tonic response dynamics, including a sensitivity to the velocity of cupular or otolithic displacement	Tonic response dynamics, resembling those expected from displacement of the cupula or otolithic membrane
High sensitivity to head rotations or linear forces (calyx units in the cristae can have low sensitivity)	Low sensitivity to head rotation or linear forces
Large excitatory responses to efferent activation, including both fast and slow response components	Small excitatory responses to efferent activation, usually including only slow response components
Large responses to externally applied galvanic currents	Small responses to externally applied galvanic currents

From Goldberg (1991)

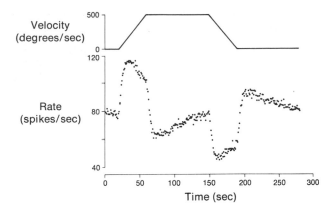

Fig. 14.11 Responses recorded in single sensory nerve fibers from the semicircular canals to stimulation by rotation (angular acceleration) of the whole animal (squirrel monkey). (From Fernandez and Goldberg, 1976)

to the angular velocity of the head. In doing this, the vestibular mechanism is performing a precise mathematical operation using mechanical and electrical components, as would an analog computer. When deceleration occurs, this is signaled in the opposite direction, by reduction of the impulse frequency below resting level.

Central Vestibular Pathways

The axons that transmit the hair cell responses to the central nervous system form part of the eighth cranial nerve. There are surprisingly few of these fibers—only about 20,000 on each side in most mammalian species, including humans. They enter the brainstem and terminate in several large cell groups, which are called the *vestibular nuclei*. From here there are three main projection systems, as indicated in Fig. 14.12. We will discuss each of these.

Vestibulospinal System

Fibers that project to the spinal cord form the vestibulospinal pathway. There are two divisions, medial and lateral. The medial tract consists of fibers that arise from cells in most of the vestibular nuclei and gather in the midline to form the *medial longitudinal fasciculus*. The descending fibers termi-

nate in anterior segments of the spinal cord, where they connect to motoneurons that control the axial muscles of the neck and trunk. By contrast, fibers to motoneurons that control limb muscles arise from the lateral vestibular nucleus and descend in the lateral tract. Although these vestibular pathways mediate reflex mechanisms that belong in Section III, we will discuss them here because they are essential for understanding the sensory functions of the central vestibular pathways.

A basic approach to analyzing vestibular control of body muscles has been to deliver single electrical shocks to the vestibular nerve or vestibular nuclei and record from single spinal neurons. This tells the neurophysiologist whether a synaptic action is excitatory or inhibitory, and whether the action is monosynaptic or mediated through interneurons. Much of this work has been carried out by Victor Wilson and his colleagues at Rockefeller University. A summary of recent studies using these methods is shown in Fig. 14.13. The important points to be made in connection with this diagram are as follows:

1. The macula projects mainly, but not exclusively, to the lateral tract, and the cristae of the canals to the medial tract.

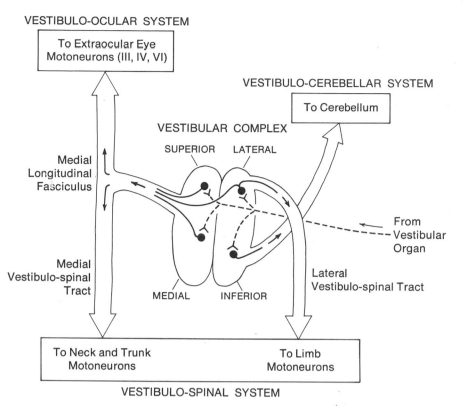

Fig. 14.12 Divisions of the vestibular nucleus, and their output connections to different parts of the brain. (Modified from Carpenter, 1976)

2. The lateral tract is mainly concerned with controlling limb motoneurons, while the medial tract is exclusively concerned with controlling neck and trunk muscles.

3. Lateral tract fibers are exclusively excitatory, whereas the medial tract contains both excitatory and inhibitory fibers.

4. Within the spinal cord, lateral tract fibers control motoneurons to the limbs through excitatory or inhibitory interneurons; in other words, the pathway is disynaptic. Medial tract fibers, by contrast, make direct monosynaptic excitatory or inhibitory connections onto neck and trunk motoneurons.

The pathways indicated in Fig. 14.13 provide an extensive system through which the vestibular input can make precise adjustments in body musculature to maintain orientation in space. If we recall the analogy between the vestibular receptor sheet and the palm of the hand, the central system depicted in Fig. 14.13 can be thought of as the internal neural representation of the palm, allowing the body muscles to "feel" the changes occurring in the receptors and adjust their states of contraction accordingly.

Figure 14.13 makes a distinction between the muscles of the head (dorsal neck) and the body, and this is important in understanding vestibular reflexes. The main point is that the head, which contains the vestibular apparatus, is connected to the neck, the neck to the trunk, and the trunk to the limbs. Neck reflexes are thus the link between movements of the head and the rest of the body. These reflexes were first studied in a systematic way by Rudolf Magnus of the Netherlands in the

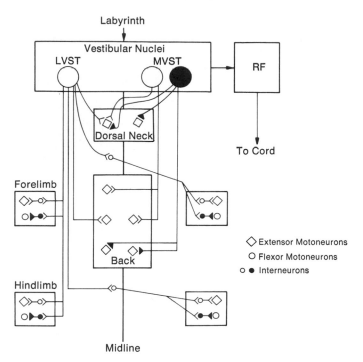

Fig. 14.13 Circuits formed by the medial and lateral vestibulospinal tracts. Open profiles indicate excitatory neurons, closed profiles indicate inhibitory neurons. LVST, lateral vestibulospinal tract; MVST, medial vestibulospinal tract; RF, reticular formation. For explanation, see text. (From Wilson and Melvill-Jones, 1979)

early part of this century. Modern studies have led to the concept of these reflexes as servosystems for stabilizing the head in space. The neck reflexes act as a closed-loop negative feedback system; since the neck muscles are attached to the head, their responses to labyrinth stimulation tend to restore the head to its normal position and reduce the stimulation. These reflexes are generally obligatory and unconscious, which is consistent with the direct synaptic pathways shown in Fig. 14.13. By contrast, body and limb muscles have variable relations to head position, through the linkage of the neck. This may be why the vestibulospinal system has both excitatory and inhibitory fibers, and the synaptic linkages are through both excitatory and inhibitory interneurons (see Fig. 14.13). The interacting effects of labyrinthine and neck reflexes on the limbs are summarized in Fig. 14.14.

Vestibulo-ocular System

The vestibular system plays a central role in controlling the eyes. This is necessary in order to maintain a stable image of the visual field on the retina despite movements of the body. The importance of a stable image is vividly exhibited in the way a bird walks with its head appearing to move with exaggerated jerks. In fact, film analysis of these movements has shown that the head is actually held almost perfectly still relative to the surrounding space, while the body moves along smoothly beneath it! This enables the animal to maintain maximal visual sensitivity to movement in its visual field while moving about itself. The stabilization is believed to be brought about by the neck reflexes mentioned above. Presumably, similar though more fluid mechanisms are present in mammals.

Movements of the eyes are controlled by a set of six extraocular muscles; their

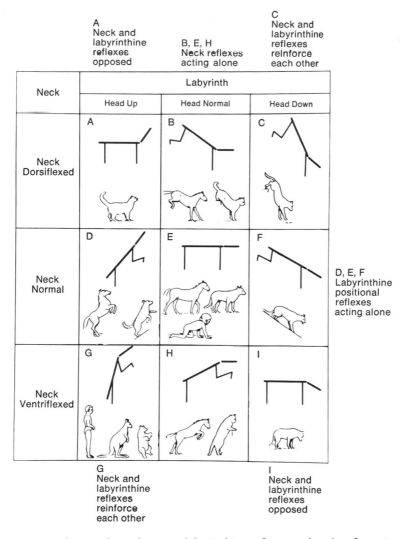

Fig. 14.14 Diagrams showing how the static labyrinthine reflexes and neck reflexes interact in the control of the limbs under different conditions of body posture. (From Roberts, in Wilson and Melvill-Jones, 1979)

arrangement is indicated in the diagram of Fig. 14.15A. Coordinated movements of both eyes together are called *conjugate movements;* it is obvious that this requires fine coordination of both sets of extraocular muscles.

We can obtain a grasp of the principles involved in this coordination by considering, in a simplified fashion, movements only in the horizontal plane. Consider that we fix our gaze on a point, and then shift our gaze to another point to the left, without moving our head. A recording of this

mechanical shift of our eyes would look like the trace in Fig. 14.15B. In the horizontal plane, this movement is brought about by contraction of the lateral rectus muscle and relaxation of the medial rectus muscle of the left eye, and the opposite activation of muscles in the right eye. This type of jumping *(saccadic)* eye movements with the head held still is characteristic of activities like reading or examination of near objects. These movements are controlled by fibers descending from the frontal eye fields of the cerebral cortex to the

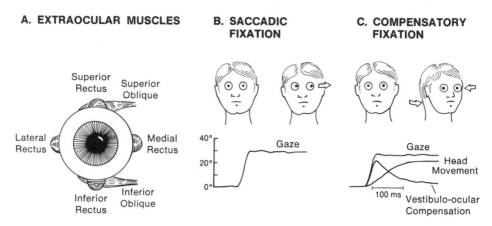

Fig. 14.15   Organization of the extraocular muscles. **A.** Arrangement of the six muscles around the right eyeball. **B.** Jumping saccadic movement, brought about through activity of the descending motor system. **C.** Compensatory eye movements, to maintain visual fixation after movement of the head; this is brought about by the vestibulo-ocular reflex. (Based in part on Morasso et al., in Wilson and Melvill-Jones, 1979)

extraocular motoneurons, and do not involve the vestibular system.

Now consider the case in which the gaze of the eyes is directed to a new point in the visual field and the head moves in order to maintain the point at the center of the visual field, with the eyes in their normal centered positions. In this case, the gaze can remain fixed in the new position only be means of compensatory eye movements which are in the opposite direction to the rotation of the head. The relative movements of eye and head are indicated in the traces of Fig. 14.15C.

Compensatory eye movements are brought about by the *vestibulo-ocular reflex.* For movements in the horizontal plane, the circuit that connects the input from the horizontal canals to the motoneurons controlling the medial and lateral rectus muscles is shown in Fig. 14.16. This circuit operates without benefit of visual or other sensory feedback; in systems terminology, one says it is an *open-loop reflex,* in contrast to the closed-loop neck reflexes described previously.

The circuit shown in Fig. 14.16 constitutes a disynaptic pathway from vestibular afferents to extraocular motoneurons. This pathway is supplemented by polysnaptic connections through the *reticular formation,* as is also true for the vestibulospinal system. The disynaptic pathway is consistent with the rapid transmission necessary in controlling eye movements.

One might suppose that a pathway of this kind would be "hard-wired" and immutable. However, much evidence indicates that the connections in this system have an extraordinary degree of plasticity. For example, after removal of one labyrinth, the ability of monkeys to perform a complicated locomotor task is completely lost and then slowly recovers; the recovery presumably depends on the establishment of new or more effective synaptic connections. More dramatically, when people wear spectacles with prisms that invert the visual field, they are initially disoriented and helpless but gradually learn to move about in a near-normal fashion. This has implied a considerable rearrangement of the synaptic circuits involved in vestibular reflexes. Microelectrode studies of single-neuron activity carried out in several laboratories have documented plastic changes in the synaptic connections of the central vestibular pathways after ablations of different components of the system (cf. Ito, 1985).

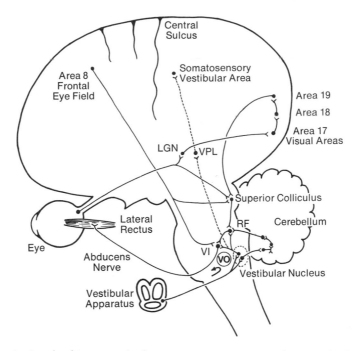

Fig. 14.16 Circuits involved in control of eye movements. For simplicity, only the control of the lateral rectus muscle is illustrated. Similar connections are involved in control of the other extraocular muscles. LGN, lateral geniculate nucleus; RF, reticular formation; VI, trochlear nucleus giving rise to the abducens nerve; VO, vestibulo-ocular reflex; VPL, ventroposterior lateral nucleus of thalamus. (Adapted from Robinson, Gouras, Schmidt, Ito, and other sources)

Vestibulocerebellar System

The importance of the vestibular system for sensorimotor coordination is further shown by the close relations of vestibular pathways to the cerebellum (see Fig. 14.16). We will discuss these relations and functions further in Chap. 21.

Vestibular Cortex

A small number of fibers ascend from the vestibular nuclei to terminate in a small part of the ventral posterior nucleus of the thalamus. From here there is a projection to a small region of cortex within the face area of somatosensory cortex (see Fig. 14.16; also 12.19). It is attractive to suppose that this vestibular area of cortex is involved in conscious perceptions of balance and movement arising from vestibular inputs. Other studies have suggested that other areas of cortex also receive vestibular thalamic inputs, including the somato-

sensory arm area and the temporal lobe, depending on the species of animal investigated. In addition to these sensory areas, there is a well-known area of the frontal lobe which contains cells that control the voluntary eye movements mentioned above. This is called the frontal eye field; electrical stimulation of this area causes conjugate eye movements to the opposite side. Figure 14.16 summarizes these cortical sensory and motor areas in the human (see also Chap. 30).

The Vestibular System and Weightlessness

We live in the space age. The most dramatic sensory effect of space travel is the effect of weightlessness on the sense of balance. Dr. Joseph Kerwin, who was aboard Skylab 2 in 1973 as the first U.S. physician astronaut in space, describes the experience in the following words (1977):

I would say there was no vestibular sense of the upright whatsoever. I certainly had no idea of where the Earth was at any time unless I happened to be looking at it. I had no idea of the relationship between one compartment of the spacecraft or another in terms of a feeling for "up or down." . . . What one thinks is up, is up. After a few days of getting used to this, one plays with it all the time. . . . It's a marvellous feeling of power over space—over the space around one. Closing one's eyes made everything go away. And now one's body is like a planet all to itself, and one really doesn't know where the outside world is.

As the above quotation indicates, the astronauts adapted quickly, over the course of a few days, to the condition of weightlessness, and found the experience (apart from episodes of motion sickness) rather pleasant and intriguing. This adaptability is remarkable if one considers that the vestibular system and the other systems contributing to balance evolved over millions of years without ever having been exposed to these conditions.

Some indication of the neural processes of adaptation that take place during flight is seen in the behavior of the astronauts after return to earth. In a simple experiment carried out by J. L. Homick and his colleagues (1977), the astronauts were tested on their ability to stand, one foot in front of the other, on narrow rails of different widths, from 3/4 in. to 2 1/4 in. (1.9–5.7 cm), as well as on the floor. In a typical result, the pilot of Skylab 3 showed a deficit on the day after splashdown in his ability to stand on relatively narrow rails with eyes open, with recovery to normal on sub-

sequent testing days. Even with the thickest rail, there was a long-lasting deficit in the ability to maintain standing balance with the eyes closed; in fact, on the day after splashdown, he could barely stand on the floor with eyes closed, but he improved to normal on the subsequent test.

These data give clear evidence that postural mechanisms are affected by prolonged periods (8–10 weeks) of weightlessness. Probably several systems contribute to these results. First, the leg muscles undergo a degree of disuse atrophy during flight, which probably affects the ability to stand in the immediate recovery period. Second, muscle tendon reflexes (see Chap. 19) are hyperactive; this could cause incoordination of the muscles used in standing. Finally, it has been postulated that a "pattern center" in the central nervous system undergoes a process of habituation during flight, and that this process must be reversed on return to earth. This pattern center integrates inputs from the vestibular organ, the muscle and tendon receptors (Chap. 13), and the tactile receptors in the skin and deep tissues (Chap. 12). It represents the distributed system for the maintenance of standing posture. The ability of this entire system to adapt between the conditions of normal gravity and zero gravity is in accord with the remarkable degrees of plasticity that are present in nervous circuits in general, and have already been demonstrated in parts of the vestibular system, as mentioned above in connection with the vestibulo-ocular system.

15

Hearing

The senses we have considered thus far are widespread throughout the Animal Kingdom. Information about molecules and chemicals, the physical environment, muscular states, and spatial orientation may be considered obligatory for any multicellular organism that carries on an active life. We now consider a type of information that is less widespread. *Audition*—the sense of hearing—is largely limited to the insects and the vertebrates. This, of course, does not mean it is any less important; on the contrary, those species that have this sense use it with great effectiveness in various ways—for example, to escape predators, find mates, and communicate socially. And in humans, there can be little doubt that hearing has been the key to development of language and, through it, much of our culture.

The Nature of Sound

Audition in its broadest interpretation is the sense of sound. Generally speaking, sound consists of waves of compression of air or water. Sound waves are produced by a wide variety of natural phenomena, including many kinds of animal movements—the beating of wings, for example,

or the clatter of hooves. These sounds are important; indeed, they may be matters of life or death, such as a warning of an approaching predator. In addition, audition is of more specific interest as the reception of sounds made explicitly by one member of a species to communicate to another; it is here that the full potentialities of the auditory sense are realized.

Animals have developed many mechanisms for producing sounds for communication (see Chap. 23). The sounds vary in their frequency over a wide range, as illustrated for a number of different animals in Fig. 15.1. Some species have relatively narrow ranges (some crickets, frogs, birds). Some ranges extend to extremely low frequencies (fish, whales and dolphins, humans). The lowest frequencies, in fact, involve vibrations that act as stimuli for various kinds of mechanoreceptors in the body, including the vestibular organs as well as auditory receptors. On the other hand, some species are able to sense extremely high frequencies, as high as 100 kHz (various mammals and insects). It is interesting to compare the wavelength at this frequency (a few millimeters) with that at the extremely low frequencies (almost 10 m). Finally, some species have very broad

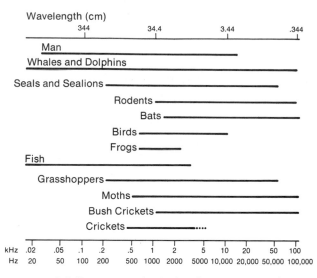

Fig. 15.1 Hearing ranges of different animals. Scales: kHZ, 1000 cycles per second; wavelength, centimeters. (After Lewis and Gower, 1980)

ranges (especially whales and dolphins, and humans). Within these different ranges, the receptors and circuits of many species are turned to certain frequencies of special importance for their behavior, as we shall see.

Modern studies of audition employ sophisticated instruments and rigorous mathematical methodologies, and many of the results are expressed in terms of decibels, power spectra, Bode plots, and other concepts that are difficult for the beginning student to grasp. We will focus our attention on the structures and properties that enable the hair cell to be adapted for reception in the acoustic range and the central nervous circuits for processing auditory information.

Lateral Line Organs and Electroreception

The sense of hearing in the vertebrates has an interesting and rather complicated phylogenetic history (see Table 15.1) that is closely related to the development of the vestibular apparatus. Both senses are commonly regarded as adaptations of the *lateral line organ*. Lateral line organs are found in all aquatic vertebrates, from primitive cyclostomes like the lamprey up through the amphibia. They consist of either pits *(ampullae)* or tubes along the sides of the body, head, or snout of the animal. At the base of the pit, or at intervals along the tubes, are right clusters of cells, called *neuromasts*. The cells are a type of hair cell, with the hairs embedded in a gelatinous mass that projects into the fluid-filled pit or tube.

The specific stimulus for the hair cells varies in different species. In some it is *mechanical vibrations* set up by the turbulence of the water around the fish. In others it may be changes in *hydrostatic pressure*. The receptors in some species are also very sensitive to *temperature,* or to the *salinity* (saltiness) of the water. Finally, in certain species of fish, these receptors are sensitive to *electrical signals*.

Although electroreception is not found in higher vertebrates, it is a fascinating sensory capability that deserves mention here. The most sensitive electroreceptors are in pits called *ampullae of Lorenzini,* which are distributed on the head and snout of certain species of fish. The threshold for eliciting a response in an individual receptor can be as low as 1 μV/cm (that is, an electric field with a gradient of one-millionth of a volt [1μV] for each centimeter of distance; this is equivalent to just a

Table 15.1 Evolution of organs for electroreception and hearing in the vertebrates

	Lateral line	Electrolocation	Electroreception for communication	Hearing	Echolocation	Hearing for communication
Cyclostomes						
Cartilaginous fish						
Bony fish						
Amphibians						
Reptiles						
Birds						
Mammals						

1-V difference over a distance of 10 km!). The threshold for a behavioral response is 10–100 times higher. The electric fields may be set up by electric organ discharge in the same fish; nearby objects produce distortions of the field, which are sensed by the electroreceptors. Alternatively, the electroreceptors may sense the fields set up by the electric organ discharges of other fish.

The Ear of Mammals

Hearing reaches its highest development in birds and mammals; in fact, the organ of hearing in these animals is one of the most complex of all sensory organs. We will describe this organ and the central auditory pathways in the mammal, with brief mention of lower forms.

The organ of hearing—the ear—has three main parts (see Fig. 15.2). The *outer ear* aids in the collection of sound and funnels it through the external ear canal to the tympanic membrane. The *middle ear* contains a system of small bones—hammer, anvil, and stirrup—that conveys the vibrations of the tympanic membrane to the inner ear. The *inner ear* consists of a fluid-filled bag, the *cochlea,* which developed as an outpouching from the vestibular labyrinth (see Fig. 14.3). Through the middle of the cochlea stretches the *basilar membrane,* containing the hair cells which are the auditory sensory receptors.

In order for sound to stimulate the hair cells, it must first be transmitted mechanically to the inner ear, and then stimulate the hair cells in an appropriate manner. The first step requires changing the sound waves from vibrations in air to vibrations of the perilymph. This is achieved through the intermediary movements of the middle ear bones. Since air is highly compressible but perilymph is incompressible, the bones must provide for a matching of the forces in the two media, a process called *impedance matching.* They do this by absorbing energy from the large area of the tympanic membrane and concentrating it in the small area of stirrup where it fits against an opening (the *oval window*) in the bone onto the membrane surrounding the cochlea (Fig. 15.2B).

The Basilar Membrane

The next step is to induce appropriate vibrations in the basilar membrane that contains the hair cells. The cochlea in humans is a coiled structure; we can understand its function best if we imagine it straightened out, as in Fig. 15.2C. We can then see that the cochlea tapers in size like a cone toward a tip, so that there is a *base,* at the oval window, and an *apex,* at the tip. In contrast, the basilar membrane is *narrower* at the base and *widens* at the apex. The student has to remember, therefore, that the basilar membrane gets wider as the cochlea gets narrower.

When the early anatomists first examined the basilar membrane under the microscope, they observed cross-striations that reminded them of the strings of a

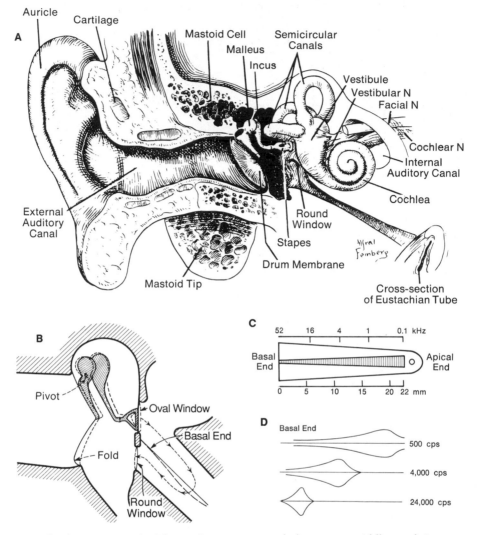

Fig. 15.2 The human ear. **A.** The main structures of the outer, middle, and inner ear. **B.** Transmission of sound vibrations through the middle ear to the inner ear (cochlea). **C.** Diagram of the cochlear partition and the basilar membrane *(above)*, and the traveling waves and their external envelopes induced by sound at different frequencies. (A, B from Davis and Silverman, 1970. C, D based on von Békésy, 1960)

piano. They imagined that the short strings (at the base) would resonate in response to high notes, and the long strings (at the apex) to low notes. Helmholtz, the great physiologist and physicist of the late nineteenth century, formulated these ideas as the *resonance theory of hearing,* which states that different frequencies of sound are encoded by their precise position along the basilar membrane. This attractive the-

ory fell victim, however, at least in part, to a stubborn fact. A string will not vibrate unless it is under tension; when Georg von Békésy (1960) tested this by making tiny slits in the basilar membrane of cochleas obtained from human cadavers, he found that the edges of the slit did not pull apart, as they would if the membrane were under tension.

By direct observation, Békésky found

that vibratory movement is transmitted as a *traveling wave* along the basilar membrane from the round window to the apex. You can make this kind of wave yourself by fastening a rope at one end and pulling lightly on the other end while moving it up and down. A ripple emanates from your end that travels through the rope until it is dampened at the far fixed end. As shown in Fig. 15.2D, the wave has its largest amplitude at a specific site along the membrane, depending on the frequency. Thus, although the wave itself travels, the envelope of the wave is stationary for a given frequency. The peak displacements for high frequencies are toward the base (where the basilar membrane is narrowest) and for low frequencies are toward the apex, just as Helmholtz postulated, but the envelope of the traveling wave is broader than he envisaged.

These experiments appeared to lend support to Helmholtz's theory, but in fact they posed a difficult problem which has only been solved in recent years. The problem was that the envelope of the traveling wave, as measured in human cadavers, is relatively broad, which contrasts with psychophysical data, as well as with common experience, which demonstrate that we are very good at hearing tones, that is, the ear is sharply tuned for different sound frequencies. To deal with this, Békésy proposed that sharpening occurs by means of lateral inhibition in the neural connections of the auditory pathway (see below). However, this was proved inadequate when recordings from single auditory nerve fibers gave clear evidence of sharp tuning already at this peripheral level.

This discrepancy has been resolved by experiments in which movements of the basilar membrane could be observed directly in the living animal by use of sensitive instruments to detect responses to small stimuli. These showed that the movements of the basilar membrane closely parallel the frequency selectivity of the neural discharge (see Fig. 15.3). There are thus properties of living cells within the cochlea

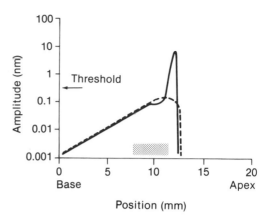

Fig. 15.3 Simple model of the effect of the cochlear amplifier of the guinea pig. The dashed line shows the amplitude of passive movement of the basilar membrane in the absence of the cochlear amplifier. The solid line shows the peak alteration added by the outer hair cells, which act over approximately half an octave on the basal (low-frequency) side of the peak of the traveling wave (shaded region). (From Ashmore, 1992)

which, added to the traveling wave, produce frequency selectivity of the hair cell responses. In order to gain insight into these properties, we must next consider transducer mechanisms in the hair cells.

Hair Cell Transducer Mechanisms

How are movements of the basilar membrane transduced into electrical responses of the hair cells? As indicated in Fig. 15.4A, the hair cells rest on the basilar membrane, with their raised tips covered by a thin flap of tissue called the tectorial membrane. The key fact about hair cells is that there are two types, depending on their relation to the organ of Corti (see Fig. 15.4A). Herein lies much of the mystery about sound reception in the cochlea: why are there two types, and what is the specific function of each?

Outer hair cells are by far the most numerous type, numbering about 20,000 arranged in three rows. Their hairs are arranged in a characteristic V-shape, as seen from above (Fig. 15.4B, C). *Inner hair cells*

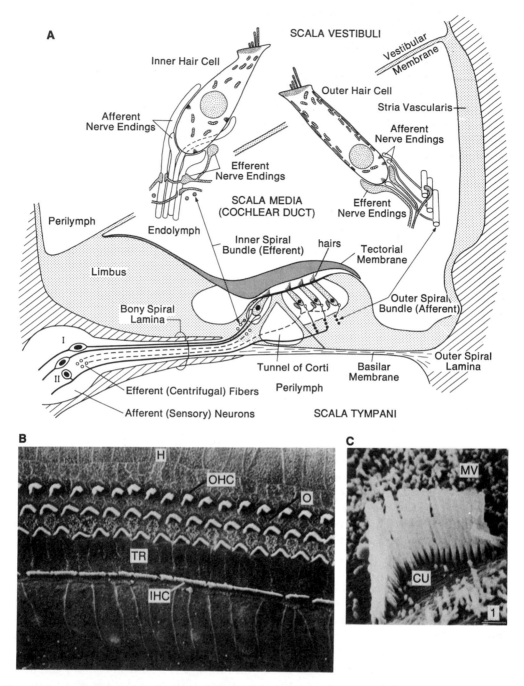

Fig. 15.4 A. Cellular organization of the cochlea (organ of Corti) of the guinea pig. Enlarged diagrams (see arrows) summarize the fine structure of the inner and outer hair cells. **B.** Scanning electron micrograph, looking down on hair cells, showing differences in arrangements of the inner (IHC) and outer (OHC) hair cells and their stereocilia. H, Henson's cell; TR, tunnel rod. **C.** Scanning electron micrograph of stereocilia of an outer hair cell. CU, cuticular plate; MV, microvilli (stereocilia). (From Smith, in Eagles, 1975; A redrawn from Smith in Brodal, 1981)

number about 3500, and are aligned in a single row (Fig. 15.4B). As in the case of the vestibular organ, the hair cells transmit their sensory responses by means of synapses onto the terminal dendritic knobs of second-order cells, the sensory neurons, which transmit the response as an impulse discharge to the central nervous system. The puzzle here is that the ratio of nerve fibers connecting to the two hair cell types is the reverse of the ratio of the number of hair cells: some 20,000 nerve fibers connect only to the 3500 inner hair cells, whereas only 1000 or so connect to the much more numerous outer hair cells. There is thus convergence of many nerve fibers onto a single inner hair cell, in contrast to divergence from one fiber onto many outer hair cells. These relations are depicted schematically in Fig. 15.5. The great preponderance of fibers to the inner hair cells means that these are the main sites of auditory transduction; the fibers connecting to them are the main auditory pathway and the main sources of single-unit recordings in studies of neural responses.

Intracellular recordings have shown that sound stimuli produce responses in which the oscillations of the membrane potential closely parallel the applied mechanical oscillations. An example of a typical inner hair cell (IHC) response is shown in Fig. 15.5B. When the entire frequency range is swept at different sound intensities, a family of tuning curves is generated, as shown at the bottom of Fig. 15.5B. For a given cell, there is a best frequency, shown as a peak in the tuning curve. Outer hair cells (OHC) have qualitatively similar response properties, but there are quantitative differences; as shown in Fig. 15.5C, the receptor potentials are somewhat smaller in amplitude, and their tuning is not as sharp.

The generation of the receptor potentials appears to involve a process of mechanoelectric transduction that is essentially similar to that which occurs in vestibular hair cells; the reader should review those mechanisms in the previous chapter. Vestibular hair cells have been much more accessible to experimental analysis for this purpose. Recently, culture systems for the organ of Corti have permitted whole-cell patch recordings of the outer hair cell transduction currents, revealing that all the main features of transduction and adaptation described earlier in vestibular hair cells are also present in auditory hair cells.

A summary of these mechanisms as they are found in the two types of auditory hair cells is shown in Fig. 15.6. Note in both types the initial depolarization due to the influx of K^+ down the steep gradient from high extracellular K^+ in the endolymph. In outer hair cells, this leads to activation of a voltage-gated Ca^{2+} conductance, which provides for modulation of a Ca^{2+}-sensitive K^+ conductance. In inner hair cells, there are at least two types of voltage-gated K^+ conductances in the basolateral membrane and a voltage-sensitive Ca^{2+} conductance in the synaptic region. The interplay of K^+ and Ca^{2+} conductances produces an oscillating potential which generates an electrical resonance, as in vestibular cells. Note the predominantly afferent innervation of inner hair cells, contrasting with the predominantly efferent innervation of outer hair cells.

Exploration of the membrane currents generated by these cells, carried out using patch-recording techniques, has begun to reveal the molecular mechanisms underlying transduction. Robert Fettiplace and his colleagues in Wisconsin have found that in some cases it has been possible to record single-channel events. As shown in Fig. 15.7A, these can be recorded in the whole-cell mode when the extracellular Ca^{2+} is reduced so that the number of active channels is reduced. They have a single-channel conductance of approximately 100 pS (100 picosiemens, or 100×10^{-12}S), which is much larger than the conductances of the channels found in the basolateral membranes of the cell. The fact that they can be activated by mechanical movement of the cilia suggests that they are indeed the sensory channels in the cilia tips. Fettiplace (1992) estimates that, since the total trans-

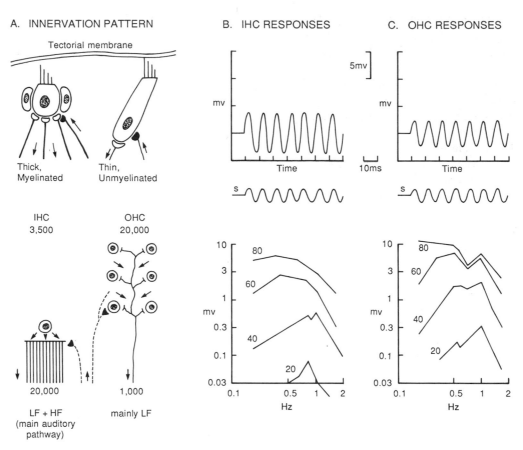

Fig. 15.5 Comparison of the organization and properties of inner hair cells (IHC) and outer hair cells (OHC). **A.** Innervation pattern. In the lower diagrams, the numbers of hair cells are given above and the numbers of auditory nerve fibers below. HF, high-frequency fibers; LF, low-frequency fibers. **B.** IHC response properties. *(Above)* Intracellular responses to tone stimulus (s). *(Below)* Response magnitude (as above) for different stimulus frequencies (abscissa) and at different stimulus intensities (20–80 dB). **C.** OHC response properties. (A based on Spoendlin, 1969; B,C based on Dallos, 1985)

ducer conductance in normal saline is of the order of 5 nS (5 nanosiemens, or 5×10^{-9}S), there may be a total of some 50 channels per cell, which is close to the estimate of approximately one channel per cilia arrived at by the studies of Howard and Hudspeth (1988).

Active Processes in Hair Cells

The ability of the hair cell to generate an oscillating membrane potential is of interest for two reasons. First, this property can add to the sensory response of the cell; by increasing the response at the best frequency, it can be an electrical mechanism for sharpening the tuning curve within the hair cell. Second, this property can provide the means for the hair cell to produce a mechanical output. The possibility that hair cells might function in this mode was first suggested by the finding that following a brief weak "click" delivered to the ear, there was a weak brief vibration of the eardrum, called a "Kemp echo" after its discoverer (Kemp, 1978). This and subsequent experiments have indicated that the

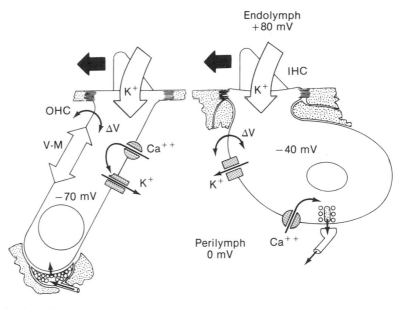

Fig. 15.6 Diagrams of the functional organization of outer hair cells (OHC) and inner hair cells (IHC). V-M, voltage to mechanical converter. See text. (From Dallos, 1992)

echo is due to output from the hair cells, and is closely related to the properties that generate the frequency selectivity of the basilar membrane and hair cells. Dramatic evidence for this property is the finding that the stereocilia can undergo spontaneous beating movements, associated with oscillatory changes in the membrane potential.

The mechanisms underlying this activity have been analyzed in outer hair cells. In these studies, on-cell patches have been used to record the tiny currents generated by the voltage-activated motor mechanisms in the basolateral membrane. As shown in Fig. 15.7B, summation of many traces revealed a transient (300 μsec) current of some 2-pA (2 picoamps, or 2×10^{-12} ampere) amplitude, which corresponds to approximately 3000 elementary charges transferred across the membrane. These charges are responsible for activation of the motor mechanisms which shorten and lengthen the outer hair cell in response to the rapidly oscillating receptor potential. The movements are so fast that they can

follow up to high frequencies: Jonathan Ashmore has produced an entertaining film of an isolated outer hair cell dancing energetically to the tune of "Rock Around the Clock"!

It is presumed that this mechanism contributes to the active tuning of hair cell responses as shown in Fig. 15.3. However, the manner in which this comes about is still the subject of debate. It must first be recognized that all outer hair cells are not alike. As summarized in Fig. 15.8, outer hair cells near the basal end of the cochlea are short and stubby, only some 30 μm in length, whereas at the apex they can reach lengths of 50 μm. Many other properties vary in accord with these changes, including a decrease in input conductance, an increase in time constant, a decrease in channel density, and an innervation ratio that increases from approximately 10 to 50. These differences are likely to be correlated with the differing frequencies that are processed along the cochlea, but the significance of these changes is still not understood.

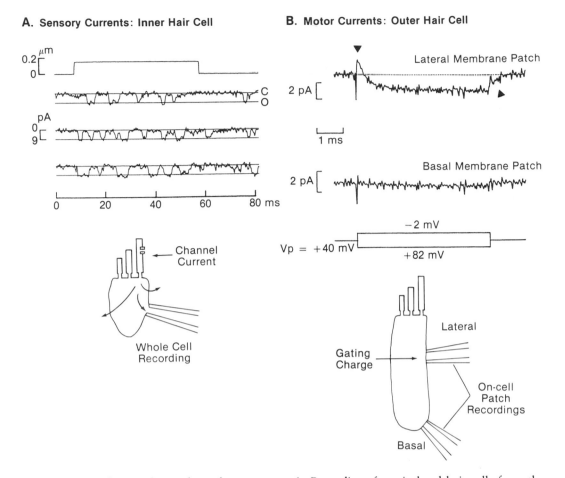

Fig. 15.7 Patch recordings of membrane events. **A.** Recordings from isolated hair cells from the basilar papilla of the turtle (see diagram below). Mechanical stimuli were applied to the hair bundle, as shown by the monitor above. Below are shown successive traces of whole-cell recordings of single-channel openings after exposing the cell to 1-μM Ca^{2+} and then returning it to normal saline containing 2.8 mM Ca^{2+}. C, closed; O, open. **B.** Gating current in the outer hair cell. In the cell-attached recording mode (see diagram below), summed responses (2000 traces) to alternating voltage steps of 42 mV revealed a rapid outward followed by slow inward membrane current in a patch from the lateral but not from the basal membrane. (A from Fettiplace, 1992; B from Ashmore, 1992)

Another subject of debate is the nature of the coupling between inner and outer hair cells. Figure 15.9 summarizes some of the ways of interpreting the feedback pathway that is believed to be involved. The essential system consists of perturbations of the basilar membrane and tectorial membrane (BM-TM), which generates the receptor potential in the IHC and the OHC. In the scheme in (a), the transducer potential in the OHC activates the cycle-

by-cycle motor force (ac motor) that is fed back to sum with the BM-TM displacement to enhance it (positive feedback) or diminish it (negative feedback), depending on the frequency. In (b), the receptor potential activates a dc potential which feeds back a steady force that alters the operating set point of the BM-TM system. In (c), the activation by the IHCs of efferent activity from the CNS is incorporated into the system to provide for additional control of the

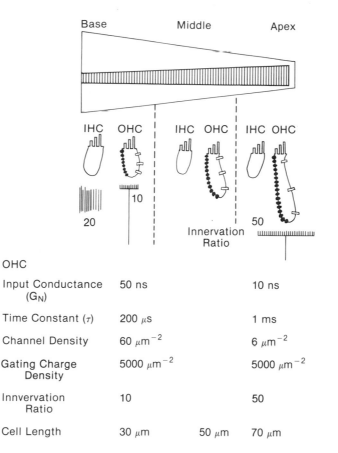

Fig. 15.8 Schematic diagram illustrating how the structure of outer hair cells (OHC) varies with location along the cochlea. *(Below)* Variations in the functional properties of outer hair cells associated with the anatomical variations. (Based on Ashmore, 1992)

operating set point and other properties of the response. Current experiments are directed toward testing these models.

Centrifugal Fibers

In addition to afferent fibers, the auditory nerve also contains efferent (centrifugal) fibers, which arise from cells in the brainstem. These fibers make synapses mainly onto the dendritic knobs of the afferent fibers connecting to inner hair cells, but they make synapses directly onto outer hair cells (see above). For many years, it has been known that stimulation of these fibers suppresses sensory responses in the auditory nerves. This was taken to be an example of descending control of sensory systems. It was believed to protect the hair cells from overstimulation, but beyond this the mechanism and its functional significance were unknown.

Our understanding of these problems has been helped considerably by the intracellular analyses of hair cells. These have shown that the centrifugal fibers inhibit hair cells by hyperpolarizing the hair cell membrane, as shown in Fig. 15.10. The transmitter at these synapses is ACh. These actions are believed to be relatively specific for outer hair cells, which, it will be recalled, centrifugal fibers contact directly (cf. Figs. 15.4A, 15.5, and 15.6). In view of the recent studies discussed previously, the significance of this inhibitory input is clear; by reducing

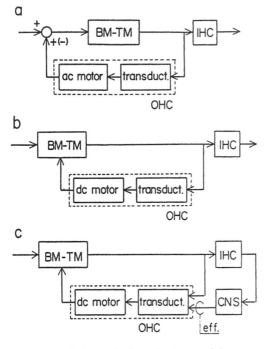

Fig. 15.9 Schematic flow diagrams of the operations involved in the cochlear feedback system. a. System for feedback of oscillatory (ac) force. b. System for feedback of steady (dc) force. c. System incorporating efferent control from the CNS. ac motor, oscillatory motor system of the basolateral membrane; BM-TM, basilar membrane-tectorial membrane system; CNS, central nervous system; dc motor, steady motor system; eff., efferent fibers from the brainstem; IHC, inner hair cell; OHC, outer hair cell; transduct., sensory transduction mechanism. See text. (From Dallos, 1992)

bipolar ganglion cell, with its cell body in the cochlea. The peripheral fiber of this cell receives the synapses of the hair cells. There are only about 25,000 auditory nerve fibers in mammals, including humans. It is sobering to think that human language, and so much of our society and culture, depends on these fibers. It recalls the words of Winston Churchill: "Seldom has so much been owed by so many to so few."

Single-unit recordings from individual

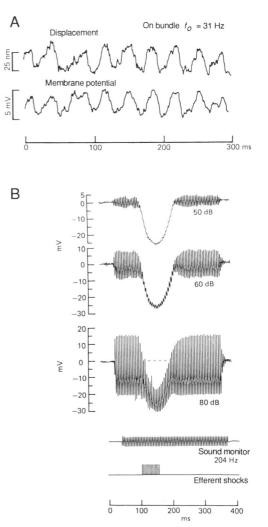

Fig. 15.10 Motor function of the hair cell, as exemplified by spontaneous mechanical *(above)* and electrical *(below)* oscillations. These recordings were made in hair cells of the turtle cochlea. (From Crawford and Fettiplace, 1985)

the motor *output* of the outer hair cells, it reduces the movement of the tectorial and basilar membranes and the sensory response of the inner hair cells. The original interpretation that these fibers provide a protection against overstimulation is therefore supported by this evidence.

Auditory Nerve Fibers

Since vertebrate hair cells lack axons, the auditory signals are transmitted to the central nervous system, as we have already noted, by a second-order neuron. This is a

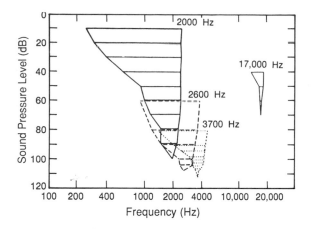

Fig. 15.11 Tuning curves for single auditory nerve fibers in the cat. (From Galambos and Davis, 1943)

auditory nerve fibers provided some of the earliest and most essential information about the encoding of auditory signals. Hallowell Davis and David Galambos in the 1940s first showed that each fiber has a characteristic tuning curve (see Fig. 15.11). As already mentioned, it is now known that the hair cells themselves have similar tuning curves (Fig. 15.5), so the synaptic coupling between hair cells and nerve fibers provides for faithful transmission of sound stimuli. An important result of the single-fiber recordings was also the finding that the nerve impulses fired in synchrony with low-frequency vibrations, but only up to about 1 kHz (1000/sec). The fibers cannot fire above this frequency because the refractory period associated with each impulse lasts about 1 msec.

These results proved that auditory frequency could not be encoded solely by the frequency of impulse firing. Rather, frequency is encoded primarily by position along the basilar membrane, referred to as *tonotopic organization*. Impulse frequency may contribute to coding at low frequencies, but in general it is primarily involved in coding stimulus *intensity*. It may be noted that in the absence of stimulation, the auditory fibers show considerable spontaneous activity. This means that hair cells, synapses, and auditory fibers are primed

to respond to threshold stimuli and small changes in stimulation, just as in most other sensory systems.

Brainstem Auditory Pathways

Since the auditory nerve contains so few fibers, it might be wondered if the auditory input would simply get submerged when it enters the central nervous system. Our acute sense of hearing tells us that this is not so, and a consideration of the synaptic organization at the first central relay shows how this comes about.

Cochlear Nucleus

There are two specializations that ensure that auditory acuity is preserved despite the small number of input channels. The first is that each auditory nerve fiber, on entering the brainstem and reaching the collection of cells known as the *cochlear nucleus*, divides into a large number of terminal branches. Within the nucleus, the branches form a rigid geometrical array. In this way, each fiber projects to several terminal regions in an orderly fashion. In addition, the tonotopic sequence along the basilar membrane is projected by the array of auditory fibers onto the separate regions of the cochlear nucleus. Thus, the single cochlea

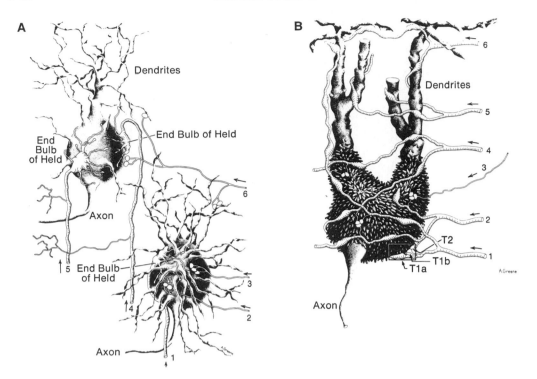

Fig. 15.12 A. Synaptic connections onto a "bushy" neuron in the anterior ventral cochlear nucleus of cat. Numbers indicate types of auditory nerve axons: (1), (4), and (5) are thick axons making large terminals (end bulbs of Held) on a bushy cell body; (2), (3), and (6) are thin axons making small boutons on cell body [and dendrites (6)]. **B.** Synaptic connections onto an octopus cell in the posterior ventral cochlear nucleus of the cat. Axon types: (1), (2), and (4) are thick axons making large, mostly en passant (in passage) contacts with cell body of octopus cell; (3) is a thin axon making a small bouton on cell body; (5) and (6) are thick axons making large en passant contacts on octopus cell dendrites [see also ascending branch of (1)]. Heterogeneity of terminals from a single axon is shown by large terminal (T1a) and small terminals (T1b and T2) from single thick axons (1) and (2). (From D. K. Morest, in Eagles, 1975)

is mapped into *multiple representations* within the cochlear nucleus.

The second specialization is seen in the types of synapses and cells in the different parts of the nucleus. The main parts have large and distinctive relay neurons, each of which receives a distinctive type of synaptic terminal. Two examples are shown in Fig. 15.12. The significance of these specializations is twofold: they provide for a very secure coupling of input to output, and they provide the morphological basis for different kinds of input–output processing of the auditory signals.

The nature of the processing that takes

place at these different relay sites has been revealed by single-cell recordings. A useful scheme of classification was introduced in 1966 by Ross Pfeiffer, on the basis of the type of response to a pure-tone stimulus. Figure 15.13 shows the main response types, together with diagrams of the cells and simplified circuits which may account for the response properties. The legend of this figure may be consulted for details. The important general conclusion from these data is that the bifurcations of the primary axons and their synaptic connections with different types of relay cell provide the basis for the conversion of the

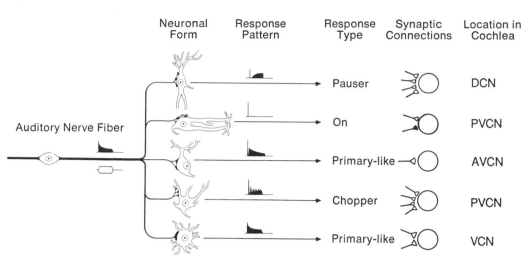

Fig. 15.13 Correlation of synaptic connections and neuron types with response properties in the cochlear nucleus. Responses to a tone burst (50 msec duration) are displayed as poststimulus time histograms. Some postulated circuit connections are indicated to the right (open terminals are excitatory, filled terminals are inhibitory). AVCN, anterior ventral cochlear nucleus (end bulbs of Held); DCN, dorsal cochlear nucleus (pyramidal-fusiform cell); PVCN, posterior ventral cochlear nucleus (On: octopus cell; Chopper: globular-bushy cell); VCN, ventral cochlear nucleus (multipolar stellate cell). (Adapted from Kiang, in Eagles, 1975)

single response envelope in the primary axons into different output patterns. The different synaptic microcircuits create, in effect, new classes of responses, each class representing an abstraction of one particular feature of the input. For example, the simple ON response is ideally suited to transmitting high-frequency stimulation. The "primary-like" response obviously preserves the envelope of the auditory input signal. The "pauser" and "buildup" types provide for a differentiation of the onset and ensuing phases of a tone, similar to the dynamic and static phases of responses we have noted in other sensory systems.

Ascending Pathways

The organization depicted in Fig. 15.13 provides a means by which different properties of the auditory stimulus are given their own private (or semiprivate) channels. This is an expression of the same general principle found in other sensory systems, which we may summarize as follows:

Different functional properties are processed and transmitted in parallel pathways. Each channel has its unique set of microcircuits, mediating a set of functions extracted from the input.

There are extraordinarily rich and complex relations between each part of the cochlear nucleus and various brainstem centers (see Fig. 15.14). However, if we keep in mind the principle of parallel pathways, some of the relations can be seen to be logical. The spherical and globular cells make ipsilateral and contralateral connections to a cell group known as the *olivary nuclear complex*, a center that is essential for the binaural localization of sounds in space. The octopus cells project to cells within the same complex, which, in turn, project through the olivocochlear bundle back to the cochlea, to provide for centrifugal control of the hair cells, as previously mentioned. Cells of the dorsal cochlear nucleus do not share these close relations with lower brainstem centers; instead, their outputs are destined for higher centers, in

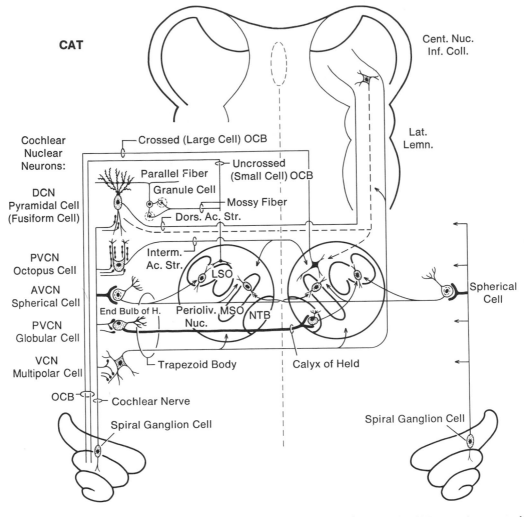

Fig. 15.14 Organization of the central auditory pathways in the cat. AVCN, anterior ventral cochlear nucleus; Cent. Nuc. Inf. Coll., central nucleus of inferior colliculus; Dors. Ac. Str., dorsal accessory stria; DCN, dorsal cochlear nucleus; End Bulb of H., end bulb of Held; Interm. Ac. Str., intermediate accessory stria; Lat. Lemn., lateral lemniscus; LSO, lateral superior olive; MSO, medial superior olive; NTB, nucleus of trapezoid body; OCB, olivocochlear bundle; Perioliv. Nuc., periolivary nucleus; PVCN, posterior ventral cochlear nucleus; VCN, ventral cochlear nucleus. (From Moore and Osen, 1979)

the midbrain (inferior colliculus) and thalamus. The general organization of these pathways thus involves certain kinds of information processing at lower levels and other kinds at higher levels.

Auditory Cortex

The main thalamic relay nucleus for auditory information is the medial geniculate nucleus (MGN). As in the somatosensory and visual systems, the projection area of the thalamic relay cells onto the cortex defines the primary auditory cortex.

Cortical Areas

Traditionally, it has been thought that there is a single area of primary cortex, but here, as in the other systems, recent work has provided evidence for multiple divi-

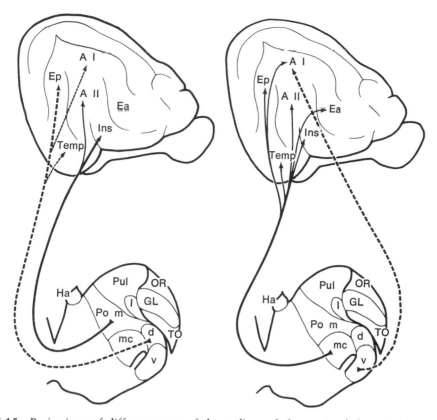

Fig. 15.15 Projections of different parts of the auditory thalamus (medial geniculate nucleus) to the cerebral cortex of the cat. A I, auditory area I; A II, auditory area II; d, dorsal division of the medial geniculate body; Ea, anterior ectosylvian gyrus; Ep, posterior ectosylvian gyrus; GL, dorsal lateral geniculate body; Ha, habenula; I, inferior division of the pulvinar complex; Ins, insular area; mc, magnocellular division of the medial geniculate body; OR, optic radiations; Po m, medial division of the posterior nuclear group; Pul, pulvinar complex; Temp, temporal field; TO, optic tract; v, ventral division of the medial geniculate body. (From Diamond, 1979)

sions within both thalamic relay nucleus and cortex, and parallel pathways connecting them. For example, HRP injections were made by Irving Diamond and his colleagues at Duke University into different areas, and the labeled cells in the thalamus were identified. As summarized in Fig. 15.15, these studies showed that, in the cat, the most specific thalamic projection is from a ventral subdivision of the MGN to primary auditory cortex. By contrast, the magnocellular subdivision projects not only to the primary cortex, but to a number of surrounding cortical areas as well. Several other subdivisions have multiple, but more specific, projections. Diamond has suggested that the multiple and diffuse projections are phylogenetically older and the

single specific system projection is more recent, in analogy with the presumed phylogeny of the two ascending pathways in the somatosensory system (Chap. 12).

Multiple auditory areas are also found in primates (these were depicted in the diagram of sensory cortical areas in Fig. 12.18). The auditory areas in humans are localized on the dorsal aspect of the temporal lobe (see Chap. 30). We will discuss their relations to other cortical regions, including the language and speech areas, in Chap. 30.

Tonotopic Representation

The discovery of a topographical representation within the primary somatosensory cortex suggested that there might be a cor-

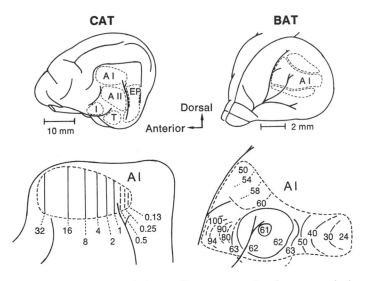

Fig. 15.16 Tonotopic organization of the auditory cortex in the cat and the mustache bat *Pteronotus*. Note the regular tonotopic representation in the cat and the distortion of this regularity by the large representation of 61–62 kHz in the bat (the frequency of the echolocating signal in this species). EP, posterior ectosylvian gyrus; I, insula; T, temporal field. Frequency bands in cortical maps are indicated in kHZ. (Based on data from Woolsey, Merzenich et al., and Suga; in Suga, 1978)

responding representation in auditory cortex of the cochlea *(cochleotopy)* or of the localization of tones along the cochlea *(tonotopy)*. The first evidence for this was obtained in experiments on dogs by Archie Tunturi in Oregon in 1944. This fit very well with other evidence showing precise tonotopy in lower centers like the inferior colliculus and medial geniculate nucleus. However, subsequent workers had difficulty in replicating the results in the cortex, and for many years the matter was unresolved. One of the problems appears to have been that tonotopic organization is more evident in anesthetized than in unanesthetized animals. In 1975, Michael Merzenich and his colleagues in San Francisco clearly demonstrated tonotopic organization in the auditory cortex of anesthetized cats. A typical map is shown in Fig. 15.16 (cat).

These results have been followed up and confirmed in other species. In most mammals there is an area in primary auditory cortex that has a large and detailed map; this is called A I. There are additional representations in several other areas. There are also some areas receiving thalamic input, as discussed above, which are not organized tonotopically; for example, A II in the figure.

The regular progression of frequency bands in the tonotopic representation of animals like cats and humans reflects the broad spectrum of sounds to which we are responsive. What about an animal like the bat, which uses special vocal signals as a kind of radar for echolocation? Let us discuss one example, the mustache bat. The echolocating sound emitted by this animal is an almost pure tone of 61 kHz (far above the hearing range of humans; see Fig. 15.1). The specializations of the bat for sensing this signal begin in the periphery, where auditory nerve fibers with relatively broad tuning curves may be found across the whole frequency spectrum, but those with best frequencies of 61 kHz have extremely narrow tuning curves. In the cortex, there is a tonotopic organization within A I, but it is distorted by an extremely large representation of 61 kHz (see Fig. 15.16 [bat]).

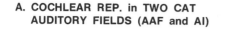

**A. COCHLEAR REP. in TWO CAT
AUDITORY FIELDS (AAF and AI)**

B. BINAURAL BANDS within AI

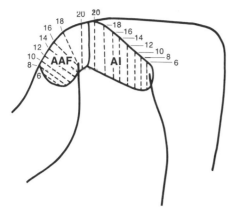

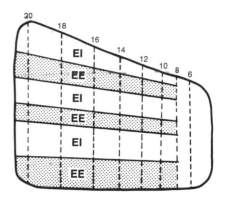

Fig. 15.17   Relation of tonotopic bands to binaural bands in cat auditory cortex. In this schematized view, isofrequency bands are at right angles to binaural bands (EE, neurons excited by both ears; EI, neurons excited by stimulation in contralateral ear and inhibited by ipsilateral ear). The combination of a binaural band and an isofrequency band forms a hyperband, or hypercolumn. AI, primary auditory area; AAF, anterior auditory field. Frequencies indicated in kHZ. (Based on Middlebrooks et al., in Merzenich and Kaas, 1980)

One can consider this as analogous to the large thumb area in somatonsensory cortex (Fig. 12.15), or the large area for the fovea in visual cortex (Chap. 16), all places where the highest acuity occurs. The acuity in this auditory area is used to detect Doppler shifts between the emitted sound and the echo, to tell the bat whether a moth or other prey is approaching or receding.

Functional Units

In view of the difficulty in demonstrating the tonotopic organization of auditory cortex, it is perhaps not surprising that there has been less clear evidence for columnar functional organization than in the somatosensory or visual systems. However, in penetrations perpendicular to the surface, an electrode characteristically records units in different layers that respond to the same frequency, which would appear to represent the same principle of functional organization as in the other systems. The isofrequency bands, of course, are in the form of slabs, not columns. In addition, recent experiments indicate that, when tested with tones in either ear, cortical cells show either summation of excitation from both ears (EE), or excitation by the contralateral ear and inhibition by the ipsilateral ear (EI). The EE and EI responsive cells are organized into slabs which cut across the isofrequency bands at right angles, as shown in Fig. 15.17. This overlay of slabs with different functional properties is also present in the visual cortex, as we shall see.

16

Vision

The earth is bathed in a constant flow of energy from the sun and the rest of the universe. This energy is in the form of electromagnetic radiation that has both the properties of waves and of particles, called photons. The radiation travels at the speed of light (300,000 km/sec), but it has different wavelengths, as indicated in Fig. 16.1. Radiation with short wavelengths (and correspondingly high frequencies) has high energies that disrupt molecular bonds. Although these waves played a role in early chemical evolution of this planet (see Chap. 2), they are deleterious to life as we know it. Fortunately, they are absorbed by the protective blanket of ozone in the atmosphere, or else life as we know it would not be possible. Radiation with long wavelengths, by contrast, has very low energy, for which there are few known receptors in living organisms. However, there is a narrow band of wavelengths with neither too much energy nor too little, and this we call *light*. Given the crucial role that this radiation plays in sustaining life on this planet, it is not surprising that plants and animals have developed special mechanisms for sensing it, and for using these signals to control a variety of behaviors.

The simplest kind of sensitivity to light is the ability to perceive different *intensities* of diffuse illumination. This ability is present in many plants and most animals. We can refer to it as the basic property of *photosensitivity*. This property is found in single-cell animals, in the skin of many simple organisms, as well as in specialized visual organs. Sensitivity to different levels of light underlies the daily rhythms of activity that govern the lives of most animals (see Chap. 25).

Most complex organisms have evolved mechanisms for sensing changes in illumination that are more rapid in time and more localized in space. These abilities constitute what we call the sense of *vision*. A remarkable feature of vision is the large number of submodalities, which means that there are correspondingly a number of different functions that vision may subserve. As summarized in Table 16.1, the simplest function, beyond the sensing of intensity, is the ability to detect *motion* in the visual field; this capacity is widespread among animals, attesting to its usefulness in detecting both predators and prey. This function requires that the receptors be arranged in a sheet in order to portray the movement across the visual field. Discrimination of the *form* of objects, so they can

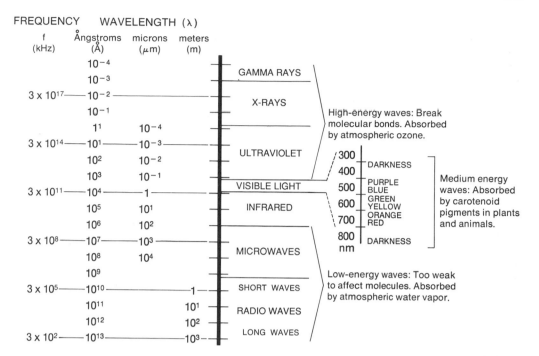

Fig. 16.1 The electromagnetic spectrum. (Modified from Gordon, 1970)

be recognized and manipulated, requires focusing of the visual image and therefore the development of a number of accessory structures for accomplishing this. Some animals can discriminate the *polarization* of diffuse light and use it for orientation and navigation. Since most animals are organized bilaterally, the two images from the eyes need to be combined, and this is used in some animals for *depth perception*. Finally, a few species, including humans, have mechanisms for discriminating different wavelengths within the visible spectrum so that they can perceive *colors*. Thus, it is readily apparent that parallel processing of multiple submodalities is a basic principle in the organization of the visual system, as it is in the other sensory systems. The subsystems act in parallel, and the neural elements within each subsystem also act in parallel, as we shall see.

The neuronal mechanisms underlying these functions have received a great deal of attention from neurobiologists, for several reasons. First, although other senses, such as the olfactory, play the dominant role in providing the cues that initiate feeding and mating behavior in most animals, vision is

Table 16.1 Submodalities of vision

Function	Mediated by
Potosensitivity (diffuse light)	Photosensitive molecule (rhodopsin) in microvilli/cilia
Form discrimination (spatial localization)	Sheet of photoreceptors (continuous or cartridges) plus focusing mechanism
Movement discrimination	Sheet of photoreceptors
Binocular vision and depth perception	Fusion of images (extraocular muscles, etc.)
Polarized light	Cellular organization and orientation
Color vision	Different photopigments

often crucial in carrying out the behavior. Second, vision becomes increasingly important in the higher invertebrates and vertebrates, particularly insects and mammals. Third, vision is overwhelmingly important in the life of humans; the more we learn about vision, the more we learn about ourselves, and also, hopefully, the more we learn about ways to prevent or cure the diseases that can cause blindness. Finally, light is a stimulus that can be controlled easily and accurately, which gives the experimenter a great advantage in analyzing neural mechanisms. The work in the visual system thus not only has given us insight into vision, but also has provided some of the best models for the functional organization of nervous systems.

We will begin by noting some basic properties of photoreception, and then discuss the functional organization of the visual pathway in mammals.

Molecular Basis of Photoreception

Rhodopsin

In a number of places in this book we learn about a particular mechanism that is so effective that it has been adapted for use across a wide range of different cells and organisms. Such, for example, is the Krebs cycle for energy metabolism, the actin–myosin complex for muscle contraction, and the synapse for neuronal communication. The same applies to photoreception. We have noted (Fig. 16.1) that radiation in the narrow band of visible light has energy just sufficient to be absorbed by molecules but not so much that it disrupts them. What is needed as a sensor for light is a molecule that will convert light energy into a maximum possible amount of chemical free energy. The molecules that do this most efficiently belong to the class called *carotenoids,* of which vitamin A is a member.

As explained by George Wald, who received a Nobel Prize for his studies of photopigments, Vitamin A is converted by metabolism into Vitamin A aldehyde, also called retinal (see Fig. 16.2), whose carbon chain of alternating single and double bonds can take on a number of different geometrical shapes. The critical shape involves a bend at carbon 11 to form 11-*cis*-retinal, which is a very unstable configuration. When 11-*cis*-retinal absorbs a photon, it quickly undergoes isomerization to the straight-chain shape of all-*trans* retinal, with release of free energy. This energy can then be used as the signal for photoreception.

In order to utilize this energy, two conditions are necessary. First, the retinal must be restrained in the unstable 11-*cis* shape, and second, the energy released when the molecule changes shape must be transferred to another molecule. Both of these conditions are met by the binding of retinal to a colorless protein called *opsin;* the two together are called *rhodopsin* (see Fig. 16.3). Rhodopsin, in various forms, is the nearly universal mediator of photoreception in animals. Opsin has been cloned and sequenced (se Fig. 16.3A). It is an integral membrane protein belonging to the superfamily of seven transmembrane receptor proteins, which includes the receptors for certain neurotransmitters (acetylcholine, dopamine, serotonin) and neuropeptides (Chap. 8), and odor molecules (Chap. 11). The human opsins that comprise the three color pigments in cones have been cloned and sequenced. These show a high degree of sequence differences from the opsin in rods, consistent with the view that the color genes (probably initially for the blue pigment) arose early in evolution. By comparison, the differences between the red and green pigment opsins are limited to a few key residues (see Fig. 16.3A). It is interesting to realize that these few specific amino acids are responsible for much of our world of color. Genetic defects in the genes for these pigment molecules cause the common types of color blindness for red and green in humans.

Although it may seem strange that a protein involved in vision should be similar to proteins involved in reception of mem-

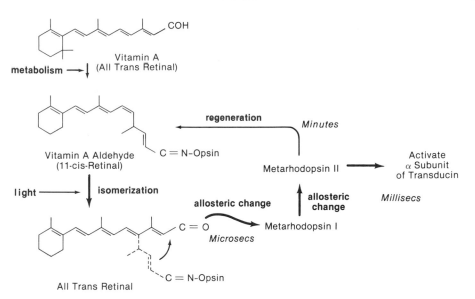

Fig. 16.2 Diagrams showing the metabolic pathways by which retinal is synthesized and the steps through which it interacts with opsin to mediate the response to light. See text.

brane signal molecules, this becomes understandable when one realizes that retinal functions as a ligand in relation to opsin. Normally, it nestles within the pocket formed by the transmembrane segments (see Fig. 16.3B), covalently bound by its terminal aldehyde group to a nitrogen on amino acid 296. This binding restrains 11-*cis*-retinal in its unstable conformation. Absorption of light overcomes this binding, allowing the molecule to change to all-trans. This in effect gives a "micro-kick" to opsin, sending it into a little dance in which it assumes a series of shapes (called allosteric conformations). The shape called metarhodopsin II is the critical one, enabling it to activate an attached molecule called transducin (Fig. 16.2), thereby initiating a second messenger cascade which results in the electrical response of the cell, as we will describe later.

At the end of their allosteric dance, retinal and opsin split apart (see Fig. 16.3). This causes loss of color of the molecule, an effect called *bleaching*. Regeneration of rhodopsin occurs by enzymatic synthesis, requiring energy in the form of ATP.

To maximize the probability of capture of photons, rhodopsin is packed very tightly into the disc membranes (see below) of the photoreceptors. It is estimated that the several hundred disc membranes contain a total of 10^9 rhodopsin molecules, accounting for over 80% of the disc membrane protein at a density of approximately 20,000 rhodopsin molecules per square micrometer. This is one of the highest densities of a specific type of membrane protein anywhere in the body, higher than for Na channels at the node of Ranvier (Chap. 6) or receptor proteins at synapses (Chap. 7).

Phototransduction in Modified Cilia

In photoreception we find a second expression of a universal mechanism; in most species, the rhodopsin is contained in fine hairlike processes. In some cases these are cilia, or modifications of cilia; in other cases they are microvilli, or modifications thereof. Richard Eakin of the University of California at Berkeley reviewed the varieties of these structures in different species, and suggested that there are two main lines in the evolution of photoreceptors. As

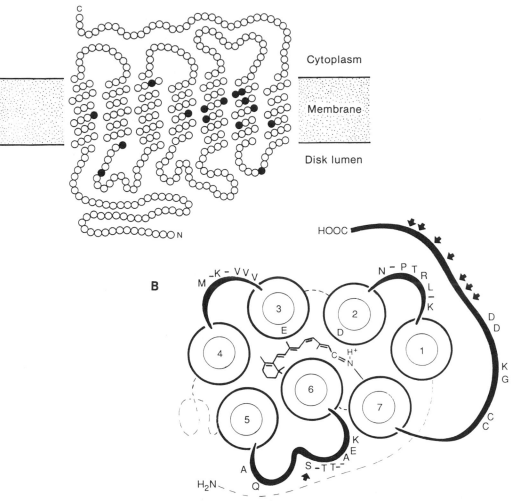

Fig. 16.3 A. The amino acid sequence and secondary folding structure of human opsin. The receptor consists of a single polypeptide. The folding structure is interpreted from analysis of hydropathic (intramembrane) and hydrophilic domains, supported by other studies. The seven transmembrane segments proceed from left to right in the diagram. The filled circles indicate differences between red and green opsins. The cytoplasmic loops (on the C-terminus side) are believed to be sites of interaction with the G protein, transducin. (Adapted from Nathans et al., 1986, in Kandel et al., 1992) **B.** The rhodopsin molecule as viewed from the cytoplasmic surface, showing the seven transmembrane segments and associated loops. Black arrows indicate potential phosphorylation sites (serine and threonines) on the cytoplasmic side. The 11-*cis*-retinal chromophore nestles in the interior of the opsin, covalently attached as a Schiff base to lysine residue 296 (segment 7). (From Baehr and Applebury, 1986)

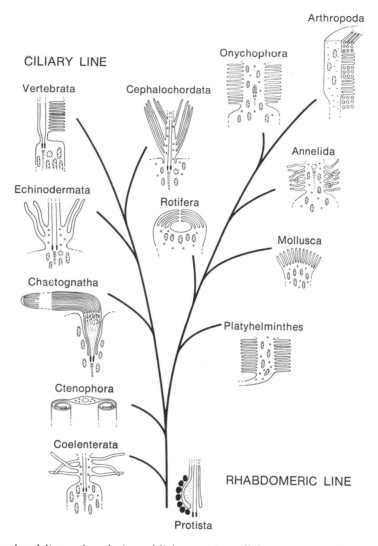

Fig. 16.4 Postulated lines of evolution of light-sensitive cellular structures. There are two main lines, one utilizing modified cilia, the other the elaboration of the rhabdome. (From Eakin, 1965)

summarized in Fig. 16.4, there is a line through flatworms–annelids–arthropods which uses the microvillus, arranged in a rhabdome, as the site of rhodopsin and photoreception, and a line through coelenterates–echinoderms–chordates which uses modified cilia at these sites. Although there are exceptions in these lines (as always in biology), the schema is a useful way of summarizing the diversity across species, and it also serves to highlight the importance of tiny hairlike processes as sites of sensory transduction. Eakin suggested that membranes of cilia or microvilli "provide a planoarrangement of the molecules of photopigment for the most effective absorption of photons of light."

Types of Eyes

The simplest organ specialized for sensing light is a group of cells in a shallow surface pit. This is called an *ocellus* (see Fig. 16.5A). It is present in the coelenterates

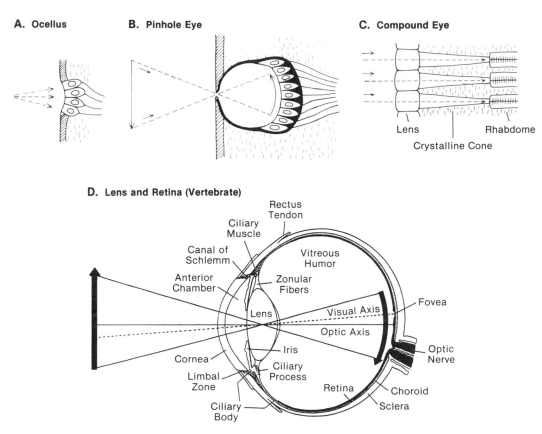

A. Ocellus

B. Pinhole Eye

C. Compound Eye

Lens   Rhabdome

Crystalline Cone

D. Lens and Retina (Vertebrate)

Rectus
Tendon

Ciliary
Muscle

Canal of
Schlemm

Anterior
Chamber

Vitreous
Humor

Zonular
Fibers

Lens

Visual Axis

Fovea

Optic Axis

Cornea

Iris

Limbal
Zone

Ciliary
Process

Optic
Nerve

Ciliary
Body

Retina

Choroid

Sclera

Fig. 16.5 Different types of eyes. **A–C.** Invertebrates. See text. **D.** The mammalian eye. The image of an object is inverted and focused by the lens on the retina, which lines the back wall of the eye. The cornea adds slightly to the refraction by the lens. The lens is focused on nearby objects by contraction of the ciliary muscle, which relaxes the zonular fibers, reducing their tension on the lens, permitting the lens to bulge, thereby increasing its refractive power. The elasticity of the lens decreases with age (presbyopia), so that lens correction is needed to permit focusing of a near object on the retina. For greatest acuity the image is focused on the fovea, where there is the highest density of photoreceptors (see text); this gives the eye a visual axis that is slightly displaced from the optical axis determined by the orientation of the lens. (A–C adapted from Knowles and Dartnall, 1977; D modified from Eckert and Randall, 1983)

and is thus, together with the neuromast, one of the first special sense organs to appear in phylogeny. Its main function is sensing light intensity. It apparently does this quite effectively, for ocelli are present in many invertebrate organisms, including the social insects.

In order for an eye to mediate true vision, it must have a way of forming an image. There are three main ways to do this. The simplest uses the principle of the pinhole camera, in which the image is formed by narrow rays passing from the object through the pinhole. The eye of the mollusc *Nautilus* is constructed on this principle (see Fig. 16.5B). Because of the small pinhole, the eye can work effectively only in bright light.

A much more efficient way to form an image is to funnel the light through multiple channels. If the channels are arranged so that they point out at different angles, a large visual field will be projected onto a small sheet of receptors, thus producing a magnification effect. By virtue of many separate channels, called *ommatidia,* this is referred to as a *compound eye* (see Fig. 16.5C). It is the characteristic eye of arthropods. As noted by A. Knowles and J. A. Dartnall (1977), the advantages of this type of eye are that:

(1) its great depth of focus makes it sensitive to movement at any distance; (2) the relatively short path length for light entering the ommatidium minimizes the loss of UV radiation, and this, coupled with freedom from the chromatic aberration of lens systems, allows it to operate over a very wide range of wavelengths and (3) the receptor cells can be arranged to make the eye sensitive to the plane of polarization of light.

In order to achieve its advantages, the compound eye sacrifices resolving power, that is, high acuity—the ability to discriminate very fine detail. This, by contrast, is maximized in the *refracting eye,* in which the image is formed by refraction of the light rays through a lens. The image formed is brought into focus on a receptor sheet, called the *retina.* This, of course, is similar to the lens system and film of a modern camera. The refracting eye is found in certain molluscs, such as the octopus, and is characteristic of all vertebrates (see Fig. 16.5D).

The importance of vision in vertebrate life is paralleled by the attention given by neurobiologists to this subject. Studies of vertebrate visual systems, in all their aspects, may well exceed those for any other part of the nervous system. Some of these aspects, such as the optics of the eye, are subjects for special study. We wish to concentrate on cellular properties and synaptic circuits, in order to understand the principles that underlie visual processing, and for the insights they may give us into the neural basis of visual perception.

Photoreceptors and Phototransduction

The vertebrate eye works on the principle of the refracted image, as already mentioned. The photoreceptors, together with the neurons involved in the first two levels of synaptic processing, are arranged in the thin *retina* at the back of the eye where the image is formed (see Fig. 16.5).

Photoreceptors

The receptor cells are arranged along the outer (posterior) surface of the retina. This means that light must pass through all the nerve layers of the inner part of the retina to reach the receptor sites. This peculiar arrangement is due to the embryological origins of the retina as an outpouching from the brain. There is little loss of light in reaching the receptors, because the retina is transparent.

The receptor cells are of two types, *rods* and *cones* (Fig. 16.6A). In both types the *outer segments* are modified cilia. They contain stacks of *disc membranes,* which are formed by inpouchings of the plasma membrane. The disc membranes contain the photopigment molecules: rhodopsin in rods, and related molecules sensitive to red, green, and blue wavelengths in cones. The receptors continually shed disc membranes from their distal tips and synthesize new membranes proximally. The location of the receptors against the outer surface of the retina facilitates the removal of the discarded membranes. The cones of reptiles and birds contain oil droplets of different colors in their inner segments, a filtering mechanism that contributes to color vision; this is an additional use of the deep placement of the receptors mentioned above.

The Second Messenger Mechanism

The mechanism of transduction of photons into the electrical response of the photoreceptor has been discussed in Chap. 10, where we saw that a second messenger is the link between the activation of rhodopsin in the disc membrane and the closing of Na^+ channels in the plasma membrane of the outer segment. Let us consider this mechanism more closely.

When we left the activation of rhodopsin by light (see above), the rhodopsin had been converted to an active intermediate,

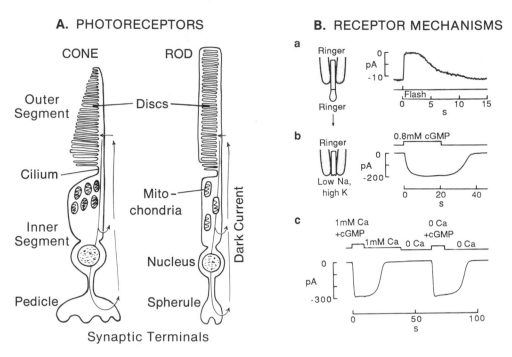

Fig. 16.6 A. Diagrams of vertebrate cone and rod photoreceptors, showing main cellular features and pathways for flow of dark current. **B.** Experiments providing evidence that the second messenger for phototransduction is cGMP. a. Single rods from the toad retina are inserted into the electrode tip and placed in a Ringer solution in the dark. A flash of light on the rod produces complete suppression of the inward "dark" current, as shown in the recording on the right. b. The inner segment and part of the outer segment are then broken off, leaving the interior of the remaining outer segment exposed to the bath; the bath is changed to a low-Na, high-K solution, approximating the intracellular medium. Addition of 0.6 mM cGMP to the bath results in a large inward current. c. Changing the concentration of Ca^{2+} in the bath had no effect on the cGMP response, showing that Ca^{2+} is not essential as a second messenger for the responses. Current is given in picoamperes (pA; 10^{-12} A); time in seconds. (From Yau and Nakatani, 1985)

metarhodopsin II. Beginning around 1970, Lubert Stryer, then at Yale, reported that there is an intermediate associated with the membrane that, when activated by meta-rhodopsin, binds GTP in exchange for GDP and then diffuses within the membrane to stimulate a phosphodiesterase (PDE), an enzyme that hydrolyzes active cGMP to inactive 5'-guanylate monophosphate. Since it functions to transduce the response of rhodopsin into the response of the cell, it was named *transducin*. This was the first of the GTP-binding proteins (G proteins) to be identified. It has been cloned and sequenced, and it shares a high degree of

homology with other G proteins linked to neurotransmitter (Chap. 8) and olfactory (Chap. 11) receptors.

At first the role transducin plays in phototransduction was not clear, because during the 1970s the leading candidate for a second messenger that gates the membrane channel to generate the electrical response to light stimuli was Ca^{2+}. However, in 1985, E. Fesenko and his colleagues in Moscow obtained inside-out patches of outer segment membrane and showed that the membrane conductance is activated directly by cGMP, in the absence of phosphates (ATP) in the bathing medium and

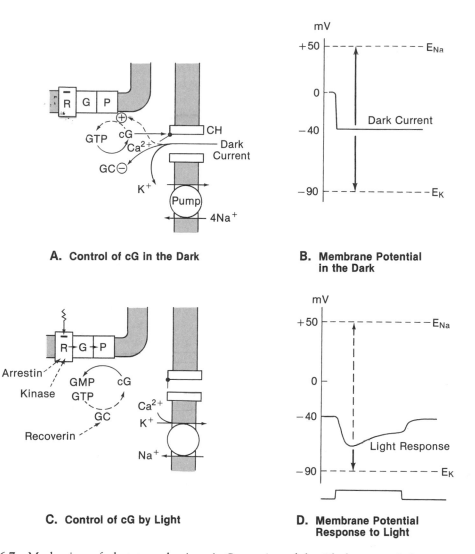

Fig. 16.7 Mechanism of phototransduction. **A.** Generation of the "dark current," due to opening of the membrane channel by the action of cGMP. **B.** The membrane potential is held relatively depolarized in the dark by the simultaneous Na$^+$ current offsetting the K$^+$ current. **C.** Light causes hydrolysis of cGMP, closing the channel. Mechanisms of adaptation are indicated by dashed lines. **D.** Membrane potential during light stimulation hyperpolarizes toward E_K. For full explanation, see text. (A, C adapted from Baylor, 1992)

in the absence of Ca^{2+}. This was soon confirmed and extended by K.-W. Yau and his colleagues at Galveston (see Fig. 16.6B).

This work, together with related biochemical and physiological studies, provided a complete account of the second messenger system involved in phototransduction. *In the dark* (Fig. 16.7A), guanylate

cyclase makes cGMP, which holds the channel open, maintaining the "dark current." This is a nonspecific cation current carried primarily by Na$^+$ but also including Ca^{2+} and K$^+$. *During light stimulation* (Fig. 16.7C), metarhodopsin II activates the subunit of transducin to bind GTP in exchange for GDP. The activated subunit diffuses in the membrane, bumping into PDE molecules and activating them at a rate of about 10^6 (1 million) per second. Thus, there is a tremendous amplification of the signal from rhodopsin through transducin to PDE. By hydrolyzing cGMP, PDE reduces the concentration of cGMP, thereby reducing the dark current during the light stimulus.

We are now in a position to understand the generation of the electrical response to light in the photoreceptor. *In the dark,* the increased conductance to Na^+ caused by cGMP produces a net inward current through the membrane channel which partially depolarizes the resting membrane potential to about 40 mV (see Fig. 16.7B). *Light stimulation* decreases this conductance by reducing the concentration of cGMP. This removes the depolarizing effect of Na^+, so that the membrane hyperpolarizes toward the equilibrium potential set by K^+ conductances in other parts of the cell (Fig. 16.7D). Thus, the response to light in the vertebrate photoreceptor is a hyperpolarization of the membrane. This of course is in contrast to the depolarizing response that is characteristic of most other types of receptors to their specific sensory stimulus.

Visual Adaptation

We have seen that most sensory receptors change their responsiveness during maintained stimulation, a property known as *adaptation*. In photoreceptors, at least two types of adaptation have been identified.

The first type is the decrease of the response of a photoreceptor during a maintained stimulus (see Fig. 16.7C, D). The studies of second messenger mechanisms have suggested that Ca^{2+} plays a key role in this process. As shown in Fig. 16.7B, Ca^{2+} entering the cell through the cGMP-gated conductance has a negative effect on the activity of guanylate cyclase and a positive effect on the PDE. When this inflow is interrupted during light stimulation, there is a decrease in guanylate cyclase

suppression, leading over time to increased cGMP production and consequent reopening of the membrane channel. This tends to depolarize the membrane, thus reducing the response from its initial phasic peak, as shown in Fig. 16.7D. Other related adaptation mechanisms include the actions of rhodopsin kinase and arrestin on activated rhodopsin and the action of recoverin on guanylate cyclase (see Fig. 16.7C).

The second main type of visual adaptation is the well-known experience of entering a darkened room and having to wait a minute or so before being able to see objects in the dark. This is due to the difference in sensitivity between cones and rods. In the light, the more sensitive rods are saturated; they are maximally stimulated and therefore cannot relay any information. Cones, on the other hand, have lower sensitivities, and thus function very well in the light. With increasing darkness the dim light becomes insufficient to stimulate them, whereas the rods become unsaturated and can be activated over their operating range. However, it takes time for them to regenerate rhodopsin from the bleached condition induced by bright light, so it is this process that determines the time it takes for us to begin to see in the dark.

The Dark Current Channel

Molecular Structure of the Channel. The cGMP-gated channel has been cloned and sequenced by the groups of Benjamin Kaupp in Germany and Randall Reed at Johns Hopkins. It shares high sequence homology with the olfactory nucleotide–gated channel (Chap. 11). As shown in Fig. 16.8, the hydropathicity plots suggest that the molecule consists of six transmembrane domains, with the cGMP-binding region in the long intracellular carboxyl terminal tail. In addition to this ligand-binding site, the molecule also contains a so-called S4 region that dips into the membrane between segments 3 and 4. This is similar to S4 regions that are believed to confer voltage sensitivity in voltage-gated channels

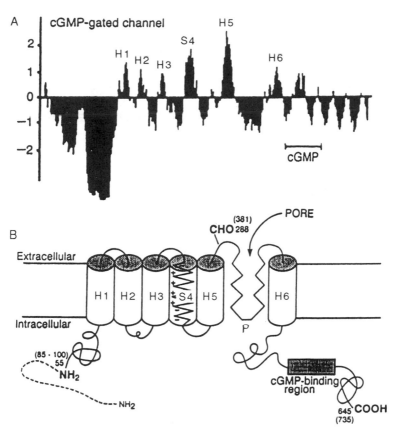

Fig. 16.8 A. Hydropathicity plot of the sequence of amino acids that has been identified for the cGMP-gated channel of the rod photoreceptor. The successive hydrophobic segments H1–H6 are indicated, and also a segment with homology to the presumed voltage sensor of voltage-sensitive channels (see S4 segment, Chap. 5) and the presumed cGMP binding site. **B.** Tertiary structure of the channel protein inferred from the hydropathicity plot, in which hydrophobic segments are presumed to lie within the membrane and the hydrophilic segments to lie in the extracellular and intracellular compartments. The S4 segment may lie within the membrane to perform its voltage-sensor function (though this is very weak in the cGMP-gated channel); it may represent a vestige of the core structure of the ancestral form of this channel family (see Chap. 5). (Figure and legend adapted from Kaupp, 1991 and Bönigk, 1993)

(see Chap. 5). It has been suggested that this S4 region was present in an ancestral form; during evolution, it became adapted as a voltage sensor for voltage-gated channels, and it serves some structural role in cyclic nucleotide–gated channels.

How Many Channels? It is interesting to calculate how many channels are present in the outer segment of a rod. Following Yau and Baylor (1989), we start with the estimate that a single channel passes approximately 4 fA of current (4 femtoamperes, or 4×10^{-15} amperes), and the dark current of a salamander rod is approximately 40 pA (40×10^{-12} A). Since in the dark the maximum number of channels is open, we can divide 4 fA into 40 pA to find that there must be approximately 10,000 channels activated at any given moment.

Experiments have shown that, even in the dark, only 1–2% of the channels are open; therefore, the total number of channels must be around 0.5–1 million $(0.5–1 \times 10^6)$. Since the surface area is some 1000 μm^2, this means that the density of channels in the plasma membrane is around 500–1000 μm^{-2}. This is at an intermediate level, more than the density of Na channels in the squid axon, but less than rhodopsin in disc membranes or ACh receptors at muscle endplates (see above and Chap. 7).

Enhancing the Signal relative to Noise. Normal levels of Ca^{2+} in the external fluid block the channel, reducing the current it can pass from 1 pA in calcium-free solution to the value of 4 fA mentioned above. This seems paradoxical: why would the cell want to reduce its ability to respond to light? Yau and Baylor (1989) explain that this is probably related to the need of photoreceptors, like all sensors, to enhance the *signal-to-noise ratio,* that is, the size of a given response above the background noise in the system (see Chap. 10). This comes from application of simple Poisson statistics, which state that for a large population of independent events, each with a very low probability of happening, the standard deviation of the spontaneous fluctuations is $N^{1/2}$, where N is the mean. This was originally applied to such phenomena as radioactive decay, in which there are many radioactive molecules, each with a very small probability of decay; it is also applied to the release of neurotransmitter quanta (Chaps. 6 and 29). For photoreceptors, if there are a lot of channels, each with a tiny unitary current, then the noise will be relatively low (that is, 10,000 channels, each with 4 fA current, gives $100 \times 4 = 0.4$ pA standard deviation of current noise). By contrast, a few channels, each with large currents, would be very noisy (25 channels, each with 1 pA current, gives $5 \times 1 = 5$ pAs, or more than 10 times the noise). This helps to explain why it is easier to detect the occurrence of a single photon if there

are many tiny channels than if there are a few large ones.

Comparisons between Invertebrates and Vertebrates. Although rhodopsin is the universal light receptor, the response to illumination is hyperpolarizing in vertebrates but depolarizing in invertebrates. This difference is believed to be correlated with the different types of eyes (see Fig. 16.5 above) and the different retinal circuits into which the photoreceptors send their output. In most other respects the photoreceptors have similar properties. These include mediation by a cGMP-sensitive channel and a variety of adaptation mechanisms.

Quantal Responses. The smallest electrical response of the photoreceptor cell is due to the reduction in cGMP molecules caused by one photon, referred to as the *quantal response.* Quantal responses were recorded by Denis Baylor and his colleagues (1979b) at Stanford using an elegant technique. Figure 16.9 shows how the rods are teased apart and drawn into the pipette (A) and stimulated with small bars of light (B). The resulting response of the rod, due to the closing of the dark current channels, changes the amount of current flowing across the membrane, as discussed previously. This current has to flow electrotonically along the rod to complete the electrical circuit. The tight seal of the electrode tip against the rod places a high resistance in this longitudinal current path, which generates voltages that can be recorded by the electrode.

Use of this method with very weak illumination made it possible to record small voltage fluctuations, as shown in Fig. 16.9C (bottom trace). Each fluctuation is a quantal event, due to the photoisomerization of a single rhodopsin molecule by a single photon. There is a quantal current of 1 pA, and also a corresponding conductance change very similar to that of a single acetylcholine-sensitive channel in the neuromuscular junction. The photoreceptor

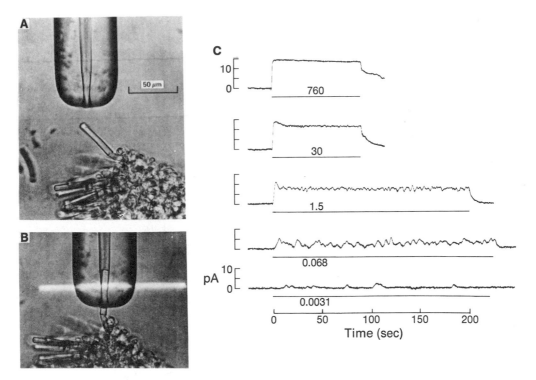

Fig. 16.9 Recordings of responses from single isolated rod photoreceptors of the toad. **A.** Suction electrode approaching the outer segment of a receptor protruding from a piece of retina. **B.** Outer segment is sucked up into electrode. Light bar is shone on small parts of the outer segment, while membrane current, proportional to longitudinal current flowing along the outer segment, is recorded by the electrode. **C.** Receptor responses, showing quantal events at low illumination (bottom), merging to a smooth graded response at higher illumination (upper traces). Note that these are recordings of membrane current (in pA: 10^{-12} A); the upward deflections signal the current flows associated with the membrane hyperpolarization that is characteristic of vertebrate photoreceptors. Intensity of light stimulation in photons/μm^2/sec. (From Baylor et al., 1979a,b)

quantal response has a rounded shape and a duration of a few seconds. It is believed that this represents the action of cGMP on a number of channel sites. Patch recordings have shown that the light-sensitive channel has a unitary conductance of only 25 fS (25 femtosiemens, or 25×10^{-15} S), one of the lowest values known for any membrane channel.

With stronger illumination, the quantal events merge, and the response becomes a smoothly graded waveform (Fig. 16.9C).

Retinal Circuits

How is the sensory information contained in photoreceptors transmitted to the brain? This is in fact a daunting task because, as we noted in Table 16.1, visual information is of many different submodalities. Each requires some initial processing before it is in a form that can be deciphered by the brain. This is the task of the circuits in the retina.

A Brief Historical Perspective

Analysis of this problem began with the studies of Keffer Hartline and his colleagues in the 1930s that showed the contrast enhancement properties of cells in the eye of *Limulus* (see Chap. 10). This approach was extended to the mammal by Stephen Kuffler (1953) in Baltimore and Horace Barlow (1953) in Cambridge. Kuffler revealed the center–surround nature of the receptive fields of ganglion cells in the cat, and Barlow showed the sensitivity of some ganglion cells to direction of motion. Responses of some ganglion cells to complex shapes were demonstrated in the frog by H. R. Maturana and colleagues (1959) at M.I.T. These results showed that considerable processing of visual stimuli can take place in the retina.

The earliest attempts at intracellular recordings were made by Gunnar Svaetichin in Stockholm in the early 1950s, but the modern era of intracellular studies began with the recordings from photoreceptors by Tomita (1965) in Japan. At the same time, electron microscopical studies by Brian Boycott and John Dowling (1965) provided clear evidence regarding the fine structure of retinal neurons and their synaptic connections. In 1969, Frank Werblin and Dowling, then at Johns Hopkins, obtained the first systematic intracellular recordings from all cell types. This kind of work at the single-cell level requires interpretation at the systems level, which was greatly stimulated by the work of David Marr (1981) on computational operations underlying visual perception.

Anatomical and Functional Cell Types

The main cell types in the mammalian retina are shown in Fig. 16.10. In the upper diagram the cells are shown in their true relative size, to emphasize how small they are and how short are their processes; below, the diagrams are magnified to bring out better the different types. The retina is barely 200 μm thick, scarcely the diameter of the cell body of a large pyramidal cell in the cortex. The functional consequence is that synaptic potentials can spread rapidly through a cell by electrotonic means (see Chap. 5) without need of an impulse, and indeed that is the mode of operation of all the cell types except the ganglion cells, which must generate impulses to transmit the output of the retina to the brain.

A striking feature of the retina is that the main cell types are divided into many subtypes, by virtue of different shapes and branching patterns. For example, Fig. 16.10 illustrates two pairs of subtypes of bipolar cells, each pair related to a subtype of ganglion cells. At least 10 distinct types of bipolar and ganglion cells have been described, as have some 20 types of amacrine cells. Correlations of intracellular recordings with staining of the recorded cells have shown that different morphological types have different synaptic connections and different physiological properties. As Peter Sterling (1990) of the University of Pennsylvania comments, ". . . there is confidence that to delineate a distinctive morphology for a . . . cell is to predict for it a distinctive physiology." Thus, although the architect Louis Sullivan maintained that "form follows function," neuroscientists argue that function can also follow form!

Given this plethora of information at the cellular level, it is not surprising that the retina has become one of the main models for understanding the neural basis of functional circuits in the brain. We will focus on several of the basic circuits related to the cones and the rods. This will by no means exhaust the types of information processed by the retina (see Table 16.1). However, we will not only get a grasp of the principles underlying the functional organization of retinal circuits, but we will also have a foundation for following this information as it is processed through higher levels within the visual pathway to form the basis for visual perception.

Center–Surround Organization of Receptive Fields

The basic physiological tool used in building our present concept of retinal circuits has been the analysis of the center–surround organization of receptive fields, as exploited by Kuffler in his pioneering studies. As summarized in Fig. 16.11, a ganglion cell may respond to either the onset or the offset of a light stimulus at the center of its receptive field (top diagrams); conversely, it shows the opposite responses

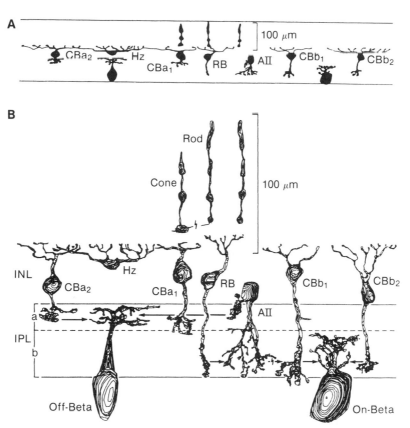

Fig. 16.10 **A.** The main neuronal elements of the mammalian retina, as exemplified by the cat. **B.** The same neurons enlarged in the vertical scale to show more clearly the cell types and synaptic layers. *Photoreceptors:* rods and cones. Hz, *horizontal cell. Bipolar cells:* CB, cone bipolars; subscripts indicate subtypes that are depolarizing or hyperpolarizing and ON or OFF; RB, rod bipolars. A II: *amacrine cell,* type II (rod amacrine). *Ganglion cells:* Off- and On-β (X in monkey). INL, inner nuclear layer; IPL, inner plexiform layer, with sublaminae a and b. (From Sterling, 1990)

to stimulation of its surround (middle diagrams). Kuffler showed that ganglion cells are most sensitive to these types of stimuli; by contrast, diffuse stimulation of both centers and surrounds gives only weak responses (bottom diagrams). The interpretation of these properties is that circuits between the photoreceptors and the ganglion cells abstract and enhance information about spatial contrast and suppress information about diffuse illumination. These results made it clear that the visual scene is not transmitted faithfully like a photograph to the brain; rather, only the specific information essential for perception is selected. From that time on, the task for

visual scientists has been to identify the mechanisms and circuits that perform this selective enhancement and transmission to the brain.

Basic Response Properties of Retinal Cells

Intracellular recordings have been essential to this task. The early experiments were carried out in the retina of *Necturus,* the mudpuppy. The cell bodies in this retina are large, making inviting targets for the probing microelectrode. A summary of the results is shown in Fig. 16.12. The diagram is arranged to show, on the left, the responses of each cell to a spot of light and, on the right, to a surround.

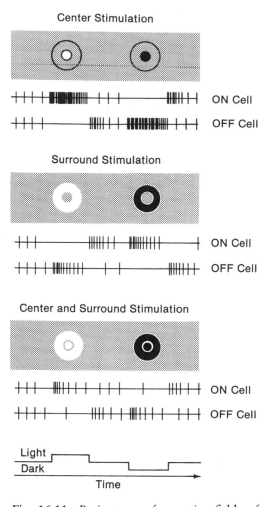

Fig. 16.11 Basic types of receptive fields of ganglion cells, in relation to center–surround antagonism and responses to ON and OFF of light stimuli. Schematic spike responses are shown as they would be recorded with an extracellular microelectrode near the ganglion cell body. (Adapted from Schiller, 1992)

There are three main points to take away from this diagram. First, in addition to the receptors, the horizontal cells and bipolar cells show only graded responses to stimulation. These are among the clearest examples of nonspiking neurons in the vertebrate nervous system. Amacrine cells show mostly graded potentials also, though they do generate a few small spikes, which may help to boost transmission through their long dendritic processes. Only the ganglion cells generate large action potentials, which is consistent with their role as the output neurons of the retina.

Second, there are marked differences between the responses in the center and the surround. The bipolar cell potentials are of opposite polarities, and ganglion cells (G_1) show excitation at the center on the left and inhibition in the surround on the right. This expresses the fundamental property of center–surround antagonism in the organization of receptive fields of ganglion cells. Since the receptors show only a graded

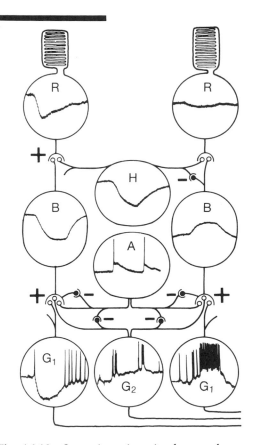

Fig. 16.12 Synaptic actions in the vertebrate retina, as recorded intracellularly from neurons in *Necturus* (mudpuppy). *(Left)* Responses recorded at the center of a spot of light (bar above). *(Right)* Responses in the surround. A, amacrine cell; B, bipolar cell; $G_{1,2}$, ganglion cells; H, horizontal cell; R, receptor. (From Dowling, 1979)

reduction of their responses in the surround, these results demonstrate that the center–surround antagonism is the result of processing by the synaptic circuits in the retina, principally those through the laterally oriented elements, the horizontal and amacrine cells.

Third, the ganglion cell (G_2) depicted in Fig. 16.12 shows transient responses at ON and OFF of the stimulus. This type of response is especially tuned to transmitting information about moving stimuli. This property can be seen to be due to synaptic circuits, principally through complex interactions of the amacrine cells.

Simple Versus Complex Retinas. Different species differ in the amount of synaptic processing that takes place in their retinas. One of the most complex retinas in this respect is that of the frog or toad. As shown in the pioneering studies of H. R. Maturana, Jerome Lettvin, and their colleagues at M.I.T. in 1960, frog ganglion cells may be tuned to one of several features of a visual stimulus, including a stationary edge, a convex edge, a moving edge, or a dimming or brightening of illumination. Some cells are so narrowly tuned to particular shapes that they seem to function virtually as "bug detectors"!

An interesting note is that complex retinas are found in some mammals (such as rabbits) as well as in lower vertebrates, so there is not a consistent phylogenetic progression. The fact that higher mammals, like cats and primates, have relatively simple retinas is taken as an expression of *encephalization of nervous control;* that is, the tendency for complex processing to be shifted from peripheral to central sites, where interactions with other circuits can take place.

Ganglion Cell Types. In the mammal, the ganglion cells are differentiated into several distinct morphological types, each with special functional properties. The main distinction is between the majority of cells (X cells) located mostly near the fovea at high density where they are responsible for high-acuity vision, and a smaller number of scattered cells (Y cells) located in the periphery where they subserve motion detection. This demonstrates how different submodalities of vision are mediated in parallel by different subsets of neurons. These and other properties of ganglion cells are summarized in Table 16.2. We will see that these properties continue to be transmitted in separate channels as far as the visual cortex.

How We See in Daylight, Dusk, and Starlight

We have noted that our ability to see over an enormous range of light is aided by covering the range with two types of photoreceptors, rods and cones. This implies that the retina must have two sets of circuits in relation to these ranges.

Daylight Vision: The Cone Pathway. The way the retina is organized to do this has been worked out in several species; we will summarize the findings in the cat made by Peter Sterling of the University of Pennsylvania and the primate by Nigel Daw at Washington University and Yale. As illustrated in Fig. 16.13, in bright light, information passes mainly from cones to two subsets of bipolar cells: ON and OFF cone bipolars. These depolarize and hyperpolarize, respectively, in response to a brief light stimulus. Since hyperpolarization of the cone in response to a light flash leads to hyperpolarization of the OFF bipolar, the synapses are called *sign-conserving;* by contrast, the fact that cone hyperpolarization leads to depolarization of the ON bipolars means that the synapses are *sign-inverting.* These terms can be more useful than the traditional terms, excitatory and inhibitory, to describe synaptic actions in the retina.

In accord with the close relation between form and function, these cone subsets have distinctive branching patterns in the inner plexiform layer, where they connect to subsets of ON and OFF ganglion cells (see also Fig. 16.10). The ganglion cells (β cells in

Table 16.2 Summary of different morphological types of retinal ganglion cells and their functional properties, in the cat

	X	Y	W
Morphology			
ganglion cell size	medium (β)	large (α)	varied
number	many; most near fovea	few; most in periphery	few
axons	medium conduction rate	fast conduction rate	varied conduction rate
projection sites	lateral geniculate nucleus	lateral geniculate nucleus and superior colliculus (and medial interlaminar nucleus)	lateral geniculate nucleus and superior colliculus
Function			
spatial summation	linear	nonlinear	mixed
movement sensitivity	+	+ + +	±
directional selectivity	no	no	yes (a few cells)
center–surround antagonism	yes	yes	±
color coded	yes (in primates)	no	?

the cat, X cells in the primate) are small and tightly packed at or near the fovea. As Sterling puts it, ". . . the beta cell's most important function is to create a fine-grained neural image using the lowest contrasts that it possibly can." To do this, in the cat 36 cones collect light and converge onto 9 cone bipolar cells, which connect to one β ganglion cell. In the primate, there is an even tighter relation, from one cone through one cone bipolar to one ganglion cell. Note that the pathway involves only one cell link, through the bipolar cells, to get from visual transduction to retinal output to the brain. Several mechanisms enable the signals to be enhanced relative to noise: inputs from neighboring cones tend to be highly correlated; convergence amplifies the signal; and electrical synapses between cone synaptic terminals allow for spread of the signal, thereby averaging it to reduce noise. Under these conditions of bright light, the rhodopsin in the rods is saturated, so the rods are unresponsive.

Evening and Night Vision: The Rod Pathway. As evening comes and darkness falls, the signal contrast falls, so there is a loss of visual acuity. Vision is critically important under these conditions, however, for this is the time when many animals search for

food, including prey. The lower level of illumination brings the rods into their operating range, and they begin to become activated in the dim light, while the cones become inactive. Paradoxically, responses to light can still be recorded in cones. This is because (as shown in the cat) each cone synaptic terminal collects signals from 50 surrounding rods, to which it is connected by electrical synapses. In the dimmest light, under the stars, the cones are completely inactive, and the rod terminals become uncoupled electrically so that the signal within them from single photons is as large as possible, to ensure transmission to the bipolar cells and on to the ganglion cell. This is helped both by convergence, of 1500 rods onto 100 bipolar cells and from there onto one β ganglion cell, and by divergence, one rod connecting to two bipolar cells and from there onto two ganglion cells (Sterling, 1990). This rod circuit thus enhances transmission of single-photon stimuli.

As shown in Fig. 16.13, the rod pathway is less direct than the cone pathway. Rods connect to a subset of rod bipolar cells; the hyperpolarizing rod response leads to a depolarizing bipolar response. By the definition explained above, this is a sign-inverting synapse. The rod bipolar cell does

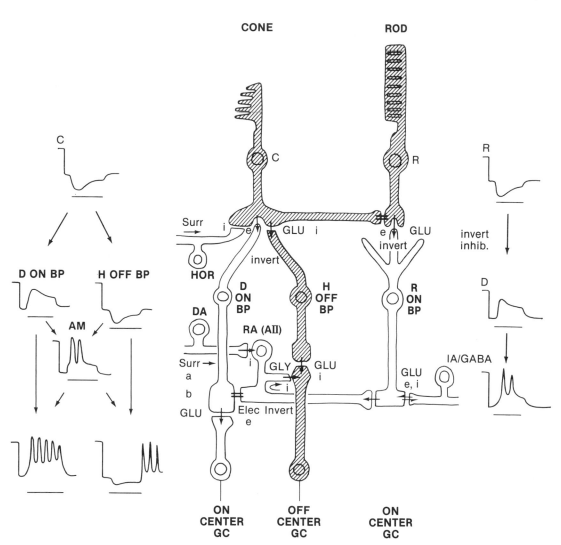

Fig. 16.13 Diagram of cone and rod circuits in the mammalian retina showing characteristic intracellular recordings from each cell type in response to center stimulation. (Modified from Daw et al., 1990) The diagram illustrates how retinal circuits shift from cone-related to rod-related as vision shifts from conditions of daylight to dusk and starlight. See text. (Based on Sterling, 1990)

not connect directly to ganglion cells; instead, it has excitatory synapses on a specific type of rod amacrine cell (A II), which makes two further types of connections: gap junctions (sign-conserving) onto the depolarizing cone bipolar terminals (which leads to excitation of ON center ganglion cells), and inhibitory chemical synapses (sign-inverting) onto β (X) ganglion cells (which inhibits OFF center ganglion cells).

In summary, rods and cones have their

distinct parallel pathways through the retina, with sharing of some elements. The pathways interconnect and interact in different ways at different levels of illumination, by virtue of their dynamic properties in relation to different activity states.

Neurotransmitters. The retina is rich in transmitters (Daw et al., 1990). Some of these are indicated in Fig. 16.13. Photoreceptors use glutamate for their synapses onto bipolar cells. The action of glutamate can be depolarizing or hyperpolarizing onto the different subsets of cone bipolars; it is depolarizing onto all rod bipolars. Bipolar cells are also glutamatergic. Amacrine cells are divided biochemically into several subtypes, for example, cholinergic, GABAergic, or dopaminergic. The dopaminergic type of amacrine cell is particularly interesting, because it is the only amacrine cell with an axon; this recurs to the horizontal cell layer, where it terminates among the synaptic connections and can modulate electrical coupling between the horizontal cells.

Color Vision

Color vision is limited in most mammals, reflecting the fact that the early mammals were probably nocturnal (i.e., rod) animals. Color vision reemerged mainly in the line leading to primates, in association with the adoption of diurnal habits. As Timothy Goldsmith of Yale has pointed out (1980), our color vision system has been reconstructed, in an evolutionary sense, from a less capable retina. It therefore lacks the specializations, such as oil droplets and sensitivity to ultraviolet light, that make the avian retina the supreme instrument for both day and night vision.

Cone Mechanisms. The rhodopsin present in rods has different sensitivities to different wavelengths of light, giving a spectral sensitivity curve as shown by the dashed line in Fig. 16.14A. Despite this varying sensitivity, a rod receptor containing only rhodopsin, or a cone receptor containing only one pigment such as green (G in Fig. 16.14A), cannot signal a specific wavelength (color), because it cannot distinguish between wavelengths in ascending and descending limbs of the curve that produce equal responses, and it cannot distinguish color from brightness; that is, a smaller response in the receptor could be due to lower place in the spectral sensitivity curve (compare objects 1 and 2 in Fig. 16.14A) or to less light, and the central nervous

system receiving these signals could not tell the difference.

This ambiguity is overcome partly by the presence of two types, and completely by the presence of three types, of cone receptors, each type having a visual pigment with a different spectral sensitivity curve. These three curves peak in the blue, green, and red regions. The key fact about these curves is that they overlap. Because of this, a particular wavelength of light gives a unique combination of degrees of activation of the three cone populations. Thus, as shown in Fig. 16.14B, the light reflected by object 1 at 450 nm produces a small response in G cones, a large response in B cones, and none at all in R cones. By contrast, object 2, reflecting light at 600 nm, produces a different pattern of responses (see Fig. 16.14B). Because the patterns are unique for different wavelengths, by comparing the information coming from the three sets of receptors the central nervous system can distinguish which wavelength is being signaled, no matter what the level of brightness. This mechanism, comparing overlapping spectra across elements acting in parallel, is a basic model for discrimination in other sensory systems (see taste and olfaction, Chap. 11; audition, Chap. 15).

Color Coding. From these considerations it can be seen that perception of color depends on comparisons between information arising in different cone systems. How is this information incorporated into the receptive field organization of retinal cells? Comparison between the signals in the three different types of cones does not take place immediately. Instead, the circuits in the retina provide first for an enhancement of the contrast between color-coded cells that is similar in principle to enhancement of spatial contrast. Although the mechanism of color coding is not understood in detail, certain principles have been identified. The dual ON and OFF cone system indicated in Fig. 16.13 is subdivided into three parallel subsystems for each of the cone types: ON and OFF systems for green,

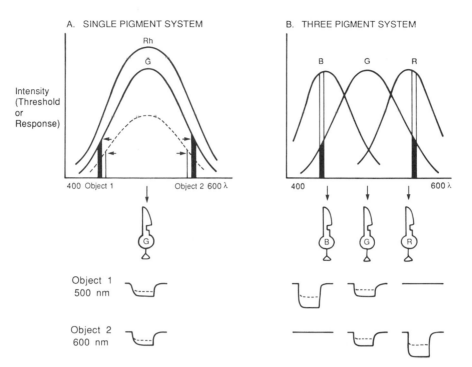

Fig. 16.14   Color-coding mechanisms at the photoreceptor level, illustrating the necessity for more than one visual pigment. **A.** A single pigment (G, for maximum sensitivity in green) gives a receptor different sensitivities to different wavelengths (λ), but the receptor could not distinguish between objects reflecting wavelengths of 450 nm and 600 nm (which have identical sensitivities). If the luminosity is decreased (dashed line), the receptor could not distinguish between the change in luminosity and a change in wavelength (arrows). Sample recordings in the G receptor under these conditions are shown below. **B.** A three-pigment system can distinguish wavelength independently of intensity. The pigments must have overlapping spectra. The two objects stimulate the three photoreceptors (B, blue; R, red) in different amounts. Each object stimulates the receptors to different degrees, so that the color code for each object is unique, and maintained despite a reduction in luminosity (dashed lines in recordings). (Modified from Gouras, 1985)

red, and blue. The green ON and OFF system and red ON and OFF system have been described by Peter Schiller of M.I.T. and are illustrated in Fig. 16.15A. These systems do not simply transmit their particular information directly to the brain. Instead, there are antagonistic interactions between them, mediated by the lateral connections through horizontal and particularly amacrine cells. These interactions set up opponent responses; Fig. 16.15B summarizes the types of responses given by red and green ON and OFF ganglion cells. Note that these responses are not fixed but are sensitive to the ambient lighting. When stimulated

against a dark background (left), the ON cells give strong responses and the OFF cells give weak responses. However, when stimulated under brightly lit conditions (right), the ON cells give no ON response but a weak OFF response to the opponent color; by contrast, the OFF cells now give weak ON responses to their preferred color but intense OFF responses to the opponent color. According to Schiller (1992), these differences arise

because the red spot is made visible as a result of the greatest amount of light decrement in the green portion of the spectrum, and the green spot is visible due to the greatest amount of

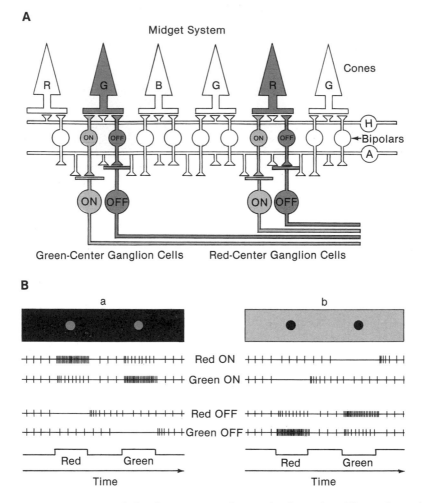

Fig. 16.15 **A.** Organization of the three cone pathways in the retina. The main pathways are organized to generate ON and OFF responses in specific subsets of green-center and red-center ganglion cells; blue OFF bipolar and ganglion cells are rare. **B.** Summary of main types of ON and OFF responses in red-center and green-center ganglion cells. With a dark background, the cells show a simple opponency pattern of responses to red and green. With a light background, there is a shift in the firing patterns, though the basic opponency characteristic is retained. See text. (Adapted from Schiller, 1992)

light decrement in the red portion of the spectrum. Thus, the sensation of "redness" is produced by the activity of red ON cells and green OFF cells, while the sensation of "greenness" is produced by the activity of the green ON and red OFF cells. Convergent input from these cells giving rise to the center mechanism of so-called "double opponent" cortical cells could lead to color integration, so as to discount background levels and thereby provide constant color sensations.

In summary, these circuits enable us to discriminate different colors despite wide variations in the level of light.

Central Visual Pathways

The ganglion cell axons run along the inner surface of the retina and gather together to form the optic nerve. This is the second cranial nerve. By embryonic origin it is a part of the central nervous system. In lower vertebrates, the main projection of the optic nerve is to the optic tectum of the midbrain, where the retina is mapped in an orderly, *retinotopic* manner as discussed in Chap. 9. Single-unit studies have shown that in the frog or toad, tectal cells are exquisitely tuned to detect particular types of movements and spatial patterns that allow them to discriminate prey (such as a worm or fly), which elicits prey-catching behavior, or predator, which elicits escape reactions. The bug detector cells in the retina thus are matched by bug detector cells in the tectum.

In mammals, the optic nerve projects primarily to the lateral geniculate nucleus (LGN) in the thalamus (Fig. 16.16). The input to the superior colliculus, the homologue of the optic tectum, still mediates midbrain reflexes important in prey and predator behavior. Most of this input is carried by the Y axons (see Table 16.2), which arise from ganglion cells in the peripheral parts of the retina and are tuned to detecting movements in the peripheral visual field. By contrast, the input to the LGN is carried in both X and Y axons. As noted previously, X axons arise mostly from ganglion cells near the *fovea*, the center of the visual field, where visual acuity is highest.

In addition to the visual projections to the midbrain and thalamus, there is a projection to the hypothalamus. This is important in the control of circadian rhythms, as will be discussed in Chap. 25.

In lower vertebrates, the two optic nerves cross (decussate) and supply the tectum and thalamus on the opposite side. In most of these animals the eyes are set on the side of the head, and there is little overlap of the two visual fields. In most mammals, however, the eyes are set forward in the head, and the two visual fields partially overlap. Associated with this is usually only a partial decussation of the optic nerve fibers. The situation in the human is depicted in Fig. 16.16. As can be seen, the ipsilateral fibers are those that arise from

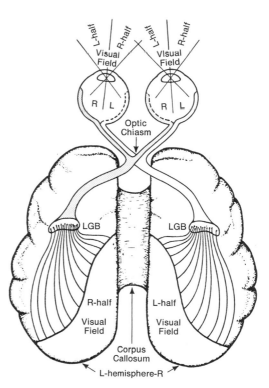

Fig. 16.16 Schematic diagram of the central visual pathway in the human. Note the projections of the visual fields onto the retinae, the partial decussation of the optic tracts, and the orderly projection from the lateral geniculate body (LGB) in the thalamus to the primary visual cortex in the occipital lobe. (From Popper and Eccles, 1977)

the outer (temporal) half of the retina, which receives stimuli from the inner (nasal) half of the visual field.

In the LGN, the inputs from the two eyes are kept separated from each other in a series of layers before being relayed to the visual cortex. Figure 16.17 shows how this comes about. The LGN in primates consists of six layers. Ipsilateral retinal input is transmitted through layers 2, 3, and 5, whereas contralateral input is transmitted through layers 1, 4, and 6. Together, the six layers transmit a complete representation of the contralateral (i.e., temporal) visual hemifield, as is illustrated in Fig. 16.17. The lamination of the LGN appears to be necessary, for reasons not fully under-

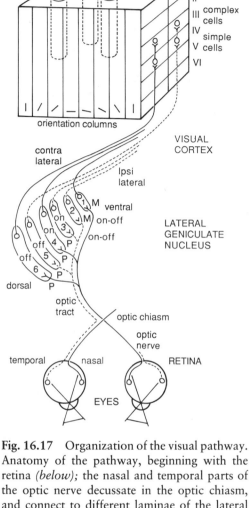

Fig. 16.17 Organization of the visual pathway. Anatomy of the pathway, beginning with the retina *(below)*; the nasal and temporal parts of the optic nerve decussate in the optic chiasm, and connect to different laminae of the lateral geniculate nucleus in the thalamus as shown (uncrossed fibers project to layers 2, 3, and 5, whereas crossed fibers project to layers 1, 4, and 6). The thalamic relay cells project in turn to the primary visual cortex *(above)*, with segregation of projections from ipsilateral and contralateral eyes to form ocular dominance columns (500 μm diameter). Cortical cells responding to different edge orientations of a light stimulus are organized into narrower columns (50 μm). A cortical unit containing both types of column is sometimes called a hypercolumn. A third type of column is the cortical peg, which stains richly for cytochrome oxidase; it is concerned with the processing of color information. See text.

stood, to keep channels of information from the retina separate before they are combined in the cortex. An example of this principle is that layers 1 and 2 contain large cells (magnocellular) and transmit mainly inputs from Y cells, whereas layers 3–6 have small cells (parvocellular) and transmit inputs mainly from X cells. The Y cell axons have large terminal ramifications correlated with their wide fields for movement detection, whereas the X cell axons have small terminal fields correlated with their role in mediating high visual acuity.

Visual Cortex

Figure 16.16 indicates that in humans the LGN projects to the occipital lobe of the cerebral cortex. This projection defines the primary visual cortex, called visual I (VI) by physiologists and area 17 by anatomists. As in other sensory systems, the visual pathway retains a precise topographical order, in which the map of the retina, and hence of the visual field, is projected onto the cortex. The fovea, where acuity is highest, occupies a large part of the cortical representation, similar to the way that regions of highest acuity dominate the maps in other sensory cortical areas (see for example the large areas for face and thumb in the somatosensory homunculus in Chap. 12).

Cortical Processing

The discovery by Stephen Kuffler in 1953 of the center–surround organization of ganglion cells not only was the basis for our understanding of the retina, but also provided the key to unlocking the mysteries of the cortex, which until then had seemed too complex to yield to single-unit analysis. Armed with this tool, David Hubel and Torsten Wiesel, in Kuffler's laboratory, took the first step centrally by recording from single cells in the LGN. They found little difference from the properties of ganglion cells. Emboldened, they tackled the visual cortex. By careful control of the visual stimuli, they were able to elucidate a logical sequence of processing of visual signals, and suggest some simple ways the cortex could be organized to accomplish this. Their work called forth an enormous outpouring of papers, which have extended and modified the original findings and concepts in many ways. For a generation the Hubel and Wiesel approach was the touchstone for virtually all work in the field. It not only focused speculations on the visual mechanisms underlying perception, but also has been an inspiration to those working in other parts of the nervous system as well, in showing that a complicated system can be made to be understandable. This was recognized by the awarding of the Nobel Prize to them in 1981.

The basic findings of Hubel and Wiesel begin with the fact that, in the primary visual cortex, the terminals of LGN input fibers are the only elements with center–surround antagonism. The simplest properties of cortical cells are responsiveness to a bar or edge of light, with a particular orientation, in a particular position in the visual field. Cells with these properties are called *simple cells* (Fig. 16.18). More complicated are responses to a bar or edge, with a specific orientation, but placed anywhere in the visual field. Cells with this property, of signaling orientation independently of position, are called *complex cells* (Fig 16.18). Responses to bars of specific length and width are another type, initially called *hypercomplex;* most workers consider these to be variations of the other two types. In addition to these properties, cells can be classified as to whether they are driven by one eye or the other *(ocular dominance)* and by their sensitivity to movement.

With the same technique of making vertical electrode penetrations used by Mountcastle in somatosensory cortex (see Chap. 12), Hubel and Wiesel found that cells encountered in a single microelectrode penetration all tend to be driven by one eye or the other. This suggested that cells are organized in alternating *ocular dominance* columns. Similarly, it was found that all

the cells in a penetration tend to be tuned to the same orientation of an edge or bar. These are called *orientation columns.*

On the basis of these findings, it was predicted that the two types of columns would combine to form a *hypercolumn,* as depicted in the top diagram of Fig. 16.17. Each orientation column is actually a thin slab or wall, about $50 \times 500 \mu m$, which receives input mainly from one eye. An ocular dominance column is actually a larger slab or wall, $500~\mu$ wide and extending through an indefinitely long repeated sequence of orientation columns. A hypercolumn is a pair of ocular dominance columns, containing all the input from both eyes in all possible orientations. A hypercolumn also contains other types of organizational units for other aspects of the visual stimulus; for example, as shown in Fig. 16.17, cells processing color information are grouped in columns called cortical pegs. The cells in these pegs have the types of single- and double-opponent color receptive fields illustrated previously in Fig. 16.15 (see also next section).

The overlapping patterns of orientation columns and ocular dominance columns have been visualized by recording from the visual cortex using video-imaging after exposure to voltage-sensitive dyes. The dyes bind to the membranes and change their emission properties when the cells are depolarized during activity. Using this approach, Gary Blasdel and his colleagues showed that the activation pattern set up in the cortex reflects the orderly arrangement of orientation columns (slabs) within the thicker stripes related to ocular dominance. The pattern is not nearly so rigid as in the formal diagram of Fig. 16.17, but nonetheless confirms the basic concept. We will consider further aspects of this level of cortical organization in Chap. 30.

Visual Cortical Areas

In the traditional view, the only precise retinotopic map was contained in area 17, and the surrounding bands of cortex (areas 18 and 19) were given over to nontopo-

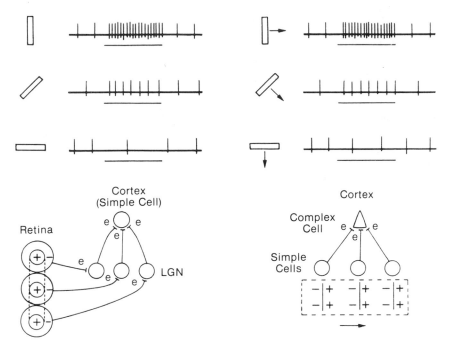

Fig. 16.18 The basic types of simple and complex cell responses identified by Hubel and Wiesel in cat primary visual cortex; their suggestions of simple connectivity rules that could generate these responses are shown below. See text. (Based on Hubel and Wiesel, 1962)

graphical "association" functions. As in the auditory and somatosensory systems, recent studies indicate that there are more representations of the peripheral fields than previously suspected. Over 20 specific visual areas have been identified in the monkey; Fig. 16.19 provides a current view.

The first evidence as to the functions of these areas came from Hubel and Wiesel. They extended their study of cell responses in area 17 in the cat to surrounding areas, and found an area in which cells were predominantly sensitive to motion. Semir Zeki in London extended this work in the monkey, showing that cells in area V5 (also called the MT area) are primarily sensitive to motion, whereas cells in area V4 are primarily sensitive to color. This suggested that these areas are *functionally specialized,* and that the visual cortical areas are organized as a mosaic, within which the different submodalities, such as form, motion, and color, undergo separate but parallel processing.

This and other work stimulated a great number of studies of the subdivisions of the visual cortex, resulting in the identification of the areas shown in Fig. 16.19. The primary physiological method for identifying these areas has been single-cell recordings of responses to different submodalities of visual stimuli. Anatomical studies using tract tracing methods (such as transport of HRP and phaseolus vulgaris [PHAL] from microinjections in specific areas) have demonstrated that the areas are richly interconnected by both feedforward and feedback connections. The physiological results have been extended to the human by use of positron emission tomography (PET). These studies have shown that a subject viewing a colored Mondrian-like painting (consisting of differently colored rectangles and stripes) has peak activity in a part of the cortex corresponding to area V4, whereas exposure to a scene of moving black and white squares gives peak activity in an area corresponding to V5 (see Zeki,

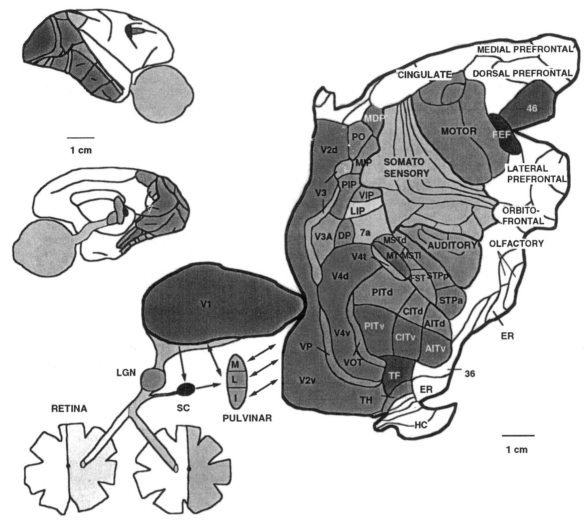

Fig. 16.19 The visual cortical areas and their relations to other cortical areas of the macaque monkey. In the upper left are lateral (above) and medial (below) views of the brain with the visual areas shaded. In the lower left is shown the two retinae and their projections through the lateral geniculate nucleus (LGN) to the primary visual cortex (V1), as well as smaller projections to the superior colliculus (SC) and the pulvinar (medial (I), lateral (L) and inferior (I)), which has reciprocal connections with many visual areas. In the large map the cortex has been unfolded and rolled out flat. **Visual areas:** *occipital lobe:* V1; V2 (v, ventral, d, dorsal); V3 and V3A; V4v, V4d, V4t (transitional); VP (ventral posterior); VOT (Ventral occipitotemporal); MT (middle temporal); *temporal lobe:* FST (floor of superior temporal); PIT (posterior inferotemporal, d and v); CIT (central inferotemporal, d and v); AIT (anterior inferotemporal, d and v); STP (superior temporal polysensory, a, anterior, p, posterior); TF, TH; *parietal lobe:* MST (medial superior temporal, d and l, lateral); PO (parieto-occipital); PIP, LIP, VIP, MIP (posterior, lateral, ventral, medial intraparietal); MDP (medial dorsal parietal); DP (dorsal prelunate); 7a; *frontal lobe:* FEF (frontal eye field); 46. Also indicated are the areas designated as **somatosensory; auditory; hippocampal complex** (HC) (Entorhinal cortex, ER); 36; **olfactory; orbitofrontal; prefrontal** (lateral, dorsal, and medial); **motor;** and **limbic cortex.** (From van Essen et al., 1992)

1993). Thus, the concept of functional specialization obtained from animal studies appears to apply to the human; in fact, it seems likely that the human may contain even more different areas than shown for the monkey.

An extremely interesting area is one that is specialized for faces. Cells in this area are particularly sensitive to faces or complex shapes that resemble facial patterns. This is seen in cats (area 7) and monkeys. Recent studies have shown that there is a modular organization within this area in the monkey.

The significance of these different areas is that they provide for abstraction, enhancement, and mixing of specific visual submodalities. As earlier emphasized in Table 16.1, vision is not one sense but many, and the different cortical areas give expression of this. Some of the best understood areas and their submodalities are summarized in the diagram of Fig. 16.20. We see, first, that color information arrives from parvocellular LGN and is processed through V1 and V2 to V4. Stimulus orientation, however, arrives in V1 from both parvo- and magnocellular LGN, and is processed through V2 and V3 to V4 and the region referred to as MT. A similar sequence applies to binocular disparity information. Information about direction of movement, by contrast, arrives in V1 from magnocellular LGN, and is transferred directly, as well as through V2, to MT above.

It can be seen that, in general, a given area has cells preponderantly tuned to a given submodality, such as color in V4 or motion in MT, but other submodalities are integrated with that information as well. Thus, each area provides for a unique combination of specific submodalities. We may note that the concept of integration of primary and secondary submodalities within each area is similar to the integration of primary and secondary types of odor ligands that is believed to occur within olfactory glomeruli (see Chap. 11). This parallel array of abstracted information is then fed into the higher visual centers. The two main streams for this information are into the inferotemporal and parietal regions, where complex processing related to the qualities and the spatial relations of objects in visual space is carried out. We will discuss these mechanisms of higher level vision in Chapter 30.

The Neural Basis of Visual Perception

Our ideas about the neural basis of visual perception have evolved through several stages related to the advance of anatomical and physiological knowledge summarized above (see Marr, 1981; Livingstone and Hubel, 1988; Churchland and Sejnowski, 1992; Martin, 1992; and Zeki, 1993).

A traditional view, dating from the early anatomical studies around 1900, was that the retina forms a cameralike image, which then is projected through the lateral geniculate onto the primary visual cortex. Perception then would consist in this image being "read," so to speak, by the "association areas," which would combine this image with other sensory images to form our perceptual world.

The studies of Kuffler, Hubel, and Wiesel in the 1950s and 1960s showed that this "camera hypothesis" is incorrect. Instead, beginning in the retinal circuits, key elements of the visual image, such as borders, contrasts, and motion, are abstracted and sent to the visual cortex. Within the primary cortex there begins a process of combining these elementary features in order to reconstruct key features of the image. This process was presumed to continue through further steps of "higher level processing" in the visual association areas to form the perceived image. An extreme view is the idea that this processing culminates in a "grandmother cell," so called because of the implication that the processing steps would converge onto a final recognition unit formed by previous visual experience.

In the 1970s, David Marr brought new concepts to bear on this problem from the perspective of computational systems. This approach emphasized the abstract proper-

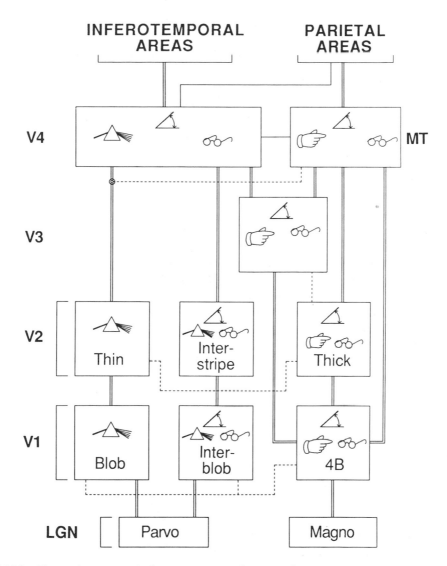

Fig. 16.20 The major anatomical connections and neuronal response selectivities in early visual areas of the macaque monkey. Icons are placed in each compartment to symbolize a high incidence of cells showing selectivity for stimulus wavelength (prisms), orientation (angle symbols), direction (pointing hands), or binocular disparity (spectacles). The lateral geniculate nucleus (LGN) is indicated with its parvocellular and magnocellular subdivisions. Subdivisions of cortical area V1 include layer 4B plus the dark "blobs" and pale "interblobs" of the superficial layers revealed by cytochrome oxidase histochemistry. The same technique reveals a tripartite subdivision of area V2, including a set of thick and thin dark stripes and an interposed set of pale "interstripes." Double-line connections denote robust pathways; single-line connections denote weaker pathways; dashed lines denote pathways only tentatively identified. Higher level processing in inferotemporal and parietal cortical areas is discussed in Chap. 30. (From DeYoe and van Essen, 1988)

ties of information processing in breaking down our visual world into a one-dimensional representation in the retina, and then reconstructing the three-dimensional world by a sequence of steps presumed to occur in the visual cortical areas. This approach stimulated the construction of a wide variety of computational models of visual processing that could carry out such tasks as pattern recognition, motion detection, and the detection of shapes against diffuse backgrounds (the so-called "shape from shading" problem). In computer terminology, this approach is "top-down," concerned only with the software programs for simulating the processing, with little regard for the actual "hardware" that might be used by the brain. These approaches have had the merit of emphasizing the importance of parallel processing for difficult tasks of image recognition. However, the relevance of these simulations to the actual mechanisms used by the brain to carry out its processing of visual images has been uncertain (see Chap. 1 for a discussion of the problem of deducing brain mechanisms from arbitrary models). The incorporation into these network simulations of more realistic models of neurons and neural connections, as discussed elsewhere in this book, should give this approach much more relevance.

We are presently in a phase when the evidence for multiple areas specialized for processing different submodalities dominates our thinking. As is obvious from Fig. 16.20, the abstracted visual "image" is distributed between many different areas. There must therefore be mechanisms for combining the different operations in order to reconstruct some kind of coherent representation of the visual image. This is sometimes referred to as the "binding problem." Since this must take place over some period of time while we perceive an object, there must be continuing interactions between the different areas, presumably mediated by the rich reciprocal connections between them. This process of reciprocal reexcitation is sometimes referred to as "reentry." One possibility is that these reciprocal interactions occur in rhythmic fashion; oscillatory waves at specific frequencies, such as 40 Hz, may be involved.

In conclusion, visual processing obviously involves the analytical abstraction and separation of submodalities as well as the integration of the abstract representations of those submodalities into a unified percept. The unified percept, however, is not confined to one location; it is distributed across the different processing areas, bound together by the multiple anatomical connections between them. Visual perception thus illustrates very well the principle of distributed processing that was introduced in Chap. 1 and will be further illustrated in the human in Chap. 30.

III

MOTOR SYSTEMS

17

Introduction: The Nature of Motor Function

Animal experience begins with information about the world that flows in through sensory organs and sensory pathways, as discussed in the previous section. However, the behavior of an animal depends on how it combines that information with its internal states and drives in order to do something. Doing something requires motor organs and the nervous circuits to control them, which together form what is called *motor systems*.

Motor systems bring about movement, and the importance of movement has always been apparent to students of animal life. The early Greek philosophers recognized that the ability to move is the essence of being alive. Furthermore, the ability of an organism to move itself about and perform actions on its environment, under control of a nervous system, is one of the crucial features that distinguishes animals from plants. In the evolution of animal life, different motor abilities have been among the chief agents for the diversity of adaptations that characterize different species. Motor abilities have been no less important in human evolution. The making of fire, development of tools, invention of the wheel, and use of weapons for hunting and implements for farming—all involved the

elaboration and extension of the human motor apparatus. Finally, the capacities for speech, writing, and artistic expression are all motor activities.

The study of this amazing array of activities is a fascinating one for neurobiologists. In one respect, the study is easy, because motor actions are observable and can be measured. To get much further than this, however, is very difficult. One of the main problems has been to identify what are the basic units of nervous organization relative to the basic units of motor function. When reflexes were recognized in the course of the nineteenth century, it was hoped that the reflex arc might serve as a basic functional unit. This has been true to some extent, as we shall see in Chap. 19.

A related problem in approaching motor systems is that they may seem quite different from the sensory systems discussed in the preceding section. After all, information flows *into* the organism (through *afferent* pathways) in sensory systems, whereas it flows *out* of the organism (through *efferent* pathways) in motor systems. Furthermore, the peripheral motor organs—glands and muscles—appear to be radically different from the receptors that receive sensory stimuli. Nor have the neural systems related to motor control appeared to share obvious similarities with the organization of sensory systems.

Because of these difficulties, there has been skepticism for many years among motor physiologists that a set of principles can be formulated that would apply generally to the organization of motor control, equivalent to the kinds of principles (receptive fields, contrast enhancement, abstraction of sensory information, parallel processing of multiple submodalities) that have been deduced for sensory systems. Here we will take a more optimistic view, building on what you learned about molecular and cellular properties of nerve cells in Section I and the organization of sensory systems in Section II. We will see that, at the *cellular* level, glands and muscles represent logical variations on the basic cellular plan discussed in Chap. 3. The *junctions* between these motor organs and the nerve fibers that innervate them are similar in principle to synapses between neurons, and between neurons and receptors (Chaps. 2 and 6, and Section II). And motor pathways are organized into *circuits* along many of the same general principles as are sensory pathways. In sum, modern studies are in fact leading to the view that motor organization shares many of the basic principles which characterize the general organization of nerve circuits throughout the nervous system.

These similarities mean that we can approach the study of motor systems using the same logical approach to underlying mechanisms that was applied to sensory systems. The basic mechanisms operate at similar levels: the peripheral *organs,* the *neural circuits,* and the *behavior* of the whole organism.

In this chapter we will first review briefly the different types of effector organs. We will then consider the skeletal muscle fiber in some detail, because it is the best understood type of effector organ and because it is the target for control by the largest proportion of motor systems in the nervous

system. Gland cells and other types of muscle cells (smooth muscle and cardiac muscle), and their nerve supplies, will follow in Chap. 18. We then move into the realm of neural systems, starting with the simplest reflex pathways and motor pattern generators (Chap. 19), the organization of circuits involved in locomotor activity (Chap. 20), and the hierarchy of systems at higher levels of the brain that provide for complex motor behavior and willed movements (Chap. 21). In recent years it has been possible to study skilled movements in awake, behaving animals, a study that is important for human motor activity, which is the subject of Chap. 22. The most special motor activity for humans is, of course, speech, which is covered in Chap. 23.

Effector Organs

In sensory systems there are different types of sensory receptors. Similarly, in motor systems there are different types of motor organs. Because their actions have effects, we also refer to them as *effectors,* or *effector organs.*

Among higher metazoans the two main types of effector organs are *glands* and *muscles.* Corresponding to these are the two main types of effector (motor) actions, *glandular secretion* and *muscle contraction.* It has been said (somewhat irreverently) that the only things an animal can do are squeeze a muscle and squirt a juice! It is true that glands and muscles are the characteristic effector organs of most invertebrates and vertebrates. However, as Table 17.1 shows, this simple statement hides a wealth of diversity. To begin with, glands are divided into two main classes (endocrine and exocrine). Within these two classes is an enormous variety of specialized types; Table 17.1 lists only a few examples. Similarly, muscles are divided into two main types, smooth and striated, and these are deployed in many different ways. Finally, there are other types of effectors; these include, for example, the elec-

Table 17.1 Types of motor organs

	Examples	
	Invertebrate	Vertebrate
Glands		
endocrine	neuroendocrine cells	neuroendocrine cells
		hypothalamus
	neurohemal organs	pituitary gland
	endocrine organs	endocrine organs
exocrine		
internal	goblet cells (mucus)	goblet cells (mucus)
	digestive glands	digestive glands
	salivary glands (enzymes, silk, jelly)	—
external (integument)	slime-secreting cells	sweat, sebaceous glands
	adhesive-secreting cells	—
	pheromone-secreting cells	pheromone-secreting cells
	toxin-secreting cells	toxin-secreting cells
	ink glands	—
	chromatophores	chromatophores
Muscles		
smooth muscle	(rare)	visceral (involuntary)
striated muscle		
cardiac	—	heart
skeletal	gut, trunk, appendages	trunk and appendages
modified muscle	—	electroplaque organ
Cilia	numerous small organisms	lining of various organs
Self-contained organs	nematocysts	

troplaque organ of certain fish, composed of modified muscle cells. Thus, as Table 17.1 illustrates, animals can send shocks, spin silk, change color, and do a number of things not adequately accounted for by the traditional functions of simple glandular secretion or muscular contraction.

If we focus our attention on muscles, we find that they show a range of properties reflecting the incredible diversity of motor activities among different animals. Figure 17.1 summarizes just one property, the speed of contraction. The fastest animal movements known are the beating of wings in certain insects, which can reach as high as 1000 Hz; however, this is by a special "click" mechanism of mechanical resonance in the wing joints, which does not require muscle contraction at that rate. The fastest known muscle contractions are involved in direct activation of muscles of insect wings, which beat up to about 200 Hz. Perhaps, in a remote jungle or desert, there is a species of insect or other animal— one that we do not yet know about—capa-

ble of faster muscle contractions. Identifying its mechanism might greatly enlarge our understanding of the molecular basis of muscle movements; this is one reason why preserving the world's ecosystems and saving endangered species is so important for biologists and for society.

As Fig. 17.1 shows, muscles have rates of contraction that range over 4 orders of magnitude, from several milliseconds to several tens of seconds. The muscles of the limbs in vertebrates can be seen to lie in the midrange of 40–100 msec. These muscles have received the most attention because of their roles in locomotion, but the data in the figure provide an important reminder that even in mammals the muscles show extraordinary diversity, from the extremely rapid contractions of the skeletal muscles involved in movements of the eyes and the larynx, to the very slow contractions of the smooth muscles of the intestines. These differences in contraction times are associated with a number of properties related to the differing metabolic demands,

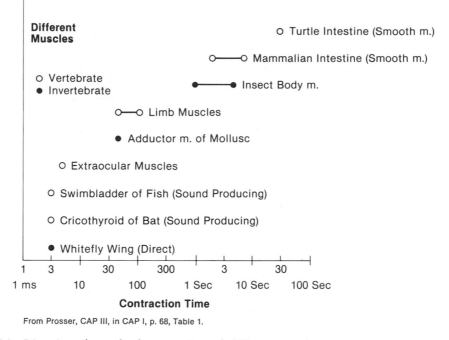

Different Muscles

○ Turtle Intestine (Smooth m.)

○———○ Mammalian Intestine (Smooth m.)

○ Vertebrate
● Invertebrate

●———● Insect Body m.

○———○ Limb Muscles

● Adductor m. of Mollusc

○ Extraocular Muscles

○ Swimbladder of Fish (Sound Producing)

○ Cricothyroid of Bat (Sound Producing)

● Whitefly Wing (Direct)

| 1 | 3 | 30 | 300 | 3 | 30 |
| 1 ms | 10 | 100 | 1 Sec | 10 Sec | 100 Sec |

Contraction Time

From Prosser, CAP III, in CAP I, p. 68, Table 1.

Fig. 17.1 Diversity of speed of contraction of different muscles. Note that the abscissa is a logarithmic scale, in order to cover the 10,000-fold differences. (Based on data in Prosser, 1991)

forces generated, and neural innervation, as we shall see.

Skeletal Muscle

Muscle cells owe their properties to specializations of their cytostructure. These structures are part of the basic equipment of all cells, but they are developed to a higher degree. In the case of muscle, the properties reside principally in two filamentous proteins, *actin* and *myosin*. These proteins are widely distributed in body cells and subserve a variety of functions related to cell movements, such as formation of pseudopodia and the cell movements that occur during development (Chap. 9). Actin falls into the class of microfilaments discussed in Chap. 3. In muscle, these proteins and their mechanism of interaction have become specialized for the specific tasks of producing movement, not just of individual cells, but of whole organs.

We know most about these proteins and their mechanisms in *skeletal muscles*, those that move bones and joints. Skeletal muscle is also called *striated*, or *striped*, muscle, because under the light microscope the individual muscle fibers have a banded appearance. The bands reflect the division of each fiber into a series of *sarcomeres*, which are the contractile units of the fiber. Each sarcomere is composed of a set of myofilaments; the myofilaments are actin or myosin, organized in a precise overlapping array. These features are depicted in Fig. 17.2. Note how this diagram of the hierarchical organization of skeletal muscle expresses our general theme of the hierarchical organization of the nervous system introduced in Chap. 1.

Sliding-Filament Model

The explanation for how the overlapping actin and myosin molecules bring about contraction is embodied in the *sliding-*

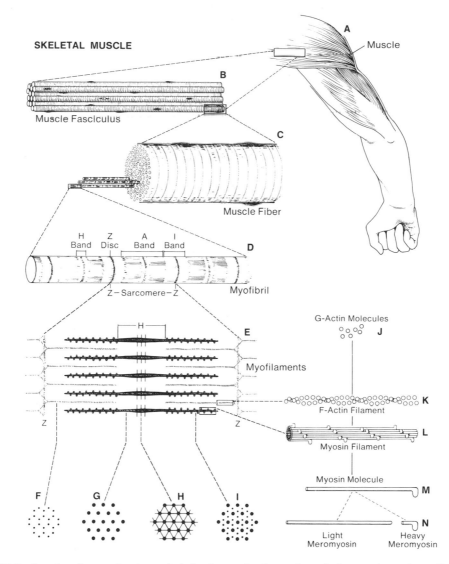

Fig. 17.2 Levels of organization of skeletal muscle, from the whole muscle to its cellular and molecular constituents. (From Bloom and Fawcett, 1975)

filament model. This was first conceived in 1954 by two independent groups of workers, Hugh Huxley and Jean Hanson in London, and Andrew Huxley and Robert Niedergerke in Cambridge, England. Their work, initially based on observations in the light microscope, has since been confirmed and greatly elaborated by many studies.

The essence of the model is illustrated in Fig. 17.2. It had previously been speculated that contraction might come about by a shortening or crumpling of the individual filaments, but this was replaced by the idea that the actin and myosin filaments slide along between each other. This sliding brings about changes in the widths of some of the bands (D), without changing the lengths of the filaments themselves.

The molecular mechanism which has been postulated to cause the sliding of the actin filaments past the myosin filaments is illustrated in Fig. 17.3. Actin filaments

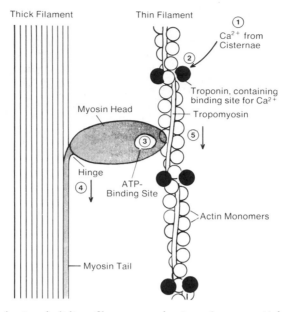

Fig. 17.3 Molecular basis of sliding filament mechanism. See text. (Adapted from Huxley, in Lehninger, 1975)

contain tropomyosin and troponin, in addition to actin. In the relaxed state, tropomyosin inhibits the myosin attachment sites on the G-actin filaments. Also, in the relaxed state, free Ca^{2+} is very low around the filaments. Muscle activation begins with release of Ca^{2+} ①, which binds to the troponin. This induces a conformational change in the troponin ②, which exposes the myosin binding sites of the β-actin subunits. The attachment of myosin ③ forms a *force-generating complex*. This induces a conformation change in the heavy meromyosin head, and a consequent rotation at the hinge between the head and the rest of the myosin molecule ④. The rotation generates the *power stroke* that causes the actin to be displaced ⑤.

The energy for these movements is supplied by ATP. The myosin heads have binding sites for ATP. It is believed that the ATP may be bound in the form of ADP + P with the energy of the phosphate bond transferred in some way to the myosin head to hold it in the *energized* conformation (Fig. 17.3). At the end of the power stroke, the myosin head assumes its *deenergized*

form when the ADP + P are released; they are replaced by new ATP, which causes the myosin head to return to its energized conformation.

Excitation–Contraction Coupling

There remains to account for the initial release of Ca^{2+} that sets this molecular machinery in motion. The full sequence of these events is depicted in Fig. 17.4. In some muscles, usually smaller and more slowly contracting fibers, the graded endplate potential (EPP) ② may be the only electrical response. In large and rapidly contracting muscles, an action potential ③ is set up by the EPP. The action potential propagates along the muscle membrane in essentially the same manner as the action potential in a nerve. The depolarization spreads rapidly into the interior of the muscle fiber through a special membrane system called the *T-tubule system* ④. The T-tubules are in close apposition (30 nm) to the membrane of the sarcoplasmic reticulum. Depolarization of the tubules brings about depolarization of the cisternae ⑤, which causes release of free calcium ⑥

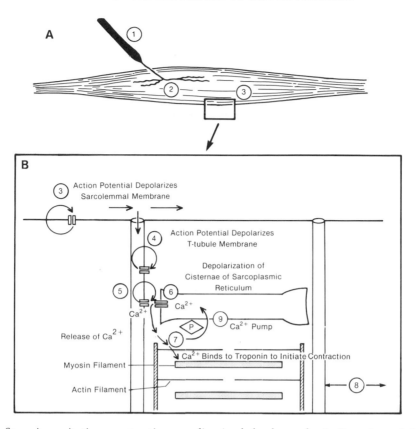

Fig. 17.4 Steps in excitation–contraction coupling in skeletal muscle. **A.** Overview of the muscle and its nerve supply. The nerve action potential ① invades the nerve terminals, activating the neuromuscular junction to release acetylcholine and set up the endplate potential (EPP) ②. This triggers the muscle action potential ③. **B.** Steps in the muscle include depolarization of T-tubules ④, activating voltage-sensitive channels ⑤ that activate Ca^{2+} channels in the sarcoplasmic reticulum ⑥. This increases Ca^{2+} flux into the sarcoplasm, where it binds to troponin ⑦ to initiate contraction ⑧, with pumping of Ca^{2+} back into the sarcoplasmic reticulum ⑨. (Based in part on Lamb and Stephenson, 1992)

into the sarcoplasm surrounding the muscle myofilaments. The Ca^{2+} binds to troponin ⑦ and initiates the contractile events ⑧ depicted in Fig. 17.3. The Ca^{2+} is restored to the cisternae by a very active *Ca pump* ⑨.

The whole sequence of events shown in Fig. 17.4 brings about a transformation from the electrical signal in the muscle membrane to a mechanical change in the myofilaments. This sequence is referred to as *excitation–contraction coupling*. It is analogous to the sequence of *excitation–secretion coupling* that takes place in gland cells and at synapses.

Depolarization of the cisternae of the sarcoplasmic reticulum is obviously critical in bringing about release of Ca^{2+}. The molecular basis has been the subject of considerable interest. Embedded in the T-tubule membrane is a protein that has been found to bind a blocking substance named dihydropyridine: this protein is therefore termed the *dihydropyridine receptor* (DHPR; see Fig. 17.5). Molecular cloning has shown that DHPR has considerable homology to the voltage-sensitive Na^+ channel (see Chap. 5)—perhaps not surprising, because DHPR is also voltage sensitive. It is presently surmised that depolar-

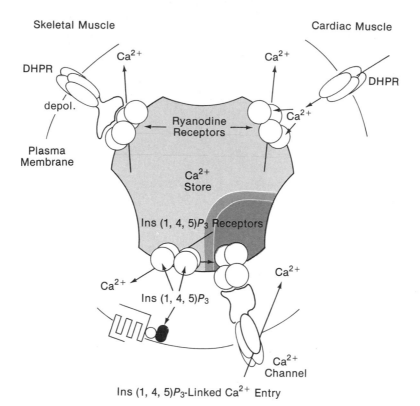

Fig. 17.5 Schematic diagram of the relations between the dihydropyridine receptor (DHPR) and the ryanodine receptor in skeletal muscle. The diagram illustrates the direct relation between the two receptors in skeletal muscle, the indirect relation between the two receptors in cardiac muscle, and the close homology with the IP_3 receptor. (Modified from Taylor and Marshall, 1992; see also Takamura et al., 1989, and Prosser 1991)

ization of the T-tubule membrane causes a voltage-dependent change in the loop between the second and third transmembrane domains of DHPR, which is transmitted by a direct mechanical coupling to a second type of protein embedded in the membrane of the sarcoplasmic reticulum called the *ryanodine receptor*. Activation of the ryanodine receptor mobilizes Ca^{2+}, causing its release from the sarcoplasmic reticulum into the cytoplasm, where it can bind to troponin and initiate muscle contraction. The ryanodine receptor also expresses the principle that muscle cells adapt general biological molecules to special uses. It has been cloned and sequenced, and it was found to belong to the same family as the *IP_3 receptor,* which, as we discussed in Chap. 8, is responsible for mobilizing Ca^{2+}

from the endoplasmic reticulum in neurons and other cells. This reflects the way that the sarcoplasmic reticulum is an adaptation of the endoplasmic reticulum for the special demands of delivering Ca^{2+} to the contractile apparatus in the depths of the muscle fiber. Figure 17.5 summarizes these molecular mechanisms and homologies (Taylor and Marshall, 1992).

Muscle Properties

How is this basic mechanism of contraction of individual skeletal muscle fibers adapted for the range of motor functions required in the mammalian body? As you already know, the body muscles of mammals are generally of two types, red and pale, or dark and white. Anatomists and physiologists have agreed on this but added a third

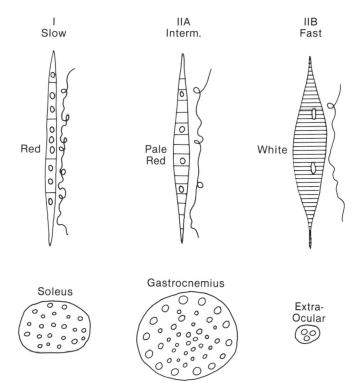

Fig. 17.6 The three main types of muscle fiber found in vertebrate skeletal muscles. (Based on Goldspink, in Eckert and Randall, 1983; Burke, 1981; and Binder, 1989; see also Table 17.2)

category intermediate between the two. These three types are illustrated by the diagrams of Fig. 17.6, and their properties are summarized in Table 17.2. As can be seen, red fibers are thin, packed with mitochondria, and richly supplied with blood vessels. The red color is due to high levels of oxygen-storing myoglobin. As you would expect from these attributes, these muscles have a high level of aerobic metabolism, that is, ATP production is dependent on oxygen delivered in the blood. When stimulated directly, they give rise to slow contractions. They are therefore called *slow phasic fibers* (sometimes slow tonic fibers). They are found in muscles that

Table 17.2 Summary of the three main types of muscle fiber in skeletal muscle and the differences in their properties

Property	I (slow)	II A (intermediate)	II B (fast)
Diameter	small	medium	large
SR and T-tubules	least	intermediate	most
Mitochondria	abundant	intermediate	sparse
Capillary supply	abundant	abundant	sparse
ATP generation	aerobic	aerobic/anaerobic	aerobic/anaerobic
Glycogen	low	high	high
Myosin	type a	type b	type b
Sarcomeres (Z bands)	wide	narrow	narrow
Myoglobin	abundant		sparse

Adapted from Burke (1981) and Binder (1989)

show sustained activity, as in maintaining upright posture or in long-distance running.

At the other extreme are the thick pale muscle fibers. When stimulated directly they give rapid contractions, reflecting a rapid turnover rate of their myosin ATPase, but they fatigue rapidly. ATP production is from glycolysis; because of their sparse mitochondria and blood supply, they rapidly build up an energy debt during contraction; the debt must therefore be repaid by oxygen from the blood. They are referred to as *fast phasic glycolytic fibers*. These fibers make up muscles that are used to generate large forces for short periods of time, as in sprinting or jumping. Finally, an intermediate type shows intermediate properties between red and pale; they are called *fast phasic oxidative fibers*.

Motor Units

Thus far our focus has been on each muscle cell as an individual unit. However, to be useful in generating movements, cells must be brought together in functional groups and provided with appropriate nervous control. As indicated in Fig. 17.2, the simplest arrangement of muscle fibers is in a fasciculus, and many fasciculi together make up a given muscle.

The lowest level of nervous control of muscles is concerned with the connections of the motoneurons to the muscle cells. A motoneuron, along with the population of muscle fibers it innervates, is called a *motor unit*. In many invertebrate muscles, a single fiber may receive endplates from more than one motoneuron, and multiple innervation also occurs during early development of vertebrate muscle. However, in the adult, the rule is that a muscle fiber receives an endplate from only one motoneuron.

Because of this rule, the size of a motor unit is determined by the extensiveness of branching of the motoneuron axon. This in turn is matched to the particular functional demands in a given muscle. Motor unit size varies widely, in accord with the range of functions of different muscles in the body.

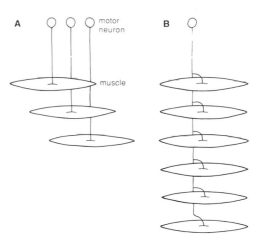

Fig. 17.7 The organization of motor units. **A.** Each motoneuron innervates only one muscle; this is the smallest possible motor unit size. **B.** A larger motor unit with an innervation ratio of 6.

At one extreme are very small muscles involved in controlling very fine movements. This is exemplified by the extraocular muscles to the eye (Chap. 14; Fig. 14.15). In these muscles, the motor unit may approach the limit of a single muscle cell for each motoneuron (Fig. 17.7A). At the other extreme are muscles that are large, and whose functions are to generate large forces (such as the gastrocnemius) or sustained contractions (such as the soleus). These properties have been discussed above (Fig. 17.4). For these kinds of muscles, the innervation ratios may be several hundred, even up to 1000 (Fig. 17.7B).

Motor Unit Types

We are now in a position to ask whether the different types of muscle fibers are organized into different types of motor units. To test this, the experimenter anesthetizes an animal and carefully dissects a peripheral motor nerve and the muscle it innervates (Fig. 17.8A). Electrical stimuli can be delivered to the nerve while the contractions of the muscle are recorded. Parallel experiments can be done in the human, as indicated in Fig. 17.8B.

A Animal Experiments   **B Human Experiments**

C Motor Unit Types (Gastrocnemius)

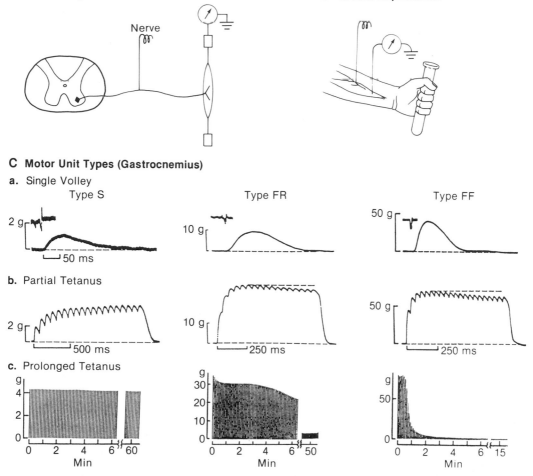

Fig. 17.8   The main types of motor units as expressed by their contraction properties. **A.** Setup for stimulation and recording in animal experiments. **B.** Setup for stimulation and recording in the human. **C. a.** Responses to a single nerve impulse in Type S (slow), Type FR (fast fatigue resistant), and Type FF (fast fatiguing). **b.** Responses to partial tetanus. Dashed lines show amount of sag of FR and FF responses. **c.** Responses to prolonged complete tetanus. (Modified from Burke, 1973, in Binder, 1989)

Figure 17.8 summarizes the results of a classical set of experiments carried out by Robert Burke and his collaborators at NIH, who found that the motor units could be divided into three types. Type S (slow) responded to a single nerve volley with a small contraction (C, a). They then stimulated the nerve repetitively and found that there was summation of the contraction to a higher and maintained level. This is because the individual muscle twitches, be-

ing slower than the nerve impulses, build up on top of each other (C, b). This maintained contractile state is called a *muscle tetanus*. Although it is artificially induced by the electrical shocks, it in fact mimics the way muscles are activated by impulse discharges in the motoneurons. Even over long periods of time, the tetanic contraction in an S motor unit is maintained (Fig. 17.8C, e).

By contrast, at the other extreme is the

type of motor unit that gives a large contraction in response to a single nerve volley (C, a), almost 50 times as large as the response to an S motor unit (C, 1). However, when this unit is activated tetanically, it starts to fade (C, b), and over longer periods (C, c) it declines rapidly to practically no contractile force at all. This type is called fast fatiguing (FF). Between these is a third type that shows intermediate properties: a modest contractile response to a single volley (C, b), and only modest fading with tetanic stimulation (C, b, e). This type is therefore termed fast fatigue–resistant (FR).

Thus, there is a very close correlation between the types of muscle fibers we previously described and the types of motor units. This implies that a given motoneuron axon supplies muscle fibers belonging to the same type. This is a logical way to organize the innervation of muscle fibers because it means that the motoneurons can selectively activate different muscle types, and, in turn, the higher centers in the brain can selectively activate those muscles by selectively activating the appropriate motoneurons. These selective connections are laid down during development and fine-tuned by experience. They provide the basis for extremely delicate control of different types of muscles in relation to different types of motor activity. In Chaps. 19 and 20 we will see how these properties are further matched to the properties of the motoneurons and the central neural circuits that control them.

Two further principles are of interest. First, it is clear from the responses to single volleys (Fig. 17.8C) that brief nerve impulses give rise to prolonged muscle responses. Thus, a single brief neural event can serve to trigger a longer lasting mechanical event in another cell. This could apply to actin–myosin interactions in nerve cells as well as in muscle cells, and it applies also to secretory events in nerve cells as well as in gland cells (see above). Second, during the long-lasting mechanical responses in a skeletal muscle, repeated neu-

ral inputs find the muscle fiber at different stages of contraction and relaxation. The summation that occurs is therefore complex, and very frequency-dependent. High-frequency tetanic stimulation, summed over many fibers in a muscle, is the basis for the generation of maximal muscle power. Medium-frequency contractions, providing for complex summation among subgroups of fibers with similar properties at different frequencies, are the basis for movements requiring precision. These two principles underlie much of the diversity of motor behavior which we will encounter in the ensuing chapters.

The Motor Hierarchy

How then are the motor units controlled? That is a central question of motor systems and the subject of the remaining chapters in this section. The essence of these systems is that they are organized in a hierarchical fashion. In fact, they illustrate, better than any other part of the nervous system, the principle of hierarchical organization that was introduced in Chap. 1.

The concept that motor control is embedded in successively higher levels of organization was recognized at an early stage by students of animal behavior. A schema which expresses this is shown in Fig. 17.9 It relates specifically to the reproductive behavior of the male stickleback fish, but it should be regarded as a framework for motor behavior in general. Beginning at the bottom are the muscle fibers and motor units ⑧. Above this are several levels (⑦ to ⑤) building up to the coordinated movement of a whole fin. Above this are the levels (① to ④) of increasingly larger components of behavior that involve the whole animal. For example, one goes from biting ④, to fighting ③, to territorial behavior ②, to migration ①. Although nerve circuits are not explicitly included in this diagram, they are implied for the mechanisms controlling the lower levels of organization and generating the behaviors at the higher levels.

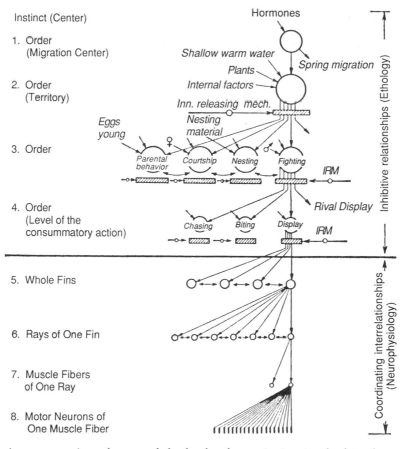

Fig. 17.9 A representation of some of the levels of organization involved in the generation of motor behavior. See text. (After Tinbergen, 1951)

The important aspect of the higher levels is that a particular behavior is characteristically elicited by a particular stimulus, called an innate releasing mechanism (IRM). As an example, Fig. 17.10 shows a simple shape that was used as a visual stimulus to test for escape responses in turkeys. When the pattern was moved to the left, it elicited avoidance by the turkeys, because the shape and orientation (short neck, long tail) mimicked a hawlike bird of prey. By contrast, when the pattern was moved in the other direction, it elicited little or no response, because it had gooselike features (long neck, short tail) which were not threatening.

This kind of result has an interesting connection with properties of sensory systems. Recall from our discussion in Chap. 10 that perception involves recognition of a pattern as a whole, called a gestalt; in the example of Fig. 10.9, the illustration tends to be perceived as a face or a vase,

Fig. 17.10 A simple pattern elicits quite different motor responses in a test animal, depending on the direction of movement. See text.

but not both. Similarly, for the turkey, the shape in Fig. 17.10 tends to be seen as predator or innocuous neighbor, and the entire motor pattern appropriate for one or the other gestalt is played out. This illustrates why our study must proceed step by step to the higher levels of circuit organization in motor systems and central systems in order to achieve an understanding of motor behavior.

18

Autonomic Functions

Despite all the variations in body form, one can usually divide the animal body into two parts: a *visceral* part containing the internal organs, and a *somatic* part consisting of the musculoskeletal apparatus. The first is concerned with maintaining the internal environment and carrying out functions within the body, the second with moving the animal about and mediating interactions with the external environment. In some simple species the division between the two is so clear that it seems as if there are two animals, the muscular animal moving about with a visceral animal on its back. In the vertebrates, a similar principle is at work, with visceral and somatic body parts clearly distinguishable (Fig. 18.1).

The nerves that supply the internal organs constitute the autonomic nervous system. "Autonomic" indicates that most of the nervous control is believed to be "autonomous." What does this term mean? One meaning is that this control is automatic and involuntary, in contrast with voluntary, willed control over skeletal muscles. Another meaning is that this system functions in an independent, "autonomous," manner with a self-contained hierarchy of functional units and internal state generators. These distinctions are useful,

but they are not hard and fast. For example, autonomic nerves control the blood supply to muscles and skin in the somatic part of the body. Moreover, activity of skeletal muscles requires the autonomic nervous system to divert blood from gut to muscles. So the line between autonomic and voluntary nervous control, between involuntary and voluntary actions, is crossed in both directions, and accordingly there needs to be a higher level of central control which coordinates these two great systems (see Chap. 24).

As commonly defined, the autonomic nervous system consists only of the nerves carrying motor impulses to the internal organs; by this definition, it is exclusively peripheral and motor. While it is useful to be able to define this component, a larger view of the nervous mechanisms involved in regulating the body requires a broader perspective. This would take account of the sensory information from the internal organs, the effects of a great many peptides and hormones, and the circuits in the central nervous system that control the peripheral ganglia and nerves.

This chapter begins with an overview of the autonomic nervous system of vertebrates. We will then examine the organi-

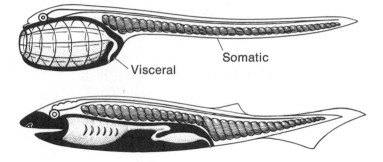

Fig. 18.1   Organization of the vertebrate body. *(Above)* Hypothetical primitive chordate. *(Below)* Representative lower vertebrate, such as the fish. Note the divisions into somatic and visceral (shaded) body parts. (After Romer and Parsons, 1977)

zation of the ganglia which transfer information from the motoneurons to the peripheral organs. Next we will consider the functional properties of gland cells as the other main type of effector organ in parallel with muscle cells. We will complete our review of muscle cells, which started with skeletal muscle in the previous chapter, by considering first the properties of smooth muscle cells and finishing with cardiac muscle and its control by the special conducting fibers of the heart.

Organization of the Autonomic Nervous System

The discovery that the nerves to the viscera constitute a distinct system was made by anatomists from observations in humans. Thomas Willis in 1664 first described the two chains of ganglia running on either side of the vertebral column. Willis also made the important distinction between nerves that subserve voluntary (somatic) and involuntary (visceral) functions. It was not until 1732 (how slow progress can be!) that Winslow, in France, described the many nerves that connect the chain to the internal organs, and speculated that these nerves bring the organs into "sympathetic" relation with each other.

Modern studies of these nerves began with Gaskell and Langley in England, around 1900; the contributions made by their early work toward laying the founda-

tion for our concepts of the chemical nature of synaptic transmission were noted in Chap. 8. The terms that are now generally used date from that time. The nerves to the internal organs constitute the *autonomic nervous system*. The system has two main divisions: the *sympathetic* nervous system (originally termed the orthosympathetic by Langley) and the *parasympathetic*. This is shown for the human in Fig. 18.2. The basic organization of these two systems appears to be similar in most vertebrate species.

In the parasympathetic division, the peripheral ganglia are located in the organs that their cells innervate. The ganglia receive their neural inputs from fibers arising from cells in certain nuclei of the brainstem and sacral spinal cord. By virtue of their positions relative to the ganglia, fibers innervating a ganglion are called *preganglionic*, and those arising from it are *postganglionic*.

Sympathetic ganglia, by contrast, are arranged in a cord along the vertebral column (Figs. 18.2, 18.3) or in the mesentery of the gut. Their output cells have long postganglionic fibers which branch and innervate the internal organs. The cells, in turn, are innervated by preganglionic fibers of cells located in the intermediolateral column of the thoracolumbar portions of the spinal cord. It can thus be seen that in both divisions of the autonomic nervous system, there is, in addition to the centrally located

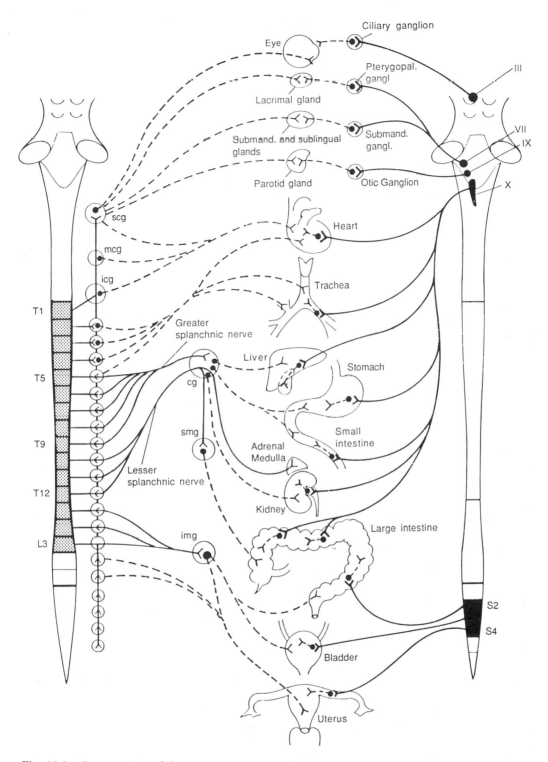

Fig. 18.2 Organization of the autonomic system. The sympathetic system is shown on the left, the parasympathetic system on the right. Abbreviations for the names of sympathetic ganglia are as follows (from top to bottom): scg, superior cervical ganglion; mcg, middle cervical ganglion; icg, inferior cervical ganglion; cg, celiac ganglion; smg, superior mesenteric ganglion; img, inferior mesenteric ganglion. T1–T12, segments of thoracic spinal cord; L3, third lumbar spinal segment; S2, S4, sacral spinal cord segments. III–X, cranial nerves. The diagram represents the human, but it applies generally to vertebrate species. (Modified from Heimer, 1983)

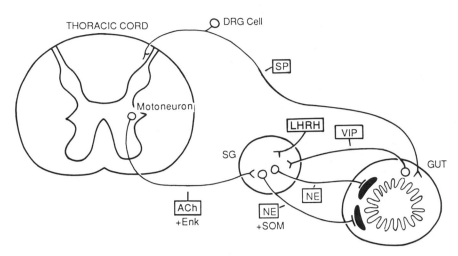

Fig. 18.3 Localization of neurotransmitters and neuropeptides in neurons of the sympathetic nervous system. ACh, acetylcholine; DRG, dorsal root ganglion; Enk, enkephalin; LHRH, luteinizing hormone releasing hormone; NE, norepinephrine; SG, sympathetic ganglion; SOM, somatostatin; SP, substance P; VIP, vasoactive intestinal peptide. (Modified from Hökfelt et al., 1980)

motoneuron within the spinal cord, a peripherally located motoneuron within a ganglion.

A special subdivision of the sympathetic system is located in the interior (medulla) of the adrenal gland. Here cells known as *chromaffin* cells are packed together. These cells contain vesicles filled with catecholamines (mostly epinephrine). Like sympathetic ganglia, the adrenal gland receives its innervation from motor cells of the intermediolateral column of the spinal cord. The chromaffin cells themselves lack axons; upon stimulation, they discharge their epinephrine into the bloodstream. Through this mechanism, the sympathetic nervous system can mediate rapid hormonal-type actions on various organs throughout the body.

Recently, the chromaffin cells have taken on a new significance for neurobiology through the experiments on brain transplants. Since the synthetic pathway for epinephrine passes through dopamine, these cells have been transplanted into the neostriatum (see Chap. 9) in the hope of inducing them to synthesize and release dopamine, and thereby replace the dopamine that has been lost through the degeneration of dopaminergic cells in Parkinson's dis-

ease. Animal experiments have showed that these experiments are feasible, and the first operations on human patients have given encouraging results in producing relief of the motor incapacities of these patients.

The main functions of the parasympathetic and sympathetic nerves are summarized in Table 18.1. Several general conclusions about the relations between the two divisions can be drawn from the table. First, either system may be stimulatory or inhibitory to a given organ. Second, when an organ is innervated by both systems, they are usually (but not always) opposed to each other in their actions. Third, some organs are predominantly or exclusively controlled by one system or the other.

The different patterns of innervation of the two divisions depicted in Fig. 18.2 are important to understand in assessing the contrasting actions of the two divisions on the behavior of the organism. The ganglion cells in the sympathetic system have wide fields of peripheral innervation, and this means that their activity tends to have widespread effects and evoke mass responses. Traditionally, it has been believed that the overall effect of the sympathetic system is to decrease activity in the visceral organs and stimulate the heart and somatic

Table 18.1   Effects of autonomic nervous activity on different body organs

Organ	Effect of sympathetic stimulation	Effect of parasympathetic stimulation
Eye		
pupil	dilated	contracted
ciliary muscle	none	excited
Glands		
nasal	vasoconstriction	stimulation of thin, copious secretion
lacrimal		containing many enzymes
parotid		
submaxillary		
gastric		
pancreatic		
Sweat glands	copious sweating (cholinergic)	—
Apocrine glands	thick, odoriferous secretion	—
Heart		
Muscle	increased rate	slowed rate
	increased force of contraction	decreased force of atrial contraction
coronaries	vasodilates	constricted
Lungs		
bronchi	dilated	constricted
blood vessels	mildly constricted	—
Gut		
lumen	decreased peristalsis and tone	increased peristalsis and tone
sphincter	increased tone	decreased tone
Liver	glucose released	—
Gallbladder and bile ducts	inhibited	excited
Kidney	decreased output	—
Ureter	inhibited	excited
Bladder		
detrusor	inhibited	excited
trigone	excited	inhibited
Penis	ejaculation	erection
Systemic blood vessels		
abdominal	constricted	—
muscle	constricted (adrenergic)	—
	dilated (cholinergic)	
skin	constricted (adrenergic)	dilated
	dilated (cholinergic)	
Blood		
coagulation	increased	—
glucose	increased	—
Basal metabolism	increased up to 100%	—
Adrenal cortical secretion	increased	—
Mental activity	increased	—
Piloerector muscles	excited	—
Skeletal muscle	increased glycogenolysis	—
	increased strength	

From Guyton (1976)

muscles, and that these effects prepare the whole organism for "fight-or-flight" behavior. By contrast, the ganglion cells of the parasympathetic system, being located in their target organs, have narrow fields of innervation. It has been generally thought that their effects are thereby local, and related to facilitating the activities of their respective organs. While the patterns of innervation in the two systems are indeed clearly different, and while it is true that sympathetic activity has more global effects, Table 18.1 makes it clear that there are exceptions to the traditional generalizations about the functions of the systems.

We will discuss further the relation of autonomic activity to behavior below and in Chaps. 24–28. For the remainder of this chapter, we will consider the organization within a ganglion and examples of mechanisms involved in the control of the three types of effector cells: glands, smooth muscle, and cardiac muscle.

Sympathetic Ganglion

The traditional view of the autonomic nervous system represented in the diagram of Fig. 18.2 has been greatly extended by recent studies of transmitter substances and the synaptic organization within a ganglion.

Transmitters and Modulators

A simple rule for autonomic neurotransmitters was laid down by the classical workers. In both divisions, the preganglionic motoneurons use acetylcholine (ACh); the postganglionic cells also use ACh in the parasympathetic system, but norepinephrine (NE) in the sympathetic system. Figure 18.3 summarizes this pattern of localization in the sympathetic innervation of the gut.

When neuroactive peptides began to be identified, it was soon found that they were present in the autonomic nervous system. Some of the clearest evidence for colocalization of neuropeptides with classical transmitters has come from these studies. Thus, as indicated in Fig. 18.3 in the sympathetic division, enkephalin (Enk) is found in cholinergic preganglionic nerves, and somatostatin (SOM) is found in many postganglionic noradrenergic nerves. In addition to motor fibers, peptides are also present in the sensory fibers; vasoactive intestinal peptide (VIP) has been found in sensory neurons in the gut, and substance

P (SP) in dorsal root ganglion cells (see Fig. 18.3).

Synaptic Organization

What are the actions of transmitters and neuropeptides in the autonomic pathways? Diagrams such as Figs. 18.2 and 18.3 give the impression that a sympathetic ganglion is the site of a simple relay from pre- to postganglionic nerve. However, we have already seen that neuromodulators may have complex effects on synaptic transmission (see Chap. 8). Studies of synaptic organization have shown that in fact the ganglion is a complex integration system, in which the different transmitters and neuropeptides act on specific receptors with distinct properties for controlling the transmission of information to the target organs.

The sympathetic ganglion of the frog has served as a valuable model for elucidating these principles.

As shown in Fig. 18.4, stimulation of the preganglionic fibers gives rise to a sequence of synaptic effects in a ganglion cell. The briefest action (A, in Fig. 18.4) is a *fast EPSP*, lasting 10–20 msec. It is mediated by ACh, liberated from the nerve endings and acting on nicotinic receptors (receptors that are activated by the substance nicotine; see circuit diagram in F). The mechanism underlying this response is believed to be similar to that at the neuromuscular junction, involving an increased conductance to cations (Na^+, Ca^{2+}, and K^+).

The next component of the response (B in Fig. 18.4) is a *slow IPSP,* lasting several hundred milliseconds. This appears to be mediated by a conductance-decrease synapse, which hyperpolarizes the membrane by turning off Na^+ conductance (see Chap. 7). A number of studies have been carried out to identify the transmitter for this synaptic action. One possibility has been that the nerves excite an interneuron, and that this interneuron inhibits the ganglion cells by releasing a catecholamine, either dopamine or norepinephrine. Anatomical studies have shown that many ganglia contain small cells, packed with large dense-core

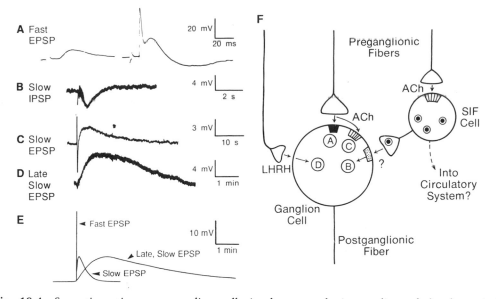

Fig. 18.4 Synaptic actions on ganglion cells in the sympathetic ganglion of the frog. **A–D.** Intracellular recordings from ganglion cells, showing different types of responses to electrical stimulation of preganglionic fibers; note the increasingly slower oscilloscope traces from A to D. (From Jan et al., 1979) **E.** Comparison of the responses in A, C, and D on the same time base and with the same amplification, to emphasize the contrasting time courses of these responses. (From Jones and Adams, 1987) **F.** Summary diagram of the synaptic organization of the sympathetic ganglion, showing sites of generation of different types of response in A–D. (Modified from Libet, in Shepherd, 1979) LHRH, luteinizing hormone releasing hormone; SIF, small intensely fluorescing.

vesicles. Since these cells fluoresce when treated with the paraformaldehyde technique, they are called *small intensely fluorescing cells,* or SIF cells. It has been proposed that the slow IPSP response of the ganglion cell involves activation of an adenylate cyclase and a second messenger system, as described in Chap. 8. However, a slow IPSP is seen in some ganglia that lack SIF cells, so other mechanisms must be considered as well.

The third response component is a *slow EPSP,* lasting several seconds (C in the figure). This is mediated by ACh acting on muscarinic receptors (receptors sensitive to activation by the substance muscarine). Forest Weight and J. Votava showed in 1970 that this is due to a decrease in a resting K conductance; the membrane depolarizes toward the Na equilibrium potential, according to the mechanism discussed in Chap. 7. Paul Adams and his colleagues

termed this resting K current the M (for *muscarinic*) current and have characterized its voltage-dependent properties (see Chap. 5). By itself, closure of M channels by ACh gives only a slow, small depolarization of a few millivolts, as indicated in Fig. 18.4C; however, the closure greatly potentiates concurrent large nicotinic EPSPs (as in A) by raising the input resistance of the cell and reducing the repolarizing effect of the K conductance. This is an example of a transmitter acting as a neuromodulator.

The final response component (D) is a *late slow EPSP,* lasting several minutes. In 1979, an elegant analysis by Lily Yeh Jan, Yuh Nung Jan, and Stephen Kuffler of Harvard implicated the polypeptide luteinizing hormone releasing hormone (LHRH) (see Chap. 8) in mediating this response. Several procedures could be carried out in this simple system, such as microionophoresis of LHRH and its analogues, and radioimmunoassay for presence of LHRH in the ganglion, which thus satisfied several of the criteria deemed necessary for positive identification of a neurotransmitter or neuromodulator (Table 8.4 in Chap. 8). Suppression of the M current is also the mechanism for this response. Thus, the same type of neuromodulation can be brought about by different transmitter and neuropeptide systems acting over different time periods.

Receptors for Different Durations, Operations, and Strengths

Why are multiple transmitters and receptors needed in this pathway through the sympathetic ganglion? Better insight into this question is obtained by replotting the responses of Fig. 18.4 A–D on a common time scale, as in E. This brings out more clearly that each receptor function is distinctive in at least three ways: duration, actions, and strength. In *duration,* each synaptic potential has its distinctive time course, though there is also overlap (cf. also Fig. 8.15). In *action,* the synaptic responses are depolarizing or hyperpolarizing, excitatory or inhibitory. Other possible actions would include biasing (cf. Chap. 7), second messenger metabolic effects (Chap. 8), or the setting of a behavioral state (Chap. 24). In *strength,* the responses vary in amplitude. At one extreme, the fast EPSP has a large amplitude, presumably with the function of generating an impulse response; at the other extreme, the late slow EPSP has a small amplitude, as would be suitable for modulating the frequency of the impulse discharge.

These considerations suggest the principle that the complexity of information processing in a neural system is likely to depend on how rich a repertoire of molecular receptor functions it has for generating responses with different durations, actions, and strengths. This helps to answer the question, why are there so many transmitters and modulators in the nervous system? The answer is that they are needed to enable a neural center to provide for different functional operations, coordinated over

different time periods, and adjusted for different strengths.

Seen in this light, the sympathetic ganglion is a local circuit system (see Fig. 18.4F). It is much more complex than the simple relay as traditionally conceived, but it probably is not as complex as some regions of the central nervous system, which have to deal with spatial processing and exquisite temporal discrimination as well. As a model system, it demonstrates admirably how circuits integrate multiple inputs by distinctive mechanisms covering a range of durations, actions, and strengths. These principles are likely to be important for the contribution of the sympathetic system to the behavior of the organism.

Target Organs

Let us move on from the site of the ganglion cell body, and examine the synaptic mechanisms whereby the ganglion cell exerts control over its target organ. We will consider the three main types of target in the autonomic nervous system: glands, smooth muscle, and cardiac muscle.

Gland Cells

As previously mentioned (Chap. 8), gland cells are divided into two main types. *Endocrine* glands manufacture hormones that are secreted into the bloodstream to act on distant cells and organs. The adrenal medulla noted above is of this type. These glands are discussed in later chapters dealing with the hypothalamus and pituitary body (see Chap. 24).

In contrast to endocrine cells is the *exocrine* type of secretory cell. Characteristically these cells are grouped together to form a gland, and the substance is carried away in a duct. Glands of this type perform a variety of functions in the body. Some are involved in nonnervous functions, such as the digestive glands of the gastrointestinal system.

As indicated in Table 18.1, a number of glands are under autonomic control.

Glands in the walls of the intestinal tract are the *tubular* type, which take the form of pits or tubules. They contain cells that secrete either mucus, to lubricate the gut wall, or enzymes, to aid in digestion.

The other main type of gland is the *acinar* type, which consists of collections of cells which discharge their contents into central ducts. This characterizes more complex glands, such as the salivary glands, pancreas, and liver.

Secretion may be stimulated in several ways: by local chemical cues from the ingested food; by nervous reflexes elicited by local chemical, tactile, or mechanical cues; by central nervous activity; or by circulating hormones. The relative importance of these mechanisms varies for different glands.

We generally think of the glands of the body as involved mostly in housekeeping chores, helping to maintain the constancy of the internal environment *(milieu interieur)* and enabling it to respond to stress. These are indeed important functions. However, as indicated in Table 17.1, many glands are parts of the motor apparatus through which the animal operates on the external world. The neural control of these glands characteristically involves sensory recognition of the appropriate releasing stimuli and precise timing of secretion within a sequence of behavioral acts.

Excitation–Secretion Coupling

The cellular basis for the secretory activity of gland cells was described in Chap. 3. It will be recalled that in secretory cells the transitional endoplasmic reticulum of the cytoplasm is grouped in stacks, called the Golgi complex, where the specific proteins are stored, and packaged into secretory granules (see Fig. 18.5). The granules are released when the cell receives its appropriate stimulus. The features vary considerably, depending on the particular type of secretory activity. For example, secretory granules may not be demonstrable in the electron microscope in some cells, particularly those that continuously secrete small

amounts. On the other hand, granules tend to be more prominent in cells that store the secretions and release them intermittently in massive amounts. Secretory cells exhibit variations in fine structure, reflecting these differences in function.

In glands under neural control, stimulation occurs by means of a neurotransmitter liberated from the terminals of the motor axon. The neurotransmitter generally brings about a depolarization of the gland cell, similar to the endplate potential in a muscle. The depolarization gives rise to an action potential in some gland cells, as was pointed out in Chap 6; in others the electrical response consists only of the graded postsynaptic potential.

What is the linkage between this depolarization and the release of secretory substances? This linkage has been termed *excitation–secretion coupling*, in analogy with excitation–contraction coupling in muscle. In the gland cell, the coupling usually involves the following sequence: depolarization of the membrane; activation of second messenger systems; rise in intracellular Ca^{2+}; movement of superficial granules to plasmalemma; fusion of vesicle membrane and plasmalemma and release of secretory substances. The sequence is similar to the steps controlling release of neurotransmitter at a chemical synapse. In fact, it seems very likely that the basic mechanism of excitation–secretion coupling became adapted in evolution for synaptic transmission. The similarities between excitation–secretion mechanisms in a prototypical gland cell and nerve cell are summarized in Fig. 18.5.

Smooth Muscle

Knowledge of the properties of smooth muscle is limited because of inherent difficulties: the cells tend to be small and thin; they form syncytial networks, often intertwined within other complex structures (like the gut or blood vessels); they characteristically undergo spontaneous contrac-

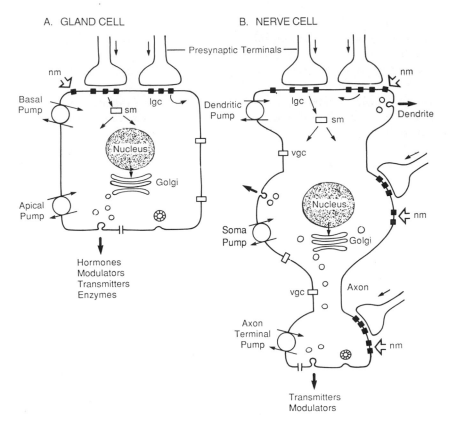

18.5   Comparison between cellular organization of a gland cell and a neuron. Note that the gland cell is polarized, with a basal (receptive) and apical (effector) region. The neuron is also polarized, from dendrite to axon, but receptive and effector sites may be distributed widely over the surface. Most of the cellular elements are common to the two types. lgc, ligand-gated channels; nm, neuromodulator; sm, second messenger; vgc, voltage-gated channels.

tions, which make a very unstable target for a probing microelectrode.

Membrane Properties

In order to circumvent the difficulties of working on smooth muscle, investigators have developed preparations of isolated muscle cells for physiological analysis. A single smooth muscle cell from the stomach wall is shown in Fig. 18.6A. In the middle panel, a pulse of ACh was ejected from the micropipette, causing contraction of the cell. ACh causes a slow depolarization of the cell, giving rise to a discharge of impulses, as shown in Fig. 18.6B. What is the mechanism of this response?

The response begins with the binding of ACh to muscarinic receptors in the mem-

brane. These are distributed widely over the cell membrane; there is not a well-defined endplate, as in the case of skeletal muscle. The slow depolarization is characteristic of the slow excitatory responses to ACh mediated by muscarinic receptors. For many years it has been believed that this slow depolarization is due to a simple increase in membrane conductance for cations (Na, Ca, K) in a relatively nonspecific manner. Recent experiments with patch recordings have indeed confirmed that this mechanism is present; ACh evokes a depolarization of the membrane that is associated with an overall increase in input conductance; under voltage clamp, this response to ACh is associated with an inward current, due to Na and Ca. However, ACh

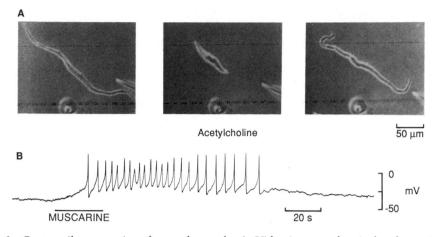

Fig. 18.6 Contractile properties of smooth muscle. **A.** Video images of an isolated gastric smooth muscle cell: *(left)* in normal Ringer's; *(middle)* 15 seconds after pressure ejection of ACh from pipette tip at lower right; note vigorous contraction; *(right)* recovery after 4 minutes. **B.** Electrical response of another cell to the cholinergic agonist muscarine. The trace shows intracellular recording of slow burst of Ca^{2+} action potentials which generate the muscle contraction (cf. Fig. 18.8). (From Sims et al., 1986)

has other actions through muscarinic receptors as well. One is to suppress spontaneous Ca^{2+}-activated K^+ channels; another is to close K^+ channels, thus turning off an M current (cf. Chap. 5). The M channels can also be modulated by peptides such as substance P and LHRH. The slow depolarization activates voltage-dependent Ca^{2+} channels, which generate the impulse discharge. This influx of Ca^{2+} then contributes to excitation–contraction coupling.

Excitation–Contraction Coupling

The mechanism of excitation–contraction coupling differs in certain respects from that in skeletal muscle. Since there is no endplate, there is no endplate potential (see Fig. 18.7). The muscarinic receptors are coupled to second messenger systems in the membrane ③. One of the main systems is the inositol phospholipids ④ discussed previously in Chap. 8. Activation of the second messenger system has two effects. One is to change the conductance of channel proteins in the membrane, leading to depolarization of the membrane and entry of Ca^{2+} ⑤, as discussed above. The second is to produce an increase in internal free Ca^{2+} through the second messenger ④.

The rise in free Ca^{2+} from both these sources activates calcium/calmodulin (Ca/CaM), which is similar to troponin C in skeletal muscle ⑦. Ca/CaM activates the enzyme myosin light-chain kinase, which phosphorylates the regulatory subunit of the myosin head, so that it binds actin ⑧, and brings about contraction.

Slow Contractile Properties

Like skeletal muscle, smooth muscle contains actin and, in smaller amounts, myosin, but they are not organized into repeating sarcomere units (see Fig. 18.7). The contractile mechanism itself is believed to involve sliding of the actin past the myosin, essentially as in the sliding-filament model. Other types of filaments are also present in smooth muscle, which may provide for additional force generation, as well as attachment to the cell wall.

By virtue of these adaptations, smooth muscle shows special contractile properties. The contractions, compared with those of skeletal muscle, are slow and sustained, lasting for seconds or minutes (or even hours or days) instead of 10–100 msec as in the case of vertebrate skeletal muscle (cf. Table 17.1 in the previous chapter). These

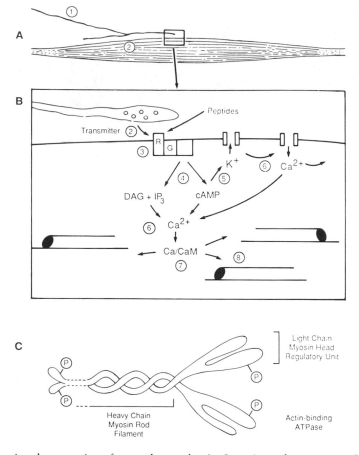

Fig. 18.7 Functional properties of smooth muscle. **A.** Overview of neuromuscular unit. Steps in excitation–contraction coupling begin with nerve impulse ①, which activates diffuse release of transmitter ②. **B.** Further steps in muscle include activation of a receptor ③, which stimulates or inhibits a second messenger system (inositol-lipid): ④. This may cause closing of M-type K^+ channels ⑤, which leads to membrane depolarization and generation of a Ca^{2+} action potential ⑥ (A. Marty, personal communication). Alternatively, the second messengers may cause an increase in free Ca^{2+} in the cytosol, activating Ca/CaM ⑦ contractile apparatus ⑧. **C.** Molecular conformation of myosin, consisting of heavy and light chains, and actin-binding sites.

slow contractile properties are important for a wide range of functions, such as maintaining the tonus of blood vessel walls. The rate of sliding of the actin and myosin filaments is 100–1000 times slower than that of skeletal muscle, although the force of contraction reached is nearly as great. The rate of energy consumption due to ATP hydrolysis is 5–10 times lower to reach the same force, so the slow contractions can be maintained for long periods of time with minimal energy demands or

fatigue. Thus, smooth and skeletal muscle can be seen to be each beautifully adapted for distinctly different motor tasks.

In summary, smooth muscles carry out slow motor functions that may be precise or powerful, and neuronal circuits that control them must be organized to control these functions. At the molecular level the interactions of actin and myosin in smooth muscle cells serve as a valuable model for understanding the contractile properties of neurons, especially with regard to the mo-

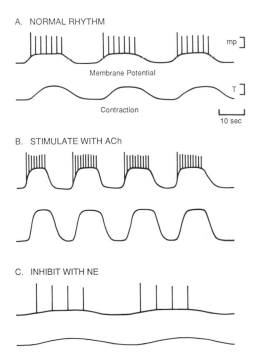

A. NORMAL RHYTHM

Membrane Potential

mp]

Contraction

T]

10 sec

B. STIMULATE WITH ACh

C. INHIBIT WITH NE

Fig. 18.8 Rhythms of contraction of smooth muscle in the gut. **A.** Normal rhythm. Membrane potential (mp) scale is 20 mV; tension (T) scale is in arbitrary units. **B.** Acetylcholine (ACh) stimulates stronger and more rapid contractions. **C.** Norepinephrine (NE) makes contractions weaker and slower. (After Golenhofen, in Schmidt and Thewes, 1983)

tility of axonal growth cones (Chap. 9), cell migration during early development (Chap. 9), axonal regeneration and sprouting in response to injury (Chap. 9), and synaptic plasticity underlying these processes and memory (Chap. 29). We may conclude that, just as the neuron can be viewed as a modified secretory cell (Fig. 18.5), so can it also, with equal validity, be viewed as a modified smooth muscle cell.

Rhythmic Activity

The smooth muscle of the gut undergoes *myogenic* spontaneous contractions. These contractions are slow, lasting several seconds, and are associated with slow depolarization of the muscle membrane giving rise to impulse discharges (see Fig. 18.8). Slow contractions induced by ACh are associ-

ated with modulation of an M current, influx of Ca^{2+}, and release of internal Ca^{2+} through second messenger systems, as discussed above. The slow contractions that occur spontaneously require a pacemaker mechanism, involving a relatively slowly inactivating channel which depolarizes the membrane and sets up a discharge of impulses. The increase in Ca^{2+} influx leads to activation of the Ca^{2+}-activated K^+ channels which repolarize the membrane and terminate the contractile wave. The steady pacemaker depolarization then initiates a new cycle.

This example demonstrates how rhythmic cell activity can arise from the interplay of membrane conductances and second messengers. Variations on this basic mechanism underlie the generation of rhythmic activity throughout the nervous system. In this case, ACh speeds up the rhythms and makes them stronger, through increasing the membrane depolarization; norepinephrine, by contrast, opposes this action, making the rhythm slower and weaker.

The neural systems which modulate smooth muscle rhythmic contractions in the gut are contained in two ganglia within the gut wall, named Meissner's and Auerbach's plexuses. From a combination of morphological, neurochemical, and electrophysiological studies, a picture of the organization of Auerbach's plexus has emerged, which is summarized in Fig. 18.9. As can be seen, the circuit has its own source of *neurogenic* rhythm, in the "burst-type oscillators" that fire periodic bursts of impulses. These drive bursting follower cells that have inhibitory noradrenergic synapses on the muscles. Tonically discharging interneurons, activated by stretch of the muscles, have inhibitory synapses on the follower cells; through these connections, they can release the circular muscles of the gut from inhibition. By sequential activation of mechanoreceptors and interneurons, a peristaltic contractile wave moves along the gut wall, as from left to right in the figure.

The slow and smoothly graded nature

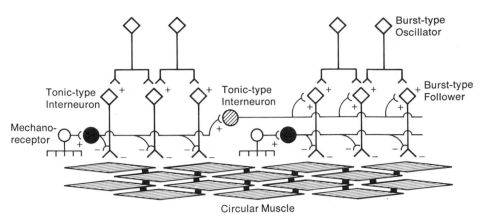

Fig. 18.9 Neuronal circuit that mediates inhibitory control of intestinal circular muscles. Nerve cells have excitatory ($+$) or inhibitory ($-$) connections. Muscle cells are interconnected by electrical synapses. Neuronal interactions, as described in text, cause peristaltic wave of contraction of the circular muscles to move from left to right. (From Wood, 1975)

of the contractions arises from the diffuse spread of neurotransmitter to the muscle, the modulatory action of neuropeptides present in the gut wall, and the electrical synapses that couple the muscles together in a functional syncytium. In 1975, J. D. Wood of the University of Kansas characterized this neural circuit as "a simple integrative system analogous to ganglia . . . of invertebrate animals" that functions to coordinate and program the various patterns of mobility of the gut. He has suggested that it shares common integrative properties with many centers in the central nervous system. It can be seen that such systems involve a hierarchy of neural control, combined with intrinsic rhythmic properties of the muscles; such systems are under multiple controls by different transmitters and modulators.

A Plethora of Peptides

The gut has become a rich ground for identifying and localizing different neurotransmitter and neuromodulatory substances. Figure 18.10 summarizes the major types of neurons that are present in the guinea pig small intestine and innervate the circular muscle. Two types are excitatory and secrete acetylcholine, substance P, and related tachykinins; of these, one type has

long forward (oral) projections and co-releases dynorphin and enkephalin; the other has short local connections and co-releases only enkephalin. By contrast, two other types are inhibitory and secrete vasoactive intestinal peptide and nitric oxide as their primary transmitters; of these, one type has long backward (anal) projections and co-releases dynorphin and gastrin-releasing peptide (also called bombesin), whereas the other has short local projections and co-releases a combination of dynorphin, galanin, and neuropeptide Y. Coordinated co-release of these substances provides for the orderly progression of peristaltic waves as indicated in Figs. 18.8 and 18.9.

Table 18.2 summarizes the locations and roles of these substances, together with a number of other bioactive peptides that have been identified in the neurons of the intestine, but for which specific roles are not yet assigned. It is this kind of evidence that has led neuroscientists to study the gut for discovering new peptides and analyzing their possible functions. The plethora of bioactive molecules has encouraged the view that "the definition of a transmitter should be broad enough to include all those substances that are released when a neuron is activated and that influence other tissues,

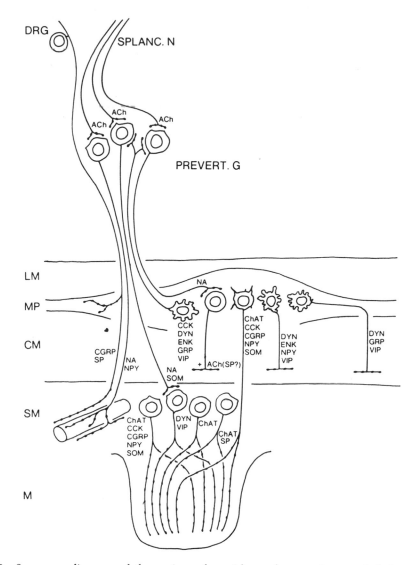

Fig. 18.10 Summary diagram of the variety of peptides and transmitters and their synthesizing enzymes in the intrinsic neurons and the extrinsic nerves of the gastrointestinal tract. The diagram depicts a lateral view of the gut, with the head to the left. ACh, acetylcholine; CCK, cholecystokinin; CGRP, calcitonin gene-related peptide; ChAT, chloramphenicol acetyltransferase; CM, circular muscle; DRG, dorsal root ganglion; DYN, dynorphin; ENK, enkephalin; GRP, gastrin-releasing peptide; LM, longitudinal muscle; M, mucosa; MP, myenteric plexus; NA, noradrenaline; NPY, neuropeptide tyrosine; PREVERT. G, prevertebral ganglion; SM, submucosa; SOM, somatostatin; SP, substance P; SPLANC. N, splanchnic nerve; VIP, vasoactive intestinal peptide. (From Costa et al., in Hökfelt, 1991)

or that modify the effectiveness of the transmission process in the short or long term" (Furness et al., 1992). Studies of gut neurons have provided some of the clearest evidence for co-release of transmitters and peptides and of short-term or long-term effects of the same transmitter on different specific target cells. Most of these substances are also found in the brain. The genes for these molecules are ancient, indi-

Table 18.2 Summary of neurotransmitters and neuropeptides that have been identified in the intestine

Substance	Location and role
Acetylcholine (ACh)	Primary excitatory transmitter to muscle, to intestinal epithelium, to parietal cells, to some gut endocrine cells, and at some neuroneuronal synapses
Adenosine triphosphate (ATP)	Probably contributes to transmission from enteric inhibitory muscle motoneurons
γ-Aminobutyric acid (GABA)	Present in different populations of neurons, depending on species and region Does not appear to be a primary neurotransmitter
Calcitonin gene-related peptide (CGRP)	Present in some secretomotoneurons and interneurons Role unknown
Cholecystokinin (CCK)	Present in some secretomotoneurons and in some interneurons May contribute to excitatory transmission Generally excites muscle
Dynorphin (DYN) and dynorphin-related peptides	Present in secretomotoneurons, interneurons, and motoneurons to muscle Does not appear to be a primary transmitter
Enkephalin (ENK) and enkephalin-related peptides	Present in interneurons and muscle motoneurons In most regions these substances probably provide feedback inhibition of transmitter release
Galanin	Present in secretomotoneurons, descending interneurons, and inhibitory motoneurons in human intestine Role unknown
Gastrin-releasing peptide (GRP) (mammalian bombesin)	Excitatory transmitter to gastrin cells Also found in nerve fibers to muscle and in interneurons, where its roles are not known
Neuropeptide Y	Present in secretomotoneurons, where it appears to inhibit secretion of water and electrolytes Also present in interneurons and inhibitory muscle motoneurons
Nitric oxide (NO)	A co-transmitter from enteric inhibitory muscle motoneurons Possible transmitter at neuroneuronal synapses
Noradrenaline	Noradrenergic nerve fibers in the intestine are not strictly enteric: they are of sympathetic origin Major roles are to inhibit motility in nonsphincter regions, to contract the muscle of the sphincters, to inhibit secretomotor reflexes, and to act as vasoconstrictor neurons to enteric arterioles
Serotonin (5-HT)	Appears to participate in excitatory neuroneuronal transmission
Somatostatin	Despite its widespread distribution in enteric neurons, no clearly defined roles have been established
Tachykinins (substance P, neurokinin A, neuropeptide K, and neuropeptide γ)	Excitatory transmitters to muscle; co-transmitters with ACh May contribute to excitatory neuroneuronal transmission
Vasoactive intestinal peptide (VIP) (and peptide histidine isoleucine [PHI])	Excitatory transmitter from secretomotoneurons Possibly a transmitter of enteric vasodilator neurons Contributes to transmission from enteric inhibitory muscle motoneurons

Note: Substances for which definite synaptic functions are known are underlined.

Adapted from Furness et al. (1992)

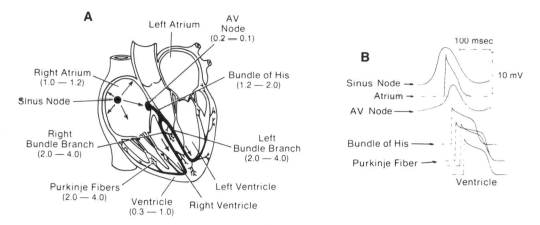

Fig. 18.11 **A.** The conducting system of the human heart. The electrical impulses begin in the sinus node, and are conducted to the AV node, through the bundle of His, to the branches and finally the Purkinje fibers, which connect to the cardiac muscle fibers. Conduction velocities (meters per second) are indicated by numbers in parenthesis. **B.** Intracellular recordings of action potentials at different sites. (From Shepherd and Vanhoutte, 1979)

cating that they have been critical elements in the evolution of nervous systems and mechanisms of nervous control. Thus, the ways that different combinations of neurotransmitters and neuropeptides achieve their subtle effects on the smooth muscle cells of the gut may provide many useful models for similar or analogous mechanisms in the brain.

Cardiac Muscle

The vertebrate heart is *myogenic,* that is, it continues to contract in the absence of innervation. Although the heart thus can beat in isolation, it nonetheless has extensive innervation, from both parasympathetic and sympathetic nerves. This is part of the nervous control of the cardiovascular system that is essential for providing the flexibility that is necessary to adapt the organism's motor performance to ongoing needs. This combined neurogenic and myogenic control of the heart is a common theme in both vertebrates and invertebrates (cf. Calabrese and Arbas, 1985).

Sequence of Activation

Rhythmic activity arises in the heart in the sinus node (see Fig. 18.11A,B). Since

activity here leads the activity in other parts of the heart, it is called the "pacemaker." The impulses spread through the cardiac muscles of the atrium to a second pacemaker site, in the atrioventricular (AV) node. From here arise *Purkinje fibers,* which course together in a bundle and then distribute themselves throughout the ventricular walls. The Purkinje fibers are large, contain few myofibrils, and are closely coupled electrically. By these specializations, they conduct impulses at 2–4 m/sec, some six times faster than normal cardiac fibers.

Action Potential Mechanisms

The action potential takes on characteristic forms in the different types of cardiac fiber (see Fig. 18.11B), and these forms reflect the differing ionic conductances involved. Analysis of these conductances has benefited greatly from the Hodgkin-Huxley model of the action potential in nerve. Beginning in 1960, Dennis Noble in England adapted the Hodgkin-Huxley equations to the specific properties of the different cardiac impulses, so that we now have quantitatively precise models incorporating information about different ionic conductances.

Let us begin with the sinoatrial (SA)

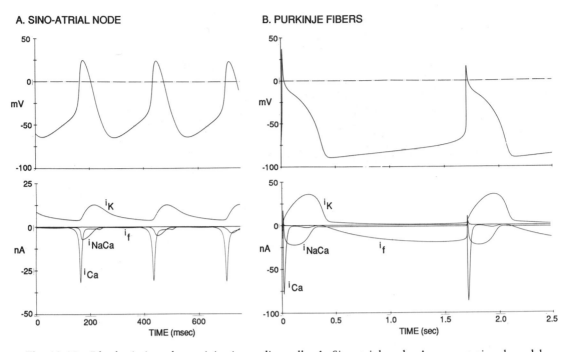

Fig. 18.12 Rhythmic impulse activity in cardiac cells. **A.** Sinoatrial node. A computational model has generated the rhythmic action potentials *(above)* and the underlying changes in membrane currents which cause them *(below)*. i_{Ca}, calcium current; i_K, delayed potassium; i_{NaCa}, Na^+–Ca^{2+} exchange current; i_f, hyperpolarizing-activated current. **B.** Purkinje fibers. Top trace shows rhythmic action potentials generated by the computational model. Bottom trace shows changes in membrane currents. Abbreviations as in A. Not shown are additional currents that are also present, including the anomalous (inward) rectifying K current. (From Noble, 1985)

node, because it serves as the pacemaker for the others. The upper trace in Fig. 18.12A shows the voltage recordings of the rhythmical firing generated by the model for the SA node. The currents in the lower trace allow us to understand the time course of each impulse, and the factors that control the frequency of firing. The rapid upstroke of the impulse is determined by a rapid inward Ca current (I_{Ca}). The slow decline from the impulse peak is the outcome of the interplay between persisting I_{NaCa} and turning on of I_K. I_K brings the membrane to a hyperpolarized level, which turns on a hyperpolarization-activated nonspecific cation current (I_f), carried by both Na and K ions. This inward current, combined with the waning I_K, underlies the slow depolarization which leads to the next impulse firing when threshold for the regenerative I_{Ca} is reached.

The mechanism for rhythmic firing of Purkinje fibers is illustrated in Fig. 18.12B. There are several interesting differences. The impulse has a more rapid upstroke to a sharper peak, due to a very fast inward calcium current (I_{Ca}). This is followed by a maintained depolarization, called a plateau, which is due to the balance between a slow inward current (I_{NaCa}), the onset of the delayed rectifier-like K current (I_K), and the turning off of a K current by the depolarizing impulse (I_{IR}, not shown). The impulse is terminated by a hyperpolarization which, as in the SA node, activates a hyperpolarization-activated nonspecific cation current (I_f). The depolarizing effect of this current is opposed by I_{IR}, however, resulting in a slower rhythm than in the SA node.

The heart therefore provides two carefully worked out examples of how the interplay of voltage-gated channels brings about generation of rhythmical activity. Some of the channels, such as I_{Ca} and I_K, have their counterparts in neuronal membranes, and Chap. 5 should be consulted for further comparisons. Some of the mechanisms appear different from nerve: for example, Ca^{2+}-activated K^+ channels, which appear so prominently in neurons, appear to have a lesser role in cardiac rhythms.

Control of Calcium Channels

It is evident from this analysis that Ca conductances play a key role in the generation of cardiac action potentials. This was recognized many years ago, before the role of Ca in neurons was suspected. Thus, heart cells have served as models for Ca mechanisms; it was here that neurochemical modulation of a voltage-gated channel and the role of cAMP in that modulation were first recognized. As Richard Tsien (1987) then of Yale, observed, "Heart cells have provided a Rosetta stone for neuromodulation."

The importance of Ca is reflected by the fact that many neurotransmitters and neurohormones converge in its control. This is summarized in Fig. 18.13, where it can be seen that the common convergence point is adenylate cyclase. The sympathetic neurotransmitters (E and NE) stimulate adenylate cyclase through a G_s protein, whereas ACh mediates parasympathetic suppression through a G_i protein. The effects of cAMP are mediated by phosphory-

lation of the Ca channel or a closely associated protein. Direct action on a cyclic nucleotide gated channel (cf. Chap. 8) has also been described.

By regulating Ca current, an exquisite control over many dynamic properties of heart muscle is achieved. As was apparent in Fig. 18.12, initiation of action potentials depends on Ca, and this in turn determines the rate of the heartbeat. Thus, the enhanced Ca current due to sympathetic stimulation increases the strength of ventricular muscle contraction (reviewed in Tsien, 1987).

Innervation and Synaptic Control

The nervous pathways that provide for modulation of the heart are shown in Fig. 18.14. The sympathetic innervation comes from postganglionic fibers of the sympathetic chain. When stimulated, these fibers release NE from their terminals; the NE acts on β_1-adrenergic receptors on the cardiac cells. As discussed above, this results in increases in heart rate, impulse conduction, and contractility.

The parasympathetic innervation of the heart comes from *pre*ganglionic fibers arising in the motor nucleus of the vagus, situated in the brainstem. The vagus is the source of most parasympathetic fibers to the viscera. As shown also in Fig. 18.2, the cardiac fibers, being preganglionic, do not innervate the heart cells directly, but instead terminate on ganglia located in the heart. The ganglia contain interneurons, and there are interactions between in-

Fig. 18.13 Convergence of excitatory (+) and inhibitory (−) neurotransmitters and neurohormones on Ca channels in cardiac muscle. ACh, acetylcholine; E, epinephrine; G_s and G_i, stimulatory and inhibitory G-binding proteins; I_{Ca}, calcium current; NE, norepinephrine. (From Tsien, 1987)

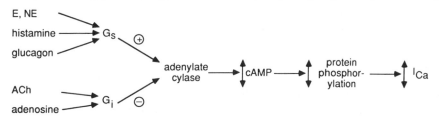

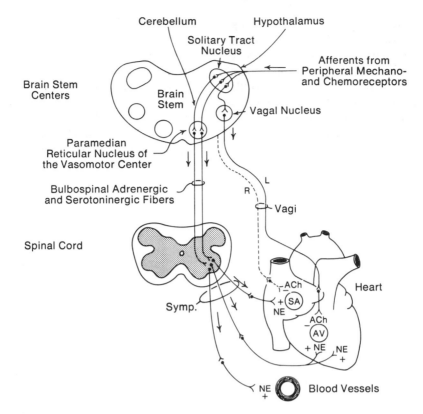

Fig. 18.14 The autonomic innervation of the heart. The right (R) and left (L) vagi mediate parasympathetic control; sympathetic control is mediated through the sympathetic (Symp.) nerves. ACh, acetylcholine; AV, atrioventricular node; NE, norepinephrine (noradrenaline); SA, sinoatrial node. (Adapted from Shepherd and Vanhoutte, 1979)

terneurons and ganglion cells. The ganglia thus appear to be complex integrative centers, like sympathetic ganglia. The studies of Kuffler and his colleagues showed that the vagal fibers have *excitatory* cholinergic synapses on the cells of the ganglion which, in turn, have *inhibitory* cholinergic synapses on the cells of the heart. Acetylcholine inhibits the heart by decreasing Ca^{2+} permeability, thereby slowing the action potential and slowing the heart rate.

These details of synaptic organization are of interest for two reasons. First, they illustrate rather nicely how a transmitter (ACh) can have an excitatory action at one synapse and an inhibitory action at another. Thus, the common statement that "the vagus inhibits the heart" actually means that the vagus *excites* the cardiac

ganglion, which then inhibits the heart. Furthermore, under natural conditions the ganglion is likely to mediate more subtle modulations of cardiac activity, through its local circuit interactions, than the simple inhibition revealed by strong electrical stimulation of the vagus. Thus, knowledge of synaptic organization gives us a much richer understanding of the peripheral neural mechanisms involved in controlling visceral functions.

Brainstem Centers

The sympathetic motoneurons of the spinal cord are under control of descending catecholaminergic and serotonergic fibers originating in the vasomotor center of the brainstem. The "vasomotor center" is, in fact, a collection of different nuclei, including the motor nucleus of the vagus and the nucleus of the solitary tract (see Fig. 18.14), which receives sensory inputs from mechanoreceptors in the walls of blood vessels and chemoreceptors such as the carotid body. This level of organization, in the brainstem, is analogous to the central ganglia controlling cardiac ganglia in invertebrates. The brainstem nuclei, in turn, are affected by higher nervous centers, especially the cerebellum, hypothalamus, and the limbic system.

19

Reflexes and Fixed Motor Responses

One of the primary objectives in the study of motor systems is to identify the *elementary units of motor behavior*. There are two main concepts that have dominated thinking about this problem. One is the idea that the simplest unit of behavior is the *reflex,* and that complex behavior is built up by a chaining together of reflexes. The other idea is that much of behavior (particularly in invertebrates and lower vertebrates) involves stereotyped sequences of actions that are either generated within the organism or triggered by appropriate environmental stimuli. These are usually referred to as *fixed-action patterns*. Figure 19.1 and Table 19.1 summarize the main features of these types of behavioral units.

This chapter first gives a brief historical background of these two concepts and then presents examples in the invertebrates and the vertebrates. The relevance of the fixed action pattern concept for more complex types of behavior will be discussed in later chapters.

Reflexes: A Brief History

The fact that there are immediate motor responses to sensory stimulation was apparent through the ages, and implicit in the writings of many scholars. However, the term "reflex" actually did not appear in the language until the eighteenth century. It comes as a surprise, for example, to realize that Shakespeare could dramatize so much of the human condition without using the word. Georg Prochaska of Vienna was one of the first to use the term in 1784 when he wrote:

The reflexion of sensorial into motor impressions . . . takes place in the sensorium commune (common sensory center). . . . This reflexion may take place either with consciousness or without. . . .

Our present use of the term dates from the earliest experimental investigations of the role of the spinal cord in mediating muscle responses to sensory stimuli. Among the important studies were those of Charles Bell in England and François Magendie in France, who first established in the 1820s that sensory fibers are contained in the dorsal roots, and motor fibers in the ventral roots, of the spinal cord. Bell stated it clearly: "Between the brain and the muscle there is a circle of nerves; one nerve conveys the influence of the brain to the muscle, another gives the sense of the condition of the muscle to the brain." The

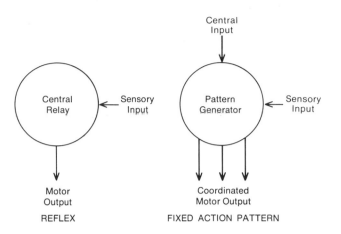

Fig. 19.1 Comparison between a reflex and a fixed-action pattern. *On the left,* there is a simple motor response (for example, a knee jerk). *On the right,* there is a coordinated sequence of motor acts, such as the tail flip of a startled fish. We will discuss both of these unitary motor actions in this chapter.

action may be all-or-nothing or graded in intensity, and it may be contingent on the type of sensory stimulus or internal state, but it maintains its basic pattern.

Marshall Hall (1790–1857) vigorously advocated the notion that spinal reflex movements are distinct from voluntary movement, dependent on the spinal cord but unconscious and independent of the rest of the brain. Hall and many other workers in the course of the nineteenth century identified and categorized a number of types of reflexes. Much of this work, however, was clouded by wrangling about whether reflexes were conscious or not, and the prevailing reticular theory of the

neuron misled many people in thinking about the nervous pathways involved.

It was at this stage, around 1890, that Charles Sherrington came onto the scene. We have already seen (Chap. 6) how his work led to the concept of the synapse. With regard to reflexes, his approach was built on two essential foundations: he carried out a careful anatomical analysis of the nerves to different muscles, and he then used this knowledge to analyze quantitatively the reflex properties of specific nerves and muscle groups. This painstaking work was the means for obtaining the first clear view of the reflex as a combined structural and functional entity, and it established the

Table 19.1 Main features of reflexes and fixed-action patterns

Reflex	Fixed-action pattern
1. A simple motor action, stereotyped and repeatable.	1. A complex motor act, involving a specific temporal sequence of component acts.
2. Elicited by a sensory stimulus, the strength of the motor action being graded with the intensity of the stimulus.	2. Generated internally, or elicited by a sensory stimulus. The stimulus acts as a trigger, causing release of the coordinated motor act. The action may be all-or-nothing or graded in intensity, and it may be contingent on the type of sensory stimulus or internal state, but it maintains its basic pattern.

reflex arc as a subject for further anatomical and physiological analysis by many twentieth-century workers. In addition, Sherrington emphasized the importance of the reflex as an elementary *unit of behavior,* and thus laid one of the cornerstones for the modern studies of animal behavior (see Sherrington, 1906).

Fixed-Action Patterns: A Brief History

There are, in general, two ways to study the behavior of an organism. One is to bring the animal into the laboratory and devise various kinds of instruments and procedures to test its abilities. This type of approach began around 1900, and gave rise to the fields of *behaviorism* and *animal psychology.* We shall discuss these fields and their methods further in Section IV.

The other approach is for the investigator to go out in the field and observe the animal in its daily life. This, of course, is as old as humankind itself, but it became a science only in the late nineteenth century. Charles Darwin's other great book, *On the Expression of the Emotions in Man and Animals* (1872), in which he attempted to demonstrate similarities in instinctual behavior between animals and humans, is often regarded as the starting point for the systematic study of naturally observed behavior. This led to the modern field of *ethology.*

How can observations of natural behavior provide evidence for basic units of behavior? A number of workers in the early part of this century contributed to this study by their careful observations of animals in the field. This culminated in 1950 with the suggestion of Konrad Lorenz that much of the repertoire of individual motor actions and motor responses of animals can be described as *fixed-action patterns.* These are acts that are instinctual, stereotyped, and characteristic of a given species. Table 19.1 summarizes some of their attributes. Since they are coordinated and purposeful,

they usually involve higher levels in the hierarchy of motor organization, which we discussed in Chap. 17 (see Fig. 17.9). They are in turn only part of a larger behavioral unit that includes at least three components: first, the "drive" to search for a certain stimulus context; second, the selective response to it—its "innate recognition"; and third, the discharge of an equally innate motor activity coping with the situation (von Holst, in Lorenz, 1981). The innate motor activity is thus fixed only in a relative sense. It can be selectively activated by different stimuli under different conditions of internal state or motivational drive. It can vary in intensity, from the slightest intentional gesture to a full-blown all-or-nothing action, such as a startle response. In fact, one of the main themes to arise from recent work is the conditional nature of most motor acts. For this reason, many workers prefer other terms, such as "response," "complex reflex," "motor pattern," or "behavior." In general, these all have in common that they describe a complete and purposeful pattern of motor response to some initiating input.

Apart from their importance for behaviorists and ethologists, the reflex and the fixed-action pattern have been useful tools in guiding the experiments of neurobiologists on the cellular mechanisms of motor systems. This has been particularly true in invertebrates, where it has been possible to isolate and study relatively simple nervous components that correspond rather closely to the circuits for specific reflexes and even simple fixed-action patterns. To the extent that reflexes and fixed-action patterns are innate, they have also been attractive for genetic experiments; for example, mutations can produce selective loss of parts of a circuit, and these parts can be correlated with the behavioral deficits produced.

In the remainder of this chapter, we will discuss several of the best known examples of reflexes and fixed-action patterns. First, we consider startle responses in invertebrates. We then consider startle responses in the vertebrate, and we finish with an overview of the reflex organization of the vertebrate spinal cord.

SIMPLE SYSTEMS IN INVERTEBRATES

The Leech: Skin Reflexes

The neurons involved in receiving stimuli in the skin of the leech and transmitting the information to the central nervous ganglia were described in Chap. 12. It will be recalled that touch, pressure, and noxious modalities have identifiable cell bodies and given locations (see Fig. 19.2). One main type of motoneuron is the L type, which innervates the longitudinal muscles. When these contract, they shorten the segment. The other main type is the AE motoneuron, which innervates the annulus erector mus-

cles. When these contract, they pucker the segment into a sharp ridge.

When touch, pressure, or noxious stimuli are applied to the skin of the intact animal, contractions are elicited in the segmental muscles. One therefore has a simple reflex pathway, and we can deduce right away that it has the three components of a classical reflex arc: a sensory inflow pathway, a central relay site, and a motor outflow pathway. The key questions for the neurobiologist are: what kinds of connections are made between the sensory and motor cells, and what are their functional properties?

Work from the laboratory of John Nicholls, of Stanford and Basel, has provided answers to these questions. Let us consider the case of the L cell. Intracellular electrodes were introduced into a sensory cell body and a motoneuron cell body in the same ganglion. When the sensory cell was

Fig. 19.2 Skin reflexes in the leech. **A.** Segmental ganglion, showing positions of touch (T), pressure (P), and noxious (N) sensory neuron cell bodies, and two motoneurons, the longitudinal (L) and annulus erector (AE). **B.** Stimulating the L motoneuron shortens the segment. **C.** Intracellular recordings from different combinations of sensory and motor neurons, to show chemical and electrical transmission, and summation of responses. **D.** Circuit diagram summarizing the pathways for the skin reflexes. Chemical synapses are indicated by small circles, electrical synapses by zigzag lines. (Based on studies of Nicholls and collaborators, in Kuffler et al., 1984)

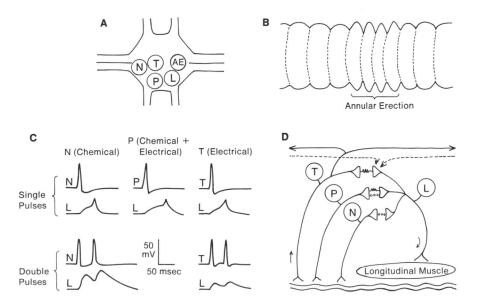

directly stimulated by an electrical pulse through the intracellular electrode, an impulse was elicited in that cell, and also a response in the L cell. The L cell response consisted of a synaptic depolarization leading up to a small spike (see Fig. 19.2). When the latencies and other properties of the responses were analyzed for the responses elicited by the three types of sensory cell, it was found that the T cell is coupled to the L cell by an electrical synapse, the N cell by a chemical synapse, and the P cell by a combination of electrical and chemical synapses. It was further found that the chemical synapses are readily modifiable, giving a much facilitated response to a second pulse, whereas the electrical synaptic response was relatiely invariant (see Fig. 19.2).

In addition to these short-term effects, there are long-term changes in the reflex pathways brought about by sustained or repeated natural stimulation of the skin. This is seen as a hyperpolarization of the sensory cells that may last for seconds or minutes after stimulation has ceased. Nicholls and his colleagues have shown that the hyperpolarization may be due either to an electrogenic sodium pump, activated by the influx of Na^+ during the impulse discharge, or to a prolonged Ca^{2+}-dependent increase in K^+ permeability. The pump mechanism predominates in T cells, the K^+ permeability mechanism in N cells, and both are present in P cells. The hyperpolarizations induced in these cells raise the thresholds for impulse generation and affect synaptic integration, but the significance of these effects for the reflex behavior of the organism is not yet understood.

The diagram in Figure 19.2 summarizes the circuits for the reflex arcs from T, P, and N cells through L cells. Since the connections are direct, without intervention of an interneuron, we say that these are *monosynaptic pathways,* and *monosynaptic reflexes.* Note that the cells in these pathways make connections with other central neurons, and in turn receive connections from other neurons; these are the means for coordinating the reflexes with other nervous activity.

Skin reflexes are virtually universal, providing the means whereby organisms withdraw from unfamiliar or harmful stimuli. It is of interest that in the leech the pathway for reflex withdrawal from a noxious stimulus is monosynaptic; the corresponding pathway in vertebrates is polysnaptic (see below).

Crayfish Escape Response

A common type of behavior in many invertebrates is a quick escape movement. This is elicited by a sensory stimulus, and it consists of a sudden synchronous contraction of special fast muscles that moves the animal away from the site of danger. Animals as diverse as the earthworm, crayfish, and squid all show this type of behavior (it is also common in vertebrates; see below). In most cases it has been found that a giant axon is a part of the nervous pathway. This is logical, because, as we learned in Chap. 5, the larger a fiber, the faster the conduction rate of the impulses—and the essence of an escape movement is speed.

The escape response is produced by the discharge of only a single impulse in the successive components of the nervous pathway. In the case of the leech reflexes, the motor responses could be seen as small pieces of larger behavioral patterns. In the case of the escape response, the motor action is an entire purposeful behavioral act in itself. There are, in fact, two pairs of giant fibers in the crayfish: medial and lateral. The medial giant fiber connects to all segments, and the resulting muscle contractions propel the animal backwards. The lateral giant fiber lacks connections in the most posterior ganglia; this tends to produce an upward movement of the animal. These connections and their correlated behaviors are shown in Fig. 19.3.

The escape response happens so quickly that there is not time for feedback information to guide or adjust the movement. Thus, there is no role for feedback informa-

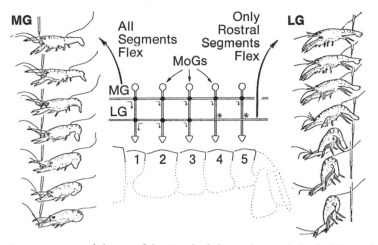

Fig. 19.3 Flexion responses of the crayfish. *On the left,* an electrical shock delivered to the medial giant fiber (MG) elicits postsynaptic responses in motoneurons (MoG) in all segmental ganglia (G_1–G_5). *On the right,* a shock to the lateral giant (LG) fiber elicits responses in motoneurons only in segmental ganglia G_1–G_3. The differences in behavioral responses are shown in the sketches of frames from high-speed cinematography; the tail flip mediated by the medial giants moves the animal backward; that by the lateral giants moves the animal upward. Large dots indicate sites of electrical synapses from giant fibers onto motor giants. These synapses are lacking at sites marked by asterisks. (From Wine and Krasne, 1982)

tion from muscle receptors, for example. It is a movement that, once triggered by peripheral stimuli, is completely under *central control.*

The ability of a single neuron or fiber, such as the giant fiber, to trigger an entire behavioral act implies that it occupies some special position in the hierarchy of motor control. From this has emerged the concept of the *command* neuron or fiber, and the idea that such a neuron or fiber has some kind of executive power to initiate or control a specific coordinated motor act. We shall discuss this concept further in Chap. 21.

Neural Circuit for the Escape Response

The neural circuit mediating the escape response has been worked out in detail in the crayfish, and the components and the sequence of events can be summarized in relation to the diagram of Fig. 19.4. Abrupt mechanical stimulation of the integument excites the hair cells (tactile afferents, TAs) to discharge an impulse, which activates electrical synapses on a giant fiber (lateral

giant command cell, LG) and chemical synapses on an interneuron (transient sensory interneuron, T); the interneuron, in turn, also has electrical synapses on the giant fiber. Although the hair cell thus has a direct connection to the giant fiber, the route through the interneuron is actually much more powerful. The impulse set up in the giant fiber excites segmental motoneurons (motor giants, MoGs) in rapid sequence as it passes through successive segmental ganglia; this occurs by means of electrical synapses. These, in fact, were the first electrical synapses to be identified, and their mechanism was described in Chap. 9. The excited motoneurons then activate their respective muscles to contract.

This response is similar to the reflexes of the leech that were discussed above, in that a sensory stimulation activates a nervous pathway which leads to an immediate motor act. It differs, however, in several important respects. Instead of one synaptic relay, it involves at least three synaptic links; we say, therefore, that it is a *polysynaptic* reflex. It has a relatively high threshold for sensory activation because it is a specialized movement which is appropriate only under specific environmental conditions. When activated in this way, it therefore has the character of an *all-or-nothing response,* in contrast to the reflexes we discussed above in the leech, whose magnitude is *graded* with the intensity of the stimulation.

The simple circuit in Fig. 19.4 represents only a beginning in understanding the principles underlying this escape response. We will discuss briefly several of the mechanisms that need to be added to this circuit. First, the reflex *habituates* with repeated stimulation. If a crayfish is tapped at intervals of 1 minute, the escape response disappears within 10 minutes. Most of this is accounted for by decreased release of neurotransmitter from the sensory afferent terminals onto the thick sensory interneurons (arrow in Fig. 19.4). We will discuss mechanisms of habituation more fully in Chap. 29.

Second, the simple circuit of Fig. 19.4 is a part of much more extensive circuits mediating control of the escape response and related motor behavior. The LG neuron not only excites the motoneurons that cause the tail flip, but also mediates feedback inhibition. This inhibition is massive and widespread, and it affects synaptic sites at virtually every level in the hierarchy of motor control. This so-called *command-derived inhibition* has several functions; for example, it can set the threshold for subsequent elicitation of the response and protect against the onset of habituation.

Third, the fact that the tail flip can be elicited by a single impulse in the lateral giant interneuron (LG in Fig. 19.4) has implied that the LG–MoG pathway is the sole pathway for the tail flip. However, further studies have shown that there is a set of nongiant motoneurons, called fast flexor (FF) motoneurons. These differ in innervating smaller groups within the flexor musculature of each segment. The giant and nongiant motoneurons are the final output sites of two parallel, linked pathways, each of which exerts different kinds of motor control (see Fig. 19.5). The giant system is specialized exclusively for producing rapid tail flips. The nongiant system, according to Krasne and Wine (1984), is specialized for "finesse of control." Synchronous stimulation of premotor neurons within this system can also elicit tail flips, though not so rapid or so vigorous; the main function of these neurons, however, is to provide for finer motor control. Stimulation of the giant system also activates the nongiant system (through a segmental giant [SG] interneuron; see Fig. 19.5), which appears to reinforce the giant-mediated tail flip. There is no connection in the reverse direction, which is reasonable, because it would be inappropriate for the nongiants to recruit the giants in their finer mechanisms of control.

Finally, recent work gives testimony to the pervasive effects of diffuse *arousal systems,* which set and modulate the level of activity in specific neural circuits. Thus, a hungry crayfish is highly aroused; its escape threshold is high, and it will fight for food until it acquires it, at which time the escape threshold abruptly lowers.

The escape response is thus not a simple reflex. It possesses many of the defining properties of a fixed-action pattern. It also exhibits many of the functional properties that are characteristic of more complex motor and central systems. Finally, it illustrates the very important principle that each synapse within a neural circuit it not just a simple relay, but rather is a site at which *multiple controls* are present. These controls enable the circuit to operate in different modes, depending on the history of use of the circuit and the behavioral state of the organism.

Neuromodulation and Behavioral States

Another approach to the analysis of stereotyped motor patterns is to analyze how they may be mediated by the actions of specific types of neurotransmitters or neuromodulators. Ed Kravitz and his col-

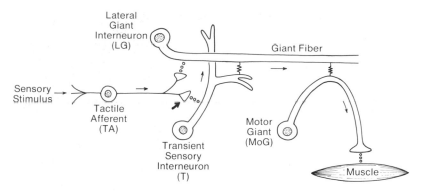

Fig. 19.4 Simplified circuit mediating the crayfish escape response. Chemical synapses shown by dots, electrical synapses by zigzag lines. Short thick arrow indicates site of habituation. (From Wine and Krasne, 1982)

leagues (see Kravitz et al., 1985) at Harvard carried out a series of experiments on the effects of biogenic amines in the lobster which nicely complement the experiments described above in the closely related crayfish. The basic behavioral finding was that after a systemic injection of serotonin, lobsters assume a doubled-up, flexed posture, resembling their posture when they are startled or incited to fight. After injection of octopamine, on the other hand, the animals lie flat, with legs extended—a submis-

sive posture such as is normally seen in a mating female.

If these substances can have these effects when injected, it suggests the hypothesis that they are released endogenously to mediate natural behavior. In order to identify the cells that might do this, serotonin (5HT) was conjugated to a large molecule (albumin); antibodies to it were prepared, and the location was visualized by a fluorescent marker. Some 100 cells throughout the nervous system were seen; each gan-

Fig. 19.5 Summary diagram of circuits controlling the tail flip. Precision tail movements are mediated by the nongiant pattern generator neurons (labeled Non-G Pattern Gen. 12,3, and other) which make excitatory chemical synapses (arrow) and electrical synapses (bar) on fast flexor (FF) motoneurons. The most rapid tail flips, on the other hand, are mediated by the medial and lateral (MG and LG) giant axons, which make electrical synapses on the giant motoneurons (moG). They also can activate the nongiant system, either indirectly, through activation of a segmental giant (SG) interneuron, or possibly directly (dotted line). Midline is indicated by dashed line. Multiple neurons are indicated by ditto marks. See text. (From Krasne and Wine, 1984)

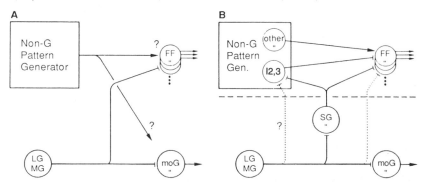

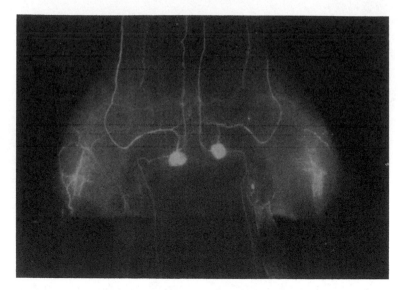

Fig. 19.6 Identification of the nerve cell type that contains serotonin in the lobster. The photomicrograph shows the T5 thoracic ganglion, with immunocytochemical staining for serotonin. The two paired cell bodies are located near the midline; their processes arborize within the ganglion and extend to neighboring ganglia. (From Kravitz et al., 1985)

glion contained at least one of these cells, such as the pair shown in Fig. 19.6. It therefore appears that these 5HT-containing cells contribute to mediation of the flexed posture. This is an excellent illustration of a correlation between a specific molecule, a specific cell type, and a specific behavior, one of the goals of neurobiology, as we discussed in Chap. 1. The next steps in the analysis are to localize the serotonin more precisely by use of monoclonal antibodies, to analyze transcription and translation of the enzymes that synthesize 5HT, and to characterize the molecular receptors for 5HT.

In further work, it was shown that the contrasting postures are not due to different actions of the two amines on the extensor and flexor muscles: both amines cause long-lasting contractures when applied directly to the muscles. This effect has been interpreted as a priming action of the amines on the muscles. Where, then, do the different actions occur?

The answer to this question emerged from experiments in which intracellular re-

cordings were made of responses of motoneurons to the two amines (Fig. 19.7). It was found that excitatory motoneurons to flexor muscles were excited by 5HT and inhibited by octopamine; by contrast, excitatory motoneurons to extensor muscles were excited by octopamine and inhibited by 5HT. The differential actions of the amines on the motoneurons were thus in accord with the behavioral findings. It was interesting that the action of an amine on the inhibitory motoneuron to a muscle was the opposite of its action on the excitatory motoneuron (thus 5HT *inhibited* inhibitory motoneurons to flexors), so that the effects were synergistic.

The interpretation of these experiments was that the amines have two sites and modes of action. In the periphery, they prime the muscles to contract more vigorously in response to the excitation by glutamate at the neuromuscular junction. In central ganglia, they have specific synaptic actions. The central actions of 5HT synapses result in activation of the central motor program for flexion, whereas octo-

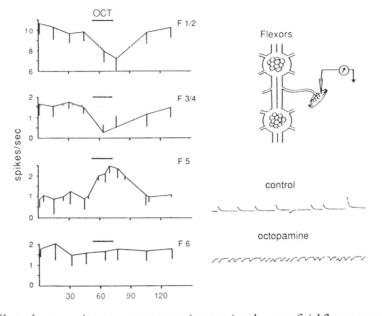

Fig. 19.7 Effect of octopamine on motoneurons innervating the superficial flexor muscles. *At upper right,* experimental setup is shown. *At lower right* are shown sample intracellular recordings of excitatory (upward deflections) and inhibitory (downward deflections) junctional potentials in a muscle fiber. Octopamine caused an increased frequency of inhibitory potentials. *On the left* are shown impulse firing rates of motoneurons following application of octopamine (OCT) (3×10^{-5} M); F5 is an inhibitory motoneuron; all others are excitatory. Octopamine inhibited impulse firing in flexor motoneurons F 1/2 and F 3/4, but enhanced firing in F 5. Vertical bars indicate ± 1 SD. (From Kravitz et al., 1983)

pamine synapses activate the central motor program for extension. Kravitz et al. (1985) concluded that "amines are interacting with the 'command neuron' circuitry in some way to trigger the readout of central motor programs for flexion and extension."

VERTEBRATE ESCAPE RESPONSES

The studies of invertebrates illustrate the advantages of being able to work on systems composed of large, identifiable cells and fibers. In the vertebrates, similar advantages are offered by Mauthner cells.

These cells are present in the medulla oblongata of many species of fish and amphibians. When the fish is exposed to a sudden vibratory stimulus (such as a tap on the aquarium), it responds with a quick flip of the tail, which displaces the animal sideways (see Fig. 19.8). It is thus essentially similar to an escape response; another term is a *startle response.*

The Mauthner cell is the key element in the startle response. Because of its large size, it has been possible to study it carefully. As indicated in Fig. 19.9, there is a single Mauthner cell on each side, situated at the level where the eighth nerve enters, carrying input from the auditory and vestibular nerves. The eighth nerve fibers make direct electrical synapses on the distal parts of the lateral Mauthner cell dendrites by means of large club endings. Some eighth nerve fibers make synapses on vestibular nucleus neurons, which then make excitatory chemical synapses on the lateral dendrite. There is thus both a monosynaptic and disynaptic pathway from the eighth nerve fibers to the Mauthner cell. Note the

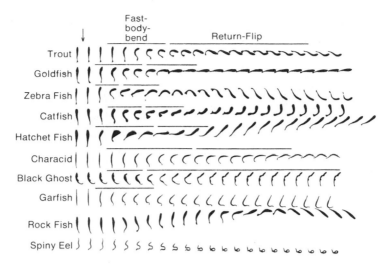

Fig. 19.8 Moving picture analysis of startle responses in different kinds of fish. Frames were taken at 5-msec intervals. Stimulus was a tap of a mallet on the aquarium, delivered at time indicated by arrow. (From Eaton et al., 1977)

close similarity of this arrangement to that in the afferent connections to the giant fiber in the crayfish escape response circuit (see Fig. 19.4).

Another set of connections mediates feedback control of the Mauthner cell. After the Mauthner cell fires its impulse, the impulse not only travels down the axon to excite the tail motoneurons, but also invades axon collaterals to excite interneurons. Through polysynaptic pathways, two kinds of interneurons are ultimately excited and feed back onto the Mauthner cell. One type has chemical inhibitory synapses on the lateral dendrite. The other has axon terminals which wind around the axon hill-

Fig. 19.9 Synaptic organization of the Mauthner cell. Electrical synapses are shown by zigzag connections, chemical synapses by small circles. Compare this circuit for a startle reflex in a vertebrate with that in Fig. 19.4, for an invertebrate. (After Furukawa, in Kuffler et al., 1984)

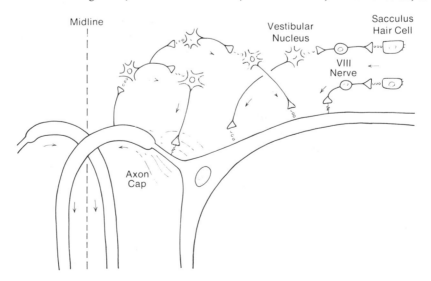

ock and initial segment of the Mauthner cell. This region is encased in a thick wrapping of glial membranes, which form an *axon cap*. The terminals make electrical synapses on the initial segment, and the axon cap increases the effectiveness of their inhibitory action by limiting the spread of extracellular current. (This mechanism was explained more fully in Chap. 7).

At a *cellular* level, the Mauthner cell illustrates the importance of dendrites in integrating different types of synaptic inputs. It also exemplifies the strategic siting of a specific type of synapse to control the axonal output of the cell at the initial segment. At a *behavioral* level, the startle response resembles the crayfish escape response in showing many of the properties of a fixed-action pattern. As in that case, there are, in addition to the Mauthner cell, other, non-Mauthner circuits that can mediate rapid responses which are similar in their pattern. The Mauthner cell discharge determines the initial C-shape bending of the body (phase 1); within 15 msec, a second group of reticulospinal neurons, situated more laterally, fires impulses. Recent studies by Robert Eaton and his colleagues (Nissanov et al., 1990) suggest that this group contributes to the C response as well as the subsequent strong flip of the body (phase 2) which propels the fish from danger. They suggest that the Mauthner cell is only one component, albeit an important one, in the overall circuit for initiating and coordinating the widespread muscle contractions that take place in this deceptively simple response.

VERTEBRATE SPINAL REFLEXES

The best known examples of reflexes, and the ones most characteristic of vertebrates, are those mediated by the spinal cord. Our knowledge about them has progressed in several stages over the past 100 years or so, being dependent on the experimental techniques available.

Sherrington's first studies, as we have mentioned, involved a correlation of anatomical tracing of sensory and muscle nerves with meticulous observations of different reflex behaviors. He introduced methods for cutting across the brainstem of a cat at the level of the midbrain (between the superior and inferior colliculi), which produced a great enhancement of tone in the extensor muscles of the limbs. This was termed *decerebrate rigidity*. The extensor muscles are the ones responsible for maintaining the animal in a standing position. In order to study the reflex basis of this activity, Sherrington and his collaborators in 1924 began to analyze the responses to passive stretch of an extensor muscle (for example, the quadriceps femoris of the thigh, which attaches to the knee cap).

Figure 19.10 illustrates the experimental setup and results. Stretch of the muscle by only a few millimeters gives rise to a large increase in tension, as measured by a strain gauge. If the muscle nerve is cut, the tension developed is small, because it results only from the passive elastic properties inherent in the muscle and its tendon. This shows that the large tension depends on a reflex pathway that passes through the spinal cord. The reflex activity produces contractions of the muscle that was stretched. Because the reflex feeds back specifically to the stretched muscle, it is a *myotatic reflex;* because it is elicited by stretch, it is also called a *stretch reflex*. This is the familiar "knee-jerk" reflex elicited by a tap on the tendon of the knee. Most muscles, invertebrate and vertebrate, show this type of reflex, though extensor muscles that work against gravity show it best. Although the feedback to the muscle stretched is excitatory, there is, in addition, an inhibitory effect on muscles with antagonistic actions at a joint; thus, when a knee flexor is stretched, some of the tension in the knee extensor melts away (see Fig. 19.10). This illustrates the principle of *reciprocal innervation* of the muscles to a joint.

The next step was to analyze the nervous pathways involved in these and other types

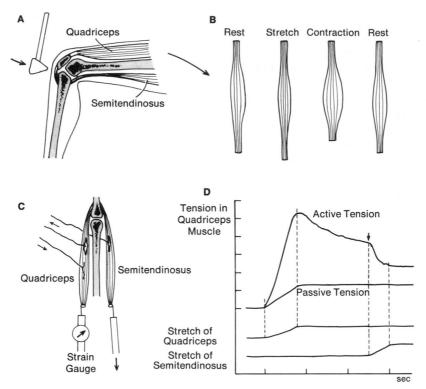

Fig. 19.10 **A.** Testing for the stretch reflex by tapping the patellar tendon of the quadriceps muscle. **B.** Different conditions of a muscle. **C.** Experimental setup for analyzing the stretch reflex in the cat. **D.** Tension of quadriceps muscle in response to stretch, before (active) and after (passive) cutting the motor nerve. Arrow indicates onset of reciprocal inhibition of quadriceps motoneurons produced by stretch of the semitendinosus muscle, an antagonist. (Based on Liddell and Sherrington, in Henneman, 1980)

of reflex activity. David Lloyd at Rockefeller University began these studies around 1940. The experiments required laborious dissection of individual peripheral nerves combined with removal of the laminae of the vertebral bones (laminectomy) to expose the spinal cord so that electrodes could be placed on the dorsal and ventral roots. A single volley could then be set up in a peripheral nerve, and the response of motoneurons could be recorded in terms of the compound action potential of their axons in the ventral root. Representative results are shown in Fig. 19.11. Stimulation of a muscle nerve produces a short-latency, brief volley in the ventral root. This shows that the input from muscles is carried over large, rapidly conducting axons, and that

there is only one, or at most two or three synaptic relays in the spinal cord. By contrast, the ventral root response to a volley in a skin nerve has a long latency, and lasts a long time. This suggests the involvement, in skin reflexes, of slower-conducting fibers, polysnaptic pathways, and prolonged activity in the neurons in these pathways. Note that this polysynaptic pathway for skin reflexes in the vertebrate spinal cord contrasts with the monosynaptic pathways in the leech (Fig. 19.2, above). The polysynaptic pathways may provide for more complex processing of the input information from the skin.

The most definitive analysis of the neural reflex circuits has been with the use of intracellular recordings. We have already

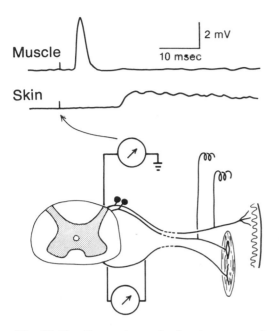

Muscle

2 mV

10 msec

Skin

Fig. 19.11 Comparison of reflex responses of motoneurons to electrical stimulation of a muscle nerve (gastrocnemius) and a skin nerve (sural) in the cat. (After Lloyd, in Henneman, 1980)

discussed the properties of synaptic excitation and inhibition in motoneurons and the mechanisms of integration (see Chap. 7). A key question has been to identify the synaptic pathways within the spinal cord for each of the main types of reflexes elicited by inputs in specific sensory afferent fibers (see Chap. 12). The first evidence from intracellular recordings, obtained by John Eccles and his collaborators in the 1950s, has been largely confirmed and much extended by subsequent studies. As shown in Fig. 19.12, inputs over the different types of muscle nerves are set up by either brief muscle stretch or a shock to a nerve. Motoneuron responses are recorded and synaptic relays estimated from delay times. This analysis assumes a delay of 0.5 msec at each synapse (see Chap. 6) and additional delays of the order of about 1 msec for impulse conduction times in the intraspinal fibers. From such measurements it has been concluded that group Ia affer-

ents make monosynaptic excitatory synapses onto their own motoneurons and disynaptic inhibitory synapses onto antagonist motoneurons. Group II afferents, by comparison, make mostly disynaptic excitatory synapses onto their own motoneurons (see Fig. 19.12). (The terminology of peripheral nerve groups was covered in Chap. 13; see Table 13.2.)

Specific Types of Reflex Pathways

We can briefly summarize the main types of reflex circuits that have been identified by this kind of analysis at the segmental level, using the simplified diagrams in Fig. 19.13. The circuits are categorized in terms of their main type of sensory input. The top three are muscle reflexes, and the bottom one is a skin reflex. The reader may review the main types of sensory fibers in Chaps. 12 and 13. Let us briefly review each circuit here; the role of each in relation to locomotion and larger patterns of motor behavior will be discussed in subsequent chapters.

Stretch Reflex

As we discussed above, the largest diameter sensory nerves, the Ia fibers from muscle spindles, make monosynaptic excitatory synapses on their own motoneurons and disynaptic inhibitory synapses onto antagonist motoneurons. This is the main pathway for the stretch reflex. Figure 19.13A shows that this reflex is present for both extensors and flexors. Classically, the prominence of the stretch reflex in antigravity extensor muscles was presumed to provide the reflex basis for maintenance of upright posture. The way this comes about is as follows. Begin with a standing posture, and let your knees start to bend. This begins to stretch your quadriceps (the large thigh muscle that extends the knee; see Fig. 19.10). This stretches the muscle spindles in the quadriceps, setting up a barrage of impulses in their axons. As can be traced in the circuit of Fig. 19.13A, this will lead to excitation of the extensor motoneurons,

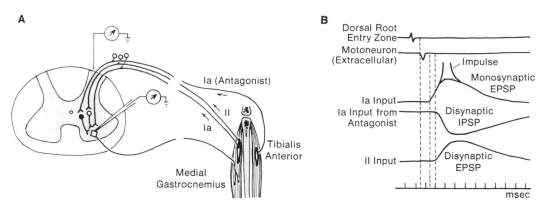

Fig. 19.12 Experimental demonstration of monosynaptic and disynaptic connections onto motoneurons of the cat. See text. (Based on Eccles, 1957, and Watt et al., 1976)

causing the extensor muscle to contract and oppose the gravity-induced stretch, thereby tending to return the body to its upright posture. At the same time, by means of reciprocal inhibition through inhibitory interneurons, the flexor motoneurons will be inhibited, further allowing the extension to take place. In addition to these segmental loops, the extensor motoneurons are under considerable excitatory control from descending fibers from higher centers (see Chaps. 20 and 21).

It may thus be appreciated that when we test for the stretch reflex by tapping the tendon at the knee, we are using a phasic stimulus (the tap) to elicit a phasic response (the knee jerk), whereas mainly this reflex circuit mediates tonic activity involved in maintaining a posture. In fact, the Ia endings are especially sensitive to the rate of change of small stretches, as was shown by Peter Matthews at Oxford using small vibratory stimuli applied to a muscle. Vibration excites a tonic vibration reflex akin to the stretch reflex.

Tension Feedback Reflex

The second circuit shown in Fig. 19.13 is that established by the large-diameter group Ib fibers from Golgi tendon organs. The connections onto motoneurons are all disynaptic. The effects on the motoneurons from a given muscle tend to be the reverse of those in the stretch reflex pathway, giving rise to the term *inverse myotatic reflex.*

The sensory input for this reflex comes from the Golgi tendon organs. As we discussed in Chap. 13, tendon organs are particularly sensitive to tension aroused by muscle contraction. When a muscle contracts, it exerts tension on the tendon organs, sending a barrage of impulses in the large Ib afferents. Within the spinal cord, the afferents activate inhibitory interneurons onto homonymous motoneurons (motoneurons to the same muscle), and excitatory interneurons onto antagonist muscles (see Fib. 19.13B).

By virtue of the synaptic organization of this reflex pathway, the effect of a muscle contraction is to decrease the amount of contraction of that muscle but increase excitation of opposing muscles. One interpretation of this arrangement is that it tends to maintain a constant tension in the muscle. Another interpretation is that the inhibitory interneuron provides a switch for reducing contraction; thus, cutaneous and joint afferents also can activate Ib inhibitory interneurons, and can switch off contraction if a limb encounters an obstacle. When combined with the stretch reflex, the inverse myotatic reflex contributes to the overall stiffness of the muscles.

It may be noted that in older textbooks, the inverse myotatic reflex is described as

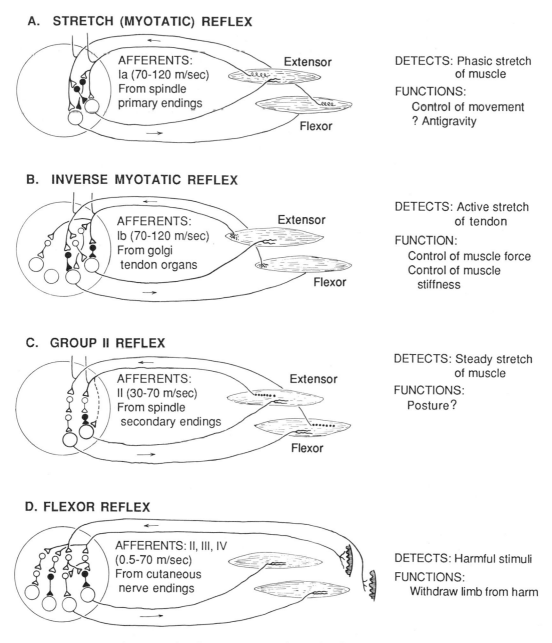

A. STRETCH (MYOTATIC) REFLEX

AFFERENTS:
Ia (70-120 m/sec)
From spindle
primary endings

Extensor

Flexor

DETECTS: Phasic stretch
of muscle
FUNCTIONS:
Control of movement
? Antigravity

B. INVERSE MYOTATIC REFLEX

AFFERENTS:
Ib (70-120 m/sec)
From golgi
tendon organs

Extensor

Flexor

DETECTS: Active stretch
of tendon
FUNCTION:
Control of muscle force
Control of muscle
stiffness

C. GROUP II REFLEX

AFFERENTS:
II (30-70 m/sec)
From spindle
secondary endings

Extensor

Flexor

DETECTS: Steady stretch
of muscle
FUNCTIONS:
Posture?

D. FLEXOR REFLEX

AFFERENTS: II, III, IV
(0.5-70 m/sec)
From cutaneous
nerve endings

DETECTS: Harmful stimuli
FUNCTIONS:
Withdraw limb from harm

Fig. 19.13 Neural circuits for the main types of spinal reflexes in the vertebrate. Each diagram shows only the minimum circuit elements involved in a reflex; for the sake of simplicity, additional interneuronal pathways contributing to modulation and control of the reflex are not shown. The spinal cord is on the left, and a flexor and extensor muscle are on the right. Within the spinal cord are flexor (F) and extensor (E) motoneurons; excitatory terminals of afferent fibers and interneurons are shown as unfilled (open) profiles, while inhibitory cells and connections are shown as filled (black) profiles. The dashed line in C indicates a weak action. See text. (Based in part on Matthews, 1972, 1982)

responding to higher levels of stretch, suggesting the interpretation that the inhibitory effect of the Golgi tendon organ afferents on homonymous motoneurons would serve as a protection against overstretch of the muscle. However, by the 1970s it was realized that stretching a passive muscle is the least physiological way of activating the Golgi tendon organs (see Fig. 13.9A). By contrast, as we saw in Chap. 13, during normal muscle activity the γ motor innervation of the spindles keeps the muscle under resting tension, and the Golgi tendon organs become exquisitely sensitive to tension changes due to active muscle contractions (see Fig. 13.9D).

Group II Reflexes

The third circuit in Fig. 19.13 is that involving the medium-size group II fibers from muscle spindles. These make mainly disynaptic connections onto motoneurons. In some experiments a weak monosynaptic connection onto homonymous motoneurons (shown by the dashed line in C), similar to that in the stretch reflex, can be demonstrated. Regardless of sensory input, disynaptic excitatory connections are directed mainly to flexor muscles and the inhibitory connections to extensor muscles; their functional role is still be debated.

Flexor Reflex

Stimulation of the skin or muscles by a noxious stimulus characteristically produces withdrawal of the affected limb. This is termed the *flexor reflex*. It can be mediated by a wide range of receptors and fibers, which are often referred to collectively as *flexor reflex afferents* (FRAs).

The bottom diagram (D) in Fig. 19.13 indicates that these fibers make polysynaptic connections which are excitatory to flexors and inhibitory to extensors. In addition, these fibers make widespread connections throughout the spinal cord which have the opposite effect: excitation of extensors and inhibition of flexors. Thus, while the hindlimb is being withdrawn,

the other limbs are being extended, for maintaining posture and participating in locomotion. Or, in more picturesque terms, the hindlimb is removed from danger, while the other three legs run away! This is another demonstration of the fact that reflexes do not take place in isolation. Here, as in the other examples of Fig. 19.13, the reflex circuit not only plays back upon the stimulated limb, but also calls forth appropriate and coordinated actions of the other limbs.

COORDINATION OF REFLEX PATHWAYS

How does the coordination of different reflex pathways come about? One possibility is that each pathway is distinct, so there would be separate and distinct sets of special interneurons that would link them together. This kind of arrangement requires multiple neuronal populations and multiple systems to control them. Another possibility is that the different pathways might share common interneurons. This would require careful interweaving of traffic through the common elements, but it would have the advantage that the control mechanisms could be overlain.

The latter possibility appears to be the case in the spinal cord reflex pathways. The work of Anders Lundberg and his colleagues in Göteborg, followed closely by Robert Burke at NIH and many others, has provided considerable evidence that individual interneurons can serve as nodal points for more than one functionally specific circuit. One of the best worked out examples of this principle is the Ia inhibitory interneuron. We saw in Fig. 19.13A that this interneuron derives its name from the fact that it mediates inhibition of antagonistic motoneurons in the stretch reflex pathway. But as shown in Fig. 19.14, this interneuron is in addition a nodal point in many other circuits. At the spinal segmental level, these include, in addition to Ia

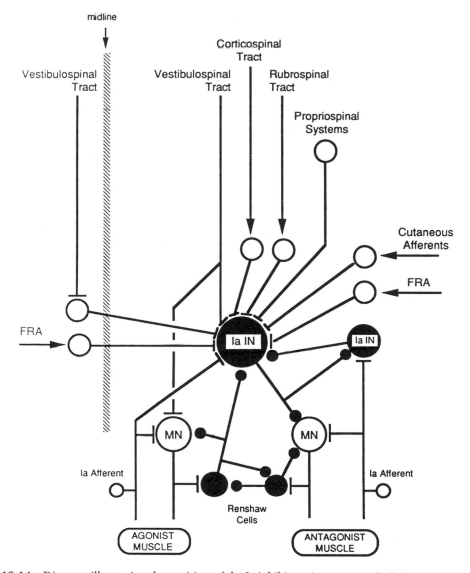

Fig. 19.14 Diagram illustrating the position of the Ia inhibitory interneuron (Ia IN) as an integrating node in a number of spinal circuits. Excitatory neurons and synaptic terminals are indicated by open profiles, inhibitory by closed profiles. FRA, flexor reflex afferents; MN, motoneuron; Ia afferent, sensory nerve from muscle spindle. See text. (From Burke, 1990)

inhibition, an inhibitory pathway from flexor reflex afferents onto motoneurons. It also provides for a longer loop in the Renshaw circuit; the student should trace this circuit through in Fig. 19.14, and realize that, when Renshaw cells inhibit this inhibitory neuron, they will actually be causing disinhibition of the effect of the Ia

interneuron on antagonist muscles. There are similar disinhibitory interactions between the Ia interneurons themselves (see right side of diagram).

The figure also indicates that the Ia interneuron is a nodal point for the descending control of the spinal cord by higher centers. These descending systems include the *vestibulospinal tract,* mediating reflexes involved in the sense of balance (Chap. 14); the *rubrospinal tract,* originating in the red nucleus; and the *corticospinal tract,* carrying impulses from the motor and related areas of the cerebral cortex. We will have ample opportunity to study the nature of these descending controls in Chaps. 22 and 23.

This principle of organization, in which functionally distinct pathways share common neural elements, is not unique to the spinal cord. Do you remember that we studied a similar organization in the retina, where distinct pathways starting at the rods and cones share common elements in their pathways through the bipolar cells to the ganglion cells? Neural network aficionados sometimes refer to this as *multiplexing,* in analogy with the way a single telephone line can carry multiple messages.

These considerations provide a framework for thinking about the organization of spinal reflex pathways. We still have to account for the relation between these pathways and the central pattern generators that are involved in coordinating the actions of these circuits during locomotion and other types of motor behavior. We turn therefore to this subject in the next chapter.

20

Locomotion

The ability to move about can be regarded as the most important characteristic of animal life. We generally refer to this overall ability as locomotion. Locomotion can be defined, in precise terms, as *the ability of an organism to move in space in purposeful ways under its own power, by efficient mechanisms suitable for the purposes of the movement.* In this chapter we will be concerned with the main types of neural mechanisms involved in this control.

Down through the ages, people have made careful observations of the movements of their fellow humans and of the animals about them, but naked-eye observations of rapid movements could be little more than fleeting impressions; of the most rapid movements, such as the wing movements of a hummingbird, there was only ignorance. Of course, ignorance is the fertilizer for the flowers of debate, and by the nineteenth century there were heated controversies about such matters as the exact positions of a horse's legs during trotting or galloping, or how a cat held upside down can right itself as it falls. (Some scholars were able to prove that the latter is theoretically impossible!)

Accurate knowledge of locomotor activity awaited the application of photographic techniques in the late nineteenth century. The pioneers in this were Emil Marey in Paris and Eadweard Muybridge in the United States. Muybridge set up a battery of 24 still cameras-side by side along a track, and triggered them in sequence while an animal walked or ran by. Examples of his results are shown in Fig. 20.1. These studies cleared up the old controversies about galloping horses and the like while providing a wealth of information about the locomotor patterns of a wide variety of animals, including humans.

Evolution of Locomotor Structures and Functions

In order to understand the neural systems that control a locomotor pattern like that shown in Fig. 20.1, we need to consider the phylogeny of invertebrates and vertebrates, for much of the basis for the evolution of different body forms is to be found in the adaptations for different types of locomotor activity in different species.

The simplest types of locomotion are found in unicellular or small multicellular organisms, which move about mainly by means of *pseudopodia* or *cilia*, aided often by *secretions* of slime. These mechanisms

Fig. 20.1 Sequences of still photographs of a galloping horse. This proved for the first time that during a gallop there is a time when all four hooves are in the air. (Photographed by Eadweard Muybridge in 1887, and republished in 1957)

are, in fact, exploited in higher organisms in many cellular functions, as we saw in Chap. 3. However, as mechanisms for moving the body through the environment they are effective only for very small organisms. Several additional specializations accompanied the development of larger and more complex organisms. A key one was the ability to develop significant amounts of *force* through muscular contractions. This depended on the development of a *skeleton,* a hydrostatic type in worms and molluscs, and a rigid type in arthropods and vertebrates. A second adaptation was the ability to carry out different *specialized functions* (for example, walking, grasping, feeding); this is greatly enhanced by the *metameric* body form, with its potential for specialization of different segments. A third specialization was the development of *appendages,* which could be adapted for many different types of locomotion. These specializations of body structure and function, and their significance for the main types of locomotor activity, are summarized in Table 20.1.

Some Common Principles in Nervous Control of Locomotion

The modern study of locomotion depends on a variety of methods. In addition to cinematographic recording of movement patterns, measurements are made of joint positions and angles and of the forces gen-

erated. The activity in individual muscles can be recorded by means of fine electrodes inserted into the muscles; the recordings are made on an *electromyograph* (EMG). Underlying any study, of course, is a thorough knowledge of the anatomy of the bones and muscles involved in the movements of interest. Recently, computer models of motor systems have become an important adjunct to these studies.

Levels of Motor Control

The key to an understanding of motor control is the concept that a pattern of locomotor activity is due to a pattern of neural activity. The task has therefore been to identify the different levels of control, the neural mechanisms at each level, and the ways that the mechanisms in the different levels are coordinated.

Historically, the first efforts, directed at *muscle reflexes,* originated in the studies of Sherrington and co-workers. One of the main concepts that came out of these studies is the idea that locomotion involves a modulation of postural reflexes. A second area, to which Sherrington also contributed, concerns the ability of the spinal cord to generate *intrinsic rhythms.* This concept owes its origin, in the early part of this century, particularly to Graham Brown in England, who studied cats with transected spinal cords. A third area of research is concerned with the control of the spinal cord by *higher motor centers.* Many work-

Table 20.1 Relations of structural specializations to modes of locomotion

Modes of locomotion	Adaptations of body structure	Significance for locomotion
swimming, creeping	hydrostatic skeleton (coelenterates)	transmission of pressure into force
swimming, creeping	coelom (annelid worms, molluscs)	more efficient transmission of pressure into force
swimming, creeping, burrowing	metamerism (annelid worms, arthropods)	1. more effective deployment of force 2. opportunity for specilization of segments 3. coordination of segments by specialized nervous system
swimming, walking, running, flying	jointed skeleton with appendages (arthropods, vertebrates)	1. extreme localization of force 2. amplification of force through limbs acting as levers 3. reduction and specialization of appendages, sometimes for multiple functions 4. more complex nervous controls

ers have contributed to this area, as we shall see in the next chapter.

At one time or another there have been claims for the predominant influence of each of these main types of mechanisms for the control of movement. One of the important developments in motor studies has been the synthesis of all these mechanisms into a general framework for nervous control. This synthesis is summarized in Fig. 20.2. The first element in this framework is the *central pattern generator* in lower centers in the spinal cord (or the relevant ganglia in the case of invertebrates) This contains the essential neural mechanisms for generating coordinated rhythmic outputs of the motoneurons. The central pattern generator is activated and controlled by descending fibers from the second key element, *higher motor centers*. There is commonly a succession of higher centers, which form the hierarchy of motor control as we discussed in Chap. 17 (Fig. 17.9). Third, there are *feedback circuits,* which feed back information from the *muscles* (proprioceptor reflexes); from the external *environment* through other sensory pathways; and, within the nervous system itself, from lower to higher *centers* (central

feedback). This internal feedback constitutes a copy of the output, which informs the higher center how faithfully the output followed the descending instructions and how much difference there is between that output and the output needed to reach some behavioral goal of the organism. It is

Fig. 20.2 The main neural components common to most motor systems.

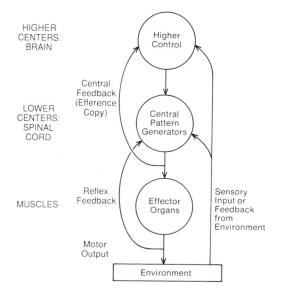

called by several names, including *corollary discharge* (because it is a corollary of the primary output discharges), *efference copy* (because it is a copy of the efferent [output] signal), and *reafference* (because it is an internal afferent ["sensory"] signal).

Despite the variety of body forms and types of locomotive activity in different species, these main components are present to some extent in the nervous systems of most higher organisms. Randy Gallistel of Philadelphia has synthesized this view of a hierarchy of motor control in *The Organization of Action* (Gallistel, 1980), which presents a persuasive case that most of the main types of motor behavior can be accounted for by motor hierarchies containing three elementary units of action: the *reflex*, the *oscillator*, and the *servomechanism*. These are all represented in the diagram of Fig. 20.2. We have dealt with the reflex in the preceding chapter, and turn now to the oscillator.

Central Pattern Generators

The generation of repeating or oscillatory patterns of muscle activity is one of the main functions of motor systems. In recent years, it has been recognized that this property rests within nerve circuits that form what is called a *central pattern generator*. This type of circuit, which may be local or distributed, has emerged from studies of both invertebrates and vertebrates. It is one of the key organizing principles for present-day concepts of mechanisms of locomotion.

The view emerging from recent studies is that, despite the diversity of body structure and locomotion patterning in the animal world, the basic types of neural circuits for generating rhythmic motor output are actually quite limited. Three of the main types that have been proposed are illustrated in Fig. 20.3.

Half-Center Model. The first model (A) was proposed by Graham Brown to account for the alternating activation of flexor and extensor muscles of the limb

of the cat during walking. Each pool of motoneurons for flexor (F) or extensor (E) muscles is activated by a corresponding "half-center," or pool, of interneurons. Another set of neurons (D) provides for a steady excitatory drive to these interneurons. Between each pool of interneurons are inhibitory connections which ensure that, when one pool is active, the other is suppressed. Graham Brown hypothesized that, as activity in the first pool progressed, a process of fatigue would build up in the inhibitory connections between the two half-centers, thereby switching activity from one half-center to the other. In more modern terms, the process of fatigue can be replaced by any process bringing about self-inhibition of the active cells.

Closed-Loop Model. A second, and related, type of model (B) conceives of the interneurons as organized in a "closed-loop" of inhibitory connections. There are corresponding pools of motoneurons activated, or inhibited, in sequence. Because of the fractionation of the pools of interneurons and motoneurons, there can be a finer differentiation in the activation of different muscles. This seems to be a more accurate description of the slightly different activation patterns of individual muscles during many locomotor acts. George Szeckely of Hungary proposed this model for the salamander in 1968, and it has been applied also by the Russian school of Shik and Orlovsky in the cat.

Pacemaker Model. In the models of Fig. 20.3A and B, rhythms result from neuronal circuit organization. By contrast, in the model of C a rhythm arises as a membrane property of a *pacemaker* cell or group of cells. This cell undergoes rhythmic excitation by intrinsic membrane mechanisms, involving the interplay of ionic currents. As discussed in Chap. 5, a common mechanism involves voltage-gated Ca channels; slow depolarization due to the entry of Ca^{2+} generates repetitive impulse firing, which is terminated by the hyperpolarizing

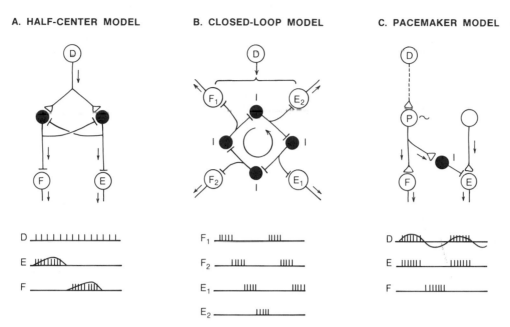

Fig. 20.3 Basic types of rhythm generators. **A–C.** Simplified diagrams of minimum number of neurons and connections for each type. Abbreviations for types of neurons: D, driver; E, extensor motoneuron; F, flexor motoneuron; P, pacemaker; I, interneuron. Neurons with excitatory actions are shown by open profiles; inhibitory, by filled profiles. Sequences of spike firing or graded potentials are shown in idealized recordings below. (For sources, see text)

action of calcium-sensitive K channels. In the model of Fig. 20.3C, the pacemaker cell drives flexor motoneurons directly, and brings about concurrent inhibition of extensor motoneurons through inhibitory interneurons. As knowledge of membrane properties generating endogenous rhythms has increased (see Chap. 6), pacemaker mechanisms have come to the fore as the primary sources of rhythmic activity, modulated by circuit organization.

For the remainder of this chapter we will consider the specific mechanisms involved in swimming and terrestrial locomotion. Swimming is important because it represents the basic pattern of alternating muscle movement that is fundamental for nearly all forms of animal movement, both invertebrate and vertebrate, and terrestrial locomotion. Terrestrial locomotion, which permitted the evolution of the higher vertebrates on land, is the underlying motor activity in our everyday lives.

Swimming

Swimming generally takes place by means of undulatory (wavelike) movements of the whole body. This type of mechanism is very widespread in the animal world, and characterizes, for example, the whiplike movements of the tail of a sperm and the locomotion of animals as diverse as worms, molluscs, fish, and snakes. Swimming is closely related to *creeping* and *burrowing*. They all involve coordinated sequences of muscle contractions, and their effectiveness (in terms of increased *speed* or *force* of movement) is enhanced by the development of a hydrostatic coelom, metameres, and a more complex organization of muscles and nervous system.

Tritonia: An Invertebrate Model

Tritonia is a sea slug that swims by making a series of alternating dorsal and ventral flexions of its body. The mechanisms that

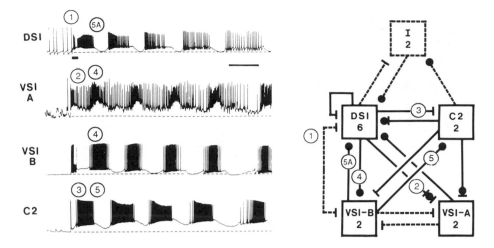

Fig. 20.4 The swimming generator in *Tritonia*. *(Left)* Intracellular recordings of the firing patterns in the main types of neurons, induced by an initial brief period of electrical stimulation ① of the dorsal swim interneuron (DSI). The other neuron types are ventral swim interneurons (VSI, class A and B) and cerebral cells (C2). Numbers ② through ⑤ indicate the sequence of impulse firing patterns. Time bar, 5 seconds. *(Right)* Diagram of connections deduced for generating the firing patterns. T-bars represent excitatory synapses; filled circles, inhibitory synapses. Note that some connections are mixed or sequential (e.g., connection ② is excitatory–inhibitory–excitatory). Dashed lines show postulated pathways. I2, unidentified neurons. (From Getting and Dekin, 1985)

generate these movements have been studied intensively by Peter Getting and his colleagues at Iowa, and as a result this has become a useful model system.

The swimming movements are produced by alternating contractions of dorsal and ventral flexor muscles. These are activated by dorsal flexion neurons (DFNA) and ventral flexion neurons (VFNs), respectively (see Fig. 20.4). Their alternating pattern of impulse discharge can be set up even when the ganglia containing the DFNs and VFNs are dissected out of the organism and placed in a recording chamber, showing that the rhythm is due to a central pattern generator located within these ganglia.

The swimming generator has been localized to several groups of premotor interneurons. The main groups are dorsal swim interneurons (DSIs) which activate the dorsal flexion neurons, and ventral swim interneurons (VSIs), which activate the ventral flexion neurons. These two groups are interconnected with each other

and with a third group of cerebral (C2) cells and a fourth group of interneurons (Is). The basic circuit for these connections is shown in Fig. 20.4 (right).

The swimming generator arises out of the membrane properties, molecular receptors, and synaptic organization of these four types of neurons. This information has come from intracellular recordings combined with intracellular current injection to test resetting of the rhythms. The way the circuit works is illustrated in Fig. 20.4. Swimming is initiated by sensory stimuli which feed into DSI and cause it to begin to fire a burst of impulses ①. DSI inhibits VSI ②, and at the same time excites C2 ③. C2 has excitatory synapses on VSI; however, the initial response of VSI neurons is delayed because of inhibition. This is due to an initial inhibitory action of the neurotransmitter followed by excitation in some neurons (VSI-A), or to initial activation of a fast transient I_A potassium current, which then inactivates so that excitation can proceed (VSI-B). The delays are important for the circuit to oscillate. VSI then fires ④, during which there is inhibition by VSI of C2 ⑤ and DSI (5A). Since this means that VSI no longer receives excitatory input from C2, its firing declines; DSI is therefore released from inhibition and is ready to fire again to initiate a new cycle.

The principles of organization of this central pattern generator can be summarized as follows. First, the rhythm is neurogenic, arising out of interactions between interneurons; the motoneurons are simply driven by the interneurons. Second, the oscillations are a network property, due to a combination of synaptic connections and membrane conductance channels, both ligand and voltage-gated. Third, the network is heavily dependent on inhibitory connections; in this respect, it resembles the type illustrated in Fig. 20.3, but it is better to consider it as a mixed excitatory–inhibitory type. Fourth, the generation of this rhythm is dependent particularly on complex postsynaptic potentials (PSPs) that include multiple components for excitation and inhibition, acting through different membrane conductances over different time scales.

From this analysis, it was concluded that the *Tritonia* swim generator cannot be described by a simple division into a pattern generator and a "command" neuron or system that turns it on and off. For example, the C2 and DSI neurons function both as command neurons (in initiating the rhythm) and as part of the pattern generator itself. Getting and Dekin (1985) therefore suggest that

the command function appears to be an emergent property of the network as a whole. It is not a function of a single cell, but a process that emerges as a consequence of multiple synaptic interactions within the network.

The same network can function in the two modes by virtue of the multifunction synapses, with different actions over different time scales, as explained above. We will return to the question of command

neurons within the context of motor hierarchies in the next chapter.

Lamprey: A Vertebrate Model

Among the vertebrates, the undulatory motions of swimming in fish or slithering in snakes are similar to those of worms. Higher speeds are achieved by increasing the frequency of the alternating contractions of muscle groups that underlie the body undulations. The waves persist after deafferentation of the spinal cord when most of the dorsal roots are cut. After a high spinal transection, interrupting all descending fibers, most fish show no spontaneous movements, but during tonic sensory stimulation (such as a pinch of the tail), wavelike contractions of the body appear. From such experiments it has been concluded that a central rhythm generator is present in the spinal cord, which in many species requires tonic input for its expression.

A key feature of vertebrate locomotion is that the vertebral column is long and flexible, so that swimming can take place by undulatory movements of the body (see Fig. 20.5). The search for the basic mechanisms underlying these movements has led from higher organisms such as the cat to the most primitive species. Sten Grillner and his colleagues in Stockholm have developed the lamprey spinal cord as a simple model for this purpose.

As shown in Fig. 20.6, the neural control systems for swimming in the lamprey spinal cord consist of the higher centers in the brainstem and the segmental systems in the spinal cord. Added to the properties of the segmental circuits we discussed in the previous chapter is the fact that each segment is capable of generating burst activity in the motoneuron axons; that is, it is a central pattern generator (CPG). Swimming occurs when the reticulospinal center sends descending excitatory inputs to the CPGs, raising their level of excitation (Fig. 20.6A). The CPGs are coupled together by intersegmental connections, so that their bursting is coordinated. Higher levels of

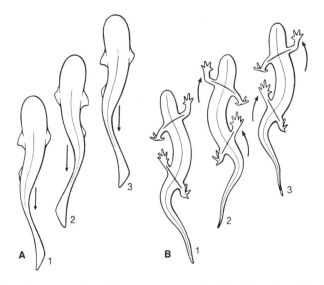

Fig. 20.5   Comparison between swimming movements of a fish (**A**) and primitive walking movements of a salamander (**B**). (From Romer and Parsons, 1977)

excitation in anterior segments lead to a slightly faster frequency, so that the anterior segments entrain the more posterior segments. There is thus a phase lag between segments that propels the animal forward (Fig. 20.6B). Higher excitation delivered to more posterior segments allows the animal to swim backward.

The basic organization for central pattern generation in each segment is summarized in Fig. 20.7. Although the circuit looks complicated, its underlying organization is relatively simple. In addition to the motoneurons (MN) there are just excitatory interneurons (E) and two types of inhibitory interneuron: those with ipsilateral (L) and contralateral (CC) connections. The descending brainstem inputs

Fig. 20.6   Outline of the principles underlying the organization of the neural systems controlling swimming motions in vertebrates. **A.** General scheme for vertebrate motor control. **B.** Undulatory movements of the body involve a sequence of activity in consecutive spinal segments; the phase lag between segments determines the rapidity of movement. See text. (From Grillner and Matsushima, 1991)

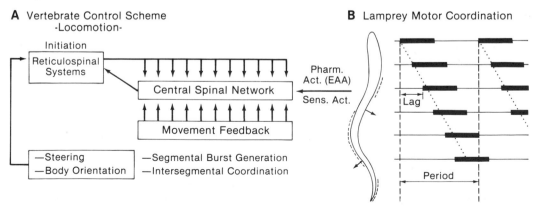

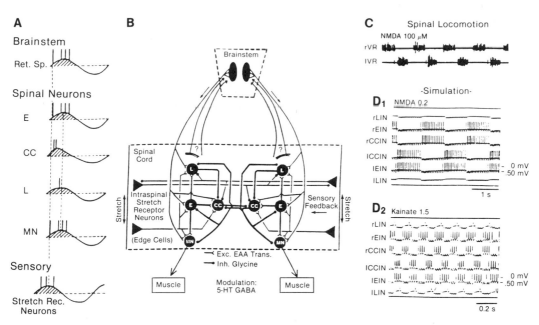

Fig. 20.7 Operation of the segmental motor circuits. **A.** Sequence of activity in the neurons of the brainstem, spinal cord, and sensory fibers. **B.** The segmental circuit, showing both sides of the spinal cord. **C.** Recordings of alternating burst activity in the right and left ventral roots (rVR and lVR, respectively) of the isolated spinal cord; this slow-burst activity was elicited by pharmacological activation of the NMDA receptors. **D.** Computer simulations of the activity in different interneurons of the segment. D_1: Simulation of activation of NMDA receptors. D_2: Simulation of more rapid bursting due to activation of kainate receptors. See text. (From Grillner and Matsushima, 1991)

have excitatory connections onto all these elements. The bursting pacemaker activity of the motoneurons is due to a sequence of voltage-dependent NMDA receptor activation and Na^+/Ca^{2+} depolarization followed by repolarization due to Ca^{2+}-activated K channels. This sequence has been studied by making intracellular recordings and activating the cells pharmacologically; all of these manipulations can be made on the isolated spinal cord. As shown in Fig. 20.7C, the NMDA receptors are important for generating a slow and steady swimming rate. The circuit has been modeled by a computational network that reproduces the slow rate related to NMDA receptor activation (D_1) and the more rapid rate due to activation of the kainate/AMPA receptors (D_2).

In conclusion, this work illustrates the use of simple animal preparations to give insight into basic properties of neural orga-

nization. Experimenters (and theoretical modelers) are thereby provided with working hypotheses which can be tested in more complicated animals.

Walking

Apart from the snakes, terrestrial vertebrates have particularly exploited the locomotory abilities of the leg in their invasion of the land. The evolution of legs from fins follows a logical sequence. In lower fish the function of the fins is mainly to provide *stabilization,* but in higher (bony) fish the fins are more specialized and contribute to *propulsion.* In the transitional forms (Crossopterygii) the pectoral and pelvic fins are more elaborate; they appear to be the basis for the fore- and hindlimbs of tetrapods.

The primitive walking movements of amphibians reflect rather strongly the basic swimming movements of fish. This is illus-

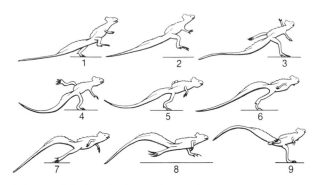

Fig. 20.8 Running of the basilisk lizard. (From Gray, 1968)

trated in Fig. 20.5. Note in B how the limbs have attachment points on either side of the pectoral (anterior) and pelvic (posterior) girdles. Forward movement is achieved by extension, placing, and thrust of the limbs, in coordination with the swimming movements of the body.

Amphibians and reptiles are constructed so that the limbs are attached laterally to the trunk, whereas in birds and mammals the legs support the body from underneath. The lateral placement has the advantage of a low center of gravity and a more stable equilibrium. The vertical placement, however, is regarded as more efficient for the purposes of locomotion. In general this is so; nonetheless, some reptiles are capable of moving very swiftly indeed. Outstanding in this regard are the lizards. The little basilisk lizard can skitter along at rates up to 7 m/sec (about 10 miles per hour). It does this by tucking in its forelimbs and running on its hindlimbs, with the heavy tail helping to maintain balance (see Fig. 20.8). The lizard thus runs with a *bipedal* gait. Several other animals, besides humans, have developed a bipedal gait (for example, kangaroos), which illustrates the principle of convergent evolution.

Control of Locomotion of the Cat

Because of its convenient size and generalized body form, the domestic cat has been an important subject for studies of neural mechanisms of locomotion in higher vertebrates ever since the pioneering experiments of Charles Sherrington in the 1890s. The patterns of movements of the feet during normal locomotion of the cat are illustrated in Fig. 20.9. The stepping sequence is: left hind leg, left front leg, right hind leg, and right front leg. Watch your cat the next time you have a chance, and see if you can identify this sequence—without the help of a photographic analysis! It has been found that this is the basic pattern for most vertebrate species, as well as for fast-moving invertebrates like the cockroach (see Fig. 20.9). The reason for this prevalence is that the pattern seems to provide for the best *stability,* which becomes increasingly important in the longer legged animals.

Gaits and Step Cycles

As the cat moves forward at increasing speeds, the steps are faster, but the same progression is maintained. However, different combinations of legs are on or off the ground at the same time as the speed increases, and these combinations are expressed as different *gaits.* The cat has several gaits, for moving to higher speeds, just as a car has several gears; these are illustrated in Fig. 20.9. For comparison, the cockroach has just one fast-moving gait. The cat can change relatively smoothly from one gait to the next *(gait conversion),* whereas the cockroach shifts more abruptly from the one to the other.

A closer analysis of the movement of one leg through a step cycle is illustrated in Fig. 20.10. As we have seen, a step consists of

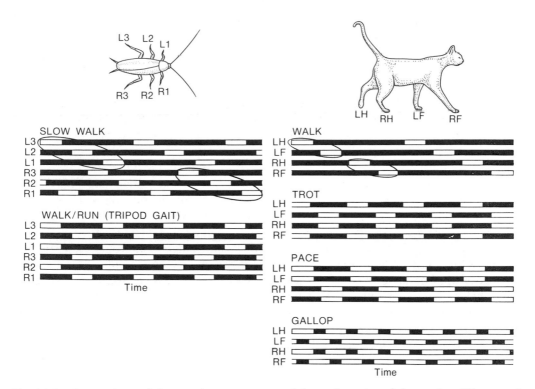

Fig. 20.9 Comparison of the stepping movements of the cockroach and the cat for different gaits characteristic of the two species. Open bars, foot lifted; closed bars, foot planted. (Adapted from Pearson, 1976)

Fig. 20.10 The step cycle, showing phases of leg flexion (F) and extension (E) and their relation to the wing and stance. *(Bottom)* Electromyograph (EMG) recordings. (Adapted from Wetzel and Stuart, 1976)

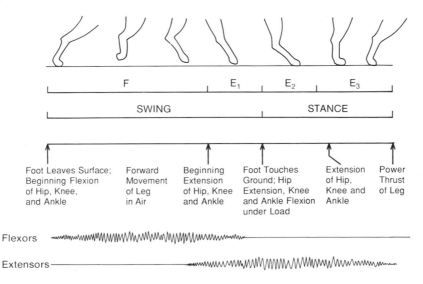

two phases, swing and stance. During the *swing* phase, the leg is lifted and brought forward, mainly by action of the leg flexor muscles (F). At the end of the swing phase (E_1), the extensors become active, and the combined activity of flexors and extensors stiffens the leg as it is planted. During the *stance* phase, extensor activity becomes dominant (E_2, E_3), providing the force that thrusts the animal forward. These phases of muscle activity seem to be quite general, and they apply to our own bipedal steps as well as to quadrupeds.

If we examine again the different gaits as depicted in Fig. 20.9, we see that the total time for a step cycle decreases as the speed goes up, but the decrease is almost entirely taken in the stance phase; the duration of the swing phase stays relatively constant. Thus, speed of locomotion comes about mainly by faster and quicker thrust. There is presumably an economy in having the swing phase be relatively stereotyped, so that the central motor circuits can change gaits by controlling muscle activity mainly during the stance phase.

Spinal Stepping

The different gaits and how they are controlled brings us back to the questions originally posed by Muybridge and his photographs. We have seen how Graham Brown's studies led to the idea of "half-centers" in the spinal cord, with intrinsic rhythmic activity controlling flexor and extensor motoneuron pools. Further analysis awaited more refined methods, in which there could be experimental control of locomotor behavior under well-defined conditions. This was first achieved by the Russian physiologists Shik, Severin, and Orlovsky, in Moscow. The experimental setup is illustrated in Fig. 20.11. The cat may have a transection at one of several levels in the neuraxis: in the brainstem (for example, a transection between the superior and inferior colliculi, producing a decerebrate animal), between brainstem and rostral spinal cord (high spinal transection), and at various levels in the spinal cord. The animal is held rigidly in a holder,

and its paws are placed on a treadmill. Stepping movements of the paws can be initiated by various procedures, such as movement of the treadmill, electrical stimulation of different parts of the brainstem below the transection, or by injection of substances into the bloodstream.

Using this preparation, Sten Grillner and his colleagues in Stockholm showed that with a high spinal transection, a cat can still generate alternating and coordinated movements of all four limbs. Even with a lower, midthoracic transection, the hindlimbs can display walking movements, and they can change to approximately simultaneous galloping movements when the speed of the treadmill is increased. Examples of recordings from the muscles of the knee, ankle, toe, and hip are shown in Fig. 20.12A, for the case of a high midbrain transection. The coordinated pattern of discharge reflects the normal sequence of activation of these muscles during walking. This basic pattern persists in some animals even after deafferentation (shown in B), although it is not as stable and can readily break down.

The conclusion from these and related studies is that locomotion depends on "a central network that generates essential features of the motor pattern and sensory feedback signals that form an integral and crucial part of the control system" (Grillner, 1985). A third element is the descending control from higher brain centers, which includes the mechanism of efference copy (see next chapter). These are the same basic elements of the swimming generator in the lamprey described previously. We thus have identified the three fundamental units of motor action mentioned at the beginning of this chapter.

Neural Mechanisms Controlling Locomotor Performance

Having identified the three kinds of neural systems involved in locomotor control, we next consider the cellular mechanisms that underlie the operational characteristics of those systems. We will first consider the

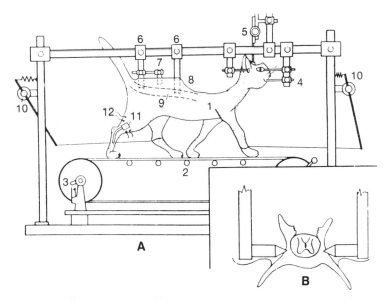

Fig. 20.11 Experimental setup for studying neuronal activity during treadmill walking in the decerebrate cat. **A.** General view. **B.** Inset showing rigid holding of vertebrae. Numbers identify the following components: 1, cat; 2, treadmill; 3, belt tachometer; 4, stereotaxic head holder; 5, electrode holder; 6, clamps to hold spine; 7, electrodes for recording unitary activity; 8, skin flap forming oil bath; 9, spinal cord; 10, detectors of longitudinal displacement of limbs; 11, joint angle detectors; 12, implanted electrodes in muscle. (From Severin et al., in Wetzel and Stuart, 1976)

way that motoneurons are responsible for the different levels of motor performance as we go from walking to running; then we will discuss how the central pattern generator controls motoneuron output during this activity.

The Concept of a Motoneuron Pool

In order to understand how different levels of activity are induced in motoneurons, we

need to introduce the idea of a motoneuron pool. This idea comes from Charles Sherrington and a brilliant group of young scientists who worked with him in his later years, during the 1920s, in Oxford. This group included John Eccles, Derek Denny-Brown, E. G. T. Liddell, John Fulton, Richard Creed, and Sybil Cooper; their work is summarized in *Reflex Activity of the Spinal Cord* (Creed et al., 1932). This work initi-

Fig. 20.12 Pattern of muscle activity in the hindlimb of a decerebrate cat placed on a moving treadmill. **A.** Control electromyograph (EMG) recordings from the knee extensor (E) quadriceps (Q) muscle, the ankle extensor lateral gastrocnemius (LG), the toe dorsiflexor extensor (EDB), and the hip flexor (F) iliopsoas (Ip). **B.** Recordings from the same muscles after bilateral transection of the dorsal roots to the hindlimbs. (Adapted from Grillner and Zangger, 1984)

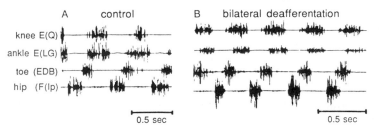

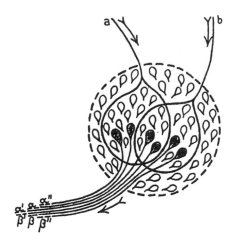

Fig. 20.13 The concept of a motoneuron pool. Sensory axons a and b have their respective fields of threshold excitation (indicated by large circles). Each separately activates its pool of motoneurons (see axons of α and β populations). Because of their overlapping fields, simultaneous activation of a and b does not give linear summation of the separate fields, a phenomenon called occlusion. See text. (From Creed et al., 1932)

ated the study of the different reflex pathways that we discussed in the preceding chapter.

In their work, these investigators stimulated different sensory nerves and recorded the reflex muscle contractions under a variety of conditions. Out of this work arose the idea of a *motoneuron pool,* that is, the population of the motoneurons connecting to a given muscle. As illustrated in Fig. 20.13, the motoneurons "which correspond to the motor units of a muscle are set inside a circle [dashed line]. This is the 'motor centre' of the muscle or its 'motoneuron pool'. The fractional portions excited to discharge by centripetal volleys in reflexes a and b are shown as small areas circumscribed by the continuation of the line representing the nerve in question, and each can be referred to as the motoneurone field of that nerve" (Creed et al., 1932). This concept helped to explain the graded activation of muscles (and motor units) by different intensities of stimulation in

different sensory nerves, as well as the graded activation of muscles by different amounts of tetanic stimulation. It also helped to explain how simultaneous excitation of two sensory nerves could induce a summed muscle contraction less than twice as strong as the sum of the separate responses to the two nerves, because of the sharing of motoneuron fields, a phenomenon called *occlusion.* The student should verify this phenomenon for herself or himself in the diagram.

The Size Principle

Study of the graded activation of muscles by graded activation of the motoneuron pool further gave rise to the concept of *recruitment,* that is, the increasing activation of motor units. It soon became evident that a motoneuron pool contains a wide variety of motoneurons, from very small cells to very large ones. What is their order of recruitment, as one goes from weak inputs to very strong? The answer is important, because it would reveal the mechanisms involved in going from low levels of activity, as in walking, to intense activity, such as running and leaping (see Fig. 20.9).

Our current understanding of these mechanisms is based on the work carried out by Elwood Henneman and his coworkers at Harvard in the 1960s. They recorded intracellularly from the alpha motoneurons to the skeletal muscles, and found that the smaller the motoneuron the more responsive it is (that is, the lower its threshold) to sensory inputs. They dubbed this the "size principle." It is illustrated in Fig. 20.14. In A are shown three representative motoneurons of increasing size. In B are shown electrotonic models of these neurons, each with a soma (cell body) and a dendritic cable. A single excitatory synapse (for example, a Ia synapse) elicits the same amount of inward synaptic current in each cell, but the resulting outward flow gives rise to larger EPSPs in the smaller motoneuron because there is less somadendritic membrane, and hence a higher input resistance giving a larger voltage change (see

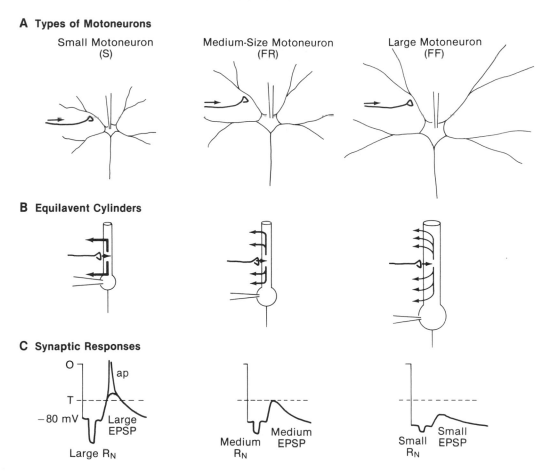

A Types of Motoneurons

Small Motoneuron
(S)

Medium-Size Motoneuron
(FR)

Large Motoneuron
(FF)

B Equilavent Cylinders

C Synaptic Responses

O
ap
T
−80 mV
Large
EPSP
Large R$_N$

Medium
EPSP
Medium
R$_N$

Small
EPSP
Small
R$_N$

Fig. 20.14 Illustration of the size principle. **A.** Different size motoneurons are correlated with S, FR, and FF motor units. **B.** For analysis of the membrane properties, the dendritic tree can be approximated by an "equivalent cylinder," according to Rall. The cell can be activated by synaptic input or by intracellular current injection. **C.** Recordings show large responses to a given synaptic input or a given amount of injected current in small cells compared with large cells. The higher input resistance (R_N) of the small cells is due to the smaller membrane area and consequently higher current density. See text. T, threshold; ap, action potential. (Based on Burke, 1981)

representative recordings in C). The student should work through these examples, because they require application of the basic principles about recordings and electrical current flows that we studied in Chaps. 5 and 6.

Subsequent work has shown that the size principle is correlated with a number of other properties. Some of these are summarized in Table 20.2. The best correlation is with the subdivisions into the three types of motor units that we characterized in the previous chapter. Thus, *small motoneurons*, in addition to small membrane areas giving high input resistances, have slow axon conduction velocities in accordance with their smaller diameters. Their motor unit properties include slow rates of contraction, small contration force, and high resistance to fatigue, which puts them in the category of *S units* supplying slow muscles (see previous chapter). By contrast, the *largest motoneurons* have the largest somadendritic size and the largest, fastest

Table 20.2 Motor unit properties

Unit types	FF	FR	S
Motoneuron properties			
input resistance	low	intermediate	high
total membrane area	largest	intermediate	smallest
axonal conduction velocity	high	high	low
Muscle unit properties			
rate of contraction	fast	fast	slow
force output	large	intermediate	small
fatigue resistance	low	high	very high
glycolytic capacity	high	high	low
oxidative capacity	low	moderate to high	high
Functional properties			
usual recruitment threshold	high	intermediate	low
output force grading by recruitment	course	moderate	fine
metabolic optimum activity	shortening	shortening	isometric
relative duty cycle	low	moderate	high
metabolic cost of maintenance	low	moderate	high
"typical" usage	gallop, jump	walk, run	posture

FF = fast contracting, large force, fatigable
FR = fast contracting, moderate force, fatigue resistant
 S = slowly contracting, small force, very resistant to fatigue

Adapted from Burke (1981)

conducting axons. Their motor unit properties include fast rates of contraction and large forces, but rapid fatigue, thus correlating with the category of *FF motor units* supplying fast muscles. Between is the third category of intermediate size motoneurons correlated with the properties of *FR motor units* with fast contractions, moderate force, and some resistance to fatigue. The student should compare these properties carefully with those of the muscles in Chap. 17.

Motor Unit Recruitment Underlying Motor Performance

The conclusion from the foregoing studies is that motoneuron properties are exquisitely matched to the properties of the motor units supplying the muscles and the properties of the muscles themselves. This has suggested some basic principles of how the synaptic circuits controlling the motoneurons are brought into play to mediate different levels of locomotor activity.

The basic concept is illustrated in Fig. 20.15. In A, the input A represents the sensory drive to the motoneuron pool arriving over Ia afferent fibers, as well as other types of input fibers with these properties. The sizes of the arrows indicate that the greatest efficacy of these inputs is to the smallest motoneurons (S type), with decreasing efficacy to the larger FR type, and the least efficacy to the FF type. Thus, at the lowest levels of input, for example, that involved in quiet standing, it is mainly the S motoneurons that are activated, as shown in the graph in Fig. 20.15B. As the level of input rises to mediate walking, recruitment of larger FR-type motoneurons occurs, and at still higher levels, FF motoneurons are recruited. Thus, larger numbers of active input fibers overcome the lower efficacy of synapses to the larger motoneurons.

It remains to account for situations in which sudden, synchronous activation of motoneurons is needed for very rapid and

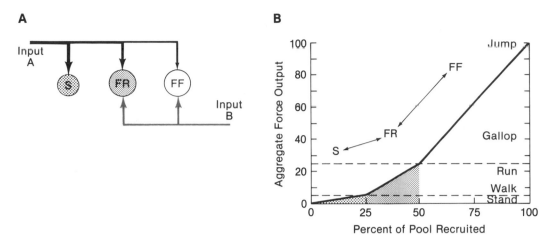

Fig. 20.15 Relation between size principle and motor performance. **A.** The three basic types of motoneurons (S, FR, and FF) receive two types of input; input A has higher efficacy for smaller motoneurons (size principle), whereas input B has high efficacy for largest motoneurons. **B.** Recruitment of motoneurons by increasing Ia inputs (representing input A) to the gastrocnemius motor units of the cat, according to the size principle, underlies the progression from standing to walking, running, and galloping. For extreme bursts of muscle force, as in jumping, input B directly activates the largest motoneurons. See text. (From Burke, 1990)

forceful movements, as in jumping away from a harmful stimulus or in extreme voluntary motor performance such as jumping. For these functions, different sets of input fibers can have more direct access to the largest motoneurons, in order to generate immediate maximum motor output. This is indicated by input B in Fig. 20.15A and by the highest force output for "jump" in the graph in Fig. 20.15B. Thus, as expressed by Robert Burke (1990) of the National Institutes of Health:

The existence of only two different orderings of synaptic efficacy. . . . provides the CNS with great flexibility in how a given motor pool might be "used" to provide the enormous range of functional demand placed on the motor systems, from sustained postural maintenance to power-ful, ballistic movements like galloping, jumping, and rapid limb shaking.

In conclusion, the examples we have studied share common principles of organization that are adapted for widely different types of locomotor activity. It expresses beautifully the importance of understanding the nature of nervous system function at the circuit level. As Meldrum Robertson and Keir Pearson (1985) eloquently phrased it:

. . . motor patterns are produced by nearly universal neuronal processes . . . the overall circuits are unique only in the way that well-described components are assembled. This is heartening for those in search of general principles of neuronal organization.

21

Motor Hierarchies

The four preceding chapters have introduced the major elements of the motor apparatus: the glands and muscles themselves, the motoneurons that innervate them, the simplest neural networks for generating rhythms and fixed-action patterns to drive the motoneurons, and the simplest reflex pathways that adjust motor performance to sensory inputs. For the most part, all of this neural machinery is at the segmental level of the nerve cord in invertebrates and the spinal cord and brainstem in vertebrates.

The next step is to ask, what are the neural mechanisms for operating this segmental apparatus, so that it serves the whole organism in purposeful ways? The two parts of this question actually relate to different levels within the hierarchy of motor organization referred to in preceding chapters (see Fig. 20.2, "higher centers"). At the first level are the descending pathways by which the segmental apparatus is controlled. That is the subject of this chapter. Above this are levels of organization that operate within the context of the whole organism. For example, is a given movement voluntary or involuntary, purposive or automatic? What is the adaptive value of a given type of movement for

the individual and for the species? These questions are dealt with in the final section of the book, on central systems. However, it is important even while dealing with lower levels of organization to realize that the most profound students of motor function, such as Charles Sherrington, have always regarded individual motor mechanisms from a perspective of their significance for the behavior of the whole organism. This demonstrates the point made at the very outset of this book (Chap. 1): "Nothing in neurobiology makes sense except in the light of behavior."

The conceptual framework of the motor hierarchy has been fundamental both for analysis of motor control in animal experiments and for the diagnosis of neurological diseases of the motor system in human patients. For more than a century, clinical neurologists have used this framework to distinguish between the *lower motor neuron,* in the spinal cord, and the *upper motor neuron,* which includes all the centers in the brainstem and motor cortex that project to the spinal cord. Our concern with the immediate descending control of the segmental apparatus is therefore with these different types of upper motor neurons.

The same general framework applies to motor control in invertebrates, where the ganglia of the nerve cord are equivalent to the spinal segments, and fibers from higher centers are equivalent to the descending fibers in the vertebrate (see Fig. 20.2). An important difference is that in the invertebrate there are characteristically giant neurons and fibers, or other kinds of individually identifiable neurons. The fact that a single such neuron or fiber can be selectively stimulated to produce a motor response has given rise to the concept of the "command neuron." This concept has been a mixed blessing; it has stimulated much of the research on invertebrate motor systems, but its validity as a concept for understanding motor control has been heatedly debated. Since one cannot study motor systems without encountering this issue, we will begin with a brief history of the development of this concept, and attempt to outline a broader definition consistent with recent research. We will then consider how decisions are made within motor control circuits as revealed by experiments in invertebrates. We will then be in a position to study the mechanisms of descending motor control in the vertebrate.

Command Neurons: A Brief History

This concept arose from the work of C. A. G. "Kees" Wiersma, at the California Institute of Technology, on motor systems in the crayfish. Beginning in the 1940s, his work was focused on the giant fiber system that controls the escape response (see Figs. 19.3–19.5), and also the circuit that controls the rhythmic beating of small abdominal appendages called phyllopodia, or *swimmerets* (Fig. 21.1A). Wiersma realized that large, identifiable fibers in the connectives to the segmental ganglia can be selectively stimulated and their individual effects on the motor output determined. In 1964, he and Kazuo Ikeda showed that the endogenous bursting of the segmental motoneurons that innervate the swimmeret muscles can be turned on and off by stimu-

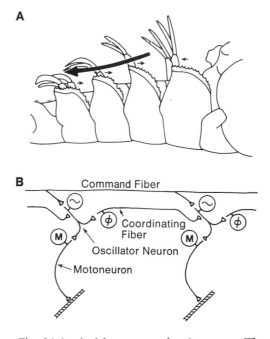

Fig. 21.1 A. Movements of swimmerets. The most anterior appendage, at right, is completing the power stroke, while the others are in successive phases of their return strokes, in the direction shown by the arrows. Movement of water is indicated by the large heavy arrow. (From Davis, in Kennedy, 1976) **B.** Rhythm generator circuits for control of swimmerets in the crayfish. Two neighboring hemiganglia are shown, connected by a coordinating fiber. In each hemiganglion, there is an oscillator neuron that receives its input from a nonrhythmic command fiber. (From Stein, in Wetzel and Stuart, 1976)

lation of specific interneurons in more rostral ganglia and connectives. They introduced the term "command fiber" for the fibers they stimulated (Fig. 21.1B). The details of the swimmeret system were worked out subsequently by Paul Stein at Washington University in St. Louis and by Donald Kennedy and his colleagues at Stanford. A summary of the circuit is shown in Fig. 21.1B.

Problems of Definition

The terms *command fiber* and *command neuron* quickly found application to a variety of systems. Very soon, however, contro-

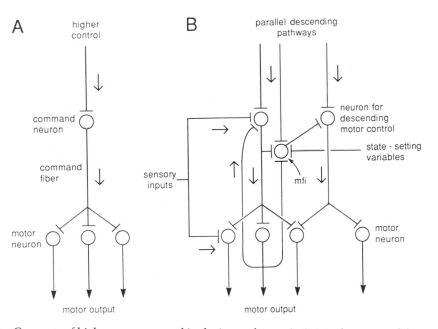

Fig. 21.2 Concepts of higher motor control in the invertebrate. **A.** Original concept of the command neuron or fiber. **B.** More complex types of circuit connections revealed by recent work. mfi, multifunction interneuron. See text.

versy arose over the definition of these terms. A suggestion, by David Bentley and Mark Konishi (1978), was as follows:

Command [neurons] can be defined as interneurons whose activation alone suffices to elicit a recognizable fragment of behavior through excitation and/or inhibition of a constellation of motoneurons.

This simple definition is represented by the diagram in Fig. 21.2A. Kupfermann and Weiss (1978), in a much quoted essay, emphasized that a command neuron or fiber should be proven to be both necessary and sufficient for eliciting a given motor output.

The problem with these and other definitions has been that the more a given system is investigated, the more complicated it becomes. We have already seen this in several of the systems we have studied thus far. For example, the escape response of the crayfish appears to be mediated by a classical command neuron. However, as we discussed in Chap. 19, recent work shows that the circuit is actually quite com-

plicated. It includes nongiant as well as giant components; feedback as well as feedforward pathways; interneurons with complex excitatory and inhibitory actions; and sensory input at several levels. In addition, virtually every type of synaptic junction in the circuit is under control of state-setting variables, such as the state of arousal, adaptation, or fatigue; we saw an example of how dramatic these variables can be in the effects of serotonin and octopamine in eliciting completely different postures in the crayfish. Motor control is thus much more complex than originally envisaged, and an expanded circuit, such as that in Fig. 21.2B, is needed to reflect the actual situation.

Another example of the problem of defining a command neuron is the swimming circuit for *Tritonia*. As we learned in the previous chapter (see Fig. 20.4), the central pattern generator (CPG) consists of three types of neuron: cerebral (C2) cells, and the dorsal and ventral swimming interneurons (DSI and VSI). By the criteria mentioned above, C2 and DSI are also command neurons, because they are both necessary and sufficient for the initiation and maintenance of the swimming rhythm (refer to Fig. 20.4B). Since these two cells function both as central pattern generators and as command neurons, they must be regarded as *multifunctional neurons*. In addition, there is a group of interneurons (I2) which coordinate C2 and DSI (see Fig. 20.4B), and thus represent *different levels* in the hierarchy of control.

New Concepts of Motor Control from Studies of Simple Systems

From these considerations it is obvious that the simple concept of the command neuron is no longer tenable. Some of the new concepts that have arisen from recent work may be summarized as follows.

1. A given neuron may display command functions, but it is likely to take part in other functions as well. This means that we cannot identify a given neuron exclusively as a command neuron, and the same applies to other cells originally identified with specific functions, such as oscillation, coordination, and feedback. Thus, one has the concept of the *multifunctional neuron*. This is a completely general concept, applying throughout the nervous system as well as to the large, morphologically identifiable command neurons in simple systems. We have already encountered several examples: bipolar cells of the retina, which process rod signals in the dark and cone signals in bright light (Chap. 16); nonspiking interneurons in the segmental ganglia of many invertebrates, whose different dendritic branches take part in different types of circuits (Chap. 20); and the Renshaw cells of the spinal cord, which are nodes in several types of pathway (Chap. 20). In addition, there are many examples in which the nonlinear *voltage-gated properties* of the soma and dendritic membrane enable single cells to operate in different firing modes at different membrane potentials (see Chaps. 5 and 25).

2. Since different functions often represent different levels of control, it is apparent that a given neuron is able to function at different levels of hierarchical control, so that we have the allied concept of the *multilevel neuron*. For example, the command function usually is interpreted as being at a higher level of control than an oscillator or coordinating function, but a given neuron may display either a command function or an oscillator function, depending on the actions of neuromodulators in setting the behavioral state or on the particular phase of motor output to alternately flexing and extending limb muscles. The ubiquitous nature of *feedback systems* helps to ensure that neurons at most levels of hierarchical control are heavily interconnected and interdependent.

3. Although some command functions are heavily dependent on one or a few large and identifiable neurons, they are the exception to the rule. More commonly in invertebrates, and almost exclusively in vertebrates, command functions are dependent on the combined actions of populations of neurons. One therefore has the concept of *command systems,* in which each neuron contributing to that function may be regarded as a *command element* (Ewert, 1980). In view of points 1 and 2 above, it is evident that the command elements and systems may shift among the neurons of a particular region, dependent on the behavioral state or phase of activity.

4. These considerations suggest that our attention should shift from commands as properties of neurons to commands as properties of networks. *A command function is a network property* that arises from specific combinations of cellular properties distributed in the neurons of the network. The neurons and their interconnections provide the substrate for the property; that substrate can shift within the network depending on different behavioral states and phases of activity. The command function is therefore a *distributed property*. Jenny Kien (1983) at Regensburg has suggested that the distributed command function is

transmitted to the locomotor output units as a kind of "across-fiber pattern," analogous to the spatial patterns of activity in many sensory systems. Thus, except for extremely specialized actions such as certain types of escape responses, motor command functions can best be thought of in terms of the actions of parallel distributed networks, similar to those involved in pattern recognition in sensory systems.

5. The shifts in cellular locus of a given function, such as the command function, can be brought about in several ways, most prominently by the actions of peripheral or descending inputs themselves, or by neuromodulation. In the previous chapter we discussed the reordering, or reconfiguring, of functional connectivity in spinal circuits by reflex activity. As Peter Getting (1989) of Iowa put it, "Anatomical connectivity defines the constraints of a network but functional connectivity determines the activity pattern." Getting further notes:

. . . input may not only activate a network but may also configure it into an appropriate mode to process that input. Descending commands and afferent inputs should be considered as both instructive and permissive in that they may organize the functional interactions within a network to be appropriate for the task at hand as well as activate the network to perform the task.

6. *Neuromodulation* has profound effects on networks. Behavioral states are commonly set by the actions of neuromodulators, which switch a neural network from one operating mode to another. Synaptic connections as well as voltage-gated conductances are prime targets for these actions. Many examples of neuromodulation are discussed in this book: see the effects of serotonin on the nervous system of the lobster (Chap. 19); the oscillatory and transmission modes of thalamic relay neurons (Chap. 25); and the sleeping and waking cycle (Chap. 25). Since neuromodulation is primarily effected through second messenger systems, the student should also thoroughly review the different neurotransmitters and neuropeptides and the mechanisms in Chap. 8.

7. If a network can shift between several operating modes, there must be *decision-making mechanisms* for selecting one behavior (such as a tail flip) over another (such as walking or feeding). An example is the giant-fiber escape system. We have already noted that this reflex pathway has a relatively high threshold for activation. Thus, the threshold alone can serve the decision-making function; if sensory stimulation is below threshold, there is a tail flip. However, the threshold itself is affected by a number of factors, such as whether, in the case of the crayfish, the animal is in or out of the water, or whether it has been injured. Sensory pathways are thus important in activating or modulating motor pathways; potentially, this may occur at all levels, from motoneurons up to command neurons, and also may involve the cells that control the command neurons.

Several systems have been analyzed in which the decision-making mechanism is more complicated. One example is the movement detector-jump system in the locust. The motor program for the jump actually has three distinct stages. First is flexion of the leg; second is cocontraction of extensor and flexor muscles; third, there is sudden relaxation of the flexors, giving rise to the jump. Thus, rather than a single neuron or gate, there is a sequence of gates; at each stage, the decision to proceed depends on the presence of several factors, including the intensity of specific sensory stimulation, the coincidence of other sensory stimuli, the state of arousal, and activity in other motor systems (summarized in Burrows, 1978). These results show that under natural conditions a given behavior is mediated not by a single command neuron, but rather by a network of such cells providing for multiple state-dependent gates.

Motor Hierarchies in Vertebrates: A Brief History

Historically, the idea that motor systems in vertebrates are organized in hierarchical fashion has a long and distinguished lineage. The idea was formulated in the latter part of the nineteenth century by the great English neurologist John Hughlings Jackson (1835–1911). Jackson, the son of a farmer, had a limited education and medical training. His genius lay in the meticulousness of his observations of patients with neurological diseases, combined with a philosophical cast of mind which enabled him to recognize the principles implied by the observations. One of his main studies was of epileptic seizures, particularly those that involve restricted parts of the body musculature. This type is now known, in his honor, as Jacksonian seizures. A tragic footnote is that his profound insights into this disease drew on his wife, who suffered from these seizures and died at an early age from cerebral thrombosis.

Jackson's study of the motor derangements associated with seizures convinced him that there are successive levels for motor control in the nervous system. In evolution, he deduced, there has been a progression from automatic toward purposive movements, and this is reflected in the nervous system in the control of automatic movements by lower levels and purposive movements by higher levels. He reasoned that the higher levels normally exert control over the lower levels, and that this control can be either excitatory or inhibitory. When upper-level function is interrupted or destroyed by disease, lower centers are "released" from higher control, and the result may be hyperactivity (such as exaggerated reflexes) if the normal descending control was inhibitory. Although knowledge of anatomy was limited in Jackson's time, he postulated that the lowest level for motor control is in the spinal cord and brainstem, the next (middle) level is in the cerebral cortex along the central (Rolandic) fissure, and the highest level is in the frontal lobe.

Motor Organization

These concepts have influenced all subsequent workers, and today they still provide a useful evolutionary framework for think-

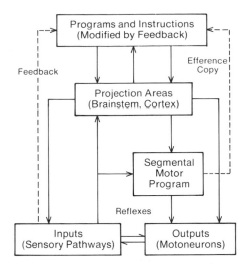

Fig. 21.3 Hierarchical organization of motor control. (Modified from Phillips and Porter, 1977)

ing about the organization of motor systems. A recent version of this framework for the mammalian nervous system is depicted in the schema of Fig. 21.3. The reader may wish to compare this with the diagrams of Figs. 19.1 and 20.2, which were mainly concerned with representing the lowest, segmental level of organization in the spinal cord. Figure 21.3 shows the immediate control of the spinal apparatus by specific motor regions of the brainstem and cerebral cortex, representing Jackson's "middle level" of control. These regions, in turn, are under the combined control of several areas, including the cerebellum and the basal ganglia (rather than exclusively the prefrontal cortex, as Jackson thought).

Several additional points may be made in reference to Fig. 21.3. One is that, as in the invertebrates, motor control in the mammal is not, in fact, strictly hierarchical. The hierarchical sequence descending from "projection areas" to "segmental motor programs" to "motoneurons" is bypassed by some direct pathways from the projection areas to the motoneurons. In other words, there are parallel pathways as well as serial ones. This recalls the studies of sensory systems, in which recent findings have emphasized the presence of both serial and parallel pathways. Thus, it appears that in both sensory and motor systems, greater adaptability and processing power are obtained by building in both types of pathways.

A second point is that there are feedback connections from every level of the motor system. Some of this feedback is from sensory pathways, conveying information from the periphery, information that alerts the animal, or tells it the effect of its motor actions on the environment. The sensory pathways themselves are under the influence of the motor outflow. In addition, the motor pathways not only carry information to the motoneurons, but also send a copy of that information back to the higher levels. As discussed in Chap. 20, this internal feedback is referred to as *reafference, corollary discharge,* or *afference copy.* Through it higher levels are kept informed of what lower levels are doing. It seems to express a principle of good management, that presidents and admirals should not put their trust entirely in their vice-presidents and colonels, but should see for themselves what is actually happening in the offices, factories, and battlefields. We have already discussed reafference as a type of circuit component which, together with the reflex and the rhythm generator, forms one of the three fundamental units of action for the organization of motor control systems (Chap. 20).

Brainstem Centers

It will be helpful to keep Fig. 21.3 in mind as our discussion moves to higher control of motor function in the vertebrate nervous system. First, the projection areas that constitute the middle level will be identified. These include the regions that are sources of descending fibers that terminate in the spinal cord. The main regions are shown in Fig. 21.4.

Reticular Nucleus

The reticular system is distributed diffusely through the brainstem. As we learned in the discussion of sensory systems (Chaps.

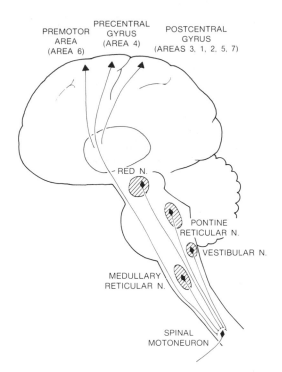

Fig. 21.4 Principal regions of the human brain involved in immediate motor control through descending fibers to motoneurons.

10–16), it receives collaterals from sensory projection pathways, and as one of its functions it contributes to states of arousal. This involves fibers that ascend toward the thalamus and descend toward the spinal cord. The functions of the reticular system, considered as part of central systems, are discussed in Chaps. 24 and 26.

The descending fibers are the ones directly involved in motor control. In fish and amphibians, a few cells in the reticular substance of the medulla are differentiated into *giant cells,* with large dendrites and a long axon which descends into the spinal cord to make synaptic connections with segmental interneurons and motoneurons. These are known as *Mueller cells* (6–8 cells in lamprey, for example) or *Mauthner cells* (a single pair, as in many teleost fish and amphibians).

From a phylogenetic viewpoint, the significance of these large, reticular neurons appears to be mainly in relation to control of the *tail*. In most vertebrates the tail is an important motor organ. It serves as an organ of locomotion, and it is also used in maintaining balance (as in swimming fish or running lizards). Part of the importance of the tail lies in the power it can generate; this is why it is so important in escape and startle responses. As we discussed in Chap. 19, the Mauthner cell, with its giant axon, provides a fast pathway for initiating quick and powerful movements of the body and tail, and we noted that its function in this respect is closely analogous to that of giant fiber systems in invertebrates.

In higher vertebrates, reticular neurons in the medulla and in the pons also send fibers to the spinal cord. Some of these fibers come from giant cells, though the cells are not uniquely identifiable. Numerous studies have indicated that the reticulospinal neurons constitute the main brainstem system for immediate control of the segmental spinal apparatus as it is employed in standing and stepping. We will discuss more details of this control below.

Vestibular Nucleus

The vestibular nuclear complex is closely associated with the reticular nuclei. This nucleus receives the afferents from the vestibular canals (Chap. 14) and also fibers from the cerebellum, and in turn provides one of the main inputs to the Mauthner cell in lower vertebrates (Chap. 19). Of the four main parts of the vestibular complex (see Fig. 14.12) in the mammal, the *lateral nucleus* is the main source of fibers to the spinal cord. The lateral nucleus contains *giant cells*, named *Deiters cells* after the early German histologist who first described them in 1865 (see the illustration from Deiters' work in Fig. 3.2); the nucleus is accordingly referred to as *Deiters nucleus*.

The *vestibulospinal tract*, composed of fibers from the giant cells as well as other cells of the lateral nucleus, descends the length of the spinal cord. The lateral nucleus is organized somatotopically, and this order is maintained in the projections to different levels of the cord. Electrical stimu-

lation of the lateral nucleus produces polysynaptic EPSPs in *extensor* motoneurons of the limbs. This has suggested that, through the vestibulospinal tract, the cerebellum exerts a facilitatory control over muscle tone in the extensor limb muscles, and thereby contributes to the standing posture.

It may be noted that the *medial vestibular nucleus* sends fibers to the upper part of the spinal cord by way of the *medial longitudinal fasciculus*. The role of these fibers in the control of eye movements has been discussed in Chap. 14. The *inferior* and *medial vestibular nuclei* project fibers to the cerebellum (see below). Through all these connections the signals from the vestibular canals are widely dispersed throughout the brainstem and spinal cord.

Red Nucleus

The red nucleus is a prominent structure in reptiles, birds, and mammals. It has a slightly pinkish color in fresh specimens, hence its name. Part of the nucleus contains *giant cells;* these, as well as cells of other sizes, project their axons to brainstem centers and to the spinal cord. In higher mammals, including humans, there are fewer giant cells, and the spinal tract is less prominent; other pathways become more important, especially the corticospinal tract (see below).

Electrical stimulation of the red nucleus in the cat causes *flexion* of the limbs. Intracellular recordings have shown that such stimulation produces polysynaptic EPSPs in flexor motoneurons and polysynaptic IPSPs in extensor motoneurons to the limbs. The excitatory effect may be mediated through segmental interneurons or by the activation of γ motoneurons in the γ loop (Chap. 13). The facilitative effect of the red nucleus on limb flexion thus contrasts with the facilitative effect of the vestibular nucleus on limb extension.

The Control of Walking

The significance of the three brainstem centers discussed above for the control of

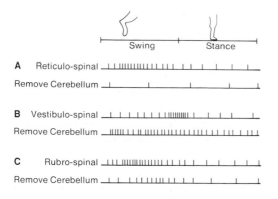

Fig. 21.5 Single-cell activity in brainstem motor centers in relation to stepping. See text for explanation. (Based on Orlovsky, in Wetzel and Stuart, 1976)

walking has been assessed by G. N. Orlovsky of Russia, who used the preparation of the cat depicted in Fig. 20.11. Some of his results are summarized in Fig. 21.5. Chief among the three centers in the control of walking is the reticular formation. In unit recordings, the reticulospinal cells increase their discharge in order to facilitate flexor motoneurons, as is necessary to bring the foot up and forward during the swing phase of stepping. After the removal of the cerebellum, the spontaneous activity of reticulospinal neurons falls precipitously, associated with a loss of muscle tone and loss of coordination of the limbs (see A in Fig. 21.5). From these results it has been concluded that the maintained discharge of reticulospinal neurons is involved in "switching on" the stepping generator, and the modulated discharge facilitates flexor activity in the swing phase.

By contrast, vestibulospinal neurons (B in Fig. 21.5) show peak firing that phase-leads the onset of the stance phase, thereby facilitating extensor motoneurons as the limb is extended to support the body against gravity. Finally, the activity of rubrospinal neurons (C) resembles that of reticulospinal neurons in facilitating flexor motoneurons during the swing phase. Neither the vestibular nor the red nucleus is essential for stepping; their roles thus appear supportive in relation to the reticular formation.

From these and related studies it has been concluded that brainstem centers control the spinal rhythm generator by two mechanisms: a *generalized activation* by means of ongoing spontaneous discharge, and *modulated discharges* that facilitate specific phases of the stepping cycle. The level of activation appropriate for switching on the generator depends on the cerebellum, while modulation is due to sensory feedback from the limbs relayed through the cerebellum. As expressed by Mary Wetzel and Douglas Stuart (1976) of Arizona:

. . . afferent modulation of efferent output by way of the cerebellum provides for selection from a larger array of corrective signals than if the modulation was achieved by directly ascending tracts, such as the spinoreticular and spinovestibular systems. The cerebellum's importance is becoming evident in the building of large-scale neural ensemble activity from stepping signals that arrive from many different parts of the nervous system.

Cerebellum

It is obvious from the foregoing remarks that a key center for sensorimotor control at the level of the brainstem is the cerebellum. The cerebellum lacks direct connection to the spinal cord, and thus stands higher in the motor hierarchy than the middle level, as indicated in Fig. 21.3. However, it is so intimately involved in brainstem mechanisms that it is appropriate to consider it here. The student should review the development of the cerebellum in Chap. 9.

The cerebellum is an outgrowth of the pons. It arises early in vertebrate phylogeny. In brains of different vertebrate classes, the cerebellum varies from a mere nubbin in some species to a large, much convoluted structure in others. It is often stated that the cerebellum increases in size during phylogeny, but there are numerous exceptions to this generalization. Most notable is the enormous expansion in certain

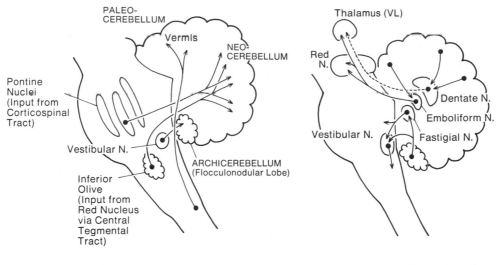

Fig. 21.6 Main input (A) and output (B) pathways of the mammalian cerebellum.

species of electric fish, which contrasts with the generally small size in most amphibians and reptiles. The large size of the cerebellum in mammals is believed to be related to several important functional roles. Three roles have traditionally been identified; they involve control of (1) *muscle tone;* (2) *balance;* and (3) *sensorimotor coordination.* Let us consider the organization of the cerebellum in terms of its input and output connections, and see how they relate to these functions.

Like Caesar's Gaul, the cerebellum is divided into three parts, which reflects its phylogenetic history. First is the *archicerebellum,* consisting of the small flocculonodular lobe. Next is the *paleocerebellum,* consisting of the anterior lobe. Third is the *neocerebellum,* formed by the great expansion of the lateral hemisphere.

Inputs to the Cerebellum

The *inputs* to the cerebellum are specific for these three parts. As shown in Fig. 21.6A, *vestibular* fibers make connections to the flocculonodular node. Fibers from the *spinal cord* ascend in the spinocerebellar tract and terminate mainly in the anterior lobe. The ventral portion of this tract

is present in fish, but the dorsal tract is present first in reptiles. These fibers carry information from muscle receptors. Finally, input to the neocerebellum comes mainly from the large masses of cells in the *pons* and from the *inferior olivary nucleus* in the medulla.

Outputs from the Cerebellum

The *output* fibers from the cerebellar cortex project to the midbrain in fish and urodele amphibians; the midbrain in these animals is one of the main centers for sensorimotor coordination. In higher vertebrates, however, the output from the cortex goes to a set of *deep cerebellar nuclei* (which may represent specializations of the more primitive midbrain center).

The relations between cortex and deep nuclei are somewhat more complicated than the simple tripartite division mentioned above. Thus, as shown in Fig. 21.6B, the *fastigial* nucleus receives fibers from the midline vermal zone of the cortex, and projects in turn to the lateral vestibular nucleus; there is also a direct connection to this nucleus as well. The *emboliform* nucleus receives fibers from a strip of cortex next to the midline; it projects in turn to the red nucleus, and also has some terminals in the ventrolateral nucleus of the thalamus. The *dentate* nucleus is by far the largest of the deep nuclei, reflecting the fact that it receives the output fibers of the large cerebellar hemisphere. The dentate nucleus projects some fibers to the red nucleus, but in higher mammals most of them go to the ventrolateral nucleus of the thalamus.

The connections to the vestibular nucleus account for the powerful control exerted by the cerebellum on mechanisms of balance. The connections to the red nucleus and reticular nucleus mediate control over reflexes and muscle tone, through the projections of these nuclei to the spinal cord. The role of the lateral cerebellar hemisphere→dentate→ventrolateral thalamus circuit in sensorimotor coordination is more subtle; it is expressed through the cerebral cortex, as we will see below.

Microcircuits of the Cerebellar Cortex

We have identified the main cerebellar pathways and the overall effects they mediate, but what actual operations are carried out within the cerebellum itself? For this we must consider the microanatomy and microphysiology of the cerebellar cortex. When we do, we find that this is one of the most astonishing pieces of cellular machinery in the animal body. It consists of a convoluted sheet, divided into a deep and a superficial layer. The deep (granule) layer is packed with tiny *granule* cells; the best estimates put their numbers at 10–100 billion, which is more than all the other cells in the nervous system combined! The granule cell axons ascend to the superficial (molecular) layer and bifurcate into two *parallel fibers,* which run for several millimeters in opposite directions. The molecular layer also contains the large *Purkinje cells,* of which there are about 7 million. Each has a widely branching dendritic tree that is flattened into a two-dimensional plane and oriented at right angles to the parallel fibers. These relations are diagramed in Fig. 21.7 (for the development of these relations, see Chap. 9). The Purkinje cell dendrites are covered with *spines,* which are the sites of synapses from the parallel fibers

Fig. 21.7 Neuronal organization of the cerebellar cortex. Inputs: mossy fibers (MF) and climbing fibers (CF), parallel fibers (pf), fibers from locus ceruleus (LC), and raphe nucleus (DR). Principal neuron: Purkinje cell (P), with recurrent collateral (rc). Intrinsic neurons: granule cell (Gr); stellate cell (S); basket cell (B); Golgi cell (Go). Histological layers are shown at the right: molecular layer (MOL), Purkinje cell body layer (PCL), granule layer (GrL). (Modified from Shepherd, 1979)

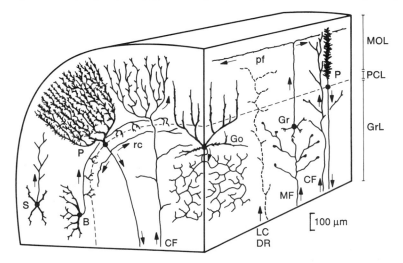

as they pass through successive Purkinje cell dendritic arbors. The Purkinje cell axons carry the output of the cerebellar cortex to the deep cerebellar nuclei.

There are three main types of fiber that carry inputs to this neuronal machinery. One type is the *mossy fiber,* which arises from the pontine nuclei, as well as several other sites, and terminates on granule cell dendrites in large endings that, to some early anatomists, seemed to have a "mossy" appearance. This input is relayed through the granule cells to the Purkinje cells, and in so doing is subjected to considerable convergence and divergence, due to branching of the parallel fibers and overlap of their connections (see diagram). The second type is the *climbing fiber,* which arises mainly from the inferior olivary nucleus. This fiber ends in an extensive arborization which literally climbs over the Purkinje cell dendritic tree and makes synapses onto it. This input thus has a one-to-one, direct relation to the Purkinje cell. A third type of fiber ramifies widely throughout the cortex; these have been identified as norepinephrine- and serotonin-containing fibers arising from the brainstem (see Chap. 24).

Basic Circuit of Cerebellar Cortex

The basic circuit put together from anatomical and physiological studies is summarized in Fig. 21.8. We will focus on four essential features of this circuit. First, the direct input pathway, through the climbing fibers, and the indirect pathway, through the mossy fiber→granule cell→parallel fiber relay, are both excitatory. Second, all other connections, by the various types of interneurons present (Golgi, basket, stellate), are inhibitory. This means that the responses of cerebellar cortical neurons to their excitatory inputs are brief, being rapidly terminated by inhibition. Third, the connections of Purkinje cell axons onto their target neurons in the deep nuclei are also inhibitory. This was surprising when first discovered, and it means that one of

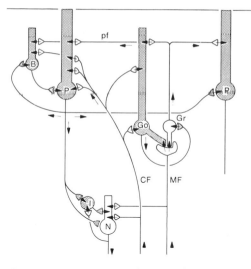

Fig. 21.8 Basic circuit diagram for the mammalian cerebellum. Note deep cerebellar nuclear cells: principal neuron (N) and intrinsic neuron (I). Other abbreviations in Fig. 21.7. (Modified from Shepherd, 1979)

the basic input–output operations of the cerebellar cortex is to convert its excitatory inputs into inhibitory outputs. Fourth, the deep nuclei, as shown in Fig. 21.8, also receive inputs from the climbing and mossy fibers. The cerebellar cortex can therefore be regarded as a highly sophisticated interneuronal system, mediating feedforward inhibition to control the input–output operations of the deep nuclei. Finally, both deep cerebellar neurons and Purkinje cells have high rates of resting discharge (50–100 impulses per second). One function of this high rate is to act as a high *set point,* so that the cerebellum will be maximally sensitive to both decreases and increases in firing rate caused by changes in input activity.

Functional Operations of the Cerebellum

How does the exquisite neuronal machinery of the cerebellum generate functional operations that mediate sensorimotor coordination? This has turned out to be a diffi-

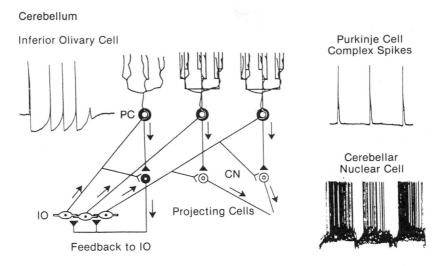

Fig. 21.9 Closed-loop generation of rhythmic activity by interactions between inferior olivary (IO) cells, Purkinje cells (PCs), and deep cerebellar nuclear (CN) cells. IO output is excitatory; PC and CN feedback is inhibitory (filled profiles). See text. (From Llinás and Walton, 1990)

cult question to answer, because of the complexity of cerebellar organization and its many input and output connections with the rest of the nervous system. We consider two properties and their functional implications: rhythmic activity and plasticity.

Rhythmic Activity. Rodolfo Llinás and his collaborators have analyzed this property by correlating the functional properties of the cerebellar cortex, deep nuclei, and the main input pathways (see Fig. 21.9).

A key finding is that inferior olivary cells (1) are electrically coupled and have strong pacemaker activity, so that their axons transmit synchronous and rhythmic excitatory synaptic input to both the cerebellar nuclear cells and, through climbing fibers, the Purkinje cells of the cerebellar cortex. The climbing fibers thus activate bands of Purkinje cells in a powerful and synchronous manner. The Purkinje cells (2) respond with intensely depolarizing burst responses (also called "complex spikes"), which are transmitted as intense phasic inhibitory inputs to the cerebellar nuclear cells. The nuclear cells (3) are driven by the excitation from the inferior olivary cells

followed by the inhibition from the Purkinje cells; the inhibition leads to rebound excitation. The nuclear cells thus rhythmically excite their target cells in the red nucleus and thalamus (see Fig. 21.7). There is also a GABAergic inhibitory feedback connection to the inferior olive, where it reduces the electrotonic coupling between the cells to restrain their output.

The effect of this closed loop of rhythmic activity, according to Llinás and Walton (1990), is that

the olivocerebellar system would serve as an oscillatory circuit capable of generating timing sequences such as observed in tremor and in the organization of coordinated movements. . . . The mossy fiber–parallel fiber system provides a continuous and delicate regulation of the excitability of the cerebellar nuclei, brought about by the tonic activation of simple spikes in Purkinje cells, that ultimately generates the fine control of movement known as motor coordination. The fact that the mossy fibers inform the cerebellar cortex of both ascending and descending messages to and from the motor centers in the spinal cord and brainstem gives an idea of the ultimate role of the mossy fiber system; it informs the cortex of the place and rate of movement of limbs and puts the motor intentions generated by the brain into the context of the status of the body at the time the movement is to be executed.

Plasticity. We have seen that neural circuits are characterized by considerable plasticity, and cerebellar circuits are no exception. As pointed out by Masao Ito (1991) of Riken, Japan, cerebellar plasticity is manifested in three main ways: (1) activity-dependent changes related to motor learning; (2) axonal regeneration and synaptic sprouting in response to injury; and (3) remodeling of neurons and connections during development. Here we discuss briefly changes related to learning.

One of the changes that is of most interest as a possible basis for learning in the cerebellum is a long-lasting depression of parallel fiber synapses following repeated excitation by climbing fibers of Purkinje cells. This is referred to as long-term depression (LTD), in contrast to long-term potentiation (LTP) (see Chaps. 8 and 29). Current evidence points to glutamate as the transmitter at parallel fiber synapses. Climbing fiber activation elicits EPSPs, which lead to large depolarizations due to dendritic Ca^{2+} spikes and plateau potentials, as recorded by microelectrodes and by Ca^{2+}-sensitive dyes. As shown in Fig. 21.10, the raised intracellular Ca^{2+} is believed to activate second messenger mechanisms, which lead to desensitization of the AMPA receptors in the spine synapses that receive the parallel fiber terminals. Several possible second messenger mechanisms are presently under study. One is the activation of metabotropic glutamate receptors coupled to IP_3 production, which would contribute to increased internal Ca^{2+} levels. Another is activation of NO synthase, production of NO, and subsequent activation of guanylate cyclase (presumably in neighboring cells; see Chap. 8) to produce the second messenger cGMP; the cGMP might then activate cGMP-dependent protein kinase (PKG) to desensitize the AMPA receptors. There might also be direct gating by cGMP of a membrane conductance, similar

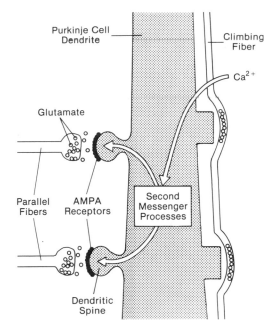

Fig. 21.10 Simplified diagram summarizing a proposed mechanism of long-term depression (LTD) in the cerebellum. Repeated climbing fiber activation induces increased intracellular Ca^{2+}, which activates second messenger mechanisms leading to desensitization of AMPA receptors for glutamate at parallel fiber synapses onto Purkinje cell spines. See text. (From Ito, 1991)

to its action in olfactory and retinal cells (see Chaps. 8, 11, and 16). As in the case of LTP, there are contributions by both presynaptic and postsynaptic mechanisms to LTD.

Motor Cortex: A Brief History

A crucial component of the middle level of motor control is the *motor cortex.* The identification of the motor part of the cerebral cortex is one of the most dramatic chapters in the history of neurophysiology. During the American Civil War in the early 1860s, Weir Mitchell, a Philadelphia neurologist, noted that one side of the brain appeared to be related to the opposite side of the body. In the Prusso-Danish War of 1864, the German physician Theodor Fritsch observed, in dressing a head wound, that irritation of the brain caused twitching in the opposite side of the body. He took this information to Eduard Hitzig, then in medical practice in Berlin, and they set about to test this finding experimentally. According to the legend (Haymaker, 1953):

At that time there were no laboratories available at the Physiological Institute in Berlin for work on warm-blooded animals, and as a consequence Hitzig and Fritsch did their first studies on dogs in Hitzig's home, operating on Frau Hitzig's dressing table.

This was one of the great discoveries in neurobiology, for it demonstrated both the *electrical excitability of the brain* and the *localization of motor function* in the cerebral cortex. The importance of the finding of localization was not lost on David Ferrier of London, who proceeded in the 1870s to map very thoroughly the excitable cortex in a variety of mammals, including monkeys. One of his aims was "to put to experimental proof of the views entertained by Dr. Hughlings Jackson." Ferrier found that the most excitable area for eliciting movements in the monkey is a strip of cortex on the precentral gyrus, the gyrus of Rolandi (where Jackson had expected it to be). He showed an orderly progression of focal areas along the gyrus for eliciting movements of the leg, hand, and face. This was dramatic confirmation indeed of Jackson's postulate that there must be an orderly sequence of discharging foci in the cortex to explain the "march of spasm" of the muscles that is so characteristic of focal epileptic convulsions (the kind now known as Jacksonian seizures).

Studies of localization of the motor cortex were extended to the great apes by Sherrington in the early 1900s, and finally to humans by Wilder Penfield, a student of Sherrington's, in the 1930s. Penfield, with his colleagues at the Montreal Neurological Institute, focally stimulated the cortex of patients during neurosurgical operations. They found, as had many workers before them in other species, a great deal of over-

lap between the areas for different parts of the body. The areas for the hand and face had the lowest thresholds and widest fields (Fig. 21.11A). Penfield and his colleagues summarized their findings in a homunculus, as shown in Fig. 21.9B. This is the motor counterpart to the sensory map of the body surface previously shown in Chap. 12.

These findings left little doubt that the precentral gyrus (cytoarchitectonic area 4 of Brodmann) is the cortical region most closely involved in immediate control of motoneurons in the spinal cord. This control is mediated by the *corticospinal tract* which provides a direct connection between the cortex and the spinal cord. Because its fibers are funneled through the pyramids on the ventral surface of the medulla in the brainstem, it is also called the *pyramidal tract*. The traditional view, that this tract mediates voluntary control of movement, seemed to many to be almost self-evident. However, this view concealed several difficult problems that have required many years of work to resolve, and have led to a much expanded view of "motor cortex."

The Organization of Motor Cortex

The first question to be asked is, which cells give rise to the fibers in the pyramidal tract? The fibers arise from pyramidal-shaped neurons in cortical layer V, which is an obvious source of terminological confusion; it must be kept in mind that the pyramidal tract derives its name from the fact that it passes through the medullary pyramids, not because its fibers arise from pyramidal-shaped neurons! The layer V cells include some giant neurons, called Betz cells after their discoverer (see Fig. 21.12). For many years it was thought that all corticospinal fibers arise from Betz cells, but now it is known that only about 3% of the tract fibers can be accounted for on this basis. Betz cells are found mainly in the leg area, and thus their large size appears to be correlated with the greater length of

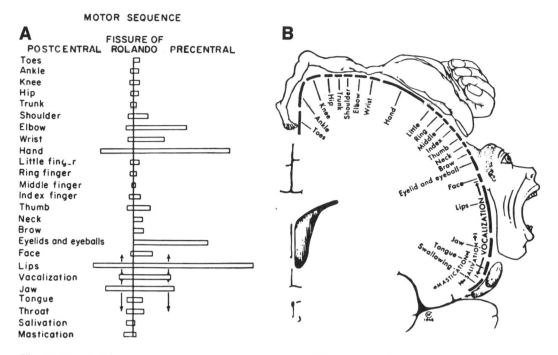

Fig. 21.11 **A.** The relative proportions of movements elicited by stimulation anterior and posterior to the central (Rolandic) sulcus in the series of patients of Penfield and Rasmussen, 1950. **B.** The motor homunculus of Penfield and Rasmussen. (From Phillips and Porter, 1977)

axon they send to caudal parts of the spinal cord. Nonetheless, it is of interest that, like the brainstem centers, the motor cortex of the middle level of the motor hierarchy exerts some of its control through giant neurons.

The second question is, does the corticospinal tract arise only from area 4? Both anatomical and physiological studies have shown that area 6 (just anterior to area 4) and the postcentral gyrus (3, 1, and 2 of the somatosensory area) also contribute fibers. Thus, as shown in Fig. 21.4, the corticospinal tract arises from extensive areas of cortex. Movements elicited from area 6 (also called the premotor area) and areas 3, 1, and 2 by electrical stimulation are less precise and have higher electrical thresholds than those elicited from the motor strip in area 4. There is thus *multiple representation* of the motor map in the cerebral cortex (Fig. 21.13).

The third question is, does the corticospinal tract mediate direct (monosynaptic) control of motoneurons? Phylogenetic comparisons provide a useful perspective on this question (see Fig. 21.14). In lower mammals, such as the rabbit, the corticospinal tract barely reaches the anterior segments of the spinal cord, so that further connections within the cord have to be by propriospinal interneurons. In mammals like the cat, the tract reaches most of the cord, but the fibers make synaptic connections only onto segmental interneurons, which provide a polysynaptic pathway onto motoneurons. It is only in primates that a monosynaptic pathway from the tract fibers to spinal motoneurons exists.

This direct connection from the cortex to spinal motoneurons is very important for the motor capacities of primates, including humans, as will be discussed in the next chapter.

Parallel Motor Pathways

The main types of motor centers and their descending pathways are summarized in

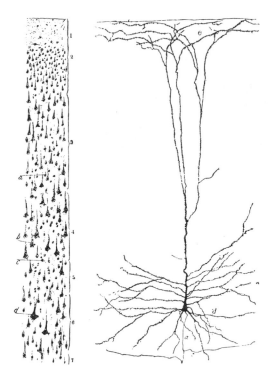

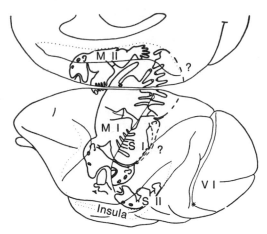

Fig. 21.13 The motor areas of the monkey cortex. M I, precentral motor area (equivalent to areas 4 and 6 of the human cortex); M II, supplementary motor area; S I, primary somatic sensory area. All these areas send fibers to the pyramidal tract as well as extrapyramidal centers. Also shown are the secondary somatic area (S II) and primary visual area (V I). (From Woolsey, in Henneman, 1980)

Fig. 21.12 *(Left)* Histological laminae of the human motor cortex. In this preparation only the cell bodies have been stained. There are seven laminae in Cajal's classification. The laminae are identified on the basis of their relative numbers of large cell bodies (pyramidal cells) and small cell bodies (pyramidal or granule [stellate] cells). The significance of the laminae will be discussed further below (cf. Fig. 21.15). *(Right)* Single Betz cell of the motor cortex, impregnated by the Golgi method. Note the thick apical dendrite which divides into three ascending branches that ramify at the cortical surface, the many shorter basal dendrites, and the axon (a) which arises from the cell body and gives off two horizontal axon collaterals. (From Cajal, 1911)

Fig. 21.15. First are the brainstem centers. The reticular, vestibular, and red nuclei were discussed as representative of these, to which can be added others such as the superior colliculus and its tectospinal tract. Present in all higher vertebrates, these constitute the middle-level mechanisms that set the main patterns of control of motor behavior. This is referred to as the *extra-*

pyramidal system, because the fibers all lie outside the medullary pyramids.

Studies by Henricus Kuypers at Cambridge have indicated that the brainstem pathways are organized into two main groups, according to where they terminate within the gray matter of the spinal cord. One is the *ventromedial* group, which includes the reticulospinal and vestibulospinal tracts. This is the basic system for overall control of movements, pertaining to "maintenance of erect posture, to integrated movements of body and limbs such as orienting movements, to synergistic movements of the individual limbs and to directing the course of progression" (Kuypers, 1985). This is in accord with our above discussion. The other is the *lateral* group, which includes the rubrospinal tract. This supplements the ventromedial control, but especially adds the ability for fine control of the distal extremities. A lesion, for example, of the red nucleus has little effect on overall motor control (which by contrast is severely affected by vestibu-

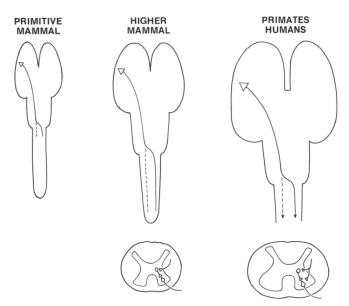

Fig. 21.14 Comparison of pyramidal tract pathways and terminations in the spinal cord of a primitive mammal (opossum), a higher mammal (cat), and a primate. Dashed lines indicate small uncrossed portion of the tract.

lospinal lesion), but impairs the ability of a monkey to use its hand and fingers. The corticospinal tract is a part of the lateral group, and adds further refinement to independent movement of the fingers, as we shall see in the next chapter.

The motor cortex, which gives rise to the corticospinal tract, may be regarded to some extent as a component of the middle level of control because of its direct connections to the spinal cord. However, it is at a higher anatomical level, and also assumes a higher level function relative to the brainstem. The corticospinal fibers not only project directly to the spinal cord, but also give off collaterals to the brainstem centers. In addition to these inputs, there are fibers from other cortical cells which terminate only in the brainstem and do not proceed further. Many of these connections have been established by making horseradish peroxidase (HRP) injections into subcortical nuclei and identifying the labeled cells in the cortex. These studies, in addition, have shown that the fibers to these nuclei arise from neurons in layer V. By contrast,

motor cortical cells projecting to other areas of cortex arise from several layers. This is summarized in Fig. 21.15.

Traditionally it has been believed that the corticobulbar cells are found in the premotor and postcentral areas, as mentioned above. Recent experiments, however, have shown that the origin of these fibers is much more extensive. For example, when HRP injections are made into the superior colliculus, retrogradely labeled cells are found in the visual and auditory cortex, as well as somatosensory. The cells are pyramidal-shaped cells in layer V, the same as the motor cells of the motor cortex. The superior colliculus, in turn, has direct connections to the spinal cord through its tectospinal tract, as well as indirect connections through other brainstem connections.

These findings suggest several important points regarding the organization of motor control. First, the pyramidal and extrapyramidal systems provide to some extent separate and parallel pathways for the control of the spinal cord. Second, the two systems are interrelated by interconnections at all

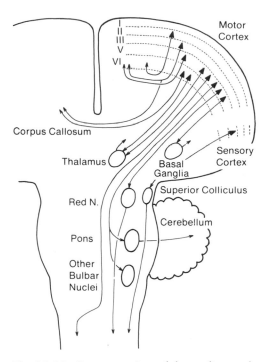

Fig. 21.15 Summary view of descending pathways, showing multiple projections and different laminar locations of cortical output neurons. Note that superficial cortical laminae (II, III) project to other cortical areas, whereas deep cortical laminae (V, VI) give rise to descending projections, both pyramidal and extrapyramidal. Lamina IV (granule layer) is poorly developed in most parts of primary and secondary motor cortex. (Based in part on Jones, 1981)

the main levels: cortex, brainstem, and spinal cord. Third, the extrapyramidal centers in the brainstem are influenced by connections from extensive areas of cortex. Together, these form what has been termed the *cortically originating extrapyramidal system,* to distinguish it from the *cortically originating pyramidal system.*

The Basal Ganglia

Now that we have identified the middle level of the motor hierarchy, we need to finish by discussing where the "Programs and Instructions" are found that represent the highest level in Fig. 21.3. One of the main regions is the basal ganglia of the forebrain.

Anatomy

The forebrain includes all of the nervous system above the diencephalon. It consists of two kinds of structure, the outer layer, called the cortex, and an inner mass, called the basal ganglia (*basal* because they seem to be within the base of the cerebral hemispheres, and *ganglia* because this was the term applied by nineteenth-century histologists to large groups of neurons).

The positions of the basal ganglia are shown in Fig. 21.16. Of the several ganglia, the caudate is an elongated extension of the putamen; the two have a similar neuronal structure, and together are called the *striatum* (this term comes from the fact that there are bands of fibers passing through, which produce a striated appearance). As indicated in the diagram, the striatum receives widespread inputs from the cerebral cortex; its output, in turn, is directed to another of the basal ganglia, the *globus pallidus,* and to the *substantia nigra.* The substantia nigra is actually located in the midbrain, but its main connections are with the striatum and globus pallidus, and it is therefore functionally linked with the basal ganglia. The output of the globus pallidus is directed to the thalamus, to the same nuclei which receive inputs from the cerebellum and project widely to the cerebral cortex. The output of the substantia nigra is directed both to the thalamus and back to the striatum. These connections can all be traced in Fig. 21.16.

Until the 1960s, almost nothing was known about the functions of these large masses of cells; they were the great silent interior of the cerebral subcontinent. The only clues were that pathological changes were found in these regions in patients with certain striking movement disorders. Lesions in the putamen and globus pallidus were associated with slow, writhing movements (athetosis). Degeneration of cells in the striatum was found in patients with Huntington's chorea, characterized by in-

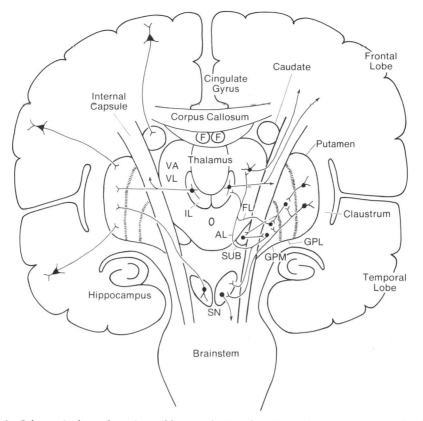

Fig. 21.16 Schematic frontal section of human brain, showing main connections and relations of the basal ganglia (shaded). Globus pallidus, lateral segment (GPL); medial segment (GPM). SUB, subthalamic nucleus; SN, substantia nigra; AL, ansa lenticularis; FL, fasciculus lenticularis. Thalamic nuclei: VA, ventral anterior; VL, ventral lateral; IL, intralaminar nucleus; F, fornix; O, cerebral aqueduct. Connections of caudate are similar to those shown for putamen. (Modified from Shepherd, 1979)

voluntary jerking movements. Lesions of a small region called the subthalamic nucleus were associated with violent flinging movements of the extremities (ballismus).

The most interesting correlation was the finding of degeneration of the dopaminergic input from the substantia nigra to the striatum in patients with Parkinson's disease. This was established in the 1960s, and was the first correlation of a neurotransmitter deficiency with a neurological disease. It provided the basis for the use of levodopa (L-dopa), a dopamine precursor, to treat these patients (see Chap. 24). It has also provided the precedent for hoping that other diseases, such as schizophrenia, may also be linked to a defect in a specific

transmitter or substance. The use of brain transplants of dopamine-containing or dopamine-synthesizing cells to alleviate the symptoms of Parkinson's disease was discussed in Chap. 9.

Local Circuits and Microcircuits

All this work indicated that the basal ganglia are crucially involved in the control of movement and in sensorimotor coordination. Spurred by these findings, neuroscientists have directed their attention to the cellular properties and microcircuits within each region. Figure 21.17 illustrates some of these results. In the striatum, Stephen Kitai and his colleagues, then at Michigan State University, injected single projection

A. CAUDATE NUCLEUS

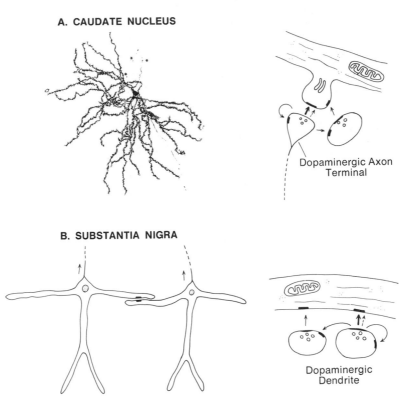

Dopaminergic Axon
Terminal

B. SUBSTANTIA NIGRA

Dopaminergic
Dendrite

Fig. 21.17 Synaptic organization and interactions in the basal ganglia. **A.** Caudate–putamen. *On the left,* the profuse dendritic tree of an HRP-filled neuron. *On the right,* the schematic diagram shows some possible synaptic actions of dopaminergic fibers from the substantia nigra. Thick arrow, brief synaptic action; dotted arrow, slow synaptic action; other arrows, actions of dopamine on autoreceptors and on other synaptic terminals. **B.** Substantia nigra. *On the left,* two dopaminergic output neurons. *On the right,* some possible actions of dopamine, analogous to those in A. (HRP-filled neuron in A is from Kitai, 1981; diagrams are based on Aghajanian and Bunney, 1976; Nowycky and Roth, 1978; Groves et al., 1979; and Glowinski et al., 1984)

neurons with HRP and showed that most have dendrites covered with spines (A). The axon gives off an enormous number of collateral branches, providing for multiple local circuit connections to neighboring neurons.

On the input side, neuroanatomists have found that fibers from the cortex, thalamus, and substantia nigra make synapses onto the spines. Most of these synapses have a type I morphology (see Chap. 6), suggesting an excitatory action. However, the action of the input fibers has been a matter of keen debate. For the case of the dopaminergic fibers, some studies have suggested a rapid, brief excitatory action,

while others have suggested a slower, inhibitory action. Biochemical studies have identified a dopamine-sensitive adenylate cyclase in the striatum, presumably activated by the dopaminergic fibers. In addition to its specific synaptic action, dopamine (DA) may act by diffusing from varicosities to postsynaptic sites, and it may also act on presynaptic terminals, as indicated in B in Fig. 21.17.

These studies indicate that a single type of fiber secreting a single transmitter can have several kinds of actions at several sites. This is in accord with similar results obtained under more precisely controlled conditions in invertebrate neurons. At each site, a specific combination of cellular properties is brought into play, to integrate information, transmit it to neighboring sites, and exert local feedback and modulatory control, actions that in addition may be use-dependent and lead to plastic changes. David Smith and his colleagues at Oxford have speculated that the combination of excitatory and inhibitory synaptic inputs onto individual spines functions as an AND-NOT gate in processing information in the striatum (Freund et al., 1984).

Similar results have been obtained in the substantia nigra. Phillip Groves and his colleagues in San Diego have postulated that the output firing of the dopaminergic cells is controlled by feedback of DA, similar to the presynaptic control in the striatum. DA is contained within the dendrites of these cells. Its release from the dendrites has been shown by Jacques Glowinski and his colleagues (1984) in Paris. They inserted a push–pull cannula into the substantia nigra, which allowed them to infuse various substances that activate the nigral neurons and then collect the released DA. During natural activity it is postulated that released DA may act on autoreceptors in the same dendrites, or diffuse to receptors on neighboring dendrites, or act at dendro-dendritic synapses; such synapses have been identified in the electron microscope (see Fig. 21.17B).

Mosaic Organization of the Striatum

An important step forward in understanding the anatomical and functional organization of the striatum was the discovery by Patricia Goldman and Walle Nauta at M.I.T. that the striatum is not a homogeneous collection of cells, but is organized in compartments (Goldman and Nauta, 1977). The present picture is of two main divisions, cell islands, called *striosomes,* or *patches,* and the surrounding regions, called *matrix.* The medium spiny neurons respect these boundaries, with their dendrites confined either to the patches or to the matrix (see Fig. 21.18), whereas the interneurons have dendrites that span the boundaries, and their axons also ramify across the boundaries. Thus, the output neurons appear to process information in at least two independent parallel paths, with the interneurons presumably providing for coordination or antagonistic interactions between the two. There is evidence in the rat that there are differential cortical inputs to the two compartments, with the deeper layers of cortex projecting to the patches and the more superficial layers projecting to the matrix (Gerfen, 1992). However, this pattern is not so clear in primates. In fact, in primates there is a third compartment, the so-called *matrisomes,* discrete zones within the matrix which receive complex combinations of inputs from ipsilateral and contralateral M I and S I cortical areas (Flaherty and Graybiel, 1993).

A summary of how these compartments of the striatum relate to its input and output connections is provided in Fig. 21.19. It is simplest to begin with the striosomes (S), where medium spiny cells receive excitatory input from deep-layer cells in the prefrontal cortex, and project to the pars compacta of the substantia nigra, where they have inhibitory actions on the dopaminergic cells which provide a feedback loop primarily ending on matrix cells. A second distinct system is the medium spiny cells in the matrix that receive excitatory inputs from superficial-layer cells in sensorimotor cortex and send their outputs to the pars reticulata of the substantia nigra. This inhibits the output from this region to the superior colliculus (SC) and the return pathway to the cortex by the way of the ventral anterior (VA) and mediodorsal (MD) nuclei of the thalamus; since those outputs are inhibitory, the effect is to disinhibit those target cells. A third system involves the medium spiny cells in the matrix that project to the globus pallidus; these are subdivided into cells which project to the pars externa (GPe) and and pars interna (GPi). The cells projecting to the GPe have been shown by various methods, including

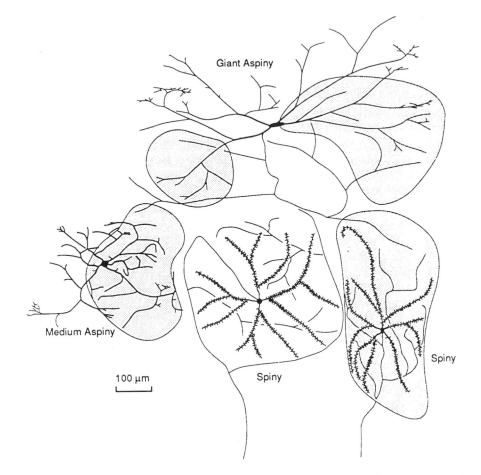

Fig. 21.18 Relationship of the dendritic fields of different neuron types to the mosaic organization of the neostriatum. Striosomes are shaded, matrix is unshaded. Note that spiny neurons tend to be confined within either patches or matrix, whereas aspiny neurons have dendrites that cross the boundaries. (From Wilson, 1990)

in situ hybridization, 2-deoxy-glucose monitoring of neural activity, and pharmacological lesions of the dopamine pathway using 6-hydroxydopamine, to express enkephalin in addition to the neurotransmitter GABA, as well as D2 receptors for dopamine (see Chap. 8). By contrast, the cells projecting to GPi express GABA, substance P, and dynorphin, as well as D1 dopamine receptors. The two parts of the globus pallidus have distinct outputs, as can be traced in the diagram.

These parallel systems, each with a unique combination of neurotransmitters and neuropeptides and set of input and output connections, provide for fine tuning of the activity of the cells in each of the centers. Note, for example, that the GPe, GPi, and SNpr (substantia nigra, pars reticulata) all receive different combinations of excitatory and inhibitory inputs and project to different targets. In this way, according to Charles Gerfen at the National Institute of Mental Health,

the relative responsiveness of striatonigral and striatopallidal neurons to cortical input determines the pattern of activity of the output neu-

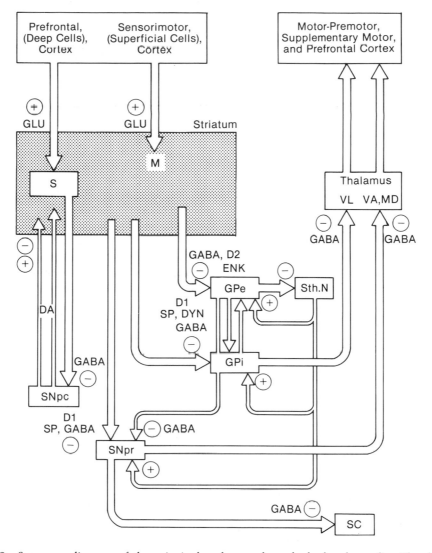

Fig. 21.19 Summary diagram of the principal pathways through the basal ganglia. The different regions include striosomes (S); matrix (M); substantia nigra, pars compacta (SNpc); substantia nigra, pars reticulata (SNpr); globus pallidus, pars externa (GPe); globus pallidus, pars interna (GPi); subthalamic nucleus (Sth.N); superior colliculus (SC); and ventrolateral (VL), ventroanterior (VA), and meriodorsal (MD) thalamic nuclei. The different neurotransmitters and neuromodulators in these pathways include glutamate (GLU); 8-aminobutyric acid (GABA); enkephalin (ENK); and substance P (SP). Excitatory synapses, +; inhibitory synapses, −. (From Graybiel, 1991)

rons of the basal ganglia in the substantia nigra . . . dopamine plays a pivotal role in modulating this responsiveness. (Gerfen, 1992)

A final level of organization of the striatum is topographical; there is an orderly progression in the projection from different areas of the cortex to the rostrocaudal extent of the striatum. The convergence and divergence patterns in relation to patch and matrix are complex, however, and vary with species, posing what Gerfen has

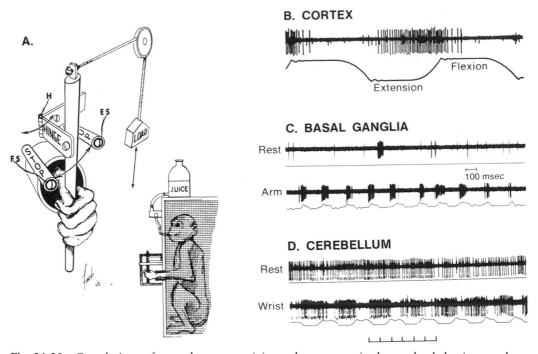

Fig. 21.20 Correlations of central neuron activity and movement in the awake, behaving monkey. **A.** Experimental setup for training a monkey to perform an extension or flexion of the wrist; it received a reward of fruit juice if the movement was performed within a brief specified period of time (400–700 msec). **B.** Activity of a single pyramidal tract neuron in the motor cortex in relation to movement. This unit was active prior to and during flexing of the wrist to a steady flexed position (but not during the maintained flexion). It increased its activity when flexion was performed against a flexion load (as in A), but was silent when flexion was performed against an extension load; this indicates that the motor cortex encodes force of movement as well as direction, displacement, and speed. **C.** Activity of a single neuron in the globus pallidus. The unit showed regular bursting in relation to push–pull movements of the arm, but not in relation to similar movements of the leg. Note the low rate of impulse firing at rest. **D.** Activity of a single Purkinje cell in the cerebellum. This cell showed regular bursting in relation to alternating extension and flexion of the wrist. Note the high rate of firing at rest, compared with the low rate of the basal ganglia cell in C. Time bar: 100-msec divisions. (A, B from Evarts, in Phillips and Porter, 1977; C from DeLong and Georgopoulos, 1981; D from Thach, in Brooks and Thach, 1981)

termed "a complex functional mapping problem." Presumably, different regions of the striatum function locally as well as globally in relation to specific regions of the cortex, but this is currently under study.

Functional Organization for Control of Movement

To complete the picture of organization, we need to consider the basal ganglia as a system, and the way it relates to other

systems. The basal ganglia feed into two thalamic nuclear groups, the ventral lateral–ventral anterior group, and the intralaminar group. These, in turn, project to the cortex, and thus complete a loop through cortical cells back onto the striatum, as shown in Fig. 21.19. This organization into closed loops is one of the outstanding characteristics of basal ganglia circuits, at all levels: it is seen in the several forms of feedback in microcircuits, in the reciprocal relations between striatum and substantia nigra, and in the loop through thalamus and cortex. In the thalamus, the input from the basal ganglia is integrated with input from the cerebellum. The cerebellum is embedded in its own loop, through thalamus to cortex and back through the pons. The organization of these circuits into loops apparently confers the advantages of tight feedback at all levels in the control of movement. It appears, however, to have the disadvantage of uncontrollable oscillations when any part is damaged, as in the disorders seen in neurological diseases involving these regions.

The extensively related circuits involve many regions of the brain, and are a good example of a *distributed system* (see Chap. 24). In order to gain insight into how activity throughout this system is coordinated, single-unit recordings have been made in several of the regions, using the methods of Evarts (see next chapter). Figure 21.20 summarizes the results obtained from three regions—cortex, basal ganglia, and cerebellum—in relation to specific movements performed by awake monkeys. These experiments have shown that in both the cerebellum and the basal ganglia, single cells begin their discharges in advance of the onset of a volitional movement. Furthermore, cerebellar cells have been found to change their activity in advance of cortical cells during the state of "readiness" that precedes a motor act, as will be discussed in the next chapter.

From these results, the remarkable conclusion emerges that the distributed system containing the highest motor programs includes centers (cerebellum and basal ganglia) that are anatomically at lower levels. This means that when we try to formulate concepts of motor control, we need to free ourselves of the idea that higher functions are lodged exclusively in the cortex—that the cortex is the chief executive and all other regions are its minions. The cortex is the necessary instrument of higher function, but it is an instrument played by programs fashioned from the interactions between centers throughout the central nervous system. The multiple interconnections between centers both provide for the higher levels of abstraction underlying volition and purpose and ensure that those functions emerge from the fabric of the organism as a whole. We will consider how these interactions are involved in voluntary control in the next chapter.

22

Manipulation

The success of an animal in thriving and procreating depends on more than simply moving about in its environment; what matters is its ability to *operate on the environment,* to extract from it the means of sustenance, to attack or defend itself, and to find mates and engage in cooperative behavior with others of its kind. These abilities usually depend on special motor organs suited for the purpose. In general, these organs, which an animal uses to operate on the environment, are complex, and they require nervous control by more complex mechanisms than those mediating locomotion.

Some representative animals with their specialized motor organs are portrayed in Fig. 22.1. It can be seen that these organs generally take one of two forms: They are either modifications of the *limbs* (usually the forelimb), or they are modifications of the *head* (usually the face). The forelimb commonly has, as an elaboration to its distal part, an apparatus for *grasping,* which may take the form of *claws* or *hands.* The facial apparatus, characteristically elaborated in relation to the mouth, consists mainly of a set of jaws, or a quite elongated and complicated *snout* or *proboscis.* These various organs can carry out

a range of operations, and it is difficult to find one word to characterize them all; somewhat arbitrarily we will class them all as one form or another of *manipulation,* to distinguish them from forms of locomotion.

The motor mechanisms we have considered in previous chapters have mainly involved repetitive events, such as the heartbeat or stepping, or single events, such as a tail flip. Specialized organs for manipulation involve more complex sequences of events. In addition, they involve higher levels of control in the motor hierarchy.

It can therefore be appreciated that the reductionist strategy of analyzing simple systems, which has been so successively applied to lower levels of motor control, has to be modified for the task of analyzing the more complex movements and the higher levels of control that are involved in manipulation. In general, the strategy must be to analyze a series of movements into its separate parts, and to identify the neural mechanisms at different levels in the hierarchy not only in relation to each part, but also in relation to the series as a whole. We will see that the neural mechanisms for manipulation incorporate principles for lower levels of motor control that we have

478

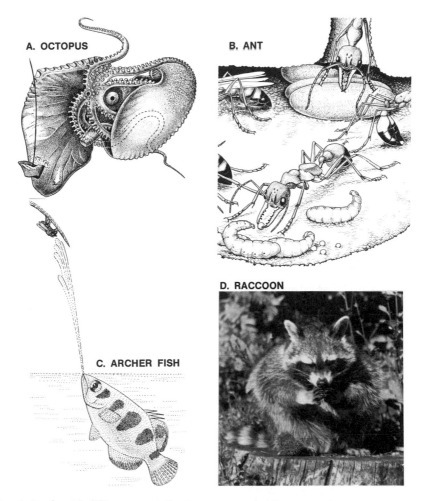

Fig. 22.1 Animals with different specialized motor organs. The octopod (**A**) performs manipulations with its tentacles, which are modified limbs, the ant (**B**) with its pincers, which are modified jaws (mandibles). The archer fish (**C**) "manipulates" water with its mouth and pharynx, in order to shoot droplets of water at airborne insects; this actually represents a form of tool use. The raccoon (**D**) performs dextrous manipulations of foodstuffs with its handlike forepaws. (A from Young, 1964; B from Wilson, 1975; C from Alcock, 1979)

already studied, as well as new principles involved in the initiation, maintenance, and termination of sequences of movements (sometimes called "motor programs") that arise at higher levels of control (see Fig. 21.3).

A critical element in these kinds of movement is that they are carried out under *voluntary control:* that is, these movements are not automatic or reflexive, as in much of the control of locomotion, but rather are willed, expressing the desire of the or-

ganism to operate on its environment. This in turn depends on a number of additional properties such as consciousness, attention, and motivation; these variables of behavioral state are primarily centrally generated within the organism, and are therefore the subject of "Central Systems" (Section IV).

In recent years the study of mechanisms of manipulation has acquired a practical relevance with growing interest in constructing artificial limbs and robotic arms for a variety of applications in industry

Table 22.1 Various specialized motor operations

Limbs (legs, arms)	Face (proboscis, jaws)
grasping	grasping
pinching	pinching
tearing	tearing
clasping	clasping
holding	holding
squeezing	squeezing
crushing	crushing
exploring	exploring
feeling	feeling
	sucking
	grinding
	beating

and space exploration. The computational problems involved in controlling an arm in three-dimensional space, guiding it to a target under visual control, and performing a manipulation on a target object turn out to be formidable. As a background, it will be desirable to acquaint ourselves with the range of strategies that have evolved in both invertebrates and vertebrates before considering the mechanisms in the human.

Octopus Tentacle

Invertebrates have evolved many kinds of organs for carrying out a variety of manipulative functions. As previously noted, the invertebrate has achieved its flexibility in this regard by virtue of the fact that the manipulative appendages evolved independently of the limbs for locomotion, and could therefore be individually adapted to their specific tasks. Table 22.1 gives at least a partial idea of the range of tasks carried out by limbs and facial organs.

Let us consider the octopus tentacle as an example of a manipulative organ that illustrates both the complexities and limitations of nervous control mechanisms in invertebrates.

The motor apparatus of the octopus consists of muscles controlling *gland cells* (chromatophores for protective color changes, and cells for discharging ink),

muscles for rapid movements (jet propulsion and giant fiber systems), and muscles for moving the arms, or *tentacles*. The tentacle is one of the most versatile manipulative organs devised in Nature. It is used to catch prey (either while the octopus is lurking quietly among the rocks and crannies of the seafloor or while it is swimming by jet propulsion); to convey the prey to the mouth; to explore the environment and test possible sources of food or danger by means of tactile and chemosensitive receptors in the arms; and to defend against predators.

Although one tends to think of the tentacle as being like an arm, it actually differs markedly from its vertebrate counterpart. The lack of bones and joints is an obvious difference; the advantage gained is that the muscle movements have great degrees of freedom, but the disadvantage is that this places heavy demands on their neural control mechanisms. The octopus has evolved a hierarchy of nervous centers for this purpose. At the lowest level are the motoneurons that innervate the muscles that move the individual suckers, and the muscles that move the arm. These motoneurons are located in ganglia within the arm. Chemosensory and tactile receptors in the skin and suckers, as well as muscle receptors, send their axons to these ganglia and establish connection with motoneurons and interneurons (see Fig. 22.2). By this means, reflex pathways for exploratory feeling or for withdrawal are present within the arm itself. A severed arm continues to show coordinated, purposeful movements such as conveying food in a mouthward direction, and, similarly, in animals whose brains have been removed, the tentacles still perform movements of grasping or withdrawal.

The tentacle thus contains its immediate motor control mechanism, and in this respect the nerve ganglia of the tentacle are analogous to the axial nerve cords of annelids and arthropods and the spinal cords of vertebrates. It is as if the octopus has attached to it eight wormlike appendages,

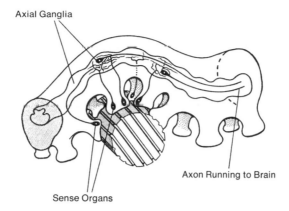

Axial Ganglia

Axon Running to Brain

Sense Organs

Fig. 22.2 Diagram of part of an arm of an octopus touching a plastic cylinder. The pathways from the receptors are inferred from physiological and degeneration experiments. The diagram illustrates an experiment in which the octopus was tested for its ability to discriminate between rods with different numbers of grooves. (After Wells and Wells, in Barrington, 1979)

each with its self-contained motor ganglia for carrying out simple movements. There is thus a *decentralization* of motor control. The degree of this decentralization is dramatically reflected by the fact that, of the total of about 500,000 nerve cells in the octopus nervous system, over half of them (300,000) are found within the ganglia of the tentacles.

Some limitations of this degree of decentralization in the motor control hierarchy have been revealed in behavioral tests. The experiment illustrated in Fig. 22.2 showed that the octopus can distinguish between different amounts of roughness of the surface of a cylinder but is poor at distinguishing patterns or shapes. Other experiments showed an inability to learn to discriminate between objects with different weights. These limitations appear to be due to the fact that sensory information is largely confined to local reflex pathways within the arm; it is not made sufficiently available to the rest of the nervous system to serve as a basis for more sophisticated spatial discriminations, or learning. Martin Wells, of England, who carried out these studies, has suggested that all of this is a consequence of the completely flexible nature of the cephalopod body and tentacles, and the enormous amounts of sensory informa-

tion about limb positions that this generates. It is more economical of nervous tissue to monitor this sensory information through peripheral circuits, but this limits the learning capacity, which can be obtained only through central associative pathways. Thus, the tentacle as a manipulative organ attains its extraordinary flexibility at the expense of limitation in discrimination abilities, and the inability of the organism to learn new manipulations.

The Hand

In our discussion of locomotion, we noted the change that occurred when vertebrate animal life became adapted from aquatic to terrestrial environments. Some of these changes involved a relatively direct adaptation of preexisting structures and functions. Two pairs of limbs were elaborated, the forelimbs from the pectoral girdle and fins, the hindlimbs from the pelvic girdle and fins. The segmental neural apparatus was adapted for generating rhythmic alternating activity in the limbs, coordinated with the undulatory movements of the trunk; higher centers were elaborated for descending control.

These adaptations can be summarized by the principle that "Form ever follows

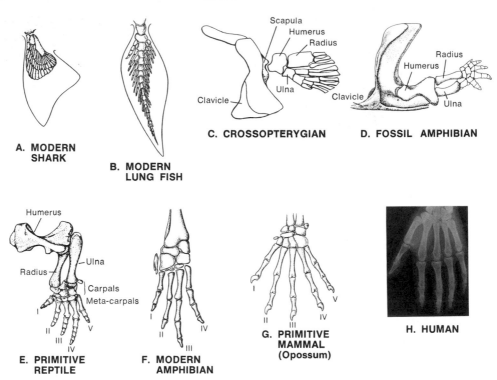

Fig. 22.3 Evolution of the bony structure of the forelimbs. (A–G from Romer and Parsons, 1977; H from Warwick and Williams, 1973)

function," a statement made by the architect Louis Sullivan many years ago to express his belief that buildings should have forms that facilitate their functional uses. This statement can serve as well to express the principle that underlies the relation between structure and function in the evolution of the nervous system. Many of these modifications of structure reflect the economical tendency in phylogeny toward adaptation of a preexisting structure to different functions in a new environment, rather than development of an entirely new structure. Some changes, however, have been more radical, and among these were the development of elongated *limbs* and the development of complicated distal appendages. These changes are illustrated in the diagrams of the skeletal elements in Fig. 22.3. Our interest here is in the forelimbs, because they gave rise to the *arm* and the *hand*. The hand is one of the most effective

and adaptable appendages to emerge along the phylogenetic scale, and its manipulative abilities were crucial to the evolution of primates. The nervous mechanisms involved in the control of the primate hand are the subject of the rest of this chapter.

The control of the hand is not simply a matter of controlling the distal digits. For most functions, many major muscle groups are brought into play. First are the muscles of the trunk that insert on the humerus, and move the whole arm at the shoulder joint. Second are the muscles that arise from the humerus and insert on the radius and ulna; these move the lower arm and the elbow. Third are the muscles that arise from the lower humerus and insert in the hand; these flex or extend the whole hand or the individual digits. Fourth are the intrinsic muscles of the hand that spread or close the fingers; of those, the muscles that enable the thumb to oppose the fingers are especially important. Although individual functions can be assigned to individual muscles (for example, the abductor policus moves the thumb toward the fingers), *a given muscle always functions as a part of the whole ensemble of muscles.* It can be readily appreciated, for example, that swinging a hammer or a tennis racquet involves coordinated activity of all the muscle groups mentioned above.

One of the main manipulative functions of the hand is *prehension,* that is, the ability to *grasp.* One sees this as a primitive reflex in a human baby; a finger placed in the baby's hand is always immediately enclosed by warm tiny fingers. One sees it revealed also in disease in the adult; a stroke commonly damages motor cells or fibers in the cortex, resulting in "release" of reflexes from their normal descending inhibitory control. Stretch reflexes become hyperexcitable, and more primitive reflexes such as the grasp reflex may be revealed. As previously noted, Hughlings Jackson's study of these release phenomena was important for the development of his concepts of motor hierarchies.

In our normal, everyday activities we tend to take it for granted that the manipulative positions of the hand are infinitely variable. However, to some extent they all involve some degree of either power or precision. A *power grip,* for example, is used by a monkey when it swings from one branch to another, or by us when we swing a tool or weapon, or lift a heavy object. A *precision grip,* in contrast, is used by a raccoon when it is cleaning and eating its food, or when a monkey picks up seeds or grooms its mate, or when we are writing with a pencil. Most manipulations involve degrees of both power and precision (Fig. 22.4). And most express the principle stated above, that any given movement depends on the coordinated action of an ensemble of muscles. For example, we know from our own experience that a power grip requires not only flexion of the digits but also extension of the hand itself (remember the childhood trick of making someone

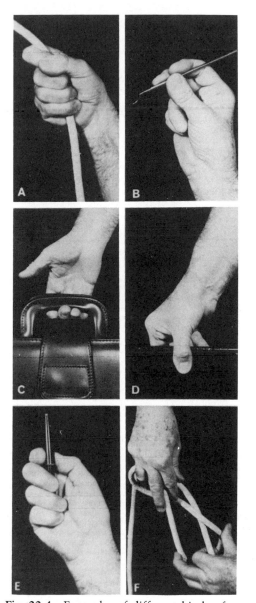

Fig. 22.4 Examples of different kinds of manipulative functions of the human hand. **A.** Power grip. **B.** Precision grip. **C.** Power (hook) grip. **D.** Power (pinch) grip. **E.** Combined power and precision grip. **F.** Complex posture and manipulation. (From Warwick and Williams, 1973)

drop an object by pressing down their hand?); thus, it is the synergy of finger flexion plus wrist extension that gives the greatest power. Similarly, threading a needle involves a delicate counterpoise be-

Table 22.2 Neural mechanisms for fine
motor control of the hand

Cortical
1. Large cortical representation of movements of
   hand and individual digits
2. Low threshold for cortical activation
3. Complex intracortical organization
4. Transcortical sensory feedback loops

Spinal
1. Strong corticospinal projection to spinal cord
2. Convergence of many fibers onto motoneurons
3. Monosynaptic connections to motoneurons
4. Facilitation of repetitive synaptic potentials

tween extensor and flexor muscles at every
level of the hand, arm, shoulder, and indeed the whole body.

Neural Mechanisms

The axial and proximal limb muscles that
contribute to movements of the hand are
the same as those that take part in maintaining posture and providing for locomotion. We therefore turn our attention to the
muscles of the distal forearm and hand,
and the neural mechanisms that relate to
their control.

Properties of the Corticospinal Tract. The
neural mechanisms that are used in the
fine control of the hand are summarized in
Table 22.2. We have already discussed
some of them in the previous chapter, such
as the importance of the direct connection
between the cortex and the spinal cord
through the corticospinal tract. Among the
pioneers in studying the physiological
properties of this connection in the monkey
were Charles Phillips and his colleagues at
Oxford. They recorded intracellularly in
the cervical region of the cord from motoneurons that send their axons into the median nerve of the forearm and thence to
muscles controlling the hand (see Fig.
22.5A). These motoneurons respond to
stimulation of the peripheral Ia fibers from
the muscle spindles with monosynaptic
EPSPs, which remain about the same amplitude during a repetitive train (Fig.

22.5B). They also respond to cortical stimulation with a monosynaptic EPSP to cortical stimulation; however, with repetitive
stimulation the EPSPs quickly undergo a
marked *facilitation* of their amplitude (Fig.
22.5C, traces 2 and 4).

This property of facilitation is not unique
to the corticospinal synapses; it is also seen
with stimulation of the rubrospinal pathway, for example, or with activation of
polysynaptic pathways through segmental
interneurons. This means that when any of
these inputs is activated, the higher the
frequency and the longer the burst, the
more potent that input becomes in com-
manding the motoneurons. Also, as Phillips
and Porter observed in their definitive
monograph on *Cortico-Spinal Neurones*
(1977), this enables the corticospinal tract
"to adjust, by alteration in frequency, the
power of its excitatory action on targets."

Cortical Organization. What of the neural
mechanisms at the cortical level? The maps
of cortical representation (Fig. 21.11) are
the starting point for investigation at this
level. One of the most interesting questions
is whether the intracortical circuits at a
given site are organized on the basis of
modules similar to those of the olfactory
glomeruli and the barrels and columns of
sensory cortex (see Chaps. 11, 12, and 16).
A motor cortex module has been difficult
to define; to begin with, there is no highly
ordered thalamic input which imparts to
sensory cortex much of its modular structure. The technique of *intracortical microstimulation* has given insight into this problem. This technique, introduced in the
1960s by Hiroshi Asanuma of the Rockefeller University, involves selective stimulation of corticospinal neurons through a
microelectrode introduced into the cortex.
The results have provided evidence that
corticospinal neurons activated along a
given electrode track through the cortex
tend to be related to the same muscle, and
to receive inputs from the corresponding
part of the limb. This has suggested that
there is a radial organization within the

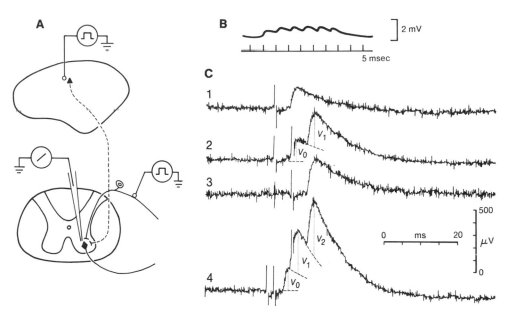

Fig. 22.5 **A.** Experimental setup for intracellular recording of responses of motoneurons to stimulation of input pathways in the baboon. **B.** Nonsummating EPSPs elicited by a train of six volleys in Ia afferents. **C.** Facilitating EPSPs elicited by 1, 2, and 3 shocks to the hand area of the motor cortex. Each trace is the average of 156 repetitions. The trace in 3 was obtained by subtracting 1 from 2, showing the facilitation of the second response compared with the first. (From Muir and Porter, in Phillips and Porter, 1977)

cortex for input–output functions related to the same muscle. It is possible that this provides the basis for columnar modules. In fact, anatomical studies have shown that, after HRP injections into lower motor centers, the retrogradely labeled output cells in the cortex are grouped in clusters (see Fig. 22.6). However, the physiological studies to date indicate that the modular organization is less distinct than its sensory counterpart; there is variation in the sizes of modules and overlap of neighboring modules. This probably reflects the point we stressed at the outset of this section, that muscles are not controlled in isolation, but always as part of an ensemble of muscles with complementary or antagonistic actions.

Synthesis of Motor Circuits

A summary of the traditional view of the circuits that link the cortical and spinal levels in control of the hand is given in Fig. 22.7. Voluntary movements begin with central programs ① which activate, in appropriate pattern and sequence, the modules of the motor cortex. The corticospinal fibers ② activate the motoneurons to the muscles ③, by the mechanisms we have discussed. Through collaterals, the corticospinal fibers also activate central sensory pathways and other ascending central systems ④ that feed back information to the cortex about the signals that have been sent; this is the "reafference," or "corollary discharge," mentioned previously (Chap. 21). Sensory input from the muscles ⑤ provides information about the state of contraction of the muscles and the extent of movement that has actually taken place. Some of this information reaches the motor cortex through direct connections from the somatosensory relay nuclei in the thalamus (see Chap. 12), while some is relayed from the somatosensory cortical areas. Connections between the somatosensory and mo-

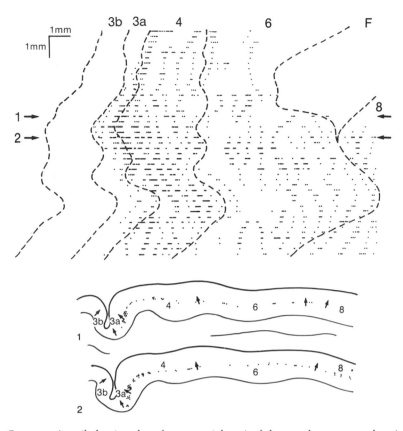

Fig. 22.6 Cross section *(below)* and surface map *(above)* of the monkey cortex, showing clustering of corticospinal neurons as revealed by injection of HRP into the red nucleus. Arrow to left of map indicates location of the cross sections. Arrows in cross sections below indicate boundaries of different numbered areas. HRP injections into other subcortical motor centers, including the spinal cord, give similar results. (From Jones and Wise, 1977)

tor areas thus provide for a "transcortical reflex loop," that can function as part of a servomechanism by which the nervous system can assess errors in accuracy of movements and correct them.

More precise information about the relation between the organization at the cortical level and the organization at the spinal level has been obtained in studies of awake, behaving animals. These experiments have built on the methods of Evarts (see below) in combining single-cell recordings and single-cell microstimulation within motor cortex with electromyographic (EMG) recordings from individual muscles. Using these methods, Eberhard Fetz and his colleagues at Washington (see Cheney et al.,

1985) have studied the precise ways that cells in the cortex control fine movements of the wrist and fingers. In one set of experiments, intracortical microstimulation of single pyramidal tract cells was used to elicit activation of individual flexor and extensor muscles of the wrist and fingers. These findings were then correlated with recordings from single cells during these same movements.

Three basic patterns of cortical cell influence on wrist flexor and extensor motoneurons were found. The hypothesized circuit connections mediating these influences are summarized in Fig. 22.8. One pattern is pure facilitation of agonist muscles (related muscles with the same physiological action,

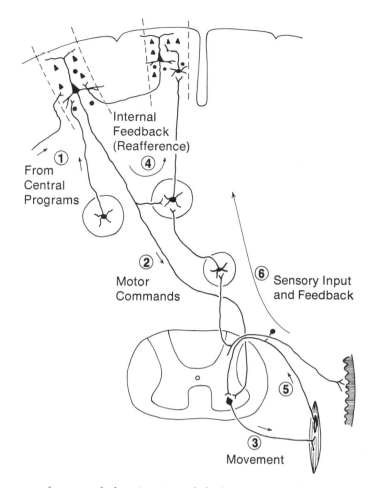

Fig. 22.7 Summary of some of the circuits and functions involved in voluntary control of movements of the hand. Most of these elements are similar in principle to those involved in control of locomotion, and are adapted for fine control of the hand, as discussed in the text.

of extension or flexion). This is mediated by cell groups A and C in Fig. 22.8. The second pattern is facilitation of agonists with suppression of antagonists, which can be mediated by cell groups B, E, and F. The third pattern is pure suppression of agonists, such as can be mediated by cell group D. For each of these cell groups, the pathways can be traced in the figure to the connections with specific types of extensor and flexor motoneurons and inhibitory interneurons that give rise to the corresponding physiological pattern.

This scheme is useful in relating the pathways for descending control of fine move-

ments to the principles of organization we have already studied, which relate to the cortical and spinal levels. At the cortical level, the pyramidal tract cells to the wrist motoneurons are found in the wrist area of the motor map within the motor strip (Figs. 21.11 and 21.13). Within that area, cells with a similar distribution of output effects appear to be grouped into clusters. These clusters reflect the columnar organization of cells revealed in the anatomical studies (Fig. 22.6, above). At the spinal level, Fig. 22.8 makes clear that the descending connections are incorporated into the segmental circuits for motoneuron con-

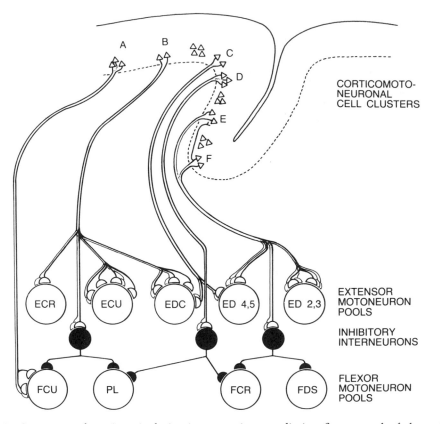

CORTICOMOTO-
NEURONAL
CELL CLUSTERS

EXTENSOR
MOTONEURON
POOLS

INHIBITORY
INTERNEURONS

FLEXOR
MOTONEURON
POOLS

Fig. 22.8 Summary of corticospinal circuit connections mediating fine control of the wrist and fingers, as indicated by experiments in awake, behaving monkeys. A F, clusters of pyramidal tract neurons in the motor cortex. Groups (pools) of motoneurons to different types of muscles: the six extensors were extensors carpi ulnaris (ECU), digitorum communis (EDC), digitorum 2 and 3 (ED2,3), digitorum 4 and 5 (ED4,5), carpi radialis longus (ECR-L), and carpi radialis brevis (ECR-B); the six flexors were flexors carpi radialis (FCR), digitorum profundus (FDP), carpi ulnaris (FCU), digitorum sublimis (FDS), palmaris longus (PL), and pronator teres (PT). Open profiles, excitatory actions; filled profiles, inhibitory actions. See text. (From Cheney et al., 1985)

trol. An expression of this is the fact that reciprocal organization, which we have seen underlies the segmental control of locomotion (Fig. 19.13), also underlies the descending control of fine movements. The same circuit elements may be involved in both kinds of control. It seems likely, for example, that the inhibitory interneuron mediating reciprocal inhibition from the cortex (Fig. 22.8) is the same interneuron mediating Ia inhibition from the muscle spindles as a part of the stretch reflex (Fig. 19.13). Thus, the evolution of the higher control of fine movements in our distal

extremities has arisen by adaptation of preexisting circuits for locomotion.

Adaptive Properties of Motor Circuits

Motor circuits have not only been adapted during evolution for different forms of control, but they can be adapted for different functions over millisecond time periods during active movements. In discussing locomotion in Chap. 20, we noted that cutaneous stimuli normally bring about flexion responses in a stimulated limb at rest and during the swing phase of walking, but this changes to extension responses during the

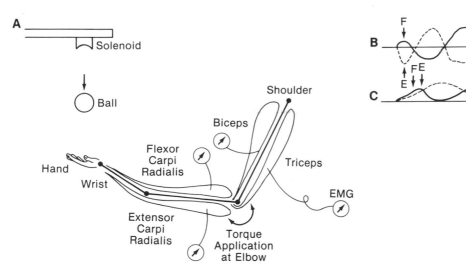

Fig. 22.9 An experiment demonstrating the adaptability of the stretch reflex in the human in response to loading changes. **A.** The experimental setup, in which the subject rests the elbow on a platform and catches a ball released at a signal. A torque motor attached at the elbow to the sleeve encasing the arm applies a perturbing small jolt to the arm as the ball drops. The subject must compensate for the jolt in order to catch the ball. Electrodes record the electrical activity of different flexor and extensor muscles that control the arm movements. **B.** Under normal conditions, the flexor and extensor muscles act reciprocally, reflecting the organization of spinal reflexes (diagram on the right). **C.** When the arm is perturbed, flexors and extensors are initially activated simultaneously in order to reposition the arm, reflecting centrally generated coactivation. See text. (From Lacquaniti et al., 1991)

stance phase, this change of effect being appropriate to maintenance of the body during this phase.

This ability to mediate *different output functions in relation to different tasks* is one of the key features of motor circuits. There has been particular interest in this property in relation to the question of how spinal circuits are adapted for the many different functions that must be performed by the arm and hand. An example of this property has been demonstrated by Francesco Lacquaniti and his colleagues in Milan. They tested the ability of subjects to catch a ball under conditions in which compensations have to be made for transient dislocations of the arm. In the experimental setup (Fig. 22.9A), the subject rests an elbow on a platform and catches a ball released by a solenoid. Electrodes placed on the skin register the electromyographic ac-

tivity in specific flexor and extensor muscles of the upper and lower arm. Under normal conditions (Fig. 22.9B), the activity in flexor and extensor muscles is reciprocally related: contraction in extensors is accompanied by relaxation in flexors, according to the law of reciprocal innervation in the spinal reflex pathways, as we learned in Chap. 13. When a small deflection of the arm is produced near the time of impact (Fig. 22.9C), there are immediate flexor and extensor muscle movements to adjust the hand position to catch the ball. The interesting features of these compensatory movements are, first, that they are extremely short latency (several tens of milliseconds), indicating that they are mediated over the fastest (Ia) muscle reflex pathways; second, they involve coactivation of flexors and extensors, which overrides the reciprocal spinal reflex pathways; and third, they occur before the ball reaches the hand, that is, they are in anticipation of where the ball will hit, not a reflex response to its impact.

The authors conclude that changes in the efficacy of spinal synapses, at both the presynaptic and postsynaptic level, must be involved in these adaptive effects. The ability of specific types of inputs to have special access to motoneurons to circumvent the usual recruitment order according to the size principle (see Chap. 20) must contribute to this process. The changes in synaptic efficacy may resemble those seen in adaptive and learning processes in the vestibulo-ocular reflex (see Chap. 14).

These kinds of results have offered insights into the nature of motor circuits mediating control of precision movements. Lacquaniti and his colleagues (1991) conclude that, in a multijoint system like the arm, the relations between inputs and outputs are not fixed but vary with the task and the system state, so that hardwired servocontrol mechanisms are not applicable.

Rather, the operation of the reflex circuit is best understood within the context of adaptive control, that is, a control process capable of estimating and modifying state and output variables on the basis of internal models of expected behavior.

Lacquaniti (1992) furthermore concludes that, in contrast to the traditional view that global planning of movements takes place at higher levels and local automatic control of muscles at lower levels, it is now recognized that some of the global mechanisms are also present at the lower levels. Thus, there is a distributed control, both within different levels and across different levels, of many types of kinematic and dynamic variables, together with the ability to switch between different modes in a distributed system in relation to different system or behavioral states.

Active Touch and Precision Grip

Similar principles emerged from analysis of hand movements in relation to sensory information flowing in from the skin.

Let us begin by referring back to Fig. 12.1, which indicates the information that flows in from other somatosensory receptors in the skin and tissues of the hand and the fingers. As we discussed in Chap. 12, the fingers have a high density of sensory receptors. This is correlated with their large cortical representation, which parallels that for the muscles that move the fingers. We employ this sensory capacity to gather information about the environment, such as the shapes of small objects or the texture of surfaces. This use of the hand is called *active touch* (see Chap. 12). It reminds us that the hand is a sense organ, as well as a motor organ.

The diagrams in Figs. 22.7 and 22.8 indicate the direct, corticospinal pathway to the spinal cord, but it should be emphasized that the corticospinal fibers have collaterals to all the brainstem motor nuclei we mentioned earlier in this chapter (reticular, vestibular, red). These are involved in the control of all the axial and proximal muscles during movements of the hand. In addition, motor control involves pathways through the cerebellum (in the brainstem) and basal ganglia (in the telencephalon), as we discussed in the previous chapters.

The way that sensory input from the hand is used in control of hand movements has been analyzed by Roland Johansson and his collaborators in their studies of precision gripping in humans. In these experiments, the subject grasped a small weighted disc between thumb and forefinger and, at a signal, raised it a couple of centimeters, held it there, and then lowered it to rest (see Fig. 22.10A). The surfaces of the disc were covered with different materials; this gave them different degrees of slipperiness, and required the subject to adjust the amount of *grip force* to produce a vertical lifting force (called *load force*) that would raise the disc without slipping.

The force coordination related to three different surface materials is shown in Fig. 22.10B. The load force required to raise the weighted disc was the same for all trials, as was the position to which it was

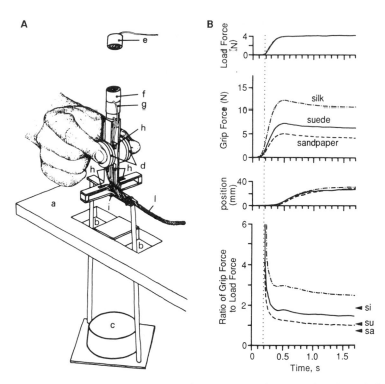

Fig. 22.10 Experiment on precision gripping in humans. **A.** Schematic drawing of the apparatus. a, table; b, holes in table; c, exchangeable weight shielded from the subject's view by the table; d, exchangeable discs; e and f, vertical position transducer with an ultrasonic receiver (e) and an ultrasonic transmitter (f); g, accelerometer; h, strain-gauge force transducers for measurement of grip force and load force (vertical lifting force); i, peg with a hemispherical tip on which the object rests while standing on the table; 1, electric line. **B.** Force coordination during the initial part of lifting trials with three different surface structures (silk, suede, and sandpaper). From top, the graphs show load force, grip force, vertical position, and ratio between grip and load force as a function of time. Weight constant at 400 g. Vertical lines indicate the beginning of the loading phases. Force ratio not shown for the preload phase. Time scale with an arbitrary origin. *(Top)* 16 sample trials superimposed (single subject). *(Bottom)* Data averaged from a total of 120 trials by nine different subjects. Arrows indicate mean slip ratios for the three surface structures, respectively. (Figure and legend from Johansson and Westling, 1984)

moved. However, the grip force was greatest for the smoothest material (silk), intermediate for suede, and lowest for the roughest material (sandpaper). This meant that the ratio of grip force to load force was highest with silk and lowest with sandpaper. There was an initial period of high ratio, followed by a period of lower ratio while the disc was held maintained in the raised position. The subject was then asked to lower gradually the gripping force until the disc slipped; this defined the slip ratio (see arrowheads in Fig. 22.10B).

These experiments demonstrated several principles in goal-directed manipulation involving precision gripping. First, sensory receptors provide exquisitely refined information about the frictional condition of the gripped surface. This comes not from a single type of receptor, but from the entire ensemble of receptors in the skin and connective tissue of the finger tips. Second, the motor system uses this information to generate a critical balance between grip force and load force that is adequate for raising and maintaining the lift under the different surface frictional conditions. This grip-to-load ratio is set at a level above the slip ratio. Third, this grip–load force combination occurs automatically; according to Johansson and Westling (1984), this

agrees with the notion that a particular pattern of afferent information might trigger release of a particular set of preprogrammed motor commands or update certain parameters of the currently executed motor programmes.

The authors speculate that these motor commands are mediated by the motor cortex and by the cerebellum, which is well-known to be involved in vestibular and postural reflexes (see Chap. 14), and in the initiation of certain movements (see Chap. 21). A final point is that force coordination in precision gripping is learned; the motor command level therefore contains in its circuit organization the memory for adjusting grip-to-load force to the appropriate frictional condition, and this memory is constantly updated for changes that occur in the surface or load.

Voluntary Movements

The motor cortex is partly at an intermediate level in the hierarchy of motor control as we have seen, but is, nonetheless, at the highest level to which we can trace pathways that can be labeled "motor." Above this, the inputs to the motor cortex come from central systems, to which the terms "motor" and "sensory" do not apply. It is these central systems that provide the "central programs" (see ① in Fig. 22.7) that control the motor cortex.

We are only beginning to understand the nature of these central motor programs. Part of the answer will come from more precise mapping of complex central circuits. But higher motor control in humans is not just more complex than in other animals; its essence lies more in the fact that what we do seems "voluntary," and is performed with "purpose." We cannot ethically study this dimension of motor control in humans at the cellular level with

present techniques, but we can approach it through experiments on awake, behaving monkeys. The results of these experiments have given us insights unobtainable by other means.

This approach to the study of motor behavior was pioneered by Edward Evarts at the National Institutes of Health in the 1960s. A typical experimental setup is illustrated in Fig. 22.11. A monkey is first prepared for chronic recording by placement of a closed recording chamber over an opening in the skull. For the behavioral study, the animal sits under light restraint in a chair and performs a task. A microelectrode inserted into the motor cortex records the activity of a single cell in relation to performance of the task. This basic setup has been used for experiments illustrated and discussed in previous chapters (Chaps. 13, 21).

Motor Operations: Preparatory Set

We have seen that there is not a tidy sequence in the onset of neuronal activity in different areas of the motor system. Neurons at apparently lower levels in the motor hierarchy begin to fire in relation to a learned movement about the same time as at higher levels, indicating that motor control involves parallel as well as serial processing (see Fig. 21.20). This raises the question of whether one could go even further, and identify activity related to the earliest mental operations involved in preparing to carry out an intended movement.

By setting up an appropriate behavioral paradigm, it is in fact possible to show that there are neurons in the motor cortex that have this type of activity. In the example of Fig. 22.11, the task was first to hold a handle in a given position for a few seconds. Then either a red or green lamp was lit. A red light meant "get ready to pull" on the handle, a green light meant "get ready to push." The handle was then displaced automatically and the task of the monkey was to give it either a pull or a push, depending on what the instruction had been. The results are shown in Fig.

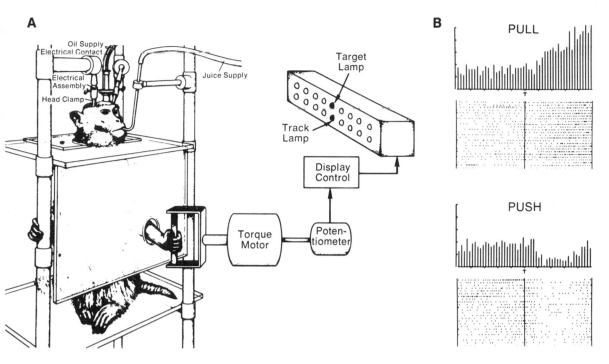

Fig. 22.11 A. Experimental setup for recording from single cells in the monkey motor cortex during visual pursuit tracking. Movements of the handle produced shifts in the lighting of the lower (tracking) row of lamps. The task was to keep the tracking lamp aligned with changing positions of the target lamps. **B.** Discharge of a corticospinal neuron 1 sec before and 1 sec before and 1 sec after the appearance of a red light (signaling "get ready to pull") and a green light (signaling "get ready to push"). *(Top)* Histograms of number of impulses in 40-msec bins. *(Bottom)* Raster displays of occurrences of impulses in successive trials. See text. (From Tanji and Evarts, in Evarts, 1981)

22.11B. The impulse discharge of this neuron is displayed for the period of one second before and one second after the onset of the red or green light (arrow). As can be seen, the cell increased its discharge after the instruction to pull, but decreased it after the instruction to push. These changes in cell activity occurred in advance of the actual motor response to the displacement of the handle.

From these kinds of results Evarts (1981) concluded that the warning light "set up preparatory states for a particular direction of centrally programmed movement." This state is reflected in the specific activity of a single cortical cell. Remarkably, this motor activity is entirely internal, as far as the behavior is concerned; there is no overt motor movement during the warning pe-

riod, and no EMG discharge in the muscles. Thus, the central motor program drives the cortical motor cells, but the motor output stays within the brain. States of readiness, such as this, can also be detected as "readiness potentials" in the electroencephalogram, recorded by gross electrodes on the scalp (see Chap. 25).

This earliest activity is related to several types of mental operations. The expectation of target appearance in these motor experiments is sometimes referred to as "perceptual set." This requires the monkey to remember where the light and target are located, and therefore overlaps with the operation referred to as "spatial memory." This in turn overlaps with spatial attention, which is required for carrying out the task. Thus, neuronal activity may be involved in

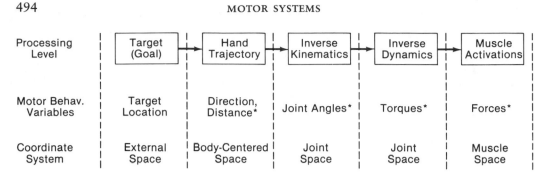

*(and Time Derivatives)

Fig. 22.12 Summary of the conceptual scheme for moving the arm to a target. The mechanisms are indicated in relation to concepts of motor processing, motor behavioral variables, and coordinate space. This theoretical scheme is in terms of the sequence of operations that is implied in going from the preparation for movement, beginning on the left, to the activation patterns of the muscles on the right. The intermediate steps of inverse kinematics and inverse dynamics indicate the operations that are implied in which nervous mechanisms control the angles and torques at the joints to achieve the arm positions that will allow the hand to reach the target. See text. (From Alexander and Crutcher, 1990b)

several categories of mental and motor operations.

Motor Operations: Preparation and Execution

After these earliest stages of preparation, movement of the arm, wrist, or fingers to a target can be analyzed into a number of distinct operations. There is now general agreement on a basic set of these operations in relation to goal-directed movements. As shown in Fig. 22.12, these can be arranged conceptually in a serial order. They start with location of the target or goal in space and end with the specific pattern of muscle activity required to reach that goal. Between are computations necessary for controlling the trajectory of the hand, which in turn depends on setting the joint angles, which in turn depends on the torques (rotational forces) at the joints by the muscles. Implied in this sequence is a hierarchical order, from the highest level of setting the target to the lowest level of moving the muscles.

There have been numerous studies of neuronal activity in different parts of the motor hierarchy in relation to these different operations. Many of these have tested the critical question of whether these operations occur in serial sequence, as implied in Fig. 22.12, or whether they occur in parallel. This is an important question, stimulated by the previous findings of simultaneous activity at different hierarchical levels (Fig. 21.20) and the increasing evidence from neural network models of parallel distributed processing in carrying out various kinds of sensory and motor operations.

A recent set of experiments by Garrett Alexander and Michael Crutcher at Johns Hopkins has tested this question in a systematic way by correlating recordings of single-cell activity in three motor regions: the motor cortex (MC), the supplementary motor area (SMA), and the basal ganglia, or putamen, (PUT). The student should review the locations of these regions in Fig. 21.13. The monkey learned a visuomotor tracking task similar to that illustrated in Fig. 22.11. With this paradigm, it was possible to analyze neural activity in the three regions and ask if a given region was more closely related to the preparatory operations or to those involved in motor execution.

The results showed that neurons with

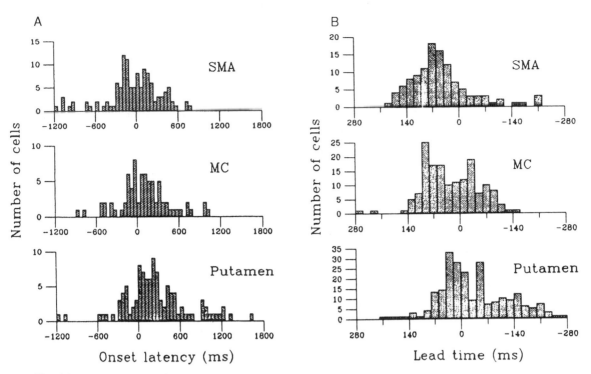

Fig. 22.13 Experimental analysis comparing the roles of the supplementary motor area (SMA), motor cortex (MC), and basal ganglia (putamen) in moving the arm to a target. **A.** Graphs of the onset latencies of impulse activity recorded by a microelectrode in the three regions, in relation to preparation for carrying out the motor task. Note the earlier onsets in SMA and MC, but also the considerable overlaps between the three regions. **B.** Graphs of the onset latencies of impulse activity in relation to movement of the arm in executing the motor task. Note the similar progression of latencies, and similar overlaps, as in B. See text. (A from Alexander and Crutcher, 1990a; B from Crutcher and Alexander, 1990)

directionally selective preparatory discharge were found in substantial numbers in all three areas; the largest proportion showing only preparatory activity were in SMA, and the smallest in MC. As shown in Fig. 22.13A, the cells tended to fire earlier in relation to the anticipated task in the SMA and MC than in the PUT, suggesting that some of the activity in the PUT may result from SMA and MC inputs. However, it can be seen in the graphs that there is a large degree of overlap in the onsets of firing, indicating simultaneous activity in the three areas, due perhaps in part to feedback loops from the PUT to the SMA and MC.

Differentiation between preparatory and movement-related properties was carried

out by noting discharges affected by loading of the arm to resist the movement and thus change the pattern of muscle activation. These results showed that neurons with movement-related discharges were also present in all three regions. This was believed to represent a neural correlate of joint torque or muscle activation pattern (see Fig. 22.12). As shown in Fig. 22.13B, there was again a progression in the onset of the discharges, with the earliest activity in the SMA and the latest in the PUT, but with considerable overlap in the firing patterns.

Summary of Movement Processing

From the various studies that have been reviewed here, a consensus view of the

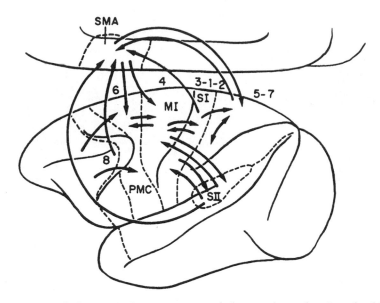

Fig. 22.14 Diagram of the cortical motor areas of the monkey, showing the high degree of interconnectedness between them. Note that most of the connections between the areas are reciprocal; these are organized in a topographic fashion, reflecting the orderly progression of representation of the muscles of the body, as in Fig. 21.13. PMC, premotor cortex; SMA, supplementary motor area; MI, primary motor area; SI, primary somatosensory area; SII, secondary somatosensory area. The numbers indicate architectonic areas. (From M. Wiesendanger, in Towe and Luschei, 1981)

neural mechanisms involved in arm movements has emerged which can be summarized as follows. The earliest activity is independent of the intended movement and appears to reflect a "perceptual set," the anticipation that a given sensory stimulus will soon occur. Next is the "motor set," which anticipates the intended direction of the limb movement. These preparatory stages are believed by theorists of motor behavior to set up a "motor equivalence" of the intended goal-oriented task. This leads finally to the "patterns of muscle activation," which control the joints to bring about the intended movement to the target.

In terms of processes, this means that there is high-level planning of a smooth movement in space to bring the hand to the target, leaving the processes controlling joint angles, joint torques, and muscle activation patterns to achieve that movement to lower levels of processing. However, the experimental evidence shows that different

levels within the *hierarchy of processing mechanisms* may occur intermixed in neurons within high levels of the *anatomical hierarchy of motor structures;* for example, both preparatory and muscle-related activity may be found in neurons within motor structures of the forebrain. This can occur because of the many cross-connections and feedback pathways that bind the regions together (see Fig. 22.14), so that each region carries out several levels of function, and a given function arises out of the action of neurons distributed in different regions. Through its unique combination of inputs, outputs, and intrinsic processing mechanisms, each region makes a unique contribution to the stages of motor processing; thus, the SMA appears to participate in the earlier stages of motor planning, whereas the PUT appears to be related to carrying out the movement. However, the great deal of overlap in the timing of this activity indicates that both serial and parallel processing occur in the distributed systems.

The principles discussed above are not unlike those involved in processing of sensory information; recall, for example, the visual system (Chap. 16), where we saw that each region of visual cortex makes its unique contribution to processing the complex set of information presented by a visual scene in both a serial and parallel manner. These considerations support the idea of common principles underlying the processing by neural circuits of sensory and motor information. In the next section, we will consider how these circuits are integrated into the central systems which control the basic behaviors necessary for the survival of the individual and the species, and which are the means for the unique capacities of the human brain.

23
Communication and Speech

Communication is an essential element of any level of organization. At the level of the cell, molecular signals and messengers are needed for accomplishing cellular functions. Similarly, the organization of individual organisms into functional groupings depends on their ability to send signals to each other. Many types of communication involve the whole organism, and draw on virtually all the motor mechanisms we have discussed thus far; for example, animals communicate with each other through different postures, different forms of walking, or by specific gestures. Other types of communication depend on specific signals, such as pheromones or tactile stimulation.

In this chapter, we will consider a type of communication that depends on special organs designed specifically for that purpose. This is communication by *sound*. Since special organs are necessary for making sound, communication by this means is limited to more complex organisms, similar to the case for manipulation. Communication by sound is particularly important in certain orders of insects, in birds, and in mammals. We have already discussed the sensory mechanisms for reception of sound signals in these organisms in Chap. 15, where we also discussed briefly the impor-

tance of this mode of communication for humans.

In this chapter, we will focus on the motor mechanisms for generating sound signals. Insect and bird song are both active fields of investigation. Many recent advances have been made in understanding neural mechanisms at all levels of the motor hierarchy, from muscles up to central programs; however, these results have received less general notice than work on locomotion.

With regard to vocalization and its neural control in mammals, neurobiologists have shown surprisingly little interest. Indeed, apart from the identification of the speech area of the human cortex, the subject does not exist as far as most textbooks are concerned. This is partly explained by the fact that the neural mechanisms seem to be too complex for successful experimental analysis, compared to those for communication in insects and birds. Even more important, speech and language are unique to humans, so there are no animal models to analyze. However, these functions are much too important to humans to allow us to ignore them. In fact, we shall see that considerable information is available on some aspects of the mechanisms. We will start with a comparison with the studies of communication in other species, which will provide some valuable perspectives on the properties that have been important for the evolution of speech as a mechanism for human communication.

Insect Song

Most rapid movements produce sounds, some of which may have a signaling value. Among higher invertebrates, crustacea and insects produce various kinds of noises. However, as specific signals these sounds are of limited value, because of their relatively coarse nature and the fact that many of these organisms appear to lack specialized auditory sensory cells. Among insects, however, certain species have specialized organs for producing and receiving specific types of sound. Best studied are crickets and grasshoppers (order Orthoptera) and cicadas (order Cicadidae). In these species the structures for producing sound are adapted from structures for locomotion. This contrasts with the evolution of flight, which in insects took place not by the adaptation of limbs to wings, but by the adding of wings.

Sound Production

The general mechanism for producing sound in insects is by scraping two parts of the exoskeleton against each other. The technical term for this is *stridulation* (meaning to scrape). It is the mechanism that a violinist uses in drawing a bow over the strings of a violin. The mechanism in the cricket is illustrated in Fig. 23.1. Across the dorsum of each wing is a structure called an *elytron*, originally a vein that has been transformed into a row of teeth, like a file. Under the median edge of each wing is a ridge, called a *pectrum*. When one wing is drawn across the other, the pectrum of one wing scrapes the file of the other, as shown in Fig. 23.1A, and induces vibrations of the wing, which make sound.

The sound produced depends on the rate of movement and the resonant properties

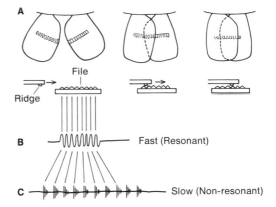

Fig. 23.1 Mechanism for producing sound in the cricket. A. Movement of the ridge (elytron) over the file (pectrum). B. Fast movement producing resonant sound. The oscillations of the wing are near the natural resonant frequency of the wing. C. Slow movement producing nonresonant sound. Each impact produces a burst of high-frequency oscillations of the wing.

of the wing. In some crickets, the teeth are small, the movement rapid, and the wing thin and flexible. In this case, each tooth impact produces an undamped oscillation, and the tooth impact rate during the wing excursion produces a sound at a frequency near the natural resonant frequency of the wing. This is called *resonant sound emission* (Fig. 23.1B). The frequency spectrum is narrow, in the range of 2–6 kHz for different species. In other crickets, the teeth are larger, the wing movement is slow, and the wing is relatively stiff. In this case each tooth impact produces a heavily dampened, rapidly decaying wave transient containing high-frequency oscillation. The slow wing excursion, with its slow tooth impact rate, produces a series of these individual complex wave forms. This is referred to as *nonresonant sound emission* (Fig. 23.1C).

Sounds produced by these two mechanisms thus differ in their frequency characteristics by virtue of both the resonant frequencies and the rates of wing beating. The main method for using these sounds to send information is by periodic interruptions of the wing beats, so that the sounds are sent

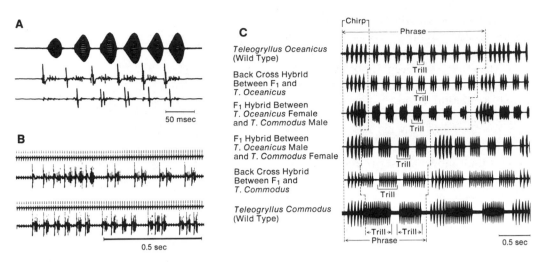

Fig. 23.2 **A.** Relation of chirp sounds in the cricket (top trace) to impulse activity in the wing-opening muscle (middle trace) and wing-closing muscle (bottom trace). **B.** Prolonged repetitive electrical stimulation of command interneuron fiber (upper trace) elicits intermittent chirping pattern of impulse discharge in motoneuron fibers to the muscles (lower trace). **C.** Song patterns of two wild cricket species (top and bottom traces), and of their hybrid offspring. (From Bentley and Hoy, 1974)

in groups, called *chirps*. By this means, a cricket is able to signal several types of behavioral states. There is a general *calling* song, a song signifying hostility or *aggression,* and a *courtship* song.

Neural Mechanisms

What are the neural mechanisms for generating these distinct motor output patterns? These have been analyzed in electrophysiological studies at several levels of the motor hierarchy. Recordings have shown that activity in the nerves to the wing muscles is closely correlated with the chirps of the cricket song (Fig. 23.2A). Synchronous volleys of impulses alternate between the nerves to the antagonistic muscles that close and open the wing. These volleys in the motoneurons arise within the thoracic ganglion; their rhythmic alternating character is relatively unaffected if the ganglion is isolated from all proprioceptive input from the muscles, and all intersegmental or descending input from other ganglia or from higher levels. This stable motor pattern generator within the thoracic ganglion

is responsible for the wing movements that produce sound.

It thus appears that communication by sound utilizes the principle of a rhythm generator at the segmental level, which is the same as that involved in generating the rhythmic movements underlying locomotion. This, in fact, is no surprise when we realize that the muscles that move the cricket's wings to produce sound are the same ones that move the wings during flight. In both cases the wing beat is neurogenic. The main difference is that, during flight, the opener and closer muscles are changed in their orientation by other muscles so that they function as depressors and elevators of the wings.

Genetic Control

The distinctiveness of the songs for a given species, and their resistance to environmental effects, are strong indications of the importance of genetic factors in determining the song pattern. This has been investigated by making crosses between males of one species and females of another. The pioneering experiments by David Bentley and Ronald Hoy at Berkeley are illustrated in Fig. 23.2C. The results show that each genotype is associated with a distinctive song, which differs from the others in its intervals between pulses within a chirp or trill, and the intervals between chirps or trills. Further experiments have indicated that the intertrill interval is controlled by genes in the X chromosome. Because of the gradual way in which the characteristics of the song patterns change with different hybrid crosses, Bentley and Hoy concluded: "The genetic system that specifies the neuronal network accounting for cricket song is therefore a complex one, involving multiple chromosomes as well as multiple genes."

Why are cricket songs so stable, so genetically dominated? Part of the answer appears to lie in the fact that there is little overlap in generations of many singing species, so that the young have no opportunity to learn the song of their species from their parents (see Chap. 27).

Hierarchical Control

The hierarchical control of the song generator follows the same principle as flight and other forms of locomotion. Stimulation of motor control fibers (command fibers) turns on chirping, and maintains it throughout the period of stimulation. Thus, as shown in Fig. 23.2B, maintained repetitive stimulation of command fibers gives rise to intermittent bursts of impulses in the motoneurons in the chirping generator. This illustrates the same principle underlying the mechanism by which a constant command generates alternating stepping movements (Chap. 20).

The upper motor neurons giving rise to these fibers are themselves under control of higher centers in the insect's brain. Repetitive stimulation at one site in the "mushroom bodies" (Fig. 23.3) elicits a "calling song" or a "rivalry song," dependent on the strength and frequency of the shocks. Stimulation at another site elicited either the calling song or a "courtship song."

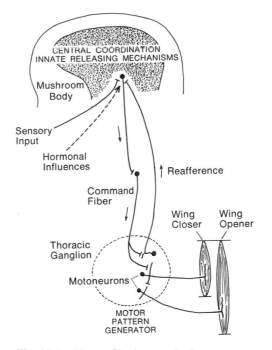

Fig. 23.3 Hierarchical control of song generation in the cricket.

The mushroom bodies represent the highest level in the motor hierarchy; their neurons interact with each other to form circuits that control lower centers in the brain, which in turn control the motoneurons in the segmental ganglia. The stimulation experiments show that the song programs are stored in these circuits and read out from them in a very stereotyped manner; there are obviously not the overlapping and flexible synaptic circuits that characterize higher control of many other types of motor output patterns, especially in vertebrates.

A schematic representation of the elements involved in higher control is shown in Fig. 23.3. The most important elements are sensory inputs, hormonal effects, and central feedback. In different species, singing is evoked by specific *sensory* cues such as a certain temperature of the air, or amount of light or dark, or the sight or song of another member of the species, or by tactile stimulation. Song production is also under close *hormonal* control; in most species only the males sing, and usually only when sexually mature, or when carrying a spermatophore prior to copulation.

This account gives only a brief introduction to the field insect song communication. The stable nature of the song pattern makes this an attractive field for analyzing the neural basis of specific motor behavior; it is thus an important meeting place for neurobiologists and neuroethologists.

Birdsong

In vertebrates, locomotion, as we have seen, depends on the adaptation of four limbs to a variety of locomotory skills, and manipulative organs have evolved as adaptations of the limbs. In contrast to insects, in vertebrates, specific organs for communicating by sound have evolved independently of the organs for locomotion. This has freed these organs from the evolutionary pressures on the limbs for locomotion, which is a factor limiting the complexity of song production by insects.

In fish, primitive kinds of sound communication occur in some species by muscular thumping of the swim bladder. In terrestrial vertebrates, however, the key strategy has been to adapt the respiratory apparatus, by expelling air from the lungs through a constricted orifice in the respiratory tract. The production of sound by this means is referred to as *vocalization*. Some amphibians and reptiles communicate by vocalization, but it is in birds especially that we see a sophisticated use of acoustic signals comparable to that in the insects.

The Syrinx

In birds, the organ for producing sound is the *syrinx*. This is located at the site where the two bronchi arise from the trachea, and is, in fact, a modification of the walls of these structures. Note that the syrinx is distinct and quite separate from the larynx, which is also present, but in birds serves merely to regulate overall air flow. The

mechanism of sound production by the syrinx is illustrated in Fig. 23.4. Within the syrinx, the bronchial walls are modified into thin *tympaniform* membranes, surrounded by air sacs. Contractions of muscles attached to the syrinx set the amount of tension on the membrane. As shown in Fig. 23.4B, when air is pressed out of the air sacs and tension is relatively slack, the tympaniform membrane bulges inward, and air flow through the bronchi sets the membranes into high-frequency oscillations.

With this relatively simple mechanism, birds are able to produce an astonishing range of songs. These serve a variety of functions, many of them similar to those in insects. Thus, there are sounds of alarm, distress, or warning. These are usually referred to as *calls*. They are usually simple in structure, and may be made by any member of the species. Contrasting with those are the elaborate vocalizations we refer to as *songs*. These are characteristically made only by males during the breeding season. The song identifies the singer and conveys information that he is defending his territory and is ready to mate. In many species, females also vocalize; these vocalizations include alarm calls and songs related to mating and nesting. In contrast to insect songs, bird songs have a rich tonal structure (as we all know from comparing the chirp of a cricket to the song of a robin), in which the different frequency components are important in conveying information.

Development of Birdsong

The songs of birds are distinct for a given species, and they have therefore been analyzed for the relative importance of genetics and environment in determining their pattern. We can surmise, to begin with, that, unlike insects, the generations overlap; parents care for their young, and thus some degree of learning by the young is possible. The degree to which this takes place is illustrated in Fig. 23.5, which summarizes

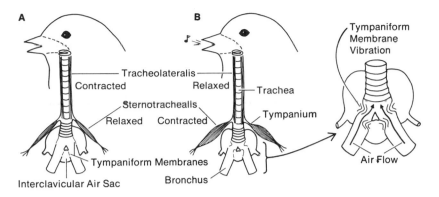

Fig. 23.4 Sound production by the syrinx in birds. **A.** The tracheolateralis muscles are contracted, while the sternotrachealis muscles are relaxed. Under this condition, the syringeal membranes are taut, and the bore from bronchi to trachea is maximally open; air flow through the syrinx generates low-frequency sounds or no sound. **B.** Tracheolateralis muscle relaxed and sternotrachealis contracted. The syringeal membranes bulge inward, and the bore leading from bronchi to trachea is maximally reduced; air flow through the syrinx generates high-frequency sounds. (From Hersch, in Nottebohm, 1975)

Fig. 23.5 Contributions of genetics and environment to the development of songs in three closely related members of the sparrow family. Each graph is a sonogram, with frequency on the ordinate and time on the abscissa. See text. (From several authors, in Marler, 1976)

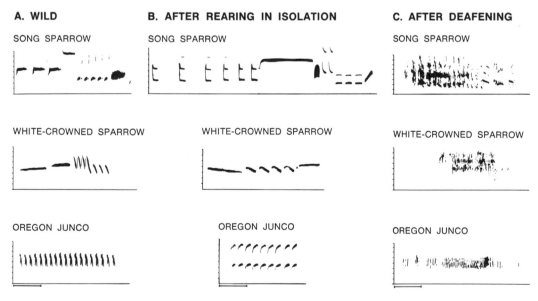

503

the results from several workers. In the wild (A of the figure), the songs of the three species are distinct. After rearing in isolation (B), the song of the song sparrow retains most of its structure; the songs of the other two lose some of their species-specific structure, though retaining some aspects of vocal control. However, when the young are deafened (C), their vocalizations as adults are coarse and scratchy, lacking in structure and species-specificity.

From these results it has been concluded that there is a sensitive period, between 10 and 15 days for the male white-crowned sparrow, during which auditory stimulation with the appropriate song pattern is necessary for the development of the ability to produce that song. This has suggested that sensory stimulation with the song pattern sets up an *auditory template* in the central auditory nerve circuits, which provides the means not only for *recognition* of the species-specific song pattern, but also for *generation* of the motor output for producing the song itself. Since the sensitive period for auditory learning precedes by several weeks the time when the bird actually begins to sing, it is as if the bird sings from memory. As a young male bird begins to sing, it makes an ever closer match between its template and its performance. Thus, genetics, sensory stimulation, sensory feedback, and, probably, internal corollary feedback, all are necessary in the ontogeny of song.

Hormones and Birdsong Circuits

A critical factor involved in the control of birdsong is the effect of hormones. As noted above, singing during courtship is characteristically done only by males (see Fig. 23.6A, top and middle). This ability in the male zebra finch is dependent on the male sex hormones; castrated males do not sing, but singing can be reinstated by administration of androgens. Injections of androgens do not induce singing in females (A in Fig. 23.6A, bottom), but they do if the females were treated at birth with

testosterone metabolites (dihydrotestosterone or estradiol; E in Fig. 23.6A, bottom). Even when reared in isolation, these females are able to produce a song that is very similar to that of the male.

Studies of the brain have revealed the neural basis for these hormonal effects. Previous stimulation and ablation studies had permitted identification of several levels in the motor hierarchy; these include the vagal motoneurons to the syringial and chest muscles, located in the brainstem; a center in the midbrain; and several centers in the telencephalon (see Fig. 23.6B). Arthur Arnold, Fernando Nottebohm, and Donald Pfaff at Rockefeller University showed that nerve cells in several of these centers are able to bind injected and radioactively labeled testosterone or its metabolites (see Nottebohm, 1980). The cells in the male are larger than in the female; they are thus *sexually dimorphic*. Mark Gurney and Mark Konishi (1980) have shown that the ability of hormone-treated females to sing is closely correlated with enlargement of brain centers in the song-producing pathways. The much smaller sizes of regions RA, HVc, and X in the female are indicated in the diagram of Fig. 23.6B. An example of the effects of hormones on one of the centers (HVc) is shown in Fig. 23.6C. These studies thus give evidence of the powerful ability of hormones to act as *organizers* of neuronal circuits (see Chaps. 8 and 27).

In addition to sexual dimorphisms, the vocal pathways in birds also demonstrate bilateral asymmetries. Each half of the syrinx is supplied by a right or left branch, respectively, of the twelfth (hypoglossal) cranial nerve. Nottebohm (1975) cut each branch separately, and found that cutting the left syringeal nerve branch severely affected the ability to sing, whereas cutting the right branch had very little effect. The basis for this difference appears to be correlated with the fact that a greater portion of the air expired during singing comes through the left bronchus from the left

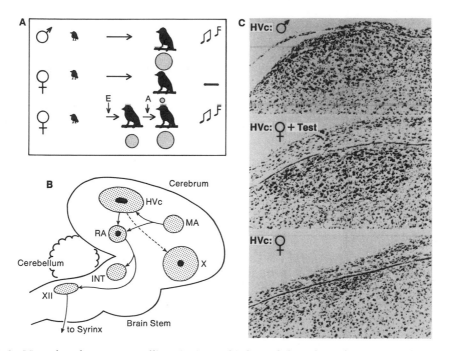

Fig. 23.6 Neural pathways controlling singing in birds, and their dependence on sex hormones. **A.** Diagrammatic representation of experiment with hormones in the finch. Normally, male finches sing but females do not (upper two rows). However, females treated at hatching with estradiol (E) and subsequently, as adults, with androgens are able to sing and show other male behavior. Relative size of the nucleus hyperstriatum ventrale (HVc) is shown by the shaded circles. **B.** The pathways for vocal control in the canary. Dots indicate sites of binding of systemically injected [³H]testosterone. Shading indicates areas more highly developed in males than in females (Area X is present only in the male). Brain regions: INT, nucleus intercollicularis; MA, nucleus magnocellularis anterior (neostriatum); RA, nucleus robustus archistriatalis; XII, twelfth cranial nerve. **C.** Histological sections, showing the HVc in normal male (upper), hormone-treated female (middle), and normal female (lower). Note the much smaller size of the HVc cell group in the female. (A from Nottebohm, 1980; B from Miller, 1980; C from Gurney and Konishi, 1980)

lung. From this work emerged the concept of left hypoglossal dominance in song production. It is somewhat surprising to realize that at the time of its discovery, around 1970, it was the only asymmetry in neural function known in a vertebrate, other than humans. Among invertebrates, there are several well-known asymmetries, such as the pincer and crusher claws of lobsters, and in recent years a number of behavioral asymmetries have been discovered in vertebrate species. We will return to this theme in discussing human speech (below and Chap. 30).

Mammalian Vocalization

Vocalization plays a prominent role in communication among members of most mammalian species. As in birds, this is correlated with a keen sense of hearing (Chap. 15). Some of this keenness is used to respond to intraspecies signals, as indicated by the complexity of many of the sounds produced. In fact, the more this question has been investigated, the more complex appear the vocalizations, their neural control, and the behavior they mediate.

A. CAT PURRING

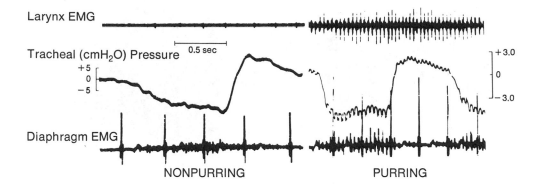

Larynx EMG

Tracheal (cmH₂O) Pressure

Diaphragm EMG

NONPURRING   PURRING

B. SONG OF THE HUMPBACK WHALE

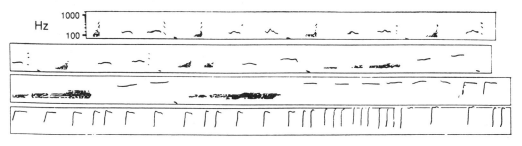

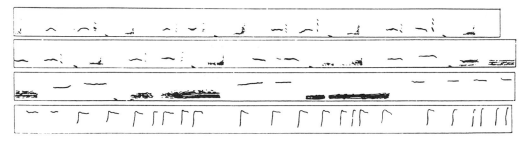

Fig. 23.7   Sound production in mammals. **A.** Purring of a cat. Upper trace: electromyograph (EMG) recordings from laryngeal muscles; middle trace: pressure in the respiratory tract; lower trace: electromyograph recordings from the diaphragm. Note correlation between impulses in EMGs and oscillations in pressure during purring. **B.** Sonograms of the song of the humpback whale, obtained from an animal in the vicinity of Bermuda. Top four strips are from a song that lasted over 10 min; bottom four strips are from a subsequent repeat of the song by the same animal. (A from Remmers and Gautier, in Doty, 1976; B from Payne and McVay, in Wilson, 1975)

The purring of a cat involves precise timing between contractions of laryngeal muscles and the diaphragm (see Fig. 23.7A). Purring is largely under central control; the purring rhythm continues in

the motoneuronal discharge despite deafferentation, or removal of the muscles.

Figure 23.7B illustrates sonograms of the song of the humpback whale. The entire song lasts 7–30 minutes; each whale sings its own song, which it repeats faithfully, as indicated in the figure. The ability to repeat a message of this complexity testifies to considerable powers of memory storage and motor readout. It has been said that this may be the most elaborate single behavioral display in any animal species.

These considerations indicate that a sophisticated apparatus for vocal communication has emerged in the course of mammalian evolution. Let us see how this apparatus has been adapted in humans for producing speech.

Human Speech

The Vocal Apparatus

Although we commonly think of speech as emanating from the larynx, our vocal apparatus is a good deal more complicated than that. Sound production is based on the principle of forced air. This requires three main components: a source of *pressure*, a set of *vibrating* elements, and a system of *resonators* and *articulators*. As shown in Fig. 23.8, each of these is a carefully coordinated system of subcomponents. *Pressure* arises by taking air into the lungs (inhalation) and expelling it (expiration). This depends on the *respiratory* muscles, principally the diaphragm, the intercostal muscles, and the abdominal muscles. The *vibrating* elements are the *vocal cords* within the *larynx;* they are controlled by a complicated set of laryngeal muscles. The larynx converts the rush of air through the trachea into a buzzing sound with many frequency components. The *resonators* and *articulators* are composed of the structures of the *upper respiratory tract;* these include the pharynx, mouth, tongue, lips, sinuses, and related structures. These provide for resonance chambers and filters that transform the laryngeal buzz into sounds with specific qualities.

These same main components are present in birds and in other mammals, except that in birds there is a syrinx instead of a larynx. This illustrates that the location of the buzz-producing element and its relation to the resonators is movable. Still another location is used when we play a musical instrument of the horn or wind family, such as a trumpet. When we blow a trumpet, we supply the pressure with our lungs; the buzz comes, not from the larynx, but from our lips pressed to the mouthpiece. The tone is formed by the resonant chambers within the tubes of the instrument.

The vibrating elements that generate the laryngeal buzz are the *vocal folds*. These are two folds of muscle tissue that have a tough ligament at their free edge and a mucous membrane cover (see Fig. 23.9). They are housed within the *thyroid* cartilage, which forms a protective shield around them. The *cricoid* cartilage forms a ring around the base of the larynx; it supports the thyroid cartilage and provides surfaces of articulation for the *arytenoid* cartilages. These are two small triangular cartilages, each of which is attached at its apex to a laryngeal fold. The base has a complex articulation with the cricoid cartilage, which allows it to rock, rotate, or slide. These movements are brought about by contractions of the intrinsic laryngeal muscles. These include the *thyroarytenoid* (which constitutes the main mass of the vocal cords), *cricoarytenoid, interarytenoid, and cricothyroid.*

Production of Speech

The individual actions of each of these muscles are too complex to detail here; suffice it to say that they provide for delicate and precise adjustments in the length, tension, and separation of the vocal cords. Figure 23.9B indicates some of the positions of the vocal folds and the way they are brought about by movements of the arytenoid cartilages. The *pitch* of the laryngeal buzz is set by the length, tension, and separation of the folds; the folds then vibrate at that frequency. The buzz is *louder* if more pressure is applied, though the muscles must counteract the pressure precisely in order to maintain pitch.

How is this buzz converted into intelligible vocal signals? This is the task of the

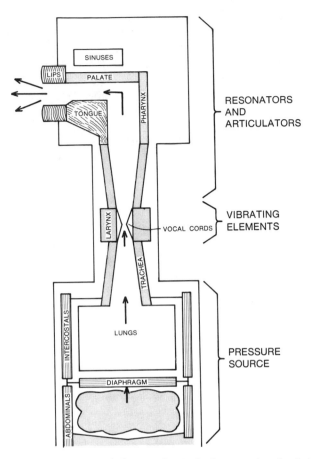

Fig. 23.8 Schematic representation of the mechanical elements involved in sound production in humans.

Fig. 23.9 **A.** Longitudinal section through the human larynx. **B.** Configurations of the glottis under different conditions. (From Zemlin, 1968)

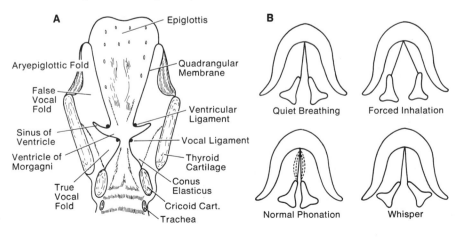

508

resonators and articulators. Each of the structures of the upper respiratory tract plays an important role, as becomes apparent when any one of them is compromised. Thus, a stuffy nose markedly changes the quality of the voice. Also, just try to say anything at all while holding the tip of your tongue! In human speech the tongue is the most important of the organs of articulation. Its complex arrangements of muscle fibers make it a most versatile motor organ, and the high density of innervation is matched by the large representation of the tongue in the motor cortex (see Chap. 21). We should also note the versatility of the tongue as an organ for manipulation and mastication of food in this regard.

All of these properties render the tongue well suited to the function of making finely graded adjustments in the configuration of the resonant chamber of the mouth. These configurations are most critical in the production of *vowel* sounds; in fact, each vowel sound is produced by a specific position of the tongue, and there is a systematic shift in these positions for the sequence of vowels (see Fig. 23.10). Consonants, by contrast, result from obstructing the air passage through the vocal tract, at the lips, teeth, hard palate, soft palate, or glottis. In English, these produce sounds that are called stops (t, p), fricatives (f, s), nasals (m), or glides (l). If you place your fingers on your face or neck while producing any of these sounds, it will become obvious that speech involves coordination of activity in most of the muscles in these regions.

Control of Pressure

We have concentrated on the muscular control of the vibrating elements and the resonators and articulators, but we must not give the impression that the pressure source acts merely as a crude bellows. In a classic study in 1959, M. H. Draper, P. Ladefoged, and David Whitteridge in Edinburgh investigated the activity of different respiratory muscles during vocalization. In addition to the EMGs of the muscles, re-corded with needle electrodes, they monitored the volume of air in the lungs and the intratracheal pressure. Figure 23.11 shows the results in an experiment in which the subject took a deep breath and slowly counted to 32. It can be seen that there is a very precise sequence of activity in these widely different muscle groups during this period of phonation. The results show that the diaphragm is relaxed through most of expiration and phonation, and that maintenance of the appropriate subglottic pressure is due to activity in the intercostal, abdominal, and latissimus dorsi muscles. Much the same sequence takes place during the singing of a single note. As Donald Proctor (1980) of Johns Hopkins observed:

. . . the production of a tone of any given intensity requires the appropriate subglottic pressure. This is accomplished by the exact blending of inspiratory and expiratory muscle effort with the elastic force associated with the lung volume at the time. This blending is largely produced through a balancing of the abdominal muscles against or with those of the chest wall across a relaxed diaphragm, occasionally supplemented by accessory expiratory muscles.

When the "appropriate subglottic pressure" is controlled by a trained and gifted singer, one of the transcendent artistic experiences of human life is produced, as illustrated in the following description of the great tenor Enrico Caruso (Scala recording 825):

. . . his voice floated on a deep and perfectly controlled column of air—something beautiful beyond description.

Birgit Nilsson (1984) described the process as follows:

I always want a very, very deep support for the breath. The whole body has to work. . . . The key to producing the sound . . . lies in . . . "support from downstairs"—that is, from the muscular area just above the pubis.

Neural Circuits for Vocalization

This delicate balancing of activity in many body muscles suggests that the motoneu-

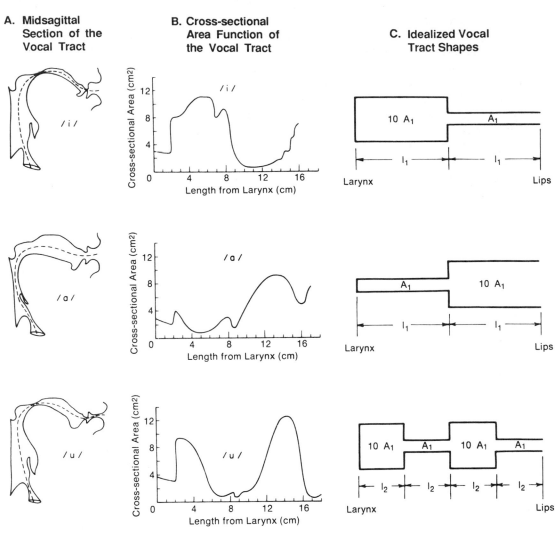

A. Midsagittal Section of the Vocal Tract

B. Cross-sectional Area Function of the Vocal Tract

C. Idealized Vocal Tract Shapes

Fig. 23.10 Some principles in the production of human speech. **A.** Positions of the tongue for forming the vowel sounds "ee" (/i/), "ah" (/a/), and "oo" (/u/). There is evidence that these are basic phonemes present in all human languages, and that other vowel sounds are variations on these. **B.** Cross-sectional areas of the supralaryngeal speech-producing spaces, from the larynx (0) through the pharynx and mouth to the lips (16), for the three different vowel sounds. **C.** Simplified computer simulations of the supralaryngeal spaces in B. Note that the considerable length of the space means that the pharyngeal and oral segments can be independently manipulated by the tongue. This is crucial for the formation of the vowels, and hence for human speech. It has been shown that the pharyngeal space is diminished or absent in chimpanzees and Neanderthal man, as well as in the human newborn (see Chap. 26). (From Lieberman et al., 1972)

rons to these muscles are under close control by descending motor pathways. In fact, it has been found that the intercostal motoneurons receive monosynaptic inputs from corticospinal fibers. These findings emphasize that it is not only the muscles of the hand that are involved in fine motor performance; muscles as different as those of the tongue and chest and abdomen may accomplish equally delicate maneuvers.

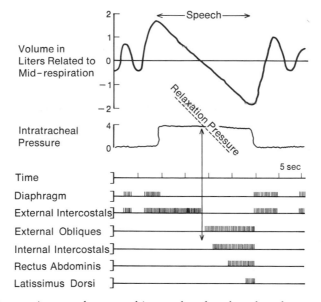

Fig. 23.11 In this experiment, a human subject took a deep breath and counted slowly from 1 to 32. The recordings show the changes in air volume in the lungs, intratracheal pressure, and impulse activity in different muscles. (From Draper et al., in Zemlin, 1968)

We can think of all the muscles of the abdomen, chest, larynx, head, and neck as being coordinated together to perform complex *manipulations of air* as it is expelled from the lungs. What are the nervous mechanisms involved in this coordination? We can begin by identifying the motoneurons involved (Fig. 23.12). For the muscles of the *pressure* apparatus, the motoneurons are located in the spinal cord: the motoneurons to the diaphragm are in cervical segments 3–5; those to the intercostals are in thoracic segments 2–5; and those to the abdominal muscles in thoracic segments

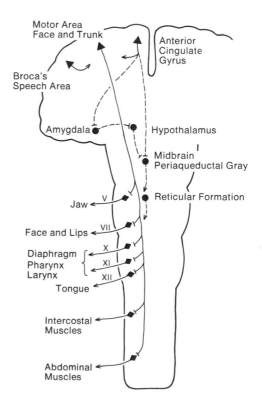

Fig. 23.12 Motor circuits and motor hierarchy involved in control of vocalization in humans. Pathways involved in precise motor control are shown by continuous lines; pathways involved in mediating emotional aspects of vocalization are shown by dashed lines. (Based in part on Jürgens and Ploog, 1981)

6–12. For the larynx, the intrinsic muscles are supplied by the vagus (X) and accessory (XI) nerves. The main branch to the larynx is the recurrent laryngeal nerve, which supplies all of the intrinsic muscles except the cricothyroid. This nerve is thus crucial to human speech; loss due to injury or infection leaves one able to speak only in a whisper. The *resonators* and *articulators* are controlled by nerves to the muscles of the pharynx (XI, X, VII), tongue (XII), and lips and face (VII). Other cranial nerves also contribute, as indicated in Fig. 23.12.

It can be seen that the motoneurons involved in vocalization constitute a complex array that is distributed along a considerable extent of the neuraxis, from the metencephalon (pons) to the lower thoracic levels of the spinal cord. The voluntary control over this array is mediated by descending fibers in the corticospinal tract. These originate in the face, neck, and trunk areas of the motor cortex (see Fig. 21.4 and Chap. 22) and make either monosynaptic or polysynaptic connections onto the motoneurons. Thus, as in the case of movements of the hand, the "intermediate" level of motor hierarchy is in the motor cortex. The mechanism of vocalization control by the motor cortex is virtually unknown, though it seems likely that it involves circuits through the basal ganglia and cerebellum, as in the case of control of manipulation. The highest level of speech control in the human includes Broca's area; the nature of the relation between Broca's area and the motor cortex in the control of speech will be discussed in Chap. 30.

IV

CENTRAL SYSTEMS

24

Introduction: The Nature of Central Systems

The analysis of sensory and motor systems has depended on several factors: they consist of localized circuits and pathways accessible to the investigator; they can be activated discretely, and they give a precise and quantifiable output. In central systems, most of these advantages are lost. The systems are deep within the central nervous system and thus relatively inaccessible to the investigator. The cells and circuits form systems that overlap and are difficult to localize, thus making selective activation difficult or impossible. The output from such systems may be too widespread to record or characterize adequately. To this is added the fact that, in the temporal domain, many of the actions of central systems last for days, months, or years. Small wonder, then, that although behaviors generated by central systems can be observed and classified, their neural substrates are extremely difficult to identify. It is, in fact, a testimonial to the power of modern methods in neurobiology that much of the experimental evidence for the molecular and cellular elements and the specific circuits of central systems has become available within the past generation or so.

The possibility of understanding the central neural substrates that govern behavior is exciting not only because it deepens our understanding of humans and of all animal life, but also because it holds the promise that we may be able to correct disorders of behavioral functions or restore functions lost by disease. However, the enthusiasm of neuroscientists for this task must always be tempered by the reminder that *identifying the neural basis for a specific behavior is one of the most difficult challenges in all biology*. As noted in Chap. 1, the history of endeavors in this area is a record of many deceptions and discouragements, because of the complexities of the systems and the difficulties in applying experimental methods noted above. This caution is all the more relevant to readers of this book, in which the accounts of many complex subjects must necessarily be brief.

As we review the different behaviors and systems in this section, remember this rule: *a given behavior may be mediated by many potential mechanisms*. The challenge for the neuroscientist is to identify the particular mechanism in a particular organism. To do this, experimental methods and theoretical models need to provide for critical tests between alternative mechanisms for a given behavior. We will see that there are attractive candidates for neural mecha-

515

nisms underlying many kinds of behavior, but we will also see that we are only at the earliest stages of devising the critical tests that can lead to a consensus on what these mechanisms are. This consensus is vitally important to everyone with an interest in the brain—students, public, clinicians, and philosophers—as we seek to understand the basis of human behavior and prevent or treat the disorders due to neurological diseases.

Definitions of Central Systems

Central systems may be defined as *cells and circuits that mediate functions necessary for the coordinated behavior of the whole organism.* The intent here is to distinguish central systems from specific sensory and motor pathways. As one proceeds more centrally, these pathways overlap with, and become incorporated in, central systems. Thus, higher levels of sensory processing clearly involve central systems involved in perception and cognition; it depends on how far centrally, through how many synaptic connections and circuits, one goes. Similarly, higher levels in the motor hierarchy involve central pattern generators and motor programs that are part of central as well as motor systems. The definition is therefore only a guide to help in identifying and characterizing systems that are less concerned with the specific tasks of sensory processing and motor control that we have discussed to this point, and more concerned with global aspects of behavior.

Central systems generally fall into two categories. One concerns systems that mediate interactions between the brain and the body. As indicated in Table 24.1 and Fig. 24.1, there are three main systems in this category. The *autonomic nervous system* was discussed in Chap. 18; it is usually regarded as a motor system, but it partially qualifies as a central system by its broad coordination of internal organ functions. The *neuroendocrine system* is responsible for coordination between the nervous system and the various endocrine

Table 24.1   Summary of central systems

Systems relating brain and body
  Autonomic nervous system
  Neuroendocrine system
  Neuroimmune system

Systems within the brain
  Specific transmitter systems
  Diffuse transmitter systems
  Distributed systems

---

organs of the body. The *neuroimmune system* mediates interactions between the nervous system and the network of organs and cells that constitute the immune system.

The other main category of central systems involves neural circuits wholly within the brain. As indicated in Table 24.1, first there is a widely projecting system of cells

Fig. 24.1 Schematic diagram of the central systems of the brain (somatic and autonomic) and their relations to sensory and motor systems, other body systems, and the environment. Solid lines indicate neural connections, dashed lines indicate humoral communication, and open arrows indicate interactions with body or environment.

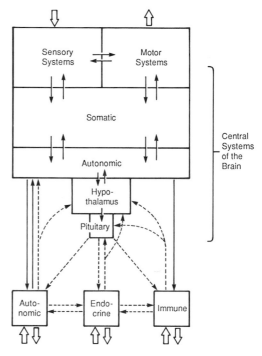

that can be regarded as a counterpart of the *autonomic nervous system* within the brain. Second, there are more specific *transmitter-identified systems;* they include central as well as sensory and motor pathways. Finally, there are systems that include circuits embracing different regions, with different transmitters, which mediate complex, higher brain functions; these are called *distributed systems.*

In this chapter, we will first discuss the neuroendocrine and neuroimmune systems. We will then summarize evidence for diffuse and specific transmitter-defined systems. The subsequent six chapters in this section will be concerned with distributed circuits underlying important types of whole-organism behavior.

Neuroendocrine Circuits

As we have seen, hormones are used in a variety of ways in the body. Those peptides and hormones that are secreted by central neurons, under central nervous control, and with actions that affect the coordinated behavior of the entire organism, may be considered to be parts of central systems.

The Pituitary Gland

Vertebrates have a master neurohemal organ, the pituitary gland. It has been traced back in phylogeny to the tunicates, the earliest chordates, where it is believed to be represented by a ciliated pit organ. An interesting fact about this pit organ is that it seems to be sensitive to pheromonal signals from other tunicates; furthermore, these pheromones have a molecular structure similar to steroid sex hormones. These similarities have supported J. B. S. Haldane's suggestion that the hormonal system of internal messengers originated in evolution from the pheromonal system of external messengers.

The pituitary consists of an endocrine and a neural part. The *endocrine* part (anterior hypophysis, or *adenohypophysis*) is derived embryologically from an outpouching of the pharynx, whereas the neural part (posterior pituitary, or *neurohypophysis*) is derived from an outpouching from the diencephalon. Both parts are under the control of the hypothalamus, but by different means.

Neurohypophysis. The *neurohypophysis* contains the terminals of axons from specific nerve cells in the hypothalamus (see Fig. 24.2). From these terminals are secreted two peptide hormones: *oxytocin,* which promotes contraction of smooth muscle in the uterus and mammary glands, and *vasopressin* (also called antidiuretic hormone, or ADH), which acts on kidney tubule membrane to promote retention of water, and on smooth muscle in arterioles of the body to raise blood pressure. We shall discuss these actions further in Chaps. 26 and 27. These hormones also have actions elsewhere within the nervous system (see below).

Adenohypophysis. The *endocrine* pituitary system, from its sites of neural control in the hypothalamus to its actions in the body, is summarized in Fig. 24.2. The identification of the pituitary hormones and their actions on target organs was largely achieved by 1950, and forms the body of classical endocrinology. The work of Geoffrey Harris and his colleagues in England thereafter showed that the anterior pituitary is controlled by the hypothalamus by means of factors transported in the hypophyseoportal system. As Harris wrote in his monograph *Neural Control of the Pituitary Gland* in 1955:

> . . . it seems likely that nerve fibers in the hypothalamus liberate some humoral substance into the primary plexus of the vessels, and that this substance is carried by the vessels to affect anterior pituitary activity. . . . If the hypothalamus . . . regulates the rate of secretion of the anterior pituitary hormones, are there as many humoral mechanisms involved as there are hormones?

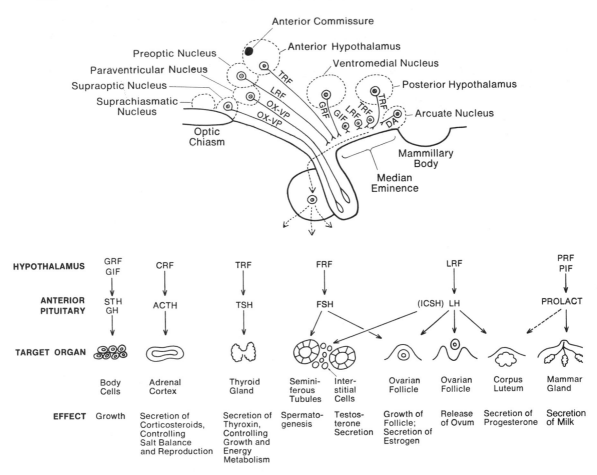

Fig. 24.2 The hypothalamic–pituitary system for neuroendocrine control in the mammal. ACTH, adrenocorticotropic hormone; CRF, corticotropin releasing factor; FRF, follicle-stimulating hormone releasing factor; FSH, follicle-stimulating hormone; GH, growth hormone; GIF, growth hormone release inhibiting factor (somatostatin); GRF, growth hormone releasing factor; ICSH, interstitial cell stimulating hormone; LH, luteinizing hormone; LRF, luteinizing hormone releasing factor (also called LHRH, luteinizing hormone releasing hormone); OX-VP, oxytocin-vasopressin; PIF, prolactin release inhibiting factor; PRF, prolactin releasing factor; STH, somatotropic hormone; TRF, thyroid hormone releasing factor; TSH, thyroid stimulating hormone. (Modified from Mountcastle, 1980)

This set the stage for an eager search for these substances, in which biochemists prepared extracts of hypothalamic tissue and identified compounds which either promote or inhibit the release and synthesis of pituitary hormones. By 1973, three of these compounds—luteinizing hormone releasing hormone (LHRH), thyrotropin releasing hormone (TRH), and somatostatin, or somatotropin-release inhibiting factor (SRIF)—had been isolated and synthesized.

These releasing factors are all peptides; their molecular structures are given in Chap. 8 (Fig. 8.11).

While this work was proceeding, other studies aimed at identifying the mechanisms for release of these factors. By electrical stimulation and local injections of hypothalamic extracts into hypothalamus, it was possible to localize the neuroendocrine cells producing the factors. More recently, these results have been extended by binding studies of receptor localization, and hybridization of mRNA probes to localize cells with genes that express different types of peptides.

Some of these sites are indicated in Fig. 24.2. The general pattern is that the cells in different regions all send axons to the median eminence, on the floor of the hypothalamus. The peptides are stored in the axon terminals, which rest on the vessels of the portal system. Discharge of the factors is controlled by neural activity within the cell and by circulating hormones. The factors act quickly (within minutes) before being inactivated in the blood. The actions on pituitary cells are to stimulate immediate release, as well as to induce long-term synthesis of hormones. Like other peptide hormones, the releasing factors act on the pituitary cells through membrane receptors and second messengers, as previously discussed in Chap. 8.

Since Harris's time, it has been clear that the nervous system is not only involved in controlling the pituitary, but is itself the target of actions of circulating hormones. If we take the case of the gonadotropins, these stimulate the production of sex hormones in the gonads (G), either testosterone or estrogen. These hormones induce the secondary sex characteristics of males and females, as well as stimulate the maturation of sperm and eggs, respectively. The levels of circulating gonadotropins are controlled by several factors; primary among them are sensory stimuli to the nervous system, eventually reaching the hypothalamus, and negative feedback by the circulating hormones onto pituitary cells and central neurons.

This scheme for the pituitary system will be complete when we have identified all the central neural connections made by the neuroendocrine cells in the hypothalamus. Recordings have been made from single hypothalamic cells that project to the median eminence (as determined by antidromic backfiring) and thus presumably secrete releasing factors. These cells can also be backfired from a number of other

sites, including the thalamus, preoptic area, amygdala, anterior hypothalamic area, and periventricular nucleus (see Fig. 24.3). Synaptic inputs, on the other hand, have been identified by orthodromic firing; these inputs have been found to arise in the amygdala, preoptic area, and hippocampus. To complete the picture on the input side, the neuroendocrine cells receive copious innervation from dopamine-containing neurons (see below). The cells are also subject to feedback regulation by local circuits. Finally, they are regulated by levels of circulating hormones and other humoral factors.

The picture that emerges in Fig. 24.3 is of a neuroendocrine cell that is under extensive control by both neural and humoral mechanisms. Through the neural connections, inputs from many central regions, including the cerebral cortex and limbic systems, reach the hypothalamus. By this means, a variety of behavioral states, such as arousal, stress, and sexual maturation, all have their influence in setting the level of discharge of releasing factors and thus the control of pituitary function.

Neuroimmune Circuits

Traditionally, the immune system has been considered to be separate from the nervous system. To begin with, its anatomical location is widely dispersed in various glands, organs, and circulating lymphocytes and immunoglobulins, in contrast to the apparently "hard-wired" nerves and brain. Its function of protecting the body against invading substances and organisms seems markedly different from the function of the nervous system of processing information. The mechanisms of the immune response involve binding of antigen, proliferation of circulating lymphocytes, and secretion of antibody molecules, mechanisms that have no obvious counterpart in the nervous system.

This traditional view is now being replaced by a new concept, which recognizes that the two systems interact in a coordi

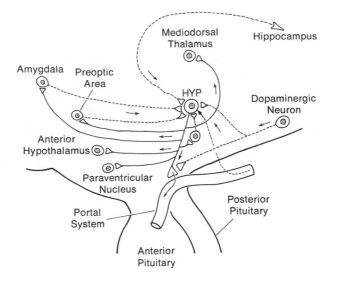

Fig. 24.3 Multiple connections of a single neuroendocrine cell in the hypothalamus. The cell and its output axon and branches are shown in continuous lines; inputs to the cell are shown in dashed lines. (After Renaud, 1977)

nated manner, and that the molecular and cellular mechanisms share many common principles. We have already seen in Chap. 3 that various types of neural adhesion molecules belong to the immunoglobulin superfamily. This new perspective is beginning to bring about revolutionary changes in our understanding of both systems. Let us summarize the evidence for how they interact, first at the systems level, and then at the molecular level.

Immunoregulatory Circuits

For many years, it has been known that psychological stress has a depressive influence on the immune system. In studies of people bereaved by the death of a spouse, or suffering severe depression, it has been shown that there is suppression of lymphocytic proliferation in response to an antigenic stimulus. Animal experiments support these findings; rats subjected to stressful situations, such as a restraining apparatus or tail shocks, also have reduced lymphocytic responses to antigen or mitogen injections.

What is the pathway by which psychological and behavioral states can affect the immune response? There is general agreement that the main pathway is through the hypothalamus (see Fig. 24.4). In some experiments, it has been shown that electrolytic lesions of the anterior hypothalamus in rats have a suppressive effect on the proliferation of lymphocytes in response to intravascular injection of a mitogen such as concanavalin A (Con A). By contrast, lesions in other brain regions, especially regions in the limbic system (see Chap. 28) such as the hippocampus which feed into the hypothalamus, lead to an increase in lymphocyte numbers.

Many experiments have documented that the hypothalamic control is exerted through the hypothalamic–pituitary–adrenocorticosteroid neuroendocrine axis. The earliest experiments showed that injections of cortisol have a suppressive effect on the immune response, implying that the influence of the brain is only suppressive and is exerted exclusively through corticosteroids. Recent experiments, however, suggest a more complicated mechanism, in which the hypothalamic–pituitary system is *immunoregulatory* rather than merely immunosuppressive, in line with the lesion

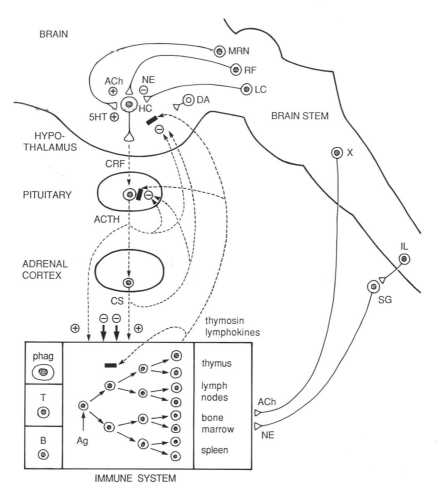

Fig. 24.4 Neuroimmune circuits. The brain regulates the immune system and the immune system modulates brain activity through interactions summarized in this diagram. A central point in the neural circuit is the hypothalamic cell (HC) secreting corticotropin releasing factor (CRF); as shown in the diagram, this cell is heavily regulated by neural inputs and by humor factors. The hypothalamic–pituitary–adrenal cortical (HYPAC) axis is the main route of immunoregulation; there is also sympathetic and parasympathetic innervation of organs of the immune system, as shown. Within the immune system organs, phagocytic (phag) and B and T lymphocytes proliferate in response to an antigenic (Ag) stimulus. The early stages of proliferation are facilitated (+) by low concentrations of corticosteroids (CS) and ACTH, but are suppressed by high concentrations (thick arrows), such as occur in behavioral states of high anxiety or stress. This is believed to be the basis of the immunosuppression that occurs in people suffering depression. The cells of the immune system secrete various substances (lymphokines, etc.) that not only regulate the immune response but also act to block the actions of other humoral substances at different levels in the HYPAC axis (see black bars). Various humoral feedback pathways in the HYPAC axis are also shown. Other abbreviations: IL, intermediolateral nucleus of the spinal cord; LC, locus ceruleus; MRN, median raphe nucleus; RF, reticular formation; SG, sympathetic ganglion; X, vagal nucleus (parasympathetic). (Based on Bulloch, 1985; Hall and Goldstein, 1985; Hall et al., 1985)

experiments above. As indicated in Fig. 24.4, corticosteroids have been found to act mainly at the earliest stages of lymphocytic proliferation, and to be stimulatory at low concentrations, changing to suppressive at high circulating levels. An interesting finding is that stress causes suppression of the immune response even in rats that are adrenalectomized and therefore secrete no corticosteroids; this has implied that pituitary hormones themselves, such as ACTH and endorphins, may participate in mediating the immunosuppressive effects.

How is the hypothalamic–pituitary pathway modulated? Proliferating lymphocytes secrete a variety of specific messenger molecules, collectively called *lymphokines,* which are crucial for coordination between B and T cell populations. Lymphokines, as well as thymokines secreted by thymus cells, have several sites of action in the brain immunoregulatory pathway: on lymphocytic corticosteroid receptors, and on cells in the pituitary body and hypothalamus. As indicated in Fig. 24.4, other feedback pathways from circulating corticosteroids and ACTH contribute to the peripheral regulation of corticotropin releasing factor (CRF) secretion from the hypothalamus. Within the brain, several transmitter systems (5HT, ACh, NE, and DA) have inputs to hypothalamic cells and can be shown to affect the neuroimmune pathway; for example, lesions of 5HT neurons in the midbrain raphe nucleus (see below) lead to increased antibody titers in response to an immunologic challenge, whereas lesions of NE neurons in the locus ceruleus lead to decreases in lymphocyte counts. Finally, there are direct nervous connections to lymphoid organs through nerves of the autonomic nervous system. These various elements can be said to constitute the *neuroimmune system.* The novel feature of the neuroimmune system is that it consists of both humoral and neural circuits, as indicated by the diagram in Fig. 24.4

Although many of the component parts of these circuits have been identified, the way they work together to achieve integrated control is not yet clear. For example, it is known that levels of circulating corticosteroids rise and fall in close association with the time course of the antibody response (see Fig. 24.5). Also correlated with the antibody response is an increased firing frequency of hypothalamic cells; these may be the cells that release CRF to bring about the rise in corticosteroids. But what is the functional significance of these correlations? A possibility is that the corticosteroids would have their greatest suppressive effect on lymphocytes with low affinity for the antigen, thereby contributing a kind of lateral inhibition to help make the immune response more specific. There is clearly much work ahead to test these and other hypotheses.

Membrane Mechanisms

If there are influences of the nervous system on the immune system, there must be membrane mechanisms to mediate them. Research at the molecular level is providing abundant and fascinating evidence that this is the case. The plasma membranes of cells of the immune system appear to be loaded with receptors for a variety of different types of molecules. Best known are the lymphokines, which serve as lymphocyte growth factors and coordinate the proliferation of T and B cells.

Receptors are also present to mediate the influence of the adrenocortical steroids, but the demonstration of receptors for ACTH and endorphins suggests that these provide the means for more direct regulation by the pituitary. In addition, an extraordinary variety of substances can affect the immune response, implying the presence of appropriate membrane receptors; some of these have been identified, as summarized in Fig. 24.6A. The surprising finding has been that these are all receptors for neurotransmitters and neuropeptides common to the nervous system.

Stimulated by this array of receptors is an arsenal of secretory products. In addi-

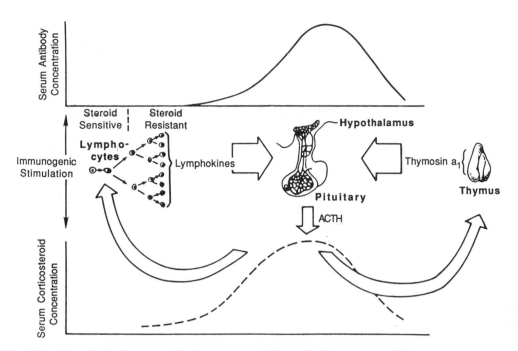

Fig. 24.5 Correlation between the time course of the antibody response of the immune system to an antigenic stimulus *(above)* and the corticosteroid response of the hypothalamic–pituitary–adrenal cortical axis *(below)*. Between the graphs, mechanisms correlating the two systems are postulated. See text. Compare with Fig. 24.4. (From Hall et al., 1985)

Fig. 24.6 Molecular properties of immune cells that are shared with nerve cells and are involved in neural–immune interactions. Abbreviations for membrane receptors in A, counter clockwise from bottom: CS, corticosteroids; LK, lymphokines; SOM, somatostatin; NT, neurotensin; SP, substance P; THY, thymosin; OXY, oxytocin; VAS, vasopressin; Arg, arginine; GH, growth hormone; ACTH, adrenocorticotropic hormone, END, endorphins; BZ, benzodiazepines; VIP, vasoactive intestinal peptide. For secretory products in A: CGT, chorionic gonadotropin. The voltage-gated channels in B were described in Chap. 5. (Based on Blalock and Smith, 1985; Cahalan et al., 1985)

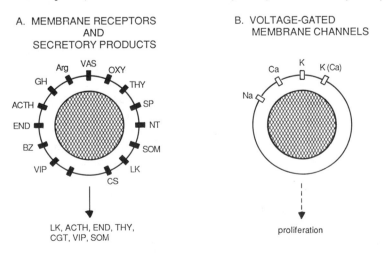

tion to the lymphokines are ACTH and β-endorphin, as in pituitary cells, as well as several neuropeptides. It should be recalled that ACTH and β-endorphin are both derived from pro-opiomelanocortin (POMC) (cf. Chap. 8). These studies indicate that virus infections elicit not only the antibody response but also the secretion of various hormones and neuropeptides, which may act as lymphocyte growth factors as well as modulate cells at other levels in the hypothalamic–pituitary immune circuit.

In addition to its many ligand-gated channels, the lymphocyte membrane contains a variety of voltage-gated channels. As indicated in Fig. 24.6B, these include channels for inward, depolarizing currents (Na^+, Ca^{2+}) and outward, hyperpolarizing currents[K^+, $K^+(Ca)$]. These have properties shared with neurons (see Chap. 5). Macrophages have in fact been shown to have the ability to generate impulses. A point of some interest is that mitogenesis in lymphocytes is accompanied by an increase in the numbers of K^+ channels (from 50 to 300 per cell in the rat). It will be interesting to test whether an increase in K^+ conductance might also accompany differentiation of neurons. This could be a mechanism shared by cells undergoing mitosis in both the nervous and immune systems, providing for control of membrane excitability during critical development periods.

The Immune System as a "Mobile Brain"

The idea that the immune system itself has brainlike properties has tantalized a number of workers in the field. The variety of studies summarized above has led to the suggestion (Blalock and Smith, 1985) that

cells of the immune system can apparently be controlled in a fashion similar to pituitary cells by a positive signal from the hypothalamus (corticotropin releasing factor) and a negative signal from the adrenal gland (a glucocorticoid hormone). Hence, "stress" as defined by an increase in circulating glucocorticoid hormone levels can, depending on the stimulus, apparently have its

ultimate origins not only in the central and peripheral nervous systems but also in the immune system itself. Such findings have led to the notion that a primary function of the immune system may be to serve as a sensory organ for stimuli such as bacteria, viruses, and tumor cells that are not recognized by central and peripheral nervous systems . . . leukocyte information then being transferred to the neuroendocrine system by peptide hormones and lymphokines . . . [thus,] certain cells of the immune system may serve as "free-floating nerve cells". Perhaps collectively, such cells represent a mobile brain.

Does the immune system constitute a "parallel brain"? Do the cells of the immune system form a network, as Niels Jerne has suggested, which has sensory, motor, and central functions equivalent to those in the nervous system? Is immunological "memory," for example, similar in any of its properties to neural memory? These questions signify that this field holds interest not only for new experimental findings, but also for conceptual advances as well.

Central Systems: Survey of Methods

In Chap. 8, the biochemistry and molecular actions of putative neurotransmitter substances were discussed. We now ask, where within the brain are the neurons that contain these substances?

Historically, the first step in answering this question was the technique for localizing monoamines introduced by Olavi Fränkö in the 1950s and refined by Falck and Hillarp in the 1960s. This technique consists of treating tissue sections with formaldehyde vapors or glyoxylic acid, which causes the monoamine compounds to fluoresce with colors characteristic for the three main types (NE, DA, and 5HT). The second step came in the 1970s with the ability to prepare antibodies to transmitter molecules. Thus, a molecule such as serotonin (5HT) is conjugated with bovine serum albumin and injected into a rabbit to produce rabbit antibodies to 5HT. This is applied to the tissue sections of interest, where the antibodies bind to the conjugated 5HT contained within cells. The sections are then treated with a fluorescence-labeled antirabbit antibody in order to visualize the cells containing the serotonin. We discussed examples of this technique in Chaps. 8 and 19. This technique is called immunocytochemistry. Its specificity has been greatly enhanced by the use of monoclonal antibodies.

The third step has been the ability to localize the genes expressing different neurotransmitters by use of mRNA hybridizing probes. These methods, described in Chap. 2, permit the unequivocal identification of the cells containing the machinery for synthesis of specific transmitter molecules.

In addition to these methods for localization of sites of transmitter synthesis, there are complementary methods for localization of sites of transmitter receptors. The traditional method is ligand binding combined with autoradiography. In this technique, tissue sections are treated with radioactively labeled molecules that bind to transmitter receptors. The sites of binding are visualized by coating the sections with a photographic emulsion so that the radioactivity makes a photographic imprint of the receptor sites, a technique called autoradiography (the section so to speak makes a radiograph of itself).

The power of this technique is twofold. First, the specific molecules used—transmitters, agonists or antagonists of transmitters, or other drugs—are employed in order to assess receptor function. Furthermore, if serial sections are made, the distribution of receptor sites may be determined throughout the entire brain; for this reason, the technique is also called *receptor mapping*.

The method of mRNA hybridization described above tests for gene expression of the mRNA for the receptor in question rather than for the presence of the receptor itself. Because of its great sensitivity for the specific nucleotide sequence of the mRNA, it is increasingly becoming the method of choice for identifying different receptor types. These studies are revealing a host of different isoforms of a given receptor family, each expressed by different cells in different regions.

In addition to neurotransmitters and their receptors, one wants to know the distribution of many other molecules of interest, including the synthesizing enzymes for neurotransmitters and second messengers and the different molecular constituents of neuronal and glial cytoplasm.

Our aim here will be to summarize the distribution in the brain of neural systems using the most common neurotransmitters and neuromodulators and related molecular constituents, whose basic mechanisms are discussed in Chaps. 6, 7, and 8.

Specific Transmitter-Defined Systems

We turn first to transmitter-defined neurons that are relatively specific in their projections. These range from neurons with long-distance projections to several regions, to intrinsic neurons whose connections are contained entirely within a region, or one lamina of a region.

Let us begin our consideration of central brain systems with specific long- and short-axon neurons. The discussion of the location of neuronal pathways will be related to a standard lateral (parasagittal) view of the rat brain, which will permit ready comparisons between different systems. In this way, the student can have a single framework for the study of the comparative neurochemistry of brain circuits.

Glutamate

Glutamate is regarded as the most prevalent neurotransmitter with a primarily excitatory synaptic action. As noted in Chap. 8, glutamate and its close relative aspartate are widespread constituents of intermediary metabolism in the body and the brain; they therefore seemed at first to be unlikely candidates for the specific actions required at a synapse. Glutamate was first established as the excitatory transmitter at the crustacean neuromuscular junction in 1959. Since then, evidence has accumulated for its excitatory action at a number of central synapses in the mammalian brain. Most of these are made by projection neurons.

At moderate levels of activity the primary action of glutamate at these synapses is a brief excitatory action. For this reason, glutamate and aspartate are also referred to as excitatory amino acids (EAAs). When the synaptic activity is intense and prolonged, glutamate activates NMDA receptors, which produce longer lasting effects in the postsynaptic neuron, such as long-term potentiation (LTP). In addition, glutamate may act on metabotropic receptors. The molecular mechanisms are explained in Chap. 8 and are further discussed in relation to mechanisms of learning and memory in Chap. 29; the role of NMDA receptors in development is described in Chap. 9.

The distribution of neurons using glutamate as a transmitter is shown in the diagram of Fig. 24.7. Of particular interest are the granule cells of the cerebellum. The granule cells are a type of short-axon projection neuron, linking the two layers of the cerebellar cortex. Because of their large numbers (10 to 100 billion, as noted in Chap. 21), we can say that there are more glutamatergic cells than all other cells combined in the nervous system! Other projection tracts using these substances include the olfactory bulb input to olfactory cortex, the entorhinal cortex input to the hippocampus and dentate fascia, and the cortical input to the caudate. At all these sites, glutamate and aspartate have excitatory actions. An intriguing note is that the mitral cells of the olfactory bulb not only project to the olfactory cortex, but also take part in dendrodendritic synapses within the microcircuits of the olfactory bulb. According to Dale's Law (Chap. 8), one might expect the transmitter at the dendritic synapses also to be glutamate or

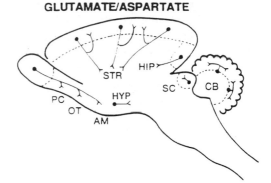

Fig. 24.7 Distribution of neurons believed to use glutamate and/or aspartate as a neurotransmitter in the rat brain. A sagittal view of the rat brain is shown for this and succeeding figures. AM, amygdala; CB, cerebellum; HIP, hippocampus; HYP, hypothalamus; OT, olfactory tubercle; PC, pyriform cutex; STR, striatum.

aspartate, and there is in fact evidence to support this.

The distribution of receptors for the NMDA receptor subunit of the glutamate receptor is revealed by mRNA in situ hybridization, as shown in the micrograph in Fig. 24.8. The regions of highest density of NMDA receptor expression include the olfactory bulb (OB), the hippocampus and dentate fascia (Hi), and the cerebellar cortex (Cb). In the olfactory bulb the laminar patterns are so distinct that it is possible to localize further the receptors to the outer beadlike glomerular layer and the thick inner granule cell layer, where the mitral/tufted cells activate inhibitory interneurons through dendrodendritic synapses; posteriorly, the accessory bulb is clearly seen. Thus, knowledge of the synaptic organization of a region enables one to analyze the functional significance of the pattern of receptor binding (see Chap. 11, Fig. 11.15). Intermediate receptor densities are seen in the cerebral cortex (note especially the most superficial layer, containing the distal dendrites of cortical pyramidal cells), striatum, superior and inferior colliculus, thala-

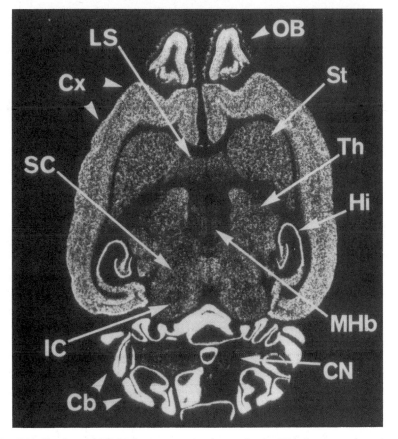

Fig. 24.8 Distribution of NMDA receptors in the rat brain. This horizontal section shows sites of in situ hybridization of probes for messenger RNA for the R1 receptor subunit. CB, cerebellum; CN, cerebellar nuclei; Cx, neocortex; Hi, hippocampus; IC, inferior colliculus; LS, lateral septal nuclei; MHb, medial habenula; OB, olfactory bulb; St, striatum; Th, thalamus. From Moriyoshi et al., 1991.

mus, and lateral septal nuclei. The student can correlate these patterns with the maps of projection neurons in Fig. 24.7.

γ-Aminobutyric Acid (GABA)

GABA is regarded as the most prevalent transmitter with a primarily inhibitory synaptic action. In mammals, GABA is found almost exclusively in the brain, where it is a constituent of intermediary metabolism. Its role as a neurotransmitter was first established in the crayfish, where it was shown to be the transmitter of the inhibitory axon to the stretch receptor cell (Chap. 7) and at the neuromuscular junction (Chap. 7). Localization of this neurotrans-

mitter in the brain has been aided by immunocytochemical methods for identifying a synthesizing enzyme, glutamic acid decarboxylase (GAD) (see Chaps. 6 and 8).

GABAergic neurons are shown in Fig. 24.9. There are short projection tracts from striatum to substantia nigra and from cerebellar cortex to deep cerebellar nuclei. The only long projection pathway thus far known has been demonstrated by immunohistochemical staining for GAD (Vincent et al., 1983). It arises from cells in the posterior hypothalamus and projects diffusely to the cerebral cortex. This could "provide a direct pathway by which limbic, emotional and visceral information can reach many

GABA

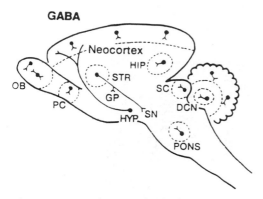

Fig. 24.9 Distribution of GABAergic neurons. DCN, deep cerebellar nuclei; GP, globus pallidus; HIP, hippocampus; HYP, hypothalamus; OB, olfactory bulb; PC, pyriform cortex; SC, superior colliculus; SN, substantia nigra; STR, striatum.

regions of cortex." It appears that this pathway shares properties with the diffuse transmitter systems discussed below.

Apart from these, most GABAergic neurons are intrinsic neurons, in such regions as cortex, olfactory bulb, hippocampus, cerebellum, and retina. Within these regions, GABA is present in high concentrations, of the order of micromoles per gram of frozen tissue, which is about 1000 times the concentrations of the monoamines. This is in accord with the powerful and specific actions of the GABAergic neurons in these regions. GABA acts on $GABA_A$ and $GABA_B$ receptors to increase Cl or K conductances, respectively (see Chap. 7); in the adult these actions are inhibitory, though at early stages of development GABA receptors may subserve excitatory actions in some regions. A long-lasting GABA action is also mediated through the metabotropic GABA receptor. These actions are exerted at axonal or dendritic output synapses, and are usually directed at controlling the output neurons. These actions are important for many functions, such as sensory processing, negative feedback, gating of rhythmic discharges, and timing and coordination of motor output. Drugs like picrotoxin and bicuculline,

which block GABA receptors (see Chap. 7), cause seizures, which has suggested that dysfunctions of GABAergic interneurons in the cortex may be critical in the development of epilepsy.

As we learned in Chap. 7, molecular cloning has revealed that the GABA receptor is a heteroligomer of five different subunits, with a number of isoforms. This tremendous combinatorial library means that there can be a different combination of subunits in the receptors of different neurons receiving GABAergic inputs. The exact combinations are being revealed by in situ hybridization of labeled oligonucleotide probes to messenger RNA, yielding micrographs of density patterns comparable to those of Fig. 24.8. The results of one such study are summarized in tabular form in Fig. 24.10. It can be seen that the patterns are exceedingly complex. However, several conclusions seem to be emerging: first, each region has its unique combination of subunits; second, each layer or subregion has its unique subunit combination; third, the number of different subunits in a layer or subregion is greater than five, implying that there is more than one type of neuron containing GABA receptors, each with its own unique composition. At present it is hypothesized that the different receptor compositions greatly increase the diversity of response properties in different neurons to the same transmitter.

Glycine

Traditionally, glycine has been localized mainly to the brainstem and spinal cord. In the spinal cord, it is believed to be an inhibitory transmitter of certain interneurons onto motoneurons.

Classically, ionophoresis experiments found little evidence for an action of glycine on neurons above the brainstem. However, the role of glycine in the brain began to undergo reevaluation in view of the evidence that glycine modulates glutamatergic synapses, and that it is present in the cerebrospinal fluid at a concentration sufficient to bring about these modulatory effects

Brain Region	α1	α2	α3	α5	α6	β1	β2	β3	γ1	γ2	δ
Olfactory bulb											
mitral cells	★★★	O	☆	O	O	★★	★★★	★	O	★★	O
granular layer	O	★	☆	☆	O	O	O	★★	☆	★	★
glomerular layer	★	☆	☆	O	O	O	★	★	☆	★	☆
accessory olf. bulb	★★	★	O	O	O	★★★	★★	★★	★	★	O
Caudate putamen	O	★	☆	O	O	☆	☆	★★	O	★	★
Cerebral cortex											
lamina I	O	O	O	O	O	O	O	O	O	O	O
lamina II + III	★★	★	★	O	O	☆	★★	★	☆	★	★
lamina IV +V	★★	☆	★	☆	O	☆	★★	★	☆	★	☆
lamina VI	★★	O	★	★	O	★	★★	★	☆	★	O
Pallidum											
ventral pallidum	★★	☆	☆	O	O	O	★	O	★	★	O
globus pallidum	★★	☆	O	O	O	O	★	O	–	★	O
insula Calleja	★★★	☆	O	O	O	O	★	O	★	☆	O
Hippocampal formation											
cells in stratum oriens	★	O	O	O	O	O	★	O	☆	☆	O
CA1	★	★	O	★	O	★	★	★★	☆	★★	☆
CA2	★	★★	☆	★	O	★★	★	★★	☆	★★	O
CA3	★	★★	☆	★	O	★	★	★★	☆	★★	O
hilus cells	★	★	O	O	O	★	☆	★★	☆	★★	O
dentate gyrus	★	★★	☆	★	O	★	★	★★	☆	★★	★
subiculum	★	☆	☆	★	O	O	☆	★	O	★	★
Thalamus	★★	O	★	O	O	O	★★★	O	☆	★	★★
Substantia nigra											
zona compacta	O	O	★	O	O	O	O	O	O	☆	O
zona reticularis	★★★	O	O	O	O	☆	★★	O	★	★	O
Cerebellum											
Bgl layer	O	☆	O	O	O	☆	O	O	☆	O	O
molecular layer	★	O	O	O	O	O	☆	O	O	☆	O
PC	★★★	O	O	O	O	O	★★	O	O	★★	O
granular layer	★★	O	O	☆	★★★	☆	★★★	★	O	★	★★★
white matter	O	O	O	O	O	O	O	O	O	O	O
Colliculus inferior	★★	O	☆	O	O	☆	★★	☆	O	★	O
Colliculus superior									SuG		
optic nerve layer	★	☆	★	★	O	★	★	★	☆	★	☆
Nucleus interpositus	★★★	O	O	O	O	O	★★	O	O	★	O
Vestibular nucleus	★	★	☆	☆	O	O	★	★	O	★★	☆
Facial nucleus	★★	★★	☆	O	O	☆	★	★	O	☆	O
Motor trigeminal nucl.	O	★	☆	☆	O	O	★	★	O	★★	O
Pontine nuclei	☆	★	O	☆	O	O	☆	☆	★	★★	O
Spinal cord											
spinal cord layers 2-3	O	O	★	☆	O	O	O	★	O	★	O
spinal cord layers 4-5	★	O	★	☆	O	O	★	★	O	★	O
spinal cord layers 6-9	★	★★	☆	☆	O	O	★	★	☆	★★	O

Fig. 24.10 Example of the diversity of receptor types for a given transmitter in the brain, as demonstrated by in situ hybridization. The table summarizes the mRNAs encoding 11 different subunits of $GABA_A$ receptors in the rat central nervous system by region, layer, and cell type. The intensity of the hybridization signals is indicated as follows: negative, open circles; weakly positive, open stars; positive, one filled star; strongly positive, two filled stars; very strongly positive, three filled stars. (From Persohn et al., 1992)

529

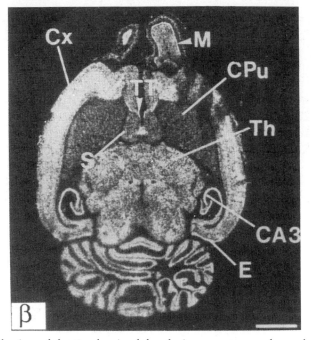

Fig. 24.11 Distribution of the β subunit of the glycine receptor, as shown by in situ hybridization autoradiography. CA3, hippocampus; CPu, caudate putamen (striatum); Cx, neocortex; E, entorhinal cortex; M, mitral cell layer; Th, thalamus. (From Betz, 1991)

(see Chap. 7). Recent years have seen a revision of the classical view with the evidence of widespread distribution of glycine receptors in the brain. An autoradiograph of in situ hybridization to messenger RNA transcripts for the B subunit of the glycine receptor is shown in Fig. 24.11. Note the high density of receptors in some of the same regions where the NMDA receptor is found (for example, olfactory bulb, hippocampus, and cerebellar cortex) as well as cerebral cortex, thalamus, and superior and inferior colliculi. In addition to this wider distribution of glycinergic receptors, the pharmacological properties of glycinergic receptors are also under revision (see Chap. 7). All in all, there seems to be much more to be learned about the functional significance of glycine as an inhibitory transmitter in the brain.

Acetylcholine

In earlier years, the localization of ACh as a neurotransmitter in the brain was inferred from a positive staining for the enzyme acetylcholinesterase, which hydrolyzes and inactivates ACh at the synapse. However, this gives many false positive results, because the enzyme is widely distributed in the brain and the body, as if on guard to chew up any stray molecules of this transmitter. The better method is to identify the presence of the enzyme choline acetyltransferase, which synthesizes ACh. To this has recently been added other methods, such as the ability to identify ligand binding sites by different types of ACh receptors (recall Chap. 8).

The locations of ACh neurons are shown in Fig. 24.12. Among central systems, ACh is specific for two cell populations ·in the limbic system, the septal-hippocampal and habenulo-endopeduncular projection neurons, and a cell population in the caudate-putamen nucleus which is involved in motor coordination (see below). ACh acts on either nicotinic receptors (producing brief synaptic potentials) or muscarinic receptors

ACETYLCHOLINE

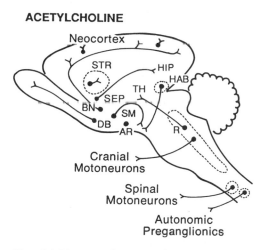

Fig. 24.12 Distribution of cholinergic cell groups and their projections in the rat brain. AR, arcuate nucleus; BN, basal nucleus; DB, diagonal band; HAB, habenula; HIP, hippocampus; R, reticular nucleus; SEP, septum; SM, stria medullaris; STR, striatum; TH, thalamus.

(producing slow synaptic potentials), as explained in Chaps. 7, 8, and 18. It produces EPSPs by both these actions on Renshaw interneurons in the spinal cord (Chap. 19). In the hippocampus, it produces both slow excitatory and inhibitory responses on pyramidal cells through muscarinic receptors (see Chaps. 7 and 29). In the caudate, it is believed to be the transmitter of short-axon cells and of axon collaterals that are involved in local circuits (Chap. 22). Degeneration of these cells occurs in certain neurological diseases in which there are uncontrollable jerky movements (Huntington's chorea) and dementia.

In recent years, much of the interest in ACh in the human brain has been motivated by the apparent link between ACh and the type of senile dementia known as Alzheimer's disease. This disease is characterized by the presence of neurofibrillary tangles within the cell bodies of cortical neurons and extracellular amyloid accumulations forming plaques in the cortical grey matter. In patients dying with Alzheimer's disease, there is also a reduction in the

ACh-synthesizing enzyme, acetylcholine acetylase, and the hydrolyzing enzyme, acetylcholinesterase; this reduction has been traced to degeneration of cholinergic cells in the nucleus basalis (see Fig. 24.12). It has been attractive to postulate that Alzheimer's disease might belong to the same category as Parkinson's disease in being due to the degeneration of a single type of transmitter-specific neuron. However, investigation of the pathology of this disease has necessarily been limited to study of brains of terminally ill patients, and there is much more work to be done on the factors that are responsible for the initial changes—genetic, viral, or otherwise—that lead to the derangements of brain functions.

Dopamine

Dopamine (DA) is the first neuroactive substance in the synthetic pathway for catecholamines (refer back to Fig. 8.9A). The locations of DA neurons are shown in Fig. 24.13. There are two main populations of projection neurons, located within the midbrain (mesencephalon). One consists of

Fig. 24.13 Distribution of dopamine-containing neurons in the rat brain. AM, amygdala; ARC, arcuate nucleus; NA, nucleus accumbens; OB, olfactory bulb; PC, pyriform cortex; SC, superior colliculus; SEP, septum; SN, substantia nigra; STR, striatum; VTA, ventral tegmental area.

DOPAMINE

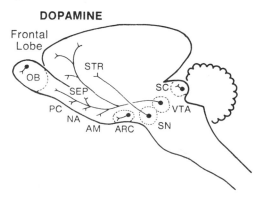

the output cells of the substantia nigra, which project to the caudate-putamen (striatum). In Parkinson's disease these cells degenerate, and the resulting loss of DA synapses in the striatum is believed to be a primary cause of the movement disorders, such as limitation of movement and the resting tremor of the hands, that are characteristic of this disease. The therapy of high doses of the compound L-dopa, the precursor of DA, is aimed at correcting this deficiency, and is successful in some, but unfortunately not all, patients. Neurosurgical operations to correct this deficiency by transplantation of dopamine-producing cells to the striatum have been discussed in Chap. 9.

The other main population of projection neurons is located in the ventral tegmental area (VTA) of the mesencephalon. These cells project to a number of sites in the forebrain, including the amygdala, olfactory tubercle, septal area, nucleus accumbens, and prefrontal cortex. Since these regions are part of what is called the limbic system, this is called the mesolimbic projection. The limbic structures served by this pathway are implicated in such functions as emotion and aggression (Chap. 28); the prefrontal cortex is crucial for some of our highest cognitive functions (Chap. 30). It is likely, therefore, that the mesolimbic pathway plays a role in coordinating these functions. There is also evidence that schizophrenia may be associated with derangements of DA metabolism and DA synaptic transmission, possibly within the mesolimbic pathway.

In addition to these projection tracts, there are also short-axon DA cells in several regions. In the hypothalamus, DA cells send axons to the median eminence, to modulate the output of releasing factors by neuroendocrine cells there (see above). Intrinsic DA cells in the retina, olfactory bulb, and optic tectum take part in local circuits in those structures. Finally, there is a system of DA cells around the fourth ventricle, extending within the core of the brainstem to the hypothalamus.

Central State Circuits

Among the central neurons that have been shown to contain specific transmitters, two in particular—norepinephrine (NE) and serotonin (5HT)—appear to be associated with neurons whose axons project widely throughout the central nervous system. This branching pattern appears to preclude functions involved in transmitting specific information about space or time, and favors the view that these neurons are more involved in slower and more global adjustments of the excitability state of neurons throughout the brain. In this respect, they appear to provide a central parallel to the actions of the peripheral autonomic nervous system.

Norepinephrine

Norepinephrine (NE), or noradrenaline (NA), is synthesized from DA by the enzyme dopamine β-hydroxylase. NE is found especially in clusters of cells in the midbrain. One of these groups is called the locus ceruleus, and it is certainly one of the most extraordinary cell populations in the entire nervous system. There are only a few hundred neurons in the locus ceruleus, yet they send axons to almost every region in the central nervous system (see A in Fig. 24.14). In order to do this, each axon branches repeatedly. There are relatively few terminals within any region, but the NE branches and terminals achieve their effects by secreting NE diffusely onto synaptic terminals in the surrounding neuropil. As noted in Chap. 8, these actions are believed to be neuromodulatory in nature. Because of their widespread ramifications and diffuse action, the NE cells are well suited for setting levels of central neural activity underlying different behavioral states. As noted above, they may in this respect act somewhat like a central autonomic nervous system, to complement the peripheral autonomic system, and function in parallel with the adrenal medulla and its release of epinephrine into the bloodstream.

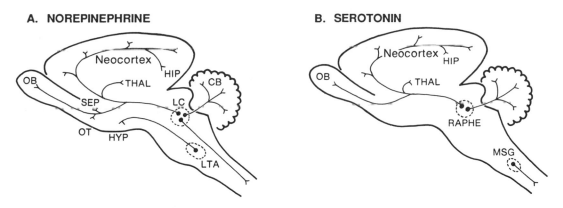

A. NOREPINEPHRINE

B. SEROTONIN

Fig. 24.14   A. Distribution of norepinephrine-containing neurons and their axonal projections. B. Distribution of serotonin-containing neurons and their projections. CB, cerebellum; HIP, hippocampus; HYP, hypothalamus; LC, locus ceruleus; LTA, lateral tegmental area; MSG, medullary serotonin group; OB, olfactory bulb; OT, olfactory tubercle; SEP, septum; THAL, thalamus.

Complementing the locus ceruleus is a nearby cluster of NE cells in the lateral tegmental area (LTA). These projections overlap those of the locus ceruleus but are directed mainly to the hypothalamus, where they are believed to participate in the regulation of releasing factors.

Epinephrine

We note that epinephrine (adrenaline) is synthesized from norepinephrine in the catecholamine pathway. It is an important hormone of the peripheral autonomic system, where it is liberated from the adrenal medulla and is crucial in preparing the body for action and stress (see Chap. 17). In the nervous system, it is present in a few clusters of cells in the lower brainstem (medulla), which project anteriorly as far as the diencephalon, and posteriorly to the spinal cord. Terminals are found particularly in the dorsal motor nucleus of the vagus (cranial nerve X) and the nuclei of the solitary tract (cranial nerves VII, IX, and X), where epinephrine may have a role in modulating motor control of the viscera and taste information from the tongue, respectively.

Serotonin

Serotonin, or 5-hydroxytryptamine (5HT), is a monoamine because it has a single terminal amine group, but it has a two-ring indole structure that differentiates it from the single-ring catecholamines. It is synthesized from the precursor tryptophan (see Fig. 8.8). Serotonin is present throughout the body, especially in blood platelets and the intestines. In the brain, it is found mainly in the midbrain, in clusters of cells called the raphe, and in the medulla. As shown in Fig. 24.14B, the fibers of these cells project widely to the forebrain, cerebellum, and spinal cord, in a pattern that resembles that of the NE fibers. Thus, like the NE system, the 5HT system appears to exert a widespread influence over arousal, sensory perception, emotion, and higher cognitive functions.

The first evidence for these effects came from experiments with the hallucinogenic agent lysergic acid diethylamide (LSD) in the 1950s. LSD was found to block 5HT receptors in muscle membranes, and it was suggested that the disastrous effects of LSD on mental states could therefore be attributed to blocking of 5HT receptors in the brain. Subsequent studies have shown that LSD depresses 5HT-containing neurons, and it has been suggested that this action may occur at receptors where raphe neurons receive synapses from each other by means of recurrent axon collaterals or dendrodendritic interactions. However, it has been impossible to incorporate this view into a view that accommodates conflicting evidence from lesion experiments and pharmacological experiments. Thus, rather than serving as a model for understanding the functions of a transmitter-defined pathway, the raphe pathway has served as an example of the difficulty of correlating central brain circuits with brain functions. It therefore illustrates the point that was emphasized in the introduction to this section on central systems. Cooper, Bloom, and Roth (1993) stated this principle eloquently:

> The student must realize that to track down all of the individual cellular actions of an extremely potent drug like LSD and fit these effects together in a jigsaw puzzlelike effort to solve the question of how LSD produces hallucinations is extremely difficult. Similar jigsaws lie just below the surface of every simple attempt to attribute the effects of a drug or the execution of a complex behavioral task like eating, sleeping, mating, and learning to a single family of neurotransmitters like 5-HT. While it is clearly possible to formulate hypothetical schemes by which divergent inhibitory systems like the 5-HT raphe cells can become an integral part of such diverse behavioral operations as pain suppression, sleeping, thermal regulation, and corticosteroid receptivity, a very wide chasm of unacquired data separates the concept from the documentary evidence needed to support it.

Gaseous Messenger Systems

Chief among the recently discovered class of gaseous messengers are nitric oxide and carbon monoxide.

As described in Chap. 8, nitric oxide is synthesized from L-arginine by the enzyme nitric oxide synthase (NOS). Identification of the distribution of NOS is therefore a clue to the sites where NO is produced and is likely to act as a diffusible messenger. This has been done by both immunohistochemistry, using antibodies to NOS, and by in situ hybridization, using oligonucleotide probes for NOS messenger RNA. A section of the brain stained with NOS antibodies is shown in Fig. 24.15. The densest staining for NOS is found in the accessory olfactory bulb, the olfactory bulb, and the cerebel-

lum. Other densely stained regions include the dentate fascia, the inferior and superior colliculi of the midbrain, and the supraoptic nucleus.

NO diffuses into processes of neighboring neurons to activate soluble guanylate cyclase, producing the second messenger cGMP which can act to modulate membrane channels or phosphorylate target proteins (see Chap. 8). In the olfactory bulb, NO has been proposed to mediate interactions between dendrites, contributing to the unitary function of the glomeruli in the main olfactory bulb and long-term changes in the dendrodendritic synapses in the accessory olfactory bulb (see Chap. 11). It has been proposed as a messenger between presynaptic and postsynaptic processes in mediating long-term potentiation in the hippocampus, but this possible role remains controversial (see Chaps. 8 and 29), particularly in view of the fact that levels of NOS in the hippocampus are low.

Peptides

As we learned in Chap. 8, there is a large number of peptides in the nervous system. Here we select several of the most prevalent to illustrate their distribution and actions.

Substance P. Substance P has the distinction of being the first neuroactive peptide isolated from the brain. In 1931, Ulf von Euler of Sweden and John Gaddum of England showed that this compound, present in both brain and gut, has a stimulating effect on smooth muscle. Around 1970, Susan Leeman and her colleagues at Harvard isolated a compound from the hypothalamus that stimulates the salivary gland, and named it a sialogog (sialo = saliva, gog = factor). When they synthesized this compound, it turned out to be identical to substance P isolated some 40 years previously (see amino acid sequence in Fig. 8.11).

Substance P is found in several specific short projection tracts, as shown in A of Fig. 24.16A. Its presence in the striatonigral pathway has suggested that it is a

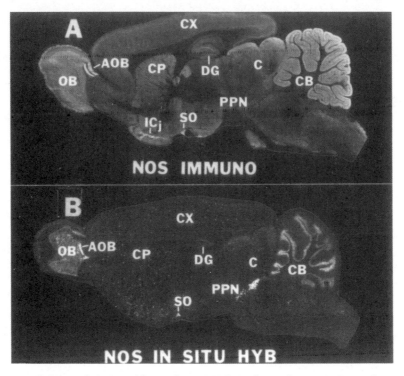

Fig. 24.15 Distribution of nitric oxide synthase (NOS) in the rat brain. **A.** Sagittal section showing localization of antibodies to NOS by immunocytochemical staining. **B.** Adjacent section showing localization of probes for NOS messenger RNA by in situ hybridization. AOB, accessory olfactory bulb; C, superior and inferior colliculi; CB, cerebellum; CP, caudate putamen; CX, neocortex; DG, dentate gyrus; ICj, islets of Calleja (olfactory tubercle); OB, olfactory bulb; PPN, pedunculopontine tegmental nucleus; SO, supraoptic nucleus. (From Bredt et al., 1991, modified in Jessell and Kandel, 1993)

Fig. 24.16 **A.** Distribution of substance P–containing neurons in the rat brain. **B.** Distribution of somatostatin-containing neurons in the rat brain. AM, amygdala; DRG, dorsal root ganglion; HAB, habenula; HYP, hypothalamus; MED, medulla; SN, substantia nigra; STR, striatum.

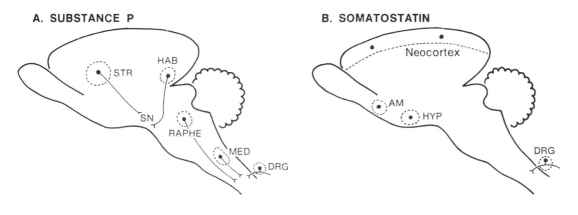

transmitter in these fibers, and it is therefore important in the motor functions of the basal ganglia (see Chap. 21). In the spinal cord, it is present in dorsal root ganglion cells. Although it is 200 times more potent than glutamate in depolarizing motoneurons when applied ionophoretically, its slow time course of action contrasts with the rapid discrete transmission of input signals in many dorsal root fibers. This is consistent with the idea that substance P has a slower modulatory action, but so far it has not been proven.

Somatostatin. This compound derives its name from its action in inhibiting the secretion of growth hormone (somatotropin) from pituitary cells. It is a tetradecapeptide (14 amino acids) (see Fig. 8.11). Like so many neuroactive peptides, it is found in autonomic fibers and other nonneural cells in the visceral organs (see Chap. 17). Within the nervous system, it is found in dorsal root ganglion cells, and, centrally, in cells of the hypothalamus, amygdala, and cerebral cortex (see B in Fig. 24.16). When injected into the cerebral ventricles, it has a depressant effect on motor activity. Like several other peptides, it has a slow inhibitory action when ionophoresed onto single neurons.

Endorphins. Among the peptides generating the most interest in recent years are the endorphins and enkephalins. This work began in 1975 with the finding of Hans Kosterlitz and Robert Hughes in Scotland that extracts of brain contain a compound which competes with opiates in assay systems and is blocked by opiate atagonists such as naloxone. The short-chain pentapeptides (5-amino acids) are called enkephalins, whereas the longer chain compounds (16- to 31-amino acids) are called endorphins. Since the enkephalin chain is contained within the endorphin chain, the term *endorphin* may be used to refer to both in general. The synthesis of these compounds from the larger molecule β-lipotropin was

discussed in Chap. 8, as were the receptor mechanisms.

Cells containing endorphins are located almost exclusively in the hypothalamus. As shown in Fig. 24.17A, their fibers project to different nuclei within the hypothalamus, and to several regions outside the hypothalamus, including the septal area and amygdala, and reach as far as the higher brainstem, where they innervate monoamine cells of the locus ceruleus and raphe nuclei. Most of the termination sites are in the core of the brain, around the ventricles.

From the moment of their discovery, it has seemed that the endorphins might act like opiate drugs, and thus function as an internal mechanism for opposing or modulating pain sensations. Part of the interest in analyzing this mechanism lies in the hope of developing a more natural substitute for morphine that is free of addictive properties and side effects. In addition, there has been much speculation on the natural functions in which the endorphins may be involved. In their enthusiasm, people have implicated them in almost every behavior imaginable, from temperature regulation to our sense of self. It is becoming clear, however, that much more work is required to establish the precise contributions of endorphins to specific behaviors. The caution expressed by Cooper, Bloom, and Roth above applies precisely to this task.

Enkephalins. Enkephalin-containing neurons are widespread in the nervous system, with a distribution that is quite distinct from that of the endorphins. The main regions in which they are found are shown in Fig. 24.17B. In the peripheral nervous system, enkephalin is found in the adrenal medulla and in fibers that innervate the smooth muscle of the gut. In the central nervous system, enkephalins are characteristically found in intrinsic neurons; in this respect they resemble GABAergic interneurons. In these cells, the enkephalin is in a position to modulate the processing of information by local circuits. A prime

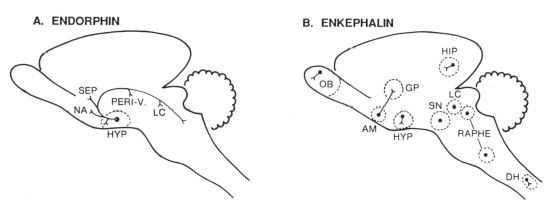

Fig. 24.17 **A.** Distribution of endorphin-containing neurons in the rat brain. **B.** Distribution of enkaphalin-containing neurons in the rat brain. AM, amygdala; DH, dorsal horn; GP, globus pallidus; HIP, hippocampus; HYP, hypothalamus; LC, locus ceruleus; NA, nucleus accumbens; OB, olfactory bulb; PERI-V., periventricular gray; SEP, septum; SN, substantia nigra.

example is the dorsal horn of the spinal cord, where it is believed that enkephalin neurons modulate the processing of input arriving over pain fibers (see Chap. 12). In addition, enkephalins are implicated in most of the functions ascribed to endorphins, as discussed above.

Closely related to the localization of endorphins and enkephalins has been the mapping of opiate-receptor binding sites. As discussed in Chap. 9, there are several types of opiate receptors, defined by binding of certain agonists and antagonists. The distribution of these receptors may be briefly summarized as follows (see Snyder, 1984; Cooper et al., 1993). The mu (μ) receptor preferentially binds morphine and its antagonist, naloxone; it is localized primarily in pain pathways in the brain, and it is believed mainly to be responsible for the painkilling and addicting properties of morphine. The delta (δ) receptor preferentially binds enkephalin; its localization in limbic parts of the brain may be related to influences on the emotions (see Chap. 28). The kappa (κ) receptor has a high affinity for dynorphin, which is believed to mediate more sedative actions at the cortical level. The sigma (σ) receptor binds a hallucinogenic drug, phencyclidine, and is especially localized in the hippocampus. The epsilon

(ϵ) receptor binds β-endorphin; it is present in neurons at the base of the brain that receive β-endorphin-containing terminals from cells in the hypothalamus and mediate a variety of behavioral effects when β-endorphin is injected into an animal, including inability to move (akinesia) and prolonged avoidance reactions.

Transmitters Across Phyla

Our attention has been focused on vertebrates and it remains to comment on the distribution of neuroactive substances across different phyla.

With regard to neurotransmitters, it should already be apparent that there is a strong tendency toward conservation across the phyla. Thus, we have seen that glutamate is an excitatory transmitter and GABA is an inhibitory transmitter (Chap. 7); indeed, the earliest evidence for the identification of these substances as transmitters was obtained in invertebrates. Similarly, ACh is widespread as a neurotransmitter in many invertebrate species; in fact, many insecticides have as their primary mode of action a blocking of cholinergic synapses. Finally, we have noted the role of 5HT in modulating behavioral states of some invertebrates (see Chap. 20). One can conclude that there is a limited set of small molecules that have similar functions across many phyla.

The situation with regard to neuropeptides is more complicated. As Greenberg and Price (1983) pointed out, invertebrate neuropeptides fall into two categories: those that are shared with vertebrates, and those that are unique to invertebrates. From a review of different phyla, they conclude that "each vertebrate peptide family is represented by one or more active structural homologues in all invertebrate animals." Unfortunately, little is known as yet about the physiological roles of these peptides in any phylum. By contrast, similar functions such as gonad stimulation, pigment dispersal, or glucose mobilization, on the evidence thus far, appear to be mediated by peptides that are different in invertebrates as compared with vertebrates. The conclusion is that peptides have diverse families with different modulatory functions within the different phyla, and that there is still much work to be done in identifying these families and characterizing their functions.

25

Biorhythms

Although motor activity is one of the cardinal features of animal life, animals are not ceaselessly in motion. Characteristically, periods of activity alternate with periods of inactivity. The inactivity can take many forms, such as simply sitting, lying, or standing still; sleeping; hibernating; or passing through stages of larval development. Underlying these periods are fluctuations in the secretions of glands, and fluctuations in many cellular functions, such as the synthesis of RNA, protein, and other molecules. Cyclical activities are thus basic characteristics of animal life, and we refer to them collectively by the term *biorhythms*.

Any cyclical activity can be defined in terms of its *period*, that is, the time it takes to complete one full cycle of activity. A cycle that lasts a day, such as the sequence of waking and sleeping, is said to have a period that is *circadian* (circa = approximately; dies = day). Longer periods (less frequent) are called *infradian* (infra = less), whereas shorter periods (more frequent) are called *ultradian* (ultra = more). Some representatives of these types are listed in Table 25.1. These biorhythms almost always involve in some way, directly or indirectly, the nervous system, and many are

directly under nervous control. We have, of course, already encountered fast ultradian rhythms in the discussion of impulse discharge, heart rate, and patterns of locomotion.

This chapter will focus on circadian rhythms, and the cellular mechanisms that lead to their generation. We will see how different neural systems interact in generating these rhythms during the course of the day and night. These rhythms provide for highly predictable neural set points that underlie control of the centrally generated behaviors that will be discussed in subsequent chapters.

A Brief History

It might be thought that the obvious relations of many plant and animal activities to day and night would have invited close study even in ancient times. However, it appears that the facts were simply too familiar. It was not until 1729 that the French geologist de Mairan did the simple experiment of placing a potted plant under constant temperature and illumination, and observed that its normal daily period of fluctuations still persisted. This showed that periodic behavior could be a function

Table 25.1　Main categories of rhythmic activity, with some examples

Infradian (longer than one day)	Circadian (approx. one day)	Ultradian (shorter than one day)
menstrual cycle	waking–sleeping	feeding
seasonal variations	body temperature	respiration
lifetime	body electrolytes	heart rate
	various hormones	nerve impulse discharge

of the organism itself. Although this finding aroused some interest, people still did not quite know what to make of it; there were suspicions that the experiment might be affected by some undetectable rays or forces.

It was not until the 1930s that biologists began to make the connection between photoperiodicity (the changing illumination during the day) and bodily rhythms. A breakthrough came in 1950, in the work of two German biologists, Gustav Kramer and Karl von Frisch. Kramer showed that birds could use the sun as a compass by virtue of the fact that they have an "internal clock" which, in effect, tells them the time of day and how much to correct for the position (azimuth) of the sun in the sky. Von Frisch came to a similar conclusion for bees. A number of biologists then initiated the search for the cellular basis of the "internal clock," a search that has continued to the present. The importance of circadian rhythms has grown in a parallel with the increasing understanding of their mechanisms, and the increasing realization of how pervasive they are in the life of the organism. As Colin Pittendrigh (1974), one of the early pioneers, put it:

. . . a circadian oscillation assumes a unique phase relationship to the 24-hour light/dark cycle that entrains it. . . . The functional significance of this is many-sided: It permits *initiation* of events in *anticipation* of the time at which their culmination most appropriately occurs, and it permits timing that cannot be entrusted to control by conditions for which the system has no adequate modality.

From the extensive work that has been carried out on circadian oscillations, we will focus on three examples: studies of simple cells and small systems in invertebrates and birds, and studies of multiple systems controlling waking and sleeping in mammals, especially humans.

Circadian Rhythms in Invertebrates

In our earlier discussions of rhythmic behavior (for example, heart rate or locomotion), we saw that there are two basic types of mechanism: there can be a pacemaker cell, which imparts its rhythmic output to other cells, or there can be a rhythmic cell group or network, in which no one cell has an intrinsic rhythmic property, but in which there is a rhythmic output by virtue of the interconnections. The same two alternatives apply in the analysis of mechanisms of circadian oscillation.

A Circadian Pacemaker

The sea hare *Aplysia* has been a useful system for studying circadian mechanisms; it was introduced for this purpose by Felix Strumwasser of the California Institute of Technology in 1965. Like most animals, *Aplysia* shows a circadian rhythm of locomotor activity; in general, it is active during the day and inactive at night. Since we do not know whether an invertebrate animal like *Aplysia* ever "sleeps," in the same sense that we apply that term to mammals, it is best to refer to the inactive periods as periods of *rest*. We then say that there is a *basic rest–activity cycle* that characterizes the overall behavior of an animal through a 24-hour day. Some animals, like *Aplysia*, are active during the day, but of course

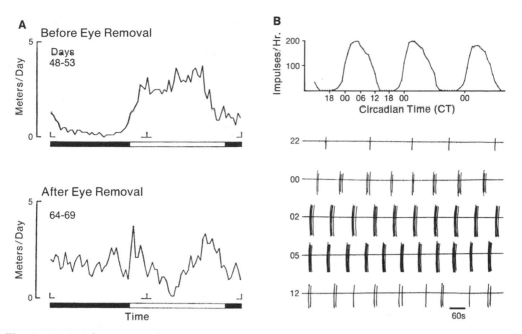

Fig. 25.1 **A.** The average locomotor activity (monitored with a video camera) of a sea hare, *Aplysia,* before *(top)* and after *(bottom)* removal of both eyes. Each trace represents the average of six days' observations. The observations for the bottom trace began three days after eye removal. (From Strumwasser, 1974) **B.** Circadian rhythm in the frequency of extracellularly recorded action potentials of the optic nerve innervating an isolated eye. *(Top)* Graph of discharge frequency (in impulses per hour), plotted against circadian time (CT). *(Bottom)* Representative recordings, at times indicated. (From Jacklet, 1981)

many animals, particularly warm-blooded predators, are active at night.

An *Aplysia* that has been entrained to a normal dark–light cycle and is then exposed to constant illumination or constant dark still shows its basic rest–activity cycle for several days. This indicates that the circadian rhythm persists in the absence of inputs from the environment, or from other ganglia. The rhythm of the ganglion maintained under constant conditions is referred to as the *free-running rhythm*. The free-running period is not quite precisely 24 hours (hence the general term circadian); we say therefore that in the intact animal the rhythm is *entrained* by the 24-hour light–dark cycle.

This result indicates that the rest–activity cycle has a circadian generator somewhere within the nervous system. The generator is not the large bursting neuron, R15, in

the abdominal ganglion; removal of the ganglia has no effect on the free-running cycle. However, removal of both eyes does abolish the locomotor cycle, in both normal and maintained conditions (Fig. 25.1A). Correlated with this is the finding that the eye of an animal maintained in constant darkness shows a free-running rhythm in the amount of spontaneous impulse activity that can be recorded from the optic nerves. A similar rhythm can be recorded from the isolated eye (see Fig. 25.1B). These experiments have therefore indicated that the neuronal mechanisms that drive the circadian rhythm of locomotor activity in *Aplysia* are found within the eye itself.

The pacemaker neurons have been localized to a group of neurons at the base of the eye. These neurons are entrained through synaptic inputs from optic nerve collaterals. They are also modulated by centrifugal fibers from the brain; these fibers are serotonergic, and their effects on the pacemaker neurons are mediated through cAMP. The centrifugal fibers are part of a widespread serotonergic system for behavioral arousal of the animal, so they provide a means whereby the state of arousal can affect the circadian rhythms.

Centrifugal Control of a Circadian Rhythm

By contrast with the pacemaker system of *Aplysia*, which is centrifugally modulated, the lateral eye of *Limulus* shows a circadian rhythm that is centrifugally driven. The mechanisms of this rhythm have been extensively analyzed by Robert Barlow and his colleagues at Syracuse. As background for this discussion the student should review the structure and function of the *Limulus* eye in Chaps. 10 and 16.

A flash of light delivered to the *Limulus* eye is transduced in the receptor cells (see Chap. 10), leading to a discharge of impulses in the optic nerve. The intensity of the discharge is not constant; it shows a circadian rhythm, being greater to the same flash during the night than during the day. The increase in impulse response is plotted in Fig. 25.2A. One might anticipate from this that there could be a similar variation in the spontaneous rate of impulse firing in the optic nerve. The rate does vary, but surprisingly in the opposite direction, showing a decrease during the night.

The mechanism underlying this disparity was analyzed by use of intracellular recordings from the photoreceptors (retinula cells). As shown in Fig. 25.2B, during the day (6 P.M.), the membrane potential is noisy, as a result of frequent small deflections representing quantal responses to single photons, and there is a relatively small response to a light flash. At night (10 P.M.), the membrane becomes quiet (which accounts for the decrease in spontaneous impulse activity in A), and there is an increase in the depolarization by a light flash (which produces the increased responses in A).

Thus, during the night there is an increased sensitivity and an increased signal-to-noise ratio, which would be advantageous to this nocturnally active animal. Because the experiments are carried out at all times in constant darkness, the differences must be due to a circadian rhythm generated either within the eye or within the brain. Tying off the optic nerve eliminates the rhythm, so, in contrast to *Aplysia,* the rhythm must be generated in the brain.

The centrifugal input brings structural as well as functional changes in the *Limulus* eye. As summarized in Fig. 25.3, these included (1) movement of pigment cells to widen and shorten the lens aperture, accompanied by (2) outward movement of photoreceptors and rearrangement of the rhabdome. These changes increase photon capture by each ommatidium. In addition to increasing sensitivity and signal-to-noise ratio, efferent activity also reduces lateral inhibition by modulating the dendrodendritic synaptic interactions that take place between the neurites of neighboring eccentric cells (refer to diagram in Fig. 10.6). This contributes to increased sensitivity at the expense of contrast enhancement.

These experiments have shown that the circadian efferent fibers exert their effects on retinular cells at the earliest stages of visual reception. The synaptic mechanisms that mediate these effects have been studied by use of microinjections of different neuroactive substances beneath the cornea of the eye. These experiments have suggested that octopamine, a biogenic amine that is common in invertebrates (see Chaps. 17 and 19), may be the transmitter for these centrifugal synapses; an injection of as little as 10 μmol octopamine can convert the daytime retina toward its nighttime structure and function. There is evidence that octopamine acts through cAMP, as summarized in the diagram and legend of Fig. 25.4.

Genetics of Circadian Clocks

Circadian pacemaker rhythms, such as those in the eye of *Aplysia,* can be reset or

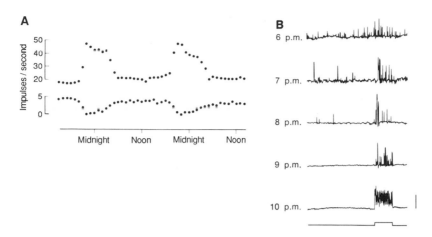

Fig. 25.2 Circadian rhythms in activity in the *Limulus* eye. **A.** Graphs of the impulse firing rate of optic nerve fibers. In the upper curve, each point gives the mean firing rate during a 6-sec light flash delivered to the eye every hour, while the animal was maintained in the dark. In the lower curve, each point gives the mean rate of spontaneous firing during a 25-sec interval in the dark each hour. **B.** Intracellular recordings from a *Limulus* retinula cell at different times indicated at left. In each trial, a 20-sec recording period preceded the 5-sec test flash. Calibration bar, 10 mV. (A from Barlow et al., 1984; B from Kaplan and Barlow, 1980)

Fig. 25.3 Structural changes that occur in the *Limulus* ommatidium as a result of the circadian clock. On the left, the structure during the day, when there is no activity impinging on the ommatidium via efferent fibers. *On the right,* a summary of the structure during the night, when activity in the efferent nerves from the circadian clock produces the changes indicated. See text. (From Barlow et al., 1984)

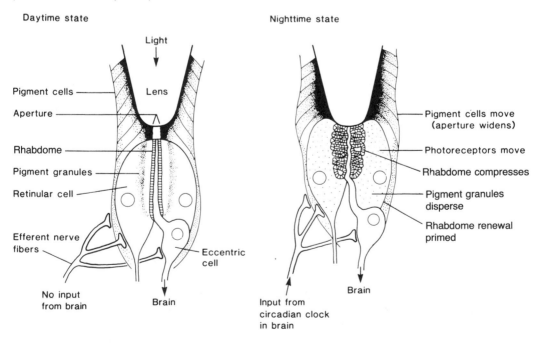

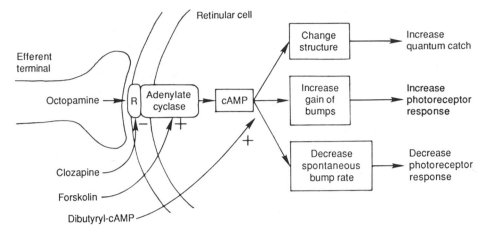

Fig. 25.4 Summary of neurochemical pathways involved in the effects produced by the circadian clock within the retinular cells of the *Limulus* ommatidium. Compare these changes with the electrophysiological recordings in Fig. 25.2 and the structural changes in Fig. 25.3. (From Barlow et al., 1984)

abolished by exposure of the cells to protein synthesis inhibitors. This has suggested that the daily synthesis of protein is a general requirement for circadian clocks (Jacklet, 1981).

Search for the genetic basis of this requirement has led to the study of certain *Drosophila* mutants with circadian defects (Konopka and Benzer, 1971). One mutant shows a shorter period of 19 hours *(per^s)*, another a longer of 29 hours *(per^l)*, and in a third circadian periodicity is abolished *(per^o)*. All of these mutations map to a single *per* locus on the X chromosome. In addition to these circadian abnormalities, the mutants show defects in other behaviors: in *per^s*, the courtship song is briefer, whereas in *per^l* it is prolonged; the *per^l* mutants are difficult to condition behaviorally.

These results indicate that there are complex interdependencies between circadian clocks, rapid (infradian) endogenous rhythms, and learning abilities. That systems with overlapping functions should share products of overlapping gene families is entirely in accord with concepts regarding the molecular diversity of neurons discussed in Chap. 2.

Is the *per* gene unique to *Drosophila?* Michael Young and his colleagues (1985) at Rockefeller University identified certain limited *per* sequences in DNA from birds, mice, and humans. One possibility was that the sequences would code for an ion channel protein, but surprisingly, the amino acid sequences in the gene products are similar to those in proteoglycans. These are molecules with a protein moiety in the plasma membrane and a large carbohydrate moiety within the extracellular space (Chap. 3). This type of molecule is important in cell–cell adhesion and recognition. Experiments are presently aimed at characterizing the way in which these extracellular molecular interactions are related to the generation of circadian rhythms of electrical activity.

Circadian Rhythms in Vertebrates

As in invertebrates, the basic rest–activity cycles are closely linked to the day–night cycle in most vertebrates. Vertebrate animals also vary in whether they are active principally during the day or night, or at specific times such as dawn or dusk. This obviously reflects the strategy of a particular species in finding food or sexual mates with the greatest success, while minimizing the risks of being preyed upon. From this perspective it can be seen that the circadian clock or clocks within an organism are essential components of the apparatus for survival, and for carving out the ecological niche of the species. The niche may in fact be very narrow in a temporal sense, requiring very strict timing of specific activities during the day–night cycle. This is seen in the fact that many species have only a few hours a day for foraging for food, and it is also seen in the exquisite coordination in timing of male and female activities at specific hours of the day or night to bring about mating in many species.

The changing illumination that occurs during the day–night cycle is as important in entraining circadian rhythms in vertebrates as it is in invertebrates. In Chap. 16, the visual pathways involved in visual perception were discussed; here we will become acquainted with the pathways that mediate this other very important visual function. As far as is known, this function is an unconscious one; this reflects the fact that sensory information may have essential roles in nervous and body functions that are separate and distinct from their roles in perception, discrimination, and consciousness. Visually entrained circadian rhythms depend simply on levels of illumination rather than on the discrimination of any particular visual pattern.

The Suprachiasmatic Nucleus

The visual pathways that are involved in control of circadian rhythms in the vertebrate are shown in Fig. 25.5A. The key pathway is made up of a small bundle of fibers which emerges from the optic nerve and terminates in a small group of cells at the anterior border of the hypothalamus. In the 1960s Curt Richter of John Hopkins showed that the anterior hypothalamic region was essential for circadian rhythms in rats. Then, in the early 1970s, a small cell group in this region was identified by independent studies of Robert Moore and

V. B. Eichler at Chicago, and F. K. Stephen and Irving Zucker at Oregon. Because of its position just over the optic chiasm, this cell group is called the *suprachiasmatic nucleus* (SCN). Studies since then, using autoradiographic tracing methods, permitted the anatomical identification of this *retinohypothalamic tract*, made up of retinal ganglion cell axons that terminate in the suprachiasmatic nucleus.

Many studies have provided evidence for the functional properties of the SCN in relation to circadian rhythms. For example, ablation of the SCN bilaterally results in disruption of circadian rhythms of many nervous and bodily functions; some of them are summarized in Table 25.2. The term "disruption" is carefully chosen; in general, the free-running rhythms are not totally abolished, and some degree of visual entrainment may persist. This has suggested that the SCN is the principal center in an extensive system including several other cell groups with weaker circadian oscillators. These groups include the lateral hypothalamic nucleus, the retrochiasmatic area, and the ventromedial hypothalamic nucleus. The relations between these regions are reciprocal and complex (Moore, 1982); together, these interconnections form a distributed system as previous defined (Chaps. 1 and 24). This system is modulated by an extraordinary array of neuropeptides. Vasopressin, somatostatin, VIP, and enkephalin have been localized in SCN neurons by immunocytochemical methods. SCN rhythms are sensitive to microinjections of neuropeptide Y (reviewed in Jacklet, 1985).

Some of the clearest evidence for the activity of the SCN has come from application of the 2-deoxyglucose (2DG) method (Chap. 8). William Schwartz and Harold Gainer at the National Institutes of Health injected [^{14}C]2DG into rats during either the day or night, and examined the autoradiographs of sections through the brain. The SCNs showed a marked circadian rhythm, with low levels of 2DG uptake indistinguishable from neighboring regions

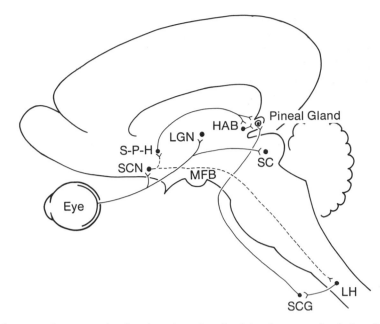

Fig. 25.5 Some pathways and related regions involved in the control of circadian rhythms in vertebrates. HAB, habenula; LGN, lateral geniculate nucleus; LH, lateral horn containing autonomic (sympathetic) motoneurons; MFB; medial forebrain bundle; SC, superior colliculus: SCG, superior cervical ganglion; SCN, suprachiasmatic nucleus; S-P-H, septal, preoptic, and hypothalamic regions.

Table 25.2 Circadian rhythms disrupted by lesions of the suprachiasmatic nucleus in rats or hamsters

---

Locomotor activity (wheel running)
Drinking
Sleep–wake rhythms
Adrenal corticosteroid levels
Estrous periods and ovulation
Temperature
Pineal N-acetyltransferase

Adapted from Menaker et al. (1978)

during the night (Fig. 25.6C: when the rats, being nocturnal animals, were most active), and high levels during the day (Fig. 25.6A: when the rats mostly slept). Since 2DG uptake is a tag for the glucose uptake needed for energy metabolism (Chap. 8), the changes in uptake presumably reflect widely differing levels of activity in the SCN during the day–night cycle.

Further work has shown that the SCN metabolic rhythm appears in the rat on the last day of gestation, before synaptic connections between the SCN neurons have been formed to any significant extent. The implications of these findings have been summarized by Robert Moore of Stony Brook (1982):

> SCN neurons are produced as genetically determined, independent circadian oscillators which become coupled and interconnected during development. Initially they are entrained by maternal influences, but postnatally this function is taken over by development of the retino-hypothalamic projection to the SCN. Within the SCN, what is initially an individual neuronal function becomes subsumed by groups of neurons functioning as interconnected networks or coupled oscillators.

Pineal Gland

In discussing visual control of circadian rhythms, we should also mention the *pineal gland*. It is, embryologically, an outpouching from the dorsal part of the diencephalon. In lower vertebrates (such as sharks, frogs, and lizards), it forms a third eye in the dorsal cranium, and functions to detect

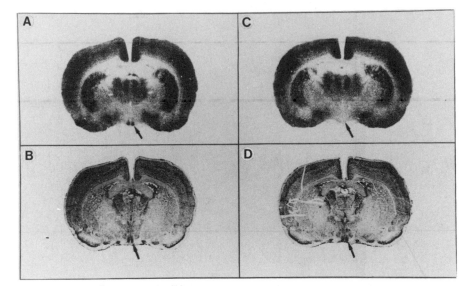

Fig. 25.6 Correlation of activity in the suprachiasmatic nuclei (SCN) with the day–night cycle in rats, with the use of the 2-deoxyglucose (2DG) method. **A.** Autoradiogram of [^{14}C]2DG patterns, showing localization of 2DG in the SCN during the day. **C.** Autoradiogram showing lack of 2DG localization in SCN at night. The circadian rhythm shown in A and C persisted even when the rats were maintained continuously in the dark. **B, D.** Histological sections used for obtaining the autoradiograms in A and C, stained with cresyl violet, to confirm the location of the SCN. (From Schwartz and Gainer, 1977)

changes in levels of illumination. In birds, it may retain some photoreceptive properties. However, in birds and especially in mammals, it is more important as a gland that secretes a hormone called *melatonin*. Melatonin is one of the indole family, and arises from metabolism of serotonin (5-hydroxytryptamine). Synthesis of melatonin has a marked circadian rhythm, as shown in Fig. 25.7. The rate-limiting enzyme is N-acetyltransferase. In mammals (though not in birds), this enzyme is controlled by norepinephrine from fibers of the sympathetic nervous system that innervate the pineal gland (cf. Fig. 25.5). The superior cervical ganglion, the source of the fibers, is believed to be influenced by fibers ultimately arising from the SCN, because lesions of either the SCN or the median forebrain bundle (containing fibers from the SCN) block the rhythm of N-acetyltransferase activity.

The norepinephrine is believed to act on a β-adrenergic receptor, which in turn activates adenyl cyclase to increase the concentration of cAMP, according to the sequence we discussed in Chap. 8. The linkage between the second messenger cAMP and the changes in N-acetyltransferase is not yet known.

Studies of pinealocytes (cultured cells prepared from the pineal gland) have shown that stimulation of N-acetyltransferase by NE (acting through a β-adrenoreceptor and cAMP) can be potentiated by stimulation of α_1-adrenoreceptors, acting through diacylglycerol and protein kinase C to enhance cAMP (and cGMP) production. This is an interesting example of a synergistic relation between the two types of second messenger systems, and indicates the complex biochemical control of this circadian rhythm (Sugden et al., 1985).

Multiple Circadian Oscillators

We thus see that many body functions have circadian rhythms, and that these are en-

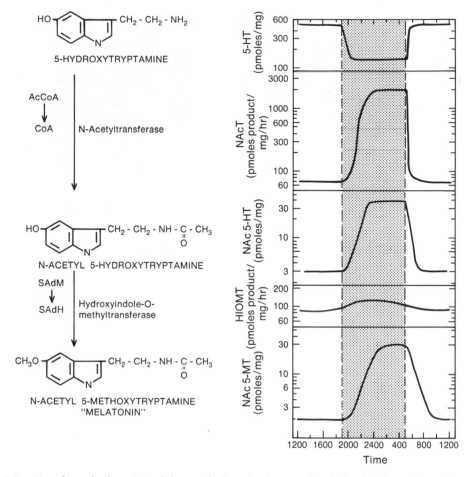

Fig. 25.7 Circadian rhythms in indole metabolism in the rat pineal gland. The pathway from 5-hydroxytryptamine (serotonin) to melatonin is shown at left. The variations in concentrations of metabolites and activities of enzymes are shown at right, in relation to dark (shaded) and light periods of the day. (From Klein, 1974)

trained by the daily light–dark cycle, acting through the visual pathway. The most important center in this pathway is the SCN, but there are other centers that also contribute to maintaining and modulating the rhythms. From this and related work has emerged the concept that circadian systems are composed of *multiple oscillators,* each with properties to some extent specific and distinct from the others. The oscillators may not only have distinct properties but may also possess different intrinsic rhythms.

A classical experiment providing evidence for multiple oscillators in humans was carried out by Jurgen Aschoff in Germany in 1969. The subject lived in an isolated chamber, deprived of any time cues. After several days he fell into a circadian sleep–wake cycle of approximately 33 hours, while his cycle of body temperature held to a period of approximately 25 hours. This suggested that these two functions are controlled by two oscillators with different free-running rhythms, which normally are both entrained to the 24-hour night–day cycle.

Recently this experiment has been repeated, by Zulley and Campbell, with Aschoff's assistance, using different mea-

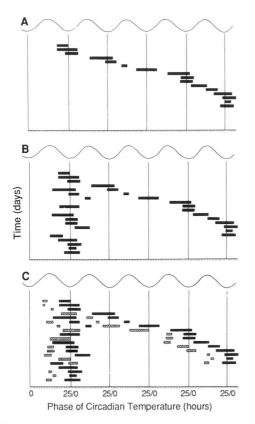

Fig. 25.8 Analysis of two circadian rhythms—body temperature and sleeping–waking—which are free-running in a human subject who lived under constant conditions. The oscillations in body temperature occurred over 25-hour periods, and are shown schematically at the top of each graph. **A.** Black bars show periods of sleep on successive days (moving from top to bottom). Note that the sleep rhythm lengthens to a period of around 30 hours. **B.** The same data are plotted a second time in relation to the first temperature cycle. **C.** Hatched bars show naps taken by subjects. Note that both sleep and naps tend to occur near temperature minima. (Based on Zulley and Campbell, in Mrosovsky, 1986)

tion, some subjects take naps, and when these are included, they tend also to occur at times of body temperature minima (C). These experiments thus indicate that the relation between different circadian oscillators may be stronger in some instances than heretofore recognized, and that new strategies of behavioral experiments will be needed to demonstrate the independent operation of individual circadian controls.

WAKING AND SLEEPING

The circadian rhythms we have discussed so far concern variations in bodily functions and locomotor activity that provide a kind of baseline for the organism during the 24-hour day. During the period of activity we say we are "awake" and "aroused"; during the period of rest we fall into a state we refer to as "sleep." Within the awake period our levels of activity and alertness vary widely, as any student sitting in a warm lecture room after eating a large lunch well knows! Similarly, sleep is a process that contains within it alternating periods of deep and light sleep, the light sleep correlated with dreaming and with a modified state of arousal.

We next consider the neurobiological mechanisms which are responsible for the waking and sleeping states, and, within the sleeping state, for the periods of arousal and dreaming.

The Electroencephalogram

The study of sleep has drawn on many disciplines. An important starting point for modern studies was the discovery that different patterns of brain activity are related to different levels of consciousness.

A Brief History

Richard Caton of England showed as early as 1875 that waves of electrical activity can be recorded from the surface of the brains of animals, but this finding lay unnoticed until the work of Hans Berger, of Germany,

sures of sleepiness and different ways of plotting the data. As shown in Fig. 25.8, this subject had indeed a 25-hour temperature rhythm (wave at top of graphs); the periods of sleep were variable in their time of occurrence, but nonetheless tended (with occasional exceptions) to occur near times of low body temperature (A, B). In addi-

in the 1920s. Berger was a psychiatrist, and also served as Rector of the University of Jena. His main research interest was determining what he called the physical basis of psychic functions. In pursuing this interest, he was led to place electrodes on the scalp of human subjects and attempt to record the electrical activity of the brain. Although the electrical activity of the heart had been recorded from skin electrodes for many years, Berger's report in 1929 that electrical waves could be recorded from the scalp (the first published recording, obtained from Berger's young son, is reproduced in Fig. 25.9) and his interpretation that they represented the activity of the brain were greeted with incredulity and even derision.

The evidence for the nervous origin of "Berger's waves" included the demonstration that the regular rhythms present in a subject resting quietly with eyes closed are replaced by low-amplitude random waves when the subject opens his or her eyes. Within a few years disbelief gave way to acceptance as many leading neurophysiologists, including Edgar Adrian of Cambridge, confirmed and extended the findings. In analogy with the electrocardiogram (ECG) recorded from the heart, "Berger's waves" came to be called the *electroencephalogram* (EEG).

Some Definitions

These early studies established that the dominant rhythm in the resting subject is 8–13 cycles per second, and is most prominent when the recording leads are over the occipital lobe of the brain (where the primary visual cortex is located; see Chap. 16). This is called the *alpha rhythm*. The replacement of these waves during arousal

was termed *alpha blocking*. It was surmised that the alpha rhythm is due to populations of brain cells acting periodically in *synchrony*, and that the low-voltage fast waves during the alpha blocking are due to *desynchronized* activity, as different cells become active in different ways during waking.

Subsequent work has enabled researchers to identify several types of EEG rhythms and correlate them with different levels of awakeness and sleep. Some characteristic patterns are shown in Fig. 25.10. Note that the largest amplitude synchronous waves are present during deepest sleep. By EEG criteria, therefore, we refer to deep sleep as *S sleep* (for synchronous or slow-wave), and light sleep as *D sleep* (for desynchronous). A desynchronized EEG can also signify *arousal* or *waking*. Thus, the EEG patterns are clues to fundamentally different levels of activity of the brain as they relate to waking and sleeping. Since observation of an animal is often not sufficient to characterize its levels of consciousness, the EEG has served as an overall monitor of brain states, especially in animals subjected to various experimental procedures.

Mechanism of the EEG

The cellular basis of the EEG has been the subject of intense study. The problem has two parts: how is the rhythm generated, and how do the potentials arise?

The Thalamic Rhythm Generator. The EEG rhythm is believed to arise mainly in the thalamus. The first comprehensive theory setting forth this concept was reported by S. Andersson of Sweden and Per Andersen of Norway in 1968. They suggested that the cortical cells are driven by cells of the thalamic nuclei, and that the rhythms arise as properties of synaptic circuits within the thalamus. These circuits provided for the gating of impulse output by means of inhibitory feedback through interneurons, in analogy with the generation of rhythmic motor output as a product of circuit properties (cf. Chap. 20).

Subsequent work has extended this con-

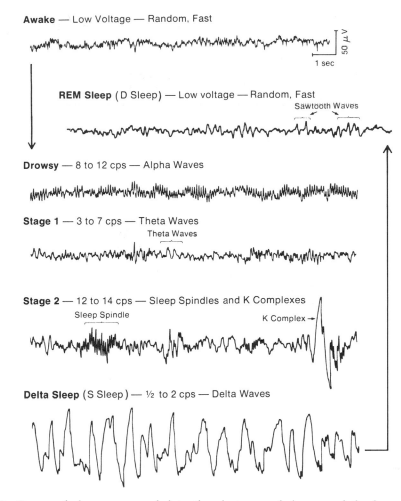

Fig. 25.10 Stages of sleep, as recorded in the electroencephalogram of the human. See text. (Modified from Hauri, 1977)

cept in two ways. First, the thalamic circuits are recognized to include both the specific relay nuclei (such as the lateral geniculate nucleus) and the nonspecific reticular thalamic nucleus. The latter receives three types of input: (1) that from the brainstem reticular system (see below); (2) collaterals of specific thalamocortical relay neurons; and (3) cortical pyramidal neurons. The intrinsic organization of the reticular thalamic nuclei consists mainly, in the primate, of inhibitory dendrodendritic and axodendritic synapses; the output is directed solely at the specific relay nuclei, and is also inhibitory. Thus, the reticular

neurons constitute a special system for inhibitory control of the thalamic relay neurons and, through them, of thalamocortical circuits.

Second, the membrane properties of thalamic neurons have been revealed by electrophysiological analysis in in vitro slice preparations. Henrik Jahnsen and Rodolfo Llinás in New York showed that the thalamic neurons have two modes of impulse firing. In one mode, the resting membrane potential is relatively depolarized, and the cell fires single impulses at a rate of about 8–12 per second. In the other mode, the resting membrane potential is relatively hy-

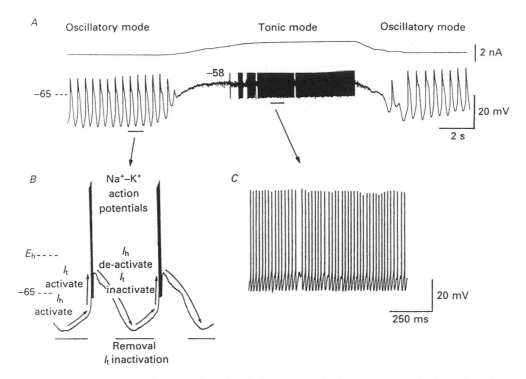

Fig. 25.11 Two different firing modes of a thalamocortical relay neuron in the lateral geniculate nucleus of the cat. **A.** Injection of current *(above)* through the recording electrode was used to shift the membrane potential between −65 mV, at which the cell displayed spontaneous oscillatory potentials, and −58 mV, at which it displayed tonic impulse firing. **B.** Proposed mechanism of generation of oscillatory potentials. **C.** Expanded sweep of tonic firing. E_h, equilibrium potential for I_h; I_h, after-hyperpolarizing current; I_t, transient Ca^{2+} current. See text, and compare with Fig. 16 in Huguenard and McCormick (1994)

perpolarized, and the cell fires in brief bursts of about 6 per second (see Fig. 25.11).

The ionic conductances that underlie these two firing modes, which were discussed in Chap. 6, will be summarized briefly here. During normal waking, the cell membrane is relatively depolarized at rest, around -65 mV. The axon hillock region generates impulses by the usual fast inward Na current followed by the delayed outward K current, as contained in the Hodgkin-Huxley model. Ca^{2+} also enters the cell during the impulse through a high-threshold (HT) Ca^{2+} conductance, leading to activation of a Ca^{2+}-dependent K current. This contributes to the afterhyperpolarization and slows the return of the mem-

brane potential to the threshold for activation of a succeeding action potential. The interplay of these several conductances generates impulses at a resting frequency of approximately 8–12 per second and is a primary determinant of the alpha rhythm (see Fig. 25.10). In addition, when such a thalamocortical relay cell is activated by sensory inputs, its frequency changes in relation to the intensity of the inputs, so that the cell is said to be in the *transmission tonic mode*.

By contrast, when the resting membrane potential is raised by only 5–10 mV, to around 65 mV, a different firing pattern is generated. Under these conditions, the hyperpolarization deinactivates the transient I_A K^+ conductance, which increases the duration of the afterhyperpolarization. The hyperpolarization also deinactivates a low-threshold (LT) Ca^{2+} conductance. As the hyperpolarization wanes, the membrane potential reaches threshold for activation of the LT Ca^{2+} conductance, producing a large slow depolarization, which generates a brief burst of Hodgkin-Huxley impulses, and the process repeats itself. The hyperpolarization is thus critical in deinactivating conductances that interact in a way to generate a pattern of intermittent impulse bursts. This *bursting oscillatory mode* prevents the accurate transmission of input information and is believed to underlie the *spindle complex* seen in deeper levels of sleep (see Fig. 25.10). You can create these two activity patterns yourself using the computer program accompanying Huguenard and McCormick (1994).

How is the cell put into the bursting mode? This is the function of modulatory systems which set the behavioral state of the brain. Muscarinic cholinergic inputs play an important role in moving thalamic relay cells from the transmission mode to the bursting mode, as the brain goes from the level of drowsiness to deeper levels of sleep (see below).

These studies thus provide evidence for the membrane properties and circuit organization that generate rhythmic inputs to the cortex. We next inquire how thalamic and cortical cells interact as a system to generate the rhythmic activity that is recorded as EEG waves.

Generation of Cortical Potentials. The first interpretation of EEG potentials was that the waves represent the envelope of summed impulses of many cells in the region of the cortex beneath a recording electrode. However, beginning in the 1950s, evidence began to accumulate in favor of the importance of synaptic activity, and it is now clear that the waves represent mostly the contributions of summed synaptic potentials in the apical dendrites of the cortical cells.

How can the activity of cells within the brain give rise to electrical potentials detectable on the surface of the scalp? The explanation is somewhat similar to the interpretation of the electrocardiogram. The rhythmic contractions of the heart are brought about by a sequence of impulses in the muscle cells and conducting fibers (Chap. 18). Each impulse discharge is so powerful and so synchronous that it gives rise to electric current that flows not only through the heart itself, but also throughout the tissues of the body. A similar argument applies to the EEG waves, except that they are much smaller in amplitude (50 μV compared with 1 mV) and have faster and more irregular rhythms. These differences reflect the fact that the populations of cells giving rise to the EEG waves are much more diverse than the muscle cells in the heart.

Figure 25.12 summarizes these various lines of evidence regarding the cellular basis of the EEG waves. The thalamic nucleus below generates the rhythmic output of the thalamocortical cells projecting to the cortex. The rhythms arise as a consequence of the intrinsic pacemaker property of cells within the nucleus and the network of excitatory and inhibitory synaptic connections as discussed above.

In the cortex, the thalamic input causes rhythmic synaptic depolarizations (EPSPs) of the apical dendrites of cortical pyramidal cells. This gives rise to current flows within the dendrites (the electrotonic spread of the EPSPs toward the cell bodies, to effect impulse generation). It also gives rise to flows of extracellular current, some of which follow a return path just outside the dendrites. The more synchronous the activity and densely packed the dendrites, the more return current will be pushed out into the surrounding tissue, including the cranium and scalp. As indicated in the diagram of Fig. 25.12, the difference between the amount of current flowing past recording electrode 1 compared with recording electrode 2 is registered as a voltage deflection in the EEG. If pyramidal cell P_1 and cell P_2 are both active in synchrony,

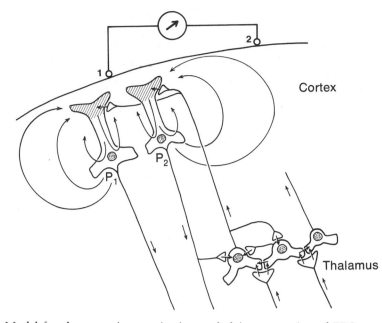

Fig. 25.12 Model for the synaptic organization underlying generation of EEG waves. The model depicts the local circuits of the thalamus, which generate a rhythmic output to the cortex. Within the cortex, the model depicts synchronous EPSPs (shaded) in the distal dendrites of cortical pyramidal cells P_1 and P_2. This generates pathways of electrical current flow as indicated by the arrows. The flow of extracellular current past electrodes 1 and 2 generates a voltage drop which is recorded as a potential difference between the two electrodes, resulting in the EEG.

their currents will summate, and the waves will be large. This is the situation in S sleep, and it also applies to alpha rhythms and other prominent waves. On the other hand, if the thalamic nucleus drives cell P_1 in a way different from that of P_2, the two cells will be asynchronous in their activity and their currents will not summate. This would be the situation during arousal or during D sleep.

This schema should be regarded as a working hypothesis for bringing together the main elements contributing to EEG waves, as an aid for understanding the application of the EEG to the study of waking and sleeping, discussed in the following section.

Early Studies of Sleep and Arousal

Sleep as a Passive State

Like circadian rhythms, sleep invited no close study until recently in modern neuro-science. The first experimental investigations of brain mechanisms were carried out by the Belgian neurologist Frederick Bremer, in the 1930s. He performed transections at many different levels of the brainstem in cats. Such lesions had been made for many years to produce different degrees of paralysis and hyperexcitability of spinal reflexes, but Bremer turned his attention to their effects on the head end of the animal. He found that with low transections, through the medulla and pons, a cat shows normal waking and sleeping cycles, but with high transections, through the midbrain, permanent somnolence ensues. These experiments pointed to the importance of the midbrain–pontine region for mechanisms of arousal.

Arousal and sleeping are behaviors involving the whole animal. As mentioned in the previous chapter, such behaviors require a central system with widespread connections. The first evidence for this

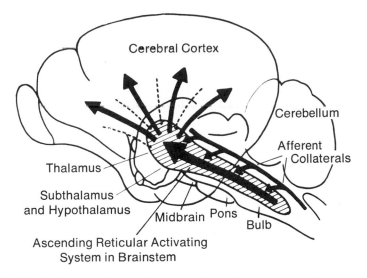

Fig. 25.13 Simplified diagram of ascending reticular activating system in the cat brain. (From Starzl et al., 1951)

came from experiments in which stimulating electrodes were placed in different parts of the thalamus. Stimulation of the specific sensory relay nuclei produced responses restricted to the appropriate sensory areas of the cortex. However, stimulation of other sites produced responses which were widespread throughout all cortical regions, and which increased in amplitude during repetitive stimulation. These sites were identified as the *nonspecific thalamic nuclei,* and the responses were termed *recruiting responses.*

The brainstem system and its linkage to the thalamic system were discovered by Guiseppi Moruzzi of Italy and Horace Magoun of Los Angeles in 1949. They mapped the brainstem using electrical shocks, and showed that high-frequency simulation in the core of the brainstem produces arousal responses in the cortex. The sites of stimulation were correlated in general with the *reticular formation,* extending from the medulla to the diencephalon, as indicated in the well-known diagram of Fig. 25.13. Further experiments showed that lesions of the reticular formation produce a state of deep sleep, and that they also block the arousal that is usually produced by somato-

sensory stimulation. It was known that specific sensory fibers send collaterals to the brainstem reticular–nonspecific thalamus region. These studies thus suggested the attractive hypothesis that arousal is mediated by the *reticular activating system,* stimulated by sensory collaterals, and activated through the nonspecific thalamic nuclei.

Sleep as an Active State

The idea that sleep is a simple state involving only a lack of arousal was soon disproven, however. In a classical study in 1953, Edward Aserinsky, a graduate student, and Nathaniel Kleitman, who had spent almost a lifetime amassing data on sleep, did a very simple experiment: they recorded the eye movements of sleeping subjects. This may seem a strange thing to do, but they explained the rationale at the beginning of their paper:

One is led to suspect that the activity of the extra-ocular musculature and the lids might be peculiarly sensitive indicators of CNS changes associated with the sleep–wakefulness cycle. The disproportionately large cortical areas involved in eye movements, the well-defined secondary vestibular pathways to the extra-ocular nuclei, and the low innervation ratio of the eye muscles point to at least a quantitative basis for their reflection of general CNS activity. A more specific relationship . . . is suggested by anatomical proximity of the oculomotor nuclei to a pathway involved in maintaining the waking state. . . .

The reader may wish to review some of these points of anatomy in Chaps. 14, 16, and 22.

People had previously observed that the eyes rotate upward and outward and that there are eye movements during sleep. However, no one had studied the eye movements carefully and correlated them with the depth of sleep throughout an entire night. It is another of those experiments which could have been done by the ancients, but had to wait until modern times. Aserinsky and Kleitman found indeed that, after falling into deep sleep, subjects went through alternating periods of *light sleep* and *deep sleep;* that light sleep is associated with *rapid eye movements,* and that subjects awakened at these times reported that they had been *dreaming.* These findings showed that sleep contains alternating periods of light and deep sleep, and that light sleep is associated with a modified state of *arousal,* in which the heart rate increases but skeletal muscles seem paralyzed. They showed that dreams occur during light sleep, and they suggested that "the rapid eye movements are directly associated with visual imagery in dreaming."

Stages of Sleep

These findings were soon independently confirmed by William Dement, then at Chicago. In 1957, Dement and Kleitman together then carefully characterized the different stages of the EEG and correlated them with the levels of sleep, as already indicated in Fig. 25.10. They could then follow the EEG patterns through a night of sleep, and relate them accurately to the occurrence of rapid eye movements, body movements, and dreaming. A typical corre-

lation is shown in Fig. 25.14. Note that sleep always begins with an initial period of deep sleep, the deepest sleep of the night; there then ensues a sequence of light and deep sleep, in which deep sleep becomes less deep and light sleep becomes more prolonged; rapid eye movements are invariably associated with light sleep and a stage I EEG.

The terminology for light and deep sleep has evolved to reflect the associated changes in the EEG, eye movements, and other behavior. Thus, deep sleep is referred to as *S sleep* (for slow-wave EEG activity). Light sleep is referred to as *D sleep* (for desynchronized EEG activity; also for dreaming). Light sleep also has other names: *REM sleep* (for its associated rapid eye movements); *emergent stage I sleep* (because it emerges in the wake of deep sleep); and *paradoxical sleep* (because the EEG activity resembles that in the awake state, but the individual is hard to arouse).

Neuronal Systems Controlling Sleep and Waking

These studies in humans showed that sleep has a complicated dynamic structure which includes periods of arousal that share some properties with arousal during waking. It became clear that sleep involves more than simply turning off arousal, and likely involves the interaction of several neuronal subsystems. A great number of animal experiments (mostly in cats) ensued in the 1960s and 1970s in the attempt to identify these subsystems. Michael Jouvet in France, in particular, carried out an extensive series of studies on the effects of lesions of specific brainstem structures. These and other studies have combined assays for different transmitter substances, and the effects of transmitter agonists and antagonists. Other experiments have involved stimulation of specific brainstem and basal forebrain regions, and unit recordings from cells in these regions.

Figure 25.15 depicts the main regions

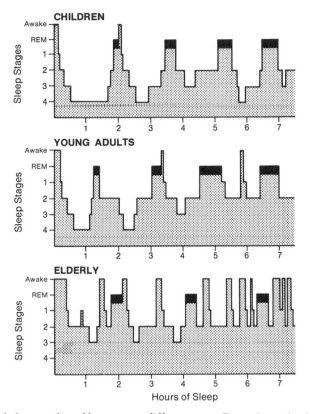

Fig. 25.14 Normal sleep cycles of humans at different ages. Dreaming episodes indicated by black bars. Note deeper sleep of children, more frequent periods of waking in the elderly. Sleep stages judged by EEG criteria. (From Kales and Kales, 1974)

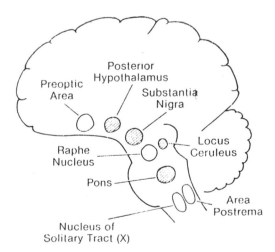

Fig. 25.15 Summary diagram showing regions of the brain that have been reported to be involved in controlling arousal or sleep. Arousal centers indicated by shaded areas, sleep centers by open areas.

that have been shown to be involved in either arousal or sleep. Let us review each region briefly.

Arousal Regions

High-frequency electrical stimulation is most effective in bringing about arousal and D sleep when it is applied to the *pontine reticular formation* (Pons, in Fig. 25.15) as in the experiments of Moruzzi and Magoun. The most effective sites are in the gigantocellular tegmental fields, whose neurons branch widely throughout the brainstem. A second arousal area is the *locus ceruleus*. As discussed in Chap. 24, this nucleus contains the noradrenergic neurons, whose axons branch widely and innervate most of the forebrain, cerebellum, and spinal cord. Jouvet (1974) showed that destruction of the locus ceruleus selectively eliminates D sleep in cats. A third center related to arousal is the dopaminergic system of fibers arising in the *substantia nigra* and nearby midbrain. Lesions of these fibers in cats render the animals comatose. However, EEG arousal can still be elicited by sensory stimulation, and the effect of the lesions is not seen in rats, suggesting that the dopaminergic fibers may be more involved in the initiation of locomotor activity than with arousal itself. Finally, there is evidence that the *basal forebrain* also contributes to arousal mechanisms; stimulation in the hypothalamus is very effective in producing arousal, and ablations of the posterior hypothalamus, as first reported by Walle Nauta in Holland in 1946, produce prolonged somnolence. However, these effects may be due in part to the fibers of the median forebrain bundle, particularly the fibers in this bundle that connect brainstem nuclei and the forebrain.

Sleep Regions

A number of studies have pointed to the importance of the serotonergic cells of the brainstem *raphe nuclei* in mediating deep sleep. Destruction of the raphe nuclei produces cats that cannot sleep (i.e., insomnia). A parallel pharmacological experiment consists of blocking serotonin synthesis with the drug *p*-chlorophenylalanine (PCPA), which inhibits tryptophan hydroxylase. This also produces insomnia, an effect which is alleviated by administration of serotonin. A second brainstem region is the *nucleus of the solitary tract*, which receives sensory fibers from the taste buds of the tongue (Chap. 11) as well as other visceral inputs. Stimulation of this region promotes a synchronization of the EEG. A related region is the *area postrema*. This area is special in that it has no blood–brain barrier, and thus can be stimulated directly by substances in the blood; toxic substances that enter the bloodstream elicit vomiting by acting on the cells in this region. Serotonin applied to this area modulates the influence of the nucleus of the

solitary tract on sleep. The mechanisms of these two regions are not understood, but they may mediate some of the effects of feeding, metabolism, and visceral activities in inducing sleep.

Finally, in the *basal forebrain*, lesions of the *preoptic region* produce insomnia, and electrical stimulation of this region induces EEG synchrony and drowsiness leading to sleep. Bremer in 1970 showed that preoptic stimulation reduces the arousal induced by stimulation of the reticular activating system.

Neuronal Mechanisms: A Synthesis

The foregoing account does not indicate the numerous controversies that have arisen in attempting to assess the contributions of each of these regions to the control of sleeping and waking. At the very least, we can say that the regions depicted in Fig. 25.15 form a distributed system involved in this control. In this system, the three brainstem centers (pons, locus ceruleus, and raphe), each using a specific transmitter, seem to be of special importance.

A coherent explanation for the neural basis of sleeping and waking states must characterize the specific changes in behavior, correlate these with different centers in the brain, and identify the circuits that connect the centers and thereby form the distributed systems for the different states. A proposal that attempts to do this for D sleep is summarized in Fig. 25.16. Each numbered site indicates a neural substrate that has been implicated in a function that is altered in D sleep. Let us consider each briefly:

① Phasic activation of motoneurons to the extraocular muscles produces rapid eye movements (REM).

② Generalized activation of the forebrain through the thalamus produces bursting activity in forebrain regions (see recordings in A); this may be associated with the production of dream images. Thalamic activity is controlled by shifts

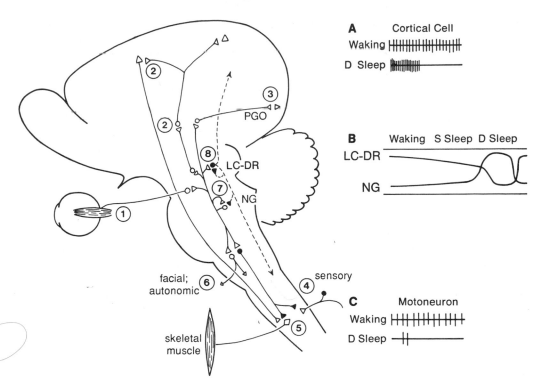

Fig. 25.16 Central circuits and neural activity involved in mediating D sleep, based on accumulated evidence from experiments in cats. The numbered sites ① through ⑧ are discussed in the text. **A.** Recordings from a cortical cell (site ②). **B.** Changes in impulse frequency of cells in locus ceruleus–dorsal raphe (LC-DR) ⑧ and nucleus gigantocellularis (NG) ⑦ during wake–sleep cycle. **C.** Recordings from a motoneuron (site ⑤). For full explanation, see text. (Based on Hobson and McCarley, 1977, and other sources)

in firing patterns by the mechanisms discussed in relation to Fig. 25.11.

③ Phasic activation of the visual pathway by means of PGO spikes (i.e., pons–lateral geniculate–occipital cortex) possibly contributes to visual imagery.

④ Inhibition of sensory input raises the threshold for arousal; this is why a person who is dreaming is hard to awaken.

⑤ Inhibition of motoneurons raises the reflex threshold and suppresses internally generated movements by cortical and brainstem motor centers (see recordings in C); this is why we lie most quietly during dreaming.

⑥ Activation and inhibition of various brainstem neurons produces phasic fine movements of facial muscles (possibly

expressing the emotional content of dreams); tonic autonomic contraction of the bladder and rectal sphincters; intermittent bursts of activity in the cardiovascular, respiratory, and other parts of the autonomic nervous system; and penile erection.

The cyclic relation between waking, S sleep, and D sleep is believed to involve interactions between the nucleus gigantocellularis (NG) neurons and neurons of the locus ceruleus (LC) and dorsal raphe (DR). In the model of Alan Hobson and Robert McCarley (1977):

⑦ NG neurons are reexcitatory to themselves; and

⑧ They also excite LC–DR neurons.

The LC–DR neurons are in turn inhibitory

to themselves and to the NG neurons. With appropriate time delays for the onset and buildup of activity in these two neuronal populations, the activity in the two populations alternates in relation to the different stages of waking and sleeping, as indicated in the graph in B.

The centers and pathways summarized in Fig. 25.16 constitute the main outlines of the distributed system that is responsible for D sleep. The diagrams together with the mechanisms detailed in Fig. 25.11, provides a working hypothesis for understanding how the different physiological functions arise from properties of brainstem and thalamic neurons and their synaptic connections, and how these properties and circuits could give rise to the sequence of waking and sleeping.

Sleep Substances

In recent years a new chapter has been opened in sleep research by the evidence that, in addition to the specific circuits described above, there are humoral factors produced by the brain or circulating in the blood that have profound effects on promoting or inhibiting sleep. We will briefly summarize these studies in relation to the summary diagram of Fig. 25.17, following the account of James Krueger (1989).

The story began in the late 1960s at Harvard Medical School, where John Pappenheimer and his colleagues found that if sheep were kept awake (sleep deprived) for several days, one could remove samples from the cerebrospinal fluid in their brains, inject it into normal rats, and induce them to increase the duration of deep slow-wave sleep. They characterized the active compound as a muramyl peptide, which is a glycopeptide that is a constituent of bacterial cell walls.

The connection between sleep and bacteria seemed remote until further research showed that muramyl peptides are in fact immune adjuvants, compounds that interact with antigens to enhance the antibody response to infection. They stimulate the production of lymphokines, which play key roles in stimulating lymphocyte proliferation (see Chap. 24). It was then found that lymphokines are even more effective than muramyl peptides in inducing slow-wave sleep when injected into rabbits. This suggested a hypothesis linking muramyl peptides to sleep: it was proposed that the breakdown of bacteria releases muramyl peptides, which stimulate lymphokines to induce sleep. One route for this effect appears to be through the production of prostaglandins (see Fig. 25.17). An attractive aspect of this hypothesis is that it accounts not only for the sleepiness that characteristically accompanies infectious diseases, but also for normal sleepiness that may involve diurnal rhythms of breakdown of intestinal bacteria; Krueger (1989) notes that each of us carries around 1 kg of bacteria in our intestines, so there is a plentiful source of potential sleep substances. One suspects that the circulating substances may act at some of the brain regions we identified in relation to Figs. 25.25 and 25.16 above.

These studies thus have provided strong evidence linking sleep to the immune system. From our previous study of the close links between the immune system and the neuroendocrine system (see Chap. 24), we might postulate further that sleep would also involve the hypothalamic-pituitary-adrenocortical system, and this has turned out to be the case. Interleukin 1 is known to stimulate the release of growth hormone (somatotropin), most of which is produced during slow-wave sleep. Growth hormone in turn has complex effects on REM and non-REM sleep. Interleukin 1 also stimulates the hypothalamo-pituitary-adrenal axis to secrete glucocorticoids, which inhibit sleep, presumably through a negative feedback loop onto interleukin. These various pathways are indicated in Fig. 25.17.

In conclusion, research on humoral substances affecting sleep has provided fascinating glimpses of powerful effects and complex reaction chains. A criticism of these effects is that many of them are seen

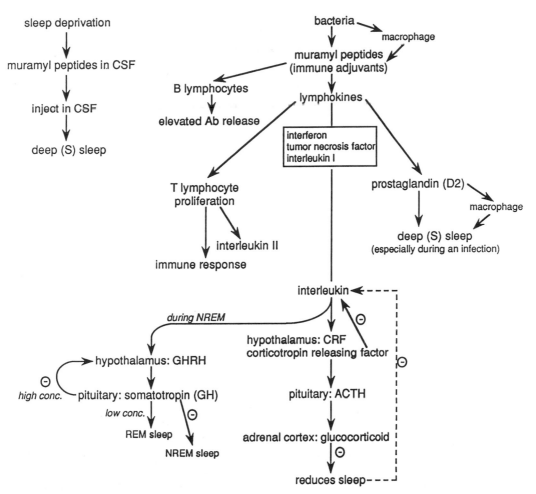

Fig. 25.17 Sleep factors and the metabolic pathways for their activation and control. The pathway for the original observations of the production of sleep-inducing substances by the sleep-deprived brain is shown in the upper left. Pathways linking sleep induction to the immune system and the neuroendocrine system are shown on the right. Compare these pathways with those linking these two systems in Chap. 24. See text. (Based on Krueger, 1989)

most clearly in sleep-deprived animals; their role in the induction and modulation of normal sleep is not as clearly established. Nonetheless, it seems evident that a comprehensive account of sleep mechanisms will require an integration of the membrane mechanisms shown in Fig. 25.11, the neural circuits illustrated in Fig. 25.16, and the humoral circuits shown in Figs. 25.17 and 24.4.

26

Visceral Brains: Feeding

Among the many types of behavior in which animals engage, two in particular stand out: feeding and mating. Feeding is necessary for the survival of the individual organism, and mating is necessary for the reproduction and propagation of the species. A good proportion of other types of behavior, such as predation and flight, or grooming and courtship, have their significance as preludes or consequences of these two fundamental activities.

A basic distinction between animal and plant life is that animals procure their food and find their mates by *active* rather than passive means. Much of the sensory apparatus is tuned to receiving stimuli from sources of food or mates, and much of the motor apparatus is adapted for moving the animal to those sources and ingesting food or copulating with a mate. The motor behavior thus has two phases: an *appetitive* phase, during which the animal has an appetite for something and seeks to find it, and a *consummatory* phase, in which the goal is achieved and the appetite is satisfied. These phases, in fact, can be recognized in even the simplest kinds of motor acts; Sherington (1906), for example, in studying the scratch reflex, noted that the cat first brings the leg to the site that itches

(the appetitive phase) and then carries out the scratching (the consummatory phase). In the cases of complex behaviors like feeding and mating, the sensory and motor mechanisms for the two phases are of course quite different, and the temporal sequence may be quite prolonged. Although it is impossible to generalize adequately across all species, the sequence usually begins with waiting or searching and proceeds through collection or capture, acceptance or rejection, and finally ingestion.

Ingested food typically enters an intestinal tract, where it is digested and the nutrient substances are absorbed into the bloodstream, in order to satisfy the needs of the body for hydration, mineral balance, and nutrition. These factors are kept in balance within the body by homeostatic mechanisms which maintain the "constancy of the internal milieu," one of the basic principles of body function first formulated by the great French physiologist Claude Bernard in the nineteenth century. The ingested foodstuffs provide the raw materials and sources of energy for these mechanisms. The whole sequence of feeding, from search to absorption, is under close control at every stage by a combination of nervous and hormonal mechanisms, and the parts of the central nervous system that mediate this control may be referred to as the "visceral nervous system" or "visceral brain." This is under control of the autonomic nervous system, as discussed in Chap. 18. The activities of the visceral brain are coordinated with the rest of the nervous system by a variety of centers which, in vertebrates, are called the limbic system. The relations between these systems will be discussed further in subsequent chapters in this section.

The fact that feeding is crucial for survival and requires a complicated series of actions means that each link in the chain is subject to the forces of natural selection. Each link can be adjusted for the greatest efficiency and advantage for a particular species, and the whole chain can undergo many permutations in different species. Therefore, when we study feeding, we are dealing with nervous mechanisms that, perhaps more than most others in the brain, directly and vividly reflect the forces of evolution acting in the daily life of the animal.

The diversity of nervous mechanisms, which we have so often noted, is therefore nowhere more evident than in the control of feeding behavior. Animals have explored a wide range of food sources, and have devised the most clever strategies for catching and consuming them. In this chapter we will consider examples that have been particularly well studied and illustrate some of the general principles of nervous control by central systems.

Humans Like to Eat!

Like every other animal species, humans must eat to provide for the basic requirements of body metabolism. But our food consumption goes far beyond meeting this need; indeed, it has a profound impact on those unique dimensions of emotional and mental activity that characterize human behavior. It has been pointed out that

the same chemicals that regulate appetite, metabolism, and weight also impact on our moods, levels of stress, physical energy, and the quality of our sex lives. "Our emotional life and state of mind," says [Sarah] Leibowitz, "will be affected by every bite we eat." (Stein, 1993)

We all know the truth of that statement with every meal we eat. The neural basis of feeding therefore has a special interest for humans. Not only does it give us insight into our normal habits of eating and our normal feelings of well being, but it also is critical for understanding the basis of feeding disorders. Most us worry at some time (and many people all the time) about being overweight, and for some young people this leads to severe disorders such as bulimia. There is continuing debate about what constitutes a healthy, balanced diet. All of us are subjected to claims for crash courses in dieting. The pharmaceutical industry has tremendous interest in drugs that can affect appetite and obesity. The government recognizes obesity as a health hazard. Thus, of all the subjects we study in this book, the neural mechanisms of feeding may be the most relevant to our everyday life.

These considerations might suggest that we should plunge directly into a consideration of human feeding. However, we will learn that the mechanisms are extraordinarily complex. Some of the mechanisms are best understood with a perspective gained from consideration of feeding mechanisms in simpler organisms. We will therefore begin with selected examples from invertebrate species before analyzing feeding in mammals and in humans.

Filter Feeders

In introductory biology, we learn that the great chain of life begins with the simple plantlike microorganisms called phytoplankton, unicellular algae that float in the sea and are capable of capturing the energy of light and using it to manufacture organic compounds, the process called photosynthesis. The phytoplankton are present in enormous numbers in most bodies of wa-

Table 26.1   Types of feeding by animals

Fluid feeding	Microphagy (small particles)	Macrophagy (large particles)
Bacterial	Filter feeding (some sponges; some arthropods)	Herbivores (cows, horses)
Planktonic		
Scavenging (some nematodes)	Deposit feeding (some annelid worms)	Carnivores (predators such as many molluscs, arthropods, vertebrates)
Predatory (some arachnids; some insects, such as flies and mosquitoes)		

ter, forming what has been called an "aquatic pasture." The simplest animals, the amebalike zooplankton, feed on this pasture. The higher animals, the metazoans, thus have three possible sources of food. One is the nutrient-rich fluids secreted or contained within other organisms; these animals are called *fluid-feeders*. Another is the microorganisms of the sea; animals that feed on them are said to be *microphagous*. Larger animals may feed on each other, in which case they are said to be *macrophagous* (see Table 26.1).

One of the most efficient methods for microphagous feeding is to let seawater flow through the animal so that the plankton can be filtered out. This method is appropriately called *filter feeding*. In many invertebrates the microorganisms are trapped in secretions and moved to the digestive tract by ciliary motion. In small aquatic arthropods, which generally lack cilia, filter feeding is achieved by ingenious use of the limbs. The swimming motions of the limbs (see the swimmerets; Fig. 21.1) create currents which cause the water to circulate through the limbs and over the opening of the mouth, before passing backward to propel the animal forward. The food is trapped by fine hairlike setules, and then passed forward to be filtered out by the mucus-covered setae around the mouth.

This provides a very nice example of how motor mechanisms serve multiple uses. We previously described in Chap. 21 the neural mechanisms for generating the orderly beating of the crustacean swimmerets. This metachronal rhythm is equally efficient for moving food and for propelling the animal. As E. J. W. Barrington pointed out in 1979:

> . . . the mechanisms involved [in filter feeding] are unexcelled for the precision and beauty of their adaptive organization. . . . The animal economically employs the limbs simultaneously for feeding as well as locomotion, while their delicate structure enables them also to serve for respiratory exchange.

Stomatogastric Nervous System of the Lobster

Filter feeding is widespread among small aquatic animals, especially crustacea and molluscs. It has been estimated that an oyster may filter up to 40 liters of water in an hour, from which it obtains less than a tenth of a gram of nutrients. Larger animals obviously need more ample sources of nourishment, and many species acquire this by consuming other large animals. This sets up one of the fundamental polarizations of animal life—the predator and its prey. Successful predation requires adequate size, speed, strength, and other capabilities, and therefore has been one of the main pressures leading to complex behaviors mediated by complex nervous systems. Since animals do not particularly enjoy being eaten, the same pressures work on animals that are prey. The behaviors must be al-

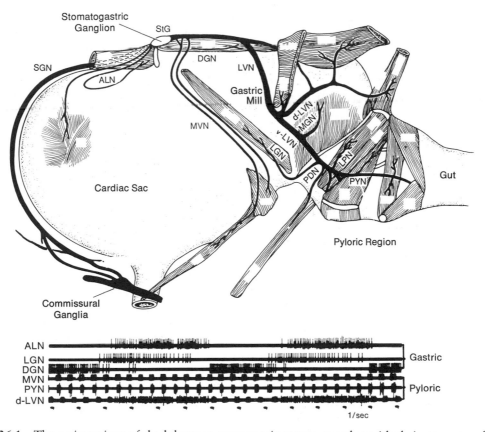

Fig. 26.1 The main regions of the lobster stomatogastric system, together with their nerve supply and some of the muscles. Recordings from nerves labeled in the diagram are shown in the traces below. These are extracellular recordings in the deafferented stomatogastric ganglion of impulse activity in motoneurons supplying their respective nerves. ALN, anterior lateral nerve; DGN, dorsal gastric nerve; LGN, lateral gastric nerve; LPN, lateral pyloric nerve; LVN, lateral ventricular nerve (d, dorsal; v, ventral); MVN, median ventricular nerve; PDN, pyloric dilator nerve; PYN, pyloric constrictor nerve; SGN, stomatogastric nerve. (Modified from Selverston, 1976)

most evenly matched, in order to maintain the ecological balance that is necessary between predators and their food supply.

Feeding on large bodies is called macrophagy (Table 26.1). It is well-exemplified by our old friend the lobster, and its freshwater cousin the crayfish. The claws of the lobster are specialized for grasping, cutting, and crushing prey, and for bringing the pieces to the mouth. The food is thus highly variable in its composition, containing many hard parts as well as soft tissue. The foregut is adapted for dealing with this mixture of ingested material. As shown in Fig. 26.1, the stomach is divided into three

parts. Foodstuff passing through the esophagus first enters the *cardiac sac,* which is mainly a large storage area. From here it passes to the *gastric mill,* which contains three calcified ossicles that function like sharp teeth. Muscles of the gastric mill wall move the teeth, to grind, macerate, and chew the food. In this way, the gastric mill functions like the vertebrate jaw. The food then passes to the *pyloric region,* where it is further churned by muscular contractions, squeezed between platelike ossicles, and subjected to the actions of digestive enzymes, before passing into the remainder of the gut where the nutrients are absorbed into the bloodstream and the waste materials are eliminated.

The muscles of the gastric mill and pyloric region are under control of nerves that originate in neurons of the *stomatogastric ganglion*. This ganglion has an unusual location, being plastered against the inside wall of the nearby ophthalmic artery that runs to the eye. There seems to be no special significance to this location other than that it is near the stomach. Experimentally, the ganglion can be exposed by making a slit in the artery from above, removing the wall beneath the ganglion, and dissecting away the periganglionic sheath. Under the dissecting microscope the nerve cell bodies can be seen to be quite large, between 40 and 90 μm in diameter; like so many invertebrate ganglia, many of the cells are identifiable by size and location, and can be recorded by using intracellular electrodes.

The attractiveness of this preparation for studying the nervous control of the stomach rhythms was recognized by Don Maynard, and has been exploited by Allan Selverston and his colleagues at San Diego, and more recently by Eve Marder and her colleagues at Brandeis, to the extent that it is one of the best understood of invertebrate neuronal systems.

There are two main rhythms in the stomach, related to the gastric mill and the pyloric region. Recordings from the nerves to the gastric mill muscles (ALN, LGN, and DGN in Fig. 26.1) show slow, prolonged impulse discharges lasting several seconds, which occur at intervals of 6–8 seconds. The discharges to antagonist muscles controlling the teeth alternate with each other, as shown in Fig. 26.1. By contrast, the nerves to pyloric muscles (MVN, PYN, and d-LVN) fire in briefer bursts of impulses, at more frequent intervals of less than a second.

A primary focus of experimental work has been on understanding the nature of these rhythms. With regard to the *gastric mill*, it has been found that there are 10 motoneurons and 2 interneurons within the stomatogastric ganglion that are involved in generating the slow rhythm of impulse firing and muscle contractions. In completely isolated ganglia, the rhythmic firing is usually abolished, indicating that it is normally dependent on excitatory input from sensory fibers. Intracellular analysis has enabled the main functional connections between the 12 cells to be identified. No spontaneous bursting cell has been found, suggesting that the gastric rhythms are a property of the whole network, of the type that we discussed in Chap. 20.

The neural system controlling the *pyloric muscles* is composed of 13 motoneurons and 1 interneuron. John Miller and Allan Selverston (1985) were able to dissect this circuit and identify the contributions of each cell type to the pyloric rhythm, by using a photoinactivation procedure. This consists of injecting a fluorescent dye, such as Lucifer Yellow, into a cell; after the dye diffuses throughout the cell, a laser microbeam is aimed through the preparation at a specific part of the neuron or its dendritic tree. In this way the effect of inactivation of an entire neuron or a dendritic branch on the generation of a circuit property such as a rhythm can be assessed.

The neural basis of the pyloric rhythm was reconstructed on the basis of this method and the known synaptic connections between the cells, as summarized in Fig. 26.2. In A, all cells but the VD and LP cells were inactivated. These two cells nonetheless generated a rhythm of alternating bursts due to their mutual inhibitory connections. This is an example of the reciprocal inhibitory half-center model (see Fig. 20.3). In B, one of the PD cells is added, so to speak, to the circuit. It fires in synchrony with VD, because it is also connected to LP with reciprocal inhibition, and has in addition weak electrical coupling with VD. In C, the system includes IC and PY neurons. The IC cell is not connected to LP and therefore fires out of phase with it, but its timing is variable because of the weaker inhibitory inputs from VD and PY (dashed lines) and the

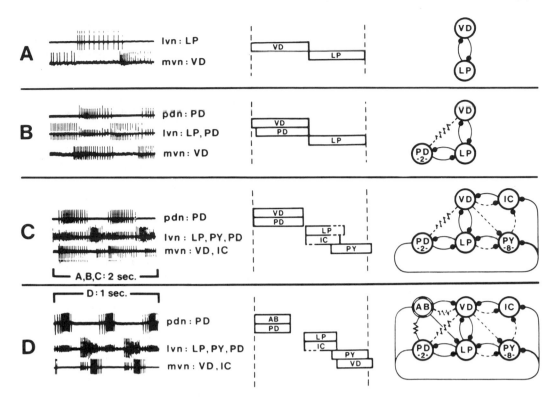

Fig. 26.2 Idealized reconstruction of the pyloric central pattern generator, based on selective inactivation of identified neurons by the dye-sensitized photoinactivation technique. **A.** System reduced to only ventricular dilator (VD) and lateral pyloric (LP) neuron (right) generates spontaneous rhythm (middle and left). **B.** System includes pyloric dilator (PD) cells. **C.** System enlarged to include IC and PY cells, which reinforce the pacemaker activity. **D.** Complete system includes anterior burster (AB) cells, which add further precision and overall control to the firing patterns. All abbreviations in Fig. 26.1 and Table 26.2. In the diagrams on the right, inhibitory chemical synapses are small filled terminals, electrical synapses are zigzag lines, weak connections are dashed lines. (From Miller and Selverston, 1982)

inhibition from the more distant PD cell. The PY cells fire slightly later than LP–IC because they do not fire until their inhibition of LP has worn off. Finally, the entire circuit includes the interneuron AB, as shown in D, which is electronically coupled to PD.

This analysis allowed three basic properties of the pyloric rhythm to be understood in terms of their neuronal basis. As summarized by Miller and Selverston (1985):

The *existence* of the pyloric pattern results from oscillatory membrane properties of the individual neurons in combination with the multiple

reciprocally inhibitory interactions within the network. The precise *phase relationships* derive from the synaptic connectivity circuit and depend on relative synaptic efficacies, postinhibitory rebound properties, and the kinetics of the plateau potential and BPP generation mechanisms [BPP = "bursting pacemaker potentials," cyclic voltage changes due to cyclic conductance changes]. The overall *cycle frequency* is determined largely by the AB interneuron via its strong intrinsic oscillatory currents and its very strong synapses with the rest of the pyloric neurons.

It is clear that the pyloric pattern produced by an isolated nervous system represents the fundamental pattern. However, the *in vivo* be-

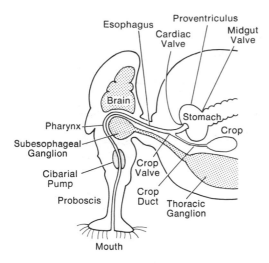

Fig. 26.3 The digestive system of the blowfly. (Modified from Dethier, 1976)

havior of the pyloric region is strongly influenced by sensory feedback and endogenous modulatory substances. The identification and physiological action of these factors on the well-characterized pyloric network promises to yield important insights into the problem of how centrally generated patterns can be modified by the environment.

Feeding in the Blowfly

One of the best understood species with respect to feeding behavior is the blowfly. The blowfly obtains all its nutrients by sucking them in through the labellum (see Fig. 26.3). The pharynx is lined with muscles which form what is called the "cibarial pump." The ingested fluid is moved through the pharynx and into the esophagus by rhythmic contractions of the cibarial muscles. These contractions are controlled by a neuronal rhythm generator located in the brain. The food passes by peristalsis from the esophagus into both the midgut and the crop, a storage region. As food is absorbed from the midgut into the blood, it is replenished by movement of stored fluid from the crop through the crop valve and cardiac valves into the midgut.

What are the mechanisms that initiate feeding behavior? This has been studied

in flies that are starved for several days. Deprivation generally enhances the locomotor activity of the fly, which engages in bursts of activity (flying about in search of food) alternating with periods of rest. This represents a balance between maximizing the chances of finding food and the need to conserve the dwindling energy stores. When food, such as a sugar solution, is encountered, excitation of chemoreceptors on the legs (tarsae) stimulates proboscis extension, and excitation of chemoreceptors on the labellum of the proboscis stimulates sucking movements. These chemosensory stimuli are believed to be the sole excitatory inputs driving feeding behavior.

What are the mechanisms involved in terminating feeding? In order to identify these factors, many experiments have been carried out in which nerves to different parts of the gut have been cut, and in which nutrients have been placed directly in different parts of the gut or removed through artificial fistulae. The outcome of these experiments appears to be that feeding is inhibited by three factors: first, the amount of peristalsis in the foregut, as sensed by stretch receptors in the gut wall; second, distention of the crop, as sensed by stretch receptors in the crop wall; and third, the amount of locomotor activity mediated by limb motoneurons in the thoracic ganglion. These factors have the effect of raising the central threshold for incoming chemosensory stimuli, so that they become less and less effective in eliciting proboscis extension and sucking. Some of these factors, and their interrelations, are indicated in Fig. 26.4.

The surprising result emerging from these studies has been summarized by Vincent Dethier (1976) in the following way:

At no stage in ingestion does the nutritive value of food regulate intake.... The initiation of feeding depends on stimulating properties; the termination of feeding depends on mechanoreceptors; the emptying of the crop is regulated by osmotic properties. In the laboratory each one of these can be varied independently of the

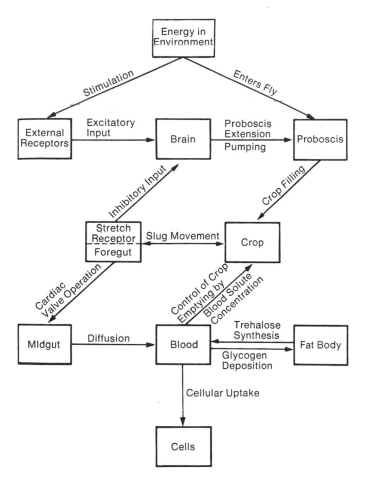

Fig. 26.4 Multiple mechanisms for regulating feeding behavior and metabolism in the blowfly. (From Gelperin, in Dethier, 1976)

other and of caloric value. In nature the different properties tend to be correlated.

In conclusion, feeding in the insect is controlled by relatively few factors, which impinge immediately and directly on the regulation of food intake. We will see that this contrasts with most vertebrates, in which multiple factors and contingencies are involved.

Feeding in Vertebrates

Vertebrates show interesting similarities and differences in feeding behavior among various species. Fish, for example, move about in a medium that contains their food; thus, their feeding is more or less continuous. Some terrestrial animals have similar habits; herbivores—cows, for example— have their food literally under their noses all the time, and spend much of their waking hours ingesting or chewing. The cow munching its way through its terrestrial pasture seems rather similar to a crustacean or mollusc filter-feeding on its aquatic pasturage. In the case of the cow, the almost continuous eating makes it difficult to characterize its feeding behavior in terms of *meal size* and *meal frequency*, two of the variables we are most familiar with in our own daily lives. Nor does a third variable, *food preferences* or *aversions*, appear to be very relevant; it seems that the cow, which consumes enormous quantities of plant material in the course of a day, may simply dilute any noxious or poisonous plants to the level of harmlessness, without needing to pick and choose the way we do. By contrast, humans, like many other terrestrial animals, especially carnivores, eat only intermittently. Thus, meal size, meal frequency, and food preferences are important variables in our eating habits; indeed, they largely determine the rhythm of daily life.

Among vertebrates, no species has been studied more intensively with regard to all aspects of feeding behavior than the laboratory rat; in fact, the studies of the rat probably outnumber those of all other species combined. Much of this is due to the possible implications of the rat as a model of mammals in general, and humans in particular. In the practical realm, the laboratories of drug companies and government agencies test rats for the possible harmful effects of food additives or substitutes, under the premise that the effects will be relevant to humans. At the research level, psychologists for almost a century have focused their concepts (and controversies) about drives and motivation on behavioral studies of the rat, much of it in relation to feeding. Since the late 1940s, neuroscientists have exerted considerable efforts to identify the main nervous pathways and mechanisms involved. We will summarize this evidence, and then discuss some implications for motivated behavior.

Feeding in the Infant Rat

Studies on the development of feeding in the infant rat have been significant for several reasons. First, it has shown that the infant is not simply a small adult; it is a very special creature, living in a very special world, during the first few weeks of life. Second, the change from the infant to the adult pattern gives insight into processes of development and learning that provide the basis for adult feeding behavior and their neuronal mechanisms. Third, suckling at a mammary gland is the defining characteristic of all mammals, and this form of feeding is therefore important for understanding this class of animals. Finally, studies in subprimates may contribute to better understanding of suckling and infant–mother relations, which are crucial in the first days and weeks of human development.

The baby rat, called a pup, is born after about three weeks' gestation. At birth the mother occupies herself with a complex series of interrelated behaviors that create the transitional environment from womb to outside world. She consumes the placenta, licks her ventrum (chest and abdomen) and nipples, and also vigorously licks the infants. Her behavior has two main functions. It imprints her ventrum and nipples with olfactory substances in her saliva, and it serves to warm, protect, and arouse the infants. The aroused infants search for and find the nipples, and begin sucking.

The act of sucking involves an extraordinary close relation between mother and infant. As shown in Fig. 26.5A, the nipple fills the mouth, and milk is ejected deep into the pharynx. Although the muscles for posture and locomotion are weak and poorly coordinated, the facial and neck muscles involved in suckling are well developed at birth; it is, in fact, the one coordinated motor act that the infant can perform on its environment. In some mammals, the sucking contractions are so strong that the infants remain attached despite locomotion of the mother, an obvious advantage if the mother has to flee from danger. The rat pup suckles virtually continuously throughout the first two weeks of life, and intermittently for the next week or two until weaning occurs. It could well be asserted that suckling must be among the most intimate and sustained (to say nothing of beneficial) relations between two organisms to be found in the animal world.

The Odor Link in Suckling Behavior

The close proximity of the nose to the mouth suggests that the sense of smell may be an important factor in suckling, and this

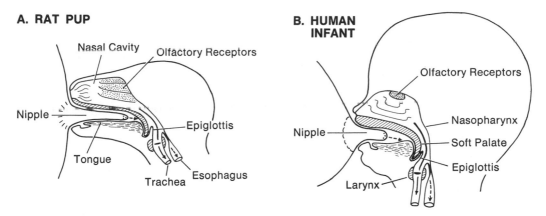

Fig. 26.5 Suckling in the rat pup (**A**) and in the human infant (**B**). Note that in both cases, the tip of the soft palate drops behind the back of the tongue and is held there by the epiglottis. This permits air flow between nose and trachea, while milk passes laterally around the epiglottis and into the esophagus. By this means, both the rat pup and the human infant can breathe while suckling. The high position of the larynx is maintained in the adult rat, an obligate nose breather. In the human, during infancy and childhood the pharynx grows in length and the larynx descends to its position in the neck in the adult. This elongation of the pharynx is crucial to our ability to form the different vowel sounds in human speech, as discussed in Chap. 24. (A based in part on J. Alberts, P. E. Pedersen, J. Laitman, and E. S. Crelin, personal communications; B adapted from Crelin, 1976)

has been found to be the case in many species. Ablation of the olfactory bulbs at birth in rats and kittens eliminates suckling. Infants must therefore be able to smell some odor in order to suckle. Following this lead, Elliott Blass and Martin Teicher at Johns Hopkins University washed the nipples of lactating mothers and found that this too eliminated attachment to the nipples by the pups (see Blass et al., 1979). Dabbing the washing fluid back on the nipples reinstated suckling, as did samples of the amniotic fluid and the mother's saliva, but not saliva of virgin females. This suggests the presence of a substance in the saliva that is dependent on the hormonal status of the mother. After suckling has begun, the pups' own saliva contains the odor cue necessary to maintain suckling.

The vital link in establishing and maintaining suckling in the rat pup is thus an odor cue on the nipple. In order to analyze the kind of activity in the olfactory pathway that mediates this response, radioactively labeled 2-deoxyglucose was injected

into suckling pups and autoradiographs of the olfactory bulbs were obtained by the Sokoloff method (see Chap. 8). As shown in Fig. 26.6, the most intense activity pattern in these olfactory bulbs was a small focus at the extreme dorsomedial margin of the main olfactory bulb, near its junction with the accessory olfactory bulb. Closer examination revealed that this is a site where the usual small, round glomeruli of the bulb are replaced by a large, irregular complex of glomeruli, an anatomically distinct region that had not previously been recognized. This was the first instance of a specific odor stimulus correlated with activity in a specific, anatomically identified, glomerulus.

As discussed in Chap. 12, olfactory stimulation gives rise to patterns of activity in the olfactory bulb characterized by distinct foci associated with more widespread, less intense patterns. The activity shown in Fig. 26.6A, 2, conforms to this pattern. One assumes that information about the suckling odor is carried both in the dense foci

A. 10 DAY OLD-SUCKLING

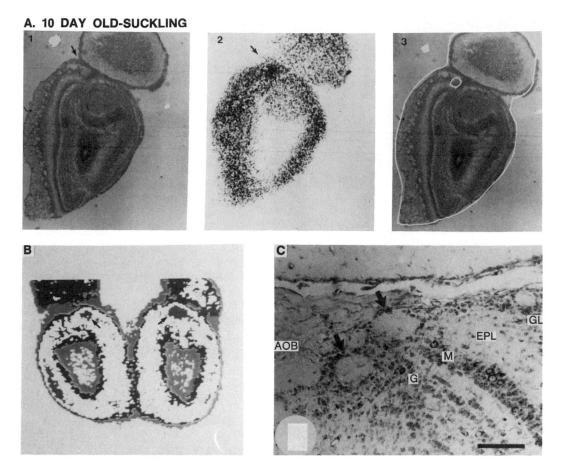

Fig. 26.6 Mapping of 2-deoxyglucose uptake in the olfactory bulb of the suckling rat pup. **A.** This 10-day-old pup was injected with [^{14}C]2DG and allowed to suckle for 45 minutes. The autoradiogram (2) shows a focus at one site in the olfactory bulb; the histological correlation (1,3) shows that the site is a modified glomerulus adjacent to the accessory olfactory bulb. **B.** Computerized image analysis of an autoradiogram of 2DG uptake in the olfactory bulbs of another suckling rat pup, age 6 days. Note the bilateral foci (arrows), shown by histological correlation to be located in the modified glomerular complex. **C.** Higher-power view of the modified glomerular region in the olfactory bulb of a 12-day-old rat pup. Arrows indicate two glomeruli within the modified complex. Accessory olfactory bulb (AOB) is to the left, layers of the main olfactory bulb are to the right. EPL, external plexiform layer; G, granule layer; GL, glomerular layer; M, mitral cell body layer. Bouin's fixation, stained with cresyl violet. Bar is 85 μm. (A from Teicher et al., 1980; B,C from Greer et al., 1982)

and the widespread pattern; by analogy, visual recognition of someone's face can be made not only from the most distinctive features (bushy eyebrows, toothy grin) but also from subtle aspects of the contours of the cheeks or wrinkles of the skin. The finding of a dense focus in the modified glomerular complex is of further interest with regard to a macroglomerular complex in the antennal lobes of insects, which is involved in processing male responses to female sexual attractant odors; in both, it is hypothesized that a crucial olfactory-mediated behavior appears to have a "labeled line" to the brain (see Chap. 11).

Further insight into the nature of this

labeled line has been gained in a study by Shinoda et al. (1989) using a polyclonal antibody against a human placental antigen X (HPAX). This antibody stains a subset of olfactory receptor neurons within the olfactory epithelium and, intriguingly, a small subset of "necklace glomeruli" along the posterior border of the main olfactory bulb, which includes the modified glomerular complex. On average, 150–200 receptor neurons project to a single glomerulus. Since the suckling cue comes from the mother's placenta, these results indicate that the neurons transmitting information about this cue in the pup express a maternal placental protein. The HPAX antibody recognizes the cytochrome P-450 portion of an enzyme with aromatase activity. (Cytochrome P-450 is a widespread enzyme which detoxifies various substrate molecules, whereas aromatase converts androgens to estrogens; see next chapter.) Other than the olfactory bulb, the HPAX antibody stains the placenta and ovary. Work in this area thus holds promise of giving further insights into the relation between reproductive behavior and the sense of smell.

Mother–Infant Interactions

Although the rat pup suckles continuously in the first two weeks, it does not receive milk continuously; milk is ejected from the nipple only intermittently. This process is called *milk letdown*. It is under the control of the hypothalamus. Neurons in the paraventricular (PV) and supraoptic nuclei synthesize *oxytocin,* an undecapeptide hormone (see Chap. 8), and transport it through their axons to their terminals in the posterior lobe of the pituitary (see Chap. 24). Single-cell recordings from PV cells in lactating mother rats show a background spontaneous discharge. Periodically, the neurons fire bursts of impulses, followed some 10–15 seconds later by ejection of milk (see Fig. 26.7). These results indicate that sensory inputs to the PV cells produce synaptic depolarization and increased impulse generation; that the impulses invade and depolarize the terminals in the posterior pituitary and cause release of oxytocin stored there; and that oxytocin acts on the smooth muscle of the mammary gland after a delay due to the time for circulation of the hormone through the blood circulation, from pituitary to nipple. Parts of this circuit are depicted in Fig. 26.7.

For their part, the rat pups are not just passive receptacles for the ejected milk. Letdown lasts only a few seconds, and the pup must be able to detect the engorgement of the nipple and respond with vigorous sucking. From these considerations it has been realized that the state of arousal of pups is very important in their ability to thrive. It appears that nipple attachment and ingestive control are exquisitely dependent on the level of arousal; a deprived pup with an empty stomach is more alert than a satiated pup, and therefore is more responsive to nipple engorgement and more active in sucking.

Arousal is mediated by the brainstem reticular system (Chap. 25), activated by the vigorous licking by the mother as described above. It is also mediated by activity in afferents from the empty stomach. The importance of arousal in initiating feeding appears to be similar to the higher activity levels of the adult blowfly after food deprivation, discussed earlier in this chapter. If rat pups are persistently aroused by experimental manipulation, they ingest milk beyond the normal capacity of the stomach. It is not until the age of 15 days that adultlike controls of ingestion begin to appear, and the rat begins to adjust intake to need.

Feeding in the Adult Rat

Theories of feeding mechanisms can be said to have evolved through two stages: the classical theory, which developed from the 1950s to the 1980s; and the modern view, based on more recent work. Since the classical theory has dominated thinking about the subject for so long and still contains

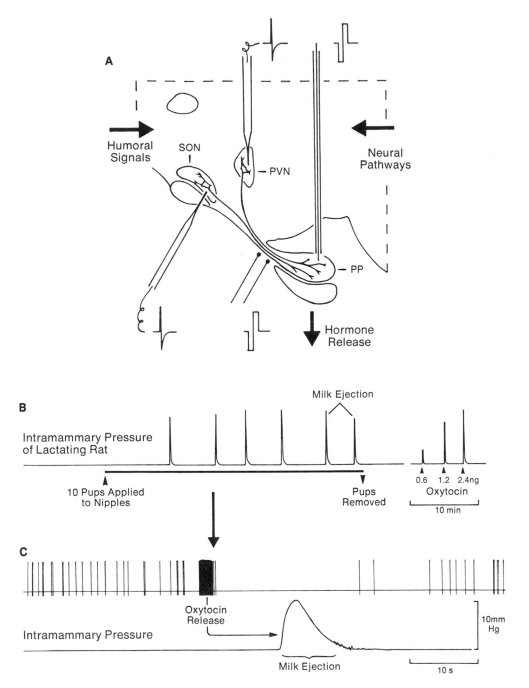

Fig. 26.7 Milk letdown in the lactating mother rat. **A.** Neurons in the paraventricular nucleus (PVN) and supraoptic nucleus (SON) in the hypothalamus and their connections to the posterior lobe (PP) of the pituitary. **B.** Intermittent pattern of milk ejection, as shown by recordings of intramammary pressure. The slow, long-duration trace shows milk ejections by an anesthetized mother rat in response to suckling by her litter. The slow, short-duration trace shows the mammary response to intravenous pulse injections of increasing concentrations of oxytocin. **C.** Unit activity recorded from a PVN neuron, showing the correlation of intense burst activity, oxytocin release, and subsequent mammary response and milk ejection. Note the much faster trace speed in C than in B. (Courtesy of D. W. Lincoln; see also Wakerly and Lincoln, 1973)

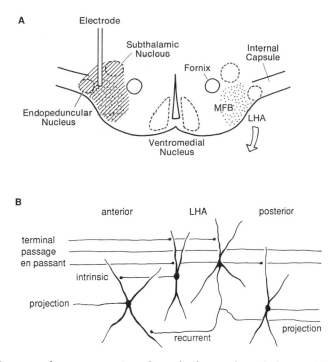

Fig. 26.8 A. Diagram of a cross section through the rat hypothalamus. Fibers of the medial forebrain bundle (MFB) are shown by dots. Note the electrode used to make electrolytic lesions. Sites of lesions of the lateral hypothalamic area (LHA) indicated by shading. **B.** Sagittal section of the rat brain, showing neurons in the lateral hypothalamic area (LHA) and some aspects of their synaptic organization. (A after Epstein, 1971; B after Millhouse, in Stellar and Stellar, 1985)

much that is relevant, it will be useful to begin with a historical perspective on how it became established and then a consideration of its shortcomings.

A Brief History

The classical theory about neural mechanisms involved in feeding had its origins in the finding by John Brobeck and his colleagues at the University of Pennsylvania in the 1940s that bilateral lesions of the ventromedial nucleus (VMN) of the hypothalamus (see Fig. 26.8) produce an animal that grossly overeats and becomes enormously fat. This implied that this nucleus normally has a restraining influence on feeding centers in the brain. This idea was supported by experiments showing that electrical stimulation in this area causes a reduction in eating.

The next step came in 1951, when Anand and Brobeck showed that small, bilateral lesions in the lateral hypothalamic area (LHA) cause a rat to stop eating (aphagia) and drinking (adipsia), leading to death within a few days. The lesions did not appear to interrupt specific sensory or motor pathways within the brain. It was therefore suggested that the LHA could be regarded as the center for mediating feeding (and drinking) behavior. The location of this area is shown in Fig. 26.8, where it can be seen that a large population of fibers, known as the medial forebrain bundle (MFB), passes through this area. As we shall see, the MFB is a major pathway uniting many regions within the core of the brain.

The evidence from these two lines of work was combined by Eliot Stellar in 1957 into the *dual-center hypothesis*. This stated that feeding is due to LHA activity, whereas cessation of feeding because of satiety is due to VMN activity; the cycle of feeding and satiety results from periodic interactions between these two centers. It was recognized that both centers receive several types of inputs, neural and humoral, so that the whole system is under *multifactorial control*.

This theory has provided an organizing principle which has stimulated and guided an enormous amount of work in the intervening time. Philip Teitlebaum and Stellar found that LHA-lesioned rats gradually recover if they are tube-fed. Since then, many variations have been practiced on the exact sites and extents of lesions, the means of producing them, and the relations of other parts of the brain to the LHA–MFB. In addition, experiments in which the LHA has been stimulated electrically have generally shown increased eating. Unit recordings from the LHA have shown increased impulse firing under conditions of deprivation. From all these results it seems clear that the LHA–MFB is intimately involved in control of feeding behavior, but there has been much disagreement on the actual pathways involved and the nature of the control.

Despite the many controversies that have arisen, two major themes have emerged. The first theme is that, despite the remarkable recovery that can take place, LHA-lesioned animals show a number of permanent deficits, not only related to feeding behavior but also profoundly affecting general behavior. The recovered rat, though able to survive, is far from normal. These defects all place serious limitations on feeding behavior; depression of salivation, for example, could by itself explain much of the inability of an animal to eat after lesioning. (Electrical stimulation of the LHA causes, on the contrary, copious salivation.) The lesion also has serious general consequences, leaving an animal that is lethargic, generally akinetic, emotionally depressed, and with limited tolerance for stress. These effects obviously contribute to the depression of feeding.

The second theme, related to the first, is that many systems, not just an isolated one, are affected by an LHA lesion, and conversely, that many parts of the nervous system have inputs to and effects on the LHA. The LHA–MFB is a locus, a common path, for many *overlapping distributed systems*.

Anatomy of the Hypothalamus–MFB System

In order to understand this key point about central systems controlling feeding behavior, we need to consider the neuroanatomy of this region. The location of the LHA has been shown in Fig. 26.8. In contrast to many other regions of the brain, the hypothalamus is not organized into a discrete region with distinct layers. This is particularly true of the lateral hypothalamic area. As indicated in Fig. 26.8A, the LHA has no clear boundaries; lesions aimed at the LHA therefore commonly impinge on neighboring areas such as the subthalamus, endopeduncular nucleus, and medial part of the internal capsule, causing more widespread and variable effects.

The intrinsic organization of the LHA is characterized by a diffuse arrangement of neurons, as illustrated in Fig. 26.8B. The triad of elements—input fibers, output neurons, and intrinsic neurons—that make up synaptic organization has been difficult for neuroanatomists to identify. There is not a clear differentiation of projection and intrinsic neurons, a feature shared with other parts of the reticular core.

A more serious problem is that the LHA lies within, and is traversed by, the fibers of the medial forebrain bundle (MFB), the massive population of fibers which serves as the main conduction pathway linking areas of the brainstem–hypothalamus–basal forebrain into a common system. Some of these fibers are afferents that have terminal boutons on the LHA dendrites and cell bodies; some are afferents that make en passant synapses, on their way to other regions; some are fibers of passage, making no synapses on their way to other regions; and, finally, some are axons arising within the LHA, either projection axons joining the MFB and destined for outside targets, or intrinsic axons within the LHA. These different types are illustrated in Fig. 26.8B.

Because of the close association of the MFB with the LHA, we must next determine the connections made by the MFB. As shown in Fig. 26.9A, in the forebrain the *inputs* are mainly from olfactory cortex, amygdala, striatum, hippocampus, and prefrontal cortex (other areas of neocortex are conspicuously absent). Caudally, the main sources of inputs are the brainstem monoamine systems (locus ceruleus [LC], dorsal raphe [DR], and ventral tegmental area [VTA]), and reticular regions. The MFB *output* targets are summarized in Fig. 26.9B. In the forebrain, these are most of the input sites mentioned above, with, in addition, most of the neocortex. Caudally, the same principle of reciprocal relations applies, with, in addition, the central taste pathway (nucleus of the solitary tract, NTS) and spinal cord.

These aspects of anatomy are important for the following reasons. The synaptic organization of the LHA shows that any lesion of this area is bound to have variable effects due to loss of intrinsic circuits, and also have widespread effects through interruption of fibers providing passage of the MFB. Similarly, electrical stimulation in the LHA must activate (or inactivate) intrinsic circuits in unpredictable ways, as well as cause orthodromic activation of MFB output fibers and antidromic activation of MFB input fibers.

Modern Concepts of Lateral Hypothalamic Function

The modern view of the function of the hypothalamus in feeding behavior began with the realization that the medial and lateral areas are involved in inhibitory feedback pathways that maintain energy balance and body weight at a specific level; this is referred to as the *set-point theory*. It arose from experiments such as those illustrated in Fig. 26.10. Normally, a rat eats at a rate which moves its weight along a curve such as A. In rats with small lesions of VMH (curve B), the satiety controls are impaired; the animal increases its food intake in relation to a higher set point. In rats with small lesions of the LHA (curve C), the animal reduces its feeding until it reaches a lower set point, which it thereafter maintains. If animals are first starved, given the same LHA lesions, and then allowed to feed, they actually gain weight initially (curve D) until they reach the same set point.

Analysis of all the mechanisms involved in the contribution of the LHA to feeding has meant, essentially, trying to track the functions of all the pathways of the MFB that pass through the LHA or make connections within it. These modern studies have employed a variety of methods, including single-unit recordings, microdialysis of transmitters and peptides, and receptor localization by ligand binding studies and by in situ mRNA hybridization. A modern synthesis of this work is shown in Fig. 26.11. Let us examine this complex system component by component, following the summary account of Hoebel (1988).

At the core of the system is the lateral hypothalamic area (LH). As already mentioned, lesions of this area cause loss of feeding (anorexia) and loss of weight until a new set point is reached, due to loss of LH cells and fibers of passage. The LH contains cells that fire impulses in response to glucose utilization. Some cells respond robustly to the taste of glucose, but gradually stop responding as a monkey becomes satiated by a glucose meal. Sensory information about taste is relayed from the tongue through the nucleus of the solitary tract (labeled ST in the diagram). Thus, LH cells integrate external and internal nutrition-related stimuli.

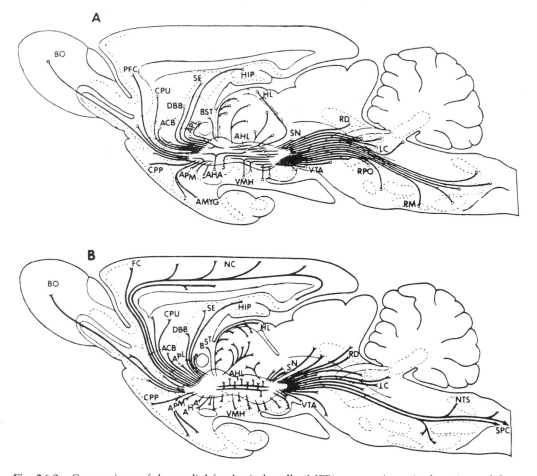

Fig. 26.9 Connections of the medial forebrain bundle (MFB), as seen in sagittal sections of the rat brain. **A.** Inputs to the MFB from different parts of the brain. **B.** Different regions of the brain that receive MFB fibers. ACB, nucleus accumbens; AHA, anterior hypothalamic area; AHL, lateral hypothalamic area; AMYG, amygdala; APL, lateral preoptic area; APM, medial preoptic area; BCA, bed nucleus of the anterior commissure; BO, olfactory bulb; BST, bed nucleus of the stria terminalis; CPA, periamygdaloid cortex; CPP, prepiriform cortex; CPU, caudate–putamen; DBB, nucleus of the diagonal band (Broca); FC, frontal cortex; HIP, hippocampus; HL, lateral habenula; LC, locus ceruleus; NC, neocortex; NTS, nucleus of the tractus solitarius; PFC, prefrontal cortex; RD, dorsal raphe; RM, raphe magnus; RPO, raphe pontis; SE, septum; SN, substantia nigra; SO, supraoptic nucleus; SPC, spinal cord; VMH, ventromedial hypothalamus; VTA, ventral tegmental area; ZI, zone incerta. (From Niewenhuys et al. in Stellar and Stellar, 1985)

The output of the LH system is important for generating the rewards for feeding. In this function it greatly overlaps with the general reward system interconnected through the MFB. This system was first discovered in 1954 by James Olds and Peter Milner, who showed that electrical stimulation of the LHA has rewarding effects. Rats readily learn to press a bar to

receive electrical stimulation; for example, they will press a lever 3000 times per hour (almost once per second) to receive half-second trains of shocks through an electrode implanted in this area.

After a large meal, self-stimulation becomes less rewarding, and rats will work to turn it off. In some experiments it has been found that LH electrodes elicit both

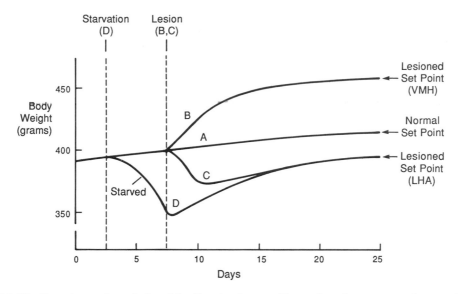

Fig. 26.10 Experimental analysis of feeding in the rat, illustrating the concept of set point for control of body weight. **A.** Curve for gradual increase of body weight in the normal adult rat, determined by a gradually increasing set point. **B.** After selective lesion of the ventral medial hypothalamus (VMH), the animal gains weight and then continues to feed at a higher set point. **C.** After selective lesion of the lateral hypothalamic area (LHA), the animal loses weight and then continues to feed at a lower set point. **D.** If the animal is initially starved, the response to an LHA lesion may be to gain weight before settling in at the new set point. (Adapted from Keesey et al., in Kandel and Schwartz, 1985)

self-stimulation and, if the stimulus train is persistent, feeding, suggesting that the "electric reward" of LH self-stimulation is related to the natural rewards of eating. This was shown in studies demonstrating covariation of LH self-stimulation and appetite. Self-stimulation was inhibited by a meal, gastric distention alone, or intravenous glucose infusion. Conversely, self-stimulation was disinhibited (that is, activated) by food deprivation, lesions of the ventral medial nuclei (VMN), or selective destruction of the ventral noradrenergic pathway from the brainstem to the hypothalamus. Thus, self-stimulation reward at a feeding site is controlled by satiety, acting by way of blood-borne factors and neural inputs to the hypothalamus. A rat that self-stimulates without eating experiences the rewards of eating without the sensation of getting full, and therefore continues to self-stimulate indefinitely. At other hypothalamic electrode sites where self-stimulation

elicits copulation by male rats, the self-stimulation has been shown to vary with castration or hormone replacement. In these experiments the electrodes appear to tap into the system for mating reward instead of feeding reward (see next chapter).

If food is accepted, there are descending feeding and reward signals from LH cells, which provide for positive feedback to neurons in the ascending pathways, including the taste centers and the sensory and motor nuclei of the brainstem involved in feeding behavior. An important finding is that these LH "reward" neurons also connect to dopaminergic (DA) cells in the midbrain tegmental area (VTA). Electrical stimulation of the LH or the VTA causes dopamine release in the projection areas of the mesolimbic dopaminergic pathway (the student should review this pathway in Chap. 24, Fig. 24.13).

Dopamine release in the nucleus accumbens (NAC) and prefrontal motor cortex

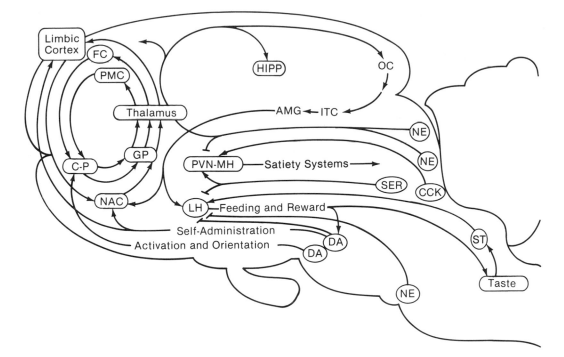

Fig. 26.11 Summary of current concepts regarding the pathways and mechanisms related to the lateral hypothalamic area and their involvement in control of feeding behavior. AMG, amygdala; CCK, cholecystekinin inputs; C-P, caudate–putamen; DA, dopaminergic inputs; FC, frontal cortex; GP, globus pallidus; HIPP, hippocampus; ITC, inferior temporal cortex; LH, lateral hypothalamic area; NAC, nucleus accumbens; NE, noradrenergic inputs; OC, occipital cortex; PMC, premotor cortex; PVN-MH, paraventricular nucleus–medial hypothalamus; SER, serotonergic inputs; ST, nucleus of the solitary tract. (Modified from Hoebel, 1988)

(PMC) is rewarding for feeding. This has been shown by the striking phenomenon of intracerebral self-injection of dopaminergic drugs, for example, amphetamine, cocaine, or dopamine itself. It is an interesting fact that almost all drugs increase synaptic dopamine in the NAC, as measured by microdialysis of the extracellular fluid. Apparently this dopamine modulates a neural loop circuit between the NAC and the PMC. A parallel loop through the striatum is modulated by the nigrostriatal dopaminergic pathway; this system is important for initiating movement and for stereotyped behavior (cf. Chap. 24, Fig. 24.14). It appears that these systems can be artificially engaged, by electrical self-stimulation in the case of the animal experiments or by drugs in the case of people, to produce

motivations toward overeating or drug abuse.

Recent microdialysis studies further suggest that release of dopamine in these systems is controlled in part by learning. A stimulus such as a novel taste was paired with intragastric injection of calorie-rich food. Later, the memory evoked by the taste alone was sufficient to release dopamine in the NAC. Conversely, a taste paired with nausea became a conditioned stimulus for inhibition of dopamine release (see one-trial learning, Chap. 29).

In summary, the LH contains cells and fiber systems that integrate information about energy supplies and send this information through descending pathways to brainstem sensory and motor centers and through ascending systems where it modulates sensorimotor forebrain systems. The forebrain systems reinforce feeding responses and engage the animal in making choices. The animal's learned history of reinforcement plays a role in this integrative and neuromodulatory process.

Modern Concepts of Medial Hypothalamic Function

In recent years, interest in the role of the medial hypothalamus in feeding has been directed toward the paraventricular nucleus. Work by Sarah Liebowitz at Rockefeller University and others has shown that a rich diversity of neurotransmitters and neuropeptides act on the neurons of this region to affect profoundly appetite, energy balance, and body weight. Analysis of these actions has suggested that they reflect separate, though closely interacting, systems that control the intake of carbohydrate, fat, and protein in the diet. We will summarize briefly the evidence relating to carbohydrate and fat (see Liebowitz, 1992).

Carbohydrate Intake Control System. The components of this system include peripheral factors, circulating in the blood, and central actions mediated by neural pathways. As indicated in Fig. 26.12, the peripheral factors include glucocorticoids, such as corticosterone, which act on organs in the body to enhance carbohydrate stores, and which act together with circulating glucose on PVN neurons to potentiate carbohydrate ingestion. Among the central inputs controlling PVN neurons is a system that functions by mean of a combination of noradrenaline (NA), neuropeptide Y (NPY), and GABA. When delivered through cannulas implanted in the PVN, these substances stimulate feeding behavior in rats and other species. The feeding behavior is selective for diets containing carbohydrate, and it involves extension of the amount and duration of feeding, suggesting that these substances attenuate the signals that normally bring about satiety and cessation of feeding.

The peripheral factors that promote satiety include the peptide cholecystokinin (CCK), released by the stomach and intestine during food digestion, as well as insulin and tryptophan. The central factors include the neurotransmitter serotonin (5-hydroxytryptamine; 5HT). Infusion of these substances into the PVN reduces feeding.

Fat Intake Control System. Control of fat ingestion and metabolism is mediated by a different system. Peripherally, the mineralocorticoids are believed to be important in maintaining fat balance, in addition to their more generally recognized function in maintaining salt and water balance. Within the PVN, the critical substances are the peptide galanin (GAL) and the opioid peptides (OP) (see Fig. 26.12), which are colocalized in neurons of the PVN. These substances are believed to act synergistically to stimulate fat intake. A single injection of GAL through a cannula implanted in the PVN leads to an increase in fat consumption lasting through the following day. In addition, it has been found that GAL gene expression and synthesis in the PVN vary in relation to a rat's appetite for fat.

A potent inhibitor of fat consumption is dopamine. Injection of dopamine through an implanted cannula in the hypothalamus reduces fat intake. This suggests that, similar to the suppressive effect of 5HT on carbohydrate consumption, dopamine may oppose the actions of GAL and the opioids to reduce fat consumption.

Temporal Aspects of Feeding Control. In addition to employing distinct neurotransmitter substances, these two systems operate at somewhat different times during feeding. As mealtime approaches, glycogen stores in the body are decreasing, so the first preference is for ingestion of carbohydrate to restore the glycogen levels (see Fig. 26.13). The carbohydrate intake control system is therefore primarily engaged early in the feeding period. In both animals and humans, preference for fat is low early in feeding but rises as feeding progresses, so

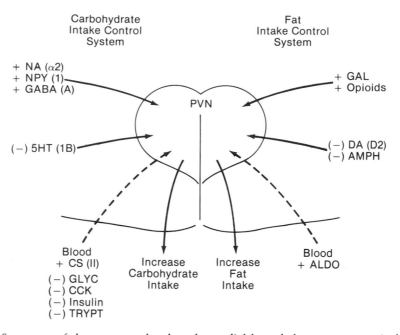

Fig. 26.12 Summary of the systems related to the medial hypothalamus–paraventricular nucleus (PVN) and its control of carbohydrate intake (on the left) and fat intake (on the right). ALDO, aldosterone; AMPH, amphetamine; CCK, cholecystekinin; DA, dopamine (type 2 receptors); GABA$_A$ receptors; GAL, galanine; GLYC, glycogen; NA, norepinephrine (α_2 receptors); NPY, neuropeptide Y (type 1 receptors); TRYPT, tryptamine. (Based on Liebowitz, 1992)

that activity in the fat intake control system is shifted later in the feeding cycle (see Fig. 26.13). The intake of fat in fact is a primary inducer of satiety.

In conclusion, work on neural mechanisms involved in feeding is moving in the same direction as studies of other systems, in showing that an apparently unitary function contains multiple subfunctions, each with its distinct neural basis. In fact, feeding should continue to be one of the most interesting behaviors for analyzing these properties, considering its critical importance for the organism and its intimate relation to emotional and motivational states. This was recognized many years ago by Jacques LeMagnen (1971) of Paris, a pioneer in the physiological and behavioral analysis of feeding, and a wise observer of both laboratory animals and his fellow countrymen:

Fig. 26.13 Circadian rhythms of neurochemicals related to the control of feeding in the rat. Solid line, substances related to carbohydrate intake; dashed line, substances related to fat intake. Active feeding period of the rat is during the night (indicated by shaded bar). For abbreviations, see legend to Fig. 26.12. (Modified from Liebowitz, 1992)

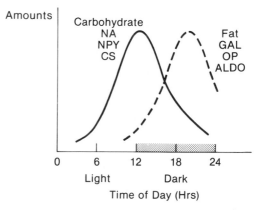

Food intake is not . . . separately regulated. . . . This field . . . represents perhaps the most advanced point in the study of a multifactorial physiological regulation. . . . A multifactorial system of regulation, in which the play of a network of positive and negative feedback mechanisms enables a body of variables to be simultaneously causes and effects, requires from the experimenter a kind of new mental program for collecting and processing data, and yet difficult to build and handle.

Drinking

Closely allied to feeding is drinking. Cells universally contain water, and metazoan animals universally consist of cells bathed in extracellular fluid. Maintenance of the fluid composition of both cells and extracellular fluid is therefore basic to animal life. As animals evolved from aquatic to terrestrial environments, homeostatic mechanisms became necessary for ensuring adequate fluid intake to offset dehydration and excretion. The need for intake is expressed as thirst, and is met by drinking.

In an animal deprived of water, the extracellular fluid begins to become more concentrated, and one says it is hyperosmotic; since the extracellular and intracellular compartments are in equilibrium across the cell wall, the cell cytoplasm also becomes hyperosmotic. In 1953, Bengt Andersson in Sweden showed that injections of hyperosmotic solutions into the hypothalamic region of goats induces drinking, suggesting that there are *osmoreceptors*, sensitive to cellular dehydration, in that region. This is a logical place to expect such receptors, near the location of the cells to the posterior pituitary (see Chap. 24) that secrete antidiuretic hormone (ADH), which stimulates the kidney to retain water. Subsequent experiments by Elliott Blass and Alan Epstein at the University of Pennsylvania in 1971 showed that the osmoreceptors are distributed rather widely throughout the hypothalamus, especially in the preoptic and lateral hypothalamic areas (see Fig. 26.14A). These and

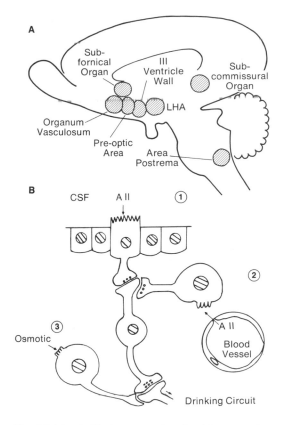

Fig. 26.14 **A.** Brain centers involved in control of drinking. **B.** Microcircuit of central receptors for angiotension (AII). Site ①, receptor in the wall of the cerebral ventricle. Site ②, receptor in circumventricular organ, where there is no blood–brain barrier. Site ③, osmoreceptor in the hypothalamus. Drinking circuit includes cells projecting to the neural lobe of the pituitary, where vasopressin (antidiuretic hormone) is released. See text (B from Phillips et al., 1977)

other workers have found that microinjections of mildly hypertonic saline into these areas readily induces drinking, as do injections into the carotid arteries which supply these areas; conversely, lesions of these areas impair drinking behavior.

In addition to cellular dehydration, a loss of extracellular fluid volume is also a stimulus to drinking. This involves the *angiotensin* system. Knowledge of this system began with the finding that a reduction of blood volume stimulates cells of the kidney to release the enzyme renin into the bloodstream. Renin acts on a circulating peptide, angiotensinogen, to make angiotensin I, which in turn is converted to angiotensin II (A II) (see Fig. 26.14B). In 1970, Epstein and his colleagues made microinjections of A II into the hypothalamus, and showed that this is a powerful stimulus for drinking. Subsequent experiments have shown that all the components of the renin–A II system are also present in the hypothalamus. The preoptic area is particularly sensitive to A II, as are several sites along the cerebral ventricles; these are indicated in Fig. 26.14B.

What are the relative contributions of osmoreceptors and A II receptors to normal drinking? In experiments carried out by Barbara Rolls, Ed Rolls, and Roger Wood at Oxford, it was found that about 65% of the drinking following water deprivation is due to cellular dehydration, and about 25% is due to blood volume reduction. This is in the rat; in the monkey, the proportions were 85% and 5%, respectively. These and other experiments have supported the idea that the osmoreceptors are more important, and that the A II receptors are supplementary, perhaps acting as an emergency system invoked under conditions of extreme deprivation or blood loss.

Water deprivation is an easy variable for the experimenter to manipulate. An animal deprived of water has a strong *drive,* or *motivation,* to correct the deficit by drinking. In behavioral experiments, the drive can be quantitated by testing how hard the animal will work, or how intense the aversive shocks it will sustain, in order to obtain water and correct the deficit. Water deprivation is thus a convenient variable for studying both physiological and behavioral mechanisms.

Among humans, much of our water intake in daily life is actually guided by other factors. We drink fluids with our meals; we drink coffee because it is time for a coffee break; we drink at cocktail hour to be sociable; we drink when our mouths feel dry; we drink because it tastes good; we drink because we are bored. These different behavioral states depend on different kinds of receptors: taste receptors in the mouth and pharynx; somatosensory receptors in the mouth, pharynx, and esophagus; stretch receptors that sense the amount of distention in the stomach and duodenum; stretch receptors and osmoreceptors in the hepatic portal veins that return blood from the liver to the heart. There is thus a complex array of signals that determines the pattern of normal drinking. This pattern is shaped by our learned habits, habits that in effect anticipate need, and keep our bodies well hydrated so that we do not have to respond in situations of need or emergency.

27

Visceral Brains: Mating

Mating is the function that is necessary for the propagation of the species. The object of mating is characteristically a copulatory union which brings the sexual organs of male and female together, so that fertilization of the female's egg or eggs by the male's sperms can occur. This requires maturation of the gonads and the copulatory organs, and timing of the preparedness of the male and receptivity of the female. In those animals in which fertilization occurs within the female, a series of internal body changes then takes place during gestation, in preparation for birth.

Mating, like feeding, takes place under a combination of nervous and hormonal control. Much of this control is mediated by parts of the nervous system within the visceral brain. In the case of feeding, we saw that this part of the brain is responsible for a variety of different control mechanisms. Similarly, in the case of mating, the visceral brain mediates the most delicate and ingenious mechanisms for bringing about copulatory unions and the mixing of genes. This part of the brain also is involved in controlling other aspects of the reproductive cycle, such as the maturation of gonads and behavior related to maternal care. Many of these functions are under combined hormonal and neural control. Modern studies, employing a variety of molecular tools, have begun to reveal the mechanisms of hormone actions at the level of the genome which mold the circuits that mediate these controls. In addition, they have begun to show a surprising overlap between hormone and synaptic actions at the level of the neuronal membrane which mediate more rapid responses contributing to these functions. In this chapter we will focus on the neural controls involved in mating.

Reproductive Strategies

To begin with, let us get a perspective on the strategies of sexual reproduction. In some organisms, each individual is a hermaphrodite, which means it contains both male and female sexual organs. This has the advantage that any two individuals can meet and reproduce. Examples include the earthworm, the sea hare *(Aplysia)*, and, among vertebrates, a number of species of fish. In these cases, nervous mechanisms appropriate for both male and female mating behavior must be present. This has the disadvantage of limiting the degree of specialized sexually related behavior that is possible, which may be part of the explanation for why this form of sexual reproduction is not more widespread.

Among vertebrates, the potential for hermaphroditism is expressed in the ability of certain species to undergo sex reversal. For example, Robert Goy and Bruce McEwen, in their book on *Sexual Differentiation of the Brain* (1980), cite studies in which genetically female fish, frogs, or salamanders grown in water containing testosterone develop into males that mate with normal females, and, conversely, males grown in water containing estradiol develop into females that mate with normal males. Even more amazing are studies on certain species of fish that live in social groups; all the members of the group develop as females, and only a few differentiate into males. If an adult male is lost, within a week or two one of the adult females differentiates into a male to take its place! A contrasting species of fish is also known, in which a group has a single female; if this female is lost from the group, a male differentiates into a female to take its place. These studies have thus provided evidence for the ability of social factors to control the differentiation of sexes, presumably through effects on nervous mechanisms and the organizational actions of sex hormones.

Among higher invertebrates and vertebrates, the common pattern is for sexual differentiation to occur early in development, and for the male and female forms to be stable throughout life. This reflects the adaptive advantages to be gained from specialization of each sex for its role in the reproductive behaviors revolving around courtship, mating, and care of the young. We have already mentioned an example of specialized nervous mechanisms in the pathways for control of singing in male birds (Chap. 23), and we will discuss several further examples below. We will also discuss the mechanisms for development of these differences in neural circuits and see that they involve a series of changes under control of the organizational effects of hormones.

Courtship

Given the importance of reproduction for the species, mating is not an activity that is entirely left to chance. It requires, first, a careful orchestration in the development of the male and female partners. During development and maturity, animals of course spend much of their time looking for food and trying to avoid being eaten themselves. As a general rule, mating is not a part of this daily business of survival. Instead, it takes place at very prescribed times. The time is set by the biorhythm generators (see Chap. 25), the maturation and readiness of the gonads and secondary sex organs, and the differentiation of the parts of the neural pathways for controlling the appropriate behavior. These are basic principles that apply more or less to all animals, invertebrates and vertebrates.

A crucial task for the neural apparatus is the control of an appropriate series of behaviors that will bring the male and female together. A basic problem is that in their daily business of survival, animals spend much of their time engaging in behaviors that are either aggressive or defensive. In order to mate, the male and female must overcome these attitudes and, at least briefly, trust each other. Furthermore, there must be an opportunity for each of the partners to test the other for its attractiveness as a source of sound genes; it is a waste to mate with a partner that is sick, weak, or maladapted, but an advantage to combine one's genes with those of a partner that is healthy, strong, and with optimal abilities to survive and flourish.

For these reasons, it is common for mating to involve a sequence of interactions between the prospective partners, during which the aggressive and defensive instincts are subdued, and attractiveness as partners is tested. Many species are every bit as meticulous in this process of choosing mates as humans are. Since we view this through human eyeglasses, we call it "courtship." Ethologists have studied these sequences in many species, and have devel-

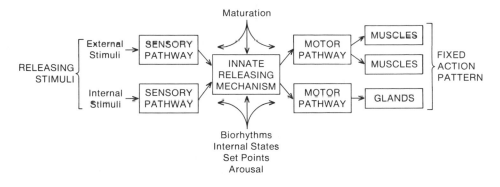

Fig. 27.1 Details of one step in a reaction train. (Adapted from Tinbergen, 1951)

oped flow charts for representing them. The whole sequence involves a number of specific motor acts and sensory stimuli that, taken together, are specific for a species, and ensure the suitability of the partners.

In the language of ethologists, sequences of this type are called a *reaction train*. Each behavior in the sequence is thought to be activated or "released" by a set of neuronal centers and their connections, called an *innate releasing mechanism*. The stimuli that activate these mechanisms are called *releasing stimuli* (see Chap. 10). It can be further realized that each behavior has the character of a *fixed-action pattern* (FAP), that is, a particular pattern of postural attitude, glandular secretion, or motor activity, such as that discussed in Chap. 19. Each of these motor acts in turn generates the stimuli which release the next act in the chain. The details of each step in the sequence are summarized in Fig. 27.1.

It should be emphasized that any given step in a reaction train is a probabilistic, not a rigidly determined, event. This, in fact, is the whole point of the sequence: any given step serves as a test for whether conditions are just right, so that the next step can take place. This is the means for assuring optimum gene transmission, as described above.

The releasing stimuli for a given behavior act on sensory receptors which in turn set up input in sensory pathways, as already discussed in previous chapters in sensory systems. In some cases it is the whole pattern of stimulation that is important, such as visual recognition of a partner or a prominent marking in a partner. In other cases, it is a conjunction of two or more sensory modalities, such as tactile and chemosensory stimuli. In many cases it involves stimulation only of a specific type of receptor.

Neural Mechanisms

The fact that mating commonly requires elaborate courtship or other types of approach behavior means that the nervous systems must be differentiated between male and female for their respective roles in these behaviors. In addition, proper differentiation of the reproductive organs must take place, together with the neuroendocrine systems for controlling them.

Our knowledge of neural mechanisms mediating mating behavior in vertebrates draws on work in many fields. There is a long tradition of biochemical work, beginning early in this century, on isolation and characterization of the *hormones* secreted by the reproductive organs and the hypothalamopituitary region. There is an enormous literature on the behavioral effects of *removal* of these organs in experimental animals, and the effects of *diseases* of these organs in human patients. In recent years, the identification of hypothalamic releasing factors and the localization of these and

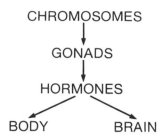

Fig. 27.2 Steps involved in sexual differentiation.

related *peptides* in many parts of the nervous system has greatly extended our concepts of the extent of neural systems involved in control of reproductive processes. Neuroanatomical studies have revealed a number of sites at which the organizational effects of hormone actions bring about *sexually dimorphic regions*. Neurophysiological studies have begun to reveal the time course and *mode of action* of nerve cells that mediate reproductive control. Finally, the *genes* for the hormones are being cloned and sequenced, and the sites of hormone *synthesis* and *reception* are being localized by in situ hybridization.

Against this background of multidisciplinary studies, let us consider first the ways by which male and female animals achieve their final adult forms, and then consider examples of mating in several species.

Sexual Differentiation

Sexual differentiation is brought about by a sequence of steps. As summarized in Fig. 27.2, the general scheme is that the chromosomes carry the code for male or female, which determines the type of gonad, male or female. The gonad synthesizes and secretes hormones which organize the secondary sex characteristics of the body. In addition, the hormones act on the brain to organize or modify circuits for appropriate nervous control of the sex organs and related behaviors.

An important generalization for under-

standing sexual differentiation is that, in higher vertebrates at least, all individuals will develop as one sex unless the chromosomal condition for the opposite sex is present and is expressed during a critical period of development. In birds, the sex chromosomes of the male are ZZ, whereas in the female they are ZO; hence birds are regarded as basically male, unless one of the sex chromosomes is missing. It is believed that ZZ inhibits the primordial ovary, allowing the testes to develop; in the case of ZO, the ovary develops while suppressing the primordial testis. By contrast, in mammals the sex chromosomes of the female are XX, whereas in the male they are XY; hence, mammals are regarded as basically female unless the Y chromosome is present. The classical idea that female development represents a passive "default condition" is being replaced by recognition that both male and female development are gene-directed processes involving active interactions between germ cells and somatic cells (see Gilbert, 1991, for further discussion).

Development of the Gonads

The differentiation of the sexes begins early in fetal life. We can illustrate the sequence of changes for the rat, a well-studied vertebrate (Fig. 27.3). The rat has a gestation period of about 21 days. By embryonic day 11 (E11), a genital ridge has formed in the lower abdomen (A). Germ cells, male or female, which have been lurking in the wall of the gut, migrate into this ridge, and bring about a proliferation of cells to form a primitive though undifferentiated gonad (B). The germ cells are grouped here into clusters, called *primary sex cords*. At about 13 days, the gonad begins to differentiate. In a genetic female, the primary sex cords migrate to the inner, medullary part of the gonad and degenerate. They are replaced in the outer, cortical part by secondary sex cords, which give rise to the egg cells *(oocytes)* (D). In the genetic male, the primary sex cords give rise to the seminiferous tubules, which contain in their walls the

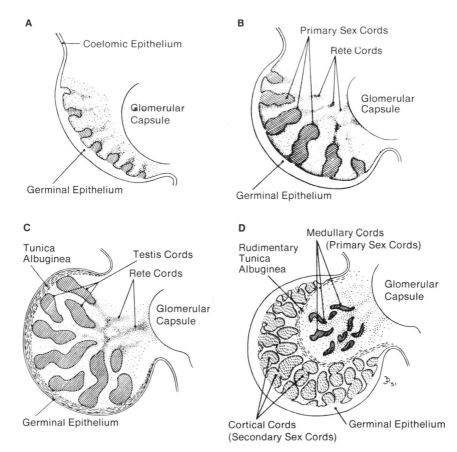

Fig. 27.3 Differentiation of gonads in vertebrates. See text. (From Burns, in Romer and Parsons, 1977)

germ cells which give rise to spermatozoa (C).

Critical Period

As soon as the male gonad begins to differentiate into a testis, interstitial cells within the testis begin to synthesize and secrete the hormone testosterone. Synthesis in the rat has been detected as early as day 13. Testosterone levels peak twice during development, just before (E18–19) and just after birth; both of these peaks are lacking in the female (reviewed by MacLusky et al., 1985). These bursts of testosterone define what is called a *critical period,* during which the message of the genetic sex is carried to the body; the testosterone acts on intracellular receptors and causes the

development of male secondary sex organs and the differentiation of certain parts of the brain responsible for male behavior patterns.

Experimental manipulations, such as castration or hormone injections, have profound and enduring effects on both the sex organs and the brain if carried out during the critical period, up to about days 5 postnatally; after this, they have little effect. This critical period is relatively brief in the rat; in species with longer gestation periods, such as the guinea pig (67 days) or monkey (160), the period may be longer, occurring mostly before birth. In some species the end of the critical period occurs at about the time postnatally when the eyes open, which has been taken to indicate that sexual differentiation takes place before neural pathways in the brain are fully functional, and hence are still more readily modifiable.

Fred Naftolin and his colleagues proposed in 1975 that much of the masculinization effects produced by testosterone are actually due to conversion of testosterone to estradiol (reviewed by MacLusky, et al., 1985). The conversion involves, biochemically, the aromatization of the hormone molecule (Fig. 27.4), and one says therefore that estrogen can act as an androgen through aromatization.

Within the brain there is a sex difference in aromatase activity during development: during the first few days of postnatal life, the levels of aromatase activity in the corticomedial amygdala and the hypothalamus are higher in males than in females. It is speculated that "early sex differences in testosterone levels could sensitize the male to subsequent androgen exposure," and that "such sensitization could occur at least in part through increased estrogen formation" in these regions (MacLusky et al., 1985). Thus, the testosterone peaks may have a "priming effect" of inducing the aromatase enzyme in the cell cytoplasm, so that these cells would be more sensitive to further testosterone actions in the male. In addition, the several pathways for metabolizing testosterone (Fig. 27.4) yield a variety of active metabolites within the target cell. As Goy and McEwen have pointed out (1980), this may be viewed

as a means to diversify the action of a single hormone by producing . . . agents able to interact at different points in intracellular metabolism. It may provide a mechanism for achieving high concentrations of specific metabolites at discrete sites and for modulating hormone action through the regulation of the activity of the metabolizing enzymes. Not surprisingly, the CNS may prove to be the target organ in which these possibilities are most extensively exploited.

These mechanisms provide means not only for diversifying the effects of testosterone, but also for protecting the male embryo from circulating estrogens from the mother.

Mechanisms of Steroid Action

We next inquire into the mechanisms of action of these hormones on their target cells. Because the steroid hormones are lipid-soluble, they readily pass through cell membranes. The traditional view was that the sex hormones bind to receptors in the cytosol, where they exert their long-term effects on sexual differentiation and related cellular changes. They thus were believed to differ markedly from neurotransmitters acting on membrane receptors to produce rapid changes in electrical activity, as we discussed in Chaps. 7 and 8. This view has given way in recent years to the evidence that hormones have both long-term *organizational* and short-term *activational* effects.

Molecular and Cellular Mechanisms

Our current view of the mechanisms of steroid action is summarized in Fig. 27.5 for the example of progesterone.

Intracellular Mechanisms. After diffusing into the cell, progesterone binds to a receptor which is also a DNA transcription factor of the type we discussed in Chap. 2. The progesterone transcription factor is a member of the large family of steroid transcription factors, which includes receptors for glucocorticoids, mineralocorticoids, estrogen, vitamin D, and retinoic acid, as well as for thyroid hormones (summarized in Hall, 1992). Each factor contains three main regions, which, from the N terminal, consist of domains for transcriptional regulation, DNA binding, and hormone binding. The hormone-binding domain is specific for a given hormone, and the DNA-binding region is specific for a given nucleotide sequence in the DNA. Binding of the hormone brings about an allosteric (conformational) change in the molecule, which leads to DNA binding and activation of transcription. DNA binding is mediated

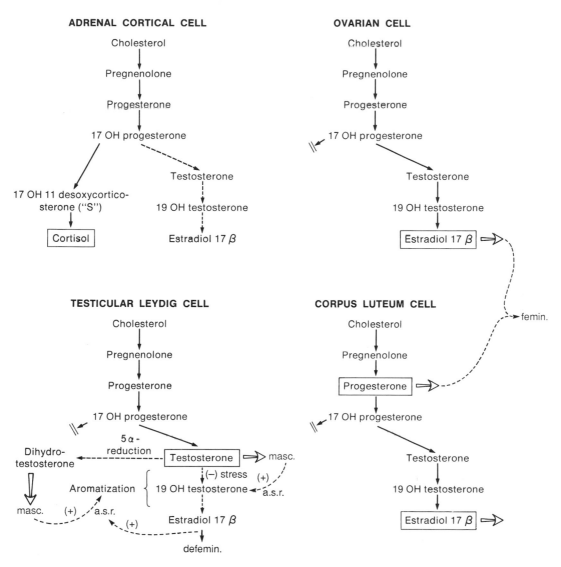

Fig. 27.4 Metabolic pathways for sex hormones. a.s.r., androgen steroid receptor; femin., feminization; defemin., defeminization; masc., masculinization. (Based on Keele and Sampson, in Gordon, 1972; N. MacLusky, personal communication)

through the zinc finger motifs discussed in Chap. 2.

Hormone actions mediated by these genomic mechanisms have effects on the differentiation and growth of cells, which occur over time scales from hours up to years, as indicated in Fig. 27.5. The very briefest actions of some hormones involve immediate early genes, such as c-*jun* and c-*fos*, that may be expressed within minutes of hormone entry into a cell.

Membrane Mechanisms. By contrast, the actions of hormones on membrane receptors have not yet been isolated and characterized because of the high degree of nonspecific binding by these lipid-soluble hormones, and by the low density of receptors. Their presence is indicated by two main lines of evidence (McEwen, 1991). Numerous biochemical studies have provided suggestive evidence for binding of steroids to membranes of various types of

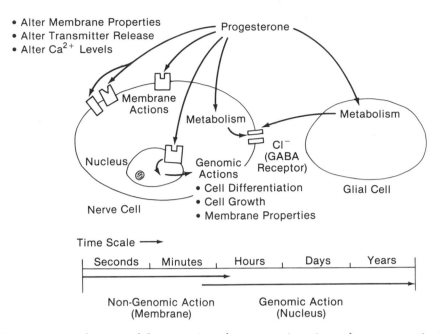

Fig. 27.5 Summary diagram of the genomic and nongenomic actions of a representative hormone (progesterone) at the cellular level. Some of the characteristic actions at the membrane (nongenomic) and nucleus (genomic) are indicated, as well as possible actions of metabolites on membrane properties. Below is shown a time scale on a logarithmic base, with arrows indicating the range of genomic and nongenomic actions; note the overlap in the range of minutes to hours. (Adapted from McEwen, 1991)

cells. And numerous electrophysiological studies have shown that hormones have effects on membrane properties that occur too quickly to be mediated by genomic actions. These nongenomic effects include alterations in Ca^{2+} entry and mobilization, modulation of GABA receptors (see Chap. 8), and changes in excitable membrane properties. They occur over seconds and minutes, as indicated in Fig. 27.5.

Both long- and short-term hormonal actions have been demonstrated on single-channel activity in isolated cell preparations (reviewed in Erulkar and Wetzel, 1987), and can be demonstrated in the hippocampal slice, as shown in Fig. 27.6 (Wong and Moss, 1992). In these experiments, intracellular recordings were obtained from pyramidal cells in the slice preparation previously illustrated in Chap. 7. The slices were prepared from female rats whose ovaries had been removed surgically (ovariectomized) two weeks previously. The recordings from slices from these animals showed a single impulse response to synaptic excitation of increasing strengths (A, a, Fig. 27.6). By contrast, in slices from animals which received a subcutaneous injection of estrogen two to three days previously, the same stimuli produced larger, longer EPSPs and repetitive impulses in a significant number of the cells (A, b, Fig. 27.6). This priming action of estrogen, lasting over several days, was presumably mediated genomically.

Short-term effects were demonstrated by introducing estrogen directly into the bath containing the slice. As shown in B of Fig. 27.6, this was followed by increases in EPSPs and impulse responses within seconds, and these effects were specific for certain glutamate receptors (AMPA, kainate, quisqualate) but not others (NMDA). The authors concluded that the estrogen

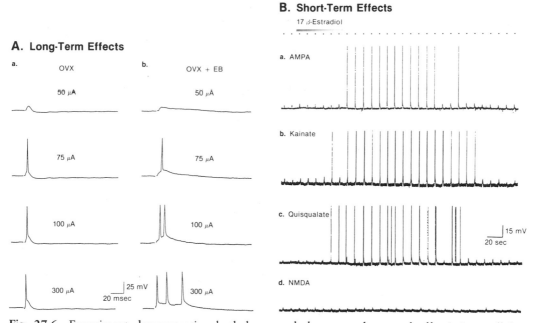

Fig. 27.6 Experiment demonstrating both long- and short-term hormonal effects. Intracellular recordings were made in the hippocampal slice from pyramidal cells responding synaptically to Shaffer collateral stimulation (same setup as shown in Chap. 7). **A.** Long-term effects investigated in ovariectomized (OVX) female rat. a. Increasing stimulus strengths (50–300 μA) gave increasing EPSPs but only a single impulse. b. In rats primed with estradiol (EB), the same strengths of stimuli gave larger EPSPs and multiple impulse responses. **B.** Short-term effects. Estradiol was introduced directly into the bath; this caused a potentiation of the non-NMDA responses (a–c) but not the NMDA response (d). (From Wong and Moss, 1992)

was probably acting postsynaptically on these non-NMDA receptors. These effects might be produced by direct binding of estrogen to a membrane protein associated with, or part of, these receptors. In other types of neurons, estrogens have been found to modulate K channels or second messenger systems such as cAMP.

In conclusion, the actions of steroid hormones now embrace both long-term organizational changes and short-term activational changes in membrane properties. One might ask why a single signal molecule should have such diverse actions. The answer may be that the special functions of these hormones are to bring about far-reaching changes in sexual differentiation which underlie the complex behaviors related to mating and reproduction, and that this requires multiple and coordinated con-

trols of many types of cells acting over many different time scales. We thus see in the actions of these single molecules the kind of spatial and temporal integration that we discussed in Chap. 8 (Fig. 8.15).

Sexual Dimorphisms

We have seen that there is considerable biochemical evidence for sexual differentiation of the brain. We next consider how this produces differences in brain structures and sexual behavior.

The behaviors of male and female organisms are necessarily different during mating. The early idea (the "peripheral hypothesis") was that these behaviors reflect the differences in the sexual organs and their hormonal control through the pituitary. In this view, the decrease in male sexual behavior indiced by castration of a newborn male rat, for example, is due to the underdeveloped penis rather than any effects on the brain. In contrast is the idea that there are differences in brain mechanisms in the two sexes (the "central hypothesis"). The first evidence for this was obtained by a closer examination of the behaviors of young animals.

It is typical of many mammalian species that the young engage in rough-and-tumble play among themselves. This play includes attempts at *mounting* and receptive displays involving downward bending of the back and exposure of the bottom, termed *lordosis*. Both sexes engage in these behaviors, but normally males tend to be more active in mounting and females in showing lordosis.

In 1959, Charles Phoenix and his collaborators at the University of Kansas reported studies in which they injected a small dose of testosterone into pregnant guinea pigs, and observed a reduction in lordotic behavior in the female progeny. This they interpreted as a *defeminizing* effect of the androgenic hormone on central brain mechanisms. In addition, guinea pigs ovariectomized at birth and subsequently injected with testosterone showed more mounting behavior, which was interpreted as a *masculinizing* effect of the hormone on the brain. Since that time, numerous studies have confirmed the generality of these two types of effects of male hormones on behavior patterns (see also Chaps. 28 and 30).

If behavior is mediated by the brain, then it should be possible to find differences in brain structure in the centers and pathways that mediate male or female behavior. At first the possibility of identifying such differences seemed remote, but the pioneering work of Geoffrey Raisman and Pauline Field, then at Oxford, in the early 1970s, showed that such differences could be found, and at the level of resolution of the electron microscope. They studied the preoptic nucleus, a region in the basal fore-brain that is involved in control of biorhythms, including the estrous cycle (see below). The nucleus receives inputs from the amygdala, a structure in the limbic system (see next chapter), as well as from other sources. After placing lesions in the amygdala, they found more nonamygdalar nondegenerating synapses on dendritic spines in the nucleus of females than in males. These results are summarized in Fig. 27.7A. This study thus showed a difference in synaptic connections in males and females, a difference that depends on exposure to sex hormones early in life. The preoptic area is involved in control of the surge in secretion of LH that underlies the estrous cycle in females (see below), and the differences in synaptic connections may be related to this control.

Stimulated by these results, research workers have looked for and found differences between males and females in several other parts of the brain. These differences in brain structure are referred to as *sexual dimorphisms*. We have already mentioned in Chap. 23 the differences in binding of sex hormones and in the size of certain centers in the pathways controlling singing in male song birds. In the mammal, various nuclei of the hypothalamus show differences in cell size and sex hormone binding. Study of Golgi-impregnated neurons in the preoptic area of the hamster has suggested that neurons in the male have dendrites more oriented toward the center of this area, whereas in the female they are more oriented toward the periphery (see Fig. 27.7B). In another part of this area, there is a nucleus of cells that is up to eight times larger in males than in females (Fig. 27.7C). In addition to these structural dimorphisms, differences in functional properties of neurons have also been found; for example, amygdalar stimulation is more effective in driving preoptic cells in males than in females.

These classical studies have been extended by much more work, which has provided evidence for sexual dimorphisms

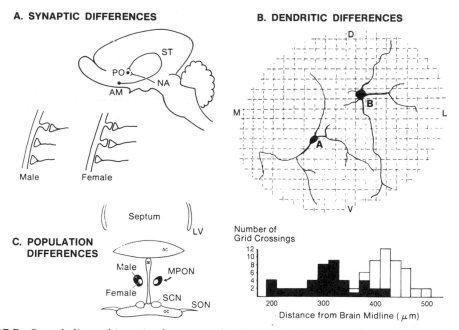

Fig. 27.7 Sexual dimorphisms in the mammalian brain. **A.** Summary of experiments in the rat, showing that nonamygdalar (NA) fibers make more synapses on dendritic spines in the preoptic (PO) nucleus of females than males. AM, amygdala; ST, stria terminalis. **B.** Two Golgi-impregnated neurons in the preoptic area of hamster brain, illustrating that neurons in males (A) have dendrites more oriented toward the center of the area, whereas in females (B) the dendrites tend to be oriented toward the periphery. **C.** Medial preoptic nucleus (MPON) in the rat, showing that this region is larger in the male than in the female. AC, anterior commissure; LV, lateral ventricle; OC, optic chiasm; SCN, suprachiasmatic nucleus; SON, supraoptic nucleus. (A based on Raisman and Field, 1971; B from Greenough et al.; C from Gorski et al., in Goy and McEwen, 1980)

in other regions and in other cell characteristics. Bruce McEwen's laboratory has reported differences in densities of dendritic spines in hippocampal neurons, implying differences in synaptic actions and dendritic integration in this region that is believed to be involved in memory storage (see Chap. 29). Patricia Goldman-Rakic's laboratory has found differences in the development of frontal lobe functions related to the performance of cognitive tasks. And Simon LeVay (1992) has provided evidence for differences in brains of homosexual and heterosexual humans. In summary, this is an active area of research, which should give us a much better understanding of what makes male and female brains differentiated for the special functions related to

mating and reproduction while united in the common nature of being human.

Brain Mechanisms in Mating

Among the behaviors that are essential for mating to occur in many vertebrates are *mounting* by the male and *lordosis* by the female. Experiments have begun to identify the specific nervous pathways that are involved in the control of these behaviors.

Mounting by the Male Frog

Mating in the frog takes place in the following manner. After suitable courtship preliminaries, the sexually mature male mounts the receptive female from behind

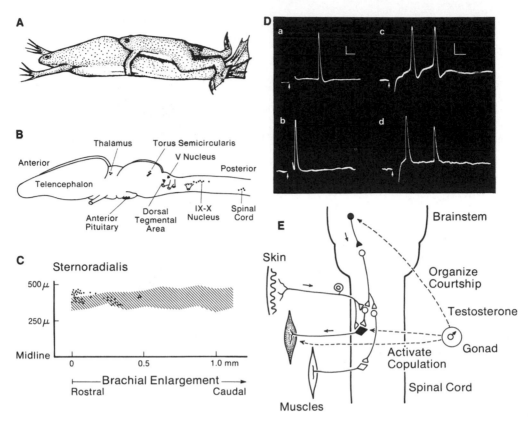

Fig. 27.8 Neural mechanisms mediating the clasp reflex of the male frog *Xenopus laevis*. **A.** Diagram of male clasping the female. **B.** Sites of binding of dihydrotestosterone to neurons in the central nervous system of the male frog. Binding in the auditory pathway (torus semicircularis) and the motor pathway controlling the larynx (dorsal tegmental area and IX–X nucleus) is related to mate calling during courtship. **C.** Localization of testosterone binding in the spinal cord. The shaded area indicates the distribution of the entire population of motoneurons innervating the sternoradialis muscle of the forelimb; dots indicate androgen-concentrating cells within that population. **D.** Intracellular recordings from a single sternoradialis motoneuron. a, b. Responses in spinal cord of castrated male to single stimulus delivered to dorsal root (a) and sternoradialis nerve (b). c, d. Response in spinal cord of a clasping male. **E.** Neural circuit mediating the male clasp reflex, showing sites of organizational and activational actions of androgen hormones. (A from Russell, 1964; B from Kelley, 1980; C, D from Erulkar et al., 1981)

so that their pelvises come together and fertilization can occur, a position known as amplexus (see Fig. 27.8A). The transfer of sperm takes many hours, which puts the male at risk because other males may come along and try to dislodge him, and the female may try to throw him over for more attractive paramours. In order to maintain his position, the male clasps the female firmly with his forelimbs. This clasp is a remarkable reflex, which can be main-

tained for up to 30 hours. The spinal nature of the reflex is shown by the fact that after decapitation a male may continue to clasp for several hours, an experimental observation that was first made in the eighteenth century.

Darcey Kelley at Princeton and Sol Erulkar at the University of Pennsylvania carried out a multidisciplinary study of the spinal circuits. The results may be summarized as follows:

Anatomy. The main muscles used by the male in clasping are the sternoradialis and flexor carpi radialis of the forelimbs. HRP injections into these muscles showed their motoneurons to be localized in the brachial enlargement of the spinal cord.

Steroid Uptake. Castrated males were injected with dihydrotestosterone (DHT), one of the active metabolites of testosterone (see above). In addition to uptake in cells of the hypothalamus, uptake was found in cells of the spinal cord, among the motoneuron populations that mediate the clasp reflex (Fig. 27.8 B, C).

Facilitation of Motoneuron Activity. Intracellular recordings were obtained from sternoradialis motoneurons in males induced to clasp by injection of gonadotropin, and compared with recordings from these motoneurons in castrated males. As shown in Fig. 27.8D, in clasping males the responses showed larger EPSPs and multiple spiking. This effect of androgens may be due to facilitation of transmitter actions, or to increased excitability of the membrane (see similar results in Fig. 27.6 above). Related experiments showed that the facilitation is blocked by cycloheximide, a protein synthesis inhibitor, suggesting that the androgen action is mediated through protein synthesis.

Enzyme Localization. The criteria for identifying neuroactive hormones parallel those for neurotransmitters (see Chap. 8). One of the criteria is the presence of appropriate enzymes. A search was therefore carried out for the presence of 5α-reductase, the enzyme that converts testosterone to DHT (see above). Segments of the spinal cord were homogenized and incubated with tritiated testosterone. Thin-layer chromatography showed substantial 5α-reductase activity in the spinal segments innervating the clasping muscles. The results are consistent with the idea that testosterone may affect clasping motoneurons by conversion to DHT.

Muscle Types. The types of muscle fibers in clasping muscles have been characterized by observing the binding of antibodies to fast myosin (present in fact twitch muscles) and slow myosin (present in slow twitch muscles). The proportion of slow myosin increases in males during the breeding season. This may be due in part to a direct action of androgens on the muscles, but Erulkar and his colleagues speculate that the changes induced in motoneuron activity by the binding of androgen, as described above, may be the main determinant of the myosin type in the muscles which these motoneurons innervate.

A summary of the neural and hormonal elements involved in male clasping behavior is shown in Fig. 27.8E. Note the two main neural targets of androgens, one in the hypothalamus to prepare the male for courtship behavior, the other in the spinal cord to prepare for clasping and copulation. In this way the spinal circuits are primed for mediating the clasp reflex; in ethological terms, the fixed-action pattern (clasping) is released by appropriate releasing stimuli (tactile stimulation). Clasping behavior is thus programmed into the spinal circuits, much as we have seen to be the case with motor activities related to posture and locomotion. The circuits thus appear to be constructed along the same principles as those controlling other motor behavior.

Lordosis in the Female Rat

Mating in the rat also involves mounting by the male, but copulation, by contrast, is extremely rapid: the entire sequence of mounting, thrusting, ejaculation, and release takes less than a second! The female's cooperation is critical, and the main motor activity which she must coordinate with mounting is lordosis. It has been known since the pioneering experiments of Frank Beach, then at Yale, in the 1940s, that lordosis is under control of female gonadal hormones. The neural circuits that mediate lordosis were extensively studied by Don Pfaff and his collaborators at Rockefeller

University. This work also demonstrated the usefulness, indeed the necessity, of a multidisciplinary approach. The results will be summarized briefly, with reference to Fig. 27.9.

Lordosis is a reflex response to tactile stimulation by the male's body against the female's rump region ①. The main tactile receptors are believed to be Ruffini endings (see Chap. 12), responding to pressure stimulation with a slowly adapting barrage of impulses in the sensory fibers to the spinal cord.

Since lordosis does not occur if the spinal cord is transected (in contrast to frog clasping discussed above), the reflex requires supraspinal connections. The ascending information is carried in fibers in the anterolateral columns ② to three sites: the lateral vestibular nucleus and the medullary and midbrain reticular formations.

Lordosis depends on estrogen; fully developed lordosis normally occurs only in mature females adequately primed by estrogens. The estrogens are primarily in neurons in the ventromedial hypothalamus and related regions (③; see also next section). The effect on these cells is to raise the level of tonic impulse activity, thereby facilitating the midbrain neurons to which these cells project.

When adequately facilitated by hypothalamic inputs, the midbrain neurons respond to the sensory inputs. Through a relay in the medullary reticular formation, this descending pathway ④ completes the reflex pathway to the spinal cord. The descending excitation is combined with sensory inputs to control the muscles of the back that execute lordosis ⑤. The lateral vestibular nucleus contributes to this control by raising the tone of postural muscles.

In this system we see again the principle of hierarchical organization; the basic circuit for controlling the pattern of motor activity is present at the spinal level, and is modulated, gated, or activated by the higher centers. It is also worth pointing out that lordosis involves primarily axial muscles; thus, as in the case of respiratory and vocal activity, a delicate control of

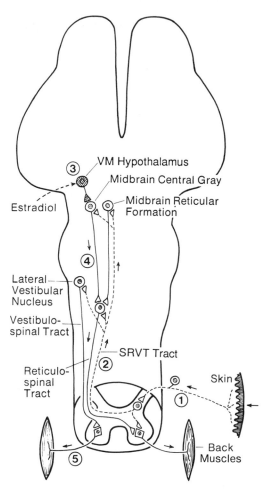

Fig. 27.9 Neural circuits mediating lordosis in the female rat. Sequence of actions (①–⑤) is described in text. SRVT, spino-reticular-vestibular tract (After Pfaff and Modianos, in Pfaff, 1980)

axial muscles is necessary, rivaling that of fine control of the extremities (cf. Chap. 22).

Neural Control of Gonadotropin Secretion in the Female Rat

It remains to consider the neural factors that affect the hormonal status of the female and its preparedness for mating.

In mammals, most females upon reaching maturity undergo cyclic changes in their preparedness to mate and produce offspring. This is termed the *estrous cycle*. It

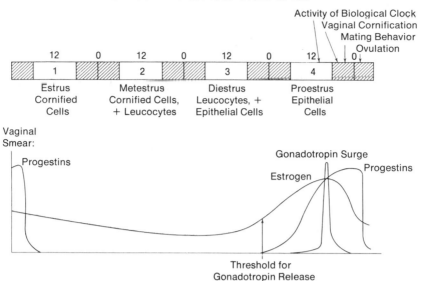

Fig. 27.10 The estrous cycle in the laboratory rat. (After Gordon, 1972)

reflects changes in the levels of the gonadal hormones, estrogens and progestins, secreted in the ovary, which in turn reflect changes in stimulation of the ovary by gonadotropin (FSH and LH) secreted by the pituitary. The way these levels change during the estrous cycle of the rat is shown in Fig. 27.10. Mating behavior must be critically timed in relation to the cycle; as indicated in the figure, mating must occur within several hours after the LH surge, and several hours before the actual time of ovulation.

The estrous cycle is thus one of the most important of the biorhythms we discussed in Chap. 25. Its duration varies considerably in different mammals. In contrast to rats, for example, are species in which the cycle lasts a year. Many primates are intermediate; in the case of humans, of course, the cycle lasts approximately 28 days, and is called the menstrual cycle.

The cyclic changes in gonadotropin secretion from the pituitary are determined by changes in levels of releasing factors secreted in the hypothalamus, as we discussed in Chap. 23. What then determines the secretion of these factors? Part of the

answer lies in the gonadal hormones themselves, which, as we have seen, are taken up and bound by specific cells in the hypothalamus. Negative feedback control is achieved in this way: increased hormone uptake suppresses production of releasing factors, decreasing the level of hormone, which in turn means less suppression and increased hormone levels, and so on. Part of the answer lies in the suprachiasmatic nucleus (SCN), which plays its role as the master oscillator in the brain; removal of the SCN abolishes the estrous cycle.

Many other regions of the brain make their contributions to the estrous cycle. These areas have been identified by a variety of studies. One approach is to stimulate electrically different regions and observe the effects on blood levels of LH; conversely, the effects of ablation of different regions can be observed. Biochemical analysis of different regions has been carried out. Localization of sex hormones has been studied using intracerebral implants of hormones, and sites of steroid uptake and steroid receptors have been identified by autoradiography. Finally may be added studies of sexual dimorphisms.

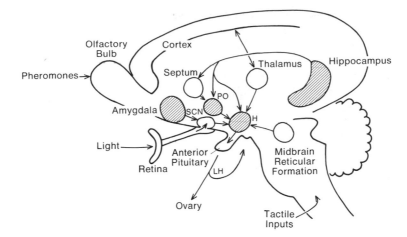

Fig. 27.11 Brain centers involved in the control of gonadotropin secretion and the estrous cycle in the female rat. Shaded area indicate regions showing sexual dimorphism. H, hypothalamus; LH, luteinizing hormone; PO, preoptic area; SCN, suprachiasmatic nucleus. (Adapted from Harlan et al., 1979)

The results of these various studies are summarized in Fig. 27.11. These regions and the pathways that connect them may be said to form a distributed system, as we defined it in Chap. 24, for the neural control of the estrous cycle. At the heart of the system is the median eminence where the releasing factors are secreted (see Fig. 24.2). Feeding into this are the mammillary body, preoptic area (PO), the dopamine neurons of the median eminence, and the SCN (see Fig. 24.3). Next are regions of the limbic system such as the amygdala, septal region, thalamus, hippocampus, and mesencephalic reticular formation (see Chap. 28). Finally are the sensory pathways, especially olfactory, tactile, and visual, that have inputs to different regions. Within the distributed system, different regions are more or less closely related to the final control of releasing factors, and thus provide multiple ways in which nervous influences can be integrated and can have an influence on circulating hormonal levels. In some species, as in the rodent, the neural controls may be relatively powerful. In other species, such as the primate, they are less powerful, and hormonal levels are more dominated by pituitary–gonadal interactions.

Variety and Adaptability of Nervous Controls

No species better illustrates the powerful influences that the nervous system can exert on reproductive processes than the common house mouse. This is one of the most adaptable and opportunistic of all mammalian species, as attested by its distribution throughout the world. A key to this dispersion is the tendency of male mice to be aggressive toward each other, an aspect of behavior which we will discuss further in the next chapter. For now, we note that the result of this aggressiveness is for dominant mice to establish their territories, and for subordinate mice to be forced to seek elsewhere for mates.

Importance of Odor Cues

One of the main ways through which mice interact is through odor cues. When you put a mouse into a new cage, the first thing it does is to go around and mark it everywhere with urine. Figure 27.12 illustrates how this can be documented by simply taking a photograph under ultraviolet light. The urine carries odor cues (pheromones) that are like a fingerprint of that mouse, conveying specific information

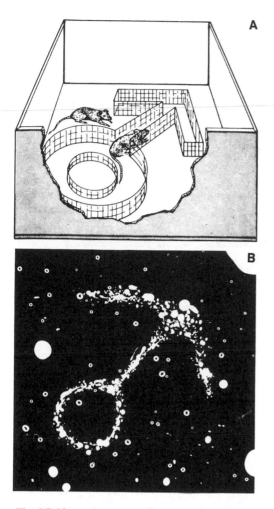

Fig. 27.12 **A.** Demonstration of urine marking by the house mouse. Diagram shows cage containing a fence in the form of the male symbol. Male mouse is inside cage, female mouse outside. **B.** Photograph of cage taken with UV light; white areas are sites of urine deposit. Picture was taken after mice had been in cage for 20 minutes. (From Bronson, 1979)

about its species, sex, sexual maturity, and social rank. By this means, a dominant male establishes his territory. The amount of pheromone in the urine is directly dependent on the high levels of circulating androgen that are present in a dominant male. The biochemical identity of the pheromone substance is still unknown.

The development of timing of ovulation in the female is critically dependent on the

pheromones in the urine. The way this comes about is summarized in Fig. 27.13. Urine deposited by other females has an inhibiting effect on the development of puberty in a young female ①. This is viewed as a protection against pregnancy while the females are crowded and still growing, and before there is maximal opportunity for dispersion. By contrast, mature male urine odors act as powerful priming pheromones, which accelerate and indeed help to organize the processes of puberty, ovulation, mating, and pregnancy. The direct stimulatory effect on LH secretion is indicated at ②. This ability of males to bring a female into ovulation is obviously an important adaptive advantage in ensuring the success of mating. It is also known that female urinary pheromones can stimulate the secretion of LH and testosterone in males ③. This effect is independent of the reproductive status of the female. By this means there is a positive feedback loop between the two sexes that ensures the greatest chances for ovulation and reproduction.

These experiments thus attest to the powerful effects of pheromones in regulating mating and reproductive behavior. These effects are found, to a greater or lesser degree, in most mammals. Many of the effects require or are enhanced by other factors, such as appropriate tactile stimulation by other animals, or adequate nutrition, or the ambient temperature.

Natural Rhythms of Ovulation

It should be stressed that most of our information comes from studies on laboratory animals living in crowded conditions, and the mechanisms of reproduction may be much modified compared with conditions in the wild. According to Frank Bronson (1979), one of the main contributors to the above account:

. . . we really have no idea whether or not the estrous cycle that has been so well delineated in the laboratory ever occurs with any regularity in the field. . . . [H]ouse mice in natural populations [may] not routinely ovulate every fourth

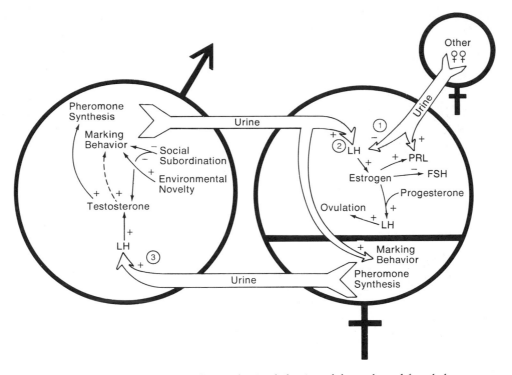

Fig. 27.13   Coordination of mating and reproductive behavior of the male and female house mouse through pheromones in the urine. ① Female–female inhibition; ② male–female stimulation; ③ female–male stimulation. FSH, follicle-stimulating hormone; LH, luteinizing hormone; PRL, prolactin; +, stimulation; –, inhibition. (From Bronson, 1979)

or fifth day. . . . They may ovulate only when the probability of a successful, ensuing pregnancy is high. Thus ovulation may be a relatively rare event in the life of a mouse. . . . It may be more adaptive for wild animals to concentrate on survival rather than reproduction unless the time for reproduction is truly propitious.

The fact that ovulation may be a relatively rare event in the mouse under some natural conditions has an interesting parallel among humans. Throughout most of human history, in many societies women have typically spent most of their childbearing years either pregnant or breastfeeding. Under these conditions, a woman ovulates only at irregular intervals, perhaps no more than a few dozen time in her entire lifetime. An interesting example of this is

seen in the present-day 'Kung tribe in Africa, in which the mother breast-feeds her babies for several years. Suckling stimulates the release of prolactin from the pituitary, which not only stimulates lactation but also suppresses ovulation. Prolactin thus acts as a natural contraceptive, and a 'Kung woman may become pregnant only a few times. Roger Short of Edinburgh, who has studied the 'Kung people, points out that we should consider infrequent and irregular ovulation and menstruation as just as normal as the modern belief in uninterrupted regular cycles. Thus, in humans as well as other mammals, there is evidence of the adaptability of the reproductive apparatus, and its ability to be shaped and modified by social and cultural factors mediated through the brain.

28

Emotion

Human beings have speculated about the nature of their emotions since earliest recorded history. The Pythagorian philosophers of ancient Greece believed that the universe is composed for four elements: fire and water, earth and air. Hippocrates and his followers deduced that the body is similarly composed of four corresponding humors: blood and phlegm (mucous discharge), black bile and yellow bile. A person's temperament was believed to express an excess of one or more of these humors. Thus, an excess of blood rendered a person sanguine: ruddy-complexioned, courageous, hopeful, amorous. An excess of phlegm made a person phlegmatic: dull, cold, even-tempered. An excess of black bile made a person melancholic and sad, whereas an excess of yellow bile made a person choleric, or angry. These ideas were so believable that they lasted until the seventeenth century.

Even though we regard this scheme now as a prescientific fairy tale, it is nonetheless sobering to realize that these terms are still used to describe human emotions. And when we try to define emotions, we find that we have made little progress, despite all our science. Shakespeare never used the word *emotion;* Hamlet, for example, praised his friend Horatio as a man

. . . whose blood and judgment are so well co-mingled . . .

Today we would say that Horatio's thoughts and emotions were in good balance, but what have we gained?

According to the *Oxford English Dictionary,* the word *emotion* is derived from the French word *mouvoir,* "to move." It came into use in the seventeenth century to describe mental feelings (pain, desire, hope, etc.) that are distinct from thoughts, or cognitions. Within this broad and rather vague definition are three types of emotion, which are illustrated in Fig. 28.1. First are complex behaviors of an animal, such as predation, feeding, copulation, and related actions, which appear to us, through our human eyes, to be *emotional actions,* but which may actually have no emotional component for the animal itself. Second are specialized motor actions—*emotional expressions*—in lower animals as well as humans, which seem to express directly some inner feelings or emotion. Third are *inner emotions* or *subjective feelings,* which are entirely felt or perceived within us, and therefore are known only to humans.

The neurobiological study of these three types of emotions has been rather limited. On the one hand, most complex behaviors are simply too complex to analyze, particu-

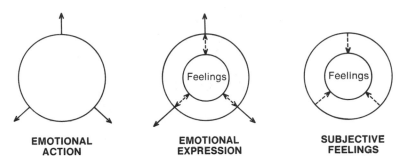

EMOTIONAL ACTION   **EMOTIONAL EXPRESSION**   **SUBJECTIVE FEELINGS**

Fig. 28.1   Types of emotional behavior.

larly when their emotional component is not obvious. On the other hand, our inner emotions are known only to ourselves. Most scientists rule out such subjective phenomena as objects of scientific study. That leaves us with only emotional expression as the type of emotion that is observable and amenable to precise analysis.

A Brief History

The subject of emotional expression was begun as a scientific study in 1806 by Sir Charles Bell of England, the same remarkable man responsible for the discovery of the sensory functions of the spinal nerves (Chap. 12), and for one of the first monographs on the structure and function of the hand (Chap 22). In his book *Anatomy and Physiology of Expression* (Bell, 1837), he discussed the detailed relations between the facial muscles and many different expressions, such as laughter or grief. Various nineteenth-century authors added their observations, but the subject was little more than popular science, a branch of phrenology, until Charles Darwin published his book *On the Expression of the Emotions in Man and Animals,* in 1872:

Sir C. Bell's view [wrote Darwin], that man had been created with certain muscles specially adapted for the expressions of his feelings, struck me as unsatisfactory. It seemed probable that the habit of expressing our feelings by certain movements, though now rendered innate, had been in some manner gradually acquired . . . expressions, such as the bristling of hair (during terror), or the uncovering of teeth (during rage), can hardly be understood, except on the belief that man once existed in a much lower and animal-like condition. . . . He who admits . . . that the structure and habits of all animals have been gradually evolved, will look on the whole subject of Expression in a new and interesting light.

Darwin based much of his study on comparisons between humans and domestic animals, and he analyzed the expressions of these animals as meticulously as he had the beaks of finches. One of his conclusions was that opposite emotions, such as hostility or affection, may be expressed by oppositely directed movements, and the illustration of Darwin's dog displaying these two attitudes (Fig. 28.2) has been often reproduced to illustrate this.

After Darwin there could be little doubt that humans express emotions by motor and muscular mechanisms that have evolved out of similar mechanisms present in ancestral forms and exemplified in present-day vertebrates, most particularly domesticated mammals. This provided the necessary rationale for scientific exploration of the neurobiological basis of emotions, using animals as experimental subjects. The techniques for doing these experiments did not become available until the 1920s and 1930s, and much of our present understanding of brain mechanisms had its origins in that era. Before discussing this work in mammals, however, we need

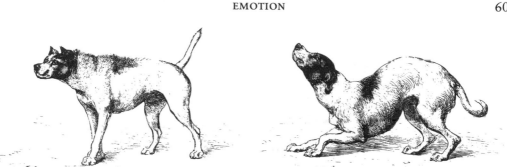

Fig. 28.2 Darwin's dog, displaying the contrasting expressions of hostility and affection (the latter to a somewhat abject degree). Can you tell which is which? (From Darwin, 1872)

first to discuss the interesting question of whether emotions occur in invertebrates and lower vertebrates.

Invertebrates and Lower Vertebrates

The question of whether lower animals have, or express, emotions has been the subject of lively debate. The view prevailing until the nineteenth century was well-summarized by Bell: "[W]ith lower creatures there is no expression but what may be referred, more or less plainly, to their acts of volition or necessary instincts." Note that Bell was willing to grant lower creatures volition while denying them emotions! These acts or instincts fall into our first category of emotional behavior (see Fig. 28.1).

By contrast, Darwin was quite willing to extend his view of emotional expression in animals to lower forms. He wrote: "Even insects express anger, terror, jealousy, and love by their stridulation." In Chap. 23 it was noted that insects use stridulation to communicate with each other, with songs that signal "calling," "hostility," and "courtship." But whether or not these are accompanied by emotions, in the sending or receiving insect, remains unanswerable.

One test for the presence of an emotion is whether a given perception or motor act is accompanied by changes in the autonomic nervous system, such as faster heart beat or increased perspiration. This, for

example, is the basis for the lie detector used to test whether suspects in a legal proceeding are telling the truth. It was stated in Chap. 18 that invertebrates have nerve ganglia that innervate their visceral organs, but that this innervation is relatively simple compared to that of vertebrates. The contrast has been summarized by Alan Epstein (1980):

. . . like all other invertebrates [the molluscs] do not have an autonomic nervous system with which so much of affective expression is achieved. They do employ the same or very similar biogenic amines in their nervous systems and their viscera is innervated by peripheral ganglia, but this is hardly an autonomic nervous system which is an anatomically widespread and highly reactive system that provides duplex and functional antagonistic innervation to the viscera, smooth muscles, and glands, and is complemented by an adrenal gland.

Epstein notes the observation of Wells, that during sexual arousal and copulation, the octopus shows no changes in heart rate.

Later in this chapter it will be shown how important the vertebrate autonomic nervous system is for the expression of emotions in mammals.

Another feature of arthropods is their rigid exoskeleton, which limits their ability to express inner states by local muscular movements on their body surface. Nonetheless, arthropods do express different behavioral states in their displays and actions related to courtship, predation, and terri-

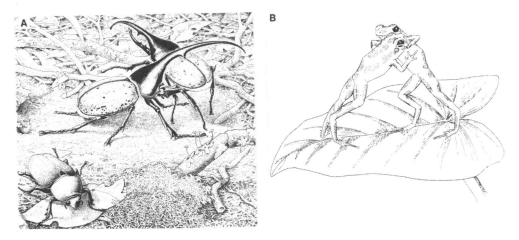

Fig. 28.3 Displays of aggressive behavior. **A.** Fight between two male Hercules beetles for dominance and access to female in lower left. **B.** Fight between two male frogs *(Dendrobates)* for possession of a territory. (A by Sarah Landry, B from Duellman, in Wilson, 1975)

tory. An example of a fierce struggle for dominance between two male beetles is depicted in Fig. 28.3A. However, these are signals conveyed by whole-body movements; there is no associated specialized system for conveying the emotional intensity or tone behind the actions. Nor is it entirely due to the limitations of an exoskeleton; frogs engaged in a struggle for territory similarly reveal little of the intensity of their efforts in their facial expressions (Fig. 28.3B).

We must conclude that a complex innervation of the internal viscera, and a complex set of muscles which can independently signal autonomic and other internal states, are two components that are necessary for the expression of emotion in animals. These two components are almost entirely lacking in invertebrates and lower vertebates. Therefore, we conclude that these animals can express behavioral states through whole-body actions, but they lack the ability to express emotions in any of the three ways defined in relation to Fig. 28.1 above.

Mammals

The fact that the expression of emotion seems to be a special ability of mammals is a clue that it may play an important role in the development of higher nervous functions in mammals. Let us consider the evidence for the neural basis of emotions, and then discuss its implications, particularly with regard to motivated behavior.

Hypothalamic Mechanisms

The success of Ferrier and his contemporaries in mapping out the motor areas of the cortex in the late nineteenth century (see Chap. 21) prompted others to investigate the effects of electrical shocks to deep structures of the brain. This early phase culminated in the demonstration by W. R. Hess of Zurich in 1928 that attack behavior, including expressions of rage, and defensive behavior, including expressions of fear, can be elicited in cats by stimulation within the hypothalamus (see Fig. 28.4).

Sham Rage. At about this time, Walter Cannon and his student Philip Bard, at Harvard, carried out a complementary series of studies in which they examined the behavior of cats after transections of the brain at different levels (Cannon, 1929; Bard, 1934). They found that removal of the forebrain (cortex, basal ganglia and thalamus, but sparing the hypothalamus)

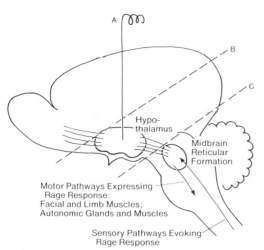

A

B

C

Hypo-
thalamus

Midbrain
Reticular
Formation

Motor Pathways Expressing
Rage Response:
Facial and Limb Muscles;
Autonomic Glands and Muscles

Sensory Pathways Evoking
Rage Response

Fig. 28.4 Diagram of brain illustrating experimental demonstrations of various emotional behaviors in the cat. Electrode (A) used to stimulate hypothalamus and produce expressions of rage or fear. Transection (B) removing the forebrain and leaving the hypothalamus (hypothalamic animal) produces sham rage; transection (C) below hypothalamus results in animal that does not display sham rage.

yields an animal that is irritable, and can be triggered into a display of rage (baring of teeth, hissing, clawing) at the slightest provocation (see Fig. 28.4). The display is accompanied by autonomic changes, such as increased heart rate and bristling of the fur. However, because of its low threshold, and uncoordinated and undirected nature, the display lacks the conscious dimension of normal attack behavior, and it was therefore termed "sham rage." This is a useful concept; after all, it is reasonable to expect a certain emptiness of meaning in the behavior of a cat that lacks its entire forebrain! When the transection occurs just below the hypothalamus, the sham rage response is lost (see Fig. 28.4).

Autonomic Changes. These two lines of evidence thus firmly established the importance of the hypothalamus for the expression of emotional behavior. This importance is seen both with respect to the somatic component (control of facial and limb muscles) and the visceral component (control of glands and muscles in the autonomic nervous system). With regard to the autonomic responses, the displays of intense emotional behaviors are one of the best ways to bring out the actions of the sympathetic, as compared with the parasympathetic, divisions, as was first described by Cannon. Thus, displays of rage or fear are accompanied by increased levels of epinephrine and norepinephrine; in addition to increased heart rate and piloerection, blood is shunted to the muscles and the brain, the eyes dilate, and so forth. These changes bring the animal to the highest level of alertness, and prepare it for the most extreme levels of physical action which may be necessary for ensuring its survival.

Aggressive Behavior. Subsequent studies have added some details to this general picture. An important advance was the series of studies which John Flynn and his associates at Yale initiated in the 1960s. By selective stimulation of sites in the hypothalamus of awake, behaving cats, combined with careful behavioral observation, they were able to distinguish between emotional and unemotional attack behaviors. The behavioral test was to put a cat and a rat into a cage and test the effects of hypothalamic stimulation on the cat. Stimulation could induce two basic types of attack. In *affective attack* (see Fig. 28.5), the animal displays most of the signs of sympathetic arousal, emotional excitement, and rage. The cat attacks the rat, with clawing and hissing, though it does not usually proceed to biting the rat unless stimulation is prolonged. In contrast, in *quiet biting,* the cat makes no sound and shows no emotion, but proceeds to capture the rat and bite it (see Fig. 28.5). This is actually similar to the normal predatory behavior of the cat. If you have seen nature films of cheetahs stalking and pursuing Thomson's gazelles in the Serengeti of Africa, you will recall the similar lack of emotional expression as they go about their business.

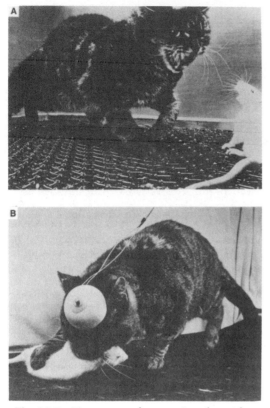

Fig. 28.5 Two types of aggression shown by a cat toward a rat. **A.** Affective attack. **B.** Quiet biting. (From Flynn, 1967)

It thus appears that one can distinguish between *predation,* which has autonomic activation but may proceed without the expressive components, and *aggressive displays,* in which some of the components of predatory behavior are displayed in heightened form and used as threats to achieve dominance or defend territory.

Overlapping Distributed Systems. We have previously seen that the hypothalamus contains neural elements and mechanisms that are involved in several types of behavior: feeding, thirst, sexual activities, and hormonal and autonomic functions. To this we now can add mechanisms that are essential for the expression of emotions.

To some extent it is logical for these mechanisms to be closely related to emo-

tions. Behaviors may be accompanied by emotions (such as attack accompanied by rage), may lead to emotions (such as eating leading to satisfaction or dissatisfaction), or emotions may be the primary motivating force leading to the behaviors (such as fear leading to flight) (see below). However, it should be emphasized that the mechanisms are to some extent distinct; thus, animals stimulated to attack and bite prey do not also eat the prey. Close analysis of these relations is difficult, because of the close proximity of different centers within the hypothalamus, and the many systems of fibers that pass through it, as we have previously noted.

For the expression of emotion, we can therefore consider the hypothalamus as one center, or group of centers; in our previous terminology, it is one node, or collection of nodes, in a distributed system. Let us now consider the rest of that system.

The "Limbic" System

The region most closely related to the hypothalamus for the expression of emotion is the *midbrain,* just posterior to it (see Fig. 28.4). Attack behavior elicited by hypothalamic stimulation is blocked by lesions of the midbrain. Midbrain stimulation by itself can elicit attack behavior, even after surgical isolation of the hypothalamus from the rest of the brain. These results have indicated that much of the neural mechanism for the control of aggressive actions is present in the midbrain and lower levels. In line with the hierarchical view of motor organization, it appears that the motor control mechanisms are delegated to the brainstem and the spinal cord, and the hypothalamus may be mainly involved in initiating and coordinating these mechanisms.

The close relations of the hypothalamus and midbrain in these respects suggested to Walle Nauta, then at Walter Reed Army Medical Center, that this part of the neuraxis has a functional unity in mediating visceral and emotional behavior. He termed it the *"septo–hypothalamic–me-*

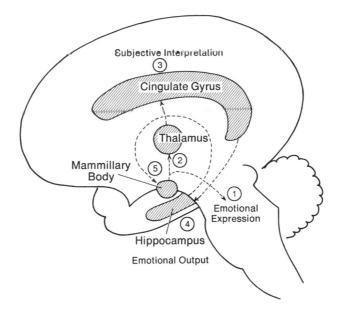

Fig. 28.6 Schematic diagram illustrating the Papez (1937) circuit for emotions. For explanation see text.

sencephalic continuum" (see Chap. 24). It is virtually identical with the regions embraced by the medial forebrain bundle.

Papez Circuit. What are the relations of this core system with other parts of the brain? The first coherent scheme of these numerous and complicated relations was put forward by James Papez of Chicago in 1937.

Papez was a neurologist who had noted reports of emotional outbursts in patients with damage to the hippocampus and to the cingulate gyrus. He carefully considered the known anatomy of the brain, and came up with a brilliant hypothesis for the neural circuit underlying emotions. The circuit starts with the mammillary body of the hypothalamus as the site of output for expression of the emotions, through its projections to the midbrain (see ① in Fig. 28.6). Collateral fibers pass to the anteroventral nucleus of the thalamus ②, where they connect to cells that project to the area of the cerebral cortex known as the cingulate gyrus ③. Here it was proposed that conscious, subjective emotional expe-

rience arises; the cingulate gyrus was considered to be the cortical receptive region for emotional input relayed through the thalamus, in analogy with the visual cortex as the receptive region for visual input relayed through the thalamus. The cingulate gyrus, in turn, projects to the hippocampus ④. The hippocampus was proposed to combine this and other inputs and organize this information for output to its main projection site (it was believed), the mammillary bodies in the hypothalamus ⑤, thus completing the circuit. The pathway from cingulate to hippocampus to hypothalamus provided a means whereby the subjective experiences at the cortical level could be combined with the emotional content of the hypothalamic output.

No one had previously conceived of how these structures of the brain might be meaningfully related, and the Papez circuit was taken up with great enthusiasm and became a powerful stimulus to further research. An attractive feature was that it gave an interesting function to the hippocampus, which up to then had been considered to be part of the rhinencephalon (nose brain) and related in some unknown way to olfaction.

Broca's Brain. It was soon recognized that Papez's circuit was reminiscent of "The Great Limbic Lobe" of Broca. In 1879, Paul Broca, the great French neurologist, had noted how the cingulate gyrus and hippocampus seem to encircle or border the base of the forebrain. He imagined that this border, "placed at the entrance and exit of the cerebral hemisphere," was like the threshold of a door; hence, the term *limbic,* the Latin for threshold being "limen." He imagined that this limbic lobe is the seat of lower faculties compared with the higher faculties in the rest of the cerebral cortex. In 1952, Paul MacLean, then at Yale, and one of the foremost workers on the visceral functions of the brain, suggested the term "limbic system" for Papez's circuit and the other regions related to it, and the term stuck.

The idea that there is a distinct "limbic system" for emotions, just as there is a visual pathway for visual perception, is of course very attractive. But over 50 years later, it increasingly appears that this is a case of a beautiful theory at the mercy of some stubborn facts. Studies have indeed upheld the role of the hypothalamus and the cingulate gyrus in emotional behavior. But the essential role of the remaining two regions in the Papez circuit, the thalamus and the hippocampus, remain uncertain. Ablation or stimulation of these regions has given variable or conflicting results in different species. In addition, several other regions have been shown to have powerful effects on emotional behavior. Foremost among these regions is the amygdala.

The Amygdala. This structure is a complex of related cells located in the cortex, at the base of the forebrain in lower mammals, and on the medial wall of the base of the temporal lobe in higher mammals. In the same year that Papez published his circuit, Heinrich Klüver and Paul Bucy at Chicago reported the results of experiments in

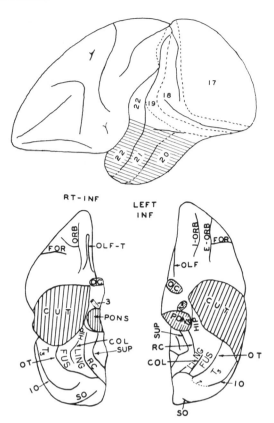

Fig. 28.7 Area of monkey temporal lobes removed (bilaterally) to produce the Klüver-Bucy syndrome. (From Bucy and Klüver, 1955)

which they bilaterally removed the temporal lobe in higher mammals (see Fig. 28.7). They noted five main effects:

1. Overattentiveness: the animals are restless; they have an urge to orient toward or respond to all stimuli.
2. Hyperorality: the animals compulsively examine all objects by putting them in their mouths.
3. Psychic blindness: the animals see but do not understand; they indiscriminately approach and examine objects even though harmful (such as a lighted match).
4. Sexual hyperactivity: the animals increase their sexual activity, also indiscriminately, even toward inanimate objects.

5. Emotional changes: monkeys previously wild and aggressive are rendered tame and placid, and can be handled easily.

This complex of features has come to be known as the *Klüver-Bucy syndrome*. The psychic blindness has subsequently been shown to be due to loss of temporal lobe neocortex (see Chap. 30). The hyperactivity may be due in part to discharging neurons on the borders of the lesion. The sexual hyperactivity (which originally was one of the most sensational aspects of the syndrome) seems to be so undirected as to be part simply of the general hyperactivity of these animals. Finally, the emotional changes in the Klüver-Bucy syndrome have been especially linked to the amygdala. However, the changes vary, depending on the species; cats, for example, are rendered savage after amygdala destruction. Some research has connected the hypersexuality with the increased aggression in these cases. However, ablation is such a crude tool, and there are so many uncertainties about the extent of damage in different studies, that no firm conclusion can be reached.

Another approach in the study of the amygdala has been to use focal electrical stimulation. Some typical results are summarized in Fig. 28.8. Although these effects are quite dramatic, and seem relatively localized to the amygdala, a number of uncertainties limit the interpretation. For example, repeated shocks are often delivered over many seconds, or even minutes, to elicit these responses, raising questions about how much spread of seizure activity there is within the brain. Or the site stimulated may actually be suppressed by excessive currents, with activity being elicited only in surrounding areas.

To understand better the functions of the amygdala, we need to know to what it is connected. By studying transport of horseradish peroxidase (HRP), dyes, and radioactively labeled amino acids, neuroanatomists have shown that the amygdala has connections with a number of brain regions. These results are summarized in Fig.

28.9. There are, first, projections from the two parts of the olfactory pathway (see Chap. 11). Then, starting from the forebrain, there are connections to cerebral cortex (frontal lobe and cingulate gyrus), thalamus (mediodorsal nucleus), septal region, hypothalamus (through a long, looping tract, the stria terminalis, as well as short, direct ventral fibers), and numerous sites in the brainstem (fibers from the taste pathway—see Chap. 11—and from the raphe and locus ceruleus). Many of these connections are reciprocal, so that the amygdala gets feedback information from the sites to which it projects.

It should be emphasized that the amygdala is actually a complex of a number of nuclei (see Figs. 28.8 and 28.9). The *cortical and medial nuclei* form one main division, concerned with olfactory and taste information. As we learned in Chap. 27, this information is used in the control of feeding, by the connections of the amygdala to the hypothalamus. The other main division of the amygdala, the *basolateral group of nuclei,* is much expanded in higher mammals. It seems likely that the connections of this division to the cortex and thalamus, as well as those to the "septo–hypothalamic–mesencephalic continuum," are involved in the expression of emotional behavior. Finally, it is likely that the microcircuits within different amygdaloid nuclei mediate special modes of local processing that the amygdala provides to the systems for emotional behavior.

In recent years there has been increasing interest in the role of the amygdala in relation to fear and anxiety, and some of the brain pathways involved in mediating these effects have been identified. This work has further refined an understanding of the functions elicited in the classical studies illustrated in Fig. 28.8 and correlated them with some of the pathways illustrated in Fig. 28.9. Figure 28.10 summarizes the results of these studies as they relate to the connections of the central nucleus of the amygdala. These functions are mediated primarily by the hypothalamus, brainstem

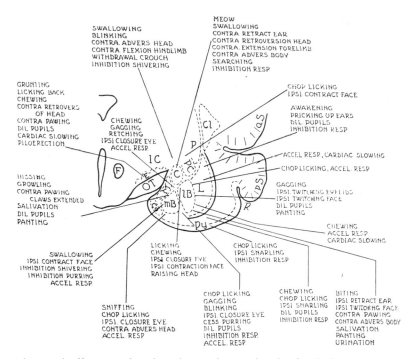

Fig. 28.8 Behavioral effects produced in the awake monkey by focal electrical stimulation at the sites shown. Abbreviations for nuclei of the amygdalar complex: C, central; PC, paracentral; M, medial; co, cortical; mB, mediobasal; lB, lateral basal; L, lateral. Other regions: IC, internal capsule; OT, optic tract; P, putamen; Cl, claustrum; aS, anterior sylvian; pS, posterior sylvian; R, rhinal sulcus; Py, pyriform cortex; F, fornix. (From MacLean and Delgado, 1953)

centers, and autonomic nervous system. This type of work holds promise for the development of pharmacological agents that can oppose the overactivity of different pathways causing harmful effects such as increased heart rate, ulcers, stress, and anxiety.

Cingulate Gyrus. Similar studies have been carried out on other regions of the limbic system. The results of studies on the *cingulate gyrus* are summarized in Fig. 28.11. These results have fully confirmed the input pathway from the mammillary bodies through the anterior–ventral nucleus of the thalamus, and the output to the hippocampus, as postulated in the Papez circuit. However, they have shown further that the cingulate gyrus has connections to many other structures. Particularly important are connections to amygdala, subiculum (a cortical region neighboring the hippocampus),

septum, and several sites within the midbrain (superior colliculi, for example, and locus ceruleus). In addition, there are connections to other areas of cortex, in the frontal, parietal, and temporal lobes. Many of the relations are reciprocal.

From these results it appears that the cingulate gyrus may well have connections to a greater variety of subcortical and cortical structures than any other region in the brain. What is the significance of this great variety? Why should there be connections, for example, to the superior colliculus, which is involved in the precise sensorimotor coordination of visual tracking, as well as to the locus ceruleus, whose diffuse projections throughout the brain are involved in biorhythms and mechanisms of consciousness (Chap. 25)? We do not know why, but part of the answer may be that emotional expression requires extensive coordination of both visual and somatic be-

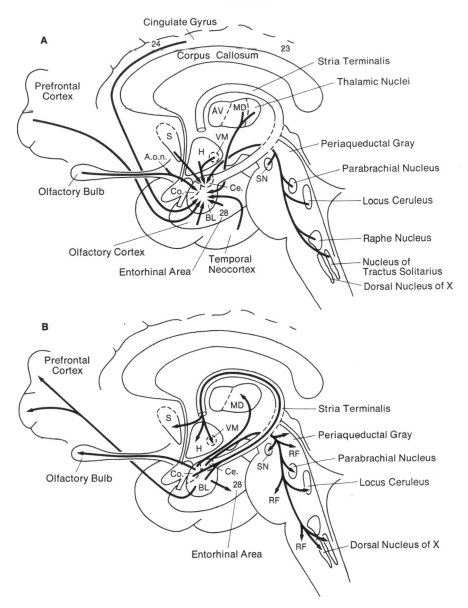

Fig. 28.9 Parts of the brain with connections to the amygdala. **A.** Inputs to the amygdala. **B.** Outputs from the amygdala. A.o.n., anterior olfactory nucleus; AV, anteroventral thalamic nucleus; BL, basolateral amygdaloid nucleus; Ce., central amygdaloid nucleus; Co., cortical amygdaloid nucleus; H, hypothalamus; MD, dorsomedial thalamic nucleus; RF, reticular formation; S, septum; SN, substantia nigra; VM, ventromedial hypothalamic nucleus. (From Brodal, 1981)

havior, and the cingulate gyrus is a center for this coordination.

Summary. In conclusion, we may summarize present concepts of the limbic system in the following manner. There is a series of structures, extending from the midbrain through the hypothalamus and into the basal forebrain, a phylogenetically ancient core system following the route of the medial forebrain bundle, that is concerned not only with visceral motor functions but also with related displays of motor behavior expressing emotion. These displays require

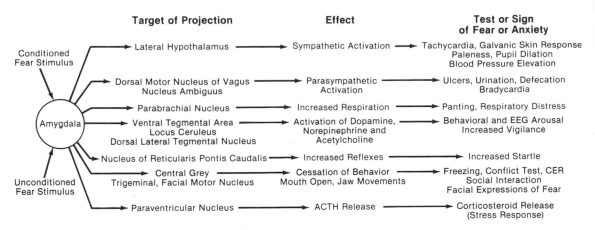

| | **Target of Projection** | **Effect** | **Test or Sign of Fear or Anxiety** |

Conditioned Fear Stimulus

Amygdala

Unconditioned Fear Stimulus

Lateral Hypothalamus → Sympathetic Activation → Tachycardia, Galvanic Skin Response Paleness, Pupil Dilation Blood Pressure Elevation

Dorsal Motor Nucleus of Vagus Nucleus Ambiguus → Parasympathetic Activation → Ulcers, Urination, Defecation Bradycardia

Parabrachial Nucleus → Increased Respiration → Panting, Respiratory Distress

Ventral Tegmental Area Locus Ceruleus Dorsal Lateral Tegmental Nucleus → Activation of Dopamine, Norepinephrine and Acetylcholine → Behavioral and EEG Arousal Increased Vigilance

Nucleus of Reticularis Pontis Caudalis → Increased Reflexes → Increased Startle

Central Grey Trigeminal, Facial Motor Nucleus → Cessation of Behavior Mouth Open, Jaw Movements → Freezing, Conflict Test, CER Social Interaction Facial Expressions of Fear

Paraventricular Nucleus → ACTH Release → Corticosteroid Release (Stress Response)

Fig. 28.10 Summary of the results of experiments on animal models of fear and anxiety, which have provided evidence on direct connections between the central nucleus of the amygdala and various hypothalamic and brainstem centers in mediating specific functions relating to these behaviors. (From Davis, 1992)

the coordinated action of other brain centers. Some of these, such as the amygdala and the cingulate gyrus, are closely related to this core system. Others, such as the hippocampus and parts of the thalamus and midbrain, are related in limited, though specific, ways. Each of these regions acts as a nodal point in a distributed system, processing unique combinations of inputs and sending this information to unique combinations of output sites. By this means is achieved the coordination of visceral and somatic motor output to the whole body that is required for the expression of emotional behavior. Thus, it is still useful to think of a limbic system in terms

Fig. 28.11 Parts of the brain with connections to the cingulate gyrus. M, mammillary body. For other abbreviations, see legend Fig. 28.9. (From Brodal, 1981)

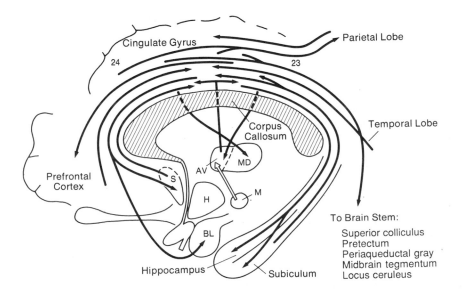

of a core with many internal and outlying integrative centers; however, the borders of this system are not sharp, and the system overlaps with many other systems in the brain. In Chap. 30, we will discuss the particular contribution of the cerebral cortex to this system.

Dermal Muscles

Most of the observations derived from studies of mammals appear to apply to the control of emotional behavior in humans as well. However, we would not want to leave the subject without discussing further the special motor apparatus that humans have for expressing emotions: the muscles of the face.

These have an interesting phylogenetic history. Snakes have small muscles that insert into their scales, and can hold the scale at different angles to help regulate how smoothly it moves; birds similarly have small muscles that regulate the angle of the feathers. Most mammals have a rela-

tively loose skin, and there is characteristically a sheet of *dermal muscles* that inserts into the skin over most of the body. When a horse twitches the skin of its back to shake off the flies, it is using dermal muscles. The dermal muscles in the face are called *facial muscles,* and they serve a variety of obvious functions, such as moving whiskers (vibrissae) on the snout, pricking up ears, moving the mouth to form different sounds, or moving the mouth during oral exploration or eating.

In higher mammals, such as the dog, and especially in primates and humans, the facial muscles have become adapted for the expression of emotions. This was recognized by the earliest writers on the subject, and the diagram in Fig. 28.12, which is taken from Bell's book, shows the detailed picture of these muscles that had been obtained by anatomists by the nineteenth century. If, as Darwin put it, "expression is the language of the emotions," then movements of the facial muscles provide the vocabulary for that language. Although

Fig. 28.12 Dermal muscles of the human face, as portrayed by Bell and Henle, in Darwin (1872). A, Occipitofrontalis, or frontal muscle; B, corrugator muscle; C, orbicular muscle of the eyelid; D, pyramidal muscle of the nose; E, medial lip-raising muscle; F, lateral lip-raising muscle; G, zygomatic muscle; K, mouth depressor muscle; L, chin muscle.

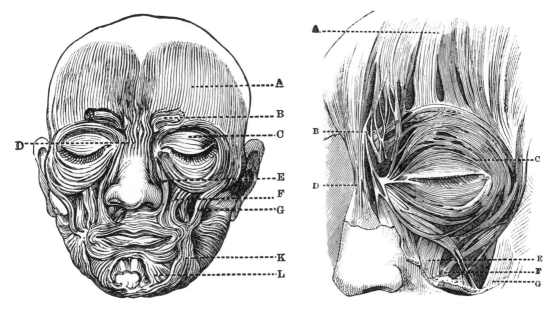

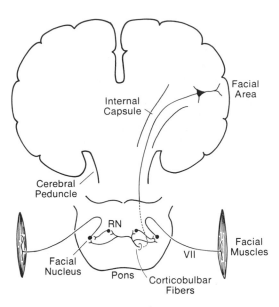

Fig. 28.13 Neural circuit mediating control of the dermal muscles of the face. RN, reticular nucleus.

whole-body movements and gestures are important in the expression of many emotions, the facial muscles represent a special apparatus for this function.

The degree of fine control exercised by the brain over these muscles is reflected in the large extent of their representation in the map of the motor cortex. It will be recalled from Chap. 21 that, in the homunculus of the body surface, the area of cortex involved in control of the facial muscles is greater even than the area for fine control of the hand. The pathways for this control are shown in Fig. 28.13. The fibers arising from pyramidal neurons in the facial area of the cortex connect bilaterally to the facial nuclei in the brainstem, there making monosynaptic and polysynaptic connections onto motoneurons that innervate the facial muscles through cranial nerve VII.

For the display of emotions by humans, we must therefore add this motor pathway to our limbic system. Detailed studies have shown that certain facial signals can be traced from primitive mammals through primates to humans. An example is shown in Fig. 28.14, for the baring of teeth combined with vocalization.

Emotion and Motivation

In our previous discussion of motivation (Chap. 26), we noted that the behavior of most animals, certainly of higher animals, is not governed simply by reflex responses to deficits or immediate needs. We saw that most behavior starts with the internal generation of behavior patterns by the brain itself. This intrinsic activity of nervous centers produces instinctual patterns of behavior, that may be triggered or guided by appropriate environmental cues and feedback circuits. The intensity of these patterns varies by a complex set of factors, including habits, incentives and rewards, learning and experience, as well as actual bodily needs. This set of factors is thought of as determining the amount of drive, or *motivation*, behind a given behavior pattern.

To this set of factors we now can add emotion. Emotion can be seen to be one of the key variables determining the kind of motor activity we engage in, and the strength or intensity of that activity. It is one of the primary factors responsible for the individuation of responses, for bringing out individual differences in behavior patterns among members of a species. We have seen that many of the neural mechanisms involved in the mediation of emotional behavior are found in the visceral part of the brain, and in the limbic system which coordinates visceral with somatic activities. Although we tend to think of these as lower functions, compared with higher functions such as learning or cognition, from an evolutionary point of view these mechanisms are among the most crucial to the success of individuals of a species, and for the adaptive survival of the species. This importance of emotions for motivated behavior has been eloquently stated by Alan Epstein (1980):

Motivated behavior is laden with affect and its performance is accompanied by overt expres-

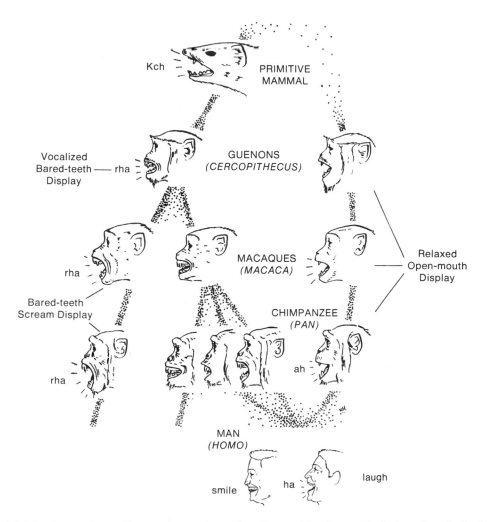

Fig. 28.14 Comparison of human expressions of smiling and laughter with the bared-teeth displays of lower primates and primitive mammals. (From van Hooff, in Wilson, 1975)

sions of internal affective states. Affect is expressed by very young animals, is often full-fledged when first exhibited, and is typically species-specific . . . it does not depend on learning. . . . These are not simply changes in limb movement or in the intensity of locomotion, but are true displays, organized into recognizable patterns, and sufficiently diversified to express a variety of internal states . . . in other words, motivated behavior is hedonic. . . . It arises from mood, is performed with feeling, and results in pleasure or the escape from pain, and although the moods, feelings, and satisfactions themselves are private and beyond our reach as

scientists, their overt expression is a necessary characteristic of motivation.

The "hedonic" properties of behavior, mentioned above, are those that relate to our conscious judgments about whether a given sensation or action gives us pleasure or pain. How does the limbic system provide for this? Papez presumed that it depends on the cingulate gyrus. Modern concepts include the contributions of other regions, especially of the cerebral cortex, as we shall see in Chap. 30.

29

Learning and Memory

The Nature of Learning and Memory

Thus far we have assembled most of the major components needed by a central system to mediate behavior. A very rough idea of how these components vary across phyla is conveyed by Fig. 29.1. It can be seen that reflexes and instincts govern the lives of most invertebrates and lower verte-

Fig. 29.1 Schematic portrayal of the relative development of different modes of adaptive behavior in phylogeny. (Modified from Dethier and Stellar, 1964)

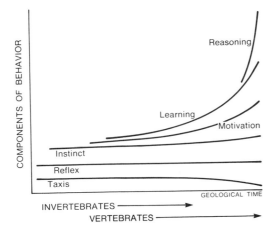

brates, with increasing contribution of motivated behavior in more complex vertebrates. We now consider an additional ability, and an essential one for most animals, the ability to learn and to remember what is learned. We shall discuss first the nature of learning and memory, and then consider what is known thus far about mechanisms.

Definition of Learning

The word "learning" shares with other words we have used, like "instinct" and "motivation," the problem of being a term in wide daily use. The ethologist S. A. Barnett (1981) maintained that

> colloquial terms with many . . . definitions, such as learning, are useful only as labels for general . . . categories of phenomena. . . . On this view it is inappropriate to ask what is the essence of the definition of learning.

This criticism notwithstanding, neurobiologists go right ahead and work on what they believe to be the problem of learning, so we need an appropriate definition for what we believe we are working on.

A very broad definition would state that *learning is an adaptive change in behavior caused by experience.* For the neurobiologist, the usefulness of this statement lies in a careful definition of each of the terms within it:

1. *Adaptive* indicates that the change must have some *meaning* for the behavior of the animal and the survival of the species.
2. By *change,* we mean that there must be a measurable *difference* between the behavior before and after some identifiable or imposed event. The change must be *selective* for the parts of the nervous system mediating the specific behavior, not just some general change in the animal, like increased metabolism or growing bigger. By the same token it must be independent of ongoing development or maturation. It should not be simply a reflection of fatigue, damage, or injury, or the normal habituation properties of receptors and nerves.
3. The *behavior* must involve *central* systems of the whole organism. It should not be confined to a part of the peripheral nervous system, or a single point in a sensory or motor pathway.

Although these qualifications may seem burdensome, we shall see that each one has meaning when we come to the experimental analysis of mechanisms.

Definition of Memory

Closely allied to learning is *memory*. Memory may be defined as *the storage and recall of previous experiences.*

This is a definition that applies as easily to computers as to animals. Memory is necessary for learning; it is the mechanism whereby an experience is incorporated into the organism, so that it can later be used to bring about adaptive changes in behavior. In lower organisms, the mechanism for storage of information may involve almost any cellular or neuronal process that can be perturbed by experience or actions of the environment. In higher vertebrates, and especially humans, we usually think of memories as those experiences that are subject to conscious recall. In many cases these recollections may be impressions of the passing world and bear no obvious relation to learning. In this sense, memory may include more than the mechanisms specific for learning.

Although everyone agrees that the mechanisms for learning and memory are found in the nervous system, there has been considerable debate about whether one could ever learn anything about them, or indeed, whether one *should* learn about them. Many cognitive psychologists, for example, believe that theories about learning and memory should be self-consistent and self-sufficient, without recourse to neuronal mechanisms. In this view, psychologists and physiologists should each have their own theories; if one tries to join the two, it can only result, in the opinion of B. F. Skinner, in "bad physiology and bad psychology." For most neurobiologists, this view is outdated, and one of the goals of modern research is to join the two levels into a coherent framework.

In Search of the Engram

An early and powerful voice in the debate on memory mechanisms was that of the psychologist Karl Lashley. In 1950, he published a paper entitled "In Search of the Engram." Engram is another word for memory trace. Lashley had spent most of a lifetime carrying out experiments on rats designed to reveal the presence of memory traces in different parts of the brain. He concluded that engrams do not exist; that memories are not localized in any one structure within the brain, but are distributed diffusely throughout the brain.

In a way, this was the reticular theory of Golgi (see Chap. 4) reincarnated with a vengeance. Just as the reticular theory made rational investigation of neuronal organization seem hopeless, so the idea of memory mechanisms spread diffusely throughout the brain seemed also to deny that they could be experimentally revealed. Lashley even observed, only half humorously, that after lifetime of studying learning he was beginning to doubt it could exist! Lashley's famous essay had a powerful influence on the field, and it took a generation to reveal the shortcomings of his experiments and interpretations, and supplant them with a more optimistic view based on modern techniques.

Cell Assemblies and Synapses

The conceptual framework for the approach of modern neurobiologists to the study of the neuronal mechanisms underlying learning and memory was laid down by two other psychologists, Donald Hebb of Montreal and Jerzy Konorski of Poland, in the late 1940s. Both drew on notions dating back to Cajal that learning and memory must involve changes in nervous circuits. Hebb, in his book on *The Organization of Behavior* (1949), hypothesized that a psychological function, like memory (or emotion or thought), is due to activity in a *cell assembly,* in which the cells are connected together in specific *circuits.* He suggested that when a cell is active, its synaptic connections become more effective (see Fig. 29.2). This effectiveness may be a relatively short-lived increase in excitability, as in short-term memory, or it may involve some long-lasting structural *change in the synapse,* as in long-term memory. Konorski's ideas were similar. The concept that brain functions are mediated by cell assemblies and neuronal circuits has become widely accepted, as will be obvious to the reader of this book, and most neurobiologists believe that plastic changes at synapses are crucial mechanisms of learning and memory.

Since the 1950s, there has been an outpouring of work by anatomists, biochemists, and electrophysiologists searching for clues to the postulated changes in synapses. Some of the main types of approaches that have been used are summarized in Table 29.1. We cannot review all this work here, and for further information the student will want to refer to textbooks and reviews in behavior, psychobiology, and physiological psychology. Much of this work was carried out before modern methods of cell biology

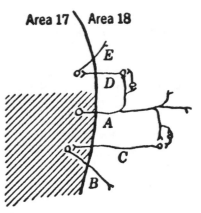

Fig. 29.2 The original concept of the "cell assembly" and the "Hebb synapse" (Hebb, 1949):

. . . perceptual integration would not be accomplished directly, but only as a slow development, and, for the purposes of exposition, at least, would involve several distinct stages, with the first of which we shall now be concerned.

The general idea is an old one, that any two cells or systems of cells that are repeatedly active at the same time will tend to become "associated," so that activity in one facilitates activity in the other. . . .

The proposal is most simply illustrated by cells *A, B,* and *C* in [the] figure. . . . *A* and *B,* visual-area cells, are simultaneously active. The cell *A* synapses . . . with a large number of cells in 18, and *C* . . . happens to lead back into 17. . . . The cells in the region of 17 to which *C* leads are being fired by the same massive sensory excitation that fires *A,* and *C* would almost necessarily make contact with some cell *B* that also fires into 18. . . . With repetition of the same massive excitation in 17 the same firing relations would recur and, according to the assumption made, growth changes would take place at synapses *AC* and *CB.* This means that *A* and *B,* both afferent neurons of the same order, would no longer act independently of each other.

At the same time, . . . *A* would also . . . synapse . . . with a cell *D* which leads back into an unexcited part of 17, and there synapses with still another cell *E.* . . . The synapse *DE,* however, would be unlikely to be traversed, since it is not, like *CB,* exposed to concentrated afferent bombardment. Upon frequent repetition of the particular excitation in area 17, a functional relationship of activity in *A* and *B* would increase much more than a relationship of *A* to *E.*

were available. Also, many of the most sensational results have been found, on reexamination, to require more modest, or alternative, interpretations.

A prime example is the recognition of

Table 29.1 Historical overview of some of the early experimental approaches to mechanisms of learning and memory

Anatomical

1. Environmental enrichment leading to bigger brains, dendrites, and synapses	Bennett et al., 1964
2. Effects of use on synapses	many authors, 1960s to present

Biochemical

1. Increased neuronal activity leading to increased RNA	Hyden, 1959
2. Memory transfer between animals	
A. Cannibalism in *Planaria*	McConnell, 1962
B. Injections of brain extracts from trained animals into untrained animals	Babich et al., Fjerdingstad et al., Reinin et al., 1960s
3. Susceptibility to antimetabolite drugs (memory consolidation requires protein synthesis)	Flexner et al., 1960s
4. Cellular biochemistry of effects of use on synapses	many authors, 1960s to present

Electrophysiological

1. Resistance to electroconvulsive shock	Duncan, 1949
2. Changes in EEG activity with learning	
A. Changes in hippocampal theta rhythms	Adey, 1960; Grastyan, 1966
B. High-frequency EEG rhythms in forebrain	Sheer, 1970
C. Slow potentials related to arousal or attention	Rowland, 1968
3. Changes in single neuron activity with learning	
A. During self-stimulation reinforcement	Olds and Olds, 1961
B. In relation to hippocampal theta rhythms	Ranck, 1973
C. In relation to amygdala and reward vs. aversive training	Fuster and Uyeda, 1971
4. Cellular electrophysiology of membrane properties at synaptic terminals	many authors (see Table 29.3)

For references, see Bennett (1977)

the crucial importance of arousal. When an animal is tested in a learning task, its level of arousal, attention, motivation, and distraction are critical determinants of its performance. Many of the findings in studies of interanimal transfer of learning (see item 2 under *Biochemical* in Table 29.1) have turned out to be attributable to nonspecific effects of the injected extracts on arousal, rather than on specific pathways in learning. Similar interpretations apply to several of the items under *Electrophysiological* in the table. This identification of the pervasive role of arousal and attention in learning is itself a valuable result, and is in accord with the importance we have placed on these functions for many aspects of behavior discussed earlier in this book.

With regard to specific mechanisms of learning, most of the approaches listed in Table 29.1 may be regarded mainly as a historical background. For the remainder of this chapter, we will focus on modern research of recent years, which has shown that the postulated mechanisms of plastic changes at synapses can indeed be investigated at the cellular and molecular level. Thus, the field of learning and memory holds out some of the best prospects for understanding how cells and synaptic circuits provide the basis for behavior.

Types of Learning and Memory

Since behavior takes a variety of forms, it should not be surprising that there are a number of different types of learning and memory. Table 29.2 lists the main categories. The plan of this chapter is to consider each of these types in turn, with the excep-

Table 29.2 Main categories of learning and memory

Types of learning	Types of memory
Simple	immediate
habituation	short-term
sensitization	long-term
	specific
Associative	
passive (classical)	
operant (instrumental)	
one-trial (aversion)	
Complex	
imprinting	
latent	
vicarious	

tion of short-term memory, which will be discussed mainly in Chap. 30. Many types of learning have their counterparts in invertebrates as well as vertebrates. We will illustrate each type with examples.

SIMPLE LEARNING

Nerve cells have a number of properties that change during or after stimulation. For example, sensory receptors adapt; their response tends to fade with continued or repeated stimulation (Chap. 10). Similarly, on the motor side, repeated stimulation of a motor nerve may cause a muscle to give either stronger (facilitation) or weaker (depression) responses. These changes are due to differences in the mobilization and release of the neurotransmitter at the neuromuscular junction. Table 29.3 lists some of the many studies of peripheral synapses that have contributed to our understanding of the plastic changes that occur at synapses as a consequence of activity.

This raises an interesting question: do we say that as a consequence of these changes the neuromuscular junction "learns" or has "memory"? The answer lies in the way we qualified our definitions in the previous section. Thus, we recognize that plastic changes may occur at many

sites in the nervous system; these changes may contribute to learning and memory, and may serve as valuable models for their mechanisms. However, learning and memory are basically properties of *central systems* that control the behavior of the whole organism, and it is therefore within the central nervous system that we must ultimately seek the underlying mechanisms.

Habituation

Closely related to the plastic properties we have just discussed are habituation and sensitization. Habituation is defined as *the decrease in behavioral response that occurs during repeated presentation of a stimulus*. It may be seen that inclusion of the term *behavioral* helps to make this response fit our definition of learning. However, habituation is a universal phenomenon, and the term gets applied to many isolated components of behavior. In addition, habituation involves only a change in the intensity of a response, not in the nature of the response itself; some workers, particularly psychologists, therefore do not consider it to be real learning. However, to the extent that habituation is of adaptive value, it is valid to consider it as an elementary form of learning.

In order to extend the studies summarized in Table 29.3 to habituation of central systems, what one wants ideally is a very simple response, a carefully controlled stimulus, and a neuron along the central pathway that can be analyzed with intracellular recordings. These conditions are met in simple organisms with large, identifiable neurons.

Aplysia

The use of *Aplysia* for this purpose has been brilliantly exploited by Eric Kandel and his colleagues at Columbia University in studies of the defensive withdrawal reflex of the siphon and gill (Fig. 29.3). The stimulus is a jet of water that activates tactile receptors in the siphon and gill and

Table 29.3 Historical overview of some of the experiments that provided the first evidence for molecular and membrane properties of synapses as a basis for plastic changes underlying learning and memory

Posttetanic potentiation at neuromuscular junction	Feng, 1941
Posttetanic potentiation in sympathetic ganglion	Larrabee and Bronk, 1947
Posttetanic potentiation in spinal cord	Lloyd, 1949
Quantal release of transmitter correlated with posttetanic potentiation and depression at the neuromuscular junction	del Castillo and Katz, 1954
Heterosynaptic interactions and presynaptic modulation	Dudel and Kuffler, 1961
Voltage dependence of transmitter release	Hagiwara and Tasaki, Takeuchi and Takeuchi, 1958–1962
Calcium dependence of transmitter release, voltage dependence of calcium channels	Katz and Miledi, 1967
$[C^{2+}]$ (internal) and synaptic plasticity	Rahamimoff, 1968
I_{Ca} and synaptic plasticity	Zucker, 1974
Membrane potential effects of synaptic potential	Shinahara and Tauc, Nicholls and Wallace, Erulkar and Weight, 1970s

For references, see Klein et al. (1980)

causes their reflex withdrawal. With repeated stimulation there is less withdrawal, and with sufficient repetition this depression of responsiveness may last for several weeks (A). Thus, both short-term and long-term habituation appear to occur.

For electrophysiological analysis, the sensory neuron mediating this response was stimulated electrically, while intracellular recordings were obtained from the motoneuron to the gill muscles (B). These experiments showed that the EPSP elicited in the motoneurons by the sensory synapse underwent a decrease in amplitude with repeated stimulation that paralleled the habituation of the reflex. By analyzing the miniature EPSPs, it was found that the decrease in EPSPs with repeated stimulation was due to a decrease in the number of transmitter quanta released by the synapse; the response of the postsynaptic membrane to each quantum was unaffected.

What causes the decrease in number of transmitter quanta released? We know that Ca^{2+} is a critical factor controlling release at the neuromuscular junction, but how can one examine this at central synapses, which are too small to record from di-

rectly? Mark Klein and Kandel used the clever tactic of analyzing the Ca^{2+} channels of the cell body of the sensory neurons, and, by careful consideration of indirect evidence, inferring the properties of the Ca^{2+} channels in the synaptic terminal of this neuron. They blocked the Na^+ channels with TTX in the medium bathing sensoryganglia, and showed that the TTX-insensitive part of the action potential, presumably mediated by Ca^{2+}, decreased in amplitude in parallel with the habituation. They concluded that depression of the Ca^{2+} current plays a critical part in short-term habituation. Recent evidence suggests that this is an N type of voltage-gated Ca^{2+} current (cf. Chap. 5). The model for this mechanism is depicted in C of Fig 29.3.

Additional changes were found to be related to long-term habituation. A large proportion of sensory neurons produced no detectable EPSPs in motoneurons, as if a functional disconnection had occurred. The basis for this finding has been studied by injecting the neurons with HRP and examining their axonal varicosities, which make the synapses, under the electron microscope after long-term habituation (Bailey

A. THE REFLEX BEHAVIOR

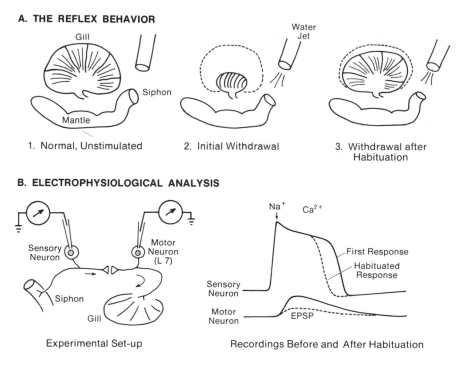

1. Normal, Unstimulated   2. Initial Withdrawal   3. Withdrawal after
                                                                    Habituation

B. ELECTROPHYSIOLOGICAL ANALYSIS

Experimental Set-up   Recordings Before and After Habituation

C. CONCEPTUAL MODELS

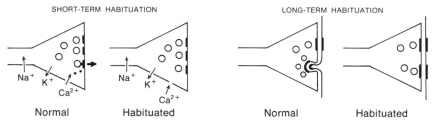

SHORT-TERM HABITUATION   LONG-TERM HABITUATION

Normal   Habituated   Normal   Habituated

Fig. 29.3 Summary of studies of habituation of the gill withdrawal reflex in *Aplysia*. **A.** The observed behavior. **B.** Intracellular recordings showing shortening of sensory action potential and decrement in motoneuron EPSP with repeated stimulation of sensory input. **C.** *(Left)* Simplified model of synapse to account for habituation by depression of inward calcium current at the terminal. *(Right)* Changes in structure of synaptic terminals during long-term habituation. (Adapted from Kandel, 1979; long-term habituation in C from Bailey and Chen, 1983)

and Chen, 1983). In normal animals, 40% of the varicosities contain synaptic active zones, whereas in the habituated animals this was reduced to 10%, and the active zones were smaller and flatter (see Fig. 29.3C). Thus, long-term habituation may involve a decrease in numbers of active synapses as well as decreased output at a given synapse.

Sensitization

Sensitization may be defined as *the enhancement of a reflex response by the introduction of a strong or noxious stimulus.* Although it appears to be the opposite of habituation, it differs in several respects. It depends on a stimulus different from that which elicits the reflex in question. Any strong stimulus activates general arousal mechanisms (Chaps. 24, 25), and so do noxious stimuli (Chap. 12). Thus, sensitization involves activation of general arousal systems, which affect the intensity of reflex response. For example, if you are startled by a loud noise, you are more sensitive to a subsequent soft sound. It is a widespread phenomenon; it alerts animals to predators and other potentially harmful stimuli, and thus is of important adaptive value.

The cellular basis for sensitization has also been investigated in the gill withdrawal reflex of *Aplysia*. Noxious stimulation can be produced by delivering a train of strong electrical shocks to the skin of the animal. The effect of this stimulation is to restore partially the original amplitude of a habituated gill withdrawal reflex response (Fig. 29.4A, B). Analysis of the quantal EPSPs shows that the restoration results from an increase in the number of transmitter quanta released by each impulse in the sensory terminal.

The nociceptors activated by these shocks make connections onto the presynaptic terminals of the sensory neurons in the gill reflex pathway. A variety of studies by James Schwartz and Kandel and colleagues have shown that sensitization via this pathway involves a series of steps at the molecular level. These include (1) activity in the nociceptor pathway, by release of serotonin and other transmitters or neuropeptides, to activate a specific receptor in the membrane of the sensory neuron terminals; (2) coupling of the receptor to adenylate cyclase, with production of cAMP; (3) activation of a cAMP-dependent protein kinase; (4) phosphorylation of a K^+ channel (S channel) to reduce K^+ currents during the impulse; (5) subsequent broadening of the impulse (as in Fig. 29.4), leading to increased Ca^{2+} influx and therefore increased transmitter release (see Fig. 29.4C).

Sensitization thus works in a manner that is opposed to habituation, to increase the amount of transmitter released, increas-ing thereby the strength of each active synapse. The essence of this mechanism is control of K^+ channels, which in turn affect Ca^{2+} influx and transmitter release. Recent studies indicate that the Ca^{2+} channels are of the N type. It is important to note that the mechanism involves interactions between several types of properties: second messenger, ionic currents, membrane potential, and neurosecretion.

Long-term training can produce long-lasting sensitization. This has been studied by electron microscopic analysis of HRP-injected sensory neurons (Bailey and Chen, 1983). In contrast to the decrease in incidence of active zones after long-term habituation (see above), long-term sensitization increases the incidence to 65%, and the active zones are larger and more complex (see Fig. 29.4C). The number of varicosities is also increased, indicating that sprouting of the terminal axonal branches has occurred. These experimental results imply that a second messenger (such as cAMP) not only has local effects in the nerve terminals underlying the short-term changes, but also stimulates protein synthesis to produce the long-lasting increases in sprouting and synaptic structure. Experiments have in fact demonstrated that treatment of the preparation with protein synthesis inhibitors can block the long-term morphological changes, and the long-term sensitization, without affecting short-term sensitization. It therefore appears that the molecular mechanisms for short- and long-term memory in *Aplysia* differ. Further studies (summarized in Kandel et al., 1991) have suggested that the increased levels of cAMP lead to increased gene transcription; the gene products mediate positive feedback to maintain persistent activation of the cAMP-dependent protein kinase as well as protein synthesis to bring about growth of the synaptic connections, as shown in Fig. 29.4C. We will discuss these molecular mechanisms further when we construct a model synapse for learning and memory at the end of this chapter.

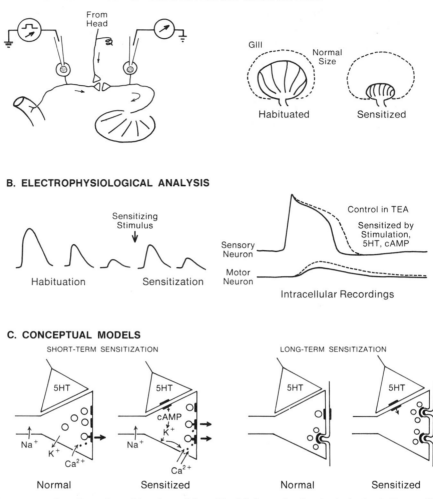

Fig. 29.4 Summary of studies of sensitization of the gill withdrawal reflex in *Aplysia*. **A.** Intracellular recordings from gill motoneuron, showing habituation to sensory input and sensitization by stimulation of nociceptors from the head. **B.** *(Left)* Amplitude of the gill response during habituation and sensitization. *(Right)* Broadening of the sensory action potential and facilitation of the motoneuron EPSP during sensitization of gill reflex. **C.** *(Left)* Simplified model of the synapse to account for sensitization (see text). *(Right)* Model of structural changes underlying long-term sensitization. (Adapted from Kandel, 1979, Shapiro et al., 1980, and Bailey and Chen, 1983)

Sensitization also occurs in the neural circuits that control feeding and heart rate in *Aplysia,* as well as in a variety of other invertebrate preparations. The three systems in *Aplysia* are similar in that sensitization appears to be mediated by serotonergic and/or peptidergic fibers, acting through cAMP to control voltage-dependent K^+ conductances, which in turn affect Ca^{2+} conductances. However, the points at which sensitization occurs are different in the three systems. The serotonergic fibers function to mediate arousal, and the different sites of connection are believed to reflect the differing significance of arousal for short-term sensitization, and for long-term plastic changes underlying learning in the case of the gill withdrawal and feeding reflexes. These arousal systems in *Aplysia* have their counterparts in vertebrates, as we discussed in Chaps. 24 and 26.

From these results, Kandel (1979) postulated that

Ca^{2+} current modulation may prove a general mechanism for learning and memory. The changes in Ca^{2+} influx can control the instantaneous level of transmitter release. . . . [M]odulation of Ca^{2+} channels . . . [might] be capable of contributing to long-term memory . . . by producing simple geometric changes in the shape of the synaptic apposition.

Similar mechanisms of synaptic modification have been proposed by others to underlie the changes that occur during neuronal development (see Chap. 9), and Kandel has speculated whether in this light one "can conceive of learning as being a late . . . stage in neuronal differentiation." Thus, Cajal's belief that development and learning share common processes of neuronal remodeling seems a step nearer realization.

The work on habituation and sensitization therefore illustrates nicely how neurobiologists, by isolating for study a very simple and elementary function, can progress through a series of experiments to arrive at results of considerable general interest. Much more work remains, of course, in applying these concepts to the behaving animal. Despite these results showing synaptic plasticity, the learning ability of *Aplysia* in behavioral tasks is in fact rather limited. We shall consider this question further below.

ASSOCIATIVE LEARNING

In associative learning, an animal makes a connection through its behavioral response between a neutral stimulus and a second stimulus that is either a reward or punishment. The best known example is the way a dog, which normally salivates when presented with a piece of meat, will salivate at the sound of a neutral stimulus like a bell,

after the bell has been paired with the presentation of the meat. This is called a *conditional reflex*. It was discovered in the early 1900s by Ivan Pavlov. By force of tradition it has come to be called *classical conditioning*.

With the advent of single-cell recording techniques, it became possible to test for the physiological mechanisms at the neuronal level. Table 29.4 summarizes some of the model systems that have been introduced for this purpose. Remarkably, a consensus is beginning to emerge on some of the cellular mechanisms that are common to classical conditioning in the different systems. For simplicity, we will consider these mechanisms by extending our discussion of *Aplysia* from the previous section.

Classical Conditioning in *Aplysia*

Tom Carew, Eric Kandel, and their colleagues have shown that the siphon–gill reflex can be classically conditioned, and the mechanism is similar to that of the underlying sensitization. The system is depicted in Fig. 29.5A. Gill withdrawal can be elicited by stimulation of the tail (middle); the siphon (below); or a nearby region, the mantle shelf (above). A strong shock to the tail was an unconditional stimulus (US in A, B), whereas a weak shock to the mantle shelf was a conditional stimulus (CS^+ in A, B). When the two were paired, the monosynaptic response recorded in the motoneurons in response to the conditioned stimulus was increased (see asterisk in C). As required in classical conditioning, the timing was critical; the conditional stimulus had to precede the unconditional stimulus (see B), and by an interval of no more than 1 second. Stimulation of another pathway (the siphon; see A, CS^-) without this temporal specificity (see B, CS^-) produced no conditioning [see C, unpaired (CS^-)].

At the molecular level, the mechanism is believed to be similar to that of sensitization. Thus, the normal sequence of events in the synaptic terminal onto motoneurons

Table 29.4   Some model central systems which have been introduced for studying of the neuronal basis of associative learning

EEG alpha blocking	Jasper and Shagass, 1941; John, 1967
Electrical stimulation of the brain	Olds and Milner, 1954
Single neurons	Olds and Olds, 1961
Flexion reflex	Buchwald et al., 1965
Motor cortex neurons	Fetz, 1961
Leg position in cockroach	Horridge, 1960s
Heart rate in pigeon	Cohen, 1969
Auditory tones	Woody, 1972
Shortening response in the leech	Henderson and Strong, 1972
Odor learning in *Drosophila* mutants	Dudai et al., 1976
Sensory stimulation in *Hermissenda*	Crow and Alkon, 1978
Sugar-odor learning in honeybees	Erber, 1978
Odor learning in *Limax*	Gelperin et al., 1978
Feeding in *Pleurobranchia*	Davis and Gillette, 1978
Withdrawal reflexes in *Aplysia*	Walters et al., 1979
Eyeblink response	Thompson et al., 1980s

For references, see Woody (1982)

in either the mantle or siphon pathway (the conditional pathway) involves invasion of the impulse into the terminal; the depolarization activates a limited Ca^{2+} influx, bringing about a low level of vesicle exocytosis, transmitter release, and postsynaptic response. During conditioning, the temporally specific pairing of action potentials (CS) with synaptic input (US) leads to a greater influx of Ca^{2+} with each action potential. The Ca^{2+} binds to calmodulin, which activates adenylate cyclase to produce cAMP; thus, the conditioned terminals produce more cAMP, leading to activation of more cAMP-dependent protein kinase, and thus to phosphorylation and closing of K^+ (S) channels; broadening of the preterminal spike; more Ca^{2+} influx at the synaptic active zone; more transmitter release; and a larger postsynaptic response. These steps may be reviewed in the diagrams of Fig. 29.4C.

As a result of this mechanism, the weak CS in effect accesses the same second messenger system employed by the strong synaptic input of the US, and replaces the US in activating this system. These changes all take place in the terminals onto the motoneurons; activity in the motoneurons is neither necessary nor sufficient for condi-

tioning to occur. It would appear therefore that classical conditioning of this reflex is based entirely on a *presynaptic* mechanism, and that this differs from a Hebb-type mechanism, in which changes are postulated to occur in both pre- and postsynaptic sites (see above). However, while the mechanism is *presynaptic* with relation to the mononeuron, it is *postsynaptic* with relation to the synaptic input from the tail. Further experiments will be necessary to determine if changes also occur in the presynaptic terminals of the tail pathway.

In comparing these results with those in other invertebrate models (see Table 29.4), Tom Carew and Chris Sahley (1986) noted some "common themes and convergent ideas [that] provide a framework for further investigations." First, experience causes changes in previously existing circuits; "in no case have novel synapses or new biophysical properties been introduced by learning." Second, modulation of K^+ conductance is involved in memory storage. Third, second messengers such as Ca^{2+} and cAMP are important in modulating the K^+ conductances. These second messengers may mediate local changes underlying short-term memory. Finally, although storage mechanisms (2 and 3

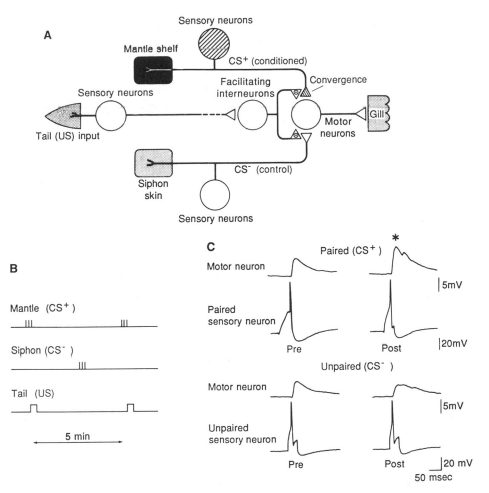

Fig. 29.5 Analysis of mechanisms of classical conditioning in *Aplysia*. **A.** Neural system controlling the gill reflex. The unconditional stimulus (US) consists of strong electrical shocks to the sensory nerves from the tail; it always elicits gill withdrawal. The conditional stimulus (CS⁺) consists of weak electrical shocks to the sensory neurons of the mantle; by itself it does not elicit a response. Weak electrical shocks to the sensory neurons to the siphon skin serve as a control conditional stimulus (CS⁻). **B.** The classical conditioning paradigm. CS⁺ is paired with US; CS⁻ is unpaired. **C.** Intracellular recordings from sensory neurons and gill motoneurons before conditioning (Pre) and one hour after (Post) a series of conditioning trials. The paired (CS⁺) trials produced a potentiation of the EPSPs in the motoneurons (asterisk), whereas the unpaired (CS⁻) trials did not. (From Kandel et al., 1991)

above) may be common, acquisition mechanisms may be variable. For example, acquisition in *Hermissenda* involves a cumulative sensory-induced depolarization that is primarily monosynaptic, whereas in *Aplysia* it involves, as we have seen, heterosynaptic enhancement.

A final point is that in most of the model systems examined thus far (see Table 29.4), the pathways are either primary sensory (e.g., *Hermissenda*) or sensorimotor reflexes (e.g., *Aplysia*). We began this chapter by noting that learning and memory need to be viewed primarily as properties of central systems. Therefore, in order to relate these model systems to learning and memory, there is a need to develop models of central systems to bridge this gap and address more directly the way central circuits mediate these functions. We turn next to some models in the mammalian brain.

Classical Conditioning in the Cerebellum

Mechanisms underlying associative learning have been pursued at the single-cell level in many parts of the mammalian brain. In reviewing recent studies, Thompson (1986) observed that memory traces do not appear to be localized to lower reflex centers in the brainstem and spinal cord; the most likely structures under current investigation are the cerebellum, hippocampus, amygdala, and cerebral cortex. We will summarize recent evidence obtained in the cerebellum and cerebral cortex, and later in this chapter consider the hippocampus.

A useful model for investigating the role of the cerebellum in learning at the cellular level has been the eyeblink reflex, introduced by I. Gormezano in 1972 and developed by Richard Thompson and his colleagues. An air puff to the cornea of a rabbit elicits an unconditional eyeblink reflex; this can be classically conditioned to an auditory tone. By making ablations of many different regions, it has been shown that the ipsilateral cerebellum is essential for learning and remembering this conditioned response, though not for the unconditioned reflex itself. The crucial site has been further localized to the lateral interpositus nucleus, one of the group of deep cerebellar nuclei (see Chap. 21).

The main afferent pathway for the unconditional reflexes is via the inferior olive, and the main efferent pathway is via the red nucleus (Fig. 29.6, diagram in upper left). The interpositus nucleus receives climbing fiber input from the inferior olive, and in turn sends its output fibers to the red nucleus (Chap. 21). Single-cell recordings from microelectrodes chronically implanted in the interpositus nucleus have documented the learning-related changes in neuronal activity. As shown in Fig. 29.6, the animal was first given separate random unpaired presentations of the conditional stimulus (CS), which elicited no significant eyelid response or change in neural activity, and the unconditional stimulus (US), which elicited the blink and a small change in neural activity. The stimuli were then paired; as training progressed through day 1 and day 2, the CS elicited an increased eyeblink response and an increased cell response (see day 2, Fig. 29.6). The cell activity is associated in time with the CS, suggesting that it represents a "model" of the conditioned, that is, learned, response.

The pathway for the conditional response has been shown to be, on the sensory side, through the auditory pathway to pontine nuclei, where mossy fibers to the cerebellum arise. Electrical stimulation of the mossy fibers is effective as a CS, producing rapid learning when paired with a US air puff. Thus, at the level of the cerebellum, the climbing fibers act as the US and the mossy fibers act as the CS. Many years ago, David Marr deduced that climbing fibers and mossy fibers could function in this way. The entire circuit identified thus far is summarized in Fig. 29.7. Although this particular circuit applies to the eyeblink response, the cerebellar components are more generally involved in learning; it has been suggested that "the cerebellum is essential for the learning of all discrete, adaptive motor responses, at least for classical conditioning with an aversive UCS" (Thompson et al., 1984).

Operant Conditioning

In classical conditioning, the animal is a passive participant. By contrast, an animal may be asked to learn a task or solve a problem, such as escaping from a box, or pressing a lever, or running a maze. This experimental method was introduced by Edward Thorndike in 1898. Since the animal learns to solve the problem and get the reward (or avoid the punishment) by

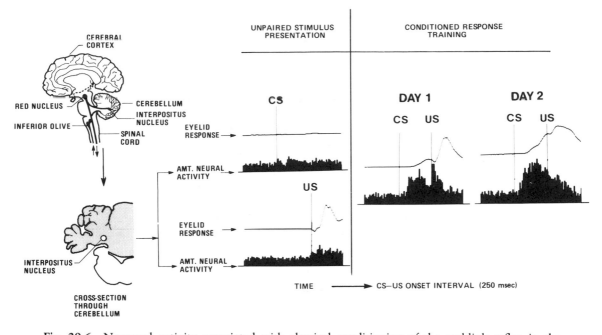

Fig. 29.6 Neuronal activity associated with classical conditioning of the eyeblink reflex in the rabbit. *(Left)* The upper diagram shows the main pathway for the unconditional eyeblink in response to an air puff to the eye; the lower diagram shows the recording site in the interpositus nucleus. *(Right)* Histograms of the neural activity together with a monitor of the eyelid movement. The pretraining control tests consisted of 104 trials with each stimulus (US air puff and CS auditory tone). Training sessions consisted of 117 trials each day. In the histograms, each bar represents the numbers of spikes in 9-msec bins. Note that the increased responses during training are related to the CS rather than the US. See text. (From McCormick and Thompson, 1984, and R. F. Thompson, personal communication)

operating on its environment, it is called *operant conditioning,* or *instrumental conditioning;* since the animal usually makes mistakes before learning the task, it is also called *trial-and-error learning.*

In both classical and instrumental conditioning, the strength of the response depends on the amount of reward or punishment. Particularly in the case of instrumental conditioning, the strength of the response can be used as a measure of the animal's "drive" to obtain the reward or escape the punishment. These experiments have thus provided the basis for studies of motivation, as discussed in Chap. 26.

One of the most surprising results of these studies has been the ability to condition single neurons in the brain by operant

techniques. This was first demonstrated by James and Marianne Olds at Michigan in 1961 (see Table 29.1). Figure 29.8 illustrates such an experiment on neurons in the motor cortex of the monkey, conducted by Eb Fetz and Mary Ann Baker (1973) of the University of Washington. The recordings were made using implanted microelectrodes in awake animals. The rate of impulse firing was registered on an illuminated dial, and when the rate exceeded a preset level, a few drops of tasty fruit juice were given as a reward. During the training period, the increase in activity took the form of bursts of impulses; as conditioning proceeded, the bursts became more frequent and more intense. When the reward was withheld, the rate of firing fell rapidly to its former level, a property called

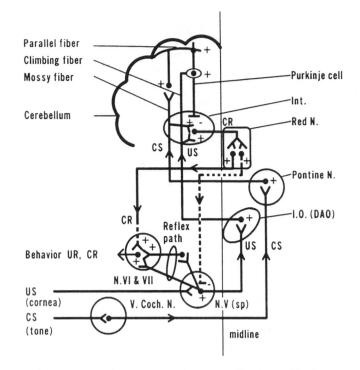

Fig. 29.7 Pathways that represent the memory trace circuit for the eyeblink response. The unconditional stimulus (US) is an air puff to the cornea. This activates the sensory fiber of the trigeminal nerve to the spinal nucleus of the trigeminal nerve [N. V (sp)]. From here the output fibers divide. Some mediate an immediate unconditional response (UR) at the brainstem level. Others ascend to the dorsal accessory portion of the inferior olive (I.O., DAO), which projects through climbing fibers to the interpositus nucleus (Int.) and to the cerebellar cortex to connect there to Purkinje cells. By contrast, the auditory conditional stimulus (CS) activates the ventral cochlear nucleus (V. Coch. N.), which projects to the pontine nuclei, from which mossy fibers arise that connect to the interpositus nucleus and cerebellar cortex. The output pathway from the interpositus nucleus goes through the red nucleus to the brainstem output nuclei. The dashed line shows an inhibitory connection from the red nucleus to the spinal nucleus of V, which is presumed to enable the red nucleus to dampen US activation of climbing fibers after the CR eyelid closure. +, synaptic excitation; −, inhibition. The cerebellar components of this circuit are believed to be involved in many other learned aversive behaviors. (From Thompson, 1986)

extinction. Occasionally, two units could be recorded from the same electrode; in some cases the firing patterns were similar, indicating coactivation; in other cases they were inversely correlated, indicating differential control of neighboring cortical cells. The relation of the neuronal activity to muscle movement was studied by recording the electrical activity of the muscles (electromyograph) and muscle movement. As shown in Fig. 29.8D, the impulse bursts tended to precede slightly the onset of muscle activity and muscle movement, suggesting that the bursts were part of the motor activity initiating motor performance during reinforcement, rather than simply a result of sensory feedback to the cortex from the moving muscles. The student should review Chaps. 12, 22, and 24 to appreciate the sensory, motor, and central circuits involved in this motor activity.

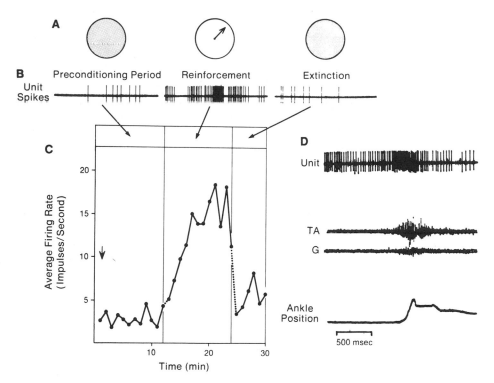

Fig. 29.8 Operant conditioning of a single neuron in the motor cortex of an awake, behaving monkey. **A.** Meter registering rate of impulse firing; when firing rate exceeded a preset level, dial became illuminated, and food reward was given. **B.** Representative records of impulse activity. **C.** Graph of impulse firing rate during successive experimental periods. **D.** Relation of muscle activity (EMG of tibialis anticus [TA] and gastrocnemius [G] muscles) and ankle movement to operant impulse burst of a reinforced motor cortical neuron. (Modified from Fetz and Baker, 1973)

Aversion Learning

For most of this century, classical and operant conditioning have dominated our concepts of how learning may occur. Each seemed to have such obvious value to an animal that it was impossible for most behavioral scientists to conceive of any other type. However, it has become clear that there are indeed other very important forms of learning. The problem first is to recognize them, and then to figure out how to study them experimentally.

An excellent example of this is aversion learning. In the 1960s, John Garcia and Robert Koelling (1966) were interested in how rats learn to associate tastes with sickness. They used the fact that strong X-irradiation of an animal damages the gastrointestinal tract and induces sickness after a period of several hours. The rats were given distinctively tasty solutions of water to drink, paired with X-irradiation. After recovering from the sickness, the rats refused to drink the tasty water. This is a laboratory demonstration of a phenomenon we all recognize: when we suspect that a food has made us sick, we lose our "taste" for the food and avoid it. In field studies of animals it is called "bait-shyness." Because it requires only one episode of sickness, it is also called "one-trial learning."

Although the pairing of an unconditional stimulus (sickness) and a conditional stimulus (taste) satisfies the criteria for associative learning, there are several differences from the situation in classical conditioning.

In classical conditioning, many trials are usually needed for transfer from the unconditional to the conditional response, and the unconditional and conditional stimuli must be timed very closely together, usually within a second or two, or else the animal cannot make the association between the two. By contrast, in aversion learning, only one trial is necessary, and the association between taste and sickness is made after a delay of several hours. These differences are so dramatic that at first few psychologists would even believe the results; one of them was quoted as commenting "These findings are no more likely than bird s____ in a cuckoo clock!" (quoted in Chance, 1979).

The basic findings have been repeatedly documented and extended to many different species (this has no implications for cuckoo clocks). Many agents can serve as the unconditional aversive stimuli; these include lithium chloride solutions, psychoactive drugs like amphetamine and apomorphine, and certain poisons. However, many agents are surprisingly ineffective; these include toxic substances like strychnine, and general factors such as stress. Electric shocks, which are so effective for classical or operant conditioning when paired with visual and auditory stimuii, have little effect. Parallel studies of the conditional stimulus have shown that one-trial learning can be demonstrated only by using stimulation of the tongue. The learning is strongest when the inputs come from the taste buds, but aversion can also be demonstrated with tactile stimulation of the tongue.

Many fascinating questions are raised by these findings, but we will consider only two here. First, what are the neural pathways for aversion learning? The taste part of the system was discussed in Chap. 11, and is reproduced in Fig. 29.9. The other pathway, mediating the aversive input, is more difficult to specify, because of the diversity of agents that may serve as unconditional stimuli. However, it is believed that much of this input is carried in visceral afferent fibers. These fibers convey sensory input from the intestines and other internal organs to the brainstem, through the tenth cranial nerve. Among the sites of termination within the brainstem is the nucleus of the solitary tract, the main sensory relay nucleus in the taste pathway. The explanation of aversion learning requires a site, or sites, at which the taste and aversive pathways meet, and the nucleus of the solitary tract appears to be a likely candidate.

In addition to the two pathways mediating the sensory inputs, a variety of other brain regions have been found to have an influence on aversion learning. These have been identified by ablation and stimulation experiments, although, as we have emphasized, these are rather crude tools. Some of the main regions identified by these means are shown in Fig. 29.9B. The diversity of these regions is remarkable. Note that many of them have been encountered before, as nodal points in limbic systems involved in mediation of feeding, arousal, and sexual functions. Each region is an integrative center that makes its special contribution to these different functions. Thus, the *nucleus of the solitary tract* is itself significant not only as the convergence point for taste and visceral inputs, but also as a region involved in arousal, as discussed in Chap. 25. As another example, lesions of the *amygdala* produce deficits in several types of learning. In the case of aversion learning, the rats seem to have difficulty in recognizing the significance of different taste stimuli, which has been interpreted as a perceptual deficit. In addition, animals with amygdalar lesions are unable to orient normally, which is believed to represent deficits in motivation.

From these considerations, the study of the neural circuits involved in aversion learning has led to a much broader perspective on learning processes in general. This view has been expressed by John Ashe and Marvin Nachman (1980) in the following way:

When an animal undergoes an experience, learning is only one of a wider complex of physiological responses that may occur. Arousal, attention, stress, and motor responses

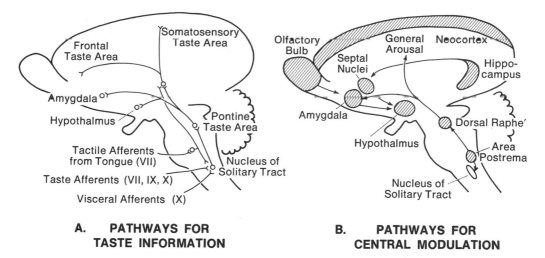

A. PATHWAYS FOR TASTE INFORMATION

B. PATHWAYS FOR CENTRAL MODULATION

Fig. 29.9 Neural pathways mediating taste aversion learning in the mammal. (Based in part on Ashe and Nachman, 1980)

are (also) . . . initiated, and it is . . . likely that these responses are coupled to associative mechanisms. . . . One of the major by-products of learned taste aversion research has been the reemphasis on the importance of understanding the total animal in elucidating principles of associative formation. In neurobiological terms . . . stimuli that initiate internal consequences that are smoothly coupled to the on-going physiology of the animal should result in robust learning and thus the animal will appear to be especially "prepared" for the acquisition of this learning.

It can thus be seen that considerable progress has been made toward understanding the neural circuits involved in aversion learning. By contrast, we know almost nothing about the cellular mechanisms that mediate these remarkably long-lasting effects. As a working hypothesis, one can speculate that the mechanisms that have been invoked for long-term changes in habituation and sensitization may be involved—that is, changes at synapses involving phosphorylation of K^+ and Ca^{2+} conductance channels, and possibly changes in DNA expression.

Invertebrates also show one-trial learning (see Table 29.4). The land slug *Limax* is a good example. *Limax* not only learns rapidly to avoid food containing a bitter substance (quinine), but also has been

shown to avoid a specific food after a single meal in which an essential amino acid (methionine) was missing from it (Delaney and Gelperin, 1986). Gelperin and his colleagues at Bell Laboratories have developed an isolated preparation of the *Limax* cerebral ganglion, buccal ganglion, and lips. They have identified several types of nerve cells that form the circuit for the feeding motor programs, and found that the program can be modulated by a serotonergic interneuron. Thus, at the cellular level there may be mechanisms that are shared with models of associative learning in other species, as we have discussed above. At the systems level, circuits for aversion learning, as well as for taste sensation, appetitive behavior (locomotion), and consummatory behavior (feeding), may be seen to be common to all animals.

COMPLEX LEARNING

The types of learning we have discussed thus far are those that have received the greatest attention from behaviorists and that have been most amenable to experimental analysis by neurobiologists. But these by no means exhaust the types of learning that occur. As indicated in Table

29.2, several additional types may be grouped under the general heading of *Complex Learning*.

Imprinting

Imprinting is the process whereby a young animal forms a behavioral attachment to a parent. It was discovered by ethologists, who found that the attachment usually depends on some special stimulus, such as the shape of the parent's body, or a particular colored spot on the plumage. A young animal can often be induced to form its attachment to an individual or object if it sufficiently resembles the specific stimulus; hence, the famous picture of young geese following Konrad Lorenz on his daily walk. Imprinting usually occurs during some early *critical period* in the young animal's life, and if it does not occur, the subsequent normal development of behavior is irretrievably lost. Thus, songbirds fail to learn their song (see Chap. 23), and animals fail to develop their adult social and sexual behaviors. Imprinting is thus an essential kind of learning in many species.

Some evidence for the brain mechanisms that are involved in imprinting has been obtained by Gabriel Horn and his colleagues at Cambridge. They have studied young chicks, which become attached to a visually conspicuous object (normally the mother hen) early in life. Experimentally, the chicks can become attached to a rotating disc. In animals trained in this manner, it was found that the incorporation of radioactive uracil into RNA was selectively increased in a part of the forebrain called the medial part of the hyperstriatum ventrale (MHV). This could reflect a growth of synaptic boutons in this region, requiring synthesis of proteins and increased RNA. Electron microscopic measurements of the synapses in the MHV showed an increase in the area of contact of about 20% over controls. Studies in other laboratories have shown changes in 2-deoxyglucose (2DG) uptake in this region during training. Thus, several lines of evidence suggest that increases in synaptic activity and synaptic effectiveness are associated with processes of imprinting in this species.

Latent Learning

Another type of complex learning occurs when an animal is introduced to an experimental environment, like a maze, and allowed to run about in it without being trained or rewarded. Although there is no evidence of learning at the time, the animal later learns an operant task in the maze much faster than an unexposed animal. This effect of experience is called latent learning.

Mechanisms of latent learning have been studied by exposing animals to a maze and then testing the effect of different brain lesions on their ability to run the maze. Daniel Kimble and his colleagues at Oregon found that animals with hippocampal lesions initially show a benefit from previous unrewarded exposure to the maze, but do not improve their performance with subsequent trials (Kimble et al., 1982). This suggests that the brain circuits that store spatial information in latent learning are distinct from those that form "cognitive maps" in learning (see Hippocampus, under Memory, below).

There is evidence that experiences such as maze running or social interactions with other animals lead to larger brains, increases in the numbers of cortical dendritic branches and spines, and even increases in the sizes of individual synapses (Table 29.1). These findings recall the observations of Darwin on the larger brains of wild compared with domesticated animals (see Chap. 30).

In the light of the perspective taken above with regard to aversion learning, familiarity with a training apparatus could facilitate learning through various ways, including alleviating stress, reducing fear, and enhancing attention and orienting mechanisms; in other words, we learn better in comfortable surroundings with a minimum of distractions.

Observational Learning

The final type of learning we will mention is called *vicarious,* or *observational, learning.* This occurs when an animal observes another animal performing a task, and then learns the task more rapidly. This is obviously extremely important in humans; it covers the way we imitate and follow examples, learn from experiences of others, and follow symbolic directions to achieve skills and attain goals. It is more difficult to demonstrate in other mammals, and, in fact, for a long time behaviorists denied that vicarious learning occurs in subhumans. There is no evidence that observational learning takes place in lower vertebrates or invertebrates.

It is believed by many behaviorists that observational learning goes beyond associative conditioning, and involves *cognitive* processes: attention, retention, and thinking. Behaviorists are only beginning to form a coherent view of these processes. Neurobiologists may soon be able to contribute their information about neural mechanisms to these emerging concepts.

MEMORY

By now we can see that *learning*—the ability to change with experience—must be intimately linked with *memory*—the capacity to store and recall those changes. Learning by classical conditioning, for example, implies that there is a memory mechanism to store and recall the effects of the conditioning. These types of mechanisms, operating over time scales from minutes up to years and lifetimes, generally fall under the category of *long-term memory,* or *associative memory.* As summarized in Table 29.5, they are involved in building up the store of information that provides for our basic knowledge, skills, and ways of thinking and behaving.

In contrast are the mechanisms that operate over very brief times of a few seconds, collectively called *working memory.* We are scarcely aware of them, but they allow us to understand what someone is saying to us as we converse, to bring to mind what we want to say, to keep in mind a telephone number to call and the myriad other kinds of information we use briefly during the business of the day and immediately forget. As noted by Patricia Goldman-Rakic (1991), "If associative memory is the process by which stimuli and events acquire permanent meaning, working memory is the process for the proper utilization of acquired knowledge."

It is obvious that these two types of memory differ distinctly in their properties, and this is likely to apply also to their mechanisms. Associative memory is closely tied to the types of learning we have discussed above, and we will therefore explore its mechanisms in the remainder of this chapter. Working memory is particularly critical for human behavior, and we will therefore discuss its mechanisms in Chap. 30 on the human cortex.

The main model that has emerged in recent years for analyzing and understanding the mechanisms underlying long-term memory is *long-term potentiation* (LTP) in the mammalian hippocampus. We will therefore review the background of this work and summarize current research on synaptic mechanisms in this system. As a preparation, the student should review Chaps. 6, 7, and 8 on the structure and function of synapses, especially glutamatergic synapses and LTP in Chap. 7; Chap. 2 on gene transcription; and Chap. 9 on neurotrophic factors in development and plasticity. This extensive review list indicates the fact that in considering the possible mechanisms of long-term memory, we will draw on most of what we have learned about the basic structure and function of synapses.

Hippocampus: Anatomy and Synaptic Organization

In order to have a model system for studying memory mechanisms, we must identify

Table 29.5 Basic memory systems

Property	Working memory	Associative memory
Contents	facts, events	facts, events
Duration	milliseconds, seconds (transient)	years, decades (archival)
Function	moment-to-moment utilization of information	acquisition of knowledge/experience
Neural localization	hippocampus, prefrontal cortex?	dentate nucleus? posterior cortex
Mechanisms	feedforward excitation; local circuits	LTP? gene expression
Productions	concepts, categorizations, plans	facts, skills, names, vocabulary, habits, rules
Colloquialisms	"scratch-pad" (Baddeley), "blackboard of the mind" (Just)	"knowing that"; "knowing how" (Squire)

From Goldman-Rakic (1991)

a brain region that has been demonstrated to be critical for memory. This criterion has been met without any shadow of doubt by the hippocampus. This was first dramatically demonstrated in the 1950s by the accidental and unfortunate discovery that bilateral removal of the hippocampus in a neurosurgical operation for relief of chronic seizures resulted in almost total loss of recent memory in the patient. Complementing this was the finding in a series of studies that brief electrical shocks applied to the hippocampal area elicit fleeting memories in awake patients undergoing neurosurgical operations for the relief of epilepsy.

The hippocampus is certainly one of the most intriguing regions in the brain. It fulfills all our criteria for a central system: a well-defined region, remote from specific sensory and motor pathways. What does it do, and how is it related to memory?

The hippocampus derives its name from its curving shape, which reminded some early neuroanatomist of a sea horse (and others of a ram's horn: Ammon's horn). This is an ancient part of the brain; as one of the first areas of the wall of the forebrain to become differentiated in primitive verte-

brates, it is called the archicortex (see Fig. 29.10A). Its functions appear to depend on the nearby septum, to which it is closely connected. As the forebrain expanded during evolution, the hippocampus got pushed and dragged around. It particularly got attached to the temporal lobe, and so as this lobe enlarged, the hippocampus was pulled along in a loop through the dorsal forebrain to the temporal lobe (B), and eventually ended up entirely within that lobe (C) in primates, snuggled up against the amygdala. It retains its primitive connection with the septum through a thick tract called the fornix, whose graceful arc describes the trajectory of hippocampal evolution.

The hippocampus is one of those regions, like the cerebellum and olfactory bulb, whose internal circuits are organized in a highly distinctive manner. This by itself is an interesting fact: stereotyped microcircuits and local circuits are used not just for processing sensory information, but also for processing information related to higher brain functions. The circuits in the hippocampus were described briefly in Chap. 7, and are shown in more detail in Fig. 29.11. The main specific inputs come

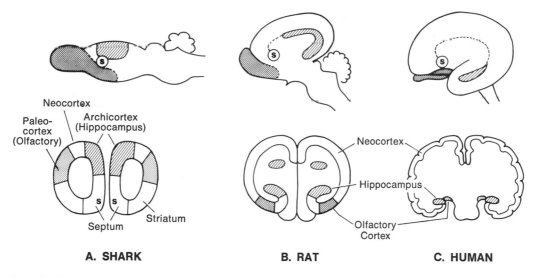

A. SHARK **B. RAT** **C. HUMAN**

Fig. 29.10 Evolution of the hippocampus, as exemplified in a lower vertebrate (shark), a mammal (rat), and the human. *(Above)* Longitudinal views. *(Below)* Cross-sectional views. (Based on Sarnat and Netsky, 1981)

Fig. 29.11 Neuronal organization of the hippocampus. A, entorhinal cortex; B, subiculum; C, hippocampus; D, dentate fascia; E, fornix; F, fibers to entorhinal cortex; G, alveus; H, periventricular gray matter; a, pyramidal cell axon; b, axon terminals; c, deep perforant fibers; d, superficial perforant fibers; e, deep perforant bundle; f, fimbria fibers; g, pyramidal neuron; h, apical dendrite of a hippocampal pyramidal neuron; i, Schaffer collateral; j, mossy fiber arising from dentate granule cell; r, recurrent axon collateral of a pyramidal neuron. (From Cajal, 1911)

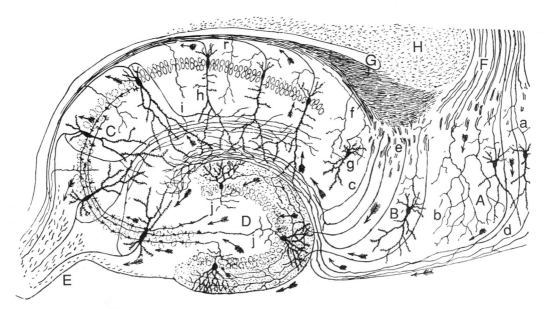

from the entorhinal cortex (A) and septum (through the fornix [E]) and the contralateral hippocampus. The inputs are excitatory, as are two pathways (the mossy fibers [j] and Schaffer collaterals [i]) for internal transfer of information. Pitted against this in the control of the output neurons are local inhibitory interneurons (see below). A delicate balance is thus set up between excitation and inhibition, a balance that can be upset by too much excitation or too little inhibition, either of which leads to uncontrolled discharges which become manifest in humans as certain kinds of epileptic seizures.

Long-Term Potentiation (LTP)

If the hippocampus is to have a role in memory, one would expect these synapses to have properties that change with use, and in fact this has turned out to be one of the outstanding characteristics of this region. In 1973, Tim Bliss and Terje Lømo in London reported studies in the intact, anesthetized rabbit, in which they recorded the field potentials evoked in the dentate fascia by a shock to the entorhinal cortex (see Fig. 29.12A). They stimulated at high frequency (tetanically) for several seconds, then tested with a single shock at various intervals after that, and found that the part of the recording due to the synaptic response of the granule cells grew to a much greater amplitude than normal (Fig. 29.12B). This phenomenon they called long-term potentiation (LTP). It is similar to, though much more powerful than, the posttetanic potentiation that is seen at the neuromuscular junction (Chap. 17). In the dentate cells, LTP persists for a surprisingly long time; as shown in Fig. 29.12, it characteristically lasts for several hours, and can even be demonstrated over periods of days and weeks. It has subsequently been demonstrated in hippocampal pyramidal cells and a number of other types of neurons as well.

Bliss and Lømo suggested several possible mechanisms for LTP, such as increased release of transmitter by the input synapses, or increased postsynaptic response. Neurobiologists were slow to assess these properties within any formal scheme until it was realized that they fit rather closely into Hebb's and Konorski's original concepts (discussed above) about the effects of activity on neural circuits. In particular, in Hebb's book of 1949 was the speculation:

When an axon of cell A . . . excite[s] cell B and repeatedly or persistently takes part in firing it, some growth process or metabolic change takes place in one or both cells so that A's efficiency as one of the cells firing B is increased.

This is the phrase that launched a thousand ships. The requirement for a conjunction of presynaptic and postsynaptic activity has come to be known as *Hebb's postulate*, or *Hebb's rule*. It has provided a formalism for guiding experiments on long-term potentiation and other candidate properties and for testing their adequacy in meeting formal criteria as the cellular basis for associative memory.

Within the hippocampus, one of the favorite models for testing Hebb's postulate has been the synaptic connections from Schaffer collaterals onto pyramidal neurons in region CA1 (see Fig. 29.12). By arranging two stimulating electrodes on these fibers and delivering weak shocks through one and strong shocks through the other, it has been possible to test whether the properties of LTP satisfy Hebb's postulate. As summarized in Fig. 29.13, weak repetitive stimulation of one input pathway may not be sufficient to elicit subsequent LTP; there must be a sufficient number of inputs, which is referred to as the criterion of *cooperativity*. Second, for weak inputs to be potentiated they must be paired with strong inputs; this is the criterion of *associativity*. Third, strong repetitive stimulation of one pathway can be sufficient to elicit LTP in that pathway, but not in any other unstimulated pathway; in other words, this is the criterion of *selectivity*.

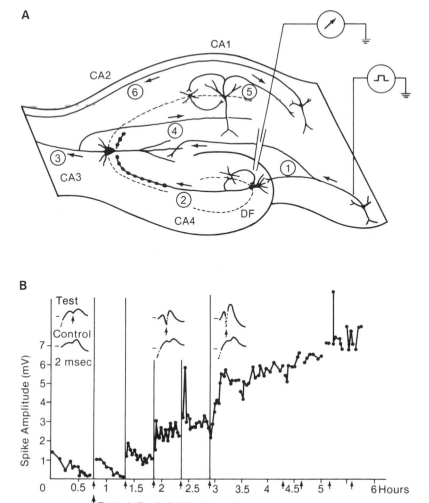

Fig. 29.12 **A.** Diagram of synaptic organization of the fascia dentata and hippocampus in the rat, showing the placement of electrodes used for stimulating the fibers of the perforant pathway and recording LTP in the dentate in the original study of Bliss and Lomø (1973). **B.** Long-lasting potentiation of the granule cell response of the dentate fascia. Electrical stimulation was applied to the perforant fibers, at a rate of 20 per second for 15 seconds; this was repeated at times marked by arrows in the graph. The amplitude of the sharp wave in the field potential (due to granule cell responses) grew, as shown in insets (arrows); the time course of the increase over a period of 6 hours is plotted in the graph. (From Bliss and Lømo, 1973)

Presynaptic or Postsynaptic Locus?

One of the key questions regarding the mechanism of LTP is whether it is primarily or exclusively based on changes in the presynaptic terminal or the postsynaptic structure. Let us consider each of these possibilities in turn, before gathering them into a synthesis of present knowledge.

Evidence for a Postsynaptic Locus

Most of the electrophysiological studies of LTP have been carried out with intracellular recordings from the target pyramidal

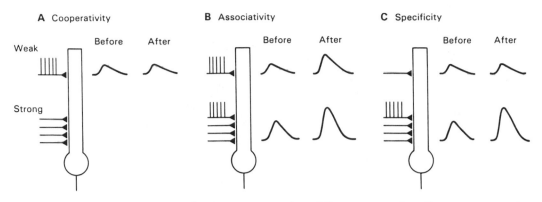

Fig. 29.13 Formal arrangements for demonstrating three different properties of long-term potentiation, by delivering weak and/or strong conditioning and testing shocks to two different subsets of Schaffer collateral fibers while recording from a pyramidal neuron in area CA1 of the hippocampus. **A.** Cooperativity is demonstrated if initial weak stimulation of one pathway (before) is sufficiently strong to produce LTP on subsequent testing (after); it does not meet that criterion in this example. **B.** Associativity is demonstrated if stimulation of the two pathways is timed appropriately. **C.** Specificity is demonstrated when strong stimulation of only one pathway produces LTP confined only to that pathway. (Based on Nicoll et al., 1988, in Kandel et al., 1991)

neurons in slice preparations of the hippocampus. These preparations enable the investigator to perform pharmacological tests of the effects of different drugs in the bathing medium on synaptic responses during LTP, as well as the effects of current injection on the responses. Using these methods it has been easiest to test the possibility that LTP depends primarily on postsynaptic mechanisms within the recorded neuron.

Among the pharmacological manipulations, the most important has been to show that blocking NMDA glutamatergic receptors with 2-amino-5-phosphonovalerate (AP5) blocks the induction of LTP. This evidence is fundamental to our present understanding of the mechanism underlying the most common type of LTP. As discussed in Chap. 7, the NMDA receptor is a channel that is blocked by internal Mg^{2+} at normal resting potentials and requires prior depolarization in order to relieve the block and allow Ca^{2+} to pass through the channel and enter the cell. This excitatory response is greatly enhanced when inhibitory inputs are blocked, which can be done

by applying picrotoxin to the bathing medium.

The electrophysiological evidence regarding LTP fits closely with the properties of the NMDA channel. If depolarization is necessary to relieve the Mg^{2+} block, it should be possible to mimic LTP by injecting depolarizing current into the cell in place of a strong excitatory input (see Fig. 29.13B). Such an experiment is shown in Fig. 29.14. The stimulating and recording setup is shown in A. The recordings of different electrophysiological tests are shown in B and C. As shown in D, LTP was elicited neither by postsynaptic depolarizing current injection alone, nor in response to a weak excitatory synaptic input (100 Hz) paired with inward current under voltage clamp (that is, no depolarization could take place). However, LTP did take place when the weak EPSPs were paired with postsynaptic depolarization. Other experiments have shown that LTP in response to strong excitatory inputs is prevented if the postsynaptic cell is injected with hyperpolarizing current, or if it is subjected to inhibitory inputs at the time

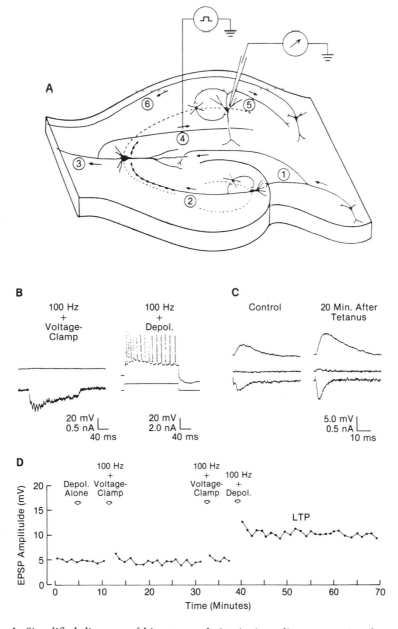

Fig. 29.14 A. Simplified diagram of hippocampal circuits in a slice preparation from the rat with stimulating and recording electrodes for demonstrating LTP. **B.** *(Left)* Voltage-clamp recording of membrane potential *(above)* and inward synaptic current *(below)* during stimulation of 100 Hz. *(Right)* Current-clamp recording of sustained postsynaptic potentials evoked by depolarizing current step with 100-Hz stimulation. **C.** Excitatory postsynaptic potential (EPSP) *(above)* and inward currents *(below)* before (control) and 20 min after 100-Hz stimulation. Middle trace shows currents under voltage clamp. **D.** Amplitude of test EPSP during application of the procedures of B and C. Only synaptic stimulation (100 Hz) paired with depolarizing current injection produces LTP. (B–D modified from Kelso et al., 1986)

of induction. Further evidence relating to postsynaptic second messenger and structural changes will be discussed below.

Evidence for a Presynaptic Locus

Some of the earliest evidence bearing on this question was obtained by measuring the release of radioactively labeled glutamate, which increases during LTP. This provides strong support for the idea that the increased effectiveness of the glutamatergic synapses during LTP is due to increased release of glutamate. Also in favor of a presynaptic locus is the analogy with the studies of sensitization in *Aplysia* mentioned earlier, where phosphorylation of ion channels may be involved. Numerous studies have reported phosphorylation of various proteins during LTP, some of which are known to be concentrated in the presynaptic terminals.

A direct way of testing for pre- and postsynaptic contributions to synaptic function is by performing a quantal analysis on the postsynaptically recorded events. This method was introduced by Bernard Katz and his colleagues in London in their analysis of synaptic transmission at the neuromuscular junction. As reviewed in Chap. 7, this work has been the foundation of our understanding of the molecular physiology of synaptic transmission. By this approach, it was shown that synaptic transmission involves the release of individual quanta, small packets of transmitter that are believed to be equivalent to the contents of an individual synaptic vesicle. As we learned, the neuromuscular junction consists of a large presynaptic terminal, with many release sites ("active zones") onto the postsynaptic muscle endplate; this situation is summarized in Fig. 29.15A. By recording from the muscle at low rates of vesicle release in response to repeated nerve stimulation, it is possible to analyze the numbers of released quanta and calculate the probability of release. This allows one, from a postsynaptic recording site, to distinguish between a presynaptic property (probability of release from the presynaptic terminal) and a postsynaptic property (the amplitude of the potentials), using well-known binomial statistics.

Application of this approach to central synapses encounters several difficulties (see Fig. 29.15B). First, individual terminals in many cases are small, with only one synapsis (= active site) per terminal. It therefore becomes problematical whether each terminal has the same number of quantal groups with the same "noise" (variance) and release probabilities. Quanta may have a postsynaptic basis, if the receptors from discrete clusters and each vesicle released saturates all the postsynaptic receptors (see Larkman et al, 1991). Finally, different terminals are situated at different sites on the dendritic tree, so that their contributions to a recording from the soma are not equivalent.

Despite these problems, several groups have been actively pursuing these types of analyses. Figure 29.15C presents an example from a study by Robert Malinow and his colleagues at Iowa. In the top graph is shown the amplitude histograms before LTP for the distribution of numbers of quantal events (excitatory postsynaptic currents) elicited by very weak stimulation of the Schaffer collaterals to a CA1 pyramidal neuron. In the bottom graph, it can be seen that LTP produced an increased separation between the peaks (note that the scale for the excitatory postsynaptic current amplitude along the abscissa is twice that for the top graph). This increased separation indicates an increased amplitude of the quantal events, signifying a postsynaptic enhancement of the response. However, there is also a reduction in the number of failures and a shift in the distribution of amplitudes toward higher values, signifying enhanced presynaptic vesicle release. Thus, these results support the idea that both presynaptic and postsynaptic factors contribute to the enhanced responsiveness that characterizes LTP.

The Role of Dendritic Spines

In the hippocampus, as in most cortical regions, excitatory synapses are made pri-

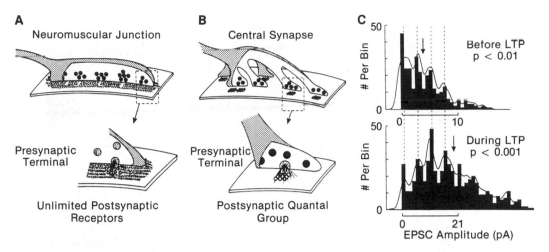

Fig. 29.15 Schematic diagrams comparing the organization of the neuromuscular junction (**A**) and central synapses (**B**). Each active zone releases a quantum of transmitter (= vesicle). Note that in this depiction the transmitter in each vesicle saturates all postsynaptic receptor (small circles in bottom diagrams). However, central synapses may not be equivalent in release probabilities or receptor numbers. (From Edwards, 1991). **C.** Analysis of quantal mechanisms at synapse between Schaffer collateral and CA3 hippocampal pyramidal neuron before (*above*) and after (*below*) LTP. Recordings were made of the unitary excitatory postsynaptic currents (EPSCs) in response to weak stimulation of the input fiber. Note that after LTP there is a decrease in failures (0 amplitude) and shift to larger quantal amplitudes (note increase in scale of EPSC amplitude, in picoamperes). (From Liao et al., 1992)

marily on dendritic spines. We were introduced to the spine as a structural unit in Chap. 6, and we now assess its structural and functional properties for their relevance to memory mechanisms.

The spine has attracted the attention of anatomists ever since it was first described by Cajal around 1890. For many years there has been a controversy about whether the spine is simply a site for receiving synaptic inputs ("spines serve only to connect") or whether spines may have special structural and functional properties that make their own contributions to information processing. A specific role in memory mechanisms was postulated by Wilfrid Rall and John Rinzel (1971), who suggested that activity-dependent changes in spine neck diameter could regulate the coupling between the synapse on the spine head and the parent dendrite, leading to increased or decreased synaptic efficacy as a basis for learning and memory. This has served as a focus for a great deal of research on spine

structure in relation to LTP and other forms of activity. It has been further noted that the equations that describe the flow of electric current between spines and dendrites are exactly equivalent to the equations that describe the diffusion of substances. Adjustments in spine stem caliber could have a critical effect on the movement of substances into and out of the spine, through which long-term control over metabolic processes in the spine could be exerted. It was therefore postulated that the spine is a device for creating a "microenvironment, whose internal composition is subject to maximal effect by synaptic action and to control through adjustments of the stem" (Shepherd, 1974).

If spines play these kinds of roles, one would expect that they would show heterogeneity in their structures and functions. Spines do indeed vary in their structure. Electron microscopic studies show that neighboring spines vary in their sizes and shapes. Some examples from the careful

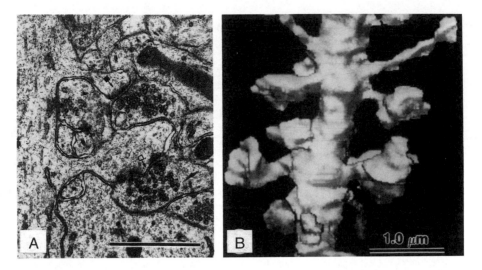

Fig. 29.16 Differences in structure of dendritic spines and synapses of pyramidal neurons in the CA1 region of the rat hippocampus. **A.** Transmission electron micrograph showing dendritic branch (on the left) giving rise to two spines which have different sizes and shapes and which receive synapses with different areas and types of postsynaptic densities (squares and arrows). **B.** Three-dimensional reconstructions of a segment of a dendritic branch with spines of varying size and shape. (Modified from Lisman and Harris, 1993)

studies of Kristen Harris at Harvard are shown in Fig. 29.16A, B. Larger spines tend to have larger synaptic areas. Harris and John Lisman at Brandeis speculate that the larger synapses "probably have more synaptic channels and are therefore likely to have greater efficacy," implying that they may be related to the kinds of activity induced by LTP. They note the "intriguing possibility that the heterogeneities in size and efficacy are the anatomists' and physiologists' views of the animal's stored memories" (Lisman and Harris, 1993). The long-term changes in synaptic size in *Aplysia* that we noted earlier in this chapter are consistent with this suggestion.

During LTP, Ca^{2+} enters through the NMDA glutamate receptor by the mechanism we discussed in Chap. 7. There is at present considerable interest in the possibility that if Ca^{2+} levels rise within the spine, they may play a critical role in the synaptic plasticity that underlies LTP. This has been directly tested in experiments in which pyramidal neurons in hippocampal slices have

been injected with the dye Fura-2, which was visualized in spines projecting from dendrites using fluorescent microscopy (Muller and Connor, 1991; Guthrie et al., 1991). Typical results are shown in Fig. 29.17 in dendrites of cells responding to stimulation of fibers making excitatory synapses on the spines. Weak stimulation resulted in focal accumulations of Ca^{2+} localized to individual identifiable spines. The focal activity did not spread to the parent dendrite or to neighboring spines. With repetitive stimulation, spine Ca^{2+} accumulation could be sustained for several minutes. In some experiments, increases in Ca^{2+} were induced in a dendritic branch; it was then observed that there was no spread from the branch into the attached spines.

These results support that idea that a spine is a semi-independent metabolic subunit, as suggested above. This independence is more than can be explained on diffusional grounds alone; it indicates that in addition there must be active mecha-

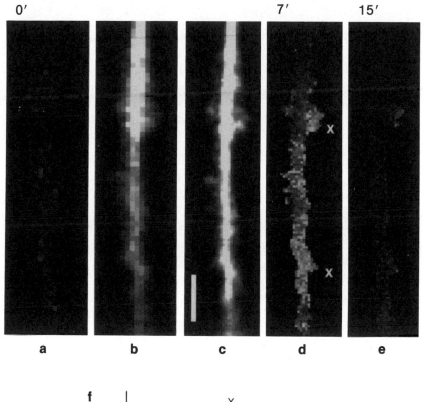

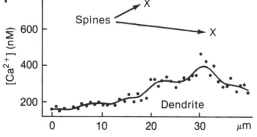

Fig. 29.17 Microscopic imaging of fluorescence of the CA^{2+}-sensitive dye Fura-2 in a dendrite at a site 200–400 μm from its parent CA3 hippocampal pyramidal cell body, which was injected with the dye through an intracellular microelectrode. a. Control, before stimulation. b, c. Fluorescence at ×20 and ×40 magnification after brief repetitive stimulation of input fibers. d. Enhanced CA^{2+} levels after 7 min repeated stimulation, showing highest levels in two spines (x). e. Return to baseline after stimulation. f. Plot of the intensity of fluorescence along the dendrite and in the two spines (x). (From Muller and Connor, 1991)

nisms for Ca^{2+} buildup and retention in the spine head. One possibility that has been suggested is local *Ca^{2+}-induced Ca^{2+} release:* the influx of Ca^{2+} could act on Ca^{2+} release channels in internal membranes, releasing more Ca^{2+} in a regenerative manner. This might be similar to the ryanodine Ca^{2+} channels in skeletal muscle (see Chap. 17). The results support the hypothesis that the spine can act as a microcompartment for NMDA receptor-mediated–Ca^{2+}-activated metabolic processes that could be involved in storage of information.

Long-Term Potentiation: A Synthesis

We can now bring together the information discussed here and in previous chapters in a model of the synaptic basis of learning and long-term memory. Since much of the evidence is preliminary, this is best regarded as a working model of possible mechanisms. This model should be considered as an expanded version of the model synapse presented in Chap. 6.

Activation of the synapse begins with impulse spread into the presynaptic terminal (Fig. 29.18, above), depolarizing the terminal membrane and activating voltage-gated Ca^{2+} channels. This leads through the steps of Ca^{2+} binding proteins and protein phosphorylation to exocytosis of synaptic vesicles and release of glutamate. The glutamate diffuses across the synaptic cleft to bind to glutamate receptors in the postsynaptic membrane. The non-NMDA (AMPA) receptors are activated to allow a net inward Na^+ current to flow, depolarizing the postsynaptic membrane (compare Fig. 7.5 in Chap. 7). The glutamate also binds to NMDA receptors, but at normal resting potentials no current flows because these channels are blocked by Mg^{2+}. However, if concurrent inputs depolarize the membrane, by the mechanisms discussed above and in Chap. 7, the Mg^{2+} block is

relieved, and a net inward current occurs which carries Ca^{2+} into the cell.

The inflow of Ca^{2+} is critical, because the Ca^{2+} acts as a second messenger to provide a link to longer lasting changes in the cell. One such action is to activate several kinds of kinases. One of these is protein kinase C (PKC), which is believed to activate NO synthase, releasing NO,

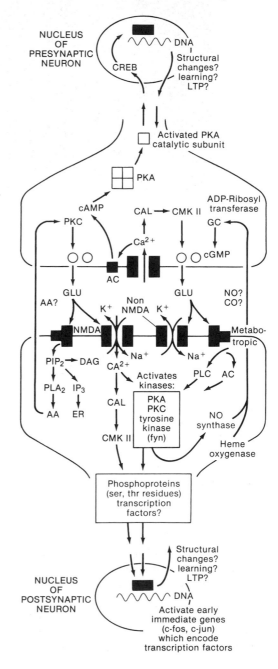

Fig. 29.18 Summary of possible synaptic mechanisms that contribute to long-term potentiation (LTP) and long-term memory. ADP, adenosine diphosphate; CAM kinase II, α-calcium-calmodulin–dependent kinase II; cAMP, cyclic adenosine monophosphate; cGMP, cyclic guanosine monophosphate; CRE, cAMP-responsive element; CREB protein, cAMP-responsive element binding protein; DAG, diacylglycerol; ER, endoplasmic reticulum; GC, guanylyl cyclase; IP_3, inositol triphosphate; PIP_2, phosphatidyl bisphosphate; PKC, protein kinase C; PLA_2, phospholipase A2; PLC, phosphoinositide-specific phospholipase C; ser, serine; thr, threonine. See text. (From Ezzell, 1992)

which diffuses to neighboring terminals, including presynaptic terminals, where it activates guanylate cyclase to produce cGMP. The target of cGMP is at present unknown; possibly it is a cyclic nucleotide channel similar to that present in olfactory neurons and photoreceptors. NO may thus provide the retrograde link that appears to be necessary for coordinating the enhancement of both pre- and postsynaptic mechanisms that contribute to LTP. Other candidates for this link are arachidonic acid (AA) and carbon dioxide (CO) (see diagram).

A second type of long-lasting mechanism occurs through binding of Ca^{2+} to calmodulin, which in turn activates Ca/calmodulin kinase II. This enzyme phosphorylates the serine and threonine groups of many target proteins over a prolonged period of time. CaM kinase II appears to play a role in LTP, because blockers of this enzyme also block LTP. Based on this finding, Alcino Silva and his colleagues at M.I.T. applied a technique known as "gene knockout" for eliminating this enzyme. In this technique, recombinant cells lacking the CaM kinase II gene are injected into mouse embryos. The chimeric mice are interbred until strains are obtained that lack the CaM kinase gene and enzyme. When hippocampal slices are prepared from these animals, LTP is lacking or greatly diminished. Furthermore, these animals take twice as long as normal animals on a spatial memory task. Parallel studies by Eric Kandel and his colleagues of knockouts of the *fyn* gene, which codes for a tyrosine kinase (see Chap. 9, Fig. 9.9), have yielded similar results in reducing both LTP and spatial learning.

On the presynaptic side, a series of studies has pointed to the possibility that long-term changes in excitability are associated with changes in gene transcription. As indicated in Table 29.1, there has been interest in increased RNA production in relation to memory dating from the 1950s, but the means to test for this adequately at the molecular level have been possible only since the advent of modern molecular tech-

niques. Among current studies, we have earlier noted that in *Aplysia* neurons long-term sensitization causes activation of adenylate cyclase in the presynaptic terminals. As indicated in Fig. 29.18, this leads to activation of protein kinase A (PKA), whose catalytic subunit is postulated to be transported retrogradely to the nucleus, where it phosphorylates a cAMP-responsive element binding protein (CREB). This in turn binds to a domain in the genome called the cAMP-responsive element (CRE) to activate transcription of proteins that may be transported back to the presynaptic terminal to bring about long-lasting changes in synaptic structure and function (see Fig. 29.4C above).

The details of these mechanisms will undoubtedly evolve with future experiments. The importance of present work is that activity-dependent changes underlying long-term memory are likely to invoke a wide range of mechanisms that overlap with those involved in development. We thus have the possibility of a unified theory of synaptic plasticity, in which gene expression and environmental influence interact in an exquisite ballet throughout the life of the individual.

Behavioral Studies

If the hippocampus has properties at the cellular level appropriate for playing a role in memory, what evidence is there for this role at the behavioral level? The most dramatic evidence was provided in 1957 by William Scoville and Brenda Milner at the Montreal Neurological Institute. They reported that one of their patients, a 27-year-old skilled mechanic known as H.M., had lost the ability to remember recent events, without any significant impairment of other intellectual abilities. This was the patient mentioned earlier in whom bilateral hippocampal removal had been performed for relief of epilepsy. Needless to say, this operation was never performed again.

H.M. has become perhaps the most studied patient in history. In a thorough reexamination 28 years after the operation, Suzanne Corkin and her colleagues in Boston reported (1981):

> He still exhibits a profound anterograde amnesia, and does not know where he lives, who cares for him, or what he ate at his last meal. . . . Nevertheless, he has islands of remembering, such as knowing that an astronaut is someone who travels in outer space. . . . A typical day's activities include doing crossword puzzles and watching television.

Corkin's studies have shown that H.M. cannot recall words that are presented to him verbally, but he can show improvement in the ability to solve puzzles and acquire perceptual skills in repeated trials. This has contributed to a variety of evidence that memory is not just one global entity, but rather consists of many subsystems, as will be discussed further below. Other interesting results in the recent studies are that H.M. has a diminished perception of pain, a lack of feelings of hunger or satiety, and an inability to identify different odors.

Learning and Memory: A Synthesis

The case of the patient H.M. stimulated workers to reproduce the symptoms in primates, in order to gain insight into the nature of the circuits that subserve memory. The first experiments in the monkey involved bilateral ablations confined to the hippocampus; these showed surprisingly limited effects. The same was true of lesions confined to the amygdala. However, H.M. lost both hippocampus and amygdala, and lesions of both these regions (the so-called medial temporal animal) in monkeys reproduced more closely the severe amnesia.

These experiments had the benefit of showing that memory is not a unitary and global process. First, memory involves distinct *steps;* there are at least three main stages: (1) a stage of *acquisition,* in which information is encoded in central circuits that include microcircuits in specific regions such as the hippocampus; (2) a stage of *storage,* in which the information is retained in those circuits, presumably dependent on changes in the synapses as postulated by Hebb; and (3) a stage of *recall,* in which the information is recalled to produce a perceptual or motor output. Stage 2 is usually recognized to consist of an early labile period of short-term memory, which then passes into a later stable period of long-term memory by a process known as consolidation.

A second important advance is the recognition of different types of information, stored as different types of memory. Among the many types that have been suggested, a useful distinction is between "knowing how" and "knowing that." "Knowing how" information is just what its name implies: knowing how to do a skilled act, such as writing or riding a bicycle, which one has learned to do successfully. "Knowing that" information, by contrast, is about specific facts or dates; it does require conscious attention and recall. These distinctions were pointed out by the philosopher Gilbert Ryle, in 1949, and have been elaborated subsequently by neuroscientists. In a formulation of Larry Squire and Neal Cohen, of San Diego (see Squire and Butters, 1984), "knowing how" memory is termed "procedural memory" and "knowing that" memory is termed "declarative memory."

These distinctions have been useful in characterizing more accurately the types of memory and identifying the types of circuits that mediate them. For instance, the learning demonstrated in *Aplysia* appears to fall in the category of "procedural memory." The reflex circuit itself is the site of the memory; there is no separate representation of the learning change elsewhere. By contrast, the amnesia of H.M. does not prevent him from learning perceptual–motor skills of the "knowing how" variety. His defect, and that of primates with similar lesions, is in retention of recent "knowing that," declarative memories. Recent tests show that this defect involves an abnormally rapid rate of forgetting that can be ascribed to inability to transfer newly acquired information to storage, or to consolidate it. In this view, the hippocampus is an essential part of the distributed circuits involved in acquisition and consolidation, but the sites of permanent memory storage must be elsewhere, because long-term memory remains and can be retrieved despite loss of the hippocampus.

The hippocampus has thus served as a useful focus for work in learning and memory. This work now extends from the level of molecular mechanisms at the synaptic level, through several levels of circuit organization within the hippocampus and be-

tween it and other regions, to the level of behavioral analysis of primate models and psychological testing in the human. The goal is to assemble a coherent framework embracing all these levels, equivalent to that for the visual task represented at the beginning of this book (Fig. 1.1), and for the many other systems that have been discussed.

Although the hippocampus is a valuable model in this respect, it must be emphasized that it is only one of many regions of the brain with specific memory functions. In the next chapter we will assess further the contributions of neocortical regions to these functions.

30

The Cerebral Cortex
and Human Behavior

In previous chapters we noted many areas of the mammalian cortex given over to specific functions related to sensory processing, control of motor outputs, and integration of centrally mediated behaviors. The diagram in Fig. 30.1 provides a summary view of the human neocortex. Specific sensory and motor areas present in lower mammals are further elaborated in primates and humans. While these specific areas are important in the evolution of human attributes, it is the other parts of the cortex that have undergone the greatest expansion. Some idea of this expansion can be grasped by realizing that the area of the cortex of a cat is about 100 cm^2, or $\frac{1}{4}$ the size of this page, whereas that of the human is about 2400 cm^2, or 6 times the size of this page. The specific areas discussed thus far account for only a small part of this total (see, for example, motor area 4, somatosensory areas 3, 1, 2, 5, visual area 17, and auditory area 41, in Fig. 30.1). It is therefore to these vast tracts of neural landscape, with billions of neurons and hundreds of billions of synapses, that we now turn to identify the circuits and mechanisms that make us uniquely human.

An understanding of higher cortical functions in humans must be built on a systematic base of knowledge. We cannot cover these areas in depth, but we can indicate what kinds of knowledge are critically important. The plan of this chapter is to use the principles you have learned in this book to construct a conceptual framework for understanding the functional organization of the cortex.

We begin by discussing the nature of human behavior and its relation to brain evolution. We then recall the *principles of neural development* that we discussed in Chap. 9, and we apply them to understanding the special properties of neocortical development in the primate. Next we use the principles of *hierarchical organization* introduced in Chap. 1 to assess how neocortical function is built up at different functional levels. We will also consider how similarities in neuronal properties and local circuits in different areas suggest a simplified view of *cortical evolution*. Finally, we consider the highest levels of cortical function, with a particular focus on *short-term working memory* and *language* and their critical roles in human cognitive functions. We shall see that a full understanding of human behavior draws natu-

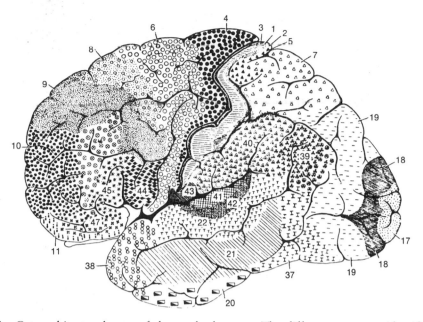

Fig. 30.1 Cytoarchitectural map of the cerebral cortex. The different areas are identified by the thickness of their layers and types of cells within them. Some of the most important specific areas are as follows. Motor cortex: motor strip, area 4; premotor area, area 6; frontal eye fields, area 8. Somatosensory cortex: areas 3, 1, 2. Visual cortex: areas 17, 18, 19. Auditory cortex: areas 41 and 42. Wernicke's speech area: approximately area 22. Broca's speech area: approximately area 44 (in the left hemisphere). (From Brodmann, in Brodal, 1981)

rally on our knowledge of neural mechanisms at all levels of organization, across all species of animals, and at all stages of development and aging.

What Makes Us Human?

It is appropriate to begin by asking, what is it that makes humans unique? The attributes that are commonly regarded as distinctive of humans are listed in Table 30.1. *Erect posture* and walking on *hind legs* allowed the *forelimbs* to be free for other functions. Recall that locomotion by the minimum number of legs expresses a trend present also in the invertebrates, as discussed in Chap. 20. The important feature was that the forelimbs did not become dedicated to other obligatory tasks, like flying birds or swinging through trees in monkeys, but rather were free to work on the environment in new and novel ways. The

Table 30.1   Human characteristics

1. Locomotion on hind legs: forelimbs free for other functions
2. Prehensile hand: making of tools and development of technology
3. Enlargement of brain relative to body size
4. Development of speech and language
5. Development of social interactions and culture: prolonged youth; division of labor in society; controls on sex and aggression
6. Individual artistic and spiritual expression

Modified from Issac and Leakey (1979)

most crucial way was provided by the *prehensile hand,* which led to *tools* and *technology.*

The parallel development of *speech* and *language* gave rise to more adaptable modes of communication, and ultimately to *symbolic thought.* These attributes involved changes in the musculoskeletal system, and associated adaptations in neural systems for sensorimotor control of posture, locomotion, throwing, grasping, vocalization, and related activities. In addition, a *prolonged childhood*, with prolonged parental nurturing (see Chap. 27), provided the basis for complex social organization and an enduring *culture*. Finally, human beings express themselves as *individuals*. The ingredients in this include emotion, motivation, and imagination; their testing ground is play; they realize themselves to the fullest in the supremely human qualities of *scientific* and *artistic expression* and *spiritual experience*.

Phylogeny of Cerebral Cortex

The human attributes listed in Table 30.1 depend on the cerebral cortex. In order to gain insight into how the human cortex has evolved, we need to consider the phylogeny of the vertebrate forebrain.

As histological stains for visualizing nerve cells and fiber tracts became available around 1900, neuroanatomists applied them across the vertebrate spectrum and attempted to identify corresponding centers and pathways in the brains of different species. This work culminated in 1936 in the monumental tome by Kappers, Huber, and Crosby entitled *The Comparative Anatomy of the Nervous System of Vertebrates, Including Man.* The view emerging from this survey was of a gradual "linear" increase in size of the forebrain and differentiation of the cortex as one ascends from fish to mammals. In fish, it seemed that the cortex was very primitive, devoted mainly to olfactory inputs (paleocortex) and higher order olfactory processing (archicortex). In amphibians, and especially in reptiles, a new cortex (neocortex) appeared; in reptiles and birds, other sensory systems began to project to the cortex through a thalamus. In mammals the neocortex, together with its thalamic inputs and closely related basal ganglia, greatly expanded. This sequence was supposed to mirror the events that lay behind the evolution of the

primate cortex. It was a satisfying view, and is summarized in Table 30.2 under "Old Ideas."

So satisfying was this view that the comparative anatomy of the nervous system came to be regarded as a rather stodgy subject, like an old exhibit in a museum. It is only since the 1970s that neuroanatomists have returned in force to rejuvenate this whole field by new studies. The surprising outcome is that most of the old ideas have had to be discarded, in the face of new evidence that significantly alters the interpretation of the evolution of the forebrain.

One of the most important new findings is that the forebrain and cortex have not undergone a gradual, linear increase, but rather there have been independent offshoots of forebrain expansion at several stages; examples of this are certain fish and the dolphin. This phenomenon should not be surprising to the reader of this book; we have already noted in Chap. 21 the enormous expansion of the cerebellum in certain species of electric fish. Similar but independently evolved structures are referred to as "homoplastic," in contrast to similar structures in a linearly evolved series, which are referred to as "homologous." The case of the dolphin brain illustrates that cortical enlargement by itself confers only limited adaptive abilities; it is the whole constellation of adaptations in Table 30.1 that is necessary for human behavior.

Another important finding has been that olfactory projections to the forebrain are much more specific than formerly believed, and that other sensory modalities are represented already in fishes by projections through thalamic relays. In studies of the visual pathway from the retina through the thalamus to the cortex, considerable variation has been found among lower vertebrates; according to R. Glenn Northcutt of San Diego, in different species the visual pathway may project ipsilaterally or bilaterally, through different central tracts, to different cortical sites. This may be re-

Table 30.2 Evolution of the forebrain (telecephalon): a comparison of old and new ideas

Old ideas	New ideas
1. There is a linear increase in size through the vertebrate series.	1. There are independent increases in size among several radiations.
2. There is early olfactory dominance and later dominance by ascending thalamic inputs.	2. There are early restricted olfactory inputs plus early restricted thalamic inputs.
3. Lower vertebrates have only a primitive cortex.	3. Main divisions of cortex are present in all vertebrates.
4. Lower vertebrates lack long descending tracts from cortex.	4. Long descending tracts are present in all vertebrate groups.
5. Functionally specific tracts are stable across different groups.	5. The same function may be mediated by different tracts in different species ("phylogenetic plasticity"). There is no truly "typical" tract or center for a given function.
6. Similar structures in different groups are homologous (evolved from a common ancestral form).	6. Similar structures in different groups are often homoplastic (they evolved independently: convergent evolution).

Based on Northcutt (1981) and others

garded as a kind of "phylogenetic plasticity," reflecting in part the flexibility of the nervous system in its routing of information through central pathways. The "New Ideas" emerging from these and related studies are summarized in Table 30.2.

Ontogeny of Cerebral Cortex

In Chap. 9, we discussed the stages involved in the development of the nervous system; we can now apply those principles to the cortex.

Neurogenesis

As we have seen in Chap. 9, during early development the birth of new neurons (neurogenesis) takes place within a proliferative zone that surrounds the central cavity (ventricle) of the embryonic brain. The neocortex in particular arises from a ventricular zone that lines the walls of the lateral ventricles of the forebrain. Within this layer the neuronal precursor cells go through repeated cycles of movements related to successive stages of cell division (mitosis), which may be summarized as follows (see Fig. 30.2).

The cycle starts with an undifferentiated stem cell containing its resting diploid

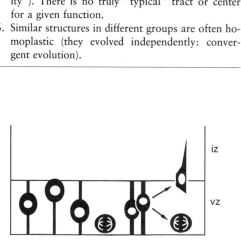

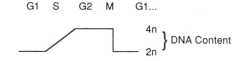

Fig. 30.2 Summary of the cell cycle in the embryonic ventricular zone of the cerebral cortex. *(Top)* Movements of the nucleus in relation to mitosis. *(Middle)* The successive stages of the cell cycle. *(Bottom)* Changes in DNA content in relation to the cell cycle. iz, intermediate zone; vz, ventricular zone. See text. (From McConnell, 1992)

DNA (stage G1). The nucleus then moves outward (away from the ventricle) as it initiates DNA replication (S phase). (It may be helpful to review the molecular mechanisms underlying replication in Chap. 2.) When the nucleus has achieved doubling of its DNA content through replication (from 2n to 4n in the diagram), it moves back toward the ventricle (G2 phase) and near the surface it initiates mitosis (phase M). Mitosis gives rise to two daughter cells, each again with the diploid content of DNA. There are two possibilities for the nature of the daughter cells. One possibility is that one of the daughter cells becomes a neuron and starts migrating along a radial glial fiber toward its final place in the cortex. The other is that it becomes another undifferentiated stem cell that reenters the cell cycle. If the result is one neuron and one stem cell, it is called an *asymmetric* cell division; this is the predominant mode during early development, because it is the necessary condition for maintaining a supply of stem cells to produce large numbers of neurons. If the result is two neurons, the division is called *symmetric;* this terminates any further genesis of neurons from this cell line.

Neuronal Migration

Neurons characteristically migrate from their site of birth (the last mitosis) to their final position, and the neocortex is no exception. As shown by studies in which animals are sacrificed shortly after radioactively labeled thymidine injections are given, all of the final mitoses take place in the proliferative zones. The neurons must therefore migrate to the primitive cortical plate. They are guided in this by the special type of cell, the radial glia, which spans the distance between the proliferative zone and the cortical plate. This mechanism has been elucidated by the elegant studies of Pasko Rakic, at Harvard and at Yale; we have already discussed this as a general mechanism (Chap. 9), and its particular role in the cortex is illustrated in Fig. 30.3. The neurons are all generated within a period of about 60 days in the monkey (100 days in the human), and the migration is complete in the monkey by day 100 (gestation lasts 165 days).

Within the cortex, the first neurons are laid down in the deepest layers, with later neurons migrating to more superficial layers; hence, there is an "inside-out" sequence of cortical development. Thus, according to the timing of its last mitosis, each neuron appears to be destined as a specific cell type in a specific layer. According to Rakic (1981):

The redistribution of such a vast number of cells during development undoubtedly provides opportunities for establishment of essential relationships and key contacts that eventually determine the radial and tangential coordinates of each neuron in the 3-dimensional map of the neocortex. Thus, the separation of proliferative centers from the final residence of neurons is of great biological significance.

Given this significance, the mechanisms of migration have come under intense scrutiny in recent years. For this purpose, preparations of isolated cells and cell slices have been developed. Analysis of the migration of granule cells in cerebellar slices from young rats has been carried out by Pasko Rakic and his colleagues using the lipophilic dye DiI, which stains the lipid cell membrane, and observing the cells under laser scanning confocal microscopy. Pharmacological blockade of N type Ca^{2+} channels was found to retard cell migration selectively (Komuro and Rakic, 1992). Ca^{2+} is known to play diverse roles in neurons (see Chap. 8), and it was suggested that its role in these migrating neurons might be essential for initiation and execution of the movements. Further studies by Komuro and Rakic have shown that migration is also selectively inhibited by blockade of NMDA receptors, and, conversely, is accelerated by the application of glutamate, providing added support for a role for Ca^{2+}. Thus, a more vivid picture is beginning to emerge of the immature neuron fighting its way through a primeval forest of nerve processes, fueled by transmitter actions and internal Ca^{2+} actions.

Although most of the migration of neurons is along radial glia, some evidence has emerged for lateral movements as well. Constance Cepko and her colleagues have marked precursor cells with a retroviral technique and mapped the clonally related cells in the cortex (Walsh and Cepko,

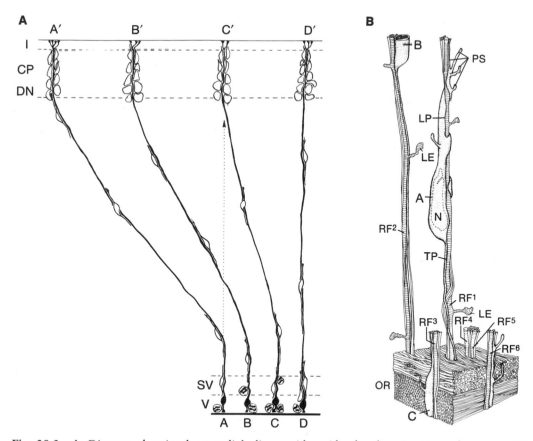

Fig. 30.3   A. Diagram showing how radial glia provide guides for the migration of neurons (N) from ventricular (V) and subventricular (SV) zones to the cortical plate (CP). Within the CP, the new neurons migrate past deeper neurons (Dn) to layer I (I). Note that despite shift of cortical position (A'–D') relative to proliferative zones (A–D), the radial glia preserve the topographical relations. **B.** Relations between migrating neuron and radial glia fiber, based on electron microscopy. Six radial fibers (RF$^{1-6}$) are shown traversing the packed fibers of the optic radiation (OR). Three neurons (A–C) are shown at various stages of migration. Neuron A is shown with nucleus (N), leading process (LP), and pseudopodia (PS). (A from Rakic, 1981; B from Rakic, in Jacobson, 1978)

1992). They found that in some cases cells from a single precursor were grouped as expected from migration of daughter cells along the radial glia, but in other cases cells were widely dispersed, indicating that these cells must have migrated laterally. In slice preparations it has also been reported that immature cells labeled by DiI may be seen migrating across radial glia. At present the consensus is that most migration is along radial glia, but some movement may also occur laterally. At stake is the question of whether cortical areas are predetermined by strictly radial migration from the ven-

tricular zone, or whether there is significant lateral postmitotic movement so that the cortex would be more similar throughout, and the different areas would have to be determined by other factors such as the ingrowth of different afferent fibers.

Neuronal Identity

An equally important question concerns the many different types of neurons within the cortex; how is neuronal identity, also called the phenotype of the neuron, determined? The reader should recall that we have already encountered this question in

Chap. 9 in discussing the mechanisms by which neural crest cells achieve their final identity as they are born and migrate through the mesoderm to their final destinations. As we observed, there are two contrasting types of mechanisms that might be involved. A cell might have its identity determined by its lineage at the time it is born, or it might be born pluripotent and have its identity determined by the environment through which it migrates.

The general conclusion from numerous studies of cortical cells is similar to the conclusion regarding neural crest cells. The phenotype of a cortical cell appears to be determined by a series of steps, starting with a broad potential at the birth of a neuron which is progressively narrowed by interactions of the developing cell with the environment through which it migrates. Susan McConnell and her colleagues at Stanford have shown that the laminar identity of neurons in the cortex is normally specified around the time of the last mitotic division (cf. Fig. 30.2); in other words, it is determined by lineage. However, experimental manipulations can alter cell fates in some cases, so it is clear that the final morphological and physiological properties of a cortical neuron are the result of progressive interactions with other elements within the developing cortex (see McConnell, 1992, for review).

Neuronal Maturation

Within the cortex, the larger pyramidal neurons (projection neurons) mature first, followed by the smaller interneurons (local circuit neurons). In comparison with the relatively brief period for cell birth, maturation is a much more prolonged process. An indication of this is given in Fig. 30.4. Note the small sizes of the cells in the newborn and their limited dendritic branches, and the subsequent growth and differentiation of the cells. It is apparent from this evidence alone that early childhood is a time of rapid and profound maturation of the cells in the cortex.

Maturation of Synapses

The maturation of cortical neurons is associated with maturation of cortical circuits. This can be documented and quantitated by counting. One might assume that the maturation of synaptic circuits would proceed linearly in parallel with the maturation of neuronal form shown in Fig. 30.4. One might have further predicted that the synaptic circuits would be laid down first in the primary sensory and motor areas (because they mediate lower levels of cortical sensory and motor processing) followed by secondary areas and lastly the association areas (which mediate higher levels of cortical processing).

Surprisingly, none of these expectations has been borne out. A direct test was carried out by Rakic, Jean-Pierre Bourgeois and colleagues, who measured the density of synapses in different cortical areas of the monkey at different ages during development (Rakic et al., 1986). This mammoth undertaking, involving the identification of over 500,000 synapses in 25,000 electron micrographs from 22 monkeys, revealed that the density of synapses follows a very similar time course during development of different regions. The basic pattern (see Fig. 30.5) is a steady increase in density during fetal life, reaching approximately adult levels at birth. The density continues to increase, however, reaching a peak at two to four months of age and then declining, at first rapidly and then slowly to the adult level at two to four years. Regardless of whether the region is primary sensory, primary motor, association, or limbic, the results are very similar, as shown in Fig. 30.5.

These results indicate, first, that synapses in cortex are formed as elsewhere in the nervous system, with an overproduction followed by elimination (cf. Chaps. 9 and 29). With synaptic maturation comes the formation of functioning cortical circuits. Second, the similar time course of overproduction and subsequent decline to a similar neuronal density in all cortical regions sug-

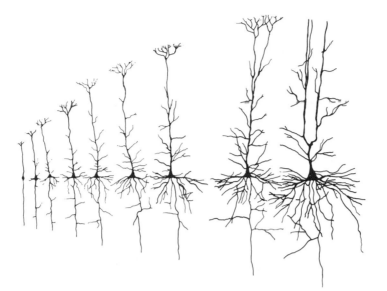

Fig. 30.4 Growth and differentiation of the dendritic trees and axon collaterals of cortical pyramidal cells in the human, from fetus to adult. (Courtesy of P. Rakic)

gests that "the cortex develops as a whole rather than regionally" (Rakic et al., 1986); in other words, there may be a basic cortical plan under common genetic control. We will explain this basic plan further below. Finally, the early development of association areas is in accord with behavioral studies showing the early development of certain cognitive functions, such as ability to perform delayed-response tests.

It is important to realize that maturation of neurons and synapses does not occur in a vacuum; in addition to the timetable prescribed by the genetic program, a neuron reaches its final form under the influence of its environment. The old debate of nature vs. nurture has been rendered obsolete, and we now recognize that both factors shape the neuron. To speak of one without the other is like asking what is the sound of one hand clapping. Furthermore, the question is not just whether a neuron by itself will mature, but whether it will survive in the competition for available nutrients, synaptic connections, and functional validation. Thus, the competition of organisms for survival in the external world mirrors a competition in the inner world among neurons to fashion the circuits that will be most effective in the external world. We shall see clear evidence of the relations between the two worlds when we discuss the organization of cortical circuits.

We may summarize this section by noting that ontogeny, like phylogeny, is not a simple linear process. It involves a series of stages, in which the organism at one stage gives rise to the next stage through the complex interaction of a number of factors. This is evident at the cellular level in the development of the cortex, and it is also evident in the behavior of the developing fetus and child. This has been summarized elegantly by Myron Hofer of Albert Einstein College of Medicine as follows (1981):

Development is characterized by transformations in which certain functional patterns come into being that are not found in previous or even in subsequent stages. Functionally as well as structurally, it is not the same creature at two different stages of its own development. The differences between stages of development in a single animal are almost as profound as those between species of vertebrates. The impli-

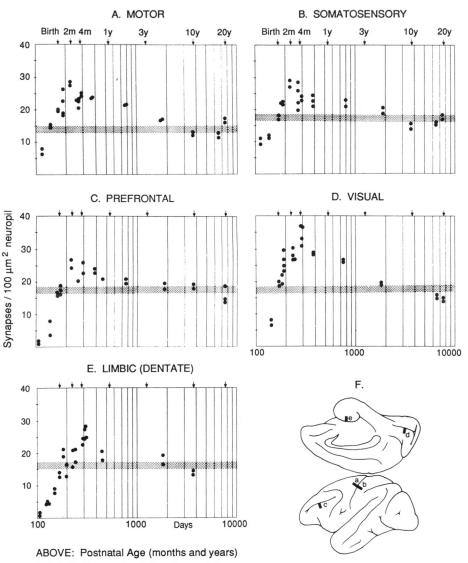

Fig. 30.5 The time course of development of synapses in five different regions of the cerebral cortex in the monkey is depicted in A–E: **A.** Motor cortex. **B.** Somatosensory cortex. **C.** Prefrontal cortex. **D.** Visual cortex. **E.** Limbic cortex (dentate molecular layer). The inset (**F**) shows lateral *(below)* and medial *(above)* views of the left cerebral hemisphere, depicting the regions examined: a, motor cortex (area 4); b, somatosensory cortex (area 1); c, prefrontal association cortex (area 9); d, visual cortex (area 17); e, molecular layers of dentate gyrus. The numbers identify the areas in the human according to Brodmann (see Fig. 30.1). In each of these areas, a track of cortical tissue that passed through all layers (I–VI) was examined with 100 serial electron micrographs in order to count all the synapses along that track. This yields a value for number of synapses/100 μm^2 of neuropil (defined as cortical tissue minus cell bodies, blood vessels). Each point is the value obtained from a single track. The horizontal stippled stripe denotes the average value in the adult as a reference. Note the logarithmic scale, spanning from conception to 20 years. Parts A–E are graphs of the data for animals of different ages before and after birth. (From Rakic et al., 1986)

cations of this fact are difficult for adults to fully grasp, accustomed as we are to relative stability in our remembered past experience. It means that we cannot generalize about the mechanism of an action, the effect of a stimulus, or the long-term impact of an experience from one stage of development to another. We have to understand exactly how the developing infant interacts with its environment separately at each age.

Levels of Cortical Organization

We are now in a position to tackle the key question: What are the cortical mechanisms that underlie our uniquely human abilities? The nineteenth-century idea was that higher human faculties actually reside in the cortex, and constitute a kind of supreme executive council which hands down orders to its neural minions. These faculties were believed to be embedded in some fashion in the billions of cells distributed throughout the cortex.

It should be obvious to the reader of this book that the modern view is radically different. Rather than being like a self-contained college of bishops and cardinals, the cortex actually contains many areas that function at multiple levels of management of the processing of signals within sensory and motor systems. Furthermore, there is increasing evidence that higher functions are mediated by distributed systems, in which the cortex is only one part. Thus, the significance of any given region in the cortex is to be found in the internal organization of its synaptic circuits and the external organization of its connections to other regions, cortical and subcortical. The integrative actions of synaptic circuits generate particular functional properties, and the external connections determine the contributions of these properties to the distributed systems of which they are a part.

In order to understand how cortical circuits function, we must use the concept of levels of organization that was introduced in Chap. 1 and has been a central theme in discussing systems throughout this book. The main levels of cortical organization

thus far characterized are summarized in Table 30.3. Each of these is a specific example of one of the levels of system organization, extending from molecule to behavior, depicted previously in Figs. 1.1 and 1.2. We will discuss each of these specific levels in turn, with the ultimate goal of obtaining insight into the relations between them and human mental activity.

Genes, Molecules and Channels

The level of the gene in relation to cortical development has already been discussed. The next two levels in the cortex correspond to the same levels in other systems. The properties of ion channels may be reviewed in Chaps. 5 and 7, and the properties of receptors for neurotransmitters and neuromodulators in Chaps. 7 and 8. The specific relation of these properties to cortical functions such as memory has already been discussed in Chap. 29, and we will discuss further examples below. As far as is known, the significance of these two levels of organization lies not in unique properties of the molecules, receptors, or channels in the cortex, but rather in their integrative context within cortical circuits.

Synapses and Spines

At the first level of circuit organization is the individual synapse and its associated pre- and postsynaptic structures. Although one may think of a single synapse as a simple link, or circuit element, we have seen (cf. Chaps. 6, 8, and 29) that it is a complex functional unit in its own right, with a variety of time- and use-dependent controls.

This is demonstrated vividly by synapses in the cortex, especially those made onto the dendritic spines of pyramidal cells. The increase in cortical synapses during development is largely associated with spines; as shown in Fig. 30.6, the spines in a 7-month-old human fetus are few in number and irregular in shape, whereas the stouter dendrite of even an 8-month-old infant is bristling with well-formed lollipop-shaped

Table 30.3 Functional units at different levels of cortical
organization

General neural systems levels	Specific cortical systems levels
System	Distributed systems
Pathway	Multiple cortical representations
	Lobes
	Hemispheres
Local region	Cortical areas
Local circuit	Basic cortical circuits;
	Cortical modules
Neuron	Complex integrative units
Microcircuit	Multiple spine units; local active regions
Synapse	Spine units, local active sites
Channels and receptors	Channels and receptors
Molecules and ions	Molecules and ions
Genes	Development; activity

Compare this table with the diagrams in Figs. 1.1 and 1.2. Each level is defined by
one or more molecular, structural, or functional units that provide the basis for the
organization of the level above.

spines. Each of these spines is the site of a
synapse, usually a type I synapse (see Chap.
6) with an excitatory action. Synapses on
spines account for up to 70% of all synapses in the cortex of rhesus monkey; during development, the ratio of type I (mostly
spine) to type II (mostly soma and dendritic
shaft) synapses increases during the overproduction phase and then decreases during the elimination phase, suggesting that
overproduction and elimination mainly involve the spine synapses (see Rakic et al.,
1986).

Cortical synapses have been shown to be
extraordinarily sensitive to a number of
environmental influences. In the visual cortex, the number of spines on dendritic
shafts of deep pyramidal cells is reduced
following removal of one eye in newborn
rodents. This could be due to several factors, including loss of spines already
formed, or failure of dendrites to mature.
A reduction in visual experience, caused by
raising animals in the dark, also leads to a
reduction in numbers of synapses in visual
areas, and to reduction in sizes of individual synapses.

Even more dramatic results are seen in
experiments comparing the cortices of ani-

mals raised in enriched environments, containing lots of toys to play with and mazes
to run in, with those raised in barren cages
(see Table 29.1 and Greenough, 1984). The
animals raised in enriched environments
have thicker cortices and bigger synaptic
contacts. Similarly, the cortex may be up to
one-third thicker in wild animals compared
with those that have been domesticated.
The fact that the brain is reduced in size
in domesticated animals was actually first
noted by Darwin. Finally, we may note
that dendritic spines are affected in certain
kinds of neurological diseases that produce
mental retardation; the derangements of
spines that occur in Patau's syndrome and
in Down's syndrome (mongolism) are
shown in Fig. 30.6A.

The fact that cortical synapses, and especially those on spines, are sensitive to environmental influences should not be surprising. Spines in the lateral geniculate nucleus
are also sensitive to visual deprivation, as
shown in Fig. 30.6B, and many examples
could be cited in other systems. Among the
most interesting are spines of neurons in
the corpora pedunculata, the highest integrative centers in the brains of insects (see
Chap. 23). Brandon and Coss (1982) made

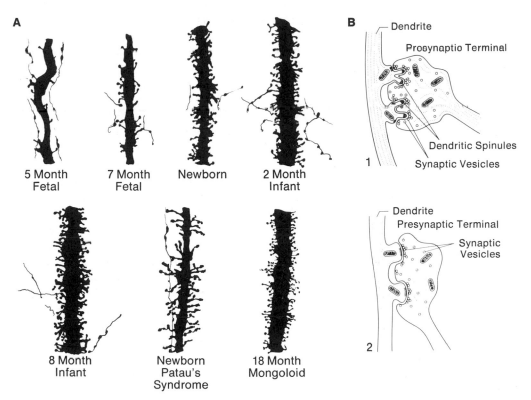

Fig. 30.6 **A.** Increase in number of spines on apical dendrites of large pyramidal neurons of layer V of human cortex, at five different ages before and after birth. In a disease known as Patau's syndrome, the spines are fewer, elongated, and irregular in form. In Down's syndrome, the spines are thin and tiny. **B.** Effects of deprivation on spines in the lateral geniculate nucleus of the dog. 1, normal; 2, animal reared in the dark. (A from Marin-Padilla, in Lund, 1979; B from Hamori, in Hofer, 1981)

careful measurements of the dimensions of these spines in honeybees just before and just after their first orientation flight from the hive; they found that the spine necks become significantly thicker after this single flight.

These findings begin to suggest why synapses are so often situated in spines. Building on our previous discussions in Chaps. 6 and 29, each spine can be conceived of as creating a microenvironment, wherein the postsynaptic response is modulated by use and exerts its first and most immediate effects on its neighbors (see Chap. 29). Each spine thus acts as a miniature input–output unit, whose properties depend on its history, its metabolic machin-

ery, its inputs, and its interactions with its neighbors.

Multiple Spine Units

The next level of organization is the microcircuit level, in which the pattern of synaptic connections and the interactions they mediate give rise to basic operations of information processing. Many instances have been cited in this book of such operations, which include spatial summation, temporal summation, and feedforward and feedback inhibition. It has been emphasized that there are two main neuronal substrates where these operations are carried out. The predominant site is within dendritic trees, with local output either through dendrodendritic output synapses or by means of intradendritic spread to control impulse output in the axon. The other site is within axonal terminals, where output can be modulated by axoaxonic synapses.

In the cerebral cortex, most of the synapses are on spines of dendritic trunks and branches, so it may be assumed that dendritic microcircuits provide the main substrate for synaptic interactions. Since much of this dendritic substrate is remote from the cell body, our knowledge of the basic properties that are involved is limited. However, information has been obtained by recording intracellularly from dendrites in isolated cortical slices (cf. Chap. 7). These experiments have supported previous evidence suggesting that cortical dendrites, like dendrites of many other types of neurons, contain ionic membrane channels that are voltage sensitive. It has been believed that these sites are located at branch points, where they serve to boost the responses of distal dendrites.

The ubiquitousness of voltage-dependent channels (see Chap. 5) has suggested that they may also be present in spines. The way that this would contribute to spine responses and spine interactions has been explored in computer simulations. A simulation consists in representing a portion of a dendritic tree with its spines by a system of compartments (see Fig. 30.7A, B), each compartment comprising the electrical properties of a dendritic segment or spine as discussed in Chap. 5 (see Fig. 30.7C, D). It was shown that an active response in a spine would boost the amplitude of the synaptic response spreading out of the spine (Fig. 30.7E). It was further shown that current spreading passively out of one spine readily enters neighboring spines, where it can trigger further active responses; thus, distal responses can be brought much closer to the cell body by a process resembling saltatory conduction in axons.

A third interesting property is that the interactions between active spines can be readily characterized in terms of logic oper-ations. Thus, an AND operation is performed when two spines must be synaptically activated simultaneously in order to generate spine responses (E). An OR operation is performed when either one spine or another can be activated by a synaptic input. Finally, a NOT-AND operation occurs when a response can be generated by an excitatory synapse if an inhibitory synapse is simultaneously active (F).

The interest of these simulations is that these three logic operations, together with a level of background activity, are sufficient for building a computer. This result of course does *not* mean that the cortex *is* a computer; rather, it helps to define more clearly the nature of the synaptic interactions that take place at the microcircuit level. Defining these interactions more clearly is a step toward identifying the basic operations underlying the functions of higher levels of circuit organization. One can speculate that interactions of this nature within dendrites may underlie some of our higher cognitive functions, such as logical and abstract reasoning. If so, the precise deployment of active ionic channels in the membranes of distal dendrites and dendritic spines of cortical neurons could be added to the list of attributes contributing to the unique qualities of humans (see Table 30.1, above.)

Cortical Neurons as Integrative Systems

Dendrites and dendritic microcircuits combine within a neuron to form a complex integrative system. There are many types of neuron in the cortex. These can be characterized simply as either pyramidal and nonpyramidal types. We have already considered the pyramidal type in several contexts (see Chaps. 6, 25, and 29, and see Fig. 30.4 above). This is the main output neuron of the cerebral cortex. The nonpyramidal type includes numerous subtypes, most of which function as interneurons with primarily inhibitory actions.

We discussed the strategy of organization of the cortical pyramidal neuron in Chap. 6. The student should review this

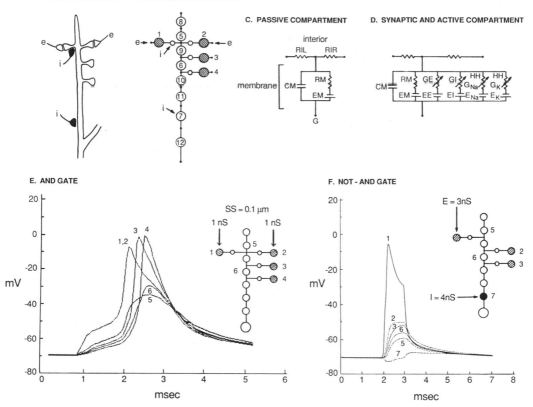

Fig. 30.7 Multiple spine interactions could generate logic operations in dendrites of cortical neurons. **A.** A distal dendritic segment of a cortical neuron is depicted in simplified form, consisting of a branch with four spines. Four possible synaptic inputs are shown: two excitatory (e), and two inhibitory (i). **B.** Compartmental representation of the system in A. Compartments with passive membrane properties are shown by open circles; those with synaptic and active properties are shaded. **C.** Equivalent electrical circuit of a passive compartment in B. CM, membrane capacitance; EM, membrane potential; G, electrical ground; RIL, RIR, left and right limbs of internal resistance; RM, membrane resistance. **D.** Equivalent circuit of a synaptic and/or active compartment in B. GE, excitatory synaptic conductance; EE, excitatory equilibrium potential; GI, inhibitory synaptic conductance; EI, inhibitory equilibrium potential; HH, Hodgkin-Huxley model for nerve impulse; G_{Na}, Na conductance; E_{Na}, Na equilibrium potential; G_K, K conductance; E_K, K equilibrium potential. **E.** Computer simulation of the response of the system when simultaneous excitatory inputs are delivered to spines 1 and 2 (see insert). Each synaptic response consists of an increase in the excitatory conductance (GE) of 1 nS. The spine stem diameter, which controls the spread of electric current between the spine head and dendritic branch, is 0.1 μm. With these parameters, spine heads 1 and 2 generate simultaneous impulses, which spread passively through the dendritic branch (compartments 5 and 6) to trigger impulses in spines 3 and 4. The system thus functions as an AND gate in logic terms. **F.** Computer simulation of the response of the system when an excitatory input to spine head 1 (with conductance changes of 3 nS) is paired with an inhibitory input (conductance change of 4 nS), sited at a distance on the dendritic branch. By itself, this excitatory input was sufficient to generate an impulse in spine head 1, which then spread passively to elicit impulses in spines 2 and 3 (not shown). The effect of the inhibition was to block the impulse in 2 and 3, but not in 1; the blocked impulse spread only passively throughout the system, as shown by the transients 2–7. This provides the basis for this system to function as a NOT-AND gate. Not shown are simulations of the system functioning as an OR gate. The simulations were run on an electrical circuit analysis program (ASTAP). (From Shepherd and Brayton, 1987)

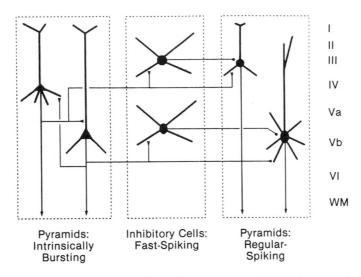

Fig. 30.8 Diagram summarizing the correlation between morphological types of cortical neurons, their location in the cortical layers, their main axonal connections, and their functional characteristics. Layers I–VI are shown at the right, surface above and white matter (WM) below. See text. (From Chagnac-Amitai and Connors, 1989). The patterns of impulse discharge may be reproduced and studied in Huguenard and McCormick (1994).

organization, to appreciate the multiple processing units in the dendrites, the division into apical and basal dendritic systems, the multiple levels of excitatory inputs within these dendritic systems, the functions of reexcitatory collaterals in feeding back excitation onto an excited cell, the flow of excitatory inputs through these dendritic systems and their gating by inhibitory inputs from local to global levels within the neuron.

Although pyramidal neurons in the different layers appear to differ mainly in size, interesting differences in structure and function have begun to emerge (see McCormick et al., 1985). Some neurons respond to prolonged excitatory inputs (injected current or synaptic excitation) by a regular series of impulses, and are therefore called regular spiking (RS) cells. By contrast, others respond with a burst of impulses, and are called intrinsically bursting (IB). In superficial layers the cells tend to be of the RS type, whereas in deeper layers both types are found (see Fig. 30.8). By contrast, interneurons in all layers are fast spiking (FS). The functional significance of these

different firing patterns is still under investigation. It has been postulated that the IB cells form a network of excitatory interactions which can initiate and coordinate synchronous synaptic activity within local populations of cortical neurons (Chagnac-Amitai and Connors, 1989). Connors and his colleagues have suggested that this may perform a crucial synchronizing role in normal cortical function. It may also underlie the development of epileptic seizure activity if the balance between excitation and inhibition is disturbed, so that too much excitation or too little inhibition leads to uncontrolled activity. Maintenance of this balance is therefore a crucial requirement for integration within dendritic systems as well as in the local circuits between neurons.

Local Circuits

As in every other region in the brain, the synaptic circuits in the cortex are built up out of the triad of neural elements: input fibers, output neurons, and intrinsic neurons. By combining methods for intracellular staining and recording with cellular identification of transmitters, it has been possible to begin to construct the basic circuits that characterize each of the three main types of cerebral cortex: palaeocortex (olfactory cortex), archicortex (hippocampus), and neocortex.

The Basic Cortical Circuit. Insight into the complexities of neocortex can be gained by considering the simple types of cortex first. We discussed olfactory cortex in Chap. 11 and hippocampus in Chaps. 7 and 29. Careful consideration of the organization of these regions shows that each consists of a basic circuit constructed of three types of connections: (1) primary afferent fibers make excitatory synapses onto distal apical dendritic spines of pyramidal neurons; (2) intrinsic axon collaterals are reexcitatory (RE) to pyramidal neurons over long distances; and (3) local interneurons are activated by afferents to give feedforward inhibition (FI) and/or axon collaterals to give lateral inhibition (LI). These connections can be identified in the basic circuit diagram of Fig. 30.9A.

In view of the fact that submammalian vertebrates have a cortex (Fig. 29.10), it is of interest to know its fundamental plan. Studies (see Kriegstein and Connors, 1986) have yielded a basic circuit which is in fact very similar to that for olfactory and hippocampal cortex. This is significant because submammalian cortex is believed to be similar to the phylogenetic precursor of neocortex (see Fig. 29.10A).

These facts suggest that the basic circuit for olfactory–hippocampal–general cortex may represent a framework that has been elaborated into the neocortex in the course of mammalian evolution. The way this might have come about is indicated in Fig. 30.9. As shown in B, agranular association cortex is similar to simple cortex. The primary input is from other parts of cortex (see Cortical Inputs), and this terminates mainly in the most superficial layers. Pyramidal cells mediate reexcitation of pyramidal cells through direct connections of their axon collaterals (RE). Feedforward inhibi-

tion (FI) and lateral inhibition (LI) are also mediated much as in simple cortex.

In granular sensory cortex, such as visual cortex, the afferent input ends mainly in a specific population of stellate cells that defines a new layer, lamina IV, midway through the cortex (see Primary Afferents in Fig. 30.9C). This population of cells can be seen to function as a kind of intracortical relay, receiving the thalamic input and transferring it to the pyramidal neurons, providing thereby a more complex preprocessing of information prior to its input to the pyramidal neurons. This is part of the evolutionary process of encephalization, in which functions carried out in peripheral centers in earlier species are transferred to central centers in order to enhance the complexity of processing by interactions with other central centers.

What Makes Neocortical Circuits Unique? We can now begin to get a perspective on the properties that are special about the neocortex. First, it is placed where it is accessible to every major sensory input, arriving either directly (from the olfactory cortex) or relayed from below through the brainstem and thalamus. Second, it is a layered structure doubled back on itself, so that inputs arrive from the depths and outputs exit through the depths. This leaves the cells in every layer potentially accessible to every input. When the local circuits through collaterals and interneurons are added to the picture, the potential ways by which information can be integrated, stored, and recombined become enormous. Such a structure is no longer dominated by the operational sequence demanded of a particular input or output. Third, rather than there being one type of output cell, as is common in so many centers, there are several types. In fact, each layer is the source of output fibers; some fibers go only to other layers (such as outputs from layer I and IV), others go to different distant targets (such as outputs from layers II, III, V, and VI; see Fig. 30.9). As a result, each layer, in effect, acts as a semi-independent

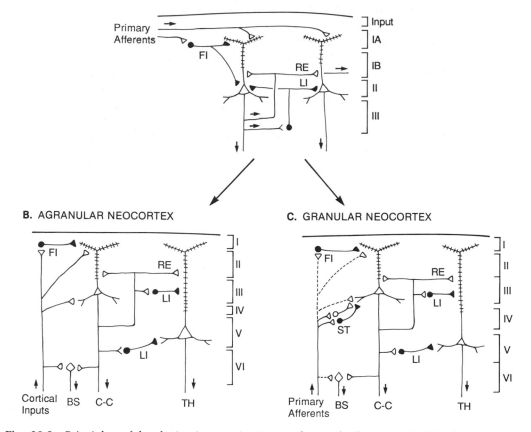

Fig. 30.9 Principles of local circuit organization in the cerebral cortex. **A.** Simple cortex, as represented by olfactory cortex, hippocampus, and reptilian dorsal (general) cortex. Layers indicated on right. FI, feedforward inhibition; LI, lateral (and feedback) inhibition; RE, recurrent (and lateral) excitation. **B.** Agranular motor and association areas of neocortex. Layers on right. Output sites below: BS, brainstem; C–C, corticocortical; TH, thalamus. **C.** Granular sensory and association areas of neocortex. Note that the main difference from the circuit in B is the predominance of primary afferents from the thalamus which project to an intracortical relay through stellate (ST) cells in layer IV. Connections to other layers are still present, but less prominent (dashed lines). Excitatory connections shown by open profiles; inhibitory, by closed profiles. (Modified from Shepherd, 1987)

unit, defined by its particular inputs, outputs, intrinsic connections, and relations to its neighboring layers. Finally, give this structure the chance to expand, through some property of its not-quite-rigid braincase, and one has the opportunity to go on enlarging individual areas, or adding on new ones, in order to combine information from new combinations of inputs or control different combinations of output targets.

Local Circuit Modules. In our studies of nervous systems, both invertebrate and vertebrate, we have seen that local circuits tend to be arranged not diffusely, but rather in discrete clusters, or modules. The ganglia of invertebrates, and the discrete regions of neuropil called glomeruli in the vertebrate olfactory bulb, are perhaps the clearest anatomical expressions of this modularization. In the cortex, modularization is expressed in many ways; some of them are summarized in Table 30.4.

Modules are not static, hard-wired entities. We have noted that individual neurons and synapses must compete for functional validation and survival, and it should not be surprising therefore to learn that the same applies to the populations within modules. This has been seen most clearly in the visual cortex.

In the monkey, the fibers from the two lateral geniculate nuclei first project during development in a diffuse and overlapping manner to the visual cortex; as Rakic has shown, it is only about two weeks before birth that segregation into columns dominated by input from one or the other eye is first seen. This process of establishment of ocular dominance columns is complete in the monkey by about three to six weeks after birth (Fig. 30.9A). We have already noted (Chap. 9) that the sorting out of the terminal fields depends on activity. Injection of TTX, which blocks inward Na^+ currents and thereby impulse activity in the axons of the visual pathway, prevents the formation of the ocular dominance domains (Shatz, 1990). The expression of specific molecules by specific types of neurons is also dependent on activity. A cell surface proteoglycan called Cat-301 is expressed selectively in cats by Y cells (equivalent to the M cells in monkeys) but is reduced by TTX injections and other deprivation procedures (reviewed in Hockfield and Kalb, 1993).

Hubel and Wiesel showed in the early 1960s that if one eye is closed at birth, the ocular dominance columns of the normal eye expand. They demonstrated this with unit recordings, and it has recently been documented by injecting radioactively labeled amino acids into the spared eye and making autoradiographs of the distribution of the transported material in the cortex.

Table 30.4 Modules in cerebral cortex

Hippocampus	transverse lamellae
Somatosensory cortex	modality-specific columns glomeruli (barrels)
Visual cortex	ocular dominance columns orientation columns hypercolumns
Motor cortex	colonies "columns"
Entorhinal cortex (fetal human)	glomeruli
Frontal association cortex	columns

References in Shepherd (1979), Goldman-Rakic (1981); for entorhinal cortex, I. Kostovic (personal communication)

As shown in Fig. 30.10B, the terminal fields of the fibers from the normal eye have expanded, at the expense of the neighboring fields from the deprived eye. Moreover, the fact that this can be demonstrated with deprivation beginning at 5½ weeks, when the normal adult ocular dominance columns are already established, suggests that the expansion of the fields from the normal eye involves an actual sprouting of the fibers. There is evidence that this plasticity after monocular deprivation depends on the presence of norepinephrine in the cortex, with the implication that this might be one of the "state-setting" functions of the NE fibers from the brainstem to the cortex.

Plasticity in cortical modules has also been implicated in the orientation selectivity of cells in the visual cortex. In kittens exposed only to a pattern of vertically oriented stripes, the recordings from the cortex show units that tend to respond maximally to stimulation with vertically oriented stripes. This implies that the circuit connections within orientation columns are dependent to some extent on visual experience.

Cortical Areas and Lobes

At the next higher level of organization are areas and lobes. The mammalian cortex consists of four main lobes: occipital, pari-

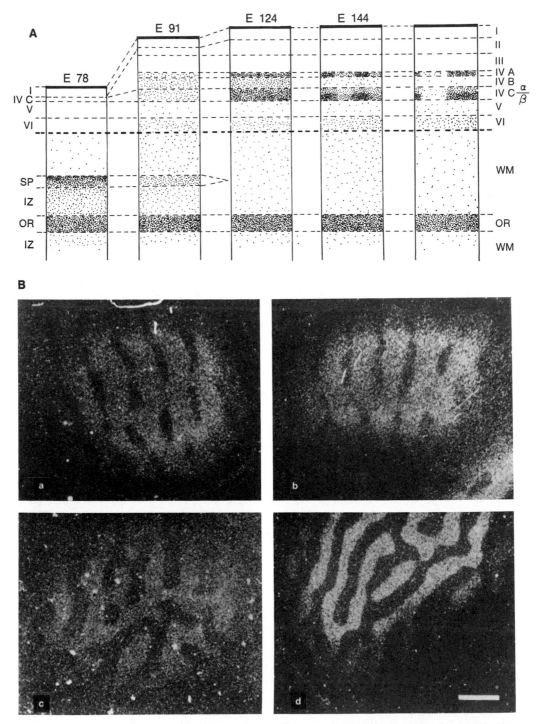

Fig. 30.10 A. Summary of establishment of ocular dominance columns in the developing rhesus monkey. The five diagrams start at embryonic day 78 and end with the adult. For each age, a monkey had been injected with radioactively labeled amino acids into one eye 14 days previously; the amino acid was incorporated into protein by the ganglion cells in the retina of that eye and transported by axonal transport to the lateral geniculate, where interneuronal transfer took place to geniculate cells that projected to the visual cortex. The diagrams show how the pattern of

670

etal, frontal, and temporal (Fig. 30.11). These terms are already familiar. You already know that the occipital lobe receives visual input, the parietal lobe receives somatosensory input, the temporal lobe receives auditory input, and the frontal lobe is the origin of many motor pathways.

The areas of each lobe that are not directly related to a specific sensory or motor function have traditionally been termed *association areas*. Since these are the areas that have undergone greatest expansion in the human brain, it has been commonly assumed that they have a large role to play in the attributes that are distinctly human. Three main functions have been ascribed to these association areas. First, a surprisingly large expanse of what appeared to be "association" cortex is actually given over to multiple representations of sensory or motor fields. This was a common theme in our discussion of visual, somatosensory, auditory, and motor cortex. Second, increasingly complex processing takes place within these multiple sensory areas; this is seen clearly, for example, in the abstraction of visual information in the occipital lobe. Third, the higher association areas are increasingly concerned with multimodal integration of information from other lobes. The capacity to integrate higher-order sensory information and use it to control different kinds of motor outputs lies at the heart of many of our higher cognitive functions.

Distributed Cortical Systems

It would be convenient if each lobe subserved one of the higher mental functions, but this is not generally the case. Most cortically mediated behavior depends on interactions between areas in different lobes. This is nicely illustrated by the very first type of behavior considered in this book, the act of reading from a page. As shown in Fig. 1.1, this involves visual input to the occipital lobe, processing of the visual information in areas of the occipital lobe and the neighboring parietal and temporal lobes, and control of eye movements and other motor output through the frontal lobe. Thus, cortical circuits in all the lobes must be engaged in a coordinated manner in order to carry out this simple act.

From these considerations the old idea that cerebral functions are organized in terms of lobes is giving way to the new idea that cerebral functions are organized in terms of distributed systems, as we noted in Chap. 1. Each area within a lobe contributes the special operational properties mediated within its local circuits, the centers being tied together by multiple long tracts, collateral branches, and feedback connections. A crucial feature is that different areas of the cortex are accessible to each other, so that there are maximal opportunities for the areas to interact. Within each area there is maximal opportunity for different inputs to tap or utilize specific properties of the local-circuit machinery, as we have already described above.

There is increasing experimental evidence for the pathways that connect the areas of different lobes to form distributed systems, and for the cognitive functions which they mediate. We will consider several of these distributed systems, involved in higher visual processing, memory for

termination within the cortex changed from diffuse (E 91) to columnar (E 144 and adult). Cortical layers 1–6 are indicated at the side. IZ, intermediate zone; OR, optic radiation; SP, subplate layer; WM, white matter. **B.** Experiments showing effects of long-term monocular deprivation on ocular dominance columns at different ages in the rhesus monkey. Deprivation was begun at 2 weeks (a), 5½ weeks (b), 10 weeks (c), and in the adult (d). In each case, the normal (nondeprived) eye was injected with radioactive amino acids. In the photomicrographs, the sites of label appear white. Note the expansion of the terminal fields of the geniculate input from the nondeprived eye up to 5½ weeks. (A from Rakic, 1981; B from LeVay et al., 1981)

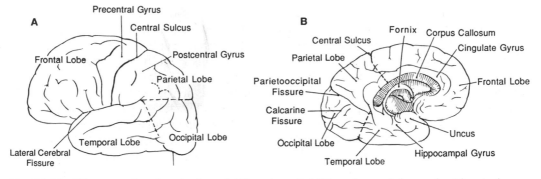

Fig. 30.11 Diagrams showing the lateral (**A**) and medial (**B**) surfaces of the cerebral hemispheres. Note the division of the lateral hemisphere into four major lobes. The main gyri and sulci are also indicated. Compare with the more detailed cytoarchitectonic map of cortical areas in Fig. 30.1. (Modified from Daube et al., 1986)

representational knowledge, personality, and language.

Higher Visual Processing

Processing of visual information is not limited to the visual association areas of the occipital lobe. By tracing connections between areas, it has been found in the monkey that the connections from the primary visual cortex diverge in two main pathways. As illustrated in Fig. 30.12A, the ventral pathway passes from primary visual cortex (OC) to successive association areas (OB, OA) and thence to the temporal lobe. Within the temporal lobe, there are connections from the posterior (TEO) to the anterior (TE) part. The reader may compare this scheme with the higher resolution map of cortical areas in Fig. 16.19.

The functions mediated by this pathway have been revealed by behavioral tests such as that illustrated in Fig. 30.12B. In this test, the animal is first familiarized with one object, and is then presented with a second object and rewarded for selecting it if it is different from the first. This requires synthesis of the physical properties of the first object (size, color, texture, shape), storing of a central representation of that combined information, and discrimination of the second object with that representa-

tion. According to Mortimer Mishkin at NIH, monkeys can carry out this type of recognition after only one exposure to the test object, even after a delay of several minutes (see Mishkin et al., 1983). Lesions of area TE render the monkey unable to perform the recognition.

By contrast, a dorsal pathway diverges within occipital cortex (OB, OA) to reach inferior parietal cortex (PG; see Fig. 30.12A). The type of visual information processed by this pathway is illustrated in Fig. 30.12C. In this test, a "landmark" object, such as a tall cylinder, is positioned randomly in successive trials between two identical rectangular food wells; the monkey is rewarded for choosing the feed well closer to the landmark. Lesions of the inferior parietal area severely disrupt this type of spatial vision. Mishkin et al. conclude that "posterior parietal cortex seems to be concerned with the perception of the spatial relations among objects," in contrast to inferior temporal cortex, which is concerned with the intrinsic qualities of an object, not its position in space (cf. Fig. 16.20). In addition, the dorsal pathway is a polysensory area, mediating "a supramodal spatial ability that subsumes both the macrospace of vision and the (tactile) microspace encompassed by the hand."

What are the mechanisms whereby visual

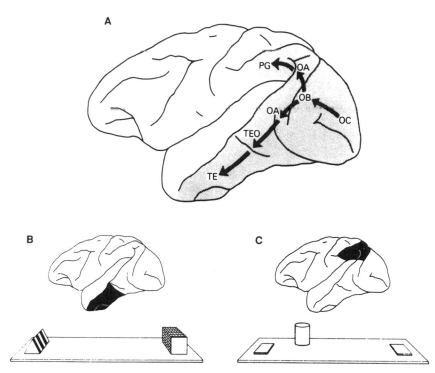

Fig. 30.12 Distributed cortical systems for higher visual processing. **A.** Diagram of lateral surface of the rhesus monkey brain. Shading indicates cortical areas involved in visual processing. Arrows show dorsal and ventral pathways that have been identified by anatomical and behavioral studies. OC, primary visual area (area 17; V1); OB, secondary visual area (V2); OA in ventral pathway represents tertiary visual areas (V_3, V_4); OA in dorsal pathway represents tertiary visual areas (area MT; see Figs. 16.19 and 16.20); PG, inferior parietal cortex; TEO, posterior temporal area; TE, inferior temporal cortex. TE connects through the amygdala to many other regions (see Chap. 28). Both PG and TE have connections with the frontal lobe (see below). (From Mishkin et al., 1983)

and tactile information are integrated in the posterior parietal area in controlling the hand? Vernon Mountcastle and his colleagues have succeeded in recording from units in this area in awake, behaving monkeys (see Fig. 30.13). Most of these neurons responded to one or another submodality of muscle, joint, or skin stimulation, similar to neurons in the primary somatosensory cortex (see Chap. 14). However, some units showed complex properties. One type was active only when the animal performed a movement of the arm or a manipulation of the hand—in other words, movement in the immediate extrapersonal space. Another type was active only when the animal visually fixated on an object in which it had an obvious interest, such as an item of

food when the monkey was hungry. The activity continued even when the object was moved in space. These experiments thus gave evidence of higher levels of abstraction, in which the mutual effects of location in space of a limb, location in space of an object, and the motivational state of the animal combine to specify the activity of a single cortical cell. Dependence of activity on multiple contingencies is one of the defining characteristics of higher abstraction. It was concluded that the kind of sensory integration displayed by this type of neuron could serve as a basis for higher abstraction, and be concerned with spatially orienting the animal toward behavioral goals (cf. Chap. 22).

In conclusion, we may note that already

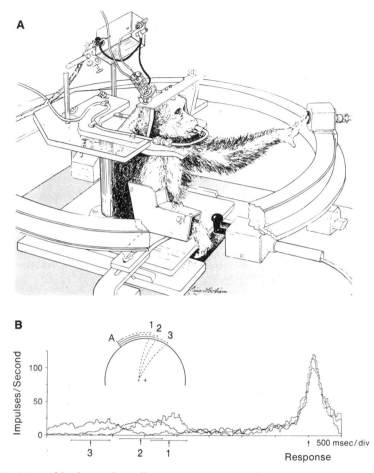

Fig. 30.13 Activity of higher-order cells in posterior parietal association cortex related to control of purposive movement. **A.** Monkey sitting in recording and test apparatus. The small box containing a signal light moves along the circular rail. The monkey starts by resting its left hand on the signal key (see black knob), then releases and projects the hand to touch the moving signal light when it is illuminated, as shown. **B.** Activity of an arm projection neuron. In this experiment, the signal light moved through three different arcs (inset); solid lines show the movement before, dashed lines the movement after, the signal came on. The trajectories of the hand, from the starting position (+) to the final points (1–3) at which the finger contacted the signal light, are also shown by dashed lines. Histograms of the firing frequency of this neuron are superimposed in the graph, aligned on the instant at which the finger touched the signal light (Response). Horizontal bars (1–3) below the abscissa indicate the detection times for the different trials. The similar firing patterns of this neuron during different arm trajectories indicate that the activity of this neuron is related to command signals for this type of purposive movement, rather than to detailed instructions for individual muscle contractions. (From Mountcastle et al., 1975)

by the eighteenth century, philosophers had postulated a "sensorium commune," where sensory information comes together to form a coherent representation of our perceptual world. The discussion above indicates that we are making progress toward

understanding the neural mechanisms involved in this process. Indeed, Mountcastle (1986) has stated the modern credo of this endeavor: "Any aspect of perception identified in the behavior of a non-human primate and brought under quantitative experimental control can be studied directly at the level of mechanism."

Internal Representations and the Frontal Lobes

If the brain constructs internal representations of the external world, the question arises of how it uses those representations to guide the behavior of the animal. Study of this problem has implicated the frontal lobes.

In 1937, Carlyle Jacobson, working in John Fulton's laboratory at Yale, studied the effects of frontal lobe lesions. The experimental paradigm was the "delayed-response" test. As illustrated in Fig. 30.14A, the monkey first watches as the experimenter puts food in one of the wells and then covers both wells. There is then a delay period (middle diagram) during which an opaque panel is lowered. The panel is then raised, and the monkey is tested for its ability to select the well with the food. Jacobson showed that lesions of the prefrontal area of the frontal lobe disrupt the ability to remember the location of the food. This loss of "delayed response" learning is quite specific; the performance on other complex learning tasks is unaffected.

A simple way to characterize these results is: "out of site, out of mind." The results imply that the frontal lobe is involved in circuits which construct an internal representation of the visual information about the place of an object, and then read out that information to control a motor response at an appropriate later time. There is a direct and interesting parallel with humans in this regard; human infants show the same inability to perform the delayed-response test as frontal-lobe–lesioned monkeys. The ability to perform the test is not manifest until around one year of age. The implication from the animal experiments is therefore that the prefrontal cortex is responsible for

the capacity to recognize that an object exists in time and space when it is not in view. In

children such a basic function may be a building block for object internalization and symbolic reasoning, crucial components of cognitive capacity. (Goldman-Rakic, 1984)

What are the pathways that mediate these crucial functions? Neuroanatomists have used a variety of tract-tracing methods to investigate the connections between the prefrontal cortex and other parts of the brain. The results of a comprehensive study by Goldman-Rakic and Carmen Cavada are summarized in Fig. 30.14B. Note first the strong connections between prefrontal and parietal association cortex; these provide the pathways whereby supramodality spatial and visual information is transferred to the frontal lobe. Second, each connection is reciprocal, so that the parietal areas receive input from the prefrontal area; this provides a means whereby prefrontal cortex can "participate in the regulation of the attentional and motor commands for the task" (Goldman-Rakic, 1987). Third, there are multiple parallel pathways between the subareas in the two lobes; this implies that each pathway is distinctive in the type of information that is processed. Here again we see an expression of the parallel architecture of the brain, even as one approaches the highest levels of information processing.

Working Memory

The delayed response test is a prime example of the use of working memory. In the previous chapter we focused on the mechanisms underlying long-term memory, leaving short-term memory for this chapter. This does not mean it is less important; as Goldman-Rakic (1991) has observed,

The significance of working memory for higher cortical function is not necessarily self-evident. Perhaps even the quality of its transient nature misleads us into thinking it is somehow less important than the more permanent archival nature of long-term memory. However, the brain's working memory function, that is, the ability to bring to mind events without direct stimulation, may be its inherently most flexible mechanism and its evolutionarily most signifi-

A. DELAYED RESPONSE TEST

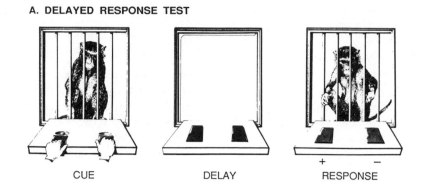

CUE   DELAY   RESPONSE

B. PREFRONTAL - PARIETAL CONNECTIONS

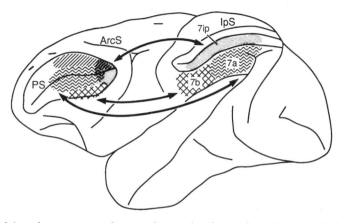

Fig. 30.14 The delayed response and its mediation by the prefrontal–parietal distributed system. **A.** The delayed response test, consisting of three stages: "cue" (the monkey is shown food being put into one of two wells); "delay" (an opaque panel hides the monkey's view of the wells for 1–10 minutes); "response" (the monkey selects the well with the food). Lesions of the prefrontal cortex selectively disrupt the ability to perform this task (see text). **B.** Lateral view of the monkey brain, showing the connections between prefrontal and parietal cortex that mediate spatial memory underlying the delayed response. Note the parallel and reciprocal connections between subareas in each cortex. ArcS, arcuate sulcus; IpS, intraparietal sulcus; PS, principal sulcus; 7ip, 7a, 7b, subareas of area 7 of parietal cortex. (From Goldman-Rakic, 1987)

cant achievement. . . . Working memory confers the ability to guide behavior by [internal] representations of the outside world rather than by immediate stimulation and thus to base behavior on ideas and thoughts.

A Neuronal Model for Working Memory. As shown above, modern research has left no doubt that the prefrontal cortex within the frontal lobe plays a critical role in working memory. In seeking to explore

the cellular mechanisms of this function, Goldman-Rakic, Charles Bruce, and their colleagues have developed an oculomotor delayed response paradigm. The reader should review at this point the vestibulo-ocular pathway which controls the rapid jumping movement of the eyes, called a saccade, by which we direct our eyes toward a target (Chap. 14). In their paradigm, impulse activity is recorded from a single cell in the frontal eye field of an awake monkey. The monkey is trained to fixate on a central visual target light while a second light is briefly flashed at different locations in the visual field. After a delay period, the animal must move its eyes to the site where the second light last appeared. The experiments show that a given neuron tends to increase its firing when the light at a particular location disappears from view, and increases it even further when the saccade is then made to that site. Similar enhanced activity during the delay period has been observed in other brain regions connected to the frontal eye fields; these include the posterior parietal cortex, basal ganglia, and the motor and premotor areas, overlapping with the areas indicated in Fig. 30.14 for spatial memory. The results (see Goldman-Rakic et al., 1992) suggest that neurons in these areas form a distributed network for mental representation, each area specialized for a particular function involved in a representation. Thus, it appears that the prefrontal cortex temporarily holds directional information about the site of the stimulus and transmits this to the motor cortex, where it is recorded as delay-period activity for the subsequent direction of the movement. The sum of activity in these neurons constitutes the way that the animal keeps information "in mind" for immediate use.

Working Memory, Human Intelligence, and Mental Disorders. Although short-term, working, memory is found in most mammals, it appears to be especially highly developed in humans. There is increasing evidence that it may be a critical factor in the evolution of human intelligence. It is closely linked to the evolution of language. When you converse with someone, you are using working memory to hold "on line" the other person's speech in segments of several seconds to scan the words and obtain their meaning, while continuing to receive the next segment of words for subsequent scanning. Some feel that this close association of working memory and language may come the closest to defining the special basis for human intelligence.

If working memory is thus critical for human mental processes, it suggests that disorders of working memory may underlie mental disorders. Neuroscientists and neuropsychiatrists have been especially interested in the relation between working memory disorders and schizophrenia. A common trait of schizophrenics is that they are easily distracted by presentation of immediate stimuli. This has been called the "tyranny of external stimuli." Not surprisingly, these patients do not do well on "delayed-response" tests of the type shown in Fig. 30.14. And brain scans show that these patients have reduced blood flow to the prefrontal cortex. Goldman-Rakic (1991) has brought together the experimental and clinical data summarized above and suggested that to the extent that the prefrontal cortex is involved in working memory, schizophrenia may involve "a breakdown in the processes by which representational knowledge governs behavior." Thus, research on working memory not only gives insight into the neural basis for human intelligence, but it may also provide leads for prevention and treatment of the most serious disorders that affect that precious ability.

Personality and the Frontal Lobes

The role of the frontal lobe in reading out internal representations of stored information is only part of a larger role involving, in humans, the entire personality of an individual. This was first indicated by the sad case of Phineas P. Gage in the early nineteenth century. Following a deep wound to his frontal lobes in a railroad construction accident, his personality changed from that of a responsible foreman to being erratic and undependable.

A century passed before the experiments of Jacobson revived an interest in the neural basis of this change. In his animals, the prefrontal lesions not only disrupted the specific ability to perform on the delayed-response test, but also induced changes such as reducing the expression of agitation in frustrating situations. This result led to the introduction of operations for frontal lobotomy (surgical undercutting of the frontal lobes) in schizophrenia patients who displayed uncontrolled aggressive behavior. Unfortunately, the results of these operations were inconsistent, and the attendant changes in personality, many of them remindful of the Gage case, were generally unacceptable.

These results made clear the need for much more research on the neural basis of frontal lobe function in relation to normal personality and to psychotic illness. A review of this large body of work, and the theories it has spawned, is beyond the scope of the present account. Currently, there is considerable interest in the roles of specific neurotransmitters and neuromodulators, especially dopamine. Abnormalities in dopamine neurotransmission have long been suspected in schizophrenia, and there is therefore great interest in the mesocortical dopaminergic innervation of the frontal lobes (see Chap. 24). A variety of neuropharmacological manipulations have documented the antipsychotic effects of neuroleptic drugs on dopaminergic systems, including the prefrontal cortex (see Bunney, 1984). These studies support the belief that psychotic behavior is due to malfunctioning of specific brain circuits, and the hope that normal function can be restored by drugs acting on the synapses of these circuits.

Hemispheres: Laterality and Dominance

All the lobes together constitute a hemisphere, and the whole forebrain thus consists of two hemispheres. Just as each hemisphere is differentiated into lobes, so are the two hemispheres differentiated from each other in mediating the highest levels of cerebral function.

The first evidence that the hemispheres are different came from the French neurologist Paul Broca; in 1863 he described a patient with the inability to speak (aphasia) who turned out to have a tumor in the left frontal lobe. Broca deduced that this is the area of cortex that controls speech; in his

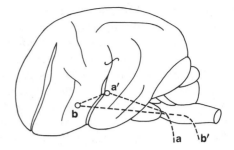

Fig. 30.15 Wernicke's original drawing in 1874, illustrating his concept of the brain circuits involved in language. Sounds received in the ears are converted into neural signals and transmitted through the auditory pathway (a) to the brain, where the sound "images" are stored in Wernicke's area (a'). These neural images are transferred to Broca's area (b), where they activate the descending pathway for motor control of speech (b'). The diagram indicates how lesions in a' give rise to sensory aphasia (an inability to understand spoken words or organize words coherently), whereas lesions in b give rise to motor aphasia (an inability to articulate spoken words). (From Kolb and Whishaw, 1985)

words, "We speak with the left hemisphere." In 1875, Carl Wernicke, a 26-year-old German neurologist, reported that aphasia can also be caused by a lesion in the temporal lobe. He showed the difference between sensory aphasia, the lack of ability to formulate words due to temporal lobe damage, and motor aphasia, the inability to produce words (speak) due to frontal lobe damage. His diagram of the way these two regions are involved in the control of speech (Fig. 30.15) was one of the first attempts to identify the brain circuits underlying specific behavioral functions.

This work thus clearly established that the left hemisphere is "dominant" for a specific function, speech. And there the matter rested. Until the 1950s, this stood as an isolated exception to what appeared to be the general equivalence of the two hemispheres in all their other functions, sensory and motor. Then R. D. Myers and

Roger Sperry carried out a series of elegant experiments in cats, in which they cut the corpus callosum, the thick band containing millions of fibers that connect the two hemispheres. Until that time, no significant function had been ascribed to these fibers. When visual stimuli were presented to both eyes, the animals behaved as competently as normal cats. However, after the decussating fibers in the optic chiasm were cut and each eye was tested separately, it was found that each hemisphere functioned independently; visual learning in one hemisphere was not transferred to the other hemisphere.

Sperry and Michael Gazzaniga then examined a series of human patients in whom the callosum had been cut to prevent spread of epileptic seizures. These studies confirmed that, with independent visual input to one hemisphere or the other, the hemispheres also functioned and learned independently of one another. This work, which earned Sperry the Nobel Prize in 1981, led to our present concepts of the laterality of higher functions in the human brain. The left hemisphere is dominant for control of speech, language, complex voluntary movement, reading, writing, and arithmetic calculations. The right hemisphere is specialized for mainly nonlinguistic functions: complex pattern recognition in vision, audition, and the tactile senses; the sense of space, spatial shapes, and direction in space; the sense of intuition (see Fig. 30.16).

There have been many attempts to read a lot into these differences, such as the idea that the left hemisphere is scientific whereas the right is artistic. A useful generalization is that the left hemisphere is specialized for certain specific kinds of motor output, whereas the right is more specialized for global relations of the body in space, something akin to the perceptual Gestalt referred to in Chap. 10. Recalling our previous discussion of different types of information and their storage as memories (Chap. 29), it further appears that the right hemisphere is specialized for handling procedural information, and the left hemisphere for declarative information. No matter how one characterizes the two hemispheres, it is important to realize that neither one is "dominant" in the absolute sense; each constellation of functions is of adaptive value, and the human brain attempts to optimize both by letting the hemispheres specialize in these two directions.

In recent years, lateralization of function has been found in a number of species, invertebrate as well as vertebrate; we have noted several instances in this book. It may well be that it is an inherent tendency in an animal with bilaterally organized body and brain.

Language

The simple circuit first proposed by Wernicke as the basis for cortical mechanisms in language still holds, but it has been considerably elaborated by recent studies. An important finding has been the discovery that there is an anatomical asymmetry associated with the functional asymmetry of language in the human brain. This simple fact had escaped most neuroanatomists and neuropathologists, in their examinations of the brains of humans obtained post mortem, until 1968.

In that year, Norman Geschwind and W. Levitsky at Harvard reported that measurements of the right and left temporal lobes showed a striking difference. In most brains, an area called the "planum temporale," located on the upper border of the temporal lobe and extending deep into the Sylvian fossa, is considerably larger on the left. This is shown diagrammatically in Fig. 30.17. The planum temporale contains Wernicke's speech area. It is tempting to conclude that the larger speech area on the left is correlated with left-hemispheric dominance for language in most humans. This, of course, does not explain *how* Wernicke's area mediates language, but it indicates that one of the mechanisms for obtaining greater complexity of information processing, that of increasing the number

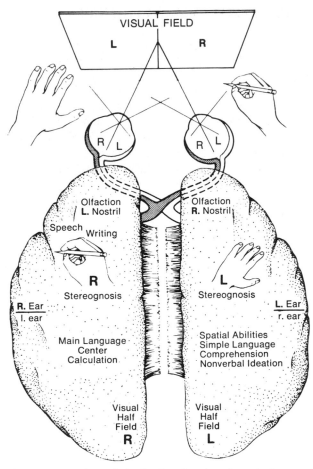

Fig. 30.16 Schematic representation of the brain, showing the relative specialization of the two hemispheres and their relations to sensory inputs and motor outputs. Cutting of the corpus collosum is shown in the midline. (From Sperry, 1974).

Fig. 30.17 Diagram of a horizontal section through the human brain, showing asymmetry of the temporal lobe upper surface, and the larger area called planum temporale on the left. (From Geschwind and Levitsky, in Geschwind, 1980)

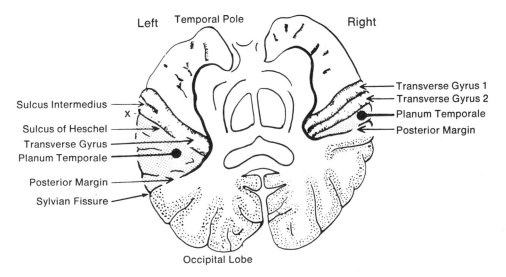

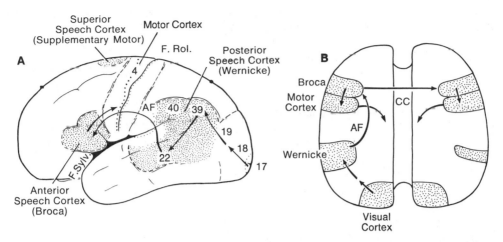

Fig. 30.18 Summary of the main pathways believed to be involved in seeing an object and saying its name. **A.** Lateral view of left hemisphere. **B.** Overhead view of both hemispheres. Bilateral inwardly pointing arrows represent the descending corticospinal motor pathway. AF, arcuate fasciculus; CC, corpus callosum. F. Rol., Rolandic fissure; F. Sylv., Sylvian fissure. (A from Popper and Eccles, 1977; B modified from Geschwind, 1980)

of neurons and extent of local circuits, has been utilized in this case.

An expanded view of the distributed system for language is shown in Fig. 30.18. This is based on the results from electrical stimulation of speech areas in human patients, as well as on anatomical studies in monkeys and humans. The diagram emphasizes the connections involved in one language task: naming a seen object. Visual information is first received in area 17, and elaborated in areas 18 and 19. From here, the perceptual image of the object is transferred to a large "posterior speech area," which includes area 39 (of parietal cortex) as well as the classical Wernicke's area. Area 39 transfers the visual representation of the object to its auditory representation in area 22. From area 22, the information is transferred to Broca's speech area, where the motor programs for speech are located. These programs are then "read out" to the face area of the motor cortex, where they control the complex spatiotemporal coordination of the muscles of speech, so that the name of the observed object can be articulated.

The Thinking Brain

The pathways described above were inferred from traditional methods of anatomy and physiology applied to animal experiments or from the study of the effects of brain lesions in human patients. Thanks to modern scanning methods, activity in these pathways can now be directly visualized. These methods, which have greatly extended the traditional recordings obtained by electroencephalography (see Chap. 25), include positron emission tomography (PET), magnetic resonance imaging (MRI), and magnetoencephalography (ME). PET utilizes radioactive isotopes, MRI is based on the spectroscopic properties of intrinsic molecules in the brain, and ME is based on the magnetic fields generated by electrical activity. The most widely used of these methods for carrying out experimental studies of brain activity has been PET. This method is based on the fact that when positrons are emitted by a radioactive isotope, they travel a few millimeters from the nucleus and come to rest, at which time they interact with an

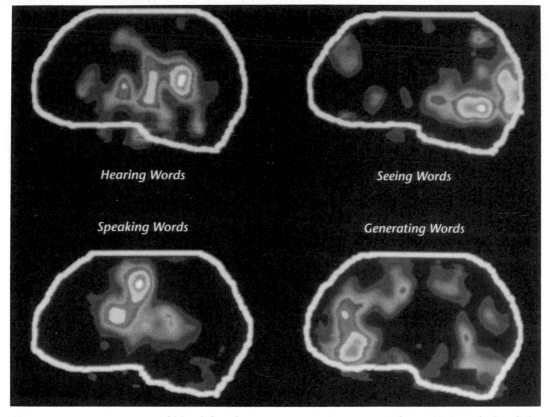

Fig. 30.19 Measurements of blood flow by positron emission tomography (PET) in right-handed normal volunteer subjects during three different cognitive tasks of increasing cognitive complexity. For each task, the blood flow pattern of the simpler task was subtracted from that of the more complex task in order to reveal the areas whose activity was specifically related to the more complex task. Task 1 (sensory reception; modality-specific word code): passive reception of a word presented auditorally (A) or visually (B), compared with simple visual fixation. Task 2 (motor output; articulatory code): speaking the presented word (C), compared with passive reception. Task 3 (semantic association): speaking a word associated with the presented word (D), compared with speaking the presented word. See text. (From M. E. Raichle, based on Petersen et al., 1988)

electron; the interaction annihilates the two particles and produces two photons traveling in opposite directions for long distances. These can be detected by electronic coincidence detectors arranged in a circle around the source organ. With this method, a radioactive isotope such as [^{18}F] fluorodeoxyglucose can be used to measure local glucose utilization or [^{15}O] water can be used to measure regional blood flow changes in the brain related to neural activity, similar to the way that 2-deoxyglucose is used in animal experiments to measure changes in local glucose utilization related to neural activity (see Chap. 8).

PET scans have been made of brain activity under many different conditions, both normal and in relation to neurological diseases. The method obviously has the potential to reveal complex patterns of brain activity underlying cognitive functions, but interpretation of the results requires careful experimental design. Such a study was carried out by Marcus Raichle and Michael Posner and their colleagues in a pioneering experiment at Washington University

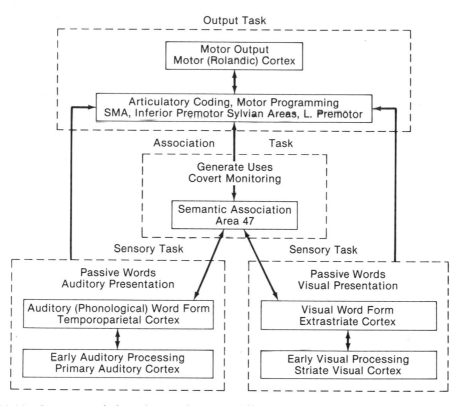

Fig. 30.20 Summary of the relations between different cortical areas and different cognitive functions involved in the processing of words, as shown by the PET study of Fig. 30.19. The dashed lines indicate the successive subtractions of the cortical activity patterns in Fig. 30.19; the solid lines indicate the anatomical areas and their levels of lexical processing. (From Petersen et al., 1988)

which applied PET scans to a carefully designed set of psychological tasks (Petersen et al., 1988). The aim was to study the processing of words and test whether this took place by successive operations in a series of brain regions, as in the model of Fig. 30.18, or by operations in parallel brain regions.

In these experiments, the subjects were asked to perform a series of cognitive tasks involving word processing that was graded in three levels of complexity. The simplest, baseline, task was to fixate on a visual target without any stimulation. The first-level task was simply passive sensory reception of the presentation of single words, either visual or spoken. The results showed that auditory presentation produced activity in primary and secondary auditory cor-

tex, whereas visual presentation produced activity in primary and secondary visual areas (Fig. 30.19). This is in accord with our expectations regarding the initial cortical processing of these modalities.

The second-level task required the subject to speak aloud each presented word. This provided the opportunity to identify areas involved in output coding and motor control. Activity was localized in the mouth area of primary sensorimotor cortex, together with nearby areas overlapping with Broca's area in the left hemisphere (Fig. 30.19). However, activity was also seen in this general region in the right hemisphere as well, suggesting that these areas are involved in general motor output rather than language-specific output. Significantly, the activation of these regions was similar with either auditory or visual word presentation, suggesting that visually received words do not have to go through a decoding operation in Wernicke's area before being forwarded for motor output.

In the third-level task, the subject was asked to respond to a noun word by speaking an appropriate verb word (for example, for "cake," say "eat"). The aim of this task was to identify areas involved in word association and semantic processing. Activity occurred in the left inferior frontal area, for both visual and auditory word presentation (Fig. 30.19). This supports the conclusion that the two modalities have parallel rather than serial pathways to this area. An additional site of activity in the anterior cingulate gyrus appeared to be related more to attention than to semantic processing per se.

The authors conclude that processing of word information proceeds more by parallel pathways, as summarized in Fig. 30.20, than by the classical serial model of Fig. 30.18. They further note (Petersen et al., 1988) that

combined cognitive and neurobiological approaches such as this give information about the functional anatomy of perception, attention, motor control, and language. As these endeavours proceed, solutions to the problem of mind–brain interactions that have intrigued us for so long should be illuminated.

If we compare these results with our initial concept of the level of brain organization in Fig. 1.1, we see that the PET scans give data about the neural mechanisms underlying cognitive functions at the highest level of distributed systems. These brain maps give important information about the extent and relations of the activated systems, much as road maps show the main transportation routes of a country. The functional units and their specific mechanisms at the underlying levels of synapses, neurons, circuits—reviewed in this chapter and in this book—give rise to these patterns; our task is to identify precisely which of those mechanisms are active for each cognitive task, so that we understand *how the brain does it*. Then the nature of cortical function, and the nature of our own being, will be a bit closer to our grasp. Our progress toward this goal will continue to depend on our ingenuity in devising new methods, our imagination in interpreting the results, and our wisdom in maintaining a balanced view of humankind among the creatures of this planet.

Appendixes

Appendix A: Study Guide for Selected Topics

This list provides a guide to topics that are covered in several different chapters. It can be used as an outline for a one-semester course.

Topic	Chapters	Topic	Chapters
Synapses		*Sensory systems*	
1 Neuromuscular junction	2, 3, 7, 9, 17	12 Olfactory system organization	6, 7, 10, 11, 24, 26, 30
2 Squid giant synapse	8, 19	13 Somatosensory system organization	9, 10, 12
3 Central synapses	6, 7, 8, 9, 17, 19, 24, 29, 30	14 Auditory system organization	15, 16, 23
4 Neurotransmitters	6, 7, 8, 17, 24, 29	15 *Limulus* eye	10, 16, 25
5 Neuropeptides	8, 17, 18, 24, 29	16 Visual system organization	6, 10, 12, 15, 16, 24, 25, 30
6 Second messengers	8, 10, 16, 17, 18, 25, 29	*Behavior*	
7 Dendritic spines	3, 6, 7, 8, 29, 30	17 Escape responses	7, 10, 13, 20
Electrophysiology		18 Rhythm-generating networks	19, 20, 25, 26
8 Action potentials	5, 18, 25	19 Motor control in awake monkey	12, 21, 22, 30
9 Synaptic potentials	7, 8, 18, 29	20 Cerebral cortex	6, 11–16, 22–25, 29, 30
Development and learning			
10 Neural development	2, 9, 17, 23, 26, 30		
11 Learning	7, 8, 29, 30		

Appendix B: Reference Table for Amino Acids

Each amino acid has a carbon atom attached to a hydrogen, an amino group, a carboxyl group, and a variable sidechain called an R group. The amino acids are classified in the table according to properties of their R groups. At right are the three-letter and one-letter abbreviations used to represent the amino acid sequences in peptides and proteins.

Amino acid classification	Abbreviations		Amino acid classification	Abbreviations	
Acidic (negatively charged R group)			Serine	Ser	S
Aspartic acid (Aspartate)	Asp	D	Threonine	Thr	T
Glutamic acid (Glutamate)	Glu	E	Tyrosine	Tyr	Y
Basic (positively charged R group)			*Nonpolar (hydrophobic R group)*		
Arginine	Arg	R	Alanine	Ala	A
Histidine	His	H	Isoleucine	Ile	I
Lysine	Lys	K	Leucine	Leu	L
Polar (uncharged R group)			Methionine	Met	M
Asparagine	Asn	N	Phenylalanine	Phe	F
Cysteine	Cys	C	Proline	Pro	P
Glutamine	Gln	Q	Tryptophan	Trp	W
Glycine	Gly	G	Valine	Val	V

Appendix C: Units of Measurement.

	$d = rt$ distance	$t = \dfrac{d}{r}$ time	$M = MW$ quantity	$E = IR$ voltage	$I = \dfrac{E}{R}$ current	$R = \dfrac{E}{I}$ resistance	$G = \dfrac{I}{E}$ conductance
10^{18}							
10^{15}							
10^{12}							
10^{9}						GΩ gigohm	
10^{6}						MΩ megohm	
10^{3}	kw kilometers					kΩ kilohm	
1	m meter	s second	M mole	v volt	A ampere	Ω ohm	s siemen
10^{-3}	mm millimeter	ms millisecond	mM millimole	mv millivolt	mA milliamp		mS millisiemen
10^{-6}	μm micrometer	μs microsecond	μM micromole	μv microvolt	μA microamp		μs microsiemen
10^{-9}	nm nanometer	ns nanosecond	nM nanomole		nA nanoamp		nS nanosiemen
10^{-12}			pM picomole		pA picoamp		ps picosiemen
10^{-15}			fm femtomole		fA femtoamp		fs femtosiemen
10^{-18}			aM attomole				

Appendix D: Greek Alphabet

	A	alpha	A	α
	B	beta	B	β
	G	gamma	Γ	γ
	D	delta	Δ	δ
short	E	epsilon	E	ϵ
	Z	zeta	Z	ζ
long	H	eta	H	η
	TH	theta	Θ	θ
	I	iota	I	ι
	K	kappa	K	κ
	L	lambda	Λ	λ
	M	mu	M	μ
	N	nu	N	ν
	X	xi	Ξ	ξ
short	O	omicron	O	o
	P	pi	Π	π
	R	rho	P	ρ
	S	sigma	Σ(C)	$\sigma\varsigma$
	T	tau	T	τ
	U	upsilon	Υ	υ
	PH	phi	Φ	ϕ
	CH	chi	X	χ
	PS	psi	Ψ	ψ
long	O	omega	Ω	ω

Literature Cited

Adrian, E. D., and Zotterman, Y. (1926). The impulses produced by sensory nerve endings. 2. The response of a single end-organ. *J. Physiol. (London)* 61:151–171.

Aghajanian, G. K., and Bunney, B. S. (1976). Dopamine "autoreceptors": pharmacological characterization by microiontophoretic single cell recording studies. *Naunyn-Schmiedeberg's Arch. Pharmacol.* 97:1–7.

Aghajanian, G. K., and Rasmussen, K. (1988). Basic electrophysiology. In *Psychopharmacology: The Third Generation of Progress* (H. Meltzer, ed.). New York: Raven Press.

Aguayo, A. J. (1985). Axonal regeneration from injured neurons in the adult mammalian central nervous system. In *Synaptic Plasticity* (C. E. Cotman, ed.). New York: Guilford Press, pp. 457–484.

Ahmad, I., Leinders-Zufall, T., Kocsis, J. D., Shepherd, G. M., Zufall, F., and Barnstable, C. T. (1994). Retinal ganglion cells express a cGMP-gated cation conductance activatable by nitric oxide donors. Neuron 12:

Aidley, D. J. (1978). *The Physiology of Excitable Cells.* Cambridge: Cambridge University Press.

Alberts, B., Bray, D., Lewis, J., Raff, M., Roberts, K., and Watson, J. D. (1989). *Molecular Biology of the Cell.* 2nd ed. New York: Garland.

Alcock, J. (1979). *Animal Behavior: An Evolutionary Approach.* Sunderland, MA: Sinauer.

Alexander, G. E., and Crutcher, M. D. (1990a).

Neural representations of the target (goal) of visually guided arm movements in three motor areas of the monkey. *J. Neurophysiol.* 64:164–178.

Alexander, G. E., and Crutcher, M. D. (1990b). Preparation for movement: neural representations of intended direction in three motor areas of the monkey. *J. Neurophysiol.* 64:133–150.

Alexander, G. E., DeLong, M. R., and Strick, P. L. (1986). Parallel organization of functionally segregated circuits linking basal ganglia and cortex. *Annu. Rev. Neurosci.* 9:357–381.

Alexander, R. McN. (1979). *The Invertebrates.* Cambridge: Cambridge University Press.

Alexandrowicz, J. S. (1951). Muscle receptor organs in the abdomen of *Homarus vulgaris* and *Palinrus vulgaris. Q. J. Microsc. Sci.* 92:163–199.

Andersen, P., and Andersson, S. A. (1968). *Physiological Basis of the Alpha Rhythm.* New York: Appleton-Century-Crofts.

Andersen, P. S., Sundberg, S. H., Sveen, O., and Wigstrom, H. (1977). Specific long-lasting potentiation of synaptic transmission in hippocampal slices. *Nature* 266:736–737.

Anderson, D. J. (1984). New clues to protein localization in neurons. *Trends Neurosci.* 7:355–357.

Armstrong, C. M. (1981). Sodium channels and gating currents. *Physiol. Rev.* 61:644–683.

Aserinsky, E., and Kleitman, N. (1953). Two types of ocular motility occurring during sleep. *J. Appl. Physiol.* 8:1–10.

Ashe, J. H., and Nachman, M. (1980). Neural

mechanisms in taste aversion leaning. *Prog. Psychobiol. Physiol. Psychol.* 9:233–262.

Ashmore, J. F. (1992). Mammalian hearing and the cellular mechanisms of the cochlear amplifier. In *Sensory Transduction* (D. P. Corey and S. D. Roper, eds.). New York: Rockefeller University Press, pp. 395–412.

Assad, J. A., Shepherd, G. M. G., and Corey, D. P. (1991). Tip link integrity and mechanical transduction in vertebrate hair cells. *Neuron* 7:985–994.

Astic, L., and Saucier, D. (1986). Anatomical mapping of the neuroepithelial projection to the olfactory bulb in the rat. *Brain Res. Bull.* 16:445–454.

Atwood, H. L. (1977). Crustacean neuromuscular systems: past, present, and future. In *Identified Neurons and Behavior of Arthropods* (G. Hoyle, ed.). New York: Plenum, pp. 9–29.

Augustine, G. J., Charlton, M. P., and Smith, S. J. (1987). Calcium action in synaptic transmitter release. *Annu. Rev. Neurosci.* 10:633–693.

Avenet, P., Kinnamon, S. C., and Roper, S. D. (1993). Peripheral transduction mechanisms. In *Mechanisms of Taste Transduction* (S. A. Simon and S. D. Roper, eds.). Boca Raton, FL: CRC Press.

Baehr, W., and Applebury, M. L. (1986). Exploring visual transduction with recombinant DNA techniques. *Trends Neurosci.* 9:198–203.

Bailey, C. H., and Chen, M. (1983). Morphological basis of long-term habituation and sensitization in *Aplysia*. *Science* 220: 91–93.

Baitinger, C., Cheney, R., Clements, D., Glicksman, M., Hirokawa, N., Levine, J., Meiri, K., Simon, C., Skene, P., and Willard, M. (1983). Axonally transported proteins in axon development, maintenance, and regeneration. *Cold Spring Harbor Symp. Quant. Biol.* 48:791–802.

Bard, P. (1934). On emotional expression after decortication with some remarks on certain theoretical views: parts I and II. *Psychol. Rev.* 41:309–329, 434–439.

Barker, D. (1974). The morphology of muscle receptors. In *Handbook of Sensory Physiology*, Vol. 3, Part 2: *Muscle Receptors* (C. C. Hunt, ed.). New York: Springer-Verlag, pp. 1–190.

Barlow, R. B., Chamberlain, S. C., and Kass, L. (1984). Circadian rhythms in retinal function. In *Molecular and Cellular Basis of Visual Acuity: Cell and Developmental Biology of the Eye* (S. R. Hilfer and J. B. Sheffield, eds.). New York: Springer-Verlag, pp. 31–53.

Barnard, E. A. (1992). Receptor classes and the transmitter-gated ion channels. *Trends Biochem. Sci.* 17:368–374.

Barnett, S. A. (1981). *Modern Ethology: The Science of Animal Behavior*. New York: Oxford University Press.

Barrington, E. J. W. (1979). *Invertebrate Structure and Function*. New York: Wiley.

Baylor, D. A. (1992). Transduction in retinal photoreceptor cells. In *Sensory Transduction* (D. P. Corey and S. D. Roper, eds.). New York: Rockefeller University Press, pp. 151–174.

Baylor, D. A., Lamb, T. D., and Yau, K.-W. (1979a). The membrane current of single rod outer segments. *J. Physiol. (London)* 288:589–611.

Baylor, D. A., Lamb, T. D., and Yau, K.-W. (1979b). Responses of retinal rods to single photons. *J. Physiol. (London)* 288:613–634.

Beal, M. F., Hyman, B. T., and Koroshetz, W. (1993). Do defects in mitochondrial energy metabolism underlie the pathology of neurodegenerative diseases? *Trends Neurosci.* 16:125–131.

Bell, C. (1833/37). *The Hand*. London: Pickering.

Bender, W. (1985). Homeotic gene products as growth factors. *Cell* 43:559–560.

Bennett, M. V. L. (1977). Electrical transmission: a functional analysis and comparison with chemical transmission. In *Handbook of Physiology*, Sec. 1: *The Nervous System*, Vol. 1: *Cellular Biology of Neurons* (E. R. Kandel, ed.). Bethesda, MD: American Physiological Society, pp. 367–416.

Bennett, T. L. (1977). *Brain and Behavior*. Belmont, CA: Wadsworth.

Bentley, D., and Hoy, R. R. (1974). The neurobiology of cricket song. *Sci. Am.* 231: 34–44.

Bentley, D., and Konishi, M. (1978). Neural control of behavior. *Annu. Rev. Neurosci.* 1:35–59.

Berridge, M. J. (1985). The molecular basis of communication within the cell. *Sci. Am.* 253:142–152.

Betz, H. (1991). Glycine receptors: heterogeneous and widespread in the mammalian brain. *Trends Neurosci.* 14:458–461.

Binder, M. D. (1989). Properties of motor units. In *Textbook of Physiology, Excitable Cells and Neurophysiology*, 21st ed. (H. D. Patton, A. F. Fuchs, B. Hille, A. M. Scher, and R. Steiner, eds.). Philadephia: W. B. Saunders, pp. 510–521.

Birnbaumer, L. (1990). Transduction of receptor signal into modulation of effector activity by G proteins: the first 20 years or so. . . . *FASEB J.* 4:3178–3188.

Blalock, J. E., and Smith, E. M. (1985). The

immune system: our mobile brain? *Immunol. Today* 6:115–117.

Blass, E. M., and Epstein, A. N. (1971). A lateral preoptic osmosensitive zone for thirst in the rat. *J. Comp. Physiol. Psychol.* 76:378–394.

Blass, E. M., Hall, W. G., and Teicher, M. N. (1979). The ontogeny of suckling and ingestive behaviors. *Prog. Psychobiol. Physiol. Psychol.* 8:243–300.

Bliss, T. V. P., and Lømo, T. (1973). Long-lasting potentiation of synaptic transmission in the dentate area of the anaesthetized rabbit following stimulation of the perforant path. *J. Physiol. (London)* 232: 331–356.

Block, S. M. (1992). Biophysical principles of sensory transduction. In *Sensory Transduction* (D. P. Corey and S. D. Roper, eds.) New York: Rockefeller University Press, pp. 1-8.

Bloom, W., and Fawcett, D. W. (1975). *A Textbook of Histology.* Philadelphia: W. B. Saunders.

Bodian, D. (1967). Neurons, circuits, and neuroglia. In *The Neurosciences: A Study Program* (G. C. Quarton, T. Melnechuk, and F. O. Schmitt, eds.). New York: Rockefeller University Press, pp. 6–24.

Bodian, D. (1972). Synaptic diversity and characterization by electronmicroscopy. In *Structure and Function of Synapses* (G. D. Pappas and D. P. Purpura, eds.). New York: Raven Press, pp. 45–66.

Boeckh, J., and Boeckh, V. (1979). Threshold and odor specificity of pheromone-sensitive neurons in the deutocerebrum of *Antheraea pernyi* and *A. polyphemus (Saturnidae). J. Comp Physiol.* 132:234–242.

Boeckh, J., Distler, P., Ernst, K. D., Hosl, M., and Malun, D. (1990). Olfactory bulb and antennal lobe. In *Chemosensory Information Processing* (D. Schild, ed.). Berlin: Springer-Verlag, pp. 201–227.

Borg, G., Diamant, H., Strom, L., and Zotterman, Y. (1967). The relation between neural and perceptual intensity: a comparative study on the neural and psychophysical response to taste stimuli. *J. Physiol. (London)* 192:13–20.

Boring, E. G. (1950). *A History of Experimental Psychology.* New York: Appleton.

Brandon, J. G., and Coss, R. G. (1982). Rapid dendritic spine stem shortening during one-trial learning: the honey bee's first orientation flight. *Brain Res.* 252:51–61.

Brazier, M. A. B. (1970). *The Electrical Activity of the Nervous System.* London: Pitman.

Bredt, D. S., Glatt, C. E., Hwang, P. M., Fotuhi, M., Dawson, T. M., and Snyder, S. H. (1991). Nitric oxide synthase protein and mRNA are discretely localized in neuronal populations of the mammalian CNS together with NADPH diaphorase. *Neuron* 7:615–624.

Breer, H., and Shepherd, G. M. (1993). Implications of the NO/cGMP system for olfaction. *Trends Neurosci.* 16:5–9.

Brodal, A. (1981). *Neurological Anatomy in Relation to Clinical Medicine,* 3rd ed. New York: Oxford University Press.

Bronner-Fraser, M. (1993). Crest destiny. *Curr. Biol.* 3:201–203.

Bronson, F. H. (1979). The reproductive ecology of the house mouse. *Q. Rev. Biol.* 54: 265–299.

Brooks, V. B., and Thach, T. T. (1981). Cerebellar control of posture and movement. In *Handbook of Physiology,* Sec. 1: *The Nervous System,* Vol. 2: *Motor Control* (V. B. Brooks, ed.). Bethesda, MD: American Physiological Society, pp. 877–946.

Brown, R. M., Crane, A. M., and Goldman, P. S. (1979). Regional distribution of monoamines in the cerebral cortex and subcortical structures of the rhesus monkey: concentrations and in vivo rates. *Brain Res.* 168:133–150.

Buck, L., and Axel, R. (1991). A novel multigene family may encode odorant receptors: a molecular basis for odor recognition. *Cell* 65:175–187.

Bucy, P. C., and Klüver, H. (1955). An anatomical investigation of the temporal lobe in the monkey *(Macaca mulatta). J. Comp. Neurol.* 103:151–252.

Bulloch, K. (1985). Neuroanatomy of lymphoid tissue: a review. In *Neural Modulation of Immunity* (R. Guillemin, M. Cohn, and T. Melnechuk, eds.). New York: Raven Press, pp. 111–142.

Bullock, T. H. (1976). *Introduction to Neural Systems.* San Francisco: W. H. Freeman.

Bunge, R. P. (1968). Glial cells and the central myelin sheath. *Physiol. Rev.* 48:197–251.

Burgess, P. R., Wei, J. Y., Clark, F. J., and Simon, J. (1982). Signalling of kinesthetic information by peripheral sensory receptors. *Annu. Rev. Neurosci.* 5:171–187.

Burke, R. E. (1981). Motor units: anatomy, physiology, and functional organization. In *Handbook of Physiology,* Sec. 1: *The Nervous System,* Vol. 2: *Motor Control,* Part 1 (V. B. Brooks, ed.). Bethesda, MD: American Physiological Society, pp. 345–422.

Burke, R. E. (1990). Spinal cord: ventral horn. In *The Synaptic Organization of the Brain,* 3rd ed. (G. M. Shepherd, ed.). New York: Oxford University Press, pp. 88–132.

Cahalan, M. D., Chandy, K. G., DeCoursey, T. E., and Gupta, S. (1985). A voltage-gated potassium channel in human T lymphocytes. *J. Physiol. (London)* 358: 197–237.

Cajal, S. Ramón y (1911). *Histologie du Système Nerveux de l'Homme et des Vertèbres.* Paris: Maloine.

Cajal, S. Ramón y (1990). *New Ideas on the Structure of the Nervous System in Man and Vertebrates* (English translation of *Les Nouvelles Idées sur la Structure des Centres Nerveux chez l'Homme et chez les Vertèbres,* by N. Swanson and L. W. Swanson). Cambridge, MA: MIT Press.

Calabrese, R. L., and Arbas, E. A. (1985). Modulation of central and peripheral rhythmicity in the heart beat system of the leech. In *Model Neural Networks and Behavior* (A. I. Selverston, ed.). New York: Plenum, pp. 69–86.

Caldwell, P. C., Hodgkin, A. L., Keynes, R. D., and Shaw, T. I. (1960). The effects of injecting "energy rich" compounds on the active transport of ions in the giant axons of *Loligo. J. Physiol. (London)* 152: 561–590.

Cameron, R. S., and Rakic, P. (1991). Glial cell lineage in the cerebral cortex: a review and synthesis. *Glia* 4:124–137.

Cannon, W. B. (1929). *Bodily Changes in Pain, Hunger, Fear, and Rage.* New York: Appleton.

Cantley, L. (1986). Ion transport systems sequenced. *Trends Neurosci.* 9:1–3.

Carew, T. J., and Sahley, C. L. (1986). Invertebrate learning and memory: from behavior to molecules. *Annu. Rev. Neurosci.* 9:435–487.

Carpenter, M. B. (1976). *Human Neuroanatomy.* Baltimore, MD: Williams & Wilkins.

Carr, W. E. S. (1986). The molecular nature of chemical stimuli in the aquatic environment. In *Sensory Biology of Aquatic Organisms* (J. Atema, R. R. Fay, A. N. Popper, and W. N. Tanolga, eds.). New York: Springer-Verlag, pp. 1–36.

Catterall, W. A. (1988). Structure and function of voltage-sensitive ion channels. *Science* 242:50–61.

Caviness, V. S., Jr., and Rakic, P. (1978). Mechanisms of cortical development: a view from mutations in mice. *Annu. Rev. Neurosci.* 1:297–326.

Chagnac-Amitai, Y., and Connors, B. W. (1989). Synchronized excitation and inhibition driven by intrinsically bursting neurons in neocortex. *J. Neurophysiol.* 62: 1149–1162.

Chance, P. (1979). *Learning and Behavior.* Belmont, CA: Wadsworth.

Changeux, J.-P., Devillers-Thiery, A., and

Chemovilli, P. (1984). Acetylcholine receptor: an allosteric protein. *Science* 225: 1335–1345.

Cheney, P. D., Fetz, E. E., and Palmer, S. S. (1985). Patterns of facilitation and suppression of antagonist forelimb muscles from motor cortex sites in the awake monkey. *J. Neurophysiol.* 53:805–820.

Churchland, P., and Sejnowski, T. (1992). *The Computational Brain.* New York: W. H. Freeman.

Claudio, T., Ballivet, M., Patrick, J., and Heinemann, S. (1983). Nucleotide and deduced amino acid sequences of *Torpedo californica* acetylcholine receptor subunit. *Proc. Natl. Acad. Sci. USA* 80:1111–1115.

Connor, J. A., and Stevens, C. S. (1971). Voltage clamp studies of a transient outward membrane current in gastropod neural somata. *J. Physiol. (London)* 213:21–30.

Constantine-Paton, M., Cline, H. T., and Debski, E. (1990). Patterned activity, synaptic convergence, and the NMDA receptor in developing visual pathways. *Annu. Rev. Neurosci.* 13:129–154.

Cooper, J. R., Bloom, F. E., and Roth, R. H. (1993). *The Biochemical Basis of Neuropharmacology,* 6th ed. New York: Oxford University Press.

Corey, D. P. (1983). Patch clamp: current excitement in membrane physiology. *Neurosci. Comment.* 1:99–110.

Corey, D. P., and Assad, J. A. (1992). Transduction and adaptation in vertebrate hair cells: correlating structure with function. In *Sensory Transduction* (D. P. Corey and S. D. Roper, eds.). New York: Rockefeller University Press, pp. 325–342.

Corey, D. P., Roper, S. D. (eds.) (1992). *Sensory Transduction.* New York: Rockefeller University Press.

Corkin, S., Sullivan, E. V., Twitchell, R. E., and Grove, E. (1981). The amnesic patient H.M.: clinical observations and test performance 28 years after operation. *Soc. Neurosci. Abstr.* 7:235.

Corwin, J. T., and Warchol, M. E. (1991). Auditory hair cells: structure, function, development, and regeneration. *Annu. Rev. Neurosci.* 14:301–333.

Coss, R. G., and Perkel, D. H. (1985). The function of dendritic spines: a review of theoretical issues. *Behav. Neural Biol.* 44: 151–185.

Crawford, A. C., and Fettiplace, R. (1985). The mechanical properties of ciliary bundles of turtle cochlear hair cells. *J. Physiol. (London)* 364:359–379.

Creed, R. S., Denny-Brown, D., Eccles, J. C., Liddell, E. G. T., and Sherrington, C. S.

(1932). *Reflex Activity of the Spinal Cord.* London: Oxford University Press.

Crelin, E. S. (1976). Development of the upper respiratory system. *Ciba Clin. Symp.* 28: 1–30.

Crutcher, M. D., and Alexander, G. E. (1990). Movement-related neuronal activity selectively coding either direction or muscle pattern in three motor areas of the monkey. *J. Neurophysiol.* 64:151–163.

Cull-Candy, S. G., and Usowicz, M. M. (1987). Multiple-conductance channels activated by excitatory amino acids in cerebellar neurons. *Nature* 325:525–528.

Dale, H. H. (1935). Pharmacology and nerve endings. *Proc. R. Soc. Med.* 28:319–332.

Dallos, P. (1985). The role of outer hair cells in cochlear function. In *Contemporary Sensory Neurobiology* (M. J. Correia and A. A. Perachio, eds.). New York: Alan R. Liss, pp. 207–230.

Dallos, P. (1992). The active cochlea. *J. Neurosci.* 12:4575–4585.

Dallos, P., and Cheatham, M. A. (1992). Cochlear hair cell function reflected in intracellular recordings in vivo. In *Sensory Transduction* (D. P. Corey and S. D. Roper, eds.). New York: Rockefeller University Press, pp. 371–393.

Darwin, C. A. (1872). *On the Expression of the Emotions in Man and Animals.* London: Murray.

Daube, J. R., Reagan, T. J., Sandole, B. A., and Westmoreland, B. F., eds. (1986). *Medical Neuroscience,* 2nd ed. Boston: Little, Brown.

Davies, A. M., Bandtlow, C., Heumann, R., Korsching, S., Rohrer, H., and Thoenen, H. (1987). Timing and site of nerve growth factor synthesis in developing skin in relation to innervation and expression of the receptor. *Nature* 326:353–358.

Davis, H., and Silverman, S. R. (1970). *Hearing and Deafness.* New York: Holt, Rhinehart and Winston.

Davis, M. (1992). The role of the amygdala in fear-potentiated startle: implications for animal models of anxiety. *Trends Pharmacol. Sci.* 13:35–42.

Daw, N. W., Jensen, R. J., and Brunken, W. J. (1990). Rod pathways in mammalian retinae. *Trends Neurosci.* 13:110–115.

De Camilli, P., and Jahn, R. (1990). Pathways to regulated exocytosis in neurons. *Annu. Rev. Physiol.* 52:625-645.

DeFelipe, J., and Jones, E. G. (1988). *Cajal on the Cerebral Cortex.* New York: Oxford University Press.

Delaney, K., and Gelperin, A. (1986). Post ingestive food-aversion learning to amino acid deficient diets by the terrestrial slug *Limax maximus. J. Comp. Physiol. [A]* 159:281–295.

DeLong, M. R., and Georgopoulos, A. P. (1981). Motor functions of the basal ganglia. In *Handbook of Physiology,* Sec. 1: *The Nervous System,* Vol. 2: *Motor Control* (V. B. Brooks, ed.). Bethesda, MD: American Physiological Society, pp. 1017–1061.

Dement, W., and Kleitman, N. (1975). Cyclic variations in EEG during sleep and their relation to eye movements, body motility, and dreaming. *Electroencephal. Clin. Neurophysiol.* 9:673–690.

Dethier, V. G. (1976). *The Hungry Fly.* Cambridge, MA: Harvard University Press.

Dethier, V. G., and Stellar, E. (1964). *Animal Behavior: Its Evolutionary and Neurological Basis.* Englewood Cliffs, NJ: Prentice-Hall.

DeYoe, E. A., and van Essen, D. C. (1988). Parallel processing streams in monkey visual cortex. *Trends Neurosci.* 11:219–226.

Diamond, I. T. (1979). The subdivisions of neocortex: a proposal to revise the traditional view of sensory, motor, and association areas. In *Progress in Psychobiology and Physiological Psychology* (J. M. Sprague and A. N. Epstein, eds.). New York: Academic Press, pp. 2–44.

Dingwall, C. (1992). Soluble factors and solid phases. *Curr. Biol.* 2:503–504.

Dodge, F. A., and Rahamimoff, R. (1967). On the relationship between calcium concentration and the amplitude of the end-plate potential. *J. Physiol. (London)* 189:90–92P.

Doty, R. W. (1976). The concept of neural centers. In *Simpler Networks and Behavior* (J. C. Fentress, ed.). Sunderland, MA: Sinauer, pp. 251–265.

Dowling, J. E. (1979). Information processing by local circuits: the vertebrate retina as a model system. In *The Neurosciences: Fourth Study Program* (F. O. Schmitt and F. G. Worden, eds.). Cambridge, MA: MIT Press, pp. 163–182.

Dowling, J. E., and Boycott, B. B. (1966). Organization of the primate retina: electron microscopy. *Proc. R. Soc. London. [B]* 166: 80–111.

Dryer, S. E. (1985). Mismatch problem in receptor binding studies. *Trends. Neurosci.* 8:522.

Dubner, R., and Ruda, M. A. (1992). Activity-dependent neuronal plasticity following tissue injury and inflammation. *Trends Neurosci.* 15:96–103.

Dunant, Y., and Israel, M. (1985). The release of acetylcholine. *Sci. Am.* 252:58–66.

Dunlap, K., and Fischbach, G. D. (1978). Neurotransmitters decrease the calcium component of sensory neurone action potentials. *Nature* 276:837–839.

Eagles, E. I., ed. (1975). *The Nervous System, Human Communication and Its Disorders.* New York: Raven Press.

Eakin, R. M. (1965). Evolution of photoreceptors. *Cold Spring Harbor Symp. Quant. Biol.* 30:363–370.

Easter, S. S., Jr., Purves, D., Rakic, P., and Spitzer, N. C. (1985). The changing view of neural specificity. *Science* 230:507–511.

Eaton, R. C., Bombardieri, R. A., and Meyer, D. L. (1977). The Mauthner-initiated startle response in teleost fish. *J. Exp. Biol.* 66:65–81.

Eccles, J. C. (1975). *The Physiology of Nerve Cells.* Baltimore, MD: Johns Hopkins University Press.

Eckert, R., and Randall, D. (1983). *Animal Physiology. Mechanisms and Adaptations.* San Francisco: W. H. Freeman.

Edelman, G. M. (1988). *Topobiology. An Introduction to Molecular Embryology.* New York: Basic Books.

Edwards, F. (1991). LTP is a long term problem. *Nature* 350:271-272.

Elson, E. (1993). Barriers to diffusion. *Curr. Biol.* 3:152–153.

Emson, P. C., and Hunt, S. P. (1981). Anatomical chemistry of the cerebral cortex. In *The Organization of the Cerebral Cortex* (F. O. Schmitt, F. G. Worden, G. Adelman, and S. G. Dennis, eds.). Cambridge, MA: MIT Press, pp. 325–345.

Epstein, A. N. (1971). The lateral hypothalamic syndrome: its implications for the physiological psychology of hunger and thirst. *Prog. Psychobiol. Physiol. Psychol.* 4: 263–317.

Epstein, A. N. (1980). A comparison of instinct and motivation with emphasis on their differences. In *Neural Mechanisms of Goal-Directed Behavior and Learning* (R. Thompson and L. Hicks, eds.). New York: Academic Press, pp. 119–126.

Erulkar, S. D., and Wetzel, D. M. (1987). Steroid effects on excitable membranes. Curr. Topics Membranes and Transport 31: 141–190.

Erulkar, S. D., Kelley, B., Jurman, M. E., Zemlan, F. P., Schneider, G. T., and Krieger, N. R. (1981). Modulation of the neural control of the clasp reflex in male *Xenopus laevis* by androgens: a multidisciplinary study. *Proc. Natl. Acad. Sci. USA* 78: 5876–5880.

Evarts, E. V. (1981). Functional studies of the motor cortex. In *The Organization of the Cerebral Cortex* (F. O. Schmitt, F. G. Worden, G. Adelman, and S. G. Dennis, eds.). Cambridge, MA: MIT Press, pp. 199–236.

Ewert, J.-P. (1980). *Neuroethology: An Introduction to the Neurophysiological Fundamentals of Behavior.* New York: Springer-Verlag.

Ezzell, C. (1992). Neuroscientists lay the groundwork for detente in the battle of learning and memory research. *J. NIH Res.* 4:60–64.

Fahrenbach, W. H. (1985). Anatomical circuitry of lateral inhibition in the eye of the horseshoe crab, *Limulus polyphemus. Proc. R. Soc. London [B]* 225:219–249.

Farbman, A. I. (1986). Prenatal development of mammalian olfactory receptor cells. *Chem. Senses* 11:3–18.

Fatt, P., and Ginsborg, B. L. (1958). The ionic requirements for the production of action potentials in crustacean muscle fibres. *J. Physiol. (London)* 142:516–543.

Felleman, D. J., and Van Essen, D. C. (1991). Distributed hierarchical processing in the primate cerebral cortex. *Cereb. Cortex* 1:1–47.

Fernandez, C., and Golberg, J. M. (1976). Physiology of peripheral neurons innervating otolith organs of the squirrel monkey. III. Response dynamics. *J. Neurophysiol.* 39: 996–1008.

Fettiplace, R. (1992). The role of calcium in hair cell transduction. In *Sensory Transduction* (D. P. Corey and S. D. Roper, eds.). New York: Rockefeller University Press, pp. 343–356.

Fetz, E., and Baker, M. A. (1973). Operantly conditioned patterns of precentral unit activity and correlated responses in adjacent cells and contralateral muscles. *J. Neurophysiol.* 36:179–204.

Firestein, S., Zufall, F., and Shepherd, G. M. (1991). Single odor sensitive channels in olfactory receptor neurons are also gated by cyclic nucleotides. *J. Neurosci.* 11: 3565–3572.

Flaherty, A. W., and Graybiel, A. M. (1993). Two input systems for body representations in the primate striatal matrix: experimental evidence in the squirrel monkey. *J. Neurosci.* 13:1120–1137.

Flynn, J. P. (1967). The neural basis of aggression in cats. In *Neurophysiology and Emotion* (D. C. Glass, ed.). New York: Rockefeller University Press, pp. 40–69.

Forbes, A. (1939). Problems of synaptic functions. *J. Neurophysiol.* 2:465–472.

Foster, M. (1897). *A Textbook of Physiology.* London: Macmillan.

Frank, E., and Fischbach, G. D. (1979). Early

events in neuromuscular junction formation in vitro: induction of acetylcholine receptor clusters in the postsynaptic membrane and morphology of newly formed synapses. *J. Cell Biol.* 83:143–158.

Fraser, S. E. (1992). Patterning of retinotectal connections in the vertebrate visual system. *Curr. Opin. Neurobiol.* 2:83–97.

Freund, T. F., Powell, J. F., and Smith, A. D. (1984). Tyrosine hydroxylase-like immunoreactive boutons in synaptic contact with identified striatonigral neurons with special reference to dendritic spines. *Neuroscience* 13:1189–1215.

Fujita, I., Tanaka, K., Ito, M., and Cheng, K. (1992). Columns for visual features of objects in monkey inferotemporal cortex. *Nature* 360:343–346.

Furness, J. B., Bornstein, J. C., Murphy, R., and Pompolo, S. (1992). Roles of peptides in transmission in the enteric nervous system. *Trends Neurosci.* 15:66–71.

Furshpan, E. J., and Potter, D. D. (1959). Transmission at the giant motor synapses of the crayfish. *J. Physiol. (London)* 145:289–325.

Fuxe, K., and Agnati, L. (1991). *Volume Transmission in the Brain: Novel Mechanisms for Neural Transmission.* New York: Raven.

Gainer, H., and Brownstein, M. J. (1981). Neuropeptides. In *Basic Neurochemistry* (G. J. Siegel, R. W. Albers, B. W. Agranoff, and R. Katzman, eds.). Boston: Little, Brown, pp. 269–296.

Galambos, R., and Davis, H. (1943). The response of single auditory-nerve fibers to acoustic stimulation. *J. Neurophysiol.* 6: 39–57.

Gallistel, C. R. (1980). *The Organization of Action: A New Synthesis.* New York: Wiley.

Galzi, J.-L., Devillers-Thiery, A., Hussy, N., Bertrand, S., Changeux, J.-P., and Bertrand, D. (1992). Mutations in the channel domain of a neuronal nicotinic receptor convert ion selectivity from cationic to anionic. *Nature* 359:500–504.

Ganong, W. F. (1985). *The Nervous System.* Los Altos, CA: Lange.

Garcia, J., and Koelling, R. A. (1966). Relation of cue to consequence in avoidance learning. *Psychonomet. Sci.* 4:123–124.

Garthwaite, J. (1991). Glutamate, nitric oxide and cell–cell signalling in the nervous system. *Trends Neurosci.* 14:60–67.

Gasser, U. E., and Hatten, M. E. (1990). CNS neurons migrate on astroglial fibers from heterotypic brain regions in vitro. *Proc. Natl. Acad. Sci. USA* 87:4543–4547.

Gerfen, C. R. (1992). The neostriatal mosaic: multiple levels of compartmental organization. *Trends Neurosci.* 15:133–139.

Geschwind, N. (1980). Some special functions of the human brain. In *Medical Physiology* (V. B. Mountcastle, ed.). St. Louis: C. V. Mosby, pp. 647–665.

Gesteland, R. D. (1986). Speculations on receptor cells as analyzers and filters. *Experientia* 42:287–291.

Getchell, T. V., and Shepherd, G. M (1978). Responses of olfactory receptor cells to step pulses of odour at different concentrations in the salamander. *J. Physiol. (London)* 282:521–540.

Getting, P. A. (1989). Emerging principles governing the operation of neural networks. *Annu. Rev. Neurosci.* 12:185–204.

Getting, P. A., and Dekin, M. S. (1985). *Tritonia* swimming: a model system for integration within rhythmic motor systems. In *Model Neural Networks and Behavior* (A. I. Selverston, ed.). New York: Plenum, pp. 3–20.

Gilbert, C. D. (1992). Horizontal integration and cortical dynamics. *Neuron* 9:1–13.

Gilbert, S. F. (1991). *Developmental Biology,* 3rd ed. Sunderland, MA: Sinauer.

Glowinski, J., Tassin, J. P., and Thierry, A. M. (1984). The mesocortico-prefrontal dopaminergic neurons. *Trends Neurosci.* 7: 415–418.

Gobel, S., Falls, W. M., Bennett, G. J., Abdelmoumene, M., Hayashi, H., and Humphrey, E. (1980). An EM analysis of the synaptic connections of horseradish peroxidase–filled stalked cells and islet cells in the substantia gelatinosa of adult cat spinal cord. *J. Comp. Neurol.* 194:781–807.

Goedert, M., Crowther, R. A., and Garner, C. C. (1991). Molecular characterization of microtubule-associated proteins tau and MAP2. *Trends Neurosci.* 14:193–199.

Goldberg, J. M. (1991). The vestibular end organs: morphological and physiological diversity of afferents. *Curr. Opin. Neurobiol.* 1:229–235.

Goldman, P. S., and Nauta, W. J. (1977). An intricately patterned prefrontocaudate projection in the rhesus monkey. *J. Comp. Neurol.* 72:369–382.

Goldman-Rakic, P. S. (1981). Development and plasticity of primate frontal association cortex. In *The Organization of the Cerebral Cortex* (F. O. Schmitt, F. G. Worden, G. Adelman, and S. G. Dennis, eds.). Cambridge, MA: MIT Press, pp. 69–97.

Goldman-Rakic, P. S. (1984). The frontal lobes: unchartered provinces of the brain. *Trends Neurosci.* 7:7–11.

Goldman-Rakic, P. S. (1987). Circuitry of the prefrontal cortex and the regulation of behavior by representational knowledge. In *Handbook of Physiology*, Sec. 1: *The Nervous System*, Vol. 5: *Higher Cortical Function* (F. Plum and V. Mountcastle, eds.). Bethesda, MD: American Physiological Society, pp. 373–417.

Goldman-Rakic, P. S. (1991). Prefrontal cortical dysfunction in schizophrenia: the relevance of working memory. In *Psychopathology and the Brain* (B. J. Carroll and J. E. Barrett, eds.). New York: Raven Press, pp. 1–23.

Goldman-Rakic, P. S., Bates, J. F., and Chafee, M. V. (1992). The prefrontal cortex and internally generated motor acts. *Curr. Opin. Biol.* 2:830–835.

Goodman, C. S., and Bate, M. (1981). Neuronal development in the grasshopper. *Trends Neurosci.* 4:163–169.

Goodman, C. S., Bate, M., and Spitzer, N. C. (1981). Embryonic development of identified neurons: origin and transformation of the H cell. *J. Neurosci.* 1:94–102.

Goodwin, G. M., McCloskey, D. J., and Matthews, P. B. C. (1972). The contribution of muscle afferents to kinesthesia shown by vibration induced illusions of movement and by the effects of paralysing joint afferents. *Brain* 95:705–748.

Gordon, M. S. (1970). *Animal Physiology: Principles and Adaptations*. New York: Macmillan.

Gormezano, I. (1972). Investigations of defense and reward conditioning in the rabbit. In *Classical Conditioning*, Vol. 11: *Current Research and Theory* (A. H. Blaek and F. Prokasy, eds.). New York: Appleton-Century-Crofts, pp. 151–181.

Goslin, K., and Banker, G. (1989). Experimental observations on the development of polarity by hippocampal neurons in culture. *J. Cell Biol.* 108:1507–1516.

Gouras, P. (1985). Color vision. In *Principles of Neural Science*, 2nd ed. (E. R. Kandel and J. H. Schwartz, eds.). New York: Elsevier, pp. 384–395.

Goy, M. F. (1991). cGMP: the wayward child of the cyclic nucleotide family. *Trends Neurosci.* 14:293–299.

Goy, R. W., and McEwen, B. S. (1980). *Sexual Differentiation of the Brain*. Cambridge, MA: MIT Press.

Gray, J. (1968). *Animal Locomotion*. New York: Norton.

Graybiel, A. M. (1991). Basal ganglia: input, neural activity, and relation to the cortex. *Curr. Biol.* 1:644–651.

Graziadei, P. P. C., and Monti-Graziadei, G. A. (1979). Continuous nerve cell renewal in the olfactory system. In *Handbook of Sensory Physiology*, Vol. 9 (M. Jacobson, ed.). New York: Springer-Verlag, pp. 5–83.

Greenberg, M. J., and Price, D. A. (1983). Invertebrate neuropeptides: native and naturalized. *Annu. Rev. Physiol.* 45:271–288.

Greenough, W. T. (1984). Possible structural substrates of plastic neural phenomena. In *Neurobiology of Learning and Memory* (G. Lynch, J. L. McGaugh, and N. M. Weinberger, eds.). New York: Guilford Press, pp. 470–478.

Greer, C. A., Stewart, W. B., Teicher, M. H., and Shepherd, G. M. (1982). Functional development of the olfactory bulb and a unique glomerular complex in the neonatal rat. *J. Neurosci.* 2:1744–1759.

Gregory, R. L. (1966). *Eye and Brain: The Psychology of Seeing*. London: Weidenfeld and Nicolson.

Grenningloh, G., and Goodman, C. S. (1992). Pathway recognition by neuronal growth cones: genetic analysis of neural cell adhesion molecules in *Drosophila. Curr. Opin. Neurobiol.* 2:42–47.

Grenningloh, G., Rienitz, A., Schmitt, B., Methfessel, C., Zensen, M., Beyreuther, K., Gundelfinger, E. D., and Betz, H. (1987). The strychnine-binding subunit of the glycine receptor shows homology with nicotinic acetylcholine receptors. *Nature* 328: 215–220.

Grillner, S., and Matsushima, T. (1991). The neural network underlying locomotion in lamprey: synaptic and cellular mechanisms. *Neuron* 7:1–15.

Grillner, S., and Zangger, P. (1984). The effect of dorsal root transection on the efferent motor pattern in the cat's hindlimb during locomotion. *Acta Physiol. Scand.* 120: 393–405.

Groves, P. M., Staunton, D. A., Wilson, C. J., and Young, S. J. (1979). Sites of action of amphetamine intrinsic to catecholaminergic nuclei: catecholaminergic presynaptic dendrites and axons. *Prog. Neuropsychopharmacol.* 3:315–335.

Grumet, M. (1991). Cell adhesion molecules and their subgroups in the nervous system. *Curr. Biol.* 1:370–376.

Gurney, M. E., and Konishi, M. (1980). Hormone induced sexual differentiation of brain and behavior in zebra finches. *Science* 208:1380–1383.

Guthrie, P. B., Segal, M., and Kater, S. (1991). Independent regulation of calcium revealed by imaging dendritic spines. *Nature* 354: 76–80.

Guyton, A. C. (1976). *Texbook of Medical Physiology*. Philadelphia: W. B. Saunders.

Hagiwara, S. (1983). *Membrane Potential-*

Dependent Ion Channels in Cell Membrane: Phylogenetic and Developmental Approaches. New York: Raven Press.

Hall, N. R., and Goldstein, A. L, (1985). Neurotransmitters and host defense. In *Neural Modulation of Immunity* (R. Guillemin, M. Cohn, and T. Melnechuk, eds.). New York: Raven Press, pp. 143–160.

Hall, N. R., McGillis, J. P., Spangelo, B. L., Healy, D. L., Chronsos, G. P., Schultz, H. M., and Goldstein, A. L. (1985). Thymic hormone effects on the brain and neuroendocrine circuits. In *Neural Modulation of Immunity* (R. Guillemin, M. Cohn, and T. Melnechuk, eds.). New York: Raven Press, pp. 179–196.

Hall, Z. W. (1992). *Molecular Neurobiology.* Sutherland, MA: Sinauer.

Hall, Z. W., and Sanes, J. R. (1993). Synaptic structure and development: the neuromuscular junction. *Cell 72/Neuron* 10:99–122.

Hanamori, T., Miller, I. J., Jr., and Smith, D. V. (1988). Gustatory responsiveness of fibers in the hamster glossopharyngeal nerve. *J. Neurophysiol.* 60:478.

Harlan, R. E., Gordon, J. H., and Gorski, R. A. (1979). Sexual differentiation of the brain: implications for neuroscience. *Rev. Neurosci.* 4:31–61.

Harris, G. W. (1955). *Neural Control of the Pituitary Gland.* London: Edward Arnold.

Harrison, R. G. (1907). Observations on the living developing nerve fiber. *Anat. Rec.* 1:116–118.

Hatten, M. E. (1993). The role of migration in central nervous system neuronal development. *Curr. Opin. Neurobiol.* 3:38–44.

Hauri, P. (1977). *The Sleep Disorders.* Kalamazoo, MI: Upjohn.

Haymaker, W., ed. (1953). *The Founders of Neurology.* Springfield, IL: Charles C Thomas.

Hebb, D. O. (1949). *The Organization of Behavior.* New York: Wiley.

Hecht, S., Shlaer, S., and Pirenne, M. H. (1942). Energy, quanta and vision. *J. Gen. Physiol.* 25:819–840.

Heginbotham, L., Abramson, T., and MacKinnon, R. (1992). A functional connection between the pores of distantly related ion channels as revealed by mutant K^+ channels. *Nature* 258:1152–1155.

Heimer, L. (1983). *The Human Brain and Spinal Cord: Functional Neuroanatomy and Dissection Guide.* New York: Springer-Verlag.

Henneman, E. (1980). Organization of the spinal cord and its reflexes. In *Medical Physiology*, Vol. 1, 14th ed. (V. B. Mountcastle, ed.). St. Louis: C. V. Mosby, pp. 762–786.

Heuser, J. E., and Reese, T. S. (1977). Structure of the synapse. In *Handbook of Physiology*, Sec. 1: *The Nervous System*, Vol. 1: *Cellular Biology of Neurons* (E. R. Kandel, ed.). Bethesda, MD: American Physiological Society, pp. 261–294.

Hille B. (1977). Ionic basis of resting potentials and action potentials. In *Handbook of Physiology*, Sec. 1: *The Nervous System*, Vol. 1: *Cellular Biology of Neurons* (E. R. Kandel, ed.). Bethesda, MD: American Physiological Society, pp. 99–136.

Hille, B. (1992). *Ionic Channels of Excitable Membranes.* Sunderland, MA: Sinauer.

Hobson, J. A., and McCarley, R. W. (1977). The brain as a dream state generator: an activation-synthesis hypothesis of the dream process. *Am. J. Psychiatr.* 134: 1335–1348.

Hockfield, S., and Kalb, R. G. (1993). Activity-dependent structural changes during neuronal development. *Curr. Opin. Neurobiol.* 3:87–92.

Hockfield, S., McKay, R. D., Hendry, S. H. C., and Jones, E. G. (1983). A surface antigen that identifies ocular dominance columns in the visual cortex and laminar features of the lateral geniculate nucleus. *Cold Spring Harbor Symp. Quant. Biol.* 48:877–889.

Hodgkin, A. L., and Horowicz, P. (1959). The influence of potassium and chloride ions on the membrane potential of single muscle fibres. *J. Physiol. (London)* 148:127–160.

Hodgkin, A. L., and Huxley, A. F. (1952). A quantitative description of membrane current and its application to conduction and excitation in nerve. *J. Physiol. (London)* 117:500–544.

Hodgkin, A. L., and Katz, B. (1949). The effect of sodium ions on the electrical activity of the giant axons of the squid. *J. Physiol. (London)* 108:37–77.

Hoebel, B. G. (1988). Neuroscience and motivation: pathways and peptides that define motivational states. In *Stevens' Handbook of Experimental Psychology*, 2nd ed., Vol. 1: *Perception and Motivation* (R. C. Atkinson, R. J. Herrnstein, G. Lindzey, and R. D. Luce, eds.). New York: Wiley, pp. 547–625.

Hofer, M. A. (1981). *The Roots of Human Behavior.* San Francisco: W. H. Freeman.

Hökfelt, T., Johansson, O., and Goldstein, M. (1984). Chemical anatomy of the brain. *Science* 225:1326–1334.

Hökfelt, T., Johansson, O., Llungdahl, A., Lundberg, M., and Schultzberg, M. (1980). Peptidergic neurones. *Nature* 284: 515–521.

Hökfelt, T. (1991). Neuropeptides in perspective: The last ten years. *Neuron* 7:867-879.

Homick, J. L., Reschke, M. F., and Miller, E. F., II (1977). The effects of prolonged exposure to weightlessness on postural equilibrium. In *Biomedical Results from Skylab* (R. S. Johnston and L. F. Dietlein, eds.). Washington, DC: Scientific and Technical Information Office, NASA, pp. 104–112.

Hopkins, W. G., and Brown, M. C. (1984). *Development of Nerve Cells and Their Connections.* London: Cambridge University Press.

Horn, R., and Marty, A. (1988). Muscarinic activation of ionic currents measured by a new whole-cell recording method. *J. Gen. Physiol.* 92:145–159.

Hoshi, T., Zagotta, W. N., and Aldrich, R. W. (1990). Biophysical and molecular mechanisms of *Shaker* potassium channel inactivation. *Science* 250:533–538.

Howard, J., and Hudspeth, A. J. (1988). Compliance of the hair bundle associated with gating of the mechanoelectrical transduction channels in the bullfrog's saccular hair cell. *Neuron* 1:189–199.

Hubbard, J. L., Llinás, R., and Quastel, D. M. J. (1969). *Electrophysiological Analysis of Synaptic Transmission.* Baltimore, MD: Williams & Wilkins.

Hubel, D. H., and Wiesel, T. N. (1962). Receptive fields, binocular interaction and functional architecture in the cat's visual cortex. *J. Physiol. (London)* 160:106–154.

Hudspeth, A. J. (1985). The cellular basis of hearing: the biophysics of hair cells. *Science* 230:745–752.

Hudspeth, A. J., and Jacobs, R. (1979). Stereocilia mediate transduction in vertebrate hair cells. *Proc. Natl. Acad. Sci. USA* 76: 1506–1509.

Huguenard, J., and McCormick, D. A. (1994). *Electrophysiology of the Neuron. An Interactive Tutorial.* New York: Oxford University Press.

Isaac, G., and Leakey, R. E. F. (1979). *Human Ancestors.* San Francisco: W. H. Freeman.

Ito, M. (1991). The cellular basis of cerebellar plasticity. *Curr. Biol.* 1:616–620.

Iversen, L. (1979). Chemistry of the brain. *Sci. Am.* 241:118–129.

Jack, J. J. B., Noble, D., and Tsein, R. W. (1975). *Electric Current Flow in Excitable Cells.* Oxford: Clarendon Press.

Jacklet, J. W. (1981). Circadian timing by endogeneous oscillators in the nervous system: toward cellular mechanisms. *Biol. Bull.* 160:199–227.

Jacob, F. (1974). *The Logic of Life.* New York: Vintage.

Jacobson, M. (1978). *Developmental Neurobiology.* New York: Plenum.

Jahr, C. E., and Stevens, C. F. (1987). Glutamate activates multiple single channel conductances in hippocampal neurons. *Nature* 325:522–525.

Jan, L. Y., Jan. Y. N., and Kuffler, S. W. (1979). A peptide as a possible transmitter in sympathetic ganglia of the frog. *Proc. Natl. Acad. Sci. USA* 76:1501–1505.

Jessell, T. M., and Kandel, E. R. (1993). Synaptic transmission: a bidirectional and self-modifiable form of cell–cell communication. *Cell 72/Neuron* 10 (Suppl.): 1–31.

Johannson, O., Hökfelt, T., Pernow, B., Jeffcoate, S. L., White, N., Steinbusch, H. W. M., Verhofstad, A. J., Emson, P. C., and Spindel, E. (1981). Immunohistochemical support for three putative transmitters in one neuron: coexistence of 5-hydroxytryptamine-, substance P- and thyrotropin-releasing hormonelike immunoreactivity in medullary neurons projecting to the spinal cord. *Neuroscience* 6:1857–1881.

Johansson, R. S., and Westling, G. (1984). Roles of glabrous skin receptors and sensorimotor memory in automatic control of precision grip when lifting rougher or more slippery objects. *Exp. Brain. Res.* 56: 550–564.

Johnson, K. O., and Hsiao, S. S. (1992). Neural mechanism of tactual form and texture perception. *Annu. Rev. Neurosci.* 15:227-250.

Jones, E. G. (1981). Anatomy of cerebral cortex: columnar input–output organization. In *The Organization of the Cerebral Cortex* (F. O. Schmitt, F. G. Worden, G. Adelman, and S. G. Dennis, ed.). Cambridge, MA: MIT Press, pp. 199–235.

Jones, E. G., and Wise, S. P. (1977). Size, laminar and columnar distribution of the efferent cells in the sensory-motor cortex of monkeys. *J. Comp. Neurol.* 175:391–448.

Jouvet, M. (1974). Monoaminergic regulation of the sleep–waking cycle in the cat. In *The Neurosciences: Third Study Program* (F. O. Schmitt and F. G. Worden, eds.). Cambridge, MA: MIT Press, pp. 499–508.

Jürgens, U., and Ploog, D. (1981). On the neural control of mammalian vocalization. *Trends Neurosci.* 4:135–137.

Kaas, J. H. (1991). Plasticity of sensory and motor maps in adult mammals. *Annu. Rev. Neurosci.* 14:137–168.

Kales, A., and Kales, J. D. (1974). Sleep disorders. *New Engl. J. Med.* 290:487–499.

Kandel, E. R. (1976). *Cellular Basis of Behavior.* San Francisco: W. H. Freeman.

Kandel, E. R. (1979). Cellular insights into behavior and learning. *Harvey Lectures* 73:19–92.

Kandel, E. R. (1985). Cellular mechanisms of learning and the biological basis of individuality. In *Principles of Neural Science* (E. R. Kandel and J. H. Schwartz, eds.). New York: Raven Press, pp. 243–255.

Kandel, E. R., and Schwartz, J. H. (1985). *Principles of Neural Science*, 2nd ed. New York: Elsevier.

Kandel, E. R., Schwartz, J. H., and Jessell, T. M. (1991). *Principles of Neural Science*, 3rd ed. New York: Elsevier.

Kaplan, F., and Barlow, R. B., Jr. (1980). Circadian clock in *Limulus* brain increases response and decreases noise of retinal photoreceptors. *Nature* 286:393–435.

Kappers, C. U. A., Huber, G. C., and Crosby, E. (1936). *The Comparative Anatomy of the Nervous System of Vertebrates, Including Man.* New York: Macmillan.

Katz, B. (1950). Depolarization of sensory terminals and the initiation of impulses in the muscle spindle. *J. Physiol. (London)* 111:261–282.

Katz, B. (1962). The transmission of impulses from nerve to muscle, and the subcellular unit of synaptic action. *Proc. R. Soc. London [B]* 195:455–477.

Kauer, J. S. (1974). Response patterns of amphibian olfactory bulb to odour stimulation. *J. Physiol. (London)* 243:675–715.

Kauer, J. S. (1987). Coding in the olfactory system. In *Neurobiology of Taste and Smell* (T. E. Finger and W. L. Silver, eds.). New York: Wiley, pp. 205–231.

Kaupp, U. B. (1991). The cyclic nucleotide-gated channels of vertebrate photoreceptors and olfactory epithelium. *Trends Neurosci.* 14:150–157.

Kelley, D. B. (1981). Social signals: an overview. *Am. Zool.* 21:111–116.

Kelso, S. R., Ganong, A. H., and Brown, T. H. (1986). Differential conditioning of associative synaptic enhancement in hippocampal brain slices. *Science* 232:85–87.

Kemp, D. T. (1978). Stimulated acoustic emissions from within the human auditory system. *J. Acoust. Soc. Am.* 64:1386–1391.

Kennedy, D. (1976). Neural elements in relation to network function. In *Simple Networks and Behavior* (J. C. Fentress, ed.). Sunderland, MA: Sinauer, pp. 65–81.

Kerwin, J. P. (1977). Skylab 2 crew observations and summary. In *Biomedical Results from Skylab* (R. S. Johnston and L. F. Dietlein, eds.). Washington, DC: Scientific and Technical Information Office, NASA, pp. 27–29.

Kien, J. (1983). The initiation and maintenance of walking in the locust: an alternative to the command concept. *Proc. R. Soc. London [B]* 219:137–174.

Kimble, D. P., Jordan, W. P., and Bremiller, R. (1982). Further evidence for latent learning in hippocampal lesioned rats. *Physiol. Behav.* 29:401–407.

Kimmel, C. B., and Varga, R. M. (1986). Tissue-specific cell lineages originate in the gastrula of the zebrafish. *Science* 231: 365–368.

Kistler, J., Stroud, R., Klymkowsky, M., Lalancette, R., and Fairclouh, R. (1982). Structure and function of an acetylcholine receptor. *Biophys. J.* 37:371–383.

Kitai, S. T. (1981). Anatomy and physiology of the neostriatum. In *GABA and the Basal Ganglia* (G. Di Chiara and G. L. Gessa, eds.). New York: Raven Press, pp. 1–21.

Klein, D. C. (1974). Circadian rhythms and indole metabolism in the rat pineal gland. In *The Neurosciences: Third Study Program* (F. O. Schmitt and F. G. Worden, eds.). Cambridge, MA: MIT Press, pp. 509–515.

Klein, M., Shapiro, E., and Kandel, E. R. (1980). Synaptic plasticity and the modulation of the Ca^{2+} current. *J. Exp. Biol.* 89: 117–157.

Klüver, H., and Bucy, P. C. (1937). "Psychic blindness" and other symptoms following bilateral temporal lobectomy in rhesus monkeys. *Am. J. Physiol.* 119:352–353.

Knowles, A., and Dartnall, J. A. (1977). The photobiology of vision. In *The Eye*, Vol. 2B (H. Dawson, ed.). New York: Academic Press, pp. 1–13.

Kolb, B., and Whishaw, I. Q. (1985). *Fundamentals of Human Neuropsychology.* San Francisco: W. H. Freeman.

Komuro, H., and Rakic, P. (1992). Selective role of N-type calcium channels in neuronal migration. *Science* 257:806–809.

Konopka, R. J., and Benzer, S. (1971). Clock mutants of *Drosophila melanogaster. Proc. Natl. Acad. Sci. USA* 8:2112–2115.

Krasne, F. B., and Wine, J. J. (1984). The production of crayfish tailflip escape responses. In *Neural Mechanisms of Startle Behavior* (R. C. Eaton, ed.). New York: Plenum, pp. 179–212.

Kravitz, E. A., Beltz, B. S., Glusman, S., Goy, M., Harris-Warwick, R. M., Johnston, M. F., Livingstone, M. S., and Schwarz, T. L. (1983). Neurohormones and lobsters: biochemistry to behavior. *Trends Neurosci.* 6:346–349.

Kravitz, E. A., Beltz, B. S., Glusman, S., Goy, M., Harris-Warwick, R., Johnston, M., Livingstone, M., Schwarz, T., and Siwicki, K. K. (1985). The well-modulated lobster: the roles of serotonin, octopamine, and proctolin in the lobster nervous system. In *Model Neural Networks and Behavior*

(A. I. Selverston, ed.). New York: Plenum, pp. 339–361.

Kreutzberg, G. W., Schubert, P., and Lux, H. D. (1975). Neuroplasmic transport in axons and dendrites. In *Golgi Centennial Symposium* (M. Santini, ed.). New York: Raven Press, pp. 161–166.

Kriegstein, A. R., and Connors, B. W. (1986). Cellular physiology of the turtle visual cortex: synaptic properties and intrinsic circuitry. *J. Neurosci.* 6:178–191.

Krueger, J. M. (1989). No simple slumber: exploring the enigma of sleep. *Am. Sci.*, pp. 36–41.

Kuffler, S. W. (1953). Discharge patterns and functional organization of mammalian retina. *J. Neurophysiol.* 16:37–68.

Kuffler, S. W. (1958). Synaptic inhibitory mechanisms: properties of dendrites and problems of excitation in isolated sensory nerve cells. *Exp. Cell Res.* (Suppl.) 5:493–519.

Kuffler, S. W., and Eyzaguirre, C. (1955). Synaptic inhibition in an isolated nerve cell. *J. Gen. Physiol.* 39:155–184.

Kuffler, S. W., and Nicholls, J. G. (1976). *From Neuron to Brain.* Sunderland, MA: Sinauer.

Kuffler, S. W., Nicholls, J. G., and Martin, A. R. (1984). *From Neuron to Brain,* 2nd ed. Sunderland, MA: Sinauer.

Kupfermann, I., and Weiss, K. R. (1978). The command neuron concept. *Behav. Brain Sci.* 1:3–39.

Kuypers, H. G. J. M. (1985). The anatomical and functional organization of the motor system. In *Scientific Basis of Clinical Neurology* (M. Swash and C. Kennard, eds.). London: Churchill-Livingstone, pp. 3–18.

Lacquaniti, F. (1992). Automatic control of limb movement and posture. *Curr. Biol.* 2:807–814.

Lacquaniti, F., Borghese, N. A., and Carrozzo, M. (1991). Transient reversal of the stretch reflex in human arm muscles. *J. Neurophysiol.* 66:939–954.

Lamb, G. D., and Stephenson, D. G. (1992). Importance of Mg^{2+} in excitation–contraction coupling in skeletal muscle. *News Physiol. Sci.* 7:270–274.

LaMotte, R. H., Shain, C. N., Simone, D. A., and Tsai, E.-F. P. (1991). Neurogenic hyperalgesia: psychophysical studies of underlying mechanisms. *J. Neurophysiol.* 66:190–211.

LaMotte, R. H., Thalhammer, J. G., and Robinson, C. J. (1983). Peripheral neural correlates of magnitude of cutaneous pain and hyperalgesia: a comparison of neural events in monkey with sensory judgments in human. *J. Neurophysiol.* 50:1–26.

LaMotte, R. H., Thalhammer, J. G., Torebjork, H. E., and Robinson, C. G. (1982). Peripheral neural mechanisms of cutaneous hyperalgesia following mild injury by heat. *J. Neurosci.* 2:765–781.

Lancet, D. (1986). Vertebrate olfactory reception. *Annu. Rev. Neurosci.* 9:329–355.

Lancet, D. (1991). The strong scent of success. *Nature* 351:275–276.

Lancet, D., Greer, C. A., Kauer, J. S., and Shepherd, G. M. (1982). Mapping of odor-related activity in the olfactory bulb by high resolution 2-deoxyglucose autoradiography. *Proc. Natl. Acad. Sci. USA* 79:670–674.

Landis, D. M. D., and Reese, T. S. (1974). Differences in membrane structure between excitatory and inhibitory synapses in the cerebellar cortex. *J. Comp. Neurol.* 155:93–126.

Landmesser, L. T. (1980). The generation of neuromuscular specificity. *Annu. Rev. Neurosci.* 3:279–302.

Larkman, A., Stratford, K., and Jack, J. (1991). Quantal analysis of excitatory synaptic action and depression in hippocampal slices. *Nature* 350:344–346.

Lashley, K. S. (1950). In search of the engram. *Symp. Soc. Exp. Biol.* 4:454–482.

Le Douarin, N. M. (1982). *The Neural Crest.* New York: Cambridge University Press.

Le Douarin, N. M. (1993). Embryonic neural chimaeras in the study of brain development. *Trends Neurosci.* 16:64–72.

Lehninger, A. L. (1982). *Principles of Biochemistry.* New York: Worth.

LeMagnen, J. (1971). Advances in studies on the physiological control and regulation of food intake. *Prog. Psychobiol. Physiol. Psychol.* 4:203–261.

Lemon, R. N., and Porter, R. (1976). Afferent input to movement-related precentral neurones in conscious monkeys. *Proc. R. Soc. London [B]* 194:313–339.

Leon, M., Coopersmith, R., Lee, S., Sullivan, R. M., Wilson, D. A., and Woo, C. (1987). Neural and behavioral plasticity induced by early olfactory experience. In *Perinatal Development: A Psychobiological Perspective* (N. Krasnegor, E. Blass, M. Hofer, and W. Smotherman, eds.). New York: Academic Press, pp. 145–167.

LeVay, S. (1992). *The Sexual Brain.* Cambridge, MA: MIT Press.

LeVay, S., Wiesel, T. N., and Hubel, D. H. (1981). The postnatal development and plasticity of ocular dominance columns in the monkey. In *The Organization of the Cerebral Cortex* (F. O. Schmitt, F. G. Worden, G. Adelman, and S. G. Dennis, eds.). Cambridge, MA: MIT Press, pp. 29–46.

Lewis, D. B., and Gower, D. M. (1980). *Biology of Communication*. New York: Wiley.

Liao, D., Jones, A., and Malinow, R. (1992). Direct measurement of quantal changes underlying long-term potentiation in CA1 hippocampus. *Neuron* 9:1089–1097.

Liddell, E. G. T. (1960). *The Discovery of Reflexes*. Oxford: Oxford University Press.

Lieberman, P., Crelin, E. S., and Klatt, D. H. (1972). Phonetic ability and related anatomy of the newborn and adult human, Neanderthal man, and the chimpanzee. *Am. Anthropol.* 74:287–307.

Liebowitz, S. F. (1992). Neurochemical-neuroendocrine systems in the brain controlling macronutrient intake and metabolism. *Trends Neurosci.* 15:491–497.

Light, A. R. (1992). *The Initial Processing of Pain and Its Descending Control: Spinal and Trigeminal Systems*. New York: Karger.

Lisman, J. E., and Harris, K. M. (1993). Quantal analysis and synaptic anatomy: integrating two views of hippocampal plasticity. *Trends Neurosci.* 16:141–147.

Livingstone, M., and Hubel, D. (1988). Segregation of form, color, movement, and depth: anatomy, physiology, and perception. *Science* 240:740–749.

Llinás, R. (1988). The intrinsic electrophysiological properties of mammalian neurons: insights into central nervous system function. *Science* 242:1654–1664.

Llinás, R., McGuiness, T. L., Leonard, C. S., Sugimori, M., and Greengard, P. (1985). Intraterminal injection of synapsin I or calcium/calmodulin-dependent protein kinase II alters neurotransmitter release at the squid giant synapse. *Proc. Natl. Acad. Sci. USA* 82:3035–3039.

Llinás, R., and Walton, K. D. (1990). Cerebellum. In *The Synaptic Organization of the Brain*, 3rd ed. (G. M. Shepherd, ed.). New York: Oxford University Press, pp. 214–245.

Lockerbie, R. O. (1987). The neuronal growth cone: a review of its locomotory, navigational and target recognition capabilities. *Neuroscience* 20:719–730.

Loewenstein, O. E. (1974). Comparative morphology and physiology. In *Handbook of Sensory Physiology*, Vol. 6: *Vestibular System*, Part 1: *Basic Mechanisms* (H. H. Kornhuber, ed.). New York: Springer-Verlag, pp. 74–123.

Loewenstein, W. R. (1971). Mechano-electric transduction in the Pacinian corpuscle: initiation of sensory impulses in mechanoreceptor. In *Handbook of Sensory Physiology*, Vol. 1: *Principles of Receptor Physiology* (W. R. Loewenstein, ed.). New York: Springer-Verlag, pp. 269–290.

Loewenstein, W. R. (1981). Junctional intercellular communication: the cell-to-cell membrane channel. *Physiol. Rev.* 61:829–913.

Lømo, T., and Rosenthal, J. (1972). Control of ACh sensitivity by muscle activity in the rat. *J. Physiol. (London)* 21:453–513.

Lund, R. D. (1978). *Development and Plasticity of the Brain*. New York: Oxford University Press.

Lund, R. D. (1979). Tissue transplantation: a useful tool in mammalian neuroembryology. *Trends Neurosci.* 3:XII–XIII.

Lynch, G. (1986). *Synapses, Circuits, and the Beginnings of the Memory*. Cambridge, MA: MIT Press.

Maclean, P. D., and Delgado, J. M. R. (1953). Electrical and chemical stimulation of frontotemporal portion of limbic system in the waking animal. *EEG Clin. Neurophysiol.* 5:91–100.

MacLusky, N. J., Philip, A., Hurlburt, C., and Naftolin, F. (1985). Estrogen formation in the developing rat brain: sex differences in aromatase activity during early post-natal life. *Psychoneuroendocrinology* 10:355–361.

Makowski, L., Casper, D. L. D., Phillips, W. C., and Goodenough, D. A. (1977). Gap junction structure. II. Analysis of the x-ray diffraction data. *J. Cell BIol.* 74:629–645.

Markl, H. (1974). The perception of gravity and of angular acceleration in invertebrates. In *Handbook of Sensory Physiology*, Vol. 6: *Vestibular System*, Part 1: *Basic Mechanisms* (H. Kornhuber, ed.). New York: Springer-Verlag, pp. 17–74.

Marks, G. S., Brien, J. F., Nakatsu, K., and McLaughlin, B. E. (1991). Does carbon monoxide have a physiological function? *Trends Pharm. Sci.* 12:185–188.

Marler, P. (1976). Sensory templates in species-specific behavior. In *Simpler Networks and Behavior* (J. E. Fentress, ed.). Sunderland, MA: Sinauer, pp. 314–329.

Marr, D. 1982. *Vision*. San Francisco: W. H. Freeman.

Martin, K. A. C. (1992). Parallel pathways converge. *Curr. Biol.* 2:555–557.

Marty, A., Evans, M. G., Tan, Y. P., and Trautmann, A. (1986). Muscarinic response in rat lacrimal glands. *J. Exp. Biol.* 124:15–32.

Marx, I. (1983). Synthesizing the opioid peptides. *Science* 220:395–396.

Masu, M., Tanabe, Y., Tsuchida, K., Shigemoto, R., and Nakanishi, S. (1991). Sequence and expression of a metabotropic glutamate receptor. *Nature* 349:760–765.

Matsumoto, S. G., and Hildebrand, J. G. (1981). Olfactory mechanisms in the moth *Manduca sexta:* response characteristics and morphology of central neurons in the antennal lobes. *Proc. R. Soc. London [B]* 213:249–277.

Matthews, B. H. C. (1931). The response of a single end organ. *J. Physiol. (London)* 71:64–110.

Matthews, P. B. C. (1972). *Mammalian Muscle Receptors and Their Central Actions.* London: Edward Arnold.

Mathews, P. B. C. (1981). Evolving views on the internal operation and functional role of the muscle spindle. *J. Physiol.* 320:1–30.

Matthews, P. B. C. (1982). Where does Sherrington's "muscular sense" originate? Muscles, joints, corollary discharge. *Annu. Rev. Neurosci.* 5:189–218.

Maturana, H. R., Lettvin, J. Y., McCulloch, W. S., and Pitts, W. H. (1960). Anatomy and physiology of vision in the front *(Rana pipiens). J. Gen. Physiol.* 4:129–175.

McConnell, S. K. (1992). The determination of neuronal identity in the mammalian cerebral cortex. In *Determinants of Neuronal Identity* (M. Shankland and E. R. Macagno, eds.). New York: Academic Press, pp. 391–432.

McCormick, D. A., Connors, B. W., Lighthall, J. W., and Prince, D. A. (1985). Comparative electrophysiology of pyramidal and sparsely spiny stellate neurons of the neocortex. *J. Neurophysiol.* 54:782–806.

McCormick, D. A., and Pape, H.-C. (1990). Properties of a hyperpolarization-activated cation current and its role in rhythmic oscillation in thalamic relay neurons. *J. Physiol. (London)* 431:291–318.

McCormick, D. A., and Thompson, R. F. (1984). Neuronal responses of the rabbit cerebellum during acquisition and performance of a classically conditioned nictitating membrane-eyelid response. *J. Neurosci.* 4:2811–2822.

McCormick, F. (1993). How receptors turn Ras on. *Nature* 363:15–16.

McCulloch, W., and Pitts, W. (1943). A logical calculus of ideas immanent in nervous activity. *Bull. Math. Biophys.* 5:115–133.

McEwen, B. S. (1991). Non-genomic and genomic effects of steroids on neural activity. *Trends Pharmacol. Sci.* 12:141–147.

Meier, T., and Bloebel, G. (1992). Noppl140 shuttles on tracks between nucleolus and cytoplasm. *Cell* 70:127–138.

Melzack, R., and Wall, P. D. (1965). Pain mechanisms: a new theory. *Science* 150:971–979.

Menaker, M., Takahaski, J. S., and Eskin, A.

(1978). The physiology of circadian pacemakers. *Annu. Rev. Physiol.* 40:501–526.

Merzenich, M. M., and Kaas, J. H. (1980). Principles of organization of sensory-perceptual systems in mammals. In *Progress in Psychobiology and Physiological Psychology* (J. M. Sprague and A. N. Epstein, eds.). New York: Academic Press, pp. 2–43.

Mignery, G. A., Sudhof, T. C., Takei, K., and De Camilli, P. (1989). Putative receptor for inositol 1,4,5-trisphosphate similar to ryanodine receptor. *Nature* 342:192–195.

Miller, J. A. (1980). A song for the female finch. *Sci. News* 117:58–59.

Miller, J. A. (1983). Molecular hardware of cell communication: describing the acetylcholine receptor. *Neurosci. Comment.* 1:89–98.

Miller, J. P., and Selverston, A. L. (1982). Mechanisms underlying pattern generation in lobster stomatogastric ganglion as determined by selective inactivation of identified neurons. IV. Network properties of pyloric system. *J. Neurophysiol.* 48:1416–1432.

Miller, J. P., and Selverston, A. L. (1985). Neural mechanisms for the production of the lobster pyloric motor pattern. In *Model Neural Networks and Behavior* (A. L. Selverston, ed.). New York: Plenum, pp. 37–48.

Miller, R. H., French-Constant, C., and Raff, M. C. (1989). The macroglial cells of the rat optic nerve. *Annu. Rev. Neurosci.* 12:517–534.

Mishkin, M., Ungerleider, L. G., and Macko, K. A. (1983). Object vision and spatial vision: two cortical pathways. *Trends Neurosci.* 6:414–417.

Moore, I. K., and Osen, K. K. (1979). The human cochlear nuclei. In *Experimental Brain Research*, Suppl. II: *Hearing Mechanisms and Speech* (O. Creutzfeld, H. Scheich, and C. Schreiner, eds.). New York: Springer-Verlag, pp. 36–44.

Moore, R. Y. (1982). The suprachiasmatic nucleus and the organization of a circadian system. *Trends Neurosci.* 5:404–407.

Moriyoshi, K., Masu, M., Ishii, T., Shigemoto, R., Mizuno, N., and Nakanishi, S. (1991). Molecular cloning and characterization of the rat NMDA receptor. *Nature* 354:31–37.

Mountcastle, V. B., ed. (1980). *Medical Physiology*, 14th ed. St. Louis: C. V. Mosby.

Mountcastle, V. B. (1986). The neural mechanisms of cognitive functions can now be studied directly. *Trends Neurosci.* 9:505–508.

Mountcastle, V. B., Lynch, J. C., Georgopolous,

A., Sakata, H., and Aeuna, C. (1975). Posterior parietal association cortex of the monkey: command functions for operation within extra-personal space. *J. Neurophysiol.* 38:871–908.

Mrosovsky, N. (1986). Sleep researchers caught napping. *Nature* 319:536–537.

Muller, W., and Connor, J. A. (1991). Dendritic spines as individual neuronal compartments for synaptic Ca^{2+} responses. *Nature* 354:73–76.

Murray, R. G. (1973). The ultrastructure of taste buds. In *The Ultrastructure of Sensory Organs* (I. Friedmann, ed.). New York: Elsevier, pp. 1–81.

Muybridge, E. (1957). *Animals in Motion.* Reprinted from *Animal Locomotion* (1887). New York: Dover.

Nakamura, T., and Gold, G. H. (1987). A cyclic nucleotide-gated conductance in olfactory receptor cilia. *Nature* 325:442–444.

Nathans, J., Thomas, D., and Hogness, D. S. (1986). Molecular genetics of human color vision: the genes encoding blue, green, and red pigments. *Science* 232:193–202.

Nathans, J., Weitz, C. J., Sung, C.-H., Davenport, C. M., Merbs, S. L., and Wang, Y. (1992). Visual pigments and inherited variation in human vision. In *Sensory Transduction.* (D. P. Corey and S. D. Roper, eds.). New York: Rockefeller University Press, pp. 109–131.

Nauta, W. J. M. (1972). The central visceromotor system: a general survey. In *Limbic System Mechanisms and Automatic Function* (C. H. Hockman, ed.). Springfield, IL: Charles C. Thomas, pp. 21–38.

Neher, F., and Steinbach, J. H. (1978). Local anaesthetics transiently block currents through single acetylcholine-receptor channels. *J. Physiol. (London)* 227:153–176.

Nestler, E. J., and Greengard, P. (1984). *Protein Phosphorylation in the Nervous System.* New York: Wiley.

Neyton, J., and Trautmann, A. (1985). Single-channel currents of an intercellular junction. *Nature* 317:331–335.

Nicholls, J. G., Martin, A. R., and Wallace, B. G. (1992). *From Neuron to Brain,* 3rd ed. Sunderland, MA: Sinauer.

Nicoll, R. A. (1982). Neurotransmitters can say more than yes or no. *Trends Neurosci.* 5:369.

Nicoll, R. A., Kauer, J. A., and Malenka, R. C. (1988). The current excitement in long-term potentiation. *Neuron* 1:97–103.

Nilius, B., Hess, P., Lansman, J. B., and Tsien, R. W. (1985). A novel type of cardiac calcium channel in ventricular cells. *Nature* 316:443–446.

Nilsson, B. (1984). Master class. *The New Yorker,* December 10, pp. 44–45.

Nissanov, J., Eaton, R. C., and DiDomenico, R. (1990). The motor output of the Mauthner cell, a reticulospinal command neuron. *Brain Res.* 5–7:88–98.

Nixon, R. A., and Sihag, R. K. (1991). Neurofilament phosphorylation: a new look at regulation and function. *Trends Neurosci.* 14:501–506.

Noble, D. (1985). Ionic mechanisms in rhythmic firing of heart and nerve. *Trends Neurosci.* 8:499–504.

Noda, M., Ikeda, T., Kayano, T., Suzuki, H., Takeshima, H., Kurasaki, M., Takahashi, H., and Numa, S. (1986). Existence of distinct sodium channel messenger RNAs in rat brain. *Nature* 320:188–192.

Norgren, R. (1980). Neuroanatomy of gustatory and visceral afferent systems in rat and monkey. In *Olfaction and Taste,* Vol. 2 (H. van der Starre, ed.). London: IRL Press, p. 288.

Northcutt, R. G. (1981). Evolution of the telencephalon in non-mammals. *Annu. Rev. Neurosci.* 4:301–350.

Nottebohm, F. (1975). Vocal behavior in birds. In *Avian Biology,* Vol. 5 (D. S. Farner and J. R. King, eds.). New York: Academic Press, pp. 287–332.

Nottebohm, F. (1980). Brain pathways for vocal learning in birds. In *Progress in Psychobiology and Physiological Psychology,* Vol. 9 (J. M. Sprague and A. N. Epstein, eds.). New York: Academic Press, pp. 86–125.

Nowak, L., Bregestovski, P., Ascher, P., Herbet, A., and Prochiantz, A. (1984). Magnesium gates glutamate-activated channels in mouse central neurones. *Nature* 307:462–465.

Nowycky, M. C., Fox, A. P., and Tsien, R. W. (1985). Three types of neuronal calcium channel with different calcium agonist sensitivity. *Nature* 310:440–443.

Nowycky, M. C., and Roth, R. H. (1978). Dopaminergic neurons: role of presynaptic receptors in the regulation of transmitter biosynthesis. *Prog. Neuropsychopharmacol.* 2:139–158.

Numa, S., Noda, M., Takahashi, H., Tanabe, T., Toyasato, M., Furatani, Y., and Kikyotani, S. (1983). Molecular structure of the nicotinic acetylcholine receptor. *Cold Spring Harbor Symp. Quant. Biol.* 48:57–69.

Olds, M. E., and Olds, J. (1961). Emotional and associative mechanisms in the rat brain. *J. Comp. Physiol. Psychol.* 54:120–126.

Olson, L. (1985). On the use of transplants to counteract the symptoms of Parkinson's disease: background, experimental models, and possible clinical applications. In *Synaptic Plasticity* (C. E. Cotman, ed.). New York: Guilford Press, pp. 485–506.

O'Rourke, N. A., Dailey, M. E., Smith, S. J., and McConnell, S. K. (1992). Diverse migratory pathways in the developing cerebral cortex. *Science* 258:299–302.

Otsuka, M., and Hall, Z. W. (1979). Preface. In *Neurobiology of Chemical Transmission* (M. Otsuka and Z. W. Hall, eds.). New York: Wiley, p. vii.

Ottoson, D., and Shepherd, G. M. (1971). Transducer properties and integrative mechanisms in the frog's muscle spindle. In *Handbook of Sensory Physiology*, Sec. 1: *The Nervous System*, Vol. 1: *Principles of Receptor Physiology* (W. R. Loewenstein, ed.). New York: Springer-Verlag, pp. 443–499.

Pace, U., Hansky, E., Salomon, Y., and Lancet, D. (1985). Odorant sensitive adenylate cyclase may mediate olfactory reception. *Nature* 316:255–258.

Palade, G. E., and Farquhar, M. G. (1981). Cell biology. In *Pathophysiology: The Biological Principles of Disease* (L. H. Smith and S. O. Thie, eds.). Philadelphia: W. B. Saunders, pp. 1–56.

Papez, J. W. (1937). A proposed mechanism of emotion. *Arch. Neurol. Psychiatr.* 38:725–743.

Patlak, J., and Horn, R. (1982). Effect of N-bromo-acetamide on single sodium channel currents in excised membrane patches. *J. Gen. Physiol.* 79:333–351.

Patterson, P. H., and Nawa, H. (1993). Neuronal differentiation factors/cytokines and synaptic plasticity. *Cell 72/Neuron 10* (Suppl.): 123–137.

Patton, H. D., Sundsten, J. W., Crill, W. E., and Swanson, P. D., eds. (1976). *Introduction to Basic Neurology*. Philadelphia: W. B. Saunders.

Pearson, K. (1976). The control of walking. *Sci. Am.* 2:72–86.

Pedersen, P. E., and Blass, E. M. (1982). Prenatal and postnatal determinants of the first suckling episode in the albino rat. *Dev. Psychobiol.* 15:349–356.

Penfield, W., and Rasmussen, T. (1952). *The Cerebral Cortex of Man*. New York: Macmillan.

Persohn, E., Malherbe, P., and Richards, J. G. (1992). Comparative molecular neuroanatomy of cloned GABA$_A$ receptor subunits in the rat CNS. *J. Comp. Neurol.* 326:193–216.

Peters, A., Palay, S., and Webster, H. deF. (1990). *The Fine Structure of the Nervous System*, 3rd ed. New York: Oxford University Press.

Petersen, O. H., and Maruyama, Y. (1984). Calcium activated potassium channels and their role in secretion. *Nature* 307:693–696.

Petersen, S. E., Fox, P. T., Posner, M. I., Minton, M., and Raichle, M. E. (1988). Positron emission tomographic studies of the cortical anatomy of single word processing. *Nature* 331:585–589.

Pfaff, D. W. (1980). *Estrogens and Brain Function*. New York: Springer-Verlag.

Pfaffman, C., Frank, M., Bartoshuk, L. M., and Snell, T. C. (1976). Coding gustatory information in the squirrel monkey chorda tympani. In *Progress in Psychobiology and Physiological Psychology*, Vol. 5 (J. M. Sprague and A. N. Epstein, eds.). New York: Academic Press, pp. 1–27.

Pfeiffer, R. R. (1966). Classification of response patterns of spike discharges for units in the cochlear nucleus: tone burst stimulation. *Exp. Brain Res.* 1:220–235.

Pfenninger, K.-H. (1986). Of nerve growth cones, leukocytes and memory: second messenger systems and growth-regulated proteins. *Trends Neurosci.* 8:562–565.

Phillips, C. G. (1977). On integration and teleonomy. *Proc. R. Soc. London [B]* 199:415–424.

Phillips, C. G., and Porter, R. (1977). *Cortico-Spinal Neurones: Their Role in Movement*. London: Academic Press.

Phillips, I. M., Felix, D., Hoffman, W. E., and Ganten, D. (1977). Angiostensin-sensitive sites in the brain ventricular system. In *Society for Neuroscience Symposia*, Vol. 2 (W. M. Cowan and J. A. Ferrendelli, eds.). Bethesda, MD: Society for Neuroscience, pp. 308–339.

Phoenix, C. H., Goy, R. W., Gerall, A. A., and Young, W. C. (1959). Organizational action of prenatally administered testosterone propionate on the tissues mediating behavior in the female guinea pig. *Endocrinology* 65:369–382.

Pickles, J. O., Comis, S. D., and Osborne, M. P. (1984). Cross-links between stereocilia in the guinea pig organ of Corti, and their possible relation to sensory transduction. *Hearing Res.* 15:103–112.

Pickles, J. O., and Corey, D. P. (1992). Mechanoelectrical transduction by hair cells. *Trends Neurosci.* 15:254–259.

Pittendrigh, C. S. (1974). Circadian oscillation in cells and the circadian organization of multicellular systems. In *The Neurosciences: Third Study Program* (F. O. Schmitt and F. G. Worden, eds.). Cambridge, MA: MIT Press, pp. 437–458.

Popper, K. R., and Eccles, J. C. (1977). *The Self and Its Brain*. New York: Springer-Verlag.

Powell, T. P. S., and Mountcastle, V. B. (1959).

Some aspects of the functional organization of the cortex of the postcentral gyrus of the monkey: a correlation of findings obtained in a single unit analysis with cytoarchitecture. *Bull. Johns Hopkins Hosp.* 105:133–162.

Powis, G. (1991). Signalling targets for anticancer drug development. *Trends Pharmacol. Sci.* 12:188–194.

Proctor, D. F. (1980). *Breathing Speech and Song*. New York: Springer-Verlag.

Prosser, C. L. (1991). Animal movement. In *Comparative Animal Physiology: Neural and Integrative Animal Physiology*, 4th ed. (C. L. Prosser, ed.). New York: Wiley-Liss, pp. 67–130.

Puia, G., Santi, M. R., Visini, S., Pritchett, D. B., Purdy, R. H., Paul, S. M., Seeburg, P. H., and Costa, E. (1990). Neurosteroids act on recombinant human GABA$_A$ receptors. *Neuron* 4:759–765.

Purves, D., and Lichtman, J. W. (1985). *Principles of Neural Development*. Sunderland, MA: Sinauer.

Raisman, G. (1969). Neuronal plasticity in the septal nuclei of the adult rat. *Brain Res.* 14:25–48.

Raisman, G., and Field, P. M. (1971). Sexual dimorphism in the preoptic area of the rat. *Science* 173:731–733.

Rakic, P. (1976). *Local Circuit Neurons*. Cambridge, MA: MIT Press.

Rakic, P. (1981). Developmental events leading to laminar and areal organization of the neocortex. In *The Organization of the Cerebral Cortex* (F. O. Schmitt, F. G. Worden, G. Adelman, and S. G. Dennis, eds.). Cambridge, MA: MIT Press, pp. 7–28.

Rakic, P. (1985). Limits of neurogenesis in primates. *Science* 227:1054–1056.

Rakic, P., Bourgeois, J.-P., Eckenhoff, M. F., Zecevic, N., and Goldman-Rakic, P. S. (1986). Concurrent overproduction of synapses in diverse regions of the primate cerebral cortex. *Science* 232:232–235.

Rakic, P., Cameron, R. S., and Komuro, H. (1994) Recognition, adhesion, transmembrane signaling and cell motility in guided neuronal migration. *Curr. Opinion in Neurobiology* 3.

Rall, W. (1967). Distinguishing theoretical synaptic potentials computed for different soma-dendritic distributions of synaptic input. *J. Neurophsiol.* 30:1138-1168.

Rall, W. (1977). Core conductor theory and cable properties of neurons. In *The Nervous System*, Vol. I: *Cellular Biology of Neurons*, Part 1 (E. R. Kandel, ed.). Bethesda, MD: American Physiological Society, pp. 39–97.

Rall, W., and Rinzel, J. (1971). Dendritic spine function and synaptic attenuation calculations. *Soc. Neurosci. Abstr.,* p. 64.

Rall, W., and Shepherd, G. M. (1968). Theoretical reconstruction of field potentials and dendrodendritic synaptic interactions in olfactory bulb. *J. Neurophysiol.* 31:884–915.

Rall, W., Shepherd, G. M., Reese, T. S., and Brightman, M. W. (1966). Dendrodendritic synaptic pathway for inhibition in the olfactory bulb. *Exp. Neurol.* 14:44–56.

Ralston, H. J. (1971). Evidence for presynaptic dendrites and a proposal for their mechanism. *Nature* 230:585-587.

Ratliff, F. (1965). *Mach Bands: Quantitative Studies on Natural Networks in the Retina*. San Francisco: Holden-Day.

Reed, R. R. (1990). How does the nose know? *Cell* 60:1–2.

Reed, R. R. (1992). Signaling pathways in odor detection. *Neuron* 8:205–209.

Reichelt, K. L., and Edminson, P. D. (1977). Peptides containing probable transmitter candidates in the central nervous system. In *Pepties in Neurobiology* (H. Gainer, ed.). New York: Plenum, pp. 171–181.

Renaud, L. P. (1977). TRH, LHRH, and somatostatin: distribution and physiological action in neural tissue. In *Society for Neuroscience Symposium*, Vol. 2 (W. M. Cowan and J. A. Ferrendelli, eds.). Bethesda, MD: Society for Neuroscience, pp. 265–290.

Ribak, C. E., Vaughn, J. E., Saito, K., Barber, R., and Roberts, E. (1977). Glutamate decarboxylase localization in neurons of the olfactory bulb. *Brain Res.* 126:1–18.

Roberts, A., and Bush, B. M. H., eds. (1981). *Neurones without Impulses: Their Significance for Vertebrate and Invertebrate Nervous Systems*. Cambridge: Cambridge University Press.

Robertson, R. M., and Pearson, K. G. (1985). Neural networks controlling locomotion in locusts. In *Model Neural Networks and Behavior* (A. I. Selverston, ed.). New York: Plenum, pp. 21–36.

Romer, A. S., and Parsons, T. S. (1977). *The Vertebrate Body*. Philadelphia: W. B. Saunders.

Ross, E. M. (1992). Twists and turns on G-protein signalling pathways. *Curr. Biol.* 2:517–519.

Rumelhart, D. E., and McClelland, J. L. (1986). *Parallel Distributed Processing: Explorations in the Microstructure of Cognition*, Vol. 1: *Foundations*. Cambridge, MA: MIT Press.

Sakmann, B., and Neher, E. (1983). *Single-Channel Recording*. New York: Plenum.

Salkoff, L., and Wyman, R. (1981). Genetic modification of potassium channels in *Drosophila Shaker* mutants. *Nature* 293:228–230.

Sallaz, M., and Jourdan, F. (1993). C-fos expression and 2-deoxyglucose uptake in the olfactory bulb of odor-stimulated awake rats. *NeuroReport* 4:55-58.

Sanes, J. R. (1993). Topographic maps and molecular gradients. *Curr. Opin. Neurobiol.* 3:67–74.

Sansom, M. S. P. (1993). Peering down a pore. *Curr. Biol.* 3:239–241.

Sarnat, H. B., and Netsky, M. G. (1981). *Evolution of the Nervous System.* New York: Oxford University Press.

Schiller, P. H. (1992). The ON and OFF channels of the visual system. *Trends Neurosci.* 15:86–92.

Schindler, J. M. (1990). Basic developmental genetics and early embryonic development: what's all the excitement about? *J. NIH Res.* 2:49–55.

Schmidt, R. F., ed. (1978). *Fundamentals of Sensory Physiology.* New York: Springer-Verlag.

Schmidt, R. F., and G. Thewes, eds. (1983). *Human Physiology.* New York: Springer-Verlag.

Schofield, P. R., Darlison, M. G., Fujita, N., Burt, D. R., Stephenson, F. A., Rodriguez, H., Rhee, L. M., Ramachandran, J., Reale, V., Glencorse, T. A., Seeburg, P. H., and Barnard, E. A. (1987). Sequence and functional expression of the $GABA_A$ receptor shows a ligand-gated receptor superfamily. *Nature* 328:221–227.

Schwartz, W. J., and Gainer, H. (1977). Suprachiasmatic nucleus: use of ^{14}C-labeled deoxyglucose uptake as a functional marker. *Science* 197:1089–1091.

Scoville, W. B., and Milner, B. (1957). Loss of recent memory after bilateral hippocampal lesions. *J. Neurol. Neurosurg. Psychiatr.* 20:11–21.

Selverston, A. (1976). A model system for the study of rhythmic behavior. In *Simpler Networks and Behavior* (J. C. Fentress, ed.). Sunderland, MA: Sinauer, pp. 82–98.

Selverston, A. I., Russell, D. F., Miller, J. P., and King, D. G. (1976). The stomatogastric nervous system: structure and function of a small neural network. *Prog. Neurobiol.* 7:215–289.

Shapiro, E., Castellucci, V. F., and Kandel, E. R. (1980). Presynaptic inhibition in *Aplysia* involves a decrease in the Ca^{2+} current of the presynaptic neuron. *Proc. Natl. Acad. Sci. USA* 77:1185–1189.

Shatz, C. J. (1990). Impulse activity and the patterning of connections during CNS development. *Neuron* 5:745–756.

Shepherd, G. M. (1972). The neuron doctrine: a revision of functional concepts. *Yale J. Biol. Med.* 45:584–599.

Shepherd, G. M. (1974). *The Synaptic Organization of the Brain.* New York: Oxford University Press.

Shepherd, G. M. (1978). Microcircuits in the nervous system. *Sci. Am.* 238:92–103.

Shepherd, G. M. (1979). *The Synaptic Organization of the Brain,* 2nd ed. New York: Oxford University Press.

Shepherd, G. M. (1981). The nerve impulse and the nature of nervous function. In *Neurones without Impulses: Their Significance for Vertebrate and Invertebrate Nervous Systems* (A. Roberts and B. M. H. Bush, eds.). Cambridge: Cambridge University Press, pp. 1–27.

Shepherd, G. M. (1985). Olfactory transduction: welcome whiff of biochemistry. *Nature* 316:214–215.

Shepherd, G. M. (1986). Neurobiology: microcircuits to see by. *Nature* 319:452–453.

Shepherd, G. M. (1988). A basic circuit for cortical organization. In *Perspectives in Memory Research* (M. S. Gazzaniga, ed.). Cambridge, MA: MIT Press, pp. 93–134.

Shepherd, G. M. (1989). Studies of development and plasticity in the olfactory sensory neuron. *J. Physiol (Paris)* 83:240–245.

Shepherd, G. M. (1990a). The significance of real neuron architectures for neural network simulations. In *Computational Neuroscience* (E. Schwartz, ed.). Cambridge, MA: MIT Press, pp. 82–96.

Shepherd, G. M., ed. (1990b). *The Synaptic Organization of the Brain,* 3rd ed. New York: Oxford University Press.

Shepherd, G. M. (1991a). *Foundations of the Neuron Doctrine.* New York: Oxford University Press.

Shepherd, G. M. (1991b). Sensory transduction: entering the mainstream of membrane signalling. *Cell* 67:845–851.

Shepherd, G. M. (1992). Modules for molecules. *Nature* 358:457–458.

Shepherd, G. M. (1994). Toward a molecular basis for perception. In *Cognitive Neuroscience* (M. S. Gazzaniga, ed.).

Shepherd, G. M., and Brayton, R. K. (1979). Computer simulation of a dendrodendritic synaptic circuit for self-and lateral-inhibition in the olfactory bulb. *Brain Res.* 175:377-382.

Shepherd, G. M., and Brayton, R. K. (1987). Logic operations are properties of computer-simulated interactions between excitable dendritic spines. *Neuroscience* 21:151–165.

Shepherd, G. M., and Firestein, S. (1991a). Making scents of olfaction. *Curr. Biol.* 1:204–206.

Shepherd, G. M., and Firestein, F. (1991b). Toward a pharmacology of odor receptors and the processing of odor images. *J. Steroid Biachem.* 39:583–592.

Shepherd, G. M., and Greer, C. A. (1988). The dendritic spine: adaptations of structure and function for different types of synaptic integration. In *Intrinsic Determinants of Neuronal Form* (R. Lassek, ed.). New York: Alan R. Liss, pp. 245–262.

Shepherd, G. M., Pedersen, P. E., and Greer, C. A. (1987). Development of olfactory specificity in the albino rat: a model system. In *Perinatal Development: A Psychobiological Perspective* (N. A. Krasnegor, E. M. Blass, M. A. Hofer, and W. P. Smotherman, eds.). New York: Academic Press, pp. 127–144.

Shepherd, G. M., Woolf, T., and Carnevale, N. T. (1989). Comparisons between active properties of distal dendritic branches and spines: implications for neuronal computations. *J. Cogn. Neurosci.* 1:273–286.

Shepherd, G. M. G., Barres, B. A., and Corey, D. P. (1989). "Bundle blot" purification and initial protein characterization of hair cell stereocilia. *Proc. Natl. Acad. Sci. USA* 86:4973–4977.

Shepherd, J. T., and Vanhoutte, P. M. (1979). *The Human Cardiovascular System.* New York: Raven Press.

Sherrington, C. S. (1906). *The Integrative Action of the Nervous System.* New Haven: Yale University Press.

Sherrington, C. S. (1935). Santiago Ramón y Cajal, 1852–1934. *Obituary Notices of the Royal Society of London* 4:425–441.

Shinoda, K., Shiotani, Y., and Osawa, Y. (1989). "Necklace olfactory glomeruli" form unique components of the rat primary olfactory system. *J. Comp. Neurol.* 284:362–373.

Sholl, D. A. (1956). *The Organization of the Cerebral Cortex.* London: Methuen.

Shull, G. E., Lane, L. K., and Lingrel, J. B. (1986). Amino-acid sequence of the β-subunit of the (Na^+/K^+)ATPase deduced from a cDNA. *Nature* 321:429–431.

Sidman, R. L., and Rakic, P. (1982). Development of the human central nervous system. In *Histology and Histopathology of the Nervous System,* Vol. 1 (W. Haymaker and R. D. Adams, eds.). Springfield, IL.: Charles C. Thomas, pp. 3–145.

Sieghart, W. (1992). $GABA_A$ receptors: ligand-gated Cl^- ion channels modulated by multiple drug-binding sites. *Trends Pharmacol. Sci.* 13:446–450.

Siggins, G. R., Hoffer, B. J., and Bloom, F. E. (1971). Studies on norepinephrine-containing afferents to Purkinje cells of rat cerebellum. III. Evidence for mediation of norepinephrine effects by cyclic $3',5'$-adenosine monophosphate. *Brain Res.* 25:535–553.

Sims, S. M., Singer, J. J., and Walsh, J. V., Jr. (1986). A mechanism of muscarinic excitation in dissociated smooth muscle cells. *Trends Pharmacol. Sci.* (Suppl.) 7:28–32.

Skoglund, S. (1973). Joint receptors and kinaesthesis. In *Handbook of Sensory Physiology,* Vol. 2: *Somatosensory System* (A. Iggo, ed.). New York; Springer-Verlag, pp. 111–136.

Smith, D., and Frank, M. (1993). Sensory coding by peripheral taste fibers. In *Mechanisms of Taste Transduction* (S. A. Simon and S. D. Roper, eds.). Boca Raton, FL: CRC Press, pp.

Smith, S. J., Osses, L. R., and Augustine, G. J. (1987). Imaging of localized calcium accumulation within squid "giant" presynaptic terminals. *Biophys. Soc. Absr.* 51:66a.

Snyder, S. H. (1984). Drug and neurotransmitter receptors in the brain. *Science* 224:22–31.

Somjen, G. (1972). *Sensory Coding in the Mammalian Nervous System.* New York: Appleton-Century-Crofts.

Sperry, R. W. (1974). Lateral specialization in the surgically separated hemispheres. In *The Neurosciences: Third Study Program* (F. O. Schmitt and F. G. Worden, eds.). Cambridge, MA: MIT Press, pp. 5–19.

Spoendlin, H. (1969). Innervation patterns in the organ of Corti of the cat. *Acta Otolaryngol. (Stockholm)* 67:239–254.

Squire, L., and Butters, N., eds. (1984). *Neuropsychology of Memory.* New York: Guilford Press.

Starzl, T. E., Taylor, C. W., and Magoun, H. (1951). Collateral afferent excitation of the reticular formation of the brain stem. *J. Neurophysiol.* 14:479–496.

Stellar, J. R., and Stellar, E. (1985). *The Neurobiology of Motivation and Reward.* New York: Springer-Verlag.

Stemple, D. L., and Anderson, D. J. (1992). Isolation of a stem cell for neurons and glia from the mammalian neural crest. *Cell* 71:973–985.

Stent, G. S. (1981). Strength and weakness of the genetic approach to the development of the nervous system. *Annu. Rev. Neurosci.* 4:163–194.

Stent, G. S., Thompson, W. J., and Calabrese, R. L. (1979). Neural control of heartbeat in the leech and in some other invertebrates. *Physiol. Rev.* 59:101–136.

Sterling, P. (1990). Retina. In *The Synaptic Organization of the Brain*, 3rd ed. (G. M. Shepherd, ed.). New York: Oxford University Press, pp. 170–213.

Stevens, C. F. (1991). Making a microscopic hole in one. *Nature* 349:657–658.

Stewart, W. B., Kauer, J. S., and Shepherd, G. M. (1979). Functional organization of rat olfactory bulb analysed by the 2-deoxyglucose method. *J. Comp. Neurol.* 185:715–734.

Storer, T. I. (1943). *General Zoology.* New York: McGraw-Hill.

Strader, C. D., Segal, I. S., and Dixon, R. A. (1989). Structural basis of β-adrenergic receptor function. *FASEB J.* 3:1825–1832.

Stretton, A. O. W., and Kravitz, E. A. (1968). Neuronal geometry: determination with a technique of intracellular dye injection. *Science* 162:132–134.

Strumwasser, F. (1974). Neuronal principles organizing periodic behaviors. In *The Neurosciences: Third Study Program* (F. O. Schmitt and F. G. Worden, eds.). Cambridge, MA: MIT Press, pp. 459–478.

Suga, N. (1978). Specialization of the auditory system for reception and processing of species-specific sounds. *Fed. Proc.* 37:2342–2354.

Sugden, D., Vanecek, J., Klein, D. C., Thomas, T. P., and Anderson, W. B. (1985). Activation of protein kinase C potentiates isoprenaline-induced cyclic AMP accumulation in rat pinealocytes. *Nature* 314:359–362.

Sulston, J., Schierenberg, E., White, J., and Thomson, N. (1983). The embryonic lineage of the nematode *Caenorhaditis elegans. Dev. Biol.* 100:64–119.

Suter, U., Welcher, A. A., and Snipes, G. J. (1993). Progress in the molecular understanding of hereditary peripheral neuropathies reveals new insights into the biology of the peripheral nervous system. *Trends Neurosci.* 16:50–56.

Szekely, G. (1968). Development of limb movements: embryological, physiological and model studies. In *Ciba Foundation Symposium on Growth of the Nervous System* (G. E. W. Wolstenholme and M. O'Connor, eds.). London: Churchill, pp. 77–93.

Takei, K., Stukenbrok, H., Metcalf, A., Mignery, G. A., Sudhof, T. C., Volpe, P., and De Camilli, P. (1992). Ca^{2+} stores in Purkinje neurons: endoplasmic reticulum subcompartments demonstrated by the heterogeneous distribution of the $InsP_3$ receptor, Ca^{2+}-ATPase, and calsequestrin. *J. Neurosci.* 12:489–505.

Tanaka, K. (1992). Inferotemporal cortex and higher visual functions. *Curr. Opin. Neurobiol.* 2:502–505.

Taylor, C. W., and Marshall, I. C. B. (1992). Calcium and inositol 1,2,5-trisphosphate receptors: a complex relationship. *Trends Biochem. Sci.* 17:403–407.

Teicher, M. H., Stewart, W. B., Kauer, J. S., and Shepherd, G. M. (1980). Suckling pheromone stimulation of a modified glomerular region in the developing rat olfactory bulb revealed by the 2-deoxyglucose method. *Brain Res.* 194:530–535.

Thomas, R. C. (1972). Electrogenic sodium pump in nerve and muscle cells. *Physiol. Rev.* 52:563–594.

Thompson, R. F. (1986). The neurobiology of learning and memory. *Science* 233:941–947.

Thompson, R. F., Clark, G. A., Donegan, N. H., Lavond, D. G., Madden, J., IV, Mamounas, L. A., Mauk, M. D., and McCormick, D. A. (1984). Neuronal substrates of basic associative learning. In *Neuropsychology of Memory* (L. R. Squire and N. Butters, eds.). New York: Guilford Press, pp. 424–442.

Thompson, W., Kuffler, D. P., and Jansen, J. K. S. (1979). The effect of prolonged, reversible block of nerve impulses on the elimination of polyneuronal innervation of new-born rat skeletal muscle fibers. *Neuroscience* 4:271–281.

Tinbergen, N. (1951). *The Study of Instinct.* Oxford: Oxford University Press.

Tombes, A. S. (1970). *An Introduction to Invertebrate Endocrinology.* New York: Academic Press.

Towe, A. L., and Luschei, E. S., eds. (1981). *Handbook of Behavioral Neurophysiology*, Vol. 5. New York: Plenum.

Trombley, P. Q., and Shepherd, G. M. (1994). Submitted.

Trotier, D., and MacLeod, P. (1986). Intracellular recordings from salamander olfactory receptor cells. *Brain Res.* 268:225–237.

Tsien, R. W. (1987). Calcium currents in heart cells and neurons. In *Neuromodulation: The Biochemical Control of Neuronal Excitability* (L. K. Kaczmarek and I. B. Levitan, eds.). New York: Oxford University Press, pp. 206–242.

Tunturi, A. (1944). Audio frequency localization in the acoustic cortex of the dog. *Am. J. Physiol.* 141:397–403.

Unwin, N. (1993). Neurotransmitter action: opening of ligand-gated ion channels. *Cell 72/Neuron* 10 (Suppl.): 31–41.

Vale, R. D., Reese, T. S., and Sheetz, M. P. (1985). Identification of a novel force generating protein (kinesin) involved in microtubule-based motility. *Cell* 41:39–50.

Vallbo, A. B., and Johansson, R. S. (1984). Properties of cutaneous mechanoreceptors in the human hand related to touch sensation. *Hum. Neurobiol.* 3:3–14.

Vallee, R. B., and Bloom, G. S. (1991). Mechanisms of fast and slow axonal transport. *Annu. Rev. Neurosci.* 14:59–92.

Van Essen, D. C. (1985). Functional organization of primate visual cortex. In *Cerebral Cortex* (A. Peters and E. G. Jones, eds.). New York: Plenum, pp. 259–330.

Van Essen, D. C., Anderson, C. H., and Felleman, D. J. (1992). Information processing in the primate visual system: an integrated systems perspective. *Science* 255:419–423.

Vincent, S. R., Hökfelt, T., Steinboll, L. R., and Wu, J.-Y. (1983). Hypothalamic alpha-aminobutyric acid neurons project to neocortex. *Science* 220:1309–1311.

Virchow, R. (1860). *Cellular Pathology.* London: Churchill.

Von Békésy, G. (1960). *Experiments in Hearing.* New York: McGraw-Hill.

Von Neumann, J. (1957). *The Brain as a Computer.* New Haven: Yale University Press.

Wakerly, J. B., and Lincoln, D. W. (1973). The milk-ejection reflex of the rat: a 20- to 40-fold acceleration in the firing of paraventricular neurones during oxytocin release. *J. Endocrinol.* 57:477.

Walsh, C., and Cepko, C. L. (1992). Widespread dispersion of neuronal clones across functional regions of the cerebral cortex. *Science* 255:434–440.

Warwick, R., and Williams, P. L. (1973). *Gray's Anatomy.* Philadelphia: W. B. Saunders.

Watson, J. D., Tooze, J., and Kurtz, D. T. (1983). *Recombinant DNA: A Short Course.* New York: Scientific American Books.

Watt, D. G. D., Stauffer, E. K., Taylor, A., Reinking, R. M., and Stuart, D. G. (1976). Analysis of muscle receptor connections by spike-triggered averaging. 1. Spindle primary and tendon organ afferents. *J. Neurophysiol.* 39:1375–1392.

Watson, S., and Girdlestone, D. (1993). Receptor Nomenclature Supplement. *Trends Pharmacol. Sci.* Vol. 14.

Welker, W. I., and Seidenstein, S. (1959). Somatic sensory representation in the cerebral cortex of the raccoon *(Procyon lotor). J. Comp. Neurol.* 111:469–501.

Wells, M. (1968). *Lower Animals.* New York: McGraw-Hill.

Werblin, F. S., and Dowling, J. E. (1969). Organization of the retina of the mudpuppy, *Necturus maculosus.* II. Intracellular recording. *J. Neurophysiol.* 32:339–355.

Wersäll, J., and Bagger-Sjöbäck, D. (1974). Morphology of the vestibular sense organ. In *Handbook of Sensory Physiology,* Vol. 6: *Vestibular System,* Part 1: *Basic Mechanisms* (H. H. Kornhuber, ed.). New York: Springer-Verlag, pp. 123–170

Wetzel, M. C., and Stuart, D. G. (1976). Ensemble characteristics of cat locomotion and its neural control. *Prog. Neurobiol.* 7:1–98.

Wiesel, T. N., Hubel, D. H., and Lam, D. M. K. (1974). Autoradiographic demonstration of ocular-dominance columns in the monkey striate cortex by means of transneuronal transport. *Brain Res.* 79:273–279.

Wilcox, G. L. (1991). Excitatory neurotransmitters and pain. In *Proceedings of the 6th World Congress on Pain* (M. R. Bond, J. E. Charlton, and C. J. Woolf, eds.). New York: Elsevier, pp. 97–117.

Williams, R. W., and Herrup, K. (1988). The control of neuron number. *Annu. Rev. Neurosci.* 11:423–454.

Willis, W. D., Jr., and Coggeshall, R. E. (1991). *Sensory Mechanisms of the Spinal Cord.* New York: Plenum.

Wilson, C. J. (1990). Basal ganglia. In *The Synaptic Organization of the Brain,* 3rd ed. (G. M. Shepherd, ed.). New York: Oxford University Press, pp. 279–316.

Wilson, E. O. (1975). *Sociobiology.* Cambridge, MA: Harvard University Press.

Wilson, V. J., and Melvill-Jones, G. (1979). *Mammalian Vestibular Physiology.* New York: Plenum.

Wine, J. J., and Krasne, T. B. (1982). The cellular organization of crayfish escape behavior. In *The Biology of Crustacea,* Vol. 4 (D. C. Sandman and H. Atwood, eds.). New York: Academic Press, pp. 241–292.

Wong, M., and Moss, R. L. (1992). Long-term and short-term electrophysiological effects of estrogen on the synaptic properties of hippocampal CA1 neurons. *J. Neurosci.* 12:3217–3225.

Wood, J. D. (1975). Neurophysiology of Auerbach's plexus and control of intestinal motility. *Physiol. Rev.* 55:307–324.

Woodbury, J. W. (1965). The cell membrane: ionic and potential gradients and active transport. In *Physiology and Biophysics* (T. C. Ruch and H. D. Paton, eds.). Philadelphia: W. B. Saunders, pp. 1–25.

Woolsey, T. A., and van der Loos, H. (1970). The structural organization of layer IV in the somato-sensory region (SI) of mouse cerebral cortex. The description of a cortical field composed of discrete cytoarchitectonic units. *Brain Res.* 17:205–242.

Yamamoto, C., and McIlwain, H. (1966). Electrical activities in thin sections from the mammalian brain maintained in chemically defined media in vitro. *J. Neurochem.* 13:1333–1343.

Yau, K.-W., and Baylor, D. A. (1989). Cyclic GMP-activated conductance of retinal photoreceptor cells. *Annu. Rev. Neurosci.* 12:289–327.

Yau, K.-W., and Nakatani, K. (1985). Light-suppressible, cyclic GMP-sensitive conductance in the plasma membrane of a truncated rod outer segment. *Nature* 307:252–255.

Young, J. Z. (1964). *A Model of the Brain.* Oxford: Oxford University Press.

Young, M. W., Jackson, F. R., Shin, H. S., and Bargiello, T. A. (1985). A biological clock in *Drosophila. Cold Spring Harbor Symp. Quant. Biol.* 50:865–875.

Zeki, S. 1993. *A Vision of the Brain.* London: Blackwell.

Zemlin, W. R. (1968). *Speech and Hearing Science: Anatomy and Physiology.* Englewood Cliffs, NJ: Prentice-Hall.

Zufall, F., Firestein, S., and Shepherd, G. M. (1994). *Annu. Rev. Biophy. Biomolec. Struc.*

Author Index

Subject Index